Basic Algebraic Topology and its Applications

Mahima Ranjan Adhikari

Basic Algebraic Topology and its Applications

Springer

Mahima Ranjan Adhikari
Institute for Mathematics, Bioinformatics,
Information Technology and Computer
Science (IMBIC)
Kolkata
India

ISBN 978-81-322-3855-3 ISBN 978-81-322-2843-1 (eBook)
DOI 10.1007/978-81-322-2843-1

Printed on acid-free paper

This Springer imprint is published by Springer Nature
The registered company is Springer (India) Pvt. Ltd.

To
my parents Naba Kumar and
Snehalata Adhikari
who created my interest in mathematics
at my very early childhood

Preface

Algebraic topology is one of the most important creations in mathematics which uses algebraic tools to study topological spaces. The basic goal is to find algebraic invariants that classify topological spaces up to homeomorphism (though usually classify up to homotopy equivalence). The most important of these invariants are homotopy groups, homology groups, and cohomology groups (rings). The main purpose of this book is to give an accessible presentation to the readers of the basic materials of algebraic topology through a study of homotopy, homology, and cohomology theories. Moreover, it covers a lot of topics for advanced students who are interested in some applications of the materials they have been taught. Several basic concepts of algebraic topology, and many of their successful applications in other areas of mathematics and also beyond mathematics with surprising results have been given. The essence of this method is a transformation of the geometric problem to an algebraic one which offers a better chance for solution by using standard algebraic methods.

The monumental work of Poincaré in "Analysis situs", Paris, 1895, organized the subject for the first time. This work explained the difference between curves deformable to one another and curves bounding a larger space. The first one led to the concepts of homotopy and fundamental group; the second one led to the concept of homology. Poincaré is the first mathematician who systemically attacked the problems of assigning algebraic invariants to topological spaces. His vision of the key role of topology in all mathematical theories began to materialize from 1920. This subject is an interplay between topology and algebra and studies algebraic invariants provided by homotopy, homology, and cohomology theories. The twentieth century witnessed its greatest development.

The literature on algebraic topology is very vast. Based on the author's teaching experience of 50 years, academic interaction with Prof. B. Eckmann and Prof. P.J. Hilton at E.T.H., Zurich, Switzerland, in 2003, and lectures at different institutions in India, USA, France, Switzerland, Greece, UK, Italy, Sweden, Japan, and many other countries, this book is designed to serve as a basic text of modern algebraic topology at the undergraduate level. A basic course in algebraic topology presents a variety of phenomena typical of the subject. This book conveys the basic language of modern algebraic topology through a study of homotopy, homology, and cohomology theories with some fruitful applications which display the great beauty of the subject. For this study, the book displays a variety of topological spaces: spheres, projective spaces, classical groups and their quotient spaces, function spaces, polyhedra, topological groups, Lie groups, CW-complexes, Eilenberg–MacLane spaces, infinite symmetric product spaces, and some other spaces. As well as, the book studies a variety of maps, which are continuous functions between topological spaces.

Characteristics which are shared by homeomorphic spaces are called *topological invariants*; on the other hand, characteristics which are shared by homotopy equivalent spaces are called *homotopy invariants*. The Euler characteristic is an integral invariant, which distinguishes non-homeomorphic spaces. The search of other invariants has established connections between topology and modern algebra in such a way that homeomorphic spaces have isomorphic algebraic structures. Historically, the concepts of fundamental groups, higher homotopy groups, and homology and cohomology groups came from such a search. The natural emphasis is: to solve a geometrical problem of global nature, one first reduces it to a homotopy-theoretic problem; this is then transformed to an algebraic problem which provides a better chance for solution. This technique has been the most fruitful one in algebraic topology. The notions initially introduced in homology and homotopy theories for applications to problems of topology have found fruitful applications in other parts of mathematics. Homological algebra and K-theory are their outstanding examples.

The materials discussed here have appeared elsewhere. Each chapter opens with a short introduction which summarizes the information that sets out its central theme and closes with a list of sources for the use of readers to expand their knowledge. This does not mean that other sources are not good. Our contribution is the selection of the materials and their presentation. Each chapter is split into several sections (and subsections) depending on the nature of the materials which constitute the organizational units of the text. Each chapter provides exercises with further applications and extension of the theory. Some exercises carry hints which should not be taken as ideal ones. Many of them can be solved in a better way. The title of the book suggests the scope and power of algebraic topology and its text is expanded over 18 chapters and two appendices displayed below.

Chapter 1 assembles together some basic concepts of set theory, algebra, analysis, set topology, Euclidean spaces, manifolds with some standard notations for smooth reading of the book.

Chapter 2 is devoted to the study of basic elementary concepts of homotopy theory with illustrative examples. A homotopy formalizes the intuitive idea of continuous deformation of a continuous map between two topological spaces. It displays a variety of phenomena and related problems such as homotopy classification of continuous maps up to homotopy equivalence introduced by Hurewicz (1904–1956) in 1935, contractible spaces, H-groups (Hopf groups) and H-cogroups through their homotopy properties. Finally, this chapter presents interesting immediate applications of homotopy in dealing with some extension problems, lifting problems, and proving "Fundamental theorem of algebra" by using homotopic concepts.

Chapter 3 continues the study of homotopy theory through the concept of fundamental groups invented by H. Poincaré in 1895 which conveys the first transition from topology to algebra by assigning an algebraic structure to the set of relative homotopy classes of loops in a functorial way. The group structure of these homotopy classes of loops is proved in Sects. 3.1 and 3.2 in two different ways. This group earlier called Poincaré group is now known as *fundamental group*. It is the first influential invariant of homotopy theory and is also the first of a series of algebraic invariants π_n, called *homotopy groups* studied in Chap. 7. This chapter computes fundamental group of the circle by using the degree of a continuous map $f : S^1 \rightarrow S^1$, and studies Brouwer fixed point theorem for dimension 2, fundamental theorem of algebra, vector field problems on D^2 and knot groups by using the tools of fundamental groups.

Chapter 4 continues the study of the fundamental groups and presents a thorough discussion of covering spaces which are deeply connected with fundamental groups. Algebraic features of the fundamental groups are expressed by the geometric language of covering spaces. This chapter is designed to utilize the power of the fundamental groups and also to establish an exact correspondence between the various connected covering spaces of a given topological space B and subgroups of its fundamental group $\pi_1(B)$, like Galois theory, with its correspondence between field extensions and subgroups of Galois groups, which is an amazing result. This chapter also studies the concepts of fibrations and cofibrations with their applications born in geometry and topology.

Chapter 5 continues the study of homotopy theory through fiber bundles, vector bundles, and K-theory. Covering spaces provide tools to study the fundamental groups. Fiber bundles provide likewise tools to study higher homotopy groups (which are generalization of fundamental groups and described in Chap. 7). The importance of fiber spaces was realized during 1935–1950 to solve several problems relating to homotopy and homology. The motivation of the study of fiber bundles and vector bundles came from the distribution of signs of the derivatives of the plane curves at each point. This chapter also discusses homotopy classification of vector bundles, Milnor's construction of a universal fiber bundle for a topological group G with homotopy classification of principal G-bundles and presents the introductory concept of K-theory born in connecting the rich structure of vector bundles over a paracompact space B with the set of homotopy classes of maps from B into the Grassmann manifold of n-dimensional subspaces in infinite-dimensional space. This theory plays a vital role in applications of algebraic topology to analysis, algebraic geometry, topology, ring theory, and number theory.

Chapter 6 builds up interesting topological spaces called *polyhedra* from simplexes followed by a study of their homotopy properties and develops some tools for computing the fundamental groups of a large class of topological spaces. The geometrical objects such as points, edges, triangles, and tetrahedra are examples of low-dimensional simplexes. Simplicial complexes provide a convenient way to study manifolds. This chapter considers how simplexes may be fitted together to construct simplicial complexes which play an important role to construct interesting topological spaces such as polyhedra for the study of algebraic topology. They form building blocks of homology theory which begins in Chap. 10. The concept of triangulation is utilized to solve extension problems and that of edge-path to show that edge-group $E(K, v)$ is isomorphic to the fundamental group $\pi_1(|K|, v)$ for any simplicial complex K. Finally, van Kampen theorem is proved by using graph-theoretic results. This chapter also proves simplicial approximation theorem given by L.E.J. Brouwer (1881–1967) and J.W. Alexander (1888–1971) around 1920 by utilizing a certain good feature of simplicial complexes introduced by Alexander. This theorem plays a key role in the study of homotopy and homology theories.

Chapter 7 continues to study homotopy theory displaying the construction of a sequence of functors π_n given by W. Hurewicz (1904–1956) in 1935 from topology to algebra by extending the concept of fundamental group invented by H. Poincaré in 1895. The basic idea of homotopy groups is to classify all continuous maps from S^n to pointed topological space X up to homotopy equivalence. To study topological spaces X of low dimension, the fundamental group $\pi_1(X)$ is very useful. But it needs refined tools for the study of higher dimensional spaces. For example, fundamental group cannot distinguish spheres S^n with $n \geq 2$. Such a limitation of low dimension can be removed by considering the natural higher dimensional analogues of $\pi_1(X)$. More precisely, this chapter defines the nth (absolute) homotopy group and generalizes it to a (relative) homotopy group of a triplet and studies algebraic, functorial and fibering properties with the exactness of homotopy sequence of fibering, Hopf maps introduced by H. Hopf (1894–1971) in 1935 for the investigation of certain homotopy groups of S^n, action of π_1 on π_n, Freudenthal suspension theorem given by H. Freudenthal (1905–1990) in 1937 for the investigation of the homotopy groups $\pi_m(S^n)$ for $0 < m < n$, weak fibration which plays a key role in the study of higher homotopy groups, and the nth cohomotopy set $\pi^n(X, A)$ on which K. Borsuk (1905–1982) endowed in 1936 with an abelian group structure under certain conditions on (X, A). This chapter also discusses some interesting applications of higher homotopy groups.

Chapter 8 continues to study homotopy theory through a suitable special class of topological spaces, called CW-complexes introduced by J.H.C. Whitehead (1904–1960) in 1949 to meet the need for further development of homotopy theory. This class of spaces is broader and has some better categorical properties than simplicial complexes, but still retains a combinatorial nature that allows for computation (often with a much smaller complex). The concept of CW-complexes is introduced as a natural generalization of the concept of polyhedra by relaxing all "linearity conditions" in simplicial complexes, instead cells are attached by arbitrary continuous maps starting with a discrete set, whose each point is regarded as a 0-cell. A CW-complex is built up by successive adjunctions of cells of dimensions $1, 2, 3, \ldots$. There is an analogy between what can be done topologically with a space, and what can be done algebraically with its chain groups. In the class of CW-complexes this analogy attains its highest strength. The category of

CW-complexes is a suitable category for a systematic study of algebraic topology. Algebraic topologists now feel that a study of CW-complex should be included in the basic course of algebraic topology, and this study should move to the theorem that every continuous map between CW-complexes is homotopic to a cellular map. This chapter studies the basic aspects of CW-complexes and relative CW-complexes with their homotopy properties and proves Whitehead theorem, Freudenthal suspension theorem (general form), and cellular approximation theorem with their applications.

Chapter 9 continues to study homotopy theory through the different products in homotopy groups such as the Whitehead product introduced by J.H.C. Whitehead in 1941, mixed product introduced by McCarty in 1964, and Samelson product. Whitehead product provides a technique at least in some cases for constructing nonzero elements of the homotopy group $\pi_{p+q-1}(X)$ of a pointed topological space X. Moreover, this chapter finds a generalization of Whitehead product, establishes certain relation between Whitehead and Samelson products, and studies mixed products corresponding to a fiber space and a topological transformation group acting on it.

Chapter 10 begins to study homology and cohomology theories. Homotopy groups are very difficult to compute. There is an alternative approach of construction of a different topological invariant, the so-called homology group, which historically came earlier than homotopy groups. Homology (cohomology) theory is a covariant (contravariant) functor from the category of topological spaces to the category of abelian groups. Homology (simplicial) invented by H. Poincaré in 1895 is one of the most fundamental influential inventions in mathematics. The basic idea of homology is that it starts with a geometric object (a space) which is given by combinatorial data (a complex). Then the linear algebra and boundary relations determined by this data are used to construct homology groups. The simplicial devices in simplicial homology theory are gradually generalized to singular homology by using the algebraic properties of the singular complex. There exist different homology theories: simplicial, singular, cellular, and Čech homology theories which are studied in this chapter. The most important homology theory in algebraic topology is the singular homology. Simplicial homology is the primitive version of singular homology. Cohomology theory is also discussed. In some sense, homology theory and cohomology theory are dual to each other. More precisely, this chapter begins with a study of the concepts of chain complex, boundary, cycle introduced by W. Mayer (1887–1947) in 1929 from a purely algebraic viewpoint. This chapter presents a construction of the homology groups of a simplicial complex in two steps: first by assigning to each simplicial complex a certain complex, called *chain complex* followed by assigning to the chain complex its homology group. This construction assigns a group structure to cycles that are not boundaries with an extension to the concept of relative simplicial homology groups and generalizes simplicial homology theory to singular homology theory. These two theories are related by the basic result that the singular homology of a polyhedron is isomorphic to the simplicial homology of any of its triangulated simplicial complexes. This chapter examines the relations between absolute homology groups of simplicial chain complexes and the relative homology groups of relative simplicial chain complexes by using the language of exact sequences and shows that the relative homology groups $H_p(K, L)$ for any pair (K, L) of simplicial complexes fit into a long exact sequence. This chapter also discusses homology groups $H_n(X; G)$ with an arbitrary coefficient group G (abelian), Mayer–Vietoris sequences in singular and simplicial homology theories, cup product, and gives the Künneth formula and Eilenberg–Zilber theorem which are used for computing homology or cohomolgy of product spaces, and Euler characteristic & Jordan curve theorem from the viewpoint of homology theory.

Chapter 11 studies a special class of CW-complexes having only one nonzero homotopy groups, called *Eilenberg–MacLane spaces* which were introduced by S. Eilenberg (1915–1998) and S. MacLane (1909–2005) in 1945. Such spaces form a very important class of CW-complexes in algebraic topology. Their importance is twofold: they develop both homotopy and homology theories. They are closely linked with the study of cohomology operations. This chapter presents Eilenberg–MacLane spaces with their construction and studies their homotopy properties. The construction process of Eilenberg–MacLane spaces $K(G, n)$ for all possible (G, n) depends on a very natural class of spaces, called *Moore spaces* of type (G, n), denoted by $M(G, n)$. This chapter also studies Postnikov towers to meet the need for construction of Eilenberg–MacLane spaces.

Chapter 12 presents an approach formulating axiomatizaton of ordinary homology and cohomology theories. These axioms, now called Eilenberg and Steenrod axioms were announced by S. Eilenberg (1915–1998) and N.E. Steenrod (1910–1971) in 1945, but first appeared in their celebrated book *Foundations of Algebraic Topology* in 1952. This approach came from the problem of comparing the various definitions of homology and cohomology given in the previous years. Eilenberg and Steenrod initiated a new approach by taking a small number of their properties (not focussing on machinery used for construction of homology and cohomology groups) as axioms to characterize a theory of homology and cohomology. This axiomatic approach simplifies the proofs of many lengthy and complicated theorems and escapes the avoidable difficulty to motivate the students who are learning homology and cohomology theories for the first time as their systematic study. This axiomatic approach classifies and unifies different homology groups on the category of compact triangulable spaces and inaugurates its dual theory called *cohomology theory*. This approach is the most important contribution to algebraic topology since the invention of the homology groups by Poincaré in 1895.

Chapter 13 continues the study of homology and cohomology theories by presenting some of their interesting properties which directly follow from the Eilenberg and Steenrod axioms for homology and cohomology theories such as homotopy equivalence in these theories, relations between cofibrations and homology theory, and finally computes the ordinary homology groups of S^n with coefficients in an arbitrary abelian group G.

Chapter 14 presents further interesting applications of the homotopy, homology, and cohomology theories. The notions initially introduced in these theories to solve problems of topology that have fruitful applications, and proves many interesting theorems such as Hopf's classification theorem, hairy ball theorem, ham sandwich theorem, Borsuk–Ulam theorem, Lusternik–Schnirelmann theorem, Lefschetz fixed point theorem, and Jordan curve theorem. It also proves some results related to graph theory, fixed point theory of continuous maps, vector fields, and applications to algebra. Moreover, this chapter indicates some applications of algebraic topology in physics, chemistry, economics, biology, and medical science with specific references.

Chapter 15 conveys the concept of a spectrum originated by F.L. Lima (1929–) in 1958 and constructs its associated spectral homology and cohomology theories, and generalized homology and cohomology theories (which have been proved to be very useful theories) to distinguish them from ordinary homology and cohomology theories. Their properties and relations to homotopy theory are also discussed. For example, the ordinary homology group of certain topological spaces X can be thought of as an approximation to $\pi_n(X)$. Moreover, this chapter constructs a new Ω-spectrum $\underline{A}$, generalizing the Eilenberg–MacLane spectrum $K(G,n)$ and also constructs its associated cohomology theory $h^*(;\underline{A})$ which generalizes the ordinary cohomology theory of Eilenberg and Steenrod. This chapter conveys K-theory as a generalized cohomology theory and also studies the Brown representability theorem, stable homotopy groups, the cohomology operations, and Poincaré duality theorem.

Chapter 16 studies a theory known as "obstruction theory" by utilizing the tools of cohomology theory to encounter two basic problems in algebraic topology such as extension and lifting problems. Obvious examples are the homotopy extension and homotopy lifting problems. The homotopy classifications of continuous maps, together with the study of extension and lifting problems, play a central role in algebraic topology. The term "obstruction theory" refers to a technique for defining a sequence of cohomology classes that are obstructions to finding solution to the extension, lifting or relative lifting problems. Obstruction theory leads to make an attempt to find a general solution. This theory originated in the classical work of Hopf, Eilenberg, Steenrod, and Postinikov in around 1940. Certain sets of cohomology elements, called obstructions, are associated with both a single map in the case of extension and with a pair of maps in the case of homotopies. These are invariants depending only on the topological spaces and their continuous mappings. In polyhedra these are the characteristics for the existence or non-existence of the desired extensions and homotopies. The underlying idea of associating cohomology elements with continuous mappings was implicitly used by H. Whitney (1907–1989) and first explicitly formulated by N.E. Steenrod (1910–1971). This chapter uses cohomology theory to yield algebraic indicators for obstacles to extension and lifting problems of continuous maps and proves Eilenberg extension theorem. It presents some applications of obstruction theory to prove a homological version of Whitehead theorem, stepwise extension of cross-section and obstruction for homotopy between relative lifts.

Chapter 17 presents some similarities and interesting relations among homotopy, homology, and cohomology. In earlier chapters, some relations between these theories have been discussed. This chapter continues to convey more relations through Hurewicz homomorphism, Eilenberg–MacLane spaces, Dold–Thom theorem, Brown's representation theorem, Hopf invariant and Adams classical theorem on Hopf invariant. Historically, L.E.J. Brouwer first connected homology and homotopy in 1912 by proving that two continuous maps of a two-dimensional sphere into itself can be continuously deformed into each other if and only if they have the same degree (i.e., if and only if they are equivalent from the view point of homology theory). Hopf's classification theorem generalizes Brouwer's result to an arbitrary dimension. The homotopy groups resemble the homology groups in many respects under suitable situations proved by Hurewicz in his celebrated "equivalence theorem". There is also a lack of similarities between these two theories essentially due to the absence in higher homotopy groups the excision property for homology and also due to the absence in higher homotopy groups a theorem analogous to van Kampen theorem for fundamental group.

Chapter 18 focuses a brief history of algebraic topology highlighting the emergences of the ideas leading to new areas of study in algebraic topology and conveys the contributions of some mathematicians who introduced new concepts or proved theorems of fundamental importance or inaugurated new theories in algebraic topology starting from the creation of fundamental group and homology group by H. Poincaré in 1895, which are the first basic and influential inventions in algebraic topology. The literature on algebraic topology is very vast. Some concepts studied now in algebraic topology had been found in the work of B. Riemann (1826–1866), C. Felix Klein (1849–1925), and H. Poincaré (1854–1912). But the foundation of algebraic (combinatorial topology) was laid in the decade beginning 1895 by H. Poincaré through the publication of his famous series of memoirs "Analysis Situs" from 1895 onwards. J.W. Alexander (1888–1971) used the word "topological" in the titles of his research papers in the 1920s. This chapter also conveys more names with their contributions in algebraic topology. The early development of homotopy theory was essentially due to H. Poincaré, L.E.F. Brouwer, H. Hopf, W. Hurewicz, H. Freudenthal, and many others. W. Hurewicz first established a connection between homology and homotopy groups for $(n-1)$-connceted spaces, when $n \geq 2$. H. Hopf pioneered a study of maps into spheres during 1926–1935 and inaugurated the homotopy theory with the discovery of the Hopf map followed by the research of W. Hurewicz, and Freudenthal. Since then homotopy theory has made a rapid progress and now plays an important role in mathematics. Homology, invented by Henry Poincaré during 1895–1901, is one of the most fundamental influential inventions in mathematics. He started with a geometric object (a space) which is given by combinatorial data (a simplicial complex), then the linear algebra and boundary relations by these data are used to construct homology groups. There are other homology theories:

(i) Homology groups for compact metric spaces introduced by L. Vietoris (1891–2002) in 1927;
(ii) Homology groups for compact Hausdorff spaces introduced by E.Čech (1893–1960) in 1932;
(iii) Singular homology groups are first defined by S. Lefschitz (1884–1972) in 1933.

All these homology theories lived in isolation. Algebraic topologists in around 1940 started comparing various definitions of homology and cohomology given in the previous years. Eilenberg and Steenrod initiated a new approach in 1945 by taking a small number of their properties (not focusing on machinery used for construction of homology and cohomology groups) as axioms to characterize a theory of homology and cohomology. This approach is the most important contribution to algebraic topology since the invention of the homology groups by Poincaré and is called the *axiomatic approach* given by a set of seven axioms announced by S. Eilenberg and N. Steenrod in 1945 and published in their book in 1952. This approach classifies and unifies different homology groups on the category of compact triangulated spaces and inaugurated its dual theory for cohomology theories. This chapter also conveys the contributions of more mathematicians, S. MacLane, J.H. Whitehead, Serre, Brown, Milnor, and Grothendieck, to name a few.

Apppendix A studies classical topological groups and Lie groups that occupy a vast territory in topology and geometry. Lie groups are special topological groups and also manifolds carrying a differential structure. For example, $GL(n, \mathbf{R}), GL(n, \mathbf{C}), GL(n, H), SL(n, \mathbf{R}), SL(n, \mathbf{C}), O(n, \mathbf{R}), U(n, \mathbf{C}), SL(n, H)$ are some important classical Lie Groups. Historically, S. Lie (1842–1899) investigated group of transformations. He developed his theory of transformation groups to solve his integration problems. Such groups are now called Lie groups after his name. The Fifth Problem of Hilbert announced at the ICM 1900, Paris, is linked to Sophus Lie theory of transformation groups which asserts that Lie groups act as groups of transformations on manifolds.

Appendix B discusses category theory through the study of categories, functors, and natural transformations with an eye to study algebraic topology which consists of the construction and use of functors from some category of topological spaces into an algebraic category. This theory plays an important role for the study of homotopy, homology, and cohomology theories, which constitute the basic text of this book in addition to *adjoint functor, representable functor, abelianization functor, Brown functor, and infinite symmetric product functor.* All constructions in algebraic topology are in general functorial. Fundamental groups, higher homotopy groups, and homology and cohomology groups are not only invariants of the underlying topological space, in the sense that two topological spaces which are homeomorphic have the isomorphic associated groups (or modules) but their associated morphisms also correspond to a continuous mapping of topological spaces an induced group (or module) homomorphism on the associated groups (modules), and these homomorphisms can be used to show non-existence (or, much more deeply, existence) of mappings. So the readers of algebraic topology cannot escape learning the concepts of categories, functors and natural transformations.

The author acknowledges the Science and Engineering Research Board (SERB) under the Department of Science & Technology of the Government of India for sanctioning of the financial support towards writing this book under the Utilization of Scientific Expertise of Retired Scientists (USERS) scheme (G.O. No HR/UR/17/2010 dated December 27, 2011) and also to West Bengal University of Technology (now renamed as Moulana Abul Kalam Azad University of Technology) for providing infrastructure towards implementing the scheme. The author expresses his sincere thanks to Springer for publishing this book. The author is very thankful to the reviewers of the manuscript and also to Dr. Avishek Adhikari of the University of Calcutta for their scholarly suggestions for improvement of the book. Thanks are also due to K. Sardar for his cooperation in dealing with the typing of one version after another of the manuscript and also to many other individuals who have helped in proofreading the book. Author's thanks are due to the institute IMBIC, Kolkata, for providing the author with the library and other facilities towards the manuscript development work of this book. Finally, the author acknowledges, with heartfelt thanks, the patience and sacrifice of long-suffering family of the author, specially author's wife Minati, son Avishek, daughter-in-law Shibopriya, and grandson little Avipriyo.

Kolkata, India — Mahima Ranjan Adhikari
March 2016

Contents

About the Author

Mahima Ranjan Adhikari, Ph.D. is the former Professor of Pure Mathematics at the University of Calcutta. His main interest lies in algebra and topology. He has published a number of papers in several international journals including the Proceedings of American Mathematical Society and five textbooks. Under his supervision, 11 students have already been awarded the Ph.D. degree. He is a member of the American Mathematical Society and serves on the editorial board of several journals and research monographs. He was the President of the mathematical science section of the 95th Indian Science Congress, 2008.

He has visited several institutions in India, USA, UK, Japan, France, Greece, Sweden, Switzerland, Italy, and in many other countries on invitation. While visiting E.T.H., Zurich, Switzerland, in 2003, he made an academic interaction with Prof. B. Eckmann and Prof. P.J. Hilton. He is currently the President of the Institute for Mathematics, Bioinformatics, Information Technology and Computer Science (IMBIC). He is also the principal investigator of an ongoing project funded by the Government of India.

This book is written based on author's teaching experience of 50 years. He is the joint author (with Avishek Adhikari) of another Springer book "Basic Modern Algebra with Applications", 2014.

Chapter 1
Prerequisite Concepts and Notations

This chapter assembles together some basic concepts and results of set theory, algebra, analysis, set topology, Euclidean spaces, manifolds with standard notations for smooth reading of the book. It is assumed that the readers are familiar with these basic concepts. However, for their detailed study, the books Adhikari and Adhikari (2014), Dugundji (1966), Herstein (1964), Maunder (1970), Spanier (1966), and some other books are referred in Bibliography.

1.1 Set Theory

This section conveys some basic concepts of set theory (naive) initiated around 1870 by the German mathematician Georg Cantor (1845–1918) which are used throughout the book. Set theory occupies an important position in mathematics. Many concrete concepts and examples are based on it. It is assumed that the readers are familiar with the sets

$\mathbf{N}$ (set of natural numbers/positive integers)
$\mathbf{Z}$ (set of integers)
$\mathbf{Q}$ (set of rational numbers)
$\mathbf{R}$ (set of real numbers)
$\mathbf{C}$ (set of complex numbers)

For precise description of many concepts of mathematics and also for mathematical reasoning the concepts of relations(functions) and cardinality of sets are very important, which are discussed first.

A binary relation ρ on a nonempty set X is a subset of $X \times X$, which is said to be an equivalence relation if ρ is reflexive, i.e., $(x, x) \in \rho$ for each $x \in X$; symmetric, i.e., $(x, y) \in \rho$ implies $(y, x) \in \rho$ and transitive i.e., $(x, y) \in \rho$ and $(y, z) \in \rho$ imply $(x, z) \in \rho$ for $x, y, z \in X$.

M.R. Adhikari, *Basic Algebraic Topology and its Applications*,
DOI 10.1007/978-81-322-2843-1_1

Definition 1.1.1 Let X be a nonempty set and ρ be an equivalence relation on X. The disjoint classes $[x]$ into which the set X is partitioned by ρ constitute a set, called the quotient set of X by ρ, denoted by X/ρ, where $[x]$ denotes the class (determined by ρ) containing the element x of X. Each element x of the class $[x]$ is called a representative of $[x]$.

Example 1.1.2 Given a positive integer n, the quotient set $\mathbf{Z}_n$ consists of all n distinct classes $[0], [1], \ldots, [n-1]$. The set $\mathbf{Z}_n$ is called the residue classes of $\mathbf{Z}$ modulo n.

Remark 1.1.3 The set $\mathbf{Z}_n$ provides very strong different algebraic structures (depending on n). The visual description of $\mathbf{Z}_{12}$ is a 12-h clock.

Definition 1.1.4 Given a nonempty set I, if there exists a set X_i for each $i \in I$, then the collection of the sets $\{X_i : i \in I\}$ is called a family of sets and I is called an indexing set of the family.

For $i \neq j$, X_i may be equal to X_j. The collection of sets $\{X_i : i \in I\}$ is finite or infinite according as I is a finite set or an infinite set.

Definition 1.1.5 A relation ρ on X is said to be antisymmetric if $(x, y) \in \rho$ and $(y, x) \in \rho$ imply $x = y$ for $x, y \in X$. A reflexive, antisymmetric and transitive relation on X is called a partial order relation.

Definition 1.1.6 If ρ is a partial order relation on a set X, then the pair (X, ρ) is called a partially ordered set or a poset. A partially ordered set in which every pair of elements is comparable, is called an ordered set and the set is called totally ordered.

Zorn's Lemma Let $(X, \leq)$ be a nonempty partially ordered set. If every subset $A \subseteq X$, which is totally ordered by $\leq$, has an upper bound in X, then X has at least one maximum element.

Remark 1.1.7 Zorn's lemma is indispensible to prove many results of mathematics.

Definition 1.1.8 A map $f : X \rightarrow Y$ is said to be

(i) injective (or an injection) if different elements of X have different images in Y, i.e., $x \neq x'$ implies $f(x) \neq f(x')$ for x, x' in X;

(ii) surjective (or onto or a surjection) if every element of Y is the image of some element of X, i.e., for every element y in Y, there is some x in X such that $y = f(x)$;

(iii) bijective (or a bijection) if f is both injective and surjective.

Definition 1.1.9 Let $f : X \rightarrow Y$ and $g : Y \rightarrow Z$ be two maps. Their composite map denoted by $g \circ f$ is the map $g \circ f : X \rightarrow Z, x \mapsto g(f(x)), x \in X$.

Proposition 1.1.10 *Let $f : X \to Y$ and $g : Y \to Z$ be two maps. Then $g \circ f$ has the properties:*

(i) *If f and g are injective, then $g \circ f$ injective;*
(ii) *If f and g are surjective, then $g \circ f$ is surjective;*
(iii) *If $g \circ f$ is surjective, then g is surjective;*
(iv) *If $g \circ f$ is injective, then f is injective;*
(v) *If f and g are bijective, then $g \circ f$ is bijective;*
(vi) *If $g \circ f$ is bijective, then f is injective and g is surjective. If in particular, if $g \circ f = 1_X$, the identity map of X, then f is injective and g is surjective.*

The bijections of sets define equivalent sets.

Definition 1.1.11 Two sets X, Y are said to be equivalent denoted by $X \sim Y$ if there exists a bijection from the set X to the set Y.

Example 1.1.12 Let $S^2 = \{(x, y, z) \in \mathbf{R}^3 : x^2 + y^2 + z^2 = 1\}$ be the unit sphere in the Euclidean space $\mathbf{R}^3$ and $\mathbf{C}_\infty = \mathbf{C} \cup \{\infty\}$ be the extended complex plane. Then the stereographic projection $f : S^2 \to \mathbf{C}_\infty$ is a bijection. Thus $\mathbf{C}_\infty$ is represented as the sphere S^2 called Riemann sphere.

Example 1.1.13 A permutation of a set X is a bijection from X onto itself.

Definition 1.1.14 A set X is said to be finite if either X is empty or $X \sim \mathbf{Z}_n$ for some integer $n > 1$. Two nonempty finite sets are equivalent if they have the same number of elements.

Proposition 1.1.15 *A finite set cannot be equivalent to a proper subset of the set.*

Definition 1.1.16 A nonempty set which is not finite is said to be an infinite set.

Example 1.1.17 **Z**, **N**, **R**, **C** are infinite sets but $\mathbf{Z}_n$ is a finite set.

Definition 1.1.18 A set X is said to be countable if either X is finite or there exists a bijection $\mathbf{N} \to X$ (in the latter case X is said to be infinitely countable and its elements can be arranged as a sequence $\{x_n\}, n \in \mathbf{N}$).

Proposition 1.1.19 *Every infinite set contains a countable set.*

Example 1.1.20 The set **Q** is countable but the set **R** is not countable.

The notion of counting is extended by assigning to every set X (finite or infinite) an object $|X|$, called the cardinal number or cardinality, defined in such a way that $|X| = |Y|$ iff there exists a bijection between the sets X and Y. If X is a finite set of n elements, then $|X| = n$. The cardinal number d or c of an infinite set X asserts that the set is countable or not. Hence $|\mathbf{N}| = |\mathbf{Q}| = d$ but $|\mathbf{R}| = c$.

Definition 1.1.21 Let α and β be the cardinal numbers of two disjoint sets X and Y, then $\alpha+\beta$, $\alpha\beta$ and β^α are defined by $\alpha+\beta=|X\cup Y|$, $\alpha\beta=|X\times Y|$ and $\beta^\alpha=|Y^X|$, where Y^X denotes the sets of all maps $f: X\to Y$.

Proposition 1.1.22 *For any cardinal numbers* α, β *and* γ,

(i) $(\alpha+\beta)+\gamma=\alpha+(\beta+\gamma)$ *(associativity for addition);*
(ii) $(\alpha\beta)\gamma=\alpha(\beta\gamma)$ *(associativity for multiplication);*
(iii) $\alpha+\beta=\beta+\alpha$ *(commutativity for addition);*
(iv) $\alpha\beta=\beta\alpha$ *(commutativity for multiplication);*
(v) $\alpha(\beta+\gamma)=\alpha\beta+\alpha\gamma$ *(distributive property);*
(vi) $(\alpha\beta)^\gamma=\alpha^\gamma\beta^\gamma$;
(vii) $\alpha<2^\alpha$ *(Cantor's Theorem);*
(viii) $\alpha\le\beta$ *and* $\beta\le\alpha$ *imply that* $\alpha=\beta$.

1.2 Groups and Fundamental Homomorphism Theorem

This section conveys some basic results of group theory which are used throughout the book. Originally, a group was defined as the set of permutations (i.e., bijections) on a nonempty set X with the property that combination (called composition) of two permutations is also a permutation on X. Earlier definition of a group is generalized to the present concept of an abstract group by a set of axioms.

Definition 1.2.1 A group G is a nonempty set G together with a binary operation (called composition), that is, a rule that assigns to each ordered pair (a, b) in $G\times G$, an element of G, denoted by ab (or $a\cdot b$ called a multiplication) such that

G(1) $ab(c)=a(bc)$ for all a, b, c in G (associative law);
G(2) there exists an element e in G such that $ae=ea=a$ for all a in G (existence of identity);
G(3) for each a in G, there is an element a' in G such that $aa'=a'a=e$ (existence of inverse).

Remark 1.2.2 In a group G, e is unique and for each a in G, a' is also unique. The element a' denoted by a^{-1}, is called the inverse of a for each $a\in G$. In additive notation, ab is written as $a+b$; e is as 0 (zero) and a^{-1} as $-a$.

A group G is said to be commutative (or abelian) if $ab=ba$ for all a, b in G. We usually use the term 'abelian group' when the composition law is in additive notation. A group G is said to be finite if its underlying set G is finite; otherwise, it is said to be infinite.

Example 1.2.3 Given a nonempty set X, let $P(X)$ denote the set of all permutations (bijective mappings) on X. Then under usual composition of mappings $P(X)$ is a group, called permutation group on X. In particular, if X contains only n elements, then $P(X)$ is called the symmetric group on n elements, denoted by S_n.

Example 1.2.4 (*General Linear Groups*) GL $(n, \mathbf{R})$ (GL $(n, \mathbf{C})$) is the group of all invertible $n \times n$ real (complex) matrices under usual multiplication of matrices and is called general linear group of order n over $\mathbf{R}(\mathbf{C})$.

Example 1.2.5 (*Circle group*) The set $S^1 = \{z \in \mathbf{C} : |z| = 1\}$ forms a group under usual multiplication of complex numbers, called the circle group in $\mathbf{C}$.

An arbitrary subset of a group forming a group, called a subgroup contained in a larger group, sometimes creates interest and plays an important role in group theory and algebraic topology.

Proposition 1.2.6 *A nonempty subset H of a group G is a subgroup of G if and only if $ab^{-1} \in H$ for all $a, b \in H$.*

Example 1.2.7 **(i)** The additive group of integers is a subgroup of the additive group of real numbers. The additive group $\mathbf{Z}$ of integers is an example of infinite group. On the other hand the group $\mathbf{Z}_n$ is a finite group.

(ii) Given a group G, the center $Z(G)$ of G, defined by $Z(G) = \{g \in G : gx = xg \text{ for all } x \in G\}$ is subgroup of G.

Definition 1.2.8 For any subgroup H of an arbitrary group G, the set $aH = \{ah \in G : h \in H\}$ is said to be a left coset of H in G for every $a \in G$. On the other hand, for every $a \in G$, the right coset Ha is defined by $Ha = \{ha \in G : h \in H\}$.

Using the concept of cosets, Lagrange theorem establishes a relation between the order of a finite group and orders of its subgroups.

Theorem 1.2.9 (Lagrange Theorem) *The order of a subgroup of a finite group divides the order of the group.*

Remark 1.2.10 The converse of Lagrange's theorem claims that if a positive integer m divides the order n of a finite group G, then G contains a subgroup of order m. But it is not true in general. For example, the alternating group A_4 of order 12 has no subgroup of order 6. However, under certain particular situations, converse of Lagrange's theorem is partially true. In its support, consider the following results.

Theorem 1.2.11 *Let G be a finite group of order n and m be a positive divisor of n.*

(i) *If G is an abelian group, then corresponding to every positive divisor m of n, G contains a subgroup of order m.*
(ii) *If $m = p$, is a prime integer, then G has a subgroup of order p.*
(iii) *If m is a power of a prime p, then G has a subgroup of order m (Sylow first theorem.).*

We now define a special map between groups, called a homomorphism which preserves compositions of the groups.

Definition 1.2.12 Let G and H be groups. Then a map $f : G \to H$ is said to be a homomorphism if $f(xy) = f(x)f(y)$ for all x, y in G.

Particular homomorphisms carry special names having interesting properties.

Definition 1.2.13 Let $f : G \to H$ be a homomorphism of groups. Then

(a) f is said to be

- **(i)** an epimorphism if f is surjective;
- **(ii)** a monomorphism if f is injective;
- **(iii)** an isomorphism if f is bijective;
- **(iv)** an endomorphism if $G = H$;
- **(v)** an automorphism if $G = H$ and f is an isomorphism.

(b) **(i)** the kernel of f, defined by $\ker f = \{x \in G : f(x) = e_H\}$ is a subgroup of G.

(ii) the image of f, defined by $\operatorname{Im} f = \{y \in H : y = f(x) \text{ for some } x \in G\}$ is a subgroup of H.

Remark 1.2.14 A homomorphism between groups preserves the identity element and inverse elements.

Proposition 1.2.15 *Let $f : G \to H$ be a group homomorphism. Then f is a monomorphism if and only if* $\ker f = \{e_G\}$.

We are now interested to present isomorphic replicas of an abstract group.

Definition 1.2.16 A subgroup H of a group is said to be normal if for all $g \in G$, $gHg^{-1} \subseteq H$, where $gHg^{-1} = \{ghg^{-1} : h \in H\}$ (i.e., $gH = Hg$ for all $g \in G$).

Remark 1.2.17 Normal subgroups form an important class of subgroups in group theory, which was first recognized by the French mathematician Evariste Galois (1811–1832).

Example 1.2.18 If $f : G \to H$ is a group homomorphism, then $\ker f$ is a normal subgroup of G.

We now construct a quotient group (or factor group) by a normal subgroup N of a group using the relation $aN = Na$ for each element a of the group.

Definition 1.2.19 Let N be a normal subgroup of a group G and G/N be the set of all cosets of N in G. Then the set G/N is a group under the composition $aH \cdot bH = (ab)H$. The group G/N is called quotient group or factor group of G by N.

Remark 1.2.20 If $|G|$ denotes the order of a finite group G and N is a normal subgroup of the group G, then $|G/N| = |G|/|N|$.

Using the fact that the kernel of a homomorphism is a normal subgroup, the following theorem is proved.

Theorem 1.2.21 (Fundamental Homomorphism Theorem or First Isomorphism Theorem) *Let $f : G \to G'$ be a group homomorphism. Then f induces an isomorphism $\tilde{f} : G/\ker f \to \operatorname{Im} f, g \ker f \mapsto f(g)$. In particular, if f is an epimorphism, then $\tilde{f}$ is an isomorphism.*

Finitely generated groups are very important in algebraic topology. Let G be a group and S be a nonempty subset of G. Then the intersection of all subgroups of G containing S is also a subgroup of G, denoted by $\langle S \rangle$. It is the smallest subgroup of G which contains S.

Definition 1.2.22 If S is a nonempty subset of a group G, then $\langle S \rangle$ is called the subgroup generated by S. If $\langle S \rangle = G$, then G is said to be generated by S. If S is a finite set and the group $G = \langle S \rangle$, then G is said to be finitely generated with S a set of generators. In particular, if $S = \{x\}$ and $G = \langle S \rangle$, then G is said to be a cyclic group and x is said to be a generator of G.

Theorem 1.2.23 *Let G be a cyclic group. Then*

(i) *G is isomorphic to the group $\mathbf{Z}$ if and only if G is infinite.*
(ii) *G is isomorphic to the group $\mathbf{Z}_n$ if and only if G is finite and $|G| = n$.*

1.3 Group Representations, Free Groups, and Relations

This section conveys the concepts of group representations, free groups, and relations which are used in subsequent chapters.

1.3.1 Linear Representation of a Group

A group representation describes an abstract group in terms of linear operators of vector spaces (see Sect. 1.8). This subsection introduces the concept of group representation.

Definition 1.3.1 Let G be a group and V be a vector space over a field F. If $\mathrm{GL}(V)$ is the general linear group on V, then a representation of G on V is a group homomorphism

$$\psi : G \to \mathrm{GL}(V)$$

such that $\psi(g_1 g_2) = \psi(g_1) \circ \psi(g_2)$ for all $g_1, g_2 \in G$. The vector space V is called the representation space and dimension of V is called the dimension of the representation. The homomorphism ψ is sometimes called a linear representation of the group G.

1.3.2 *Free Groups and Relations*

This subsection introduces the concepts of free groups and relations, which are used in computation of fundamental groups and some other groups. The free groups used in multiplication notation here are not necessarily abelian.

Definition 1.3.2 A subset $X = \{x_j\}$ of a group G with identity e is called a free set of generators of G if every element $g \in G - \{e\}$ is uniquely expressable as

$$g = x_1^{i_1} x_2^{i_2} \dots x_n^{i_n} \tag{1.1}$$

where n is a positive integer and $i_k \in \mathbf{Z}$. We assume that $x_j \neq x_{j+1}$ for any j (i.e., no adjacent x_j are equal). If $i_j = 1$ for some i_j, we write x_j^1 as x_j. Again, if $i_j = 0$ for some i_j, the term x_j^0 is dropped from the expression of g.

Example 1.3.3 The expression $g = a^5 b^{-7} c\, b^8$ is acceptable but the expression $h = a^5 a^{-7} b^0$ is not acceptable.

Definition 1.3.4 If a group G has a free set of generators, it is called a free group. Given a set X, there exists a free group G such that X is a free set of generators of G. Each element of X is called 'letter'.The cardinality of a free set of generators of G is called the rank of G.

Definition 1.3.5 The product

$$w = x_1^{i_1} x_2^{i_2} \dots x_n^{i_n} \tag{1.2}$$

is called a word, where $x_j \in X$ and $i_j \in \mathbf{Z}$. If $i_j \neq 0$ and $x_j \neq x_{j+1}$, the word is called reduced word.

Remark 1.3.6 It is always possible to reduce a word by finite steps.

Definition 1.3.7 A word with no letter is called an empty word denoted by 1.

Definition 1.3.8 The set of all reduced words forms a free group, called the free group generated by X, denoted by $F(X)$, and under the multiplication is the juxtaposition of two words followed by reduction, the unit element is the empty word and the inverse of $w = x_1^{i_1} x_2^{i_2} \dots x_n^{i_n}$ is $w^{-1} = x_n^{-i_n} \dots x_2^{-i_2} x_1^{-i_1}$.

Example 1.3.9 Let $X = \{x\}$. Then $F(X) \cong \mathbf{Z}$. An arbitrary group G is in general defined by the generators and certain conditions (constraints) R on them. If $\{x_j\}$ is the set of generators, then the conditions (constraints) commonly written as

$$R : x_{j_1}^{i_1} x_{j_2}^{i_2} \dots x_{j_n}^{i_n} = 1 \tag{1.3}$$

are called relations.

Example 1.3.10 The cyclic group of order n generated by $\{x\}$ satisfies the relation $R : x^n = 1$.

Definition 1.3.11 A group G is defined by generators $X = \{x_j\}$ and relations $R = \{R_k : R_k = 1\}$ if $G \cong F/N$, where F is free on X and N is the normal subgroup of F, generated by $\{R_k\}$. The ordered pair $(X; R)$ is called a presentation of the group G, denoted by $\langle X; R\rangle$.

Example 1.3.12 **(i)** $\mathbf{Z}_n = \langle x : x^n\rangle$ and $\mathbf{Z} = \langle x : \emptyset\rangle$.

(ii) $\mathbf{Z} \oplus \mathbf{Z} = \langle x^m y^n : m, n \in \mathbf{Z}\rangle$ is free abelian group generated by $X = \{x, y\}$. Then $xy = yx$. Since $xyx^{-1}y^{-1} = 1$, there is a relation $R = xyx^{-1}y^{-1}$. Hence the presentation of $\mathbf{Z} \oplus \mathbf{Z}$ is $\langle x, y : xyx^{-1}y^{-1}\rangle$.

Definition 1.3.13 A group G is said to be finitely presented with a set S as its generating set S and a set R of relations describing the group G if both the sets S and R are finite and then G has the presentation as $G = \langle S : R\rangle$.

Example 1.3.14 The cyclic group $\mathbf{Z}_n$ has the presentation $\langle x : x^n\rangle$.

Example 1.3.15 The dihedral group D_n is completely determined by two generators x, y and the defining relations $x^n = e$, $y^2 = e$, $(xy)^2 = e$ and $x^k \neq e$ for $0 < k < n$. Hence D_n has the presentation $\langle x, y : x^n, y^2, (xy)^2\rangle$.

Example 1.3.16 The quaternion group has the presentation $\langle x, y : x^4, x^2 = y^2, yx = x^3y\rangle$.

The concepts of direct product and direct sum of groups are frequently used in algebraic topology.

Definition 1.3.17 Let G and K be two groups. The set $P = G \times K = \{(g, k) : g \in G, k \in K\}$ is a group under pointwise multiplication defined by $(g, k)\cdot(g', k') = (gg', kk')$. This group is called the direct product of the groups G and H.

We generalize this definition as follows:

Definition 1.3.18 Let $\{G_i : i \in I\}$ be a family of groups. Then an element $f \in \prod_{i\in I} G_i = G$, is a map $f : I \to \bigcup_{i\in I} G_i$ such that $f(i) \in G_i$ for $i \in I$. The set $G = \prod_{i\in I} G_i$ forms a group under the composition defined by $(fg)(i) = f(i)\cdot g(i)$ for all $i \in I$, where the right-hand multiplication is the usual multiplication in the group G_i for all $i \in I$. This group is called the direct product of the groups $\{G_i : i \in I\}$. Let e_i be the identity element of G_i, for all $i \in I$.

We now consider the subset $\bigoplus_{i\in I} G_i$ of the product group $\prod_{i\in I} G_i$, defined by

$$\oplus G_i = \{(g_i) \in \prod_{i\in I} G_i : g_i = e_i \text{ for all } i \text{ except for a finite number of indices}\}.$$

Definition 1.3.19 $\oplus G_i$ forms a subgroup of $\prod\limits_{i\in I} G_i$, called the direct sum of the given family $\{G_i : i \in I\}$ of subgroups. In particular, if I is a finite set, then the concepts of direct product and direct sum coincide.

Free abelian groups in additive notation are used in algebraic topology.

Definition 1.3.20 Let G be an additive abelian group. G is said to be a free group with a basis B if

(i) for each $b \in B$, the cyclic group $\langle b \rangle$ is infinite; and
(ii) $G = \bigoplus\limits_{b\in B} \langle b \rangle$ (direct sum).

Theorem 1.3.21 *If G is a free abelian group with a basis $B = \{b_1, b_2, \ldots, b_n\}$, then n is uniquely determined by G, and G is then said to be free abelian group of rank n.*

Definition 1.3.22 An abelian group G is said to be a finitely generated abelian group if G has a finite basis.

Remark 1.3.23 Basis of a finitely generated abelian group is not unique.

For example, $\{(1, 0), (0, 1)\}$ and $\{(-1, 0), (0, -1)\}$ are two different bases of the group $\mathbf{Z} \oplus \mathbf{Z}$.

Theorem 1.3.24 *Any two bases of a free abelian group have the same cardinality.*

This theorem leads to the following definition.

Definition 1.3.25 Let F be a free abelian group with a basis B. The cardinality of B is called the rank of F. In particular, if F is finitely generated, then the number of elements in a basis of F is the rank of F.

Example 1.3.26 **(i)** $\mathbf{Z}$ is a free abelian group of rank 1, finitely generated by 1 (or -1).
(ii) $\mathbf{Z} \oplus \mathbf{Z}$ is a free abelian group of rank 2.
(iii) $\mathbf{Z}_2$ is finitely generated by 1 but not free, because $1 + 1 = 0$ shows that 1 is not linearly independent.

Definition 1.3.27 An abelian group G has rank r (possibly infinite) if there exists a free abelian subgroup F of G such that

(a) rank of F is r; and
(b) the quotient group G/F is of finite order.

Theorem 1.3.28 *Given a family of abelian groups $\{G_i\}_{i\in I}$, there exists an abelian group G and a family of monomorphisms $f_i : G_i \to G$ such that $G = \oplus f_i(G_i)$.*

1.3.3 Betti Number and Structure Theorem for Finite Abelian Group

This subsection states some basic concepts and theorems such as fundamental theorem of finitely generated abelian group, Betti number, and structure theorem for finite abelian group which are very key algebraic results used in algebraic topology

Theorem 1.3.29 (Fundamental theorem of finitely generated abelian groups) *Every finitely generated abelian group G (not necessarily free) can be expressed uniquely as*

$$G \cong \overbrace{\mathbf{Z} \oplus \mathbf{Z} \oplus \cdots \oplus \mathbf{Z}}^{r \text{ summands}} \oplus \mathbf{Z}_{n_1} \oplus \mathbf{Z}_{n_2} \oplus \cdots \oplus \mathbf{Z}_{n_t},$$

for some integers $r, n_1, n_2, \ldots, n_t$ such that

(i) $r \geq 0$ *and* $n_j \geq 2$ *for all* j; *and*
(ii) $n_i | n_{i+1}$, *for* $1 \leq i \leq t-1$.

Definition 1.3.30 The integer r in Theorem 1.3.29 is called the free rank or Betti number of the group G given by E. Betti (1823–1892) and the integers $n_1, n_2, \ldots, n_t$ are called invariant factors of G.

Remark 1.3.31 $\overbrace{\mathbf{Z} \oplus \mathbf{Z} \oplus \cdots \oplus \mathbf{Z}}^{r \text{ summands}}$ is a free abelian group of rank r.

Theorem 1.3.32 (Structure Theorem for finite abelian groups) *Any nonzero finite abelian group G can be expressed uniquely as $G \cong \mathbf{Z}_{n_1} \oplus \mathbf{Z}_{n_2} \oplus \cdots \oplus \mathbf{Z}_{n_t}$ such that $n_i | n_{i+1}$, for $1 \leq i \leq t-1$.*

Theorem 1.3.33 *Two finite abelian groups are isomorphic if and only if they have the same invariant factors.*

1.4 Exact Sequence of Groups

This section conveys some results of exact sequences of groups and their homomorphisms which are frequently applied in algebraic topology. For this section the book Adhikari and Adhikari (2014) is referred.

Definition 1.4.1 A sequence of groups and their homomorphisms

$$\cdots \longrightarrow G_{n+1} \xrightarrow{f_{n+1}} G_n \xrightarrow{f_n} G_{n-1} \longrightarrow \cdots$$

is said to be exact if $\ker f_n = \operatorname{Im} f_{n+1}$ for all n. Clearly, $f_n \circ f_{n+1} = 0$ for an exact sequence.

Proposition 1.4.2 **(i)** *In the short exact sequence* $0 \longrightarrow G \xrightarrow{f} K$, *$f$ is a monomorphism.*

(ii) *In the short exact sequence* $G \xrightarrow{f} K \longrightarrow 0$, *$f$ is an epimorphism.*

(iii) *The sequence* $0 \longrightarrow G \xrightarrow{f} K \longrightarrow 0$ *is exact if and only if f is an isomorphism;*

(iv) *If G is a normal subgroup of K and $i : G \hookrightarrow K$ is the inclusion map (i.e., $i(x) = x$ for all $x \in G$), then the sequence*

$$0 \longrightarrow G \xrightarrow{i} K \xrightarrow{p} K/G \longrightarrow 0$$

is an exact sequence, where 0 denotes the trivial group and p is the natural homomorphism defined by $p(x) = x + G$ for all $x \in K$.

Proposition 1.4.3 *Given exact sequences of groups and homomorphisms*

$$0 \longrightarrow G_i \xrightarrow{f_i} K_i \xrightarrow{g_i} H_i \longrightarrow 0$$

for each element $i \in I$, the sequence

$$0 \longrightarrow \bigoplus_{i \in I} G_i \xrightarrow{\oplus f_i} \bigoplus_{i \in I} K_i \xrightarrow{\oplus g_i} \bigoplus_{i \in I} H_i \longrightarrow 0$$

is also exact

Theorem 1.4.4 (The Five Lemma) *Let the diagram in Fig. 1.1 of groups and homomorphisms be commutative with two exact rows. If α is an epimorphism, λ is a monomorphism and β, δ are isomorphisms, then γ is an isomorphism.*

Proposition 1.4.5 *Given an exact sequence of abelian groups and homomorphisms*

$$0 \longrightarrow G \xrightarrow{f} K \xrightarrow{g} H \longrightarrow 0$$

if $h : H \rightarrow K$ is a homomorphism such that $g \circ h : H \rightarrow H$ is the identity automorphism of H, then $K \cong G \oplus H$.

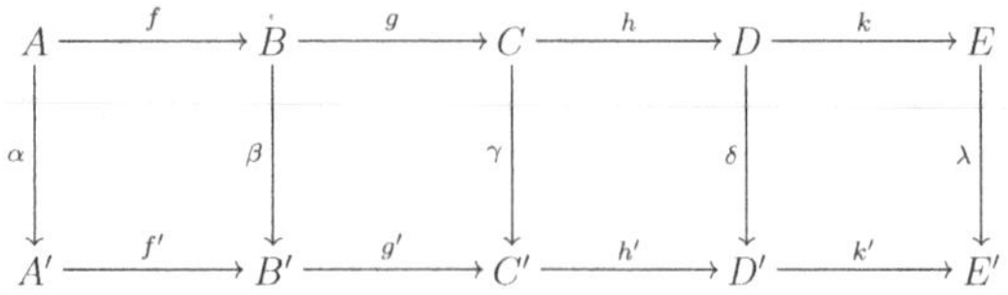

Fig. 1.1 Five lemma diagram

Remark 1.4.6 It is not true that all sequences of the form

$$0 \to G \to K \to H \to 0$$

are always exact. But if H is a free abelian group, then this is true.

Proposition 1.4.7 **(a)** *Given abelian groups G, H, K and homomorphisms f, g with $G \xrightarrow{f} K \xleftarrow{g} H$, where f is an epimorphism and H is a free abelian group, there exists a homomorphism $h : H \to G$ such that $f \circ h = g$.*

(b) *Given an exact sequence of abelian groups and homomorphisms*

$$0 \to G \xrightarrow{f} K \xrightarrow{g} H \to 0$$

where H is a free abelian group, the sequence splits and $K \cong G \oplus H$.

Remark 1.4.8 For the three lemma and four lemma of groups, see the Sect. 1.8.

1.5 Free Product and Tensor Product of Groups

This section presents the concepts of two important products such as free product and tensor product of groups which are frequently used in algebraic topology.

1.5.1 Free Product of Groups

This subsection conveys the concept of free product of groups.

Definition 1.5.1 Let G and H be groups (not necessarily abelian). Their free product denoted by $G * H$ is a group satisfying the following condition:
if there are homomorphisms i and j such that given a pair of homomorphisms $f : G \to K$ and $g : H \to K$ for any group K, there exists a unique homomorphism $h : G * H \to K$ making the diagram in Fig. 1.2 commutative.

For example, $\mathbf{Z} * \mathbf{Z}$ is a free group (of rank 2).

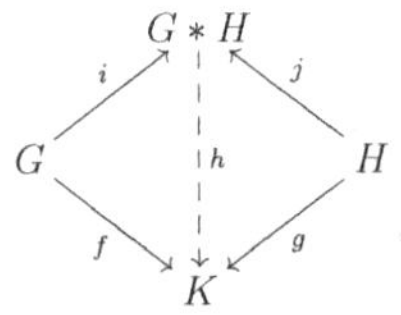

Fig. 1.2 Free product $G * H$

Remark 1.5.2 An alternative description of free product $G * H$ may be given with the help of presentations of groups G and H.

Definition 1.5.3 Let $G = \langle X : R \rangle$ and $H = \langle Y : S \rangle$ be presentations of the groups G and H in which the sets X and Y are generators (and thus the relations R and S) are disjoint. Then a presentation of $G * H$ is given by

$$G * H = \langle X \cup Y : R \cup S \rangle.$$

1.5.2 Tensor Product of Groups

This subsection conveys the concept of tensor products of groups.

Definition 1.5.4 Let G and H be two abelian groups. Their tensor product denoted by $G \otimes H$ is the group defined as the abelian group generated by all pairs of the form (g, h) with $g \in G, h \in H$ satisfying the bilinearity relations $(g + g', h) = (g, h) + (g', h)$ and $(g, h + h') = (g, h) + (g, h')$.

For example, $\mathbf{Z}_m \otimes \mathbf{Z}_n = \mathbf{Z}_{(m,n)}$, where (m, n) is the gcd of m and n: on the other hand, this tensor product is 0, if m and n are relatively prime.

1.6 Torsion Group

This section introduces the concept of torsion group. Let G be an abelian group, and p be a given prime integer. If $f : G \to G$ be the homomorphism defined by $f(x) = px$, then there exists an exact sequence

$$0 \longrightarrow ker\ f \xrightarrow{i} G \xrightarrow{f} G \xrightarrow{h} G \otimes \mathbf{Z}_p \longrightarrow 0, \tag{1.4}$$

where $h(x) = x \otimes 1$.

Definition 1.6.1 The sequence (1.4) is called a short free resolution of G and ker f is called torsion group written Tor $(G, \mathbf{Z}_p)$.

Definition 1.6.2 For each abelian group G there exists an exact sequence

$$0 \longrightarrow R \xrightarrow{i} F \longrightarrow G \longrightarrow 0 \tag{1.5}$$

with F free abelian. For any abelian group H, the torsion product of G and H is defined by Tor $(G, H) = ker(i \otimes I_B)$.

Definition 1.6.3 (*Torsion subgroup*) For any abelian group G,

$$T(G) = \{x \in G : x \text{ has finite order}\}$$

is a subgroup of G, called its torsion subgroup.

Remark 1.6.4 If for every homomorphism $f : G \to G$ of abelian groups, $T(f)$ is defined by

$$T(f) = f|_{T(G)},$$

then T defines a functor from the category of abelian groups and their homomorphisms into itself.

1.7 Actions of Groups

This section presents an action of a group G on a nonempty set X which assigns to each element of G a permutation on the set X and unifies the historical concept of the group of transformations and the modern axiomatic concept of a group. The concept of group actions is useful in the study of algebraic topology.

Definition 1.7.1 A group G with identity e, is said to act on a nonempty set X from the left (or X is said to be a left G-set), if there is a map $\sigma : G \times X \to X$, written as $\sigma(g, x) = g \cdot x$ (or gx) such that for all $x \in X$ and for all $g_1, g_2 \in G$,

(i) $e \cdot x = x$;
(ii) $(g_1 g_2) \cdot x = g_1 \cdot (g_2 \cdot x)$.

Then σ is said to be a left action of G on X and X is said to be a left G-set. Similarly, a right G-set and a right action of G on X are defined.

Remark 1.7.2 If X is a left G-set, then $x \cdot g = g^{-1} \cdot x$ defines a right G-set structure on X. As there is a bijective correspondence between left and right G-set structures it is sufficient to study only one of them.

There may exist different actions of G on X.

Example 1.7.3 Let H be a subgroup of a group G and $h \in H, x \in G$. Then

(i) $h \cdot x = hx$ (left translation);
(ii) $h \cdot x = hxh^{-1}$ (conjugation by h)

are both actions of H on G.

Theorem 1.7.4 *Let X be a left G-set. Then for any g in G, the map $X \to X, x \mapsto g \cdot x$ is a permutation on X.*

Remark 1.7.5 This theorem identifies the notion of a G-set with the notion of a representation of G by permutations on X.

Definition 1.7.6 Given a left G-set X the quotient set X/ρ by an equivalence relation ρ on X defined by $(x, y) \in \rho$ iff $g \cdot x = y$ for some $g \in G$, is called an orbit set and $[x]$ is called an orbit of $x \in X$, denoted by $orb(x)$ or Gx. The action is said to be transitive if $orb(x) = X$.

Definition 1.7.7 Let X be a left G-set. For each $x \in X$, the set $G_x = \{g \in G : g \cdot x = x\}$ is a subgroup of G, called the isotropy group or the stabilizer group of x.

Remark 1.7.8 The isotropy group G_x is a subgroup of G for every $x \in X$. For any two elements in the same isotropy group are conjugate. If $G_x = \{e\}$ for all $x \in X$, then the action of G on X is said to be free.

Theorem 1.7.9 *Let X be a left G-set. Then $|orb(x)|$ is the index $[G : G_x]$ for every $x \in X$. In particular, if G acts on X transitively, then $|X| = [G : G_x]$.*

Theorem 1.7.10 *Let G be a group. Then*

(i) *for each $g \in G$, conjugation $\sigma_g : G \to G, x \mapsto gxg^{-1}$ for all $x \in G$ is an automorphism of G.*

Definition 1.7.11 σ_g defined in Theorem 1.7.10 is called an inner automorphism of G induced by g.

1.8 Modules and Vector Spaces

This section conveys some basic concepts of modules and vector spaces needed for subsequent chapters. Modules and vector spaces play an important role in algebraic topology, specially in the study of homology and cohomology theories. The concept of modules over a ring is a generalization of the concept of abelian groups (which are modules over the ring **Z** of integers) and vector spaces are modules over a field (or a division ring).

1.8.1 Modules

This subsection begins with basic concepts of module theory. We may define a module as an action of a ring on an additive abelian group as follows.

Definition 1.8.1 Let R be a ring. A(left) R-module M is an additive abelian group M together with an action (called scalar multiplication) $\mu : R \times M \to M, (r, x) \mapsto rx$ such that for all $r, s \in R$ and $x, y \in M$,

M(i) $r(x + y) = rx + ry$;
M(ii) $(r + s)x = rx + sx$;
M(iii) $r(sx) = (rs)x$.

If R has an identity element 1, then an R-module M is said to be an unitary R-module if

M(iv) $1x = x$ for all $x \in M$.

If R is a field F (or a division ring), then the F-module M is called an F-vector space or a vector space over F.

A ring R itself may be considered as an R-module by taking scalar multiplication to be the usual multiplication of the ring R.

1.8.2 *Direct Sum of Modules*

Given R-modules $M_1, M_2, \ldots, M_t$, forget for the time being the scalar multiplication and form the direct sum of abelian groups $M_1, M_2, \ldots, M_t$. Then $M_1 \oplus M_2 \oplus \cdots \oplus M_t$ is an R-module, called the direct sum under the scalar multiplication defined by $r(m_1, m_2, \ldots, m_t) = (rm_1, rm_2, \ldots, rm_t)$. In particular, the direct sum of t-copies of R, denoted by $R^{(t)}$, is called a free R-module. Let $1 \in R$. If $e_i \in R^{(t)}$ is the t-copies having 1 in the ith place and 0 elsewhere, then every element $x \in R^{(t)}$ can be expressed uniquely as

$$x = \sum r_i e_i, r_i \in R.$$

The concepts of direct sum and direct product of a finite number of R-modules coincide.

1.8.3 *Tensor Product of Modules*

Tensor product of abelian groups and R-modules are frequently used in algebraic topology.

Let M, N, T be given R-modules. A map $f : M \times N \to T$ is said to be R-bilinear (or bilinear) if

$$f(r_1m_1 + r_2m_2, n) = r_1 f(m_1, n) + r_2 f(m_2, n)$$

and

$$f(m, r_3n_1 + r_4n_2) = r_3 f(m, n_1) + r_4 f(m, n_2)$$

for all $r_i \in R$, for all $m, m_i \in M$ and for all $n, n_i \in N$.

Let R be a commutative ring with 1 and M, N be R-modules. Consider the free R-module F on the set $M \times N$, i.e., the elements of F can be uniquely expressed as a finite linear combinations of the form: $\sum_{i,j} r_{ij}(x_i, y_j)$, where $x_i \in M$, $y_j \in N$ and $r_{ij} \in R$. Let S be the submodule of F generated by all elements of F of the form:

$(x + x', y) - (x, y) - (x', y)$;
$(x, y + y') - (x, y) - (x, y')$;

$(rx, y) - r(x, y)$ and
$(x, ry) - r(x, y)$.

Then the quotient module F/S is called the tensor product of M and N, denoted by $M \otimes N$.
If $\pi : F \to F/S$ is the canonical projection and $i : S \hookrightarrow F$, then $\psi = \pi \circ i : S \to F/S$ satisfies the relations
$\psi((x + x', y) - (x, y) - (x', y)) = 0$;
$\psi((x, y + y') - (x, y) - (x, y')) = 0$;
$\psi((rx, y) - r(x, y)) = 0$;
$\psi((x, ry) - r(x, y)) = 0$.
If we write $\psi(x, y) = x \otimes y$, then the elements $x \otimes y$ of $M \otimes N$ satisfy the following identities:
$(x + x') \otimes y = x \otimes y + x' \otimes y$;
$x \otimes (y + y') = x \otimes y + x \otimes y'$;
$(rx) \otimes y = r(x \otimes y) = x \otimes ry$.
Clearly, the map $\psi : M \times N \to M \otimes N$, $(x, y) \mapsto x \otimes y$ is R-bilinear, and $M \otimes N$ satisfies the following universal properties:

Proposition 1.8.2 *Let M and N be two R-modules. Given an R-module T and an R-bilinear map $f : M \times N \to T$, there exists a unique R-homomorphism $g : M \times N \to T$ such that the diagram in Fig. 1.3 is commutative, i.e., $g \circ \psi = f$.*

Proposition 1.8.3 *Let M, N and T be R-modules. Then there exist isomorphisms such that*

(a) $M \otimes N \cong N \otimes M$;
(b) $(M \otimes N) \otimes T \cong M \otimes (N \otimes T)$;
(c) $R \otimes M \cong M$;
(d) $(M \oplus N) \otimes T \cong M \otimes T \oplus N \otimes T$.

Theorem 1.8.4 (Structure theorem for finitely generated modules over a principal ideal domain) *Let R be a principal ideal domain and M be a finitely generated R-module. Then*

$$M \cong \overbrace{R \oplus R \oplus \cdots \oplus R}^{r \text{ copies}} \oplus R/\langle q_1 \rangle \oplus R/\langle q_2 \rangle \oplus \cdots \oplus R/\langle q_t \rangle$$

for some integer $r \geq 0$ and nonzero nonunit elements $q_1, q_2, \ldots, q_t$ of R are such that $q_1 | q_2 | \ldots | q_t$ (i.e., q_1 divides q_2, q_2 divides q_3, etc.)

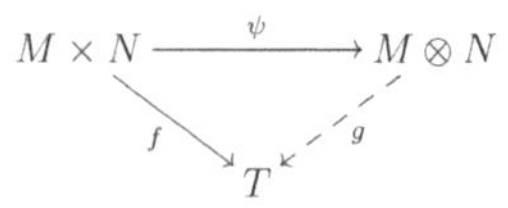

Fig. 1.3 Commutativity of the triangle for uniqueness of g

The elements $q_1, q_2, \ldots, q_t$ uniquely determined up to units are called invariant factors of M and the integer r is called the free rank or Betti number of M.

Corollary 1.8.5 (Fundamental theorem of finitely generated abelian groups) *Let G be a finitely generated abelian group. Then*

$$G \cong \overbrace{\mathbf{Z} \oplus \mathbf{Z} \oplus \cdots \oplus \mathbf{Z}}^{r\ copies}.$$

Theorem 1.8.6 (The three lemma) *Let the diagram in Fig. 1.4 of R-modules and their homomorphisms be commutative with two rows of exact sequences.*

(i) *If α, γ and f' are R-monomorphisms, then β is also an R-monomorphism.*
(ii) *If α, β and g are R-epimorphisms, then β is also an R-epimorphism.*
(iv) *If α, γ are R-isomorphisms, f' is an R-monomorphism and g' is an R-epimorphism, then β is an R-isomorphism.*

Theorem 1.8.7 (The four lemma) *Let the diagram in Fig. 1.5 of R-modules and their homomorphisms be commutative with two rows of exact sequences.*

(i) *If α, γ are R-epimorphisms and δ is an R-monomorphism, then β is an R-epimorphism.*
(ii) *If α is an R-epimorphism and β, δ are R-monomorphism, then γ is an R-monomorphism.*

Theorem 1.8.8 (The five lemma) *Let the diagram in Fig. 1.6 of R-modules and their homomorphisms be commutative with exact rows of exact sequences. If $\alpha, \beta, \delta, \lambda$ are R-isomorphisms, then γ is also an R-isomorphism.*

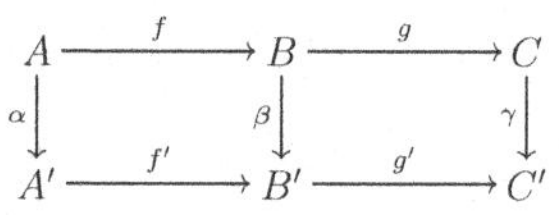

Fig. 1.4 Three lemma diagram

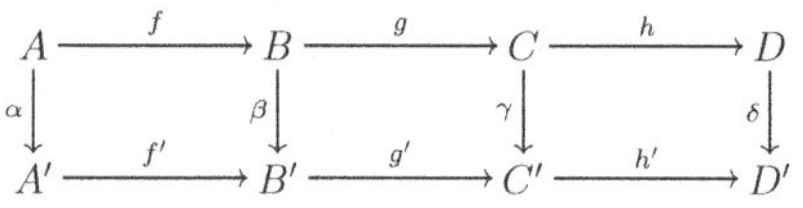

Fig. 1.5 Four lemma diagram

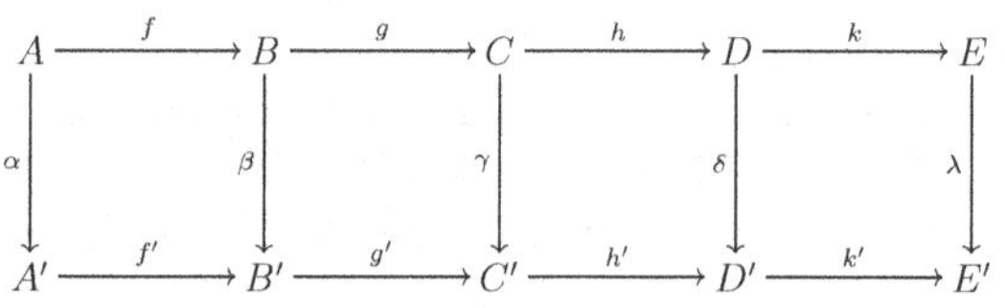

Fig. 1.6 Five lemma diagram

1.8.4 Vector Spaces

This subsection conveys some results of vector spaces. A vector space is a combination of an additive abelian group and a field (division ring) and interlinked by an external law of composition and is a module over a field.

Definition 1.8.9 A vector space or a liner space over a field F is an additive abelian group V together with an external law of composition (called scalar multiplication)

$\mu : F \times V \to V$, the image of (α, v) under μ is denoted by αv such that for all $\alpha, \beta \in F$ and $u, v \in V$

V(1) $1v = v$, where 1 is the multiplicative identity element in F;
V(2) $(\alpha\beta)v = \alpha(\beta v)$;
V(3) $(\alpha + \beta)v = \alpha v + \beta v$;
V(4) $\alpha(u + v) = \alpha u + \alpha v$.

Definition 1.8.10 A vector space V is said to be a direct sum if its subspaces $V_1, V_2, \ldots, V_n$ denoted by $V = V_1 \oplus V_2 \oplus \cdots \oplus V_n$ if every element v of V can be expressed uniquely as $v = v_1 + v_2 + \cdots + v_n$, $v_i \in V_i$, $i = 1, 2, \ldots, n$.

Definition 1.8.11 Let V be a vector space over a field F and U be a subspace of V. Then $(U, +)$ is a subgroup of the abelian group $(V, +)$. The abelian group $(V/U, +)$ is a vector space over F, under the scalar multiplication

$$\mu : F \times V/U \to V/U, (\alpha, v + U) \mapsto \alpha v + U.$$

Definition 1.8.12 The vector space V/U is called the quotient space of V by U and the map $p : V \to V/U$, $v \mapsto v + U$ is called the canonical homomorphism.

Theorem 1.8.13 (Existence Theorem) *Every vector space V over F has a basis.*

Theorem 1.8.14 *Let V be a nonzero vector space over F.*

(i) *Let S be an arbitrary linearly independent subset of V. If S is not a basis of V, then S can be extended to a basis of V.*
(ii) *If V has a finite basis consisting of n elements, then any other basis of V is also finite consisting of n elements.*
(iii) *Cardinality of every basis of V is the same.*

This theorem leads to define the dimension of a vector space.

Definition 1.8.15 Let V be a nonzero vector space over F. The cardinality of every basis of $V$is the same and this common value is called the dimension of V denoted by $\dim V$. If V has finite basis, then V is said to be finite dimensional; otherwise it is said to be infinite dimensional. If $V = \{0\}$, then V is said to be 0-dimensional.

Definition 1.8.16 Let V and W be vector spaces over the same field F. A linear transformation $T : V \to W$ is a mapping such that

L(1) $T(x + y) = T(x) + T(y),\ \forall x, y \in V$ (additive law);
L(2) $T(\alpha x) = \alpha T(x),\ \forall x \in V$ and $\forall \alpha \in F$ (homogeneity law).

Conditions **L(1)** and **L(2)** can be combined together to obtain an equivalent condition

L(3) $T(\alpha x + \beta y) = \alpha T(x) + \beta T(y),\ \forall \alpha, \beta \in F$ and $\forall x, y \in V$.

A linear transformation $T : V \to W$ is said to be an isomorphism if T is a bijection.

Proposition 1.8.17 *Given a vector space V of dimension n, the set $L(V, V)$ of all linear transformations (linear operators) $T : V \to V$ is also a vector space of dimension n^2.*

Definition 1.8.18 The group GL(V) of all nonsingular linear operators $T : V \to V$ is called a general linear group over V.

Remark 1.8.19 The group GL(V) is used to define a linear representation of a group (see Sect. 1.3).

Theorem 1.8.20 (First Isomorphism Theorem) *Let $T : U \to V$ be a linear transformation. Then the vector spaces $U/\ker T$ and $\mathrm{Im} T$ are isomorphic.*

Linear transformations and matrices are closely related.

Proposition 1.8.21 *Given an $m \times n$ matrix M over **R**, the map $T_M : \mathbf{R}^n \to \mathbf{R}^m$, $X \mapsto MX$, is a linear transformation where X is viewed as a column vector of $\mathbf{R}^m$. Conversely, if $B = \{E_1, E_2, \ldots, E_n\}$ is a set of unit column vectors of $\mathbf{R}^n$, then there is a linear transformation $T : \mathbf{R}^n \to \mathbf{R}^m$ with $T(E_j) = M_j$; where M_j is a column vector in $\mathbf{R}^n$ such that M is the matrix whose column vectors are $M_1, M_2, \ldots, M_n$.*

Definition 1.8.22 Let V be an n-dimensional vector space over F and $T : V \to V$ be a linear transformation. An element λ in F is called an eigenvalue of T if there exists a nonzero vector $v \in V$ such that $T(v) = \lambda v$. This vector v (if it exists) is called an eigenvector of T corresponding to the eigenvalue λ.

For a square matrix over F we have an analog of this definition.

Definition 1.8.23 A real division algebra is a finite dimensional real vector space with a bilinear multiplication having a both-sided identity element and satisfying the condition that each nonzero element has a both-sided multiplicative inverse.

Examples of real division algebra:

Example 1.8.24 **(i)** The real numbers **R** form a real division algebra.
(ii) The complex numbers **C** form a real division algebra.
(iii) The real quaternions **H** form a real division algebra.
(iv) The Cayley numbers form a real division algebra (an eight-dimensional non-associative algebra).

Remark 1.8.25 J.F.Adams proved that there are no other examples of real division algebra (see Chap. 17).

1.9 Euclidean Spaces and Some Standard Notations

In mathematical problems, subspaces of an n-dimensional Euclidean space arise frequently. Such spaces are used both in theory and application of topology. Some standard notations used throughout the book are given.

$\emptyset$: empty set
$\mathbf{Z}$: ring of integers (or set of integers)
$\mathbf{Z}_n$: ring of integers modulo n
$\mathbf{R}$: field of real numbers
$\mathbf{C}$: field of complex numbers
$\mathbf{Q}$: field of rational numbers
$\mathbf{H}$: division ring of quaternions
$\mathbf{R}^n$: Euclidean n-space, with $\|x\| = \sqrt{\sum_{i=1}^{n} x_i^2}$ and $\langle x, y\rangle = \sum_{i=1}^{n} x_i y_i$ for $x = (x_1, x_2, \ldots, x_n)$ and $y = (y_1, y_2, \ldots, y_n) \in \mathbf{R}^n$.
$\mathbf{C}^n$: complex n-space
I : $[0, 1]$
$\dot{I}$: $\{0, 1\} \subset I$
I^n : n-cube $= \{x \in \mathbf{R}^n : 0 \leq x_i \leq 1 \text{ for } 1 \leq i \leq n\}$ for $x = (x_1, x_2, \ldots, x_n)$
D^n : n-disk or n-ball $= \{x \in \mathbf{R}^n : \|x\| \leq 1\}$
S^n : n-sphere $= \{x \in \mathbf{R}^{n+1} : \|x\| = 1\} = \partial D^{n+1}$(the boundary of the $(n+1)$-disk D^{n+1})
$\mathbf{R}P^n$: real projective space = quotient space of S^n with x and $-x$ identified for all $x \in S^n$
$\mathbf{C}P^n$: complex projective space = space of all complex lines through the origin in the complex space $\mathbf{C}^{n+1}$
$\bigsqcup$: disjoint union of sets or spaces
$\times, \Pi$: product of sets, groups, modules, or spaces
$\cong$: isomorphism
$\approx$: homeomorphism
iff : if and only if
$X \subset Y$ or $Y \subset X$: set-theoretic containment (not necessarily proper)

1.10 Set Topology

It is assumed that the readers are familiar with the basic concepts of set topology used throughout the book. However, some of them are given in brief.

1.10.1 Topological Spaces: Introductory Concepts

Definition 1.10.1 Let X be a set. It is called a topological space if there is a family τ of subsets of X, called open sets such that

O(1) the union of any number of open sets is an open set;
O(2) the intersection of a finite number of open sets is an open set;
O(3) empty set $\emptyset$ is an open set;
O(4) the whole X is an open set.

Sometimes we write the topological space as (X, τ) to avoid any confusion regarding the topology τ.

Example 1.10.2 **(i)** (Trivial or indiscrete topology) The two subsets $\emptyset$ and the whole set X constitute a topology of X, called trivial topology.
(ii) (Discrete topology) The family of all subsets of X constitutes a topology of X, called the discrete topology of X. This topology is different from trivial topology, if X has more than one element.

Definition 1.10.3 Let (X, τ) be a given topological space. A family of open sets Ω is said to form an open base (basis) of the topology τ if every open set (relative to the topology τ) is expressible as the union of some sets belonging to Ω.

Remark 1.10.4 The unions of all sub-collections of an open base Ω of a topology τ constitute the topology τ. Thus the topology τ is completely determined by an open base Ω. But there may exists different open bases for a particular topology. For example, any topology τ always forms an open base of itself. On the other hand for the discrete topology, the subsets consisting of one point only also form an open base.

Example 1.10.5 (*Natural topology*) The null set $\emptyset$ and all open intervals (a, b), where a and b are rational numbers that form a base of a topology of the set of real numbers **R**. This topology is called the natural topology or usual topology of **R** and the set **R** endowed with this topology is called the real number space.

Definition 1.10.6 The direct product (or product) $X \times Y$ of topological spaces has a natural topology, called the product topology: a subset $U \subset X \times Y$ is open if U is the union of the sets of the form $U_1 \times U_2$, where U_1 is open in X and U_2 is open in Y.

Remark 1.10.7 Let X, Y, and Z be topological spaces. Then a function $f : X \times Y \to Z$ may be considered as a function $f(x, y)$ of two variables, with values in Z and f is continuous if it is continuous jointly in both variables x and y.

Definition 1.10.8 Let (X, τ) be a topological space. A subset V of X is said to be a neighborhood (abbreviated nbd) of a point $p \in X$ if exists an open set $U \in \tau$ such that $p \in U \subset V$.

Definition 1.10.9 (X, τ) be a topological space and A be a subset of X. A point $p \in X$ is said to be limit point of the set A if every nbd of p intersects A in at least one point other than p. The set formed by all the limit points of A is called the derived set of A, denoted by A'.

Definition 1.10.10 Let A be a subset of a topological space (X, τ). The union of all open sets of X contained in A is called the interior of A, denoted by $\text{Int}(A)$. It is sometime denoted by $\mathring{A}$ or by simply $\text{Int}A$.

Remark 1.10.11 The interior of A is the set $\text{Int}(A)$ consisting of all points $a \in A$ for which A is a nbd of a.

Definition 1.10.12 A subset A of a topological space X is said to be dense in X if for any point $x \in X$, any nbd of x contains at least one point from A (i.e., A has nonempty intersection with every nonempty open subset of X). In other words, A is said to be dense in X if the only closed subset of X containing A is X itself.

Remark 1.10.13 The real number space $\mathbf{R}$ with the natural topology has the rational numbers $\mathbf{Q}$ as a countable dense subset. This implies that the cardinality of a dense subset of a topological space may be strictly smaller than the cardinality of the space itself.

It is sometimes convenient to define a topology on a set by the axioms of closed sets.

Definition 1.10.14 A subset $F \subset X$ of a topological space is said to be closed if $X \setminus F\ (= X - F)$ is open.

Remark 1.10.15 A topological space can also be defined in terms of closed sets.

Definition 1.10.16 Let X be a set. It is called a topological space if there is a family of subsets of X, called closed sets such that

C(1) the union of finitely many closed sets is a closed set;
C(2) the intersection of any number of closed sets is a closed set;
C(3) X is a closed set;
C(4) empty set $\emptyset$ is a closed set.

Definition 1.10.17 Let (X, τ) be a topological space and A be a subset of X. The intersection of all closed sets of X containing A is called the closure of A, denoted by $\bar{A}$. If $\bar{A} = X$, then A is a dense set.

Remark 1.10.18 Let (X, τ) be a topological space. Let $A \subseteq X$. Then the closure $\bar{A}$ is given by $\bar{A} = A \cup A'$. where A' is the derived set of A.

Example 1.10.19 **(i)** The set of all rational numbers $\mathbf{Q}$ is dense in $\mathbf{R}$ with usual topology.
(ii) The set of all irrational numbers is also dense in $\mathbf{R}$ with usual topology.

Remark 1.10.20 Let (X, τ) be a topological space and A be a subset of X. Then $\text{Int}(A) = X - \overline{(X - A)}$.

Every subset Y of a topological space X is itself a topological space in a natural way.

Definition 1.10.21 Let X be a topological space and Y be a subset of X. A set $U \subset Y$ is defined to be open in Y if there exists an open set $V \subset X$ such that $V \cap Y = U$. This topology on Y is called the relative topology induced by the topology of X on Y.

1.10.2 Homeomorphic Spaces

This subsection presents the concept of a homeomorphism in topology which is a bijective function that preserves topological structure involved and is analogous to the concept of an isomorphism between algebraic objects such as groups or rings, which is a bijective map that preserves the algebraic structures involved. Its precise definition is now given.

Definition 1.10.22 Let X and Y be topological spaces. A function $f : X \to Y$ is said to be continuous if $f^{-1}(U) \subset X$ is an open set in X for every open set U in Y. A continuous function $f : X \to Y$ is said to be a homeomorphism if f is bijective and $f^{-1} : Y \to X$ is also continuous.

Remark 1.10.23 For a homeomorphism $f : X \to Y$, both f and f^{-1} are continuous means that f not only sends points of X to points of Y in a (1-1) manner, but f also sends open sets of X to open sets of Y in a (1-1) manner. This asserts that X and Y are topologically the same in the sense that a topological property enjoyed by X is also enjoyed by Y and conversely. For example, if $f : X \to Y$ is a homeomorphism, then X is compact (or connected) iff Y is compact (or connected).

Remark 1.10.24 Let X and Y be topological spaces. A bijective function $f : X \to Y$ may not be continuous. For example, let $\mathbf{R}$ be the set of real numbers in its usual topology and $\mathbf{R}_l$ be the same set in the lower topology. Then the identity function $f : \mathbf{R} \to \mathbf{R}_l, x \mapsto x$ is a bijection but not continuous.

Example 1.10.25 **(i)** The open interval $(0, 1)$ and the real line $\mathbf{R}$ with usual topology are homeomorphic.

(ii) A homeomorphism $f : (0, 1) \to \mathbf{R}$ cannot be extended to $I = [0, 1]$.

(iii) The open ball $B = \{x = (x_1, x_2) \in \mathbf{R}^2 : \|x\| < 1\}$ is homeomorphic to the whole plane $\mathbf{R}^2$.

(iv) The open square $A = \{(x, y) \in \mathbf{R}^2 : 0 < \langle x, y\rangle < 1\}$ is homeomorphic to the open ball B defined in (iii).

(v) The cone $A = \{(x, y, z) \in \mathbf{R}^3 : x^2 + y^2 = z^2, z > 0\}$ is homeomorphic to the plane $\mathbf{R}^2$.

(vi) Let S^n be the n-sphere defined by $S^n = \{x \in \mathbf{R}^{n+1} : \|x\| = 1, n \geq 1\}$, $N = (0, 0, \ldots, 1) \in \mathbf{R}^{n+1}$ be the north pole of S^n and $S = (0, 0 \ldots, -1) \in \mathbf{R}^{n+1}$ be the south pole of S^n. Then

- **(a)** $S^n - S$ is homeomorphic to $S^n - N$;
- **(b)** The stereographic projection $f : S^n - N \to \mathbf{R}^n$, defined by $f(x) = \frac{1}{1-x_{n+1}}(x_1, x_2, \ldots, x_n)$, for every $x = (x_1, x_2, \ldots, x_{n+1}) \in S^n - N$ is a homeomorphism.

(vii) A circle minus a point is homeomorphic to a line segment, and a closed arc is homeomorphic to a closed line segment.

[Hint:

(i) The function $f : (0, 1) \to \mathbf{R}, x \mapsto \log \frac{x}{1-x}$ is a homeomorphism.

(ii) Use the fact that any continuous function defined on [0, 1] must be bounded.

(iii) The function $f : B \to \mathbf{R}^2, x \mapsto \frac{x}{(1-\|x\|)}$ is a homeomorphism with its inverse $g : \mathbf{R}^2 \to B, x \mapsto \frac{x}{(1+\|x\|)}$.

(iv) Using (i) there exists a homeomorphism $f : (0, 1) \to \mathbf{R}$. Consider the functions $g : A \to \mathbf{R}^2, (x, y) \mapsto (f(x), f(y))$ and $h : \mathbf{R}^2 \to A$, $(x, y) \mapsto (f^{-1}(x), f^{-1}(y)$. Then g is a homeomorphism. Now use the result that $A \approx \mathbf{R}^2$ and $\mathbf{R}^2 \approx B$.

(v) Consider the functions $f : A \to \mathbf{R}^2, (x, y, z) \mapsto (x, y)$ and $g : \mathbf{R}^2 \to A$, $(x, y) \mapsto (x, y, \sqrt{x^2 + y^2})$. Then f is a homeomorphism with g as its inverse.

(vi)
- **(a)** Let $r : S^n - S \to S^n - N$ be the reflection map defined by $r(x_1, x_2, \ldots, x_{n+1}) = (x_1, x_2, \ldots, -x_{n+1})$. Then r is a homeomorphism.
- **(b)** Let $g : \mathbf{R}^n \to S^n - N$ be the function defined by $g(y_1, y_2, \ldots, y_n) = (ty_1, ty_2, \ldots, ty_n, 1 - t)$, where $t = \frac{2}{(1+y_1^2+y_2^2+\cdots+y_n^2)^{1/2}} \in \mathbf{R}$. Then f is a homeomorphism with g as its inverse.]

Theorem 1.10.26 *Let* $I^n = \{(x_1, x_2, \ldots, x_n) \in \mathbf{R}^n : 0 \leq x_i \leq 1\}$ *be the n-cube and* Int $I^n = \{(x_1, x_2, \ldots, x_n) : 0 < x_i < 1\}$ *be the interior of* I^n *and* $\partial I^n = \dot{I}^n = I^n -$ Int I^n *be the boundary of* I^n. *Then* I^n *is homeomorphic to the n-ball* B^n *in* $\mathbf{R}^n$ *and under this homeomorphism* $\partial I^n = I^n - \text{Int } I^n$ *corresponds to* S^{n-1}.

Proof Let $\hat{I}^n = \{(x_1, x_2, \ldots, x_n) : -1 \leq x_i \leq 1\}$. Then there exists a homeomorphism $f : \hat{I}^n \to I^n$. Under this homeomorphism ∂I^n corresponds to $\partial \hat{I}^n = \{(x_1, x_2, \ldots, x_n) : -1 \leq x_i \leq 1 \text{ and } x_i = \pm 1 \text{ for some } i\}$.
Define $f_1 : \hat{I}^n \to B^n$ and $g_1 : B^n \to \hat{I}^n$ by

$$f_1(x_1, x_2, \ldots, x_n) = \frac{\max(|x_i|)}{\|x\|}(x_1, x_2, \ldots, x_n), \text{ if } x \neq 0$$
$$= 0, \text{ if } x = 0$$

and
$$g_1(x_1, x_2, \ldots, x_n) = \frac{\|x\|}{\max(|x_i|)}(x_1, x_2, \ldots, x_n), \text{ if } x \neq 0$$
$$= 0, \text{ if } x = 0.$$

The continuity at 0 follows from the fact that $\max(|x_i|) \leq \|x\| \leq \sqrt{n} \max(|x_i|)$. Clearly, f_1 and g_1 are inverses to each other. ❑

1.10.3 *Metric Spaces*

This subsection conveys the concept of metric spaces with special reference to Banach spaces.

Definition 1.10.27 A nonempty set X is said to have a metric or a distance function $f : X \times X \to \mathbf{R}$ if for every pair of elements x, y in X

(i) $d(x, y) \geq 0$, equality holds iff $x = y$;
(ii) $d(x, y) = d(y, x)$;
(iii) $d(x, y) + d(y, z) \geq d(x, z)$ for all $z \in X$.

$d(x, y)$ is called the distance between x and y and the pair (X, d) is called a metric space or X is said to be metrized by d.

A metric space X can be made into a topological space in a natural way by defining as open sets all unions of the open balls $\beta_\epsilon(x) = \{y \in X : d(x, y) < \epsilon\}$, for $x \in X$ and $\epsilon > 0$.

We define neighborhoods and limit points in the usual way.

Definition 1.10.28 Let (X, d) be a metric space. A Cauchy sequence in X is a function $f : \mathbf{N} \to X$ such that for every positive real numbers ϵ, there exists a positive integer m such that $d(f(i), f(j)) < \epsilon$ for all $i > m$ and $j > m$.

Definition 1.10.29 A complete metric space is a metric space in which every sequence is convergent.

Example 1.10.30 $[0, 1]$ is a complete metric space but $(0, 1]$ is not so.

Definition 1.10.31 A normed linear space is a vector space X on which a real-valued function $|| \; || : X \to \mathbf{R}$ is called a norm function is defined such that

N(1) $||x|| \geq 0$ and $||x|| = 0$ iff $x = 0$;
N(2) $||x + y|| \leq ||x|| + ||y||$;
N(3) $||\alpha x|| = |\alpha| \, |x||$ for $x, y \in X$ and $\alpha \in \mathbf{R}$ or $\mathbf{C}$.

Remark 1.10.32 A normed linear space is a metric space with respect to the metric induced by the metric defined by $d(x, y) = ||x - y||$.

Definition 1.10.33 A Banach space X is a normed linear space which is complete as a metric space, i.e., every Cauchy sequence in X is convergent.

Definition 1.10.34 A Hilbert space is a complete Banach space X in which function $\langle , \rangle : X \times X \to \mathbf{C}$ is defined satisfying the following conditions:

H(1) $\langle \alpha x + \beta y, z \rangle = \alpha \langle x, z \rangle + \beta \langle y, z \rangle$;
H(2) $\overline{\langle x, y \rangle} = \langle y, x \rangle$;
H(3) $\langle x, x \rangle = ||x||^2$.

Remark 1.10.35 A Hilbert space is a complex Banach space whose norm is defined by an inner product.

1.10.4 Connectedness and Locally Connectedness

This subsection conveys the concepts of connectedness and locally connectedness.

Definition 1.10.36 A topological space X is said to be connected if the only sets which are both open and closed are $\emptyset$ and X.

Remark 1.10.37 Connectedness of topological spaces is an important topological property and is characterized by the following theorem.

Theorem 1.10.38 *A topological space X is connected iff it is not the union of two disjoint nonempty sets.*

Definition 1.10.39 Let X be a topological space, and x be a point of X. Then X is said to be locally connected at x if for every open set V containing x, there exists a connected open set U with $x \in U \subset V$. The space X is said to be locally connected if it is locally connected at x for all $x \in X$.

Example 1.10.40 The Euclidean space $\mathbf{R}^n$ is connected and locally connected for all $n \geq 1$.

Remark 1.10.41 The continuous image of a locally connected space may not be locally connected.

Definition 1.10.42 A path in a topological space X is a continuous map $f : I \to X$ from the closed unit interval I to X.

Definition 1.10.43 A topological space X is said to be path-connected, if any two points of X can be joined by a path.

Remark 1.10.44 A path-connected space is connected. A connected open subset of a Euclidean space is path-connected.

Example 1.10.45 For $n > 0$, the n-sphere S^n is path-connected.

Definition 1.10.46 A topological space X is said to be locally path-connected if for each $x \in X$, and each nbd U of x, there is a path-connected nbd V of x which is contained in U.

Example 1.10.47 The following spaces in real analysis are connected.

(i) The space $\mathbf{R}$ of real numbers;
(ii) Any interval in $\mathbf{R}$;
(iii) $\mathbf{R}^n$;
(iv) Any ball or cube in $\mathbf{R}^n$;
(v) The continuous image of a connected space is connected.

1.10.5 Compactness and Paracompactness

This subsection conveys the concept of compactness which is used throughout the book and that of paracompactness which is specially used in the classification of vector bundles.

Definition 1.10.48 An open covering of a topological space X is a family of $\{U_i\}$ of open sets of X, whose union is the whole set X.

Definition 1.10.49 A topological space X is said to be compact if every open covering of X has a finite subcovering.

Remark 1.10.50 This means that from any open covering $\{U_i\}$ of a compact space X, we can choose finitely many indices i_j, $j = 1, 2, \dots, n$ such that $\bigcup_{j=1}^{n} U_{i_j} = X$. If X is a compact space, every sequence of points x_n of X has a convergent subsequence, which means, every subsequence $x_{n_1}, x_{n_2}, \dots, x_{n_t}, \dots$, converges to a point of X. For metric spaces, this condition is equivalent to compactness.

Proposition 1.10.51 *A compact subspace of a topological space X is closed in X and every closed subspace of a compact space is compact.*

Definition 1.10.52 A topological space X is said to be locally compact if each of its points has a compact neighborhood.

Example 1.10.53 Any compact space, $\mathbf{R}^n$, any discrete space, any closed subset of a locally compact space are locally compact spaces. On the other hand, the space $\mathbf{Q}$ of rational numbers is not locally compact.

Definition 1.10.54 A topological space X is said to be a Baire space if intersection of each countable family of open dense sets in X is dense.

Example 1.10.55 Every locally compact space is a Baire space.

Definition 1.10.56 A topological space X is said be compactly generated if X is a Hausdorff space and each subset A of X satisfying the property that $A \cap C$ is closed for every compact subset C of X is itself closed.

Remark 1.10.57 If X and Y are two topological spaces such that X is locally compact and Y is compactly generated, then their Cartesian product is compactly generated.

Definition 1.10.58 A topological space X is said to be paracompact if every open covering of X has a locally finite subcovering of X.

Example 1.10.59 **(i)** $\mathbf{R}^n$ is paracompact.

(ii) Every closed subspace of a paracompact space but a subspace of a paracompact space is not necessarily paracompact.

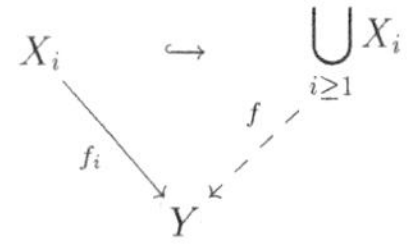

Fig. 1.7 Universal property of the union $\bigcup_{i\geq 1} X_i$ with weak topology

Definition 1.10.60 A topological space X is said to be countably compact if every countable open covering of X has a finite subcovering.

Theorem 1.10.61 (Cantor's intersection theorem) *A topological space X is countably compact iff every descending chain of nonempty closed nonempty sets of X has a nonempty intersection.*

1.10.6 Weak Topology

This subsection presents the concept of weak topology which is used in construction of some important topological spaces. Let $X_1 \subset X_2 \subset X_3 \subset \cdots$ be a chain of closed inclusions of topological spaces. Its union $\bigcup_{i\geq 1} X_i$ as the union of the sets X_i defines a topology by declaring a subset $A \subset \bigcup_{i\geq 1} X_i$ to be closed iff its intersection $A \cap X_i$ is closed in X_i for all $i \geq 1$. This topology is called union topology. It is also called weak topology with respect to the subspaces. For example, $S^\infty = \bigcup_{n=0}^{\infty} S^n$, $\mathbf{R}P^\infty = \bigcup_{n=0}^{\infty} \mathbf{R}P^n$, $\mathbf{C}P^\infty = \bigcup_{n=0}^{\infty} \mathbf{C}P^n$ have the weak topology.

Theorem 1.10.62 *The union $\bigcup_{i\geq 1} X_i$ has the following universal property:*

If a family $\{f_i : X_i \to Y : i \geq 1\}$ of continuous maps is such that $f_{i+1}|_{X_i} = f_i : X_i \to Y$, then there exists a union map $f : \bigcup X_i \to Y$ satisfying the property that $f|_{X_i} = f_i : X_i \to Y$, represented by the commutative diagram as in the Fig. 1.7.

1.11 Partition of Unity and Lebesgue Lemma

This section discusses the concept of 'partition of unity' and states Lebesgue lemma with Lebesgue number. A partition of unity subordinate to a given open covering is an important concept in mathematics.

Definition 1.11.1 Let $\mathcal{U} = \{U_j : j \in J\}$ be an open covering of a topological space X. A partition of unity subordinate to $\mathcal{U}$ consists of a family of functions $\{f_j : X \to I\}$, $j \in J$ such that $f_j|_{(X-U_j)} = 0$ for all j and each $x \in X$ has a neighborhood V with the property $f_j|_V = 0$, except for a finite number of indices j, and $\sum_j f_j(x) = 1$ for all $x \in X$.

Remark 1.11.2 The sum $\sum_j f_j(x)$ is always a finite sum.

Example 1.11.3 If $\{g_j : U_j \to \mathbf{R}\}$ is a family of continuous functions, where $\{U_j : j \in J\}$ is an open covering of a topological space X, then the function $g : X \to \mathbf{R}$ such that $g(x) = \sum_j f_j(x)g_j(x)$ is well defined and continuous.

Paracompactness of a topological space can be characterized with the help of partition of unity.

Theorem 1.11.4 *A topological space X is paracompact iff every open covering $\mathcal{U}$ of X admits a partition of unity subordinate to $\mathcal{U}$.*

Proof See (Dugundji 1966). ❑

1.11.1 Lebesgue Lemma and Lebesgue Number

Lebesgue Lemma is used to prove many important results. This lemma is also called Lebesgue Covering Lemma.

Lemma 1.11.5 (Lebesgue) *Let X be a compact metric space. Given an open covering $\{U_\alpha : \alpha \in A\}$ of X, there exists a real number $\delta > 0$ (called Lebesgue number of $\{U_\alpha\}$), such that every open ball of radius less than δ lies in some element of $\{U_\alpha\}$.*

1.12 Separation Axioms, Urysohn Lemma, and Tietze Extension Theorem

This section imposes certain conditions on the topology to obtain some particular classes of topological spaces initially used by P.S. Alexandroff (1896–1982) and H. Hopf (1894–1971). Such spaces are important objects in algebraic topology. Moreover, this section presents Urysohn Lemma and Tietze Extension Theorem which are used in this book.

Definition 1.12.1 A topological space (X, τ) is said to be a

(i) T_1-space (due to Frechet) if for every pair of distinct points p, q in X, there exist two open sets U, V such that

$$p \in U, q \in V, p \notin V, \text{ and } q \notin U.$$

In other words, every pair of distinct points is weakly separated in (X, τ): equivalently, for every pair of distinct points p, q in X, there exist a neighborhood of p which does not contain q, and a neighborhood of q which does not contain p.

(ii) Hausdorff space (due to Hausdorff) if any two distinct points are strongly separated in (X, τ):
equivalently, distinct points have disjoint neighborhoods.

(iii) Regular space (due to Vietoris) if any closed set F and any point $p \notin F$ are always strongly separated in (X, τ).

(iv) Normal space (due to Tietze) if any two disjoint closed sets are strongly separated in (X, τ), equivalently, each pair of disjoint closed sets have disjoint neighborhoods.

Remark 1.12.2 It is not true that a nonconstant real-valued continuous function can always be defined on a given space. But on normal spaces there always exist nonconstant real-valued continuous functions. Urysohn lemma characterizes normal spaces by real-valued continuous functions.

Lemma 1.12.3 (Urysohn) *A topological space (X, τ) is normal if and only if every pair of disjoint closed sets P, Q in (X, τ) are separated by a continuous real-valued function f on (X, τ), such that*

$$f(x) = \begin{cases} 0, \textit{ for all } x \in P \\ 1, \textit{ for all } x \in Q \end{cases}$$

and

$$0 \leq f(x) \leq 1 \textit{ for all } x \in X.$$

Theorem 1.12.4 (Tietze extension theorem) *If E is a closed subspace of a normal space X, then every continuous map $g : E \to I$ has a continuous extension over X.*

1.13 Identification Maps, Quotient Spaces, and Geometrical Construction

This section presents geometrical construction of some quotient spaces. The concept of quotient spaces is very important in topology and geometry to formalize the intuitive idea of 'gluing' or 'identifying' or 'pasting' mathematically.

Definition 1.13.1 Let X be a topological space and ρ be an equivalence relation on X. If $p : X \to X/\rho, x \mapsto [x]$ is the natural surjective map, then the collection Ω of all subsets $U \subset X/\rho$ such that $p^{-1}(U)$ is an open set of X, forms the largest topology on X/ρ such that p is continuous. The set X/ρ is called a quotient space of X, with the quotient topology Ω and p is called an identification map. The process of identification is sometimes called 'gluing' or 'pasting'.

Example 1.13.2 Let I be the closed unit interval and ρ be an equivalence relation on I such that $[0] = [1] = \{0, 1\}$ and $[x] = \{x\}$ for $0 < x < 1$. Then I/ρ is the quotient space homeomorphic to the circle S^1. In other words, S^1 is obtained from I by identifying the end points 0 and 1 of I.

Example 1.13.3 If we identify all the points of the circumference of a disk D^2, then the resulting quotient space is homeomorphic to the sphere S^2.

Example 1.13.4 Let X be a topological space. If we define an equivalence relation ρ on $X \times I$ by $(x, t)\rho(y, s)$ iff $t = s = 1$, then the resulting quotient space, is called the cone over X, denoted by CX. Thus $CX = X \times I/X \times \{1\}$ and this identified point $[x, 1]$ is its vertex. Thus if we consider I as a pointed space with base point 0, then the reduced cone CX over X is defined to be the quotient space obtained from $X \times I$ by collapsing $X \times 0 \cup x_0 \times I$ to a point, i.e., $CX = X \times I/(X \times 0) \cup x_0 \times I$. We use $[x, t]$ to denote the point of CX corresponding to the point $(x, t) \in X \times I$ under the identification map

$$p : X \times I \to CX.$$

The space X is embedded as a closed subspace of CX by the map

$$x \mapsto [x, 1].$$

Example 1.13.5 For topological spaces X and Y, every continuous map $f : X \times I \to Y$ satisfying the condition $f(x, 1) = y_0 \in Y$ for all $x \in X$, induces a continuous map $\tilde{f} : CX \to Y, [x, t] \mapsto f(x, t)$. In particular, if $X = S^n$ and $Y = D^{n+1}$, $f : S^n \times I \to D^{n+1}, (x, t) \mapsto (1 - t)x$, then $\tilde{f} : CS^n \to D^{n+1}, [x, t] \to (1 - t)x$ is a homeomorphism. Thus D^{n+1} is the cone over S^n with vertex $f(x, 1) = y_0$.

Example 1.13.6 Let $I \times I$ be the unit square.

(a) If we identify the point $(0, t)$ with $(1, t)$ of the square $I \times I$ for all $t \in I$, then the quotient space obtained by such identification is the Euclidean cylinder with two disjoint circles as their boundaries. If the top and bottom circles of a cylinder are glued together, the resulting quotient space is called is called the 'torus' as shown in Fig. 1.8.
Thus the torus is the quotient space obtained from unit square $I \times I$ by identifying $(t, 0)$ with $(t, 1)$ and also $(0, t)$ with $(1, t)$.

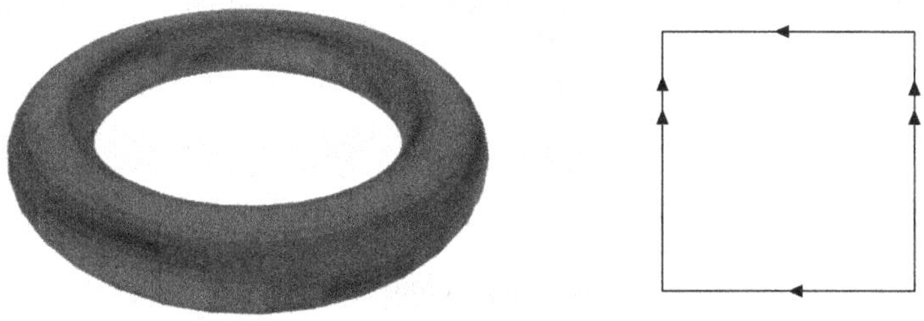

Fig. 1.8 Torus as the quotient space of unit square

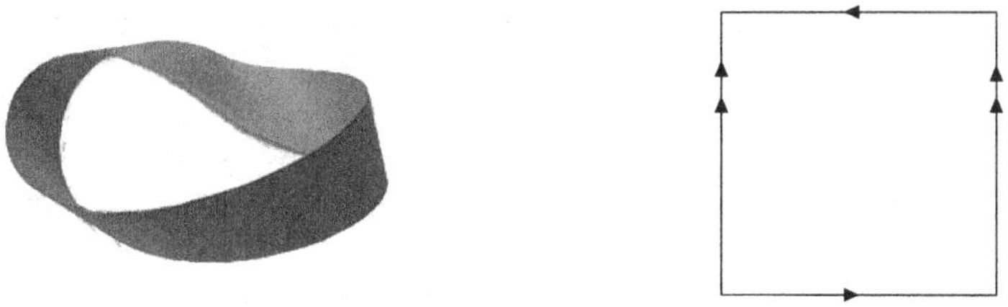

Fig. 1.9 Möbius band as the quotient space of unit square

(b) If we identify the point $(0, t)$ with $(1, 1-t)$ of the square $I \times I$ for all $t \in I$, then the resulting quotient space is homeomorphic to the space in $\mathbf{R}^3$, called Möbius band (or strip) named after A.F. Möbius (1790–1868) who invented it 1858. It is described as shown in Fig. 1.9.
Möbius band is the quotient space $M = (0, 1) \times [0, 1]/(x, 0) \sim (x, 1)$. It can be embedded in $\mathbf{R}^3$, because M is homeomorphic to a subspace of $\mathbf{R}^3$. It is a non-orientable surface with only one side and only one boundary. It can be realized as a ruled surface and is used as conveyor belts.

(c) If we identify the point $(0, t)$ with $(1, t)$ and the point $(t, 0)$ with $(1, 1-t)$ of the square $I \times I$ for all $t \in I$, then the resulting quotient space is called the Klein bottle named after F. Klein (1849–1925) who invented it in 1882. It is described as shown in Fig. 1.10.
Klein Bottle is also defined as the quotient space $K = (S^1 \times I)/(x, 0) \sim (x^{-1}, 1)$.
As Klein Bottle is the continuous image of $I \times I$, it is a compact and connected space but it cannot be embedded in $\mathbf{R}^3$, because it is not homeomorphic to any subspace of $\mathbf{R}^3$, but it can be embedded in $\mathbf{R}^4$. Klein bottle is a closed manifold in the sense that it is a compact manifold without boundary.

(d) The quotient space obtained from $I \times I$ by identifying $(t, 0)$ with $(1-t, 1)$ and also $(0, t)$ with $(1, 1-t)$ as shown in Fig. 1.11 is called the real projective plane $\mathbf{R}P^2$. It cannot be embedded in $\mathbf{R}^3$.

(e) The quotient space obtained from square by identifying its boundary to a point is the 2-sphere S^2.

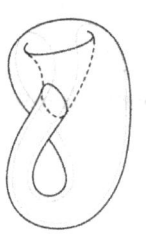

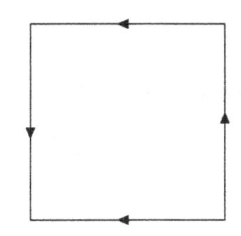

Fig. 1.10 Klein bottle as the quotient space of the unit square

Fig. 1.11 Real projective plane $\mathbf{R}P^2$

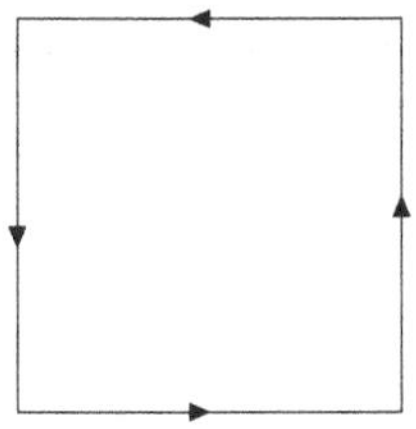

Remark 1.13.7 **(i)** The surface of a cylinder can be considered as a disjoint union of a family of line segments parametrized continuously by points of a circle. The Möbious can be considered in a similar way.

(ii) The two-dimensional torus embedded in $\mathbf{R}^3$ can be considered as a union of circle parametrized by points of another circle.

Example 1.13.8 (*Mapping Cylinder*) Let X and Y be topological spaces and $f : X \to Y$ be continuous. Let $(X \times I) \sqcup Y$ denote the disjoint union of topological spaces $X \times I$ and Y. Then both $X \times I$ and Y are open sets of $(X \times I) \sqcup Y$. If we define an equivalence relation ρ on $(X \times I) \sqcup Y$ by $(x, t)\rho y$ iff $y = f(x)$ and $t = 1$, then the quotient space $M_f = ((X \times I) \sqcup Y)/\rho$ is called the mapping cylinder of f. Thus M_f is the space obtained from Y and $(X \times I)/(x_0 \times I)$ by identifying for each $x \in X$, the points $(x, 1)$ and $f(x)$ as shown in Fig. 1.12, in which the thick line is supposed to be identified to a point (the base point of M_f). We denote $(x, t)\rho$ in M_f by $[x, t]$ and $y\rho$ in M_f by $[y]$. Then $[x] = [x, 1] = [f(x)]$, $\forall x \in X$. The space Y is embedded in M_f under the map $y \mapsto [y]$. In particular, if Y is a one-point space, then $f : X \to Y$ is a constant map and M_f is CX, the cone over X.

Example 1.13.9 Let X and Y be topological spaces with base points x_0 and y_0, respectively. Given a continuous map $f : X \to Y$, the mapping cone as shown in Fig. 1.13.

C_f is the quotient space obtained from Y and CX by identifying the point $[x, 1]$ of CX with the point $f(x)$ of Y for all $x \in X$. The base point of C_f is the point to which $[x_0, t]$ and y_0 are identical for all $t \in I$.

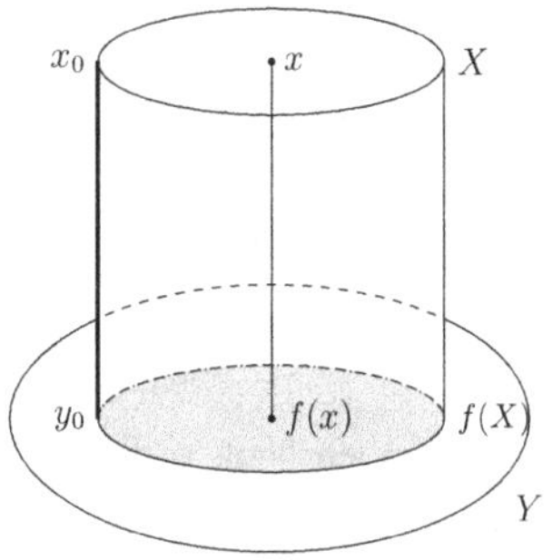

Fig. 1.12 Mapping cylinder M_f of f

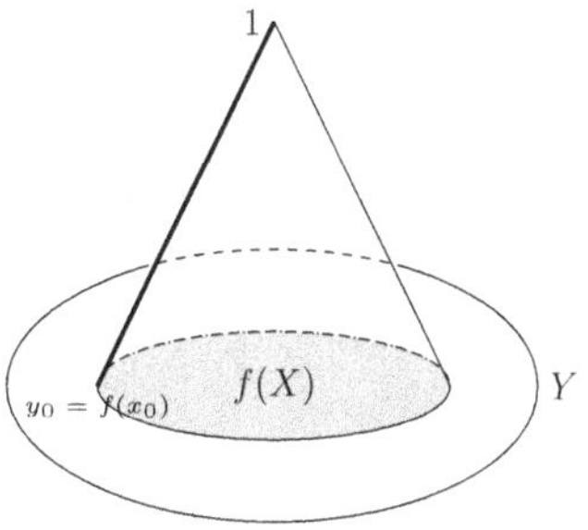

Fig. 1.13 Mapping cone C_f of f

Example 1.13.10 (*Wedge*) Let (X, x_0) and (Y, y_0) be two pointed topological spaces. Their wedge (or one-point union) $X \vee Y$ is the quotient space of their disjoint union $X \sqcup Y$ in which the base points are identified. In general, if X_i is a collection of disjoint spaces, with base point $x_i \in X_i$, then their wedge (or one-point union) $\bigvee_{i \in I} X_i$ is the quotient space X/X_0, where $X = \bigsqcup_{i \in I} X_i$ and X_0 is the subspace of X consisting of all base points x_i; the base point of $\bigvee_{i \in I} X_i$ is the point corresponding to X_0. In other words, $\bigvee_{i \in I} X_i$ is the space obtained from X by identifying together the base point x_i.

Example 1.13.11 (*Smash product*) Let X and Y be two pointed spaces with base points x_0 and y_0 respectively. Then their smash product (or reduced product) $X \wedge Y$ is defined to be the quotient space $X \times Y/(X \vee Y)$. We may think $X \wedge Y$ as a reduced version of $X \times Y$ obtained by collapsing $X \vee Y$ to a point.

Example 1.13.12 (*Reduced suspension space*) Let X be a pointed topological space with base point x_0. Then the suspension of X, denoted by ΣX, is defined to be the quotient space of $X \times I$ in which $(X \times 0) \cup (x_0 \times I) \cup (X \times 1)$ is identified to a single point. It is sometimes called the reduced suspension space. If $(x, t) \in X \times I$, we use $[x, t]$ to denote the corresponding point of ΣX under the identification map $f : X \times I \to \Sigma X$ such that $[x, 0] = [x_0, t] = [x', 1]$ for all $x, x' \in X$ and for all $t \in I$. The point $[x_0, 0] \in \Sigma X$ is also denoted by x_0. Thus ΣX is a pointed space with base point x_0 and $\Sigma X = X \wedge S^1$. In particular, $\Sigma S^n \approx S^{n+1}$ for $n \geq 0$. Moreover, if $f : X \to Y$ is a base point preserving continuous map, then $\Sigma f : \Sigma X \to \Sigma Y$ is defined by $\Sigma f([x, t]) = [f(x), t]$.

Proposition 1.13.13 *Let Y be a quotient space of X with quotient map $p : X \to Y$ and $f : Y \to Z$ be a map from Y to some space Z. Then f is continuous iff the map $f \circ p : X \to Z$ is continuous.*

1.14 Function Spaces

This section introduces the concept of function spaces topologized by the compact open topology. Function spaces play an important role in topology and geometry.

Definition 1.14.1 Let X and Y be topological spaces and Y^X(or $F(X, Y)$) be the set of all continuous functions $f : X \to Y$. Then a topology, called compact open topology, can be endowed on $F(X, Y)$ by taking a subspace for the topology all sets of the form $V_{K,U} = \{f \in F(X, Y) : f(K) \subset U\}$, where $K \subset X$ is compact and $U \subset Y$ is open.

Let $E : Y^X \times X \to Y$ be the evaluation map defined by $E(f, x) = f(x)$. Then given a function $h : Z \to Y^X$, the composite

$$\psi : Z \times X \xrightarrow{h \times 1_d} Y^X \times X \xrightarrow{E} Y$$

i.e., $\psi = E \circ (h \times 1_d) : Z \times X \to Y$ is a function.

Theorem 1.14.2 (Theorem of Exponential Correspondence) *If X is a locally compact Hausdorff space and Y, Z are topological spaces, then a function $f : Z \to Y^X$ is continuous if and only if $E \circ (f \times 1_d) : Z \times X \to Y$ is continuous.*

Theorem 1.14.3 (Exponential Law) *If X is a locally compact Hausdorff space, Z is a Hausdorff space, and Y is a topological space, then the function $\psi : (Y^X)^Z \to Y^{Z \times X}$ defined by $\psi(f) = E \circ (f \times 1_d)$ is a homeomorphism.*

Proposition 1.14.4 *If X is a compact Hausdorff space and Y is metrized by a metric d, then the space Y^X is metrized by the metric d' defined by*

$$d'(f, g) = \sup\{d(f(x), g(x)) : x \in X\}.$$

1.15 Manifolds

This section defines manifolds which form an important class of geometrical objects in topology. An n-manifold is a Hausdorff topological space which looks locally like Euclidean n-space $\mathbf{R}^n$, but not necessarily globally. A local Euclidean structure to manifold by introducing the concept of a chart is utilized to use the conventional calculus of several variables. Due to linear structure of vector spaces, for many applications in mathematics and in other areas it needs generalization of metrizable vector space, maintaining only the local structure of the latter. On the other hand, every manifold can be considered as a (in general nonlinear) subspace of some vector space. Both aspects are used to approach the theory of manifolds. Since dimension of a vector space is a locally defined property, a manifold has a dimension. Our study is confined to finite dimensional manifolds (although there are infinite dimensional manifolds).

We are familiar with curves and surfaces. Manifolds are generalizations of curves and surfaces to arbitrary dimensional objects. The concept of manifolds can be traced to the work of B. Riemann (1826–1866) on differential and multivalued functions. A curve in $\mathbf{R}^3$ is parametrized locally by a single number t as $(x(t), y(z), z(t))$, on the other hand two numbers u and v parametrize a surface as $(x(u, v), y(u, v), z(u, v))$. A curve is considered locally homeomorphic to $\mathbf{R}^1$ (real line space) and a surface to $\mathbf{R}^2$. A manifold in general, is a topological space which is locally homeomorphic to $\mathbf{R}^n$ for some n. More precisely, if every point of a topological space M has a neighborhood homeomorphic to an open subset of $\mathbf{R}^n$, we call M an n-dimensional topological manifold.

Definition 1.15.1 An n-dimensional (topological) manifold or an n-manifold M is a Hausdorff space with a countable basis such that each point of M has a neighborhood that is homeomorphic to an open subset of $\mathbf{R}^n$. An one-dimensional manifold is called a curve and a two-dimensional manifold is called a surface.

For example, S^2, torus, $\mathbf{R}P^2$ are surfaces. All manifolds M in this book are assumed to be paracompact to ensure that M is a separable metric space.

Definition 1.15.2 An n-dimensional differentiable manifold M is a Hausdorff topological space having a countable open covering $\{U_1, U_2, \ldots\}$ such that

DM(1) for each U_i, there is homeomorphism $\psi_i : U_i \to V_i$, where V_i is an open disk in $\mathbf{R}^n$;

DM(2) if $U_i \cap U_j \neq \emptyset$, the homeomorphism $\psi_{ji} : \psi_j \circ \psi_i^{-1} : \psi_i(U_i \cap U_j) \to \psi_j(U_i \cap U_j)$ is a differentiable map.

(U_i, ψ_i) is called a local chart of M and $\{(U_i, \psi_i)\}$ is a set of local charts of M.

Example 1.15.3 $\mathbf{R}^n$, S^n, $\mathbf{R}P^n$ are n-dimensional differentiable manifolds.

Example 1.15.4 $\mathbf{C}P^n$ is a 2n-dimensional differentiable manifold.

Definition 1.15.5 A Hausdorff space M is called an n-dimensional manifold with boundary ($n \geq 1$) if each point of M has a neighborhood homeomorphic to the open set in the subspace $x_1 \geq 0$ of $\mathbf{R}^n$.

Example 1.15.6 The n-dimensional disk D^n is an n-manifold with boundary.

Remark 1.15.7 Let $\mathcal{S} = \{(U_i, \psi_i)\}$ be a set of local charts of a differentiable manifold M. Then $\mathcal{S}$ is said to be a differentiable structure on M. Every subset of $\mathcal{S}$ which satisfies $M = \cup U_i$, **DM(1)** and **DM(2)**, is called a basis for the differential structure $\mathcal{S}$.

Example 1.15.8 (*Stiefel manifold*) Any ordered set of r ($r \leq n$) independent vectors in the Euclidean n-space $\mathbf{R}^n$ is called an r-frame. Let $V_r(\mathbf{R}^n)$ be the set of (orthonormal) r-frames in $\mathbf{R}^n$. Then $V_r(\mathbf{R}^n)$ is the subspace of $(v_1, v_2, \ldots, v_r) \in (S^{n-1})^r$ such that $\langle v_i, v_j \rangle = \delta_{ij}$ (Kronecker delta). Since $V_r(\mathbf{R}^n)$ is a closed subset of a compact space, it is also compact. Corresponding to each r-frame $(v_1, v_2, \ldots, v_r) \in (S^{n-1})^r$, there exists an associated r-dimensional subspace $\langle v_1, v_2, \ldots, v_r \rangle$ with a basis $\{v_1, v_2, \ldots, v_r\}$. Each r-dimensional subspace of $\mathbf{R}^n$ is of the form $\langle v_1, v_2, \ldots, v_r \rangle$. The manifold $V_r(\mathbf{R}^n)$ is called the Stiefel manifold of (orthonormal) r-frames in $\mathbf{R}^n$. It may be considered as the manifold of all orthonormal $(r-1)$ frames tangent to S^{n-1}. In particular, for $r = 2$, $V_2(\mathbf{R}^n)$ is the manifold of unit vectors tangent to S^{n-1}. The orthogonal (real) group $O(n, \mathbf{R})$ acts transitively on $\mathrm{V}_r(\mathbf{R}^n)$ with isotropy group $O(n-r, \mathbf{R})$ and hence $O(n, \mathbf{R})/O(n-r, \mathbf{R}) \approx \mathrm{V}_r(\mathbf{R}^n)$.

Example 1.15.9 (*Grassmann manifold*) An r-dimensional ($r \leq n$) linear subspace of $\mathbf{R}^n$ is called an r-plane. Let $G_r(\mathbf{R}^n)$ be the set of r-planes of $\mathbf{R}^n$ through the origin, with the quotient topology defined by, the identification map

$$V_r(\mathbf{R}^n) \to G_r(\mathbf{R}^n), (v_1, v_2, \ldots, v_r) \mapsto \langle v_1, v_2, \ldots, v_r \rangle.$$

Then $G_r(\mathbf{R}^n)$ is a compact space. Clearly, $V_1(\mathbf{R}^n) = S^{n-1}$ and $G_1(\mathbf{R}^n) = \mathbf{R}P^{n-1}$. The natural inclusion $G_r(\mathbf{R}^n) \subset G_r(\mathbf{R}^{n+1})$ gives $G_r(\mathbf{R}^\infty) = \bigcup_{r \leq n} G_r(\mathbf{R}^n)$ with the induced topology. For $g \in O(n, \mathbf{R})$ and $V \in \mathrm{G_r}(\mathbf{R}^n)$, $g(V)$ is another r-plane in $\mathbf{R}^n$. Any r-frame is taken into any other by some $g \in O(n, \mathbf{R})$, so the same is also true of r-plane. Hence $O(n, \mathbf{R})$ acts transitively on $\mathrm{G_r}(\mathbf{R}^n)$. The isotropy group of the standard $\mathbf{R}^r \subset \mathbf{R}^n$ is $O(r, \mathbf{R}) \times O(n-r, \mathbf{R})$. Hence

$$\mathrm{G_r}(\mathbf{R}^n) \approx O(n, \mathbf{R})/O(r, \mathbf{R}) \times O(n-r, \mathbf{R}).$$

Hence we get the following homeomorphisms:

$$O(n, \mathbf{R})/O(1, \mathbf{R}) \times O(n-1, \mathbf{R}) \approx \mathbf{R}P^{n-1}$$

$$U(n, \mathbf{C})/U(1, \mathbf{C}) \times U(n-1, \mathbf{C}) \approx \mathbf{C}P^{n-1}$$

and

$$S_p(n, \mathbf{H})/S_p(1, \mathbf{H}) \times S_p(n-1, \mathbf{H}) \approx \mathbf{H}P^{n-1}$$

where $O(n, \mathbf{R})$, $U(n, \mathbf{C})$ and $S_n(n, \mathbf{H})$ represent the orthogonal group over $\mathbf{R}$, the unitary group over $\mathbf{C}$ and sympletic group over quaternions $\mathbf{H}$ respectively (see Appendix A).

Example 1.15.10 (The Mbius strip or Mbius band) It is a surface with only one side and only one boundary. The Mbius strip has the mathematical property of being non-orientable.

Remark 1.15.11 Throughout the book the terms 'continuous function' and 'continuous map' (or 'map') are interchangeable in the context of topological spaces and the term 'space' means a topological space unless otherwise stated.

1.16 Exercises

1. If $f : I \to \mathbf{R}^n$ is a one-one continuous function, show that $f(I)$ and I are homeomorphic spaces.
[Hint: I is compact and $\mathbf{R}^n$ is a metric space and hence Hausdorff. The function $f : I \to f(I) \subset \mathbf{R}^n$ is onto and by hypothesis one-one continuous. Hence $f^{-1} : f(I) \to I$ exists. Let A be a closed subspace of I. Then A is also compact and $(f^{-1})^{-1}(A) = f(A)$ is a closed subset of $f(I)$, which is also Hausdorff and hence $f(A)$ is closed.]

2. Show that the family of open disks form an open base of the topology of the Euclidean plane.

3. Show that

(a) the spaces $\mathbf{R}$ and $\mathbf{R}^n (n > 1)$ are not homeomorphic.
[Hint: If possible, there exists a homeomorphism $f : \mathbf{R}^n \to \mathbf{R}$. If $x \in \mathbf{R}^n$, then the spaces $\mathbf{R}^n - \{x\}$ and $\mathbf{R} - \{f(x)\}$ must be homeomorphic. But this is not true, because $\mathbf{R}^n - \{x\}$ is connected but $\mathbf{R} - \{f(x)\}$ is not so.]

(b) the spaces I and $I^n (n > 1)$ are not homeomorphic but there exists a bijection between them.

4. Show that a space X is connected iff all continuous functions $f : X \to Y = \{0, 1\}$ are constant functions $f(x) = 0$ or $f(x) = 1$.
[Hint: Use the result that the continuous image of a connected space is connected.]

5. Show that

(i) the continuous image of a path-connected space is path-connected;

(ii) a space which is connected and locally path-connected is path-connected.

6. Let I^n be the n-cube and ∂I^n be its boundary. Show that the spaces $\mathbf{R}^n$ and $I^n - \partial I^n$ are homeomorphic.

[Hint: By Example 1.10.25(i), $I - \partial I = (0, 1)$ is homeomorphic to $\mathbf{R}^1$. Use this result to show that $I^n - \partial I^n$ and $\overbrace{(I - \partial I) \times \cdots \times (I - \partial I)}^{n\text{-products}}$ are homeomorphic.]

7. **(Topologist's sine curve)** Let $Z = \{(x, y) \in \mathbf{R}^2 : y = \sin \frac{1}{x}, 0 < x \leq \pi\}$. The map $f : (0, \pi) \to Z, x \mapsto (x, \sin \frac{1}{x})$ is continuous. Since $(0, \pi)$ is connected and Z is the continuous image of f, Z is connected. This space $T = Z \cup \{(0, y) : |y| \leq 1\}$ is known as 'Topologist's sine curve.'
Show that

(a) if a subset B of T is such that $Z \subseteq B \subseteq T$, then B is connected;
(b) in particular, if B is obtained by adjoining an additional point $(0, 1/2)$(say), or a part of the y-axis lying between the points $(0, -1)$ and $(0, 1)$ to Z, then B is connected.

8. Let X be a topological space and A be a subset of X. Show that A is connected iff its closure $\bar{A}$ is the same as A.

9. Show that every topological space with a connected dense subset is itself connected.

10. Show that

(a) $\mathbf{R}^n$ is connected;
(b) any n-ball D^n or n-cube I^n in $\mathbf{R}^n$ is connected;
(c) the n-sphere S^n in $\mathbf{R}^{n+1}$ is connected for $n > 0$.
[Hint (c): $\mathbf{R}^{n+1} - \{0\}$ is connected. Consider the map

$$f : \mathbf{R}^{n+1} - \{0\} \to S^n, (x_1, x_2, \ldots, x_{n+1})$$
$$\mapsto \left(\frac{x_1}{\sqrt{\sum x_i^2}}, \frac{x_2}{\sqrt{\sum x_i^2}}, \ldots, \frac{x_{n+1}}{\sqrt{\sum x_i^2}} \right).$$

Then f is continuous and surjective.]

11. Show that the product of two compact spaces is compact.

12. Show that

(i) $GL(n, \mathbf{R})$ is a subspace of $\mathbf{R}^{n^2}$ with the relative topology;
(ii) $GL(n, \mathbf{R})$ is not connected.

13. Let X be a metric space. Show that the following three conditions on X are equivalent:

(i) X is compact (Heine–Borel property);
(ii) X is countably compact, i.e., every infinite subset of X has a limit point (Bolzano–Weierstress property);
(iii) X is sequentially compact, i.e., every sequence in X has a convergence subsequence.

14. Show that the connectedness and compactness of topological spaces are topological properties in the sense that if $f : X \to Y$ is a homeomorphism, then

(i) X is connected iff Y is connected;
(ii) X is compact iff Y is compact.

15. Let X and Y be topological spaces and $f : X \to Y$ be a continuous and surjective. Show that if X is compact, then Y is also so.
[Hint: Let $\{V_j\}_{j\in J}$ be an open covering of Y. Then $\{f^{-1}(V_j) : j \in J\}$ is an open covering of X. Since X is compact, there exists a finite subcovering $\{f^{-1}(V_1), \dots, f^{-1}(V_r)\}$. Again since f is surjective, $f(f^{-1}(V_k)) = V_k$, for $k = 1, 2, \dots, r$ and $\bigcup_{k=1}^{r} V_k = Y$, because $\bigcup_{k=1}^{r} f^{-1}(V_k) = X$.]

16. Let $\mathbf{R}^+$ be the space of positive real numbers topologized as a subset of $\mathbf{R}^1$. The space $\mathbf{R}^+ \times S^n$ is in the product topology. Show that the map

$$f : \mathbf{R}^{n+1} - \{0\} \to \mathbf{R}^+ \times S^n, (x_1, x_2, \dots, x_{n+1})$$
$$\mapsto \left(\|x\|, \left(\frac{x_1}{||x||}, \dots, \frac{x_{n+1}}{||x||}\right)\right), \text{ where } ||x|| = \left(\sum_{i=1}^{n+1} x_i^2\right)^{1/2},$$

is a homeomorphism.

17. Let X be a regular space and A be a proper subspace of X. Show that

(i) the quotient space X/A is a Hausdorff space;
(ii) if X is a normal space and A is closed, then X/A is normal.

18. Let G be a group and H be (not necessarily abelian) subgroup of G and G/H be the set of all left cosets of H in G. Now G acts on G/H by the action

$$G \times G/H \to G/H, (g, g'H) \mapsto gg'H.$$

Show that this action is transitive and H is the stabilizer of the coset H.

19. Let a group G act on the set$X$ transitively and H be the stabilizer group of a point. Show that X is G-isomorphic to G/H, the set of all left cosets of H in G on which G acts by left translation.
[Hint Let $H = G_{x_0}$ be the stabilizer group of a point $x_0 \in X$. As the action is transitive, for each $x \in X$, $\exists$ an element $g_x \in G$ such that $g_x x_0 = x$. Define a map

$$\psi : X \to G/H, x \mapsto g_x H.$$

Then ψ is a G-isomorphism.]

20. If H and K are subgroups of a group G, show that the G-sets G/H and G/K are G-isomorphic iff H and K are conjugate subgroups in G.
[Hint: If H and K are subgroups of the group G, then they are conjugate subgroups iff $H = g^{-1}Kg$ for some $g \in G$. Let $\psi : G/H \to G/K$ be a

G-isomorphism. Then $\exists g \in G$ such that $\psi(H) = gK$. If $h \in H$, then $gK = \psi(H) = \psi(hH) = h\psi(H) = hgK$ shows that $g^{-1}hg \in K$ and $g^{-1}Hg \subset K$. Again $\psi(g^{-1}H) = g^{-1}\psi(H) = g^{-1}gK = K$ shows that $\psi^{-1}(K) = g^{-1}H$. Consequently, $g^{-1}Hg \subset H$. Hence $g^{-1}Hg = K$ shows that H and K are conjugate subgroups in G.
Conversely, let $g^{-1}Hg = K$ for some $g \in G$. Show that the map

$$\psi : G/H \to G/K, aH \mapsto agK$$

is a G-isomorphism.]

21. Let (X, A) be a normal pair (i.e., X is normal and A is closed in X) and $f : A \to S^n$ be an open map. Show that there exists an open set U of X containing A and an open extension $\tilde{f} : U \to S^n$ of f.
[Hint: Consider f as a map of A into $\mathbf{R}^{n+1}$. Then by Tietze's extension Theorem 1.12.4, there exists an extension $g : X \to \mathbf{R}^{n+1}$ of f. Let $U = X - g^{-1}(0)$, where 0 is the origin of $\mathbf{R}^{n+1}$. Define

$$\tilde{f} : U \to S^n, x \mapsto \frac{g(x)}{||g(x)||}.$$

]

22. Let (X, A) be a normal pair such that $X \times I$ is normal. Show that every map $f : X \times \{0\} \cup A \times I \to S^n$ admits a continuous extension $\tilde{f} : X \times I \to S^n$.
[Hint: Use Ex. 21 of Sect. 1.16, for the pair $(X \times I, B)$, where $B = X \times \{0\} \cup A \times I$.]

1.17 Additional Reading

[1] Artin, M., *Algebra*, Prentice-Hall, Englewood Cliffs, 1991.
[2] Chatterjee, B.C., Ganguly, S., and Adhikari, M.R., *A Textbook of Topology*, Asian Books Pvt.Ltd., New Delhi, 2002.
[3] Lang, S., *Algebra*, Addition-Wesley, Reading, 1965.
[4] Munkres, J.R., Topology, A First Course, Prentice-Hall, New Jersey, 1975.
[5] Switzer, R.M., *Algebraic Topology-Homotopy and Homology*, Springer-Verlag, Berlin, Heidelberg, New York, 1975.

References

Adhikari, M.R., Adhikari, A.: Basic Modern Algebra with Applications. Springer, New York (2014)
Artin, M.: Algebra. Prentice-Hall, Englewood Cliffs (1991)
Chatterjee, B.C., Ganguly, S., Adhikari, M.R.: A Textbook of Topology. Asian Books Pvt. Ltd., New Delhi (2002)

Dugundji, J.: Topology. Allyn & Bacon, Newtown (1966)
Herstein, I.: Topics in Algebra. Blerisdell, New York (1964)
Lang, S.: Algebra. Addition-Wesley, Reading (1965)
Maunder, C.R.F.: Algebraic Topology. Van Nostrand Reinhhold, London (1970)
Munkres, J.R.: Topology, A First Course. Prentice-Hall, New Jersey (1975)
Spanier, E.: Algebraic Topology. McGraw-Hill Book Company, New York (1966)
Switzer, R.M.: Algebraic Topology-Homotopy and Homology. Springer, Berlin (1975)

Chapter 2
Homotopy Theory: Elementary Basic Concepts

This chapter opens with a study of homotopy theory by introducing its elementary basic concepts such as homotopy of continuous maps, homotopy equivalence, H-group, H-cogroup, contractible space, retraction, deformation with illustrative geometrical examples and applications. The study of homotopy theory continues explicitly up to Chap. 9 of the present book. Its many key concepts are also applied to other chapters. The basic aim of homotopy theory is to investigate 'algebraic principles' latent in homotopy equivalent spaces. Such principles are also important in the study of topology and geometry as well as in many other subjects such as algebra, algebraic geometry, number theory, theoretical physics, chemistry, computer science, economics, bioscience, medical science, and some other subjects.

Algebraic topology flows mainly through two channels: one is the homotopy theory and other one is the homology theory. The concept of homotopy is a mathematical formulation of the intuitive idea of a continuous deformation from one geometrical configuration to other in the sense that this concept formalizes the naive idea of continuous deformation of a continuous map. On the other hand, the concept of homology is a mathematical precision to the intuitive idea of a curve bounding an area or a surface bounding a volume. Cohomology theory which is a dual concept of homology theory is also closely related to homotopy theory. The idea guiding the development of mathematical theory of homotopy, homology, and cohomology is described nowadays in the language of category theory by constructing certain functors.

Algebraic topology is one of the most important creations in mathematics which uses algebraic tools to study topological spaces. The basic goal is to find algebraic invariants that classify topological spaces up to homeomorphism (though usually classify up to homotopy equivalence). The most important of these invariants are homotopy, homology, and cohomology groups. This subject is an interplay between topology and algebra and studies algebraic invariants provided by homotopy and homology theories. The twentieth century witnessed its greatest development.

M.R. Adhikari, *Basic Algebraic Topology and its Applications*,
DOI 10.1007/978-81-322-2843-1_2

A basic problem in homotopy theory is to classify continuous maps up to homotopy: two continuous maps from one topological space to other are homotopic if one map can be continuously deformed into the other map. On the other hand, the basic problem in algebraic topology is to devise ways to assign various algebraic objects such as groups, rings, modules to topological spaces and homomorphisms to the corresponding algebraic structures in a functorial way. More precisely, although the ultimate aim of topology is to classify topological spaces up to homeomorphism, the main problem of algebraic topology is the 'classification' of topological spaces up to homotopy equivalence, the concept introduced by W. Hurewicz (1904–1956) in 1935. So in algebraic topology a homotopy equivalence plays a more influential role than a homeomorphism, because the basic tools of algebraic topology such as homotopy groups, and homology & cohomology groups are invariants with respect to homotopy equivalence.

Homotopy theory constitutes a basic part of algebraic topology and studies topological spaces up to homotopy equivalence which is a weaker relation than topological equivalence in the sense that homotopy classes of spaces are larger than homeomorphic classes. The concept of the homotopy equivalence gives rise to the classification of topological spaces according to their homotopy properties. The basic idea of this classification is to assign to each topological space 'invariants', which may be integers, or algebraic objects in such a way that homotopy equivalent spaces have the same invariants (up to isomorphism), called homotopy invariants, which characterize homotopy equivalent spaces completely. The main numerical invariants of homotopy equivalent spaces are dimensions and degrees of connectedness.

Historically, the idea of homotopy for the continuous maps of unit interval was originated by C. Jordan (1838–1922) in 1866 and that of for loops was introduced by H. Poincaré (1854–1912) in 1895 to define an algebraic invariant called the fundamental group, which is studied, in Chap. 3. The monumental work of Poincaré in 'Analysis situs', Paris, 1895, organized the subject for the first time. This work explained the difference between curves deformable to one another and curves bounding a larger space. The first one led to the concepts of homotopy and fundamental group; the second one led to the concept of homology. Poincaré is the first mathematician who systemically attacked the problems of assigning algebraic (topological) invariants to topological spaces. His vision of the key role of topology in all mathematical theories began to materialize from 1920. Of course, many of the ideas he developed had their origins prior to him, with L. Euler (1707–1783), and B. Riemann (1826–1866) above all. H. Hopf (1894–1971) introduced the concept of H-spaces and H-groups from the viewpoint of homotopy theory. Some of his amazing results have made a strong foundation of algebraic topology. Many topologists regard H. Poincaré as the founder and regard H. Hopf and W. Hurewicz as cofounders of many key concepts in algebraic topology.

Throughout this book a space means a topological space and a map means a continuous function between topological spaces; the terms: map (or continuous map) and continuous function will be used interchangeably in the context of topological spaces, unless specified otherwise.

For this chapter the books Eilenberg and Steenrod (1952), Hatcher (2002), Maunder (1970), Spanier (1966) and some others are referred in Bibliography

2.1 Homotopy: Introductory Concepts and Examples

This section is devoted to the study of the concept of homotopy formalizing the intuitive idea of continuous deformation of a continuous map between two topological spaces and presents introductory basic concepts of homotopy with illustrative examples. Homeomorphism generates equivalence classes whose members are topological spaces. On the other hand, homotopy generates equivalence classes whose members are continuous maps. The term homotopy was first given by Max Dehn (1878–1952) and Poul Heegaard (1871–1948) in 1907. It is sometimes replaced by a complicated function between two topological spaces by another simpler function sharing some important properties of the original function. An allied concept is the notion of deformation. This leads to the concept of homotopy of functions.

The relation between topological spaces of being homeomorphic is an equivalence relation. So it divides any set of topological spaces into disjoint classes. The main problem of topology is the classification of topological spaces. Given two topological spaces X and Y, are they homeomorphic? This is a very difficult problem. Algebraic topology transforms such topological problems into algebraic problems which may have a better chance for solution. The algebraic techniques are usually not delicate enough to classify topological spaces up to homeomorphism. The notion of homotopy of continuous functions defines somewhat coarser classification. This leads to the concept of a continuous deformation. The relation of homotopy of continuous functions generalizes path connectedness of a point, which is a fundamental concept of homotopy theory.

2.1.1 Concept of Homotopy

The intuitive concept of a continuous deformation is now explained with the concept of homotopy. Moreover the concept of 'flow' which is also known as one parameter group of homeomorphisms is conveyed through homotopy. Let $I = [0, 1]$ be the closed unit interval with topology induced by the natural topology on the real line $\mathbf{R}$ (sometimes written as $\mathbf{R}^1$).

Definition 2.1.1 Let X and Y be topological spaces. Two continuous maps $f, g : X \to Y$ are said to be homotopic (or f is said to be homotopic to g), if there exists a continuous map $F : X \times I \to Y$ such that $F(x, 0) = f(x)$ and $F(x, 1) = g(x)$, $\forall\ x \in X$. The map F is said to be a homotopy between f and g, written $F : f \simeq g$.

Remark 2.1.2 Geometrically, two continuous maps $f, g : X \to Y$ are said to be homotopic if f can be continuously deformed into g by a continuous family of maps $F_t : X \to Y$ defined by $F_t(x) = F(x, t)$ such that $F_0 = f$ and $F_1 = g$, $\forall\ x \in X$, $\forall\ t \in I$. By saying that the maps F_t form a continuous family, we mean that F is continuous with respect to both x and t as a function from the product space $X \times I$ to Y. Clearly, a homotopy F between two continuous maps $f, g : X \to Y$ can be considered as a special case of extension:
consider in the topological space $X \times I$ the subspace

$$A = (X \times \{0\}) \cup (X \times \{1\}) \subset X \times I$$

and consider the continuous map

$$G : A \to Y, (x, 0) \mapsto f(x), (x, 1) \mapsto g(x);$$

then a homotopy F from f to g is an extension of G from A to $X \times I$.

Definition 2.1.3 Let X be a topological space. A continuous map $f : I \to X$ such that $f(0) = x_0$ and $f(1) = x_1$, is called a path in X from x_0 to x_1. The point x_0 is called the initial point and the point x_1 is called the final or terminal point of the path f.

Definition 2.1.4 Two paths $f, g : I \to X$ are said to be homotopic if they have the same initial point x_0, the same final point x_1 and there exists a continuous map $F : I \times I \to X$ such that

$$F(t, 0) = f(t),\ F(t, 1) = g(t),\ \forall\ t \in I \tag{2.1}$$

$$\text{and } F(0, s) = x_0 \text{ and } F(1, s) = x_1,\ \forall\ s \in I \tag{2.2}$$

We call F a path homotopy between f and g as shown in Fig. 2.1 and is written $F : f \underset{p}{\simeq} g$.

Remark 2.1.5 The condition (2.1) says that F is a homotopy between f and g and the condition (2.2) says that for each $t \in I$, the path $t \mapsto F(t, s)$ is a path in X from x_0 to x_1. In other words, (2.1) shows that F represents a continuous way of deforming

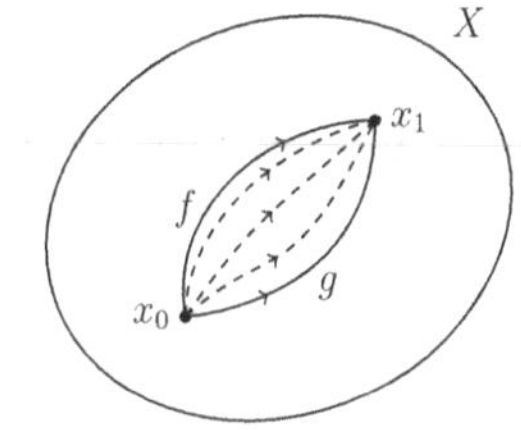

Fig. 2.1 Path homotopy

the path f to the path g and (2.2) shows that the end points of the path remain fixed during the deformation.

We now prove the following two lemmas of point set topology which will be used throughout the book.

Lemma 2.1.6 (Pasting or Gluing lemma) *Let X be a topological space and A, B be closed subsets of X such that $X = A \cup B$. Let Y be a topological space and $f : A \to Y$ and $g : B \to Y$ be continuous maps. If $f(x) = g(x), \ \forall \ x \in A \cap B$, then the function $h : X \to Y$ defined by*

$$h(x) = \begin{cases} f(x), & \forall \ x \in A \\ g(x), & \forall \ x \in B \end{cases}$$

is continuous.

Proof h defined in the lemma is the unique well-defined function $X \to Y$ such that $h|_A = f$ and $h|_B = g$. We now show that h is continuous. Let C be a closed set in Y. Then $h^{-1}(C) = X \cap h^{-1}(C) = (A \cup B) \cap h^{-1}(C) = (A \cap h^{-1}(C)) \cup (B \cap h^{-1}(C)) = (A \cap f^{-1}(C)) \cup (B \cap g^{-1}(C)) = f^{-1}(C) \cup g^{-1}(C)$. Since each of f and g is continuous, $f^{-1}(C)$ and $g^{-1}(C)$ are both closed in X. Hence $h^{-1}(C)$ is closed in X. Consequently, h is continuous. ❑

This lemma can be generalized as follows:

Lemma 2.1.7 (Generalized Pasting or Gluing lemma) *Let a topological space X be a finite union of closed subsets $X_i : X = \bigcup_{i=1}^{n} X_i$. If for some topological space Y, there are continuous maps $f_i : X_i \to Y$ that agree on overlaps (i.e., $f_i|_{X_i \cap X_j} = f_j|_{X_i \cap X_j}, \ \forall \ i, j$), then $\exists$ a unique continuous function $f : X \to Y$ with $f|_{X_i} = f_i, \ \forall \ i$.*

Proof The proof is similar to proof of Lemma 2.1.6. ❑

Theorem 2.1.8 *Let $P(X)$ denote the set of all paths in a space X having the same initial point x_0 and the same final point x_1. Then the path homotopy relation '$\underset{p}{\simeq}$' is an equivalence relation on $P(X)$.*

Proof Let $f, g, h \in P(X)$. Then $f(0) = g(0) = h(0) = x_0$ and $f(1) = g(1) = h(1) = x_1$. Let a map $F : I \times I \to X$ be defined by $F(t, s) = f(t), \ \forall \ t, s \in I$. Then F is continuous, because it is the composite of the projection map onto the first factor and the continuous map f. Hence F is a continuous map such that $F(t, 0) = f(t), F(t, 1) = f(t), \ \forall \ t \in I$ and $F(0, s) = x_0, F(1, s) = x_1, \ \forall \ s \in I$. Thus $F : f \underset{p}{\simeq} f, \ \forall \ f \in P(X)$. Next, let $f \underset{p}{\simeq} g$ and $F : f \underset{p}{\simeq} g$. Then $F : I \times I \to X$ is a continuous map such that $F(t, 0) = f(t), F(t, 1) = g(t), \ \forall \ t \in I$ and $F(0, s) = x_0, F(1, s) = x_1, \ \forall \ s \in I$. Let $G : I \times I \to X$ be the map defined by $G(t, s) =$

$F(t, 1-s)$. Since the maps $I \to I, t \mapsto t$ and $s \mapsto 1-s$ are both continuous, G is continuous. Now $G(t,0) = F(t,1) = g(t), G(t,1) = F(t,0) = f(t), \forall\, t \in I$ and $G(0,s) = F(0, 1-s) = x_0, G(1,s) = F(1, 1-s) = x_1$. Hence $G : g \underset{p}{\simeq} f$.

Finally, let $f \underset{p}{\simeq} g$ and $g \underset{p}{\simeq} h$. Then $\exists$ continuous maps $F, G : I \times I \to X$ such that $F : f \underset{p}{\simeq} g$ and $G : g \underset{p}{\simeq} h$. Consequently, for all $t, s \in I$, $F(t,0) = f(t)$, $F(t,1) = g(t), F(0,s) = x_0, F(1,s) = x_1, G(t,0) = g(t), G(t,1) = h(t)$, $G(0,s) = x_0$ and $G(1,s) = x_1$. We now define a map $H : I \times I \to X$ by the equations

$$H(t,s) = \begin{cases} F(t,2s), & 0 \le s \le 1/2 \\ G(t, 2s-1), & 1/2 \le s \le 1 \end{cases}$$

At $s = \frac{1}{2}$, $F(t,2s) = F(t,1) = g(t)$ and $G(t, 2s-1) = G(t,0) = g(t), \forall\ t \in I$ show that F and G agree at $t \times \frac{1}{2}$. Moreover, F is continuous on $I \times [0, \frac{1}{2}]$ and G is continuous on $I \times [\frac{1}{2}, 1]$. Hence by Pasting lemma, H is continuous. Now, $H(t,0) = F(t,0) = f(t), \forall\ t \in I, H(t,1) = G(t,1) = h(t), \forall\ t \in I$,

$$H(0,s) = \begin{cases} F(0,2s), & 0 \le s \le 1/2 \\ G(0, 2s-1), & 1/2 \le s \le 1 \end{cases}$$
$$= x_0, \quad \forall\ s \in I$$

and

$$H(1,s) = \begin{cases} F(1,2s), & 0 \le s \le 1/2 \\ G(1, 2s-1), & 1/2 \le s \le 1 \end{cases}$$
$$= x_1, \quad \forall\ s \in I$$

Hence $H : f \underset{p}{\simeq} h$. Consequently, '$\underset{p}{\simeq}$' is an equivalence relation on $P(X)$. ❑

Definition 2.1.9 The quotient set $P(X)/\underset{p}{\simeq}$ is called the set of path homotopy classes of paths in X.

Example 2.1.10 Let $f, g : X \to \mathbf{R}^2$ be two continuous maps. Define $F : X \times I \to \mathbf{R}^2$ by the rule $F(x,t) = (1-t)f(x) + tg(x), \forall\ x \in X, \forall\ t \in I$. Then $F : f \simeq g$. In this example, F shifts the point $f(x)$ to the point $g(x)$ along the straight line segment joining $f(x)$ and $g(x)$, as shown in Fig. 2.2. The map F is called a straight line homotopy.

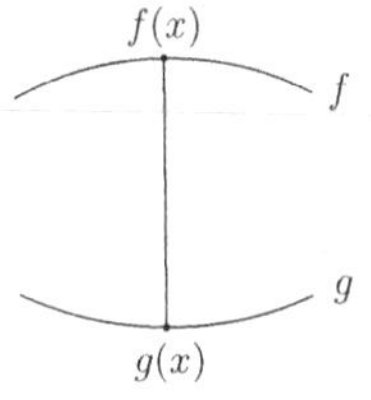

Fig. 2.2 Straight line homotopy

Example 2.1.11 If $X = Y = \mathbf{R}^n$ and $f(x) = x$ and $g(x) = 0 \equiv (0, \ldots, 0) \in \mathbf{R}^n$, $\forall\ x \in \mathbf{R}^n$, i.e., if $f = 1_X$ (identity map on X) and g is the constant map at 0, then $F : X \times I \to X$, defined by $F(x, t) = (1 - t)x$ is a homotopy from f to g, i.e., $F : f \simeq g$. Again $G : X \times I \to X$, defined by $G(x, t) = (1 - t^2)x$ is also a homotopy from f to g. These examples show that homotopy between two maps is not unique.

Remark 2.1.12 As there are many homotopies between two maps, we can deform a map f into a given map g in different ways.

Example 2.1.13 Let X denote the punctured plane $X = \mathbf{R}^2 - \{0\}$. Then the paths $f(t) = (\cos \pi t, \sin \pi t)$, $g(t) = (\cos \pi t, 2 \sin \pi t)$ are path homotopic; the straight line homotopy between them is an acceptable path homotopy.

On the other hand, the straight line homotopy between the paths $f(t) = (\cos \pi t, \sin \pi t)$ and $h(t) = (\cos \pi t, - \sin \pi t)$ is not acceptable, because it passes through 0 and hence it does not entirely lie in the space $X = \mathbf{R}^2 - \{0\}$, as shown in the Fig. 2.3. There does not exist any path homotopy in X between the paths f and h, because one cannot deform f into g continuously passing through the hole at 0.

Example 2.1.14 Let $\mathbf{D}^n = \{x \in \mathbf{R}^n : ||x|| \leq 1\}$ be the n-disk. If $f(x) = x$ and $g(x) = 0$, $\forall\ x \in \mathbf{D}^n$, then $g \simeq f$. Define $F : \mathbf{D}^n \times I \to \mathbf{D}^n$ by $F(x, t) = tx$, $\forall\ x \in \mathbf{D}^n$, $\forall\ t \in I$. Now $||tx|| = |t| \cdot ||x|| \leq 1 \Rightarrow tx \in \mathbf{D}^n$, $\forall\ t \in I$ and $\forall\ x \in \mathbf{D}^n \Rightarrow F$ is well defined. Clearly, F is continuous and $F : g \simeq f$. Similarly, $G : \mathbf{D}^n \times I \to \mathbf{D}^n$ defined by $G(x, t) = (1 - t)x$ is a continuous map such that $G : f \simeq g$.

Example 2.1.15 Let $f, g : I \to I$ be defined by $f(t) = t$ and $g(t) = 0$, $\forall\ t \in I$. Then $F : I \times I \to I$ defined by $F(t, s) = (1 - s)t$ is a continuous map such that $F : f \simeq g$.

Example 2.1.16 Let Y be a subspace of $\mathbf{R}^n$ and $f, g : X \to Y$ be two continuous maps such that for every $x \in X$, $f(x)$ and $g(x)$ can be joined by a straight line segment in Y, then $F : f \simeq g$, where $F : X \times I \to Y$ is defined by

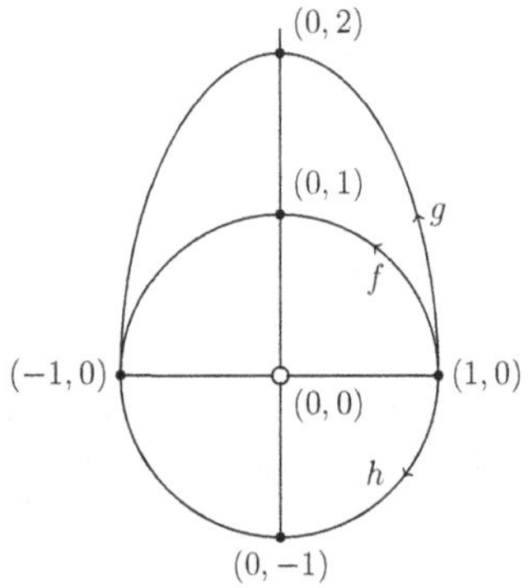

Fig. 2.3 Path homotopy

$F(x,t) = (1-t)f(x) + tg(x)$. Since $f(x)$ and $g(x)$ can be joined by a line segment in Y by hypothesis, F is well defined. To prove the continuity of F, we take $x, u \in X$ and $t, s \in I$. Then $F(u,s) = (1-s)f(u) + sg(u)$. Now $F(u,s) - F(x,t) = (s-t)(g(u) - f(u)) + (1-t)(f(u) - f(x)) + t(g(u) - g(x))$. Let $\epsilon > 0$ be an arbitrary small positive number. Then

$$||F(u,s) - F(x,t)|| \leq |(s-t)|||g(u) - f(u)|| + |(1-t)|||f(u) - f(x)|| \\ + |t|||g(u) - g(x)|| \qquad (2.3)$$

Again, f and g being continuous, $\exists$ open neighborhoods U_1 and U_2 of x in X such that for $u \in U_1 \cap U_2$, $||f(u) - f(x)|| < \epsilon/3$, $||g(u) - g(x)|| < \epsilon/3$. Then for $u \in U_1 \cap U_2$, $||g(u) - f(u)|| \leq ||g(u) - g(x)|| + ||g(x) - f(x)|| + ||f(x) - f(u)|| < c$, where c is the positive constant $||g(x) - f(x)|| + 2\epsilon/3$. Thus if $|s-t| < \epsilon/3c$, then from (2.3) it follows that

$$||F(u,s) - F(x,t)|| < \epsilon \qquad (2.4)$$

Since the set $(U_1 \cap U_2) \times (t - \epsilon/3c, t + \epsilon/3c)$ is open in $X \times I$, this shows that F is continuous. Finally, $F(x,0) = f(x)$ and $F(x,1) = g(x)$, $\forall\ x \in X$. Consequently, $F : f \simeq g$.

Remark 2.1.17 Geometrically, the above homotopy F deforms f into g along the straight line segment in Y joining the points $f(x)$ and $g(x)$ for every $x \in X$. The function F is called a straight line homotopy.

Example 2.1.18 Let X be a topological space and S^n be the n-sphere in $\mathbf{R}^{n+1}$. If $f, g : X \to S^n$ are two continuous maps such that $f(x) \neq -g(x)$ for any $x \in X$, then $f \simeq g$. To show this define the map $F : X \times I \to S^n$ by $F(x,t) = \frac{(1-t)f(x)+tg(x)}{||(1-t)f(x)+tg(x)||}$. For each $x \in S^n$ and $t \in I$, $(1-t)f(x) + tg(x) \in \mathbf{R}^{n+1}$. The given condition $f(x) \neq -g(x)$ for any $x \in X$ shows that the line segment joining $f(x)$ and $g(x)$ cannot pass through the origin $0 = (0, 0, \ldots, 0) \in \mathbf{R}^{n+1}$. In other words, $(1-t)f(x) + tg(x) \neq 0$ for any $t \in I$ and any $x \in X$. Hence $F(x,t) \in S^n$, $\forall\ (x,t) \in X \times I$ and F is well defined. We now consider f, g as $f, g : X \to \mathbf{R}^{n+1} - \{0\}$. Then by Example 2.1.16, $\exists$ a straight line homotopy $G : X \times I \to \mathbf{R}^{n+1} - \{0\}$ defined by $G(x,t) = (1-t)f(x) + tg(x)$, i.e., $G : f \simeq g$. Again consider the map $h : \mathbf{R}^{n+1} - \{0\} \to S^n$ defined by $h(x) = \frac{x}{||x||}$. Then $F = h \circ G$ is the composite of two continuous maps h and G and hence F is continuous. Finally, $F(x,0) = \frac{f(x)}{||f(x)||} = f(x)$, because $f(x) \in S^n$, $\forall\ x \in X$ and $F(x,1) = \frac{g(x)}{||g(x)||} = g(x)$, because $g(x) \in S^n$, $\forall\ x \in X$. Consequently, $F : f \simeq g$.

Example 2.1.19 Let X be a topological space and $f : X \to S^n$ be a continuous non-surjective map. Then f is homotopic to a constant map $c : X \to S^n$. By hypothesis $f(X) \subsetneqq S^n \Rightarrow \exists$ a point $s_0 \in S^n$ such that $s_0 \notin f(X)$. Define a constant map $c : X \to S^n$ by $c(x) = -s_0$, $\forall\ x \in X$. Then $f(x) \neq -c(x)$ for any $x \in X$. Hence $f \simeq c$ by Example 2.1.18.

Example 2.1.20 Let $S^1 = \{z \in \mathbf{C} : |z| = 1\} = \{e^{i\theta} : 0 \leq \theta \leq 2\pi\}$ be the unit circle in $\mathbf{C}$. Then the maps $f, g : S^1 \to S^1$ defined by $f(z) = z$ and $g(z) = -z$ are homotopic. Consider the map $F : S^1 \times I \to S^1$ defined by $F(e^{i\theta}, t) = e^{i(\theta + t\pi)}$. Clearly, F is the composite of the maps

$$S^1 \times I \to S^1 \times S^1 \to S^1, \ (e^{i\theta}, t) \mapsto (e^{i\theta}, e^{it\theta}) \mapsto e^{i(\theta+t\pi)},$$

where the second map is the usual multiplication of complex numbers. Consequently, F is a continuous map such that $F : f \simeq g$.

We now extend Theorem 2.1.8 to the set $C(X, Y)$ of all continuous maps from X to Y by extending the concept of path homotopy (obtained by replacing I by any topological space X).

Theorem 2.1.21 *Given topological spaces X and Y, the relation '$\simeq$' (of being homotopic) is an equivalence relation on the set $C(X, Y)$.*

Proof Each $f \in C(X, Y)$ is homotopic to itself by a homotopy $H : X \times I \to Y$ defined by $H(x, t) = f(x)$. Thus $f \simeq f$, $\forall\ f \in C(X, Y)$. Next suppose $H : f \simeq g$, $f, g \in C(X, Y)$. Define $G : X \times I \to Y$ by $G(x, t) = H(x, 1 - t)$. Then G is continuous, because G is the composite of continuous maps

$$X \times I \to X \times I \to Y, \ (x, t) \mapsto (x, 1 - t) \mapsto H(x, 1 - t),$$

where the first map is continuous, because the projection maps $(x, t) \mapsto x$ and $(x, t) \mapsto (1 - t)$ are continuous and the second map is H. Then $G(x, 0) = H(x, 1) = g(x)$ and $G(x, 1) = H(x, 0) = f(x)$, $\forall\ x \in X$. Thus $G : g \simeq f$. Finally, let $f, g, h \in C(X, Y)$ be such that $F : f \simeq g$ and $G : g \simeq h$. Define a map $H : X \times I \to Y$ by

$$H(x, t) = \begin{cases} F(x, 2t), & 0 \leq t \leq 1/2 \\ G(x, 2t - 1), & 1/2 \leq t \leq 1 \end{cases}$$

Then H is continuous by Pasting lemma. Finally, $H(x, 0) = F(x, 0) = f(x)$ and $H(x, 1) = G(x, 1) = h(x)$, $\forall\ x \in X \Rightarrow H : f \simeq h$. Consequently, '$\simeq$' is an equivalence relation on $C(X, Y)$. ❑

Definition 2.1.22 The quotient set $C(X, Y)/\simeq$ is called the set of all homotopy classes of maps $f \in C(X, Y)$, denoted by $[X, Y]$ and for $f \in C(X, Y)$, $[f] \in [X, Y]$ is called the homotopy class of f.

Remark 2.1.23 The set $[X, Y]$ was first systemically studied by M.G. Barratt in 1955 in his paper (Barratt 1955). This set plays the central role in algebraic topology and is used throughout the book. Some of its properties are displayed in Sect. 2.3.

We now show that composites of homotopic maps are homotopic.

Theorem 2.1.24 *Let $f_1, g_1 \in C(X, Y)$ and $f_2, g_2 \in C(Y, Z)$ be maps such that $f_1 \simeq g_1$ and $f_2 \simeq g_2$. Then the composite maps $f_2 \circ f_1$ and $g_2 \circ g_1 : X \to Z$ are homotopic.*

Proof Let $F : f_1 \simeq g_1$ and $G : f_2 \simeq g_2$. Then $f_2 \circ F : X \times I \to Z$ is a continuous map such that $(f_2 \circ F)(x, 0) = f_2(F(x, 0)) = f_2(f_1(x)) = (f_2 \circ f_1)(x), \ \forall\, x \in X$ and $(f_2 \circ F)(x, 1) = f_2(F(x, 1)) = f_2(g_1(x)) = (f_2 \circ g_1)(x), \ \forall\, x \in X$. Consequently,

$$f_2 \circ F : f_2 \circ f_1 \simeq f_2 \circ g_1 \tag{2.5}$$

Again we define $H : X \times I \to Z$ by $H(x, t) = G(g_1(x), t)$. Thus H is the composite

$$X \times I \xrightarrow{g_1 \times 1_d} Y \times I \xrightarrow{G} Z,$$

$$(x, t) \mapsto (g_1(x), t) \mapsto G(g_1(x), t).$$

Then H is a continuous map such that $H(x, 0) = G(g_1(x), 0) = f_2(g_1(x)) = (f_2 \circ g_1)(x), \ \forall \ x \in X$ and $H(x, 1) = G(g_1(x), 1) = g_2(g_1(x)) = (g_2 \circ g_1)(x)$, $\forall\, x \in X$. Hence

$$H : f_2 \circ g_1 \simeq g_2 \circ g_1 \tag{2.6}$$

Consequently, by transitive property of homotopy relation, it follows from (2.5) and (2.6) that $f_2 \circ f_1 \simeq g_2 \circ g_1$. ❑

Remark 2.1.25 Theorem 2.1.24 asserts in the language of category theory that topological spaces and homotopy classes of continuous maps form a category denoted by $\mathcal{H}tp$ called homotopy category of topological spaces (see Appendix B). Thus $\mathcal{H}tp$ is the category whose objects are topological spaces and mor (X, Y) consists of homotopy classes of continuous maps from X to Y, where the composition of maps is consistent with homotopies (see Theorem 2.1.24).

Given a topological space X the concept of a flow $\psi_t : X \to X$ $(t \in \mathbf{R})$ is closely related to homotopy.

Definition 2.1.26 A continuous family $\psi_t : X \to X$ $(t \in \mathbf{R})$ of maps is called a flow if

(i) $\psi_0 = 1_d$;
(ii) ψ_t is a homeomorphism for all $t \in \mathbf{R}$;
(iii) $\psi_{t+s} = \psi_t \circ \psi_s$.

Remark 2.1.27 It is sometimes convenient to consider a flow ψ_t as a continuous map

$$\psi : X \times \mathbf{R} \to \mathbf{R},\ (x, t) \mapsto \psi_t(x).$$

A flow is also known as one parameter group of homeomorphisms.

Proposition 2.1.28 *$\psi_t : X \to X$ is homotopic to I_X.*

Proof Consider the map

$$F : X \times I \to X, (x, s) \mapsto \psi(x, (1 - s)t).$$

This shows that every ψ_t is homotopic to 1_X. ❑

We now extend the Definition 2.1.1 of homotopy of continuous maps for pairs of topological spaces.

Definition 2.1.29 A topological pair (X, A) consists of a topological space X and a subspace A of X. If $A = \emptyset$, the empty set, we shall not distinguish between the pair $(X, \emptyset)$ and the space X. A subpair (X', A') of (X, A) is a pair such that $X' \subset X$ and $A' \subset A$.

Definition 2.1.30 A continuous map $f : (X, A) \to (Y, B)$ is a continuous function $f : X \to Y$ such that $f(A) \subset B$.

Given a topological pair (X, A), $(X, A) \times I$ represents the pair $(X \times I, A \times I)$.

Definition 2.1.31 Given pairs of topological spaces (X, A) and (Y, B), two continuous maps $f, g : (X, A) \to (Y, B)$ are said to be homotopic if $\exists$ a continuous map $F : (X \times I, A \times I) \to (Y, B)$ such that $F(x, 0) = f(x)$ and $F(x, 1) = g(x)$, $\forall\ x \in X$. Then the map F is called a homotopy from f to g and written $F : f \simeq g$.

We now consider a more restricted type of homotopy of continuous maps between pairs of topological spaces, which extends the concept of path homotopy obtained by replacing I by any topological space X and $\{0, 1\}$ by a subspace of X under consideration.

Definition 2.1.32 Let $f, g : (X, A) \to (Y, B)$ be two continuous maps of pairs of topological spaces and $X' \subset X$ be such that $f|_{X'} = g|_{X'}$ (i.e., $f(x') = g(x')$, $\forall\ x' \in X'$, which implies that f and g agree at x', $\forall\ x' \in X'$). Then f and g are said to be homotopic relative to X' if there exists a continuous map $F : (X \times I, A \times I) \to (Y, B)$ such that $F(x, 0) = f(x)$, $F(x, 1) = g(x)$, $\forall\ x \in X$ and $F(x', t) = f(x') = g(x')$, $\forall\ x' \in X'$ and $\forall\ t \in I$, and written $F : f \simeq g$ rel X'.

If $X' = \emptyset$, we omit the phrase relative to X'.

Remark 2.1.33 $f \simeq g$ rel $X' \Rightarrow f \simeq g$ rel X'' for any subspace $X'' \subset X'$.

Geometrical Interpretation: For $t \in I$, if we define $h_t : (X, A) \to (X \times I, A \times I)$ by $h_t(x) = (x, t)$, then $h_0(x) = (x, 0)$ and $h_1(x) = (x, 1)$. Thus $F : f \simeq g$ rel $X' \Rightarrow F \circ h_0 = f$, $F \circ h_1 = g$, and $F \circ h_t|_{X'} = f|_{X'} = g|_{X'}$, $\forall\ t \in I \Rightarrow$ the collection $\{F \circ h_t\}_{t \in I}$ is a continuous one parameter family of maps from (X, A) to (Y, B) agreeing on X' and satisfying the relations $f = F \circ h_0$ and $g = F \circ h_1$. Thus $f \simeq g$ rel X' represents geometrically a continuous deformation deforming f into g by maps all of which agree on X'. For example, $f \simeq g$ rel $\{0\}$ in Example 2.1.14.

Example 2.1.34 Consider $D^2 = \{z \in \mathbf{C} : z = re^{i\theta}, 0 \le r \le 1\}$ and $S^1 = \{z \in \mathbf{C} : z = e^{i\theta}, 0 \le \theta \le 2\pi\}$. Then $S^1 \subset D^2$. Let $f : (D^2, S^1) \to (D^2, S^1)$ be the identity map and $g : (D^2, S^1) \to (D^2, S^1)$ be the reflection in the origin, i.e., $g(re^{i\theta}) = re^{i(\theta+\pi)}$. Then $f \simeq g \ \text{rel}\,\{0\}$ under the homotopy $F : (D^2, S^1) \times I \to (D^2, S^1)$ defined by $F(re^{i\theta}, t) = re^{i(\theta+t\pi)}$. Moreover, $G : (D^2, S^1) \times I \to (D^2, S^1)$ defined by $G(re^{i\theta}, t) = e^{i(\theta-t\pi)}$ is also a homotopy $G : f \simeq g \ \text{rel}\,\{0\}$ (compare Example 2.1.14).

Remark 2.1.35 There may exist different homotopies from f to g relative to a subspace and thus homotopy from f to g is not unique.

Example 2.1.36 Let X be a topological space, $A \subset X$ and Y be any convex subspace of $\mathbf{R}^n$. If $f, g : X \to Y$ are two continuous maps such that $f|_A = g|_A$, then $f \simeq g \ \text{rel}\, A$ by a homotopy $G : X \times I \to Y$ defined by $G(x, t) = (1-t)f(x) + tg(x)$.

We now generalize Theorem 2.1.21.

Theorem 2.1.37 *The relation between continuous maps from (X, A) to (Y, B) of being homotopic relative to a subspace $X' \subset X$ is an equivalence relation.*

Proof **Reflexivity** Let $f : (X, A) \to (Y, B)$ be a continuous map. Define $F : (X \times I, A \times I) \to (Y, B)$ by the rule $F(x, t) = f(x)$, $\forall\ x \in X$, $\forall\ t \in I$. Then $F : f \simeq f \ \text{rel}\, X'$.
Symmetry Let $F : f \simeq g \ \text{rel}\, X'$. Define $G : (X \times I, A \times I) \to (Y, B)$ by the rule $G(x, t) = F(x, 1-t)$, $\forall\ x \in X$, $\forall\ t \in I$. For continuity of G see Theorem 2.1.21. Then $G : g \simeq f \ \text{rel}\, X'$.
Transitivity Let $f, g, h : (X, A) \to (Y, B)$ be three continuous maps such that $F : f \simeq g \ \text{rel}\, X'$ and $G : g \simeq h \ \text{rel}\, X'$. Define $H : (X \times I, A \times I) \to (Y, B)$ by the rule

$$H(x, t) = \begin{cases} F(x, 2t), & 0 \le t \le 1/2 \\ G(x, 2t-1), & 1/2 \le t \le 1 \end{cases}$$

Then H is continuous by Pasting Lemma 2.1.6. Moreover, $H : f \simeq h \ \text{rel}\, X'$ ❑

Remark 2.1.38 It follows from Theorem 2.1.37 that the set of continuous maps from (X, A) to (Y, B) is partitioned into disjoint equivalence classes by the relation of homotopy relative to X' denoted by $[X, A; Y, B]$. This set is very important in the study of algebraic topology. Given a continuous map $f : (X, A) \to (Y, B)$, $[f|_{X'}]$ represents the homotopy class in $[X, A; Y, B]$ determined by f.

We now generalize Theorem 2.1.24 for homotopies relative to a subspace.

Theorem 2.1.39 *Let $f_0, f_1 : (X, A) \to (Y, B)$ be homotopies relative to $X' \subset X$ and $g_0, g_1 : (Y, B) \to (Z, C)$ be homotopies relative to Y', where $f_1(X') \subset Y' \subset Y$. Then the composites $g_0 \circ f_0, g_1 \circ f_1 : (X, A) \to (Z, C)$ are homotopic relative to X', i.e., composites of homotopic maps are homotopic.*

Proof Let $F : f_0 \simeq f_1$ rel X' and $G : g_0 \simeq g_1$ rel Y'. Then the composite mapping

$$(X \times I, A \times I) \xrightarrow{F} (Y, B) \xrightarrow{g_0} (Z, C)$$

is a homotopy relative to X' from $g_0 \circ f_0$ to $g_0 \circ f_1$, i.e.,

$$g_0 \circ F : g_0 \circ f_0 \simeq g_0 \circ f_1 \text{ rel } X' \tag{2.7}$$

Again the composite mapping

$$(X \times I, A \times I) \xrightarrow{f_1 \times 1_d} (Y \times I, B \times I) \xrightarrow{G} (Z, C)$$

is a homotopy relative to

$$f_1^{-1}(Y') \text{ from } g_0 \circ f_1 \text{ to } g_1 \circ f_1 \tag{2.8}$$

Since $X' \subset f_1^{-1}(Y)$, (2.7) and (2.8) show that $g_0 \circ f_0 \simeq g_0 \circ f_1$ rel X' and $g_0 \circ f_1 \simeq g_1 \circ f_1$ rel X' and hence $g_0 \circ f_0 \simeq g_1 \circ f_1$ rel X' by transitivity of the relation $\simeq$.❑

2.1.2 *Functorial Representation*

This subsection summarizes the earlier discussion in the basic result from the viewpoint of category theory which gives important examples of categories, functors and natural transformations, the concepts defined in Appendix B.

Theorem 2.1.39 shows that there is a category, called the homotopy category of pairs of spaces whose objects are topological pairs and whose morphisms are homotopy classes relative to a subspace. This category contains as full subcategories the homotopy category $\mathcal{H}tp$ of topological spaces and also the homotopy category $\mathcal{H}tp_*$ of pointed topological spaces.

Theorem 2.1.40 *There is a covariant functor from the category of pairs of topological spaces and their continuous maps to the homotopy category whose object function is the identity function and whose morphism function sends a continuous map f to its homotopy class $[f]$. Moreover, for any pair (P, Q) of topological spaces there is a covariant functor $\pi_{(P,Q)}$ from the homotopy category of pairs to the category of sets and functions defined by $\pi_{(P,Q)}(X, A) = [P, Q; X, A]$ and if $f : (X, A) \rightarrow (Y, B)$ is continuous, then $f_* = \pi_{(P,Q)}([f]) : [P, Q; X, A] \rightarrow [P, Q; Y, B]$ is defined by $f_*([g]) = [f \circ g]$ for $g : (P, Q) \rightarrow (X, A)$.*

If $\alpha : (P, Q) \rightarrow (P', Q')$, then there is a natural transformation $\alpha^* : \pi_{(P',Q')} \rightarrow \pi_{(P,Q)}$. Similarly, we can define a contravariant functor $\pi^{(P,Q)}$ for a given (P, Q) of pair of topological spaces and a natural transformation $\alpha_* : \pi^{(P,Q)} \rightarrow \pi^{(P',Q')}$.

2.2 Homotopy Equivalence

This section studies the concept of homotopy equivalence introduced by W. Hurewicz (1935) to establish a connection between homotopy and homology groups of a certain class of topological spaces. The problem of classification of continuous maps from one topological space to other is closely related to the problem of classification of topological spaces according to their homotopy properties. This problem led to the concept of homotopy equivalence which is not only a generalization of the concept of homeomorphism but also gives a new foundation for the development of the combinatorial invariants of topological spaces and manifolds. The higher homotopy groups and the homology groups are invariants of the homotopy equivalent class of a topological space.

Classification of topological spaces up to homotopy equivalences is the main problem of algebraic topology. This is a weaker relation than a topological equivalence in the sense that homotopy classes of continuous maps of topological spaces are larger than their homeomorphism classes. Although the main aim of topology is to classify topological spaces up to homeomorphism; in algebraic topology, a homotopy equivalence plays a more important role than a homeomorphism. Because the basic tools of algebraic topology such as homotopy and homology groups are invariants with respect to homotopy equivalences.

Definition 2.2.1 A continuous map $f : (X, A) \to (Y, B)$ is called a homotopy equivalence if $[f]$ is an equivalence in the homotopy category of pairs. In particular, a map $f \in C(X, Y)$ is said to be a homotopy equivalence if $\exists$ a map $g \in C(Y, X)$ such that $g \circ f \simeq 1_X$ (existence of left homotopy inverse of f) and $f \circ g \simeq 1_Y$ (existence of right homotopy inverse of f). In such a situation g is unique and the map g is called a homotopy inverse of f.

Remark 2.2.2 Let f be a homotopy equivalence with g as its homotopy inverse. Then $[g] = [f]^{-1}$ in the homotopy category $\mathcal{H}tp$ which has the same objects in the category $\mathcal{T}op$ of topological spaces and their continuous maps but the morphism in $\mathcal{H}tp$ are the homotopy classes of continuous maps, so that their morphisms $\text{mor}_{\mathcal{H}tp}(X, Y) = [X, Y]$. The isomorphisms in the category $\mathcal{T}op$ are homeomorphisms and in the category $\mathcal{H}tp$ are homotopy equivalences.

Example 2.2.3 Let Y be the $(n-1)$-sphere $S^{n-1} \subset \mathbf{R}^n \subset \mathbf{R}^{n+q}$ and X be the subset of $\mathbf{R}^{n+q}$ of points not lying on the plane $x_1 = \cdots = x_n = 0$. Then the inclusion map $i : Y \hookrightarrow X$ is a homotopy equivalence.

Define $f : X \to Y$ by $f(x_1, x_2, \ldots, x_{n+q}) = (rx_1, \ldots, rx_n, 0, \ldots, 0)$, where $r = (x_1^2 + x_2^2 + \cdots + x_n^2)^{-1/2}$. Then $f \circ i = 1_Y$. Again define

$$H : X \times I \to X, (x_1, x_2, \ldots, x_{n+q}, t) \mapsto (r^{1-t}x_1, \ldots, r^{1-t}x_n, tx_{n+1}, \ldots, tx_{n+q})$$

is a homotopy from $i \circ f$ to 1_X. Consequently, i is a homotopy equivalence.

Example 2.2.4 Let D^2 be the unit disk in $\mathbf{R}^2$ and $p \in D^2$. Let $i : P = \{p\} \hookrightarrow D^2$ be the inclusion map and $c : D^2 \to P$ be the constant map. Then $c \circ i = 1_P$. Again the map $H : D^2 \times I \to D^2$ defined by $H(x, t) = (1 - t)x + tp$, being a homotopy from 1_{D^2} to $i \circ c$, i.e., $i \circ c \simeq 1_{D^2}$. Consequently, i is a homotopy equivalence.

Definition 2.2.5 If $f \in C(X, Y)$ is a homotopy equivalence, then X and Y are said to be homotopy equivalent spaces, denoted by $X \simeq Y$.

We now extend Definition 2.2.5 for pairs of topological spaces.

Definition 2.2.6 Two pairs of topological spaces (X, A) and (Y, B) are said to be homotopy equivalent, written $(X, A) \simeq (Y, B)$, if $\exists$ continuous maps $f : (X, A) \to (Y, B)$ and $g : (Y, B) \to (X, A)$ such that $g \circ f \simeq 1_X$ and $f \circ g \simeq 1_Y$, the homotopy being the homotopy of pairs.

Remark 2.2.7 Homeomorphic spaces are homotopy equivalent but its converse is not true in general.

Consider the following example:

Example 2.2.8 Let X be the unit circle S^1 in $\mathbf{R}^2$ and Y be the topological space S^1, together with the line segment I_1 joining the points $(1, 0)$ and $(2, 0)$ in $\mathbf{R}^2$, i.e., $I_1 = \{(r, 0) \in \mathbf{R}^2 : 1 \leq r \leq 2\}$. Then X and Y are of the same homotopy type but they are not homeomorphic (Fig. 2.4).

X and Y cannot be homeomorphic, because removal of the point $(1, 0)$ from Y makes Y disconnected. On the other hand, removal of any point from X leaves X connected. We claim that $X \simeq Y$. We take $f : X \hookrightarrow Y$ to be the inclusion map and $g : Y \to X$ defined by

$$g(y) = \begin{cases} y, & \text{if } y \in X \\ (1, 0), & \text{if } y \in I_1 \end{cases}$$

The continuity of g follows from Pasting lemma. Then $f \circ g, 1_Y : Y \to Y$ are two continuous maps such that $(f \circ g)(y) = f(g(y)) = g(y), \ \forall\ y \in Y$ and

$$1_Y(y) = \begin{cases} y, & \text{if } y \in X \\ (r, 0), & \text{if } y = (r, 0) \in I_1. \end{cases}$$

Since for every $y \in Y$, $(f \circ g)(y)$ and $1_Y(y)$ can be joined by a straight line segment in Y, it follows that $f \circ g \simeq 1_Y$ (see Example 2.1.16). Again $g \circ f = 1_X \simeq 1_X$. Thus f

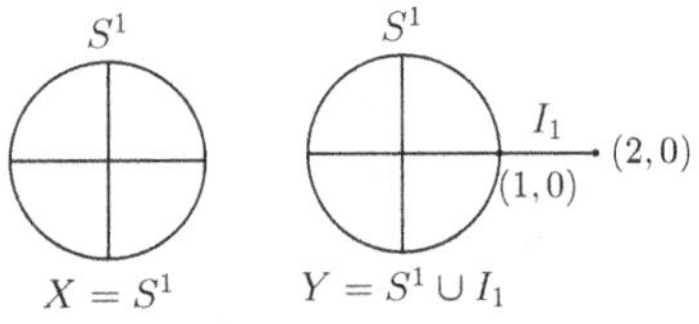

Fig. 2.4 Homotopy equivalent but non-homeomorphic spaces

and g are two continuous maps such that $f \circ g \simeq 1_Y$ and $g \circ f \simeq 1_X$. Consequently, $X \simeq Y$.

This example shows that $X \simeq Y$ does not imply $X \approx Y$.

Example 2.2.9 The topological spaces X and Y in Example 2.2.8 are homotopy equivalent. The unit disk D^2 is homotopy equivalent to a one-point topological space $\{p\} \subset D^2$ (see Example 2.2.4).

As the name suggests, the relation of being homotopy equivalent is an equivalence relation on the set of topological spaces (or pairs of topological spaces).

Theorem 2.2.10 *The relation '$\simeq$' between topological spaces (or pairs of topological spaces) of being homotopy equivalent is an equivalence relation.*

Proof **Reflexivity**: $1_X : X \to X$ is a homotopy equivalence $\Rightarrow X \simeq X$ for all X.
Symmetry: Let $X \simeq Y$. Then $\exists$ continuous maps $f : X \to Y$ and $g : Y \to X$ such that $g \circ f \simeq 1_X$ and $f \circ g \simeq 1_Y \Rightarrow g : Y \to X$ is a homotopy equivalence with homotopy inverse $f : X \to Y$. Consequently, $Y \simeq X$. Thus $X \simeq Y \Rightarrow Y \simeq X$.
Transitivity: Let $X \simeq Y$ and $Y \simeq Z$. Then $\exists$ continuous maps $f : X \to Y, g : Y \to X, h : Y \to Z$ and $k : Z \to Y$ such that $g \circ f \simeq 1_X, f \circ g \simeq 1_Y, h \circ k \simeq 1_Z$ and $k \circ h \simeq 1_Y$. Now $h \circ f : X \to Z$ and $g \circ k : Z \to X$ are continuous maps such that $(h \circ f) \circ (g \circ k) = h \circ (f \circ g) \circ k \simeq h \circ 1_Y \circ k = h \circ k \simeq 1_Z$ and $(g \circ k) \circ (h \circ f) = g \circ (k \circ h) \circ f \simeq g \circ 1_Y \circ f = g \circ f \simeq 1_X$. Consequently, $h \circ f$ is a homotopy equivalence with $g \circ k$ homotopy inverse $\Rightarrow X \simeq Z$. The equivalence relation $\simeq$ divides the set of spaces up to homotopy equivalent classes. ❑

Definition 2.2.11 The homotopy equivalent class containing X is called the homotopy type of X.

Remark 2.2.12 Two topological spaces X and Y are homotopy equivalent or of the same homotopy type if there exists a homotopy equivalence $f \in C(X, Y)$. For example, D^2 and $\{p\}$ in Example 2.2.9 are homotopy equivalent spaces. The homeomorphic spaces are said to have the same topological type. On the other hand, the homotopy equivalent spaces are said to have the same homotopy type.

Proposition 2.2.13 *Two homeomorphic spaces have the same homotopy type.*

Proof Let X and Y be two homeomorphic spaces and $f : X \to Y$ be a homeomorphism. Then its inverse $g = f^{-1} : Y \to X$ is continuous and satisfies the conditions: $g \circ f = f^{-1} \circ f = 1_X \simeq 1_X$ and $f \circ g = f \circ f^{-1} = 1_Y \simeq 1_Y$ by reflexivity of the relation $\simeq$. Consequently, f is a homotopy equivalence. Hence X and Y are of the same homotopy type. ❑

Remark 2.2.14 The converse of Proposition 2.2.13 is not true. For example, the disk D^n is of the same homotopy type of a single point $\{p\} \subset D^n$ but D^n is not homeomorphic to $\{p\}$.

Proposition 2.2.15 *Any continuous map homotopic to a homotopy equivalence is a homotopy equivalence.*

Proof Let $C(X, Y)$ denote the set of all continuous maps from X to Y and $f \in C(X, Y)$ be a homotopy equivalence. Suppose $g \in C(X, Y)$ is such that $f \simeq g$. Now f is a homotopy equivalence $\Rightarrow \exists\, h \in C(Y, X)$ such that $h \circ f \simeq 1_X$ and $f \circ h \simeq 1_Y$. Again $f \simeq g \Rightarrow f \circ h \simeq g \circ h \Rightarrow 1_Y \simeq g \circ h \Rightarrow g \circ h \simeq 1_Y$. Similarly, $h \circ g \simeq 1_X$. Consequently, g is a homotopy equivalence. ❑

Definition 2.2.16 A continuous map $f : X \to S^n$ is called inessential if f is homotopic to a continuous map of X into a single point of S^n (i.e., if f is homotopic to a constant map). Otherwise f is called essential. In general, a map $f \in C(X, Y)$ is said to be nullhomotopic or inessential if it is homotopic to some constant map.

Example 2.2.17 Let $X = Y = I$. Define $f, g : I \to I$ by $f(t) = t$ and $g(t) = 0$, $\forall\, t \in I$. Then f is the identity map and g is a constant map. Define $F : I \times I \to I$ by $F(t, s) = (1 - s)t$. Then $F : f \simeq g \Rightarrow f$ is nullhomotopic.

Remark 2.2.18 Two nullhomotopic maps may not be homotopic.

Example 2.2.19 Let X be a connected space and Y be not a connected space. Let y_0 and y_1 be points in distinct components of Y. Let $f_0(x) = y_0$ and $f_1(x) = y_1$, $\forall\, x \in X$ be two constant maps from X to Y. If possible, let $f_0 \simeq f_1$. Then $\exists$ a continuous map $F : X \times I \to Y$ such that $F : f_0 \simeq f_1$. Since $X \times I$ is connected and F is continuous, $F(X \times I)$ must be connected, which contradicts the fact that Y is not connected.

Proposition 2.2.20 *Let $f, g : (X, A) \to (Y, B)$ be pairs of continuous maps such that $f \simeq g$ as maps of pairs. Then the induced maps $\tilde{f}, \tilde{g} : X/A \to Y/B$ (corresponding quotient spaces) are also homotopic.*

Proof Let $H : (X \times I, A \times I) \to (Y, B)$ be a homotopy between f and g. Then H induces a function $\tilde{H} : (X/A) \times I \to Y/B$ such that the diagram in Fig. 2.5 is commutative, where p and q are the identification maps. Since $\tilde{H} \circ (p \times 1_d) = q \circ H$ is continuous, I is locally compact and Hausdorff, $p \times 1_d$ is an identification map, it follows that $\tilde{H}$ is a (based) homotopy between $\tilde{f}$ and $\tilde{g}$, where base points of X/A and Y/B are respectively the points to which A and B are identified. ❑

Corollary 2.2.21 *Let $f : (X, A) \to (Y, B)$ be a homotopy equivalence of pairs. Then $\tilde{f} : X/A \to Y/B$ is a (based) homotopy equivalence.*

Proof As $f : (X, A) \to (Y, B)$ is a homotopy equivalence, there exists a map $g : (Y, B) \to (X, A)$ such that $g \circ f \simeq 1_X$ and $f \circ g \simeq 1_Y$. Hence the corollary follows from the Proposition 2.2.20. ❑

$$\begin{array}{ccc} X \times I & \xrightarrow{H} & Y \\ {\scriptstyle p \times 1_d}\downarrow & & \downarrow{\scriptstyle q} \\ X/A \times I & \xrightarrow[\tilde{H}]{} & Y/B \end{array}$$

Fig. 2.5 Diagram for identification map

2.3 Homotopy Classes of Maps

This section continues the study of homotopy classes of continuous maps given in Sect. 2.2. These classes play an important role in the study of algebraic topology as depicted throughout the book. Homotopy theory studies those properties of topological spaces and continuous maps which are invariants under homotopic maps, called homotopy invariants.

Let $[X, Y]$ be the set of homotopy classes of continuous maps from X to Y: by keeping X fixed and varying Y, this set is an invariant of the homotopy type of Y, in the sense that if $Y \simeq Z$, then there exists a bijective correspondence between the sets $[X, Y]$ and $[X, Z]$. Similar result holds for pairs of topological spaces and hence for pointed topological spaces. Many homotopy invariants can be obtained from the sets $[X, Y]$ on which some short of algebraic structure is often given.for particular choice of X and Y. Most of the classical invariants of algebraic topology are homotopy invariants. Many homotopy invariants can be obtained by specializing the sets $[X, Y]$.

The following two natural problems are posed in this section but solved in Sect. 2.4.

(i) Given a pointed topological space Y, does there exist a natural product defined in $[X, Y]$ admitting the set $[X, Y]$ a group structure for all pointed topological spaces X?

(ii) Given a pointed topological space X, does there exist a natural product defined in $[X, Y]$ admitting the set $[X, Y]$ a group structure for all pointed topological space Y?

In this section we work in the homotopy category $\mathcal{H}tp_*$ of pointed topological spaces and their base point preserving continuous maps. Thus $[X, Y]$ is the set of morphisms from X to Y in the homotopy category $\mathcal{H}tp_*$ of pointed topological spaces. This set $[X, Y]$ only depends on homotopy types of X and Y. Given two continuous maps $f : X \to Y$ and $g : Y \to Z$, we can compose them and obtain $g \circ f : X \to Z$. The homotopy class of $g \circ f$ depends only on the homotopy classes of f and g. So the composition with g gives a function

$$g_* : [X, Y] \to [X, Z]$$

and the composition with f gives a function

$$f^* : [Y, Z] \to [X, Z].$$

Theorem 2.3.1 *Let X, Y, Z be pointed topological spaces and $f : Y \to Z$ be a base point preserving continuous map. Then f induces a function. $f_* : [X, Y] \to [X, Z]$ satisfying the following properties:*

(a) *If $f \simeq h : Y \to Z$, then $f_* = h_*$;*

(b) *If $1_Y : Y \to Y$ is the identity map, then 1_{Y*} is the identity function;*

(c) *If* $g : Z \to W$ *is another base point preserving continuous map, then* $(g \circ f)_* = g_* \circ f_*$.

Proof Define $f_* : [X, Y] \to [X, Z]$ by the rule $f_*([\alpha]) = [f \circ \alpha]$, $\forall\ [\alpha] \in [X, Y]$. Since $\alpha \simeq \beta \Rightarrow f \circ \alpha \simeq f \circ \beta \Rightarrow f_*([\alpha]) = f_*[\beta] \Rightarrow f_*$ is independent of the choice of the representatives of the classes. Hence f_* is well defined.

(a) Consider the functions $f_*, h_* : [X, Y] \to [X, Z]$. Then $h_*([\alpha]) = [h \circ \alpha] = [f \circ \alpha] = f_*([\alpha])$, since $f \simeq h \Rightarrow f \circ \alpha \simeq h \circ \alpha = f_*([\alpha])$, $\forall\ [\alpha] \in [X, Y]$. Hence $h_* = f_*$.

(b) $1_{Y^*} : [X, Y] \to [X, Y]$ is given by $1_{Y^*}([\alpha]) = [1_Y \circ \alpha] = [\alpha]$, $\forall\ [\alpha] \in [X, Y]$. Hence 1_{Y^*} is the identity function.

(c) $(g \circ f)_* : [X, Y] \to [X, W]$ is given by $(g \circ f) * ([\alpha]) = [(g \circ f) \circ \alpha] = [g \circ (f \circ \alpha)] = (g_* \circ f_*)[\alpha]$, $\forall\ [\alpha] \in [X, Y]$. Hence $(g \circ f)_* = g_* \circ f_*$.

❑

Corollary 2.3.2 *If* $f : Y \to Z$ *is a homotopy equivalence, then* $f_* : [X, Y] \to [X, Z]$ *is a bijection for every topological space* X.

Proof If $f \in C(Y, Z)$ is a homotopy equivalence, then $\exists\ g \in C(Z, Y)$ such that $g \circ f \simeq 1_Y$ and $f \circ g \simeq 1_Z$. Hence $(g \circ f)_* = g_* \circ f_*$ is the identity function and $(f \circ g)_* = f_* \circ g_*$ is also identity function $\Rightarrow f_*$ is a bijection with g_* as its inverse. ❑

Corollary 2.3.3 *If* $Y \simeq Z$*, then there exists a bijection* $\psi : [X, Y] \to [X, Z]$ *for every topological space* X.

Proof If $Y \simeq Z$, then $\exists$ a homotopy equivalence $f \in C(Y, Z)$. Hence $f_* = \psi : [X, Y] \to [X, Z]$ is a bijection for every X by Corollary 2.3.2. ❑

Corollary 2.3.4 *Given a pointed topological space* X*, there exists a covariant functor* π_X *from the homotopy category of pointed topological spaces to the category of sets and functions defined by* $\pi_x(Y) = [X, Y]$ *and if* $f : Y \to Z$ *is a base point preserving continuous map, then* $\pi_x(f) = f_* : [X, Y] \to [X, Z]$ *is defined by* $f_*([\alpha]) = [f \circ \alpha]$.

Proof The Corollary follows from Theorem 2.3.1. ❑

We obtain the corresponding dual results.

Theorem 2.3.5 *A base point preserving continuous map* $f : Y \to Z$ *induces a function* $f^* : [Z, X] \to [Y, X]$ *for every pointed space* X*, satisfying the following properties:*

(a) $f \simeq h : Y \to Z \Rightarrow f^* = h^*$;

(b) $1_Y : Y \to Y$ *is the identity map* $\Rightarrow 1_{Y^*}$ *is the identity function;*

(c) *If* $g : Z \to W$ *is another base point preserving continuous map, then* $(g \circ f)^* = f^* \circ g^*$.

Proof Similar to the proof of Theorem 2.3.1. ❑

Corollary 2.3.6 *If $f : Y \to Z$ is a homotopy equivalence, then $f^* : [Z, X] \to [Y, X]$ is a bijection for every pointed topological space X.*

Corollary 2.3.7 *If $Y \simeq Z$, then $\exists$ a bijection $\psi : [Z, X] \to [Y, X]$ for every pointed topological space X.*

Corollary 2.3.8 *Given a pointed topological space X, there exists a contravariant functor π^X from the homotopy category of pointed topological spaces to the category of sets and functions.*

Converses of Corollaries 2.3.2 and 2.3.6 are also true.

Theorem 2.3.9 *If $f : Y \to Z$ is a base point preserving continuous map such that*

(a) *$f_* : [X, Y] \to [X, Z]$ is a bijection for all pointed topological spaces X, then f is a homotopy equivalence.*
(b) *$f^* : [Z, X] \to [Y, X]$ is a bijection for all pointed topological spaces X, then f is a homotopy equivalence.*

Proof **(a)** In particular, $f_* : [Z, Y] \to [Z, Z]$ is a bijection (by hypothesis) $\Rightarrow \exists$ a continuous map $g : Z \to Y$ such that $f_*([g]) = [1_Z] \Rightarrow f \circ g \simeq 1_Z$. Similarly, $g_*([f]) = [1_Y] \Rightarrow g \circ f \simeq 1_Y$. Consequently, f is a homotopy equivalence.
(b) Similar to **(a)**.

❑

2.4 *H*-Groups and *H*-Cogroups

This section conveys the concept of a grouplike space, called an H-group and its dual concept, called an H-cogroup as a continuation of the study of the set $[X, Y]$ by considering the problem: when is the set $[X, Y]$ a group for every pointed topological space X (or for every pointed topological space Y)? The concepts of H-groups and H-cogroups arose through the study of such problems. These concepts develop homotopy theory. The loop spaces and suspension spaces of pointed topological spaces play an important role in the study of homotopy theory. Loop spaces of pointed spaces provide an extensive class of H-groups. On the other hand suspension spaces of pointed topological spaces form an extensive class of H-cogroups, a dual concept of H-group.

We consider topological spaces Y such that $[X, Y]$ admits a group structure for all X. There is a close relation between the natural group structures on $[X, Y]$ for all X and 'grouplike' structure on Y. Before systematic study of the homotopy sets $[X, Y]$ or $[(X, x_0), (Y, y_0)]$ by M.G. Barratt (1955) in his paper, the concept of an H-space introduced by H. Hopf in 1933 arose as a generalization of a topological group which is used to solve the above problem.

2.4.1 H-Groups and Loop Spaces

This subsection continues to study H-groups by specializing the the sets $[X, Y]$ and presents loop spaces which form an important class of H-groups. Given pointed topological spaces X and Y, we often give the set $[X, Y]$ some sort of algebraic structure. With this objective this subsection studies a grouplike space which is a group up to homotopy, called an H-group. More precisely, this subsection introduces the concepts of H-groups to obtain algebraic structures on the set of certain homotopy classes of continuous maps and introduces the concept of an H-group with loop spaces as illustrative examples. An H-group is a generalization of a topological group. Such groups were first introduced by H. Hopf in 1941 and they are named in his honor. Loop spaces of pointed topological spaces constitute an extensive class of H-groups.

The motivation of this study is to describe an additional structure needed on a pointed space P so that $\pi^P(X) = [X, P]$ is a group and for $f : X \to Y$, $f^* = \pi^P(f) : [Y, P] \to [X, P]$ is a group homomorphism. If $f : X \to Y$ and $g : X \to Z$ are continuous maps, we define $(f, g) : X \to Y \times Z$ to be the map $(f, g)(x) = (f(x), g(x))$, $\forall\ x \in X$.

If Y is a topological group, then $[X, Y]$ admits a group structure by Theorem 2.4.1.

Now the following two natural questions arise:

(i) Given a pointed topological space Y, does there exist a natural product defined in $[X, Y]$ for all pointed topological spaces X?
(ii) Given a pointed topological space X, does there exist a natural product defined in $[X, Y]$ for all pointed topological spaces Y?

We start with a topological group P (see Appendix A) followed by H-groups and H-cogroups. The essential feature which is retained in an H space is a continuous multiplication with a unit. There is a significant class of topological spaces which are H-spaces but not topological groups.

Theorem 2.4.1 *Let X be any pointed topological space and P be a topological group with identity element as base point. Then $[X, P]$ can be given the structure of a group.*

Proof Given two base point preserving continuous maps $f, g : X \to P$, let their product $f \cdot g$ be defined by pointwise multiplication, i.e., $f \cdot g : X \to P$ is defined by $(f \cdot g)(x) = f(x)g(x)$, where the right side is the group multiplication μ in P. Thus $f \cdot g = \mu \circ (f \times g) \circ \Delta$, where $\Delta(x) = (x, x)$ is the diagonal map, i.e., $f \cdot g$ is the composite

$$X \xrightarrow{\Delta} X \times X \xrightarrow{f \times g} P \times P \xrightarrow{\mu} P$$

Then $f \cdot g$ is another continuous map from X to P. Moreover, given further maps $f', g' : X \to P$ such that $f \simeq f'$ and $g \simeq g'$ then $f \cdot g \simeq f' \cdot g'$ (by Ex. 2 of Sect. 2.11)

$\Rightarrow [f]\cdot[g] = [f\cdot g] \Rightarrow$ the law of composition $f\cdot g$ carries over to give an operation '$\circ$' on $[X, P]$. Then the group structure on $[X, P]$ follows from the corresponding properties of the topological group. Consequently, $([X, P], \circ)$ is a group. ❑

Corollary 2.4.2 *If P is a topological group and $f : X \to Y$ is a base point preserving continuous map, then f induces a group homomorphism $f^* : [Y, P] \to [X, P]$ defined by $f^*([\alpha]) = [\alpha \circ f]$, $\forall$ $[\alpha] \in [Y, P]$.*

Theorem 2.4.3 *Given a topological group P, there exists a contravariant functor π^P from the homotopy category of pointed topological spaces to the category of groups and homomorphisms.*

Proof It follows from Theorem 2.4.1 and Corollaries 2.4.2 and 2.3.8. ❑

Remark 2.4.4 Given a topological group P, the group structure on $[X, P]$ is endowed from the group structure on the set of base point preserving continuous maps from X to P. We come across some situations in which $[X, P]$ admits a natural group structure, but the set of base point preserving continuous maps from X to P has no group structure. If P is a pointed topological space having the same homotopy type as some topological group P', then π^P is naturally equivalent to $\pi^{P'}$. Hence π^P can be regarded as a functor to the category of groups.

Example 2.4.5 $S^1 = \{z \in \mathbf{C} : |z| = 1\}$ is an abelian topological group under usual multiplication of complex numbers. Then $[X, S^1]$ is an abelian group and if $f : X \to Y$, then $f^* : [Y, S^1] \to [X, S^1]$ is a homomorphism of groups.

Example 2.4.6 S^3 is a topological group (the multiplicative group of quaternions of norm 1). Then $[X, S^3]$ is a group and if $f : X \to Y$, then $f^* : [Y, S^3] \to [X, S^3]$ is group homomorphism.

Remark 2.4.7 If Y is a topological group, then a product $f\cdot g : X \to Y$ is given by $(f\cdot g)(x) = f(x)\cdot g(x)\,\forall\, x \in X$. To solve the problems (i) and (ii) we search for certain other classes of pointed topological spaces, called Hopf spaces (H-spaces) and Hopf groups (H-groups).

Definition 2.4.8 A pointed topological space P with a base point p_0, is called an H-space if there exists a continuous multiplication $\mu : P \times P \to P, (p, p') \mapsto pp'$ for which the (unique) constant map $c : P \to p_0 \in P$ is a homotopy identity, i.e., each composite

$$P \xrightarrow{(c,1_P)} P \times P \xrightarrow{\mu} P \text{ and } P \xrightarrow{(1_P,c)} P \times P \xrightarrow{\mu} P$$

is homotopic to 1_P, i.e., if each of the triangles in Fig. 2.6 is homotopy commutative; sometimes it is written as an ordered pair (P, μ).

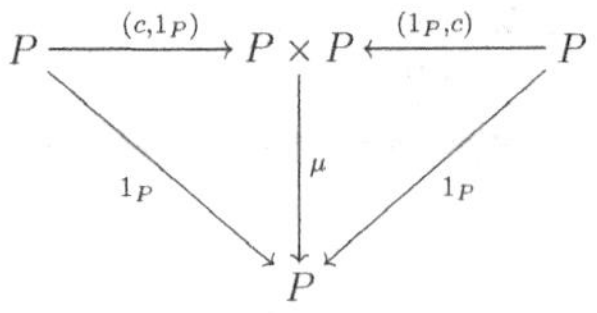

Fig. 2.6 H-space

Remark 2.4.9 The above homotopy commutativity means that the maps $P \to P$, $p \mapsto p_0p, pp_0$ are homotopic to 1_P rel $\{p_0\}$. In other words, $\exists$ homotopies L and R from $P \times I \to P$ such that $L(p, 0) = p_0p$, $L(p, 1) = p$, $L(p_0, t) = p_0$, $R(p, 0) = pp_0$, $R(p, 1) = p$ and $R(p_0, t) = p_0$, $\forall\ p \in P$ and $\forall\ t \in I$.

Definition 2.4.10 An H-space (P, μ) is said to be homotopy associative if the square in Fig. 2.7 is homotopy commutative, i.e., $\mu \circ (\mu \times 1_P) \simeq \mu \circ (1_P \times \mu)$, i.e., the two maps

$$P \times P \times P \to P, (p_1, p_2, p_3) \mapsto (p_1p_2)p_3, p_1(p_2p_3) \text{ are homotopic rel}\{p_0\}.$$

Definition 2.4.11 A continuous map $\phi : P \to P$ is said to be homotopy inverse for P and μ if each of the composites $P \xrightarrow{(1_P,\phi)} P \times P \xrightarrow{\mu} P$ and $P \xrightarrow{(\phi,1_P)} P \times P \xrightarrow{\mu} P$ is homotopic to the constant map $c : P \to P$, $p \mapsto p_0 \in P$, i.e., each of the maps $P \to P$, $p \mapsto p\phi(p), \phi(p)p$ is homotopic to c rel $\{p_0\}$.

Definition 2.4.12 An associative H-space P with an inverse is called an H-group or a generalized topological group. The point $p_0 \in P$ is called the homotopy unit of (P, μ).

Definition 2.4.13 A multiplicative μ on an H-space P is said to be homotopy abelian if the triangle in Fig. 2.8 is homotopy commutative, where $T(p, p') = (p', p)$, i.e., the two maps $P \times P \to P, (p, p') \mapsto pp', p'p$ are homotopic rel $\{p_0\}$.

Example 2.4.14 **(i)** Every topological group is an H-space with homotopy inverse. But its converse is not true in general (see (ii)). In particular, Lie groups (for

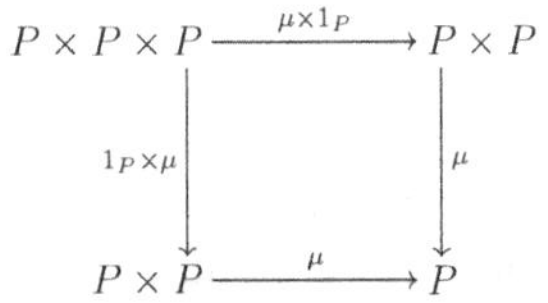

Fig. 2.7 Homotopy associative H-space

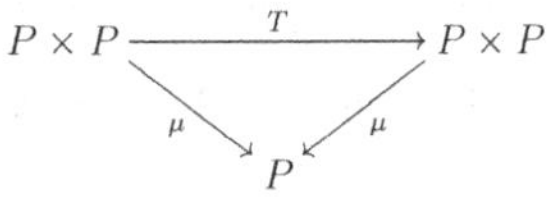

Fig. 2.8 Abelian H-space

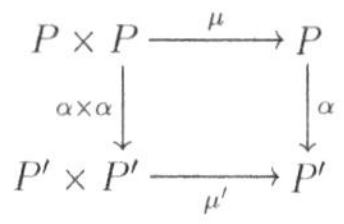

Fig. 2.9 Homotopy homomorphism

example, the general linear group, $GL(n, \mathbf{R})$ or the orthogonal group $O(n, \mathbf{R})$, see Appendix A) are H-spaces.

(ii) The infinite real projective space $\mathbf{R}P^\infty = \bigcup_{n\geq 0} \mathbf{R}P^n$ and infinite complex projective space $\mathbf{C}P^\infty = \bigcup_{n\geq 0} \mathbf{C}P^n$ are H-spaces but not topological groups.

Definition 2.4.15 Let (P, μ) and (P', μ') be two H-spaces. Then a continuous map $\alpha : P \to P'$ is called a homotopy homomorphism if the square in Fig. 2.9 is homotopy commutative.

Clearly, H-groups (H-spaces) and homotopy homomorphisms form a category.

Theorem 2.4.16 *A pointed topological space having the same homotopy type of an H-space (or an H-group) is itself an H-space (or H-group) in such a way that the homotopy equivalence is a homotopy homomorphism.*

Proof Let (P, μ) be an H-space and P' be a pointed topological space having the homotopy type of the space P. Then there exist continuous maps $f : P \to P'$ and $g : P' \to P$ such that $g \circ f \simeq 1_P$ and $f \circ g \simeq 1_{P'}$. Define $\mu' : P' \times P' \to P'$ to be the composite

$$P' \times P' \xrightarrow{g\times g} P \times P \xrightarrow{\mu} P \xrightarrow{f} P' \text{ i.e., } \mu' = f \circ \mu \circ (g \times g).$$

Then μ' is a continuous multiplication in P'. Moreover, the composites

$$P' \xrightarrow{(1,c')} P' \times P' \xrightarrow{\mu'} P' \tag{2.9}$$

$$\text{and } P' \xrightarrow{g} P \xrightarrow{(1,c)} P \times P \xrightarrow{\mu} P \xrightarrow{f} P' \tag{2.10}$$

are equal. As P is an H-space, the composite in (2.10) is homotopic to the composite $f \circ g$, because $f \circ \mu \circ (1, c) \circ g \simeq f \circ 1_P \circ g \simeq f \circ g$. Again, $f \circ g \simeq 1_{P'} \Rightarrow \mu' \circ (1, c') \simeq 1_{P'}$ by (2.9) and (2.10). Similarly, $\mu' \circ (c', 1) \simeq 1_{P'}$. Consequently, P' is an H-space with continuous multiplication μ'. Since the diagram in Fig. 2.10 is homotopy commutative, g is a homotopy homomorphism and so is f. If μ is homotopy associative or homotopy abelian, and if $\phi : P \to P$ is a homotopy inverse for P, then the composite map $f \circ \phi \circ g : P' \to P'$ is a homotopy inverse for P'. ❑

Generalizing the Theorem 2.4.1 we have the following theorem:

Fig. 2.10 Homotopy commutative square

$$\begin{array}{ccc} P' \times P' & \xrightarrow{\mu'} & P' \\ \downarrow{\scriptstyle g\times g} & & \downarrow{\scriptstyle g} \\ P \times P & \xrightarrow{\mu} & P \end{array}$$

Theorem 2.4.17 *If X is any pointed topological space and P is an H-group, then $[X, P]$ can be given the structure of a group.*

Proof Similar to the proof of Theorem 2.4.1. ❑

Theorem 2.4.18 *Let P be a pointed space with base point p_0 and $p_1, p_2 : P \times P \to P$ be the projections from the first and the second factors respectively. If $i_1, i_2 : P \to P \times P$ are inclusions defined by $i_1(p) = (p, p_0), i_2(p) = (p_0, p)$ for all $p \in P$, then the pointed space P is an H-space iff there exists a map $\mu : P \times P \to P$ such that $\mu \circ i_1 \simeq \mu \circ i_2$. Moreover, this map μ satisfies the condition $[\mu] = [p_1]\cdot[p_2]$ and if $f_1, f_2 : X \to P$ are maps then $[f_1]\cdot[f_2]$ is the homotopy class of the composite*

$$X \xrightarrow{\Delta} X \times X \xrightarrow{f_1 \times f_2} P \times P \xrightarrow{\mu} P$$

Proof It follows from hypothesis that $p_1 \circ i_1 = p_2 \circ i_2 = 1_P$ and $p_1 \circ i_2 = p_2 \circ i_1 = c$, where $c : P \to p_0$ is the constant map. If $\mu = P \times P \to P$ is a map such that $\mu \circ i_1 \simeq 1_P \simeq \mu \circ i_2$, then this μ is a multiplication admitting P the structure of an H-space.

Conversely suppose P is a pointed space and $\mu : P \times P \to P$ is a map such that $\mu \circ i_1 \simeq 1_P \simeq \mu \circ i_2$. Then given maps $f_1, f_2 : X \to P$ define $f_1 \cdot f_2 = \mu \circ (f_1 \times f_2) \circ \Delta$. This composition is compatible with homotopy, and induces a natural product in $[X, P]$. Hence P is an H-space and $p_1 \cdot p_2 = \mu \circ (p_1 \times p_2) \circ \Delta = \mu \circ 1_P = \mu$. Consequently, P is an H-space. ❑

Remark 2.4.19 For an arbitrary H-space P, the multiplication defined in $[X, P]$ may not be associative. If $f_1, f_2, f_3 : X \to P$ are maps, then $(f_1 \cdot f_2)\cdot f_3$ and $f_1 \cdot (f_2 \cdot f_3)$ are by definition the homotopy classes of the composites

$$X \xrightarrow{\Delta_3} X \times X \times X \xrightarrow{f_1 \times f_2 \times f_3} P \times P \times P \xrightarrow{\mu \times 1_P} P \times P \xrightarrow{\mu} P,$$

$$X \xrightarrow{\Delta_3} X \times X \times X \xrightarrow{f_1 \times f_2 \times f_3} P \times P \times P \xrightarrow{1_P \times \mu} P \times P \xrightarrow{\mu} P,$$

where $\Delta_3 = (\Delta \times 1_X) \circ \Delta = (1_X \times \Delta) \circ \Delta : X \to X \times X \times X$ is the diagonal map. Hence the condition$\mu \circ (\mu \times 1_P) \simeq \mu \circ (1_P \times \mu P)$ is sufficient for associativity. It is also necessary that if $X = P \times P \times P$ and f_1, f_2, f_3 are projections of $P \times P \times P$ into P, then $(f_1 \times f_2 \times f_3) \circ \Delta_3$ is the identity map.

The above discussion can be stated in the form of the following interesting result:

Theorem 2.4.20 *The set $[X, P]$ admits a monoid structure natural with respect to X iff P is a homotopy associative H-space.*

Theorem 2.4.21 *Let (P, p_0) be an H-group with multiplication μ and homotopy inverse ϕ. Then for every pointed topological space (X, x_0), the set $[(X, x_0), (P, p_0)]$ denoted $[X, P]$ can be given the structure of a group if we define the product $[f]\cdot[g]$ to be the homotopy class of the composite map*

$$X \xrightarrow{\Delta} X \times X \xrightarrow{f \times g} P \times P \xrightarrow{\mu} P,$$

where Δ is the diagonal map given by $\Delta(x) = (x, x)$. The identity element of the group is the class $[c]$ of the constant map $c : X \to p_0$ and the inverse of $[f]$ is given by $[f]^{-1} = [\phi \circ f]$. If μ is homotopy commutative, then $[X, P]$ is abelian.

Proof Define the product $[f]\cdot[g] = [\mu \circ (f \times g) \circ \Delta]$. We claim that $[f]\cdot[g]$ is well defined. To show this, let $H : X \times I \to P$ be a homotopy between f and f' and $G : X \times I \to P$ a homotopy between g and g'. Define a homotopy $M : X \times I \to P$ by $M_t(x) = M(x, t) = \mu(H(x, t), G(x, t))$. Then $M_0 = \mu \circ (f \times g) \circ \Delta$ and $M_1 = \mu \circ (f' \times g') \circ \Delta \Rightarrow \mu \circ (f \times g) \circ \Delta \simeq \mu \circ (f' \times g') \circ \Delta \Rightarrow [f]\cdot$
$[g] = [f']\cdot[g'] \Rightarrow$ the multiplication is independent of the choice of representatives of the classes. $\Rightarrow$ the multiplication is well defined.

We now prove the associativity of the multiplication. Let $h : (X, x_0) \to (P, p_0)$ be a third map. Then

$$\begin{aligned}
[f]\cdot([g]\cdot[h]) &= [\mu \circ (f \times \{\mu \circ (g \times h) \circ \Delta\}) \circ \Delta] \\
&= [\mu \circ (1 \times \mu) \circ (f \times g \times h) \circ (1 \times \Delta) \circ \Delta] \\
&= [\mu \circ (\mu \times 1) \circ (f \times g \times h) \\
&\qquad \circ (\Delta \times 1) \circ \Delta] \text{ by homotopy associativity of } \mu \\
&= [\mu \circ (\{\mu \circ (f \times g) \circ \Delta\} \times h) \circ \Delta] \\
&= ([f]\cdot[g])\cdot[h].
\end{aligned}$$

Again $[f]\cdot[c] = [\mu \circ (f \times c) \circ \Delta] = [\mu \circ (1, c) \circ f] = [1_P \circ f] = [f]$ and similarly, $[c]\cdot[f] = [f]$, $\forall\ [f] \in [X, P]] \Rightarrow [c]$ is an identity element for $[X, P]$. Finally, $[\phi \circ f]\cdot[f] = [\mu \circ ((\phi \circ f) \times f) \circ \Delta] = [\mu \circ (\phi, 1) \circ f] = [c]$ and $[f]\cdot[\phi \circ f] = [c] \Rightarrow [f]^{-1} = [\phi \circ f]$. Consequently, if (P, p_0) is an H-group, then $[X, P]$ is a group. If μ is homotopy commutative, the last part follows immediately ❑

The converse of the Theorem 2.4.21 is also true.

Theorem 2.4.22 *The set $[X, P]$ admits a group structure natural with respect to X iff P is an H-group.*

Proof It follows from the Theorems 2.4.20 and 2.4.21. ❑

Remark 2.4.23 The set $[X, P]$ can be endowed with a monoid structure natural with respect to X iff P is a homotopy associative H-space.

Theorem 2.4.24 *If $g : X \to Y$ is a base point preserving continuous map and P is an H-group, then the induced function $g^* : [Y, P] \to [X, P]$ defined by $g^*([\alpha]) = [\alpha \circ g]$ is a group homomorphism. In particular, if g is a homotopy equivalence, then g^* is an isomorphism.*

Proof Given continuous maps $f_1, f_2 : Y \to P$, we have $(f_1 \cdot f_2) \circ g = \mu \circ (f_1 \times f_2) \circ \Delta \circ g = \mu \circ (f_1 \times f_2) \circ (g \times g) \circ \Delta = \mu \circ (f_1 \circ g \times f_2 \circ g) \circ \Delta = (f_1 \circ g) \cdot (f_2 \cdot g)$. Then $g^*([f_1] \cdot [f_2]) = g^*[f_1] \cdot g^*[f_2] \Rightarrow g^*$ is a homomorphism. If g is a homotopy equivalence, then $\exists$ a continuous map $f : Y \to X$ such that $f \circ g \simeq 1_X$ and $g \circ f = 1_Y$. Then $(f \circ g)^* = g^* \circ f^* = 1_d$ and $(g \circ f)^* = f^* \circ g^* = 1_d \Rightarrow g^*$ is an isomorphism. ❑

Theorem 2.4.25 *If P is an H-group, π^P is a contravariant functor from the homotopy category of pointed topological spaces to the category of groups and homomorphisms. If P is an abelian H-group, then the functor π^P takes values in the category of abelian groups.*

Proof Define the object function by $\pi^P(X) = [X, P]$, which is a group for every pointed space X by Theorem 2.4.22. If $g : X \to Y$ is a base point preserving map, define the morphism function by $\pi^P(g) = g^*$ by Theorem 2.4.24. Then the theorem follows. ❑

The converse of the Theorem 2.4.25 is also true.

Theorem 2.4.26 *If P is a pointed topological space such that π^P takes values in the category of groups, then P is an H-group (abelian if π^P takes values in the category of abelian groups). Moreover, for any pointed space X, the group structures on $\pi^P(X)$ and on $[X, P]$ given in the Theorem 2.4.21 coincide.*

Proof Let $p_1 : P \times P \to P$ and $p_2 : P \times P \to P$ be the projections on the first and second factor respectively. Let $\mu : P \times P \to P$ be a map such that $[\mu] = [p_1] \cdot [p_2]$, where $\cdot$ is the product in the group $[P \times P, P]$. For any continuous maps $f, g : X \to P$, the induced map $(f, g)^* : [P \times P, P] \to [X, P]$ is a homomorphism and

$$[\mu \circ (f, g)] = (f, g)^*[\mu] = (f, g)^*([p_1] \cdot [p_2]) = (f, g)^*[p_1] \cdot (f, g)^*[p_2] = [f] \cdot [g]$$

implies that the multiplication in $[X, P]$ is induced by the multiplication map μ. Let X be a one-point space. The unique map $X \to P$ represents the identity element of the group $[X, P]$. Since the unique map $P \to X$ induces a homomorphism $[X, P] \to [P, P]$, it follows that the composite $P \to X \to P$, which is the constant map $c : P \to P$ represents the identity element of $[P, P]$. Hence it follows that $\mu \circ (1_P, c) \simeq 1_P$ and $\mu \circ (c, 1_P) \simeq I_P$. Consequently, P is an H-space. To prove that μ is homotopy associative, let $q_1, q_2, q_3 : P \times P \times P \to P$ be the projections. Then $[\mu \circ (1_P \times \mu)] = (1_P \times \mu)^*[\mu] = (1_P \times \mu)^*[p_1] \cdot (1_P \times \mu)^*[p_2] = [q_1] \cdot [\mu \circ (q_2, q_3)] = [q_1] \cdot ([q_2] \cdot [q_3])$. Similarly, $[\mu \circ (\mu \times 1_P)] = ([q_1] \circ [q_2]) \circ [q_3]$. Since $[P \times P \times P, P]$ has an associative multiplication, it follows that

$$\mu \circ (1_P \times \mu) \simeq \mu \circ (\mu \times 1_P).$$

Finally, we show that P has a homotopy inverse. Let $\phi : P \to P$ be the map such that $[1_P]\cdot[\phi] = [c]$. Then $\mu \circ (1_P, \phi) \simeq c$. Similarly, $\mu \circ (\phi, 1_P) \simeq c$. Hence, ϕ is a homotopy inverse for P and μ.

Consequently, P is an H-group. Moreover, if $[P \times P, P]$ is an abelian group, a similar argument shows that P is an abelian H-group. ❑

Given two H-groups P and P', we now compare between the contravariant functors π^P and $\pi^{P'}$.

Theorem 2.4.27 *Let $\alpha : P \to P'$, be a continuous map between H-groups. Then α induces a natural transformation $\alpha_* : \pi^P \to \pi^{P'}$ in the category of H-groups iff α is a homomorphism.*

Proof For each pointed topological space X, define $\alpha_*(X) : \pi^P(X) \to \pi^{P'}(X)$ by the rule $\alpha_*(X)[h] = [\alpha \circ h]$, $\forall\, [h] \in \pi^P(X)$. Then diagram in Fig. 2.11 is commutative, for every $f : Y \to X$, because, $(\pi^{P'}(f) \circ \alpha_*(X))[h] = \pi^{P'}(f)([\alpha \circ h]) = [(\alpha \circ h) \circ f]$ and $(\alpha_*(Y) \circ \pi^P(f))[h] = \alpha_*(Y)[h \circ f] = [\alpha \circ (h \circ f)]$, which are equal. Hence α_* is a natural transformation.
The converse part is left as an exercise. ❑

We now investigate the question of existence of homotopy inverses for a homotopy associative H-space.

Theorem 2.4.28 *If P is a homotopy associative H-space, then P is an H-group if and only if the shear map $\psi : P \times P \to P \times P$, given by $\psi(x, y) = (x, xy)$ is a homotopy equivalence.*

Proof Case I. First we consider the particular case when P is a topological group. Then the map ψ is a homeomorphism with inverse $\psi^{-1} : P \times P \to P \times P$ defined by $\psi^{-1}(u, v) = (u, u^{-1}v)$. Let $j = p_2 \circ \psi^{-1} \circ i_1$, where $i_1 : P \to P \times P$ is the inclusion, defined by $i_1(y) = (y, y_0)$, where y_0 is the base point of P and $p_1, p_2 : P \times P \to P$ be the projections on the first and the second factor respectively. Then $[j]$ is the inverse of the homotopy class of the identity map $1_P \in [P, P]$, so that the composites

$$P \xrightarrow{\Delta} P \times P \xrightarrow{j \times 1_P} P \times P \xrightarrow{\mu} P,$$

$$P \xrightarrow{\Delta} P \times P \xrightarrow{1_P \times j} P \times P \xrightarrow{\mu} P$$

are each nullhomotopic.

$$\begin{array}{ccc} \pi^P(X) & \xrightarrow{\alpha_*(X)} & \pi^{P'}(X) \\ {\scriptstyle \pi^P(f)=f^*}\downarrow & & \downarrow{\scriptstyle f^*=\pi^{P'}(f)} \\ \pi^P(Y) & \xrightarrow{\alpha_*(Y)} & \pi^{P'}(Y) \end{array}$$

Fig. 2.11 Natural transformation α_*

Case II. We now consider the general case. Let ψ be a homotopy equivalence with homotopy inverse ϕ. Define $j \in [P, P]$ by $j = p_2 \circ \phi \circ i_1$. Then

$$\begin{aligned}(p_1 \circ \psi)(x, y) &= p_1(\psi(x, y)) = p_1(x, xy) = x \\ &= p_1(x, y),\ \forall\ (x, y) \in P \times P \Rightarrow p_1 \circ \psi = p_1.\end{aligned}$$

Again, $(p_2 \circ \psi)(x, y) = p_2(x, xy) = xy = \mu(x, y),\ \forall\ \ (x, y) \in P \times P \Rightarrow p_2 \circ \psi = \mu$.

Hence

$$p_1 \simeq p_1 \circ \overbrace{\psi \circ \phi}^{1_d} = (p_1 \circ \psi) \circ \phi = p_1 \circ \phi,$$

$$p_2 \simeq p_2 \circ \overbrace{\psi \circ \phi}^{1_d} = (p_2 \circ \psi) \circ \phi = \mu \circ \phi.$$

In particular, $p_1 \circ \phi \circ i_1 \simeq p_1 \circ i_1 = 1_P$, since $(p_1 \circ i_1)(y) = p_1(y, y_0) = y = 1_P(y),\ \forall\ y \in P$.

Hence

$$\begin{aligned}\mu \circ (1_P \times j) \circ \Delta &= \mu \circ (p_1 \circ \phi \circ i_1 \times p_2 \circ \phi \circ i_1) \circ \Delta \\ &= \mu \circ (p_1 \times p_2) \circ (\phi \circ i_1 \times \phi \circ i_1) \circ \Delta \\ &= \mu \circ (p_1 \times p_2) \circ \Delta \circ \phi \circ i_1 \\ &= \mu \circ \phi \circ i_1 = p_2 \circ \overbrace{\psi \circ \phi}^{1_d} \circ i_1 \simeq p_2 \circ i_1 \simeq c,\end{aligned}$$

where $c : P \to p_0 \in P$ is the constant map.

Hence j is a right inverse of the identity map.

It follows from the above argument that every element of $[X, P]$ has a left inverse, and hence $[X, P]$ is a group. Conversely, if P is an H-group, then the map $\phi : (u, v) \mapsto (u, j(u)v)$ is a homotopy inverse of the shear map ψ, because $(\psi \circ \phi)(u, v) = \psi(\phi(u, v)) = \psi(u, j(u)v) = (u, uj(u)v) = (u, p_0 v)$ and since p_0 is a homotopy unit, it follows that $\psi \circ \phi \simeq 1_d$. Similarly, $\phi \circ \psi \simeq 1_d$. Consequently, ψ is a homotopy equivalence. ❑

Remark 2.4.29 Some of the techniques which apply to topological groups can be applied to H-spaces, but not all. From the viewpoint of homotopy theory, it is not the existence of a continuous inverse which is the important distinguishing feature, but rather the associativity of multiplication. If we consider S^1, S^3 and S^7 as the complex, quaternionic and Cayley numbers of unit norm, these spaces have continuous multiplication. The multiplication in the first two cases are associative but not associative in the last case. $S^!$ and S^3 are topological groups. The spheres S^1, S^3 and S^7 are the only spheres that are H-spaces proved by J.F. Adams (1930–1989) in (1962).

Remark 2.4.30 Every topological group is an H-group.

We now describe another important example of an H-group. Loop spaces form an important class of grouplike spaces, called H-groups.

Definition 2.4.31 (*Loop Space*) Let Y be a pointed topological space with base point y_0. The loop space of Y(based at y_0) denoted ΩY(or $\Omega(Y, y_0)$), is defined to be the space of continuous functions $\alpha : (I, \dot{I}) \to (Y, y_0)$, topologized by the compact open topology. Then $\Omega(Y, y_0)$ is considered as a pointed space with base point α_0 equals to the constant map $c : I \to y_0$.

The elements of ΩY are called loops in Y.

Theorem 2.4.32 *$\Omega(Y, y_0)$ is an H-group.*

Proof Define a map $\mu : \Omega Y \times \Omega Y \to \Omega Y$ by

$$\mu(\alpha, \beta)(t) = \begin{cases} \alpha(2t), 0 \leq t \leq 1/2 \\ \beta(2t-1), 1/2 \leq t \leq 1. \end{cases}$$

To show that μ is continuous, consider the evaluation map $E : \Omega Y \times I \to Y$ defined by $E(\alpha, t) = \alpha(t)$. Since I is locally compact, by Theorem of exponential correspondence (see Theorem 1.14.2 of Chap. 1) it is sufficient to show that the composite map

$$\Omega Y \times \Omega Y \times I \xrightarrow{\mu \times 1_d} \Omega Y \times I \xrightarrow{E} Y$$

is continuous.

Then the theorem of exponential correspondence and the Pasting lemma show the continuity of μ, since the above composite is continuous on each of the closed sets $\Omega Y \times \Omega Y \times [0, \frac{1}{2}]$ and $\Omega Y \times \Omega Y \times [\frac{1}{2}, 1]$.

μ **is associative**: To show this define $G : \Omega Y \times \Omega Y \times \Omega Y \times I \to \Omega Y$ by the rule

$$G(\alpha, \beta, \gamma, s)(t) = \begin{cases} \alpha(\frac{4t}{1+s}), & 0 \leq t \leq (1+s)/4 \\ \beta(4t-1-s), & (1+s)/4 \leq t \leq (2+s)/4 \\ \gamma(\frac{4t-2-s}{2-s}), & (2+s)/4 \leq t \leq 1. \end{cases}$$

The continuity of G follows from the Pasting lemma. Clearly, $G : \mu \circ (\mu \times 1_d) \simeq \mu \circ (1_d \times \mu)$

Existence of homotopy unit: If $c : \Omega Y \to \Omega Y$ is the constant map whose value is the constant loop $c : I \to Y$, $c(t) = y_0$, then $\mu \circ (\beta, c) \simeq \beta$ and $\mu \circ (c, \beta) \simeq \beta$ for every loop β.

The first homotopy is given by $F : \Omega Y \times I \to \Omega Y$, where

$$F(\beta, s)(t) = \begin{cases} \beta(\frac{2t}{1+s}), 0 \leq t \leq (1+s)/2 \\ y_0, \quad (1+s)/2 \leq t \leq 1. \end{cases}$$

The continuity of F follows from Pasting lemma. The second homotopy is defined in an analogous manner.

Existence of homotopy inverse: Let $\phi : \Omega Y \to \Omega Y$ be a map such that $\phi(\alpha)(t) = \alpha(1-t)$. Then ϕ determines homotopy inverses. The homotopy $H : \Omega Y \times I \to \Omega Y$, where

$$H(\alpha, s)(t) = \begin{cases} \alpha(2(1-s)t), & 0 \leq t \leq 1/2 \\ \alpha(2(1-s)(1-t)), & 1/2 \leq t \leq 1, \end{cases}$$

begins at $\mu \circ (\alpha, \phi(\alpha))$ and ends at c. The second homotopy is given in an analogous manner.

Consequently, ΩY is an H-group. □ ❑

Definition 2.4.33 Given a pointed space X, iterated loop spaces $\Omega^n X$ are defined inductively: $\Omega^n X = \Omega(\Omega^{n-1} X)$ for ≥ 1 and $\Omega^0 X$ is taken to be X.

Corollary 2.4.34 *For $n \geq 1$, $\Omega^n X$ is an H-group for every pointed space X.*

Theorem 2.4.35 *Ω is a covariant functor from the category of pointed topological spaces and continuous maps to the category of H-groups (Hopf groups) and their continuous homomorphisms.*

Proof If $f : X \to Y$ is base point preserving continuous maps, then $\Omega f : \Omega(X) \to \Omega(Y)$ defined by $(\Omega f)(\alpha)(t) = f(\alpha(t))$ is a homomorphisms of H-groups. The object function is given by $X \mapsto \Omega(X)$ and the morphism function is given by $\Omega f : \Omega(X) \to \Omega(X)$. Then Ω is a covariant functor. ❑

Theorem 2.4.36 *For every pointed topological space Y, ΩY is an H-group and for every pointed space X, $[X, \Omega Y]$ is a group. If $f : X \to X'$ is a base point preserving continuous map, then $f^* : [X', \Omega Y] \to [X, \Omega Y]$ is a group homomorphism.*

Proof The theorem follows from Theorems 2.4.32 and 2.4.24. ❑

2.4.2 *H-Cogroups and Suspension Spaces*

This subsection conveys the dual concepts of H-groups, called H-cogroups introduced by Beno Eckmann (1917–2008) and Peter John Hilton (1923–2010) in 1958. It involves wedge products of pointed topological spaces. Suspension spaces of pointed topological spaces form an extensive class of H-cogroups.

Recall that the wedge $X \vee X$ is viewed as the subspace $X \times \{x_0\} \cup \{x_0\} \times X$ of the product space $X \times X$. If $p_i : X \times X \to X$, for $i = 1, 2$ are the usual projections onto the first or second coordinate respectively, then define 'projections' $q_i : X \vee X \to X$, for $i = 1, 2$ by $q_i = p_i|_{X \vee X}$; each q_i sends the appropriate copy of $x \in X$, namely, (x, x_0) or (x_0, x) into itself.

Definition 2.4.37 A pointed topological space (X, x_0) is called an H-cogroup if there exists a base point preserving continuous map $\mu : X \to X \vee X$, called comultiplication, such that $q_1 \circ \mu \simeq 1_X \simeq q_2 \circ \mu$, $(1_X \vee \mu) \circ \mu \simeq (\mu \vee 1_X) \circ \mu$ (co-associativity) and there exists a base point preserving continuous map $h : X \to X$ such that $(1_X, h) \circ \mu \simeq c \simeq (h, 1_X) \circ \mu$, ($h$ is called an inverse), where $c : X \to X$ is the constant map at x_0.

Remark 2.4.38 In an H-cogroup given maps $f : X \to Z$ and $g : Y \to Z$ in $\mathcal{Top}_*$, the map $(f, g) : X \vee Y \to Z$ is defined by the characteristic property:
$(f, g)|_X = f$ and $(f, g)|_Y = g$.

H-cogroup is now defined more explicitly keeping similarity with the definition of H-group.

Definition 2.4.39 A pointed topological space X with base point x_0 is called an H-cogroup if there exists a base point preserving continuous map

$$\mu : X \to X \vee X,$$

called H-comultiplication such that the following conditions hold:
Existence of homotopy identity. If $c : X \to X$ is the (unique) constant map at x_0, then each composite map

$$X \xrightarrow{\mu} X \vee X \xrightarrow{(c, 1_X)} X \text{ and } X \xrightarrow{\mu} X \vee X \xrightarrow{(1_X, c)} X$$

is homotopic to 1_X.
Homotopy associativity. The diagram in Fig. 2.12 is commutative up to homotopy, i.e.,

$$(1_X \vee \mu) \circ \mu \simeq (\mu \vee 1_X) \circ \mu.$$

Existence of homotopy inverse. There exists a map $h : X \to X$ such that each composite map

$$X \xrightarrow{\mu} X \vee X \xrightarrow{(1_X, h)} X \text{ and } X \xrightarrow{\mu} X \vee X \xrightarrow{(h, 1_X)} X$$

is homotopic to $c : X \to X$.

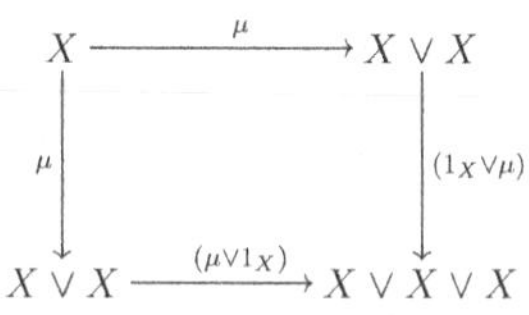

Fig. 2.12 Homotopy associativity of H-cogroups

Fig. 2.13 Diagram for abelian H-cogroup

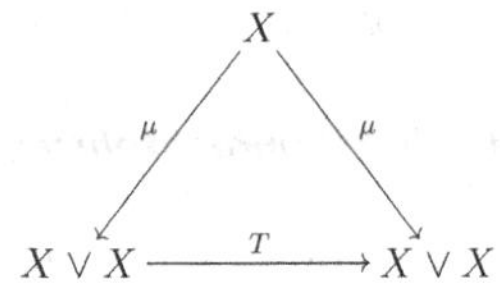

Fig. 2.14 Homotopy homomorphism of H-cogroups

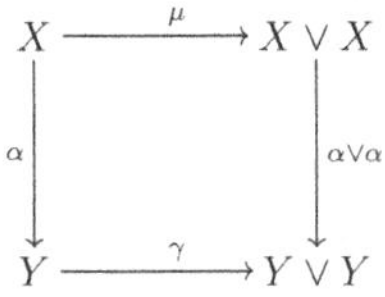

Definition 2.4.40 An H-cogroup X is said to be abelian if the triangle in Fig. 2.13 is homotopy commutative, where $T(x, x') = (x', x), \ \forall\ x, x' \in X$.

Definition 2.4.41 Let X and Y be H-cogroups with comultiplications μ and γ respectively. Then the map $\alpha : X \to Y$ is said to be a homomorphism of H-cogroups if diagram in Fig. 2.14 is homotopy commutative, i.e., $\alpha \vee \alpha \circ \mu \simeq \gamma \circ \alpha$.

Remark 2.4.42 The definition of an H-cogroup closely resembles to that of an H-group. We merely turn all the maps round and use the one-point union instead of the product.

Theorem 2.4.43 *If X is an H-cogroup and Y is any pointed space, then $[X, Y]$ can be given the structure of a group. Moreover, if $g : Y \to Z$ is a base point preserving continuous map, then the induced function $g_* : [X, Y] \to [X, Z]$ is a homomorphism in general and it is an isomorphism if g is a homotopy equivalence.*

Proof Given $f_1, f_2 : X \to Y$, define a product in $[X, Y]$ by the rule $f_1 \cdot f_2 = \nabla \circ (f \vee g) \circ \mu$, where $\nabla : X \vee X \to X$ is the folding map, defined by $\nabla(x_0, x) = \nabla(x, x_0) = x$. Proceed as in proofs of Theorems 2.4.22 and 2.4.24. ❑

Dualizing the Theorems 2.4.16, 2.4.25–2.4.27, following theorems are proved.

Theorem 2.4.44 *A pointed space having the same homotopy type of an H-cogroup is itself an H-cogroup in such a way that the homotopy equivalence is a homomorphism.*

Theorem 2.4.45 *If Q is a H-cogroup, then π_Q is a covariant functor from the homotopy category of pointed spaces with values in the category of groups and homomorphisms. If Q is an abelian H-cogroup, this functor takes values in the category of abelian groups.*

Theorem 2.4.46 *If Q is a pointed topological space such that π_Q takes values in the category of groups, then Q is an H-cogroup (abelian if π_Q takes values in the category of abelian groups). Furthermore, for a pointed topological space X the group structure on $\pi_Q(X)$ is identical with that determined by the H-cogroup Q as in Theorem 2.4.43.*

Theorem 2.4.47 *If $\alpha : Q \to Q'$ is a continuous map between H-cogroups, then there is a natural transformation from $\pi_{Q'}$ to π_Q in the category of groups if and only if α is a homomorphism.*

We now describe suspension spaces which are dual to loop spaces. Suspension spaces give an extensive class of H-cogroups which are dual to H-groups. The impact of suspension operator is realized from a classical theorem of H. Freudenthal (1905–1990) known as Freudenthal suspension theorem (see Chap. 7).

Example 2.4.48 (*Suspension Space*) Let X be a pointed topological space with base point x_0. The suspension space of X, denoted by ΣX, is defined to be the quotient space of $X \times I$ in which $(X \times 0) \cup (x_0 \times I) \cup (X \times 1)$ has been identified to a single point. This is sometimes called the reduced suspension. If $(x, t) \in X \times I$, we use $[x, t]$ to denote the corresponding point of ΣX under the quotient map $X \times I \to \Sigma X$. Then $[x_0, 0] = [x_0, t] = [x', 1]$, $\forall\ x, x' \in X$ and $\forall\ t \in I$. The point $[x_0, 0] \in \Sigma X$ is also denoted by x_0 and ΣX is a pointed space with base point x_0. Moreover, if $f : X \to Y$ is a base point preserving continuous map, then $\Sigma f : \Sigma X \to \Sigma Y$ is defined by

$$\Sigma f([x, t]) = [f(x), t].$$

Consequently, Σ is a covariant functor from the category $\mathcal{Top}_*$ of pointed spaces and continuous maps to itself.

Remark 2.4.49 If $f \simeq g : X \to Y$, then $\Sigma f \simeq \Sigma g : \Sigma X \to \Sigma Y$.

We now show that Σ is also a covariant functor from the category $\mathcal{Top}_*$ to the category of H-cogroups and homomorphisms. We define a comultiplication $\gamma : \Sigma X \to \Sigma X \vee \Sigma X$ by the formula as shown in Fig. 2.15

$$\gamma([x, t]) = \begin{cases} ([x, 2t], x_0), & 0 \leq t \leq 1/2 \\ (x_0, [x, 2t - 1]), & 1/2 \leq t \leq 1. \end{cases}$$

Clearly, γ is continuous and makes ΣX an H-cogroup.

Theorem 2.4.50 *For any pointed topological space X, ΣX is an H-cogroup. Moreover, if $f : X \to Y$ is a base point preserving continuous map, then $\Sigma f : \Sigma X \to \Sigma Y$, $[x, t] \mapsto [f(x), t]$ is a homomorphism of H-cogroups.*

Proof As ΣX and ΣY are both H-cogroups, the proof follows from the definition of Σf. ❑

Theorem 2.4.51 *For any pair of pointed spaces X and Y, $[\Sigma X \to Y]$ is a group.*

Proof As ΣX is an H-cogroup, the theorem follows from Theorem 2.4.43. ❑

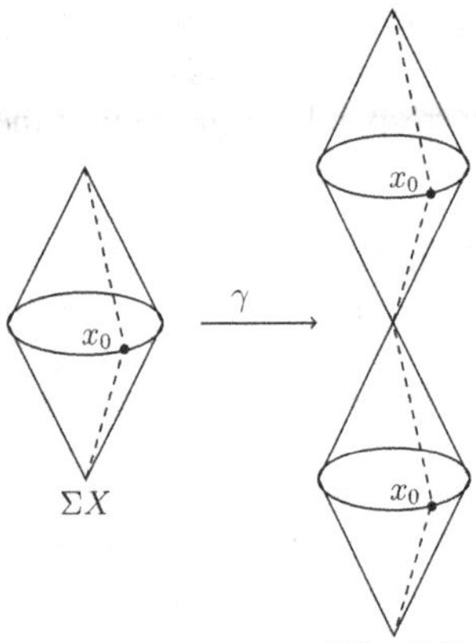

Fig. 2.15 Comultiplication γ

2.5 Adjoint Functors

This section provides an example of a special pair of functors, called adjoint functors in the language of category theory. This categorical notion of adjoint functors was introduced by Daniel Kan (1927–2013) in 1958. There is a close relation between the loop functor Ω and the suspension functor Σ in the category $\mathcal{Top}_*$.

Proposition 2.5.1 *The functors Ω and Σ defined from the category $\mathcal{Top}_*$ of pointed spaces and continuous maps to itself form a pair of adjoint functors in the sense that for pointed topological spaces X and Y in $\mathcal{Top}_*$ there is an equivalence* $\text{mor}\,(\Sigma X, Y) \approx \text{mor}\,(X, \Omega Y)$*, where both sides are the set of morphisms in the category $\mathcal{Top}_*$.*

Proof If $g : X \to \Omega Y$ is in $\mathcal{Top}_*$, then the corresponding morphism $g' : \Sigma X \to Y$ is defined by $g'[x, t] = g(x)(t)$, $\forall\ x \in X$ and $\forall\ t \in I$. Thus if $h : Y \to Y'$, then $(\Omega h \circ g)' = h \circ g' : \Sigma X \to Y'$, and if $f : X' \to X$, then $(g \circ f)' = g' \circ \Sigma f : \Sigma X' \to Y$. Then the correspondence $g \leftrightarrow g'$ gives a natural equivalence from the functor $(\Sigma -, -)$ to the functor $(-, \Omega -)$ on the category $\mathcal{Top}_*$. ❑

This natural equivalence plays an important role in the homotopy category $\mathcal{Htp}_*$ of pointed topological spaces.

Theorem 2.5.2 *There exists a natural equivalence from the functor $[\Sigma -, -]$ to the functor $[-, \Omega -]$ on the category $\mathcal{Htp}_*$.*

Proof For pointed spaces X and Y, a homotopy $G : X \times I \to Y$ maps $x_0 \times I$ into y_0. Therefore it defines a map $F : X \times I/x_0 \times I \to Y$. Since $\Sigma(X \times I/x_0 \times I)$ can be identified with $\Sigma X \times I/x_0 \times I$ by the homeomorphism $[(x, t), t'] \leftrightarrow ([x, t'], t)$, $\forall\ x \in X, t, t' \in I$, it follows that homotopies $F : X \times I/x_0 \times I \to \Omega Y$ correspond bijectively to homotopies $F' : \Sigma X \times I/x_0 \times I \to Y$. Consequently, the equivalence defined in Proposition 2.5.1 gives rise to an equivalence $[\Sigma X, Y] \approx [X, \Omega Y]$ such that if $g : X \to \Omega Y$ and $g' : \Sigma X \to Y$ are related by $g'[x, t] = g(x)(t)$, then $[g']$ corresponds to $[g]$. Hence there is a natural equivalence from the functor $[\Sigma -, -]$ to the functor $[-, \Omega -]$. ❑

Definition 2.5.3 In the language of category theory, the functors Ω and Σ are called adjoint functors in the sense of Theorem 2.5.2.

The above results are summarized in the basic Theorem 2.5.4.

Theorem 2.5.4 *The suspension functor Σ is a covariant functor from the category $\mathcal{T}op_*$ of pointed topological spaces and continuous maps to the category of H-cogroups and continuous homomorphisms. Moreover, the functor Σ preserves homotopies, i.e., if $f_0, f_1 : X \to Y_0$ are homotopic by the homotopy F_t, then $\Sigma f_0, \Sigma f_1$ are homotopic by the homotopy ΣF_t, which is a continuous homomorphism for each $t \in I$.*

Corollary 2.5.5 *The suspension functor Σ is a covariant functor from the homotopy category of pointed topological spaces and homotopy classes of continuous maps to the category of H-cogroups and continuous homomorphisms.*

We now show that for $n \geq 1$, the sphere S^n admits an extensive family of H-cogroups.

Proposition 2.5.6 *For $n \geq 0$, S^{n+1} is an H-cogroup.*

Proof To show this it is sufficient to prove that $\Sigma S^n \approx S^{n+1}$. Let $p_0 = (1, 0, \ldots, 0)$ be the base point of S^n. We consider $\mathbf{R}^{n+1}$ as embedded in $\mathbf{R}^{n+2}$ as the set of points in $\mathbf{R}^{n+2}$ whose $(n+2)$nd coordinate is 0. Then S^n is embedded as an equator in S^{n+1}. Again $S^n = \{x \in \mathbf{R}^{n+2} : ||x|| = 1 \text{ and } x_{n+2} = 0\}$ and D^{n+1} is also embedded in D^{n+2}, where $D^{n+1} = \{x \in \mathbf{R}^{n+2} : ||x|| \leq 1 \text{ and } x_{n+2} = 0\}$. Let H_+ and H_- be two closed hemispheres of S^{n+1} defined by the equator S^n. Then $H_+ = \{x \in S^{n+1} : x_{n+2} \geq 0\}$, called upper hemisphere, $H_- = \{x \in S^{n+1} : x_{n+2} \leq 0\}$, called lower hemisphere are such that $S^{n+1} = H_+ \cup H_-$ and $S^n = H_+ \cap H_-$.

The maps

$$p_+ : (D^{n+1}, S^n) \to (H_+^{n+1}, S^n), (x_1, x_2, \ldots, x_{n+1}, 0)$$
$$\mapsto \left(x_1, x_2, \ldots, x_{n+1}, \sqrt{1 - \sum_{i=1}^{n+1} x_i^2}\right)$$

and
$$p_- : (D^{n+1}, S^n) \to (H_-^{n+1}, S^n), (x_1, x_2, \ldots, x_{n+1}, 0)$$
$$\mapsto \left(x_1, x_2, \ldots, x_{n+1}, -\sqrt{1 - \sum_{i=1}^{n+1} x_i^2}\right)$$

are homeomorphisms. Again for $t \in I, x \in S^n$, the point $tx + (1-t)p_0 \in D^{n+1}$.

Clearly, the map $f : \Sigma(S^n) \to S^{n+1}$ defined by

$$f([x, t]) = \begin{cases} p_-^{-1}(2tx + (1-2t)p_0), & 0 \le t \le 1/2 \\ p_+^{-1}((2-2t)x + (2t-1)p_0), & 1/2 \le t \le 1 \end{cases}$$

is well defined and bijective. It is a homeomorphism, since $S^1 \wedge S^n$ and S^{n+1} are both compact.

Hence $f : \Sigma(S^n) \approx S^{n+1} \Rightarrow S^{n+1}$ is an H-cogroup, since $\Sigma(S^n)$ is an H-cogroup. ❑

Definition 2.5.7 Given a pointed topological space X, its iterated suspension spaces are defined inductively:

$$\Sigma^n X = \Sigma(\Sigma^{n-1} X) \text{ for } n \ge 1, \text{ and } \Sigma^0 X \text{ is taken to be } X.$$

Remark 2.5.8 The groups $[\Sigma X, \Omega Y], [X, \Omega^2 Y]$ and $[\Sigma^2 X, Y]$ are isomorphic for any pointed space Y.

Corollary 2.5.9 *For every integer $n \ge 0$, any pointed topological space X, $\Sigma^n X$ is homeomorphic to $S^n \wedge X$.*

Proof Since $S^0 = \{-1, 1\}$, it follows that $S^0 \wedge X \approx X \approx \Sigma^0 X$. Suppose $\Sigma^n X = S^n \wedge X$. Then

$$\Sigma^{n+1} X = \Sigma(\Sigma^n X) \approx \Sigma(S^n \wedge X) = S^1 \wedge (S^n \wedge X) \approx (S^1 \wedge S^n) \wedge X \approx S^{n+1} \wedge X.$$

Hence the corollary follows by induction on n. ❑

Corollary 2.5.10 *For every integer $n \ge 0$, the $(n+1)$-space S^{n+1} is an H-cogroup.*

Remark 2.5.11 The Corollary 2.5.10 shows that for $n \ge 1$, the space S^n admits an extensive family of H-cogroups.

Definition 2.5.12 (*Adjoint functors*) In the language of category theory the equivalence between $[\Sigma X, Y]$ and $[\Sigma X, \Omega Y]$ is expressed by saying that in the homotopy subcategory of pointed Hausdorff spaces of the homotopy category $\mathcal{H}tp_*$ of pointed topological spaces, the functors Ω and Σ are adjoint.

Remark 2.5.13 Recall that given a pointed topological space X, we have formed iterated loop spaces $\Omega^n X$ inductively: $\Omega^n X = \Omega(\Omega^{n-1} X)$ for ≥ 1 and $\Omega^0 X$ is taken to be X. and we have similarly formed iterated suspension spaces inductively:

$$\Sigma^n X = \Sigma(\Sigma^{n-1} X) \text{ for } n \ge 1, \text{ and } \Sigma^0 X \text{ is taken to be } X.$$

Then the groups $[\Sigma X, \Omega Y], [X, \Omega^2 Y]$ and $[\Sigma^2 X, Y]$ are isomorphic for any pointed topological space Y.

Theorem 2.5.14 **(i)** *$\Omega^n X$ is an abelian H-group for $n \geq 2$ for all pointed topological spaces X.*

(ii) *$\Sigma^n X$ is an abelian H-cogroup for $n \geq 2$ for all pointed topological spaces X.*

(iii) *For any pair of pointed Hausdorff spaces X and Y, the adjoint functors Σ and Ω give an isomorphism*

$$\psi : [\Sigma X, Y] \to [X, \Omega Y]$$

of groups. For $n \geq 2$, the isomorphisms

$$\psi : [\Sigma X, \Omega^{n-1} Y] \to [X, \Omega^n Y]$$

are of abelian groups;

Proof **(i)** Let X be a pointed topological space with base point x_0. Then for any pointed topological space Y,

$$[X, \Omega^n Y] \cong [\Sigma X, \Omega^{n-1} Y] = [\Sigma X, \Omega(\Omega^{n-2} Y)]$$

is an abelian group. Hence if $[f], [g] \in [X, \Omega^n Y]$, then $[f]\cdot[g] = [g]\cdot[f]$. This shows that

$$\mu \circ (f \times g) \circ \Delta \simeq \mu \circ (g \times f) \circ \Delta.$$

In particular, if $X = \Omega^n Y \times \Omega^n Y$, $f = p_1$, the projection on the first factor and $g = p_2$, the projection on the second factor, then we have

$$(f \times g) \circ \Delta(x, y) = (p_1 \times p_2)((x, y), (x, y)) = (x, y).$$

This implies that $(f \times g) \circ \Delta = 1_d$. On the other hand,

$$(g \times f) \circ \Delta(x, y) = (p_2 \times p_1)((x, y), (x, y)) = (y, x) = T(x, y).$$

This implies that $(g \times f) \circ \Delta = T$. Hence $\mu \simeq \mu \circ T$ shows that μ is homotopy commutative. Consequently, (i) follows from Theorem 2.4.36.

(ii) Similarly, $\Sigma^n X$ is homotopy commutative. Hence (ii) follows from Theorem 2.4.50.

(iii) It follows from Ex. 29 of Sect. 2.11. ❑

Remark 2.5.15 If X is an H-cogroup and Y is an H-group, the products available in $[X, Y]$ determine isomorphic groups which are abelian.

2.6 Contractible Spaces

This section studies a special class of topological spaces, called contractible spaces, for each of which there exists a homotopy that starts with the identity map and ends with some constant map. This introduces the concept of contractible spaces. The concept of contractible spaces is very important. Contractible spaces are in a natural sense, the trivial objects from the view point of homotopy theory, because all contractible spaces have the homotopy type of a space reduced to a single point. Such spaces are connected topological objects having no 'holes' or 'cycles' and have nice intrinsic properties. The simplest nonempty space is one-point space. We characterize the homotopy type of such spaces.

2.6.1 Introductory Concepts

This subsection opens with introductory concepts of contractible spaces.

Definition 2.6.1 A topological space X is said to be contractible if the identity map $1_X : X \to X$ is homotopic to some constant map of X to itself. If $c : X \to X$ defined by $c(x) = x_0 \in X$ is such that $1_X \simeq c$, then a homotopy $F : 1_X \simeq c$ is called a contraction of the space X to the point x_0.

Example 2.6.2 Any convex subspace X of $\mathbf{R}^n$ is contractible. Because, any continuous map $H : X \times I \to X$ defined by $H(x, t) = (1 - t)x + tx_0, x, x_0 \in X, t \in I$ is such that $H(x, 0) = x = 1_X(x), \ \forall \ x \in X$ and $H(x, 1) = x_0 = c(x), \ \forall \ x \in X$. Hence $H : 1_X \simeq c \Rightarrow X$ is contractible and H is a contraction of X to the point $x_0 \in X$. In particular, $\mathbf{R}^n$, D^n, I are contractible spaces.

Geometrical meaning: A contraction $H : 1_X \simeq c$ can be interpreted geometrically as a continuous deformation of the space X which ultimately shrinks the whole space X into the point $x_0 \in X$ and hence X can be contracted to a point of X.

Can we contract a topological space X to an arbitrary point $x_0 \in X$?

To answer this question we need the following Proposition:

Proposition 2.6.3 *A topological space X is contractible if and only if an arbitrary continuous map $f : Y \to X$ from any topological space Y to X is homotopic to a constant map.*

Proof Let X be contractible. Then the identity map $1_X : X \to X$ is homotopic to some constant map $c : X \to X, x \mapsto x_0$ (say). Let $f : Y \to X$ be any continuous map. Now $1_X \simeq c \Rightarrow 1_X \circ f \simeq c \circ f$. But $c \circ f : Y \to X, y \mapsto x_0$ is a constant map. Thus f is homotopic to a constant map. For the converse, we take $Y = X$ and $f = 1_X : X \to X$. Then by hypothesis, $1_X : X \to X$ is a constant map. Hence X is contractible. ❑

Corollary 2.6.4 *Any two continuous maps from an arbitrary space to a contractible space are homotopic.*

Proof Let X be a contractible space and $f, g : Y \to X$ be two continuous maps from an arbitrary space Y to the space X. Now $1_X \simeq c$, where $c : X \to X$ is defined by $x \mapsto x_0 \in X \Rightarrow 1_X \circ f \simeq c \circ f$ and $1_X \circ g \simeq c \circ g \Rightarrow f = 1_X \circ f \simeq c \circ f = c \circ g \simeq 1_X \circ g = g \Rightarrow f \simeq g$. ❑

Corollary 2.6.5 *If X is contractible, then the identity map $1_X : X \to X$ is homotopic to any constant map of X to itself.*

Proof If X is contractible, then by Corollary 2.6.4 it follows in particular that 1_X is homotopic to any constant map of X to itself. ❑

Remark 2.6.6 In absence of the condition of contractibility of X, the Corollary 2.6.4 fails.

Example 2.6.7 Let X be a connected topological space and $Y = \{y_0, y_1\}$, $(y_0 \neq y_1)$ with discrete topology, i.e., Y is a discrete space consisting of two distinct elements. Consider the constant maps $f, g : X \to Y$ defined by constant $f(x) = y_0$ and $g(x) = y_1$, $\forall\ x \in X$. Then f and g are not homotopic (see Example 2.2.19)

We now characterize contractible spaces.

Theorem 2.6.8 *A topological space X is contractible if and only if X is of the same homotopy type of a one-point space $P = \{p\}$.*

Proof Suppose X is contractible. Then $1_X : X \to X$ is homotopic to some constant map $c_o : X \to X, x \mapsto x_0 \in X$. Let $H : 1_X \simeq c_0$. Define maps $i : P \to X$ and $c : X \to P$ by $i(P) = x_0$ and $c(x) = p$, $\forall\ x \in X$. Then $c \circ i = 1_P$. Moreover, $H : 1_X \simeq i \circ c$, because $H(x, 0) = x$ and $H(x, 1) = c_0$. Hence $X \simeq P$. Conversely, let $X \simeq P$. then $\exists$ continuous maps $f : X \to P$ and $g : P \to X$ such that $g \circ f \simeq 1_X$ and $f \circ g \simeq 1_P$. Leth $g(p) = x_0 \in X$ and $H : 1_X \simeq g \circ f$. Since $(g \circ f)(x) = g(f(x)) = g(p) = x_0$, $\forall\ x \in X, g \circ f : X \to X, x \mapsto x_0$, is the constant map c_0. Thus $1_X \simeq c_0 \Rightarrow X$ is contractible. ❑

Corollary 2.6.9 *Two contractible spaces have the same homotopy type, and any continuous map between contractible spaces is a homotopy equivalence.*

Proof Let X and Y be two contractible spaces and P be a one-point space. Then $X \simeq P$ and $Y \simeq P \Rightarrow X \simeq Y$ by symmetry and transitivity of the relation $\simeq$. Hence $\exists$ a homotopy equivalence $f \in C(X, Y)$. Let $g : X \to Y$ be an arbitrary continuous map. Then $f \simeq g$ by Corollary 2.6.4 $\Rightarrow g$ is a homotopy equivalence by Proposition 2.2.15. ❑

Remark 2.6.10 Contractible spaces are precisely those spaces which are homotopy equivalent to a point space. Thus all contractible spaces have the homotopy type of a space reduced to a single point.

Definition 2.6.11 A topological space X is said to be contractible to a point $a \in X$ relative to the subset $A = \{a\}$ if $\exists$ a homotopy $H : X \times I \to X$ such that $H : 1_X \simeq c$ rel A, where $c : X \to X, x \mapsto a$ is a constant map.

Theorem 2.6.12 *If a topological space X is contractible to a point $a \in X$ relative to the subset $A = \{a\}$, then for each neighborhood U of a in X, $\exists$ a neighborhood V of a contained in U such that any point of V can be joined to a by a path lying entirely inside U.*

Proof Let the space X be contractible to a point $a \in X$ relative to the subset $A = \{a\}$. Then there exists a continuous map F such that $F : 1_X \simeq c$ rel $A \Rightarrow$ the line $\{a\} \times I$ is mapped by F to the point $a \in X$. We now take a neighborhood U of a. Then the continuity of $F \Rightarrow$ for each $t \in I$, neighborhoods $V_t(a)$ of a in X and $W(t)$ of t in I are such that $F(V_t(a) \times W(t)) \subset U$. Since I is compact, the open covering $\{W(t) : t \in I\}$ of I has a finite subcovering $W(t_1), W(t_2), \ldots, W(t_n)$ (say) such that $F(V_{t_i}(a) \times W(t_i)) \subset U$, for $i = 1, 2, \ldots, n$. Thus $V(a) = \bigcap_{i=1}^{n} V_{t_i}(a)$ is a neighborhood of a in X such that $F(V(a) \times I) \subset U$. Now, if $x \in V(a)$, then considering the image $F(V(a) \times I)$ in U, it follows that the point x can be joined to the point a by a path which lies inside U. ❑

Proposition 2.6.13 *Every contractible space is path-connected.*

Proof Let X be contractible to a point $x_0 \in X$ and $H : 1_X \simeq c$, where $c : X \to X, x \mapsto x_0 \in X$ is a constant map. Now $H(x, 0) = 1_X(x) = x$ and $H(x, 1) = c(x) = x_0$, $\forall\ x \in X$. Given $a \in X$ define a path $f : I \to X \times I$ by $f(t) = (a, t)$. Then $\alpha = H \circ f : I \to X$ is a continuous map such that $\alpha(0) = H(f(0)) = H(a, 0) = a$ and $\alpha(1) = H(f(1)) = H(a, 1) = x_0 \Rightarrow \alpha$ is a path from a to x_0. In other words, X is path-connected. ❑

2.6.2 *Infinite-Dimensional Sphere and Comb Space*

We now examine the contractibility of the infinite-dimensional sphere S^∞. We also study comb space which is contractible in absolute sense but not contractible in relative sense. First we describe $\mathbf{R}^\infty$, $\mathbf{C}^\infty$ and S^∞.

Definition 2.6.14 The set of all sequences $x = (x_1, x_2, \ldots, x_n, \ldots)$ of real numbers such that $\sum_1^\infty |x_n|^2$ converges, is denoted by $\mathbf{R}^\infty$. Under coordinatewise addition and scalar multiplication, $\mathbf{R}^\infty$ is a vector space over $\mathbf{R}$. Moreover, $\mathbf{R}^\infty$ endowed with a norm function defined by $\|x\| = (\sum_1^\infty |x_n|^2)^{1/2}$ is called a real Banach space. The space $\mathbf{R}^\infty$ is called infinite-dimensional Euclidean space. Similarly, the infinite-dimensional unitary space $\mathbf{C}^\infty$ is defined.

Remark 2.6.15 The space $\mathbf{C}^\infty$ is a complex Banach space. Clearly, as a topological space $\mathbf{C}^n$ is homeomorphic to $\mathbf{R}^{2n}$ and $\mathbf{C}^\infty$ is homeomorphic to $\mathbf{R}^\infty$. The space S^∞ is now defined.

Definition 2.6.16 The infinite-dimensional sphere S^∞ is the subspace of $\mathbf{R}^\infty$ (under weak topology) consisting of all real sequences $(x_1, x_2, x_3, \dots)$ such that $x_1^2 + x_2^2 + x_3^2 + \cdots = 1$ (i.e., $S^\infty = \{(x_1, x_2, x_3, \dots) \in \mathbf{R}^\infty : x_1^2 + x_2^2 + x_3^2 + \cdots = 1\}$).

As the diagram in Fig. 2.16 is commutative, we may consider S^∞ as the subspace of $\mathbf{C}^\infty$ consisting of the sequences $(z_1, z_2, \dots)$ over $\mathbf{C}$ such that $|z_1|^2 + |z_2|^2 + \cdots = 1$. We are now in a position to prove the contractibility of S^∞.

Proposition 2.6.17 *The infinite-dimensional sphere S^∞ is contractible.*

Proof Consider the map

$$\begin{aligned} F : S^\infty \times I \to S^\infty, &(x_1, x_2, x_3, \dots, t) \\ &\mapsto ((1-t)x_1, tx_1 + (1-t)x_2, tx_2 + (1-t)x_3, \dots)/N_t, \end{aligned}$$

where $N_t = [((1-t)x_1)^2 + (tx_1 + (1-t)x_2)^2 + (tx_2 + (1-t)x_3)^2 + \cdots]^{1/2}$, which is the norm of the nonzero vector of the numerator. We may parametrize F as $F_t(x_1, x_2, x_3, \dots) = F(x_1, x_2, x_3, \dots, t)$.

Then $F_0(x_1, x_2, x_3, \dots) = (x_1, x_2, x_3, \dots)$, since $N_0 = 1$ and $F_1(x_1, x_2, x_3, \dots) = (0, x_1, x_2, x_3, \dots)$, since $N_1 = 1$. Consequently, F_0 is the identity map $1_d : S^\infty \to S^\infty$, the image of F_1 is the set $X = \{x \in S^\infty : x_1 = 0\}$ and $F : F_0 \simeq F_1$.

Consider another homotopy

$$H : X \times I \to S^\infty, H(x_1 = 0, x_2, x_3, \dots, t) \mapsto (t, (1-t)x_2, (1-t)x_3, \dots)/N'_t,$$

where $N'_t = [t^2 + ((1-t)x_2)^2 + ((1-t)x_3)^2 + \cdots]^{1/2}$.

If $i : X \hookrightarrow S^\infty$ is the inclusion map, then $H : i \simeq c$, where c is a constant map.

Let $H * F : S^\infty \times I \to S^\infty$ be defined by

$$(H * F)(t) = \begin{cases} F(x, 2t), 0 \leq t \leq 1/2 \\ H(x, 2t-1), 1/2 \leq t \leq 1. \end{cases}$$

where $x = (x_1, x_2, x_3, \dots) \in S^\infty$.

Then $H * F$ is a contraction. Consequently, S^∞ is a contractible space. ❑

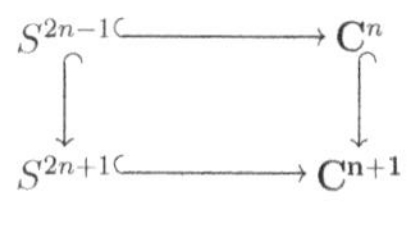

Fig. 2.16 Commutative diagram involving $\mathbf{C}^n$ and $\mathbf{C}^{n+1}$

Remark 2.6.18 The infinite-dimensional sphere S^∞ is contractible. On the other hand, the n-sphere S^n is not contractible for any integer $n \geq 0$ (see Proposition 14.1.13 of Chap. 14).

Corollary 2.6.19 *The inclusion map* $i : S^{n-1} \hookrightarrow S^n$ *is nullhomotopic.*

Remark 2.6.20 We now search for a topological space which is contractible in absolute sense but not contractible in relative sense.

Example 2.6.21 (*Comb Space*) The subspace Y of the plane $\mathbf{R}^2$ defined by

$$Y = \left\{(x, y) \in \mathbf{R}^2 : 0 \leq y \leq 1, x = 0, \frac{1}{n}(n \in \mathbf{N}) \text{ or } y = 0, 0 \leq x \leq 1\right\}$$

is called the comb space, i.e., Y consists of the horizontal line segment L joining $(0, 0)$ to $(1, 0)$ and vertical unit closed line segments standing on points $(1/n, 0)$ for each $n \in \mathbf{N}$, together with the line segment joining $(0, 0)$ with $(0, 1)$ as shown in Fig. 2.17. It is an important example of a contractible space.

Proposition 2.6.22 *Comb space* Y *is contractible but not contractible relative to* $\{(0, 1)\}$.

Proof **First part**: First we show that $L \simeq Y$. Let $p : Y \to L, (x, y) \mapsto (x, 0)$ be the projection map and $i : L \hookrightarrow Y$ be the inclusion map. Then $(p \circ i)(x, 0) = p(x, 0) = (x, 0), \ \forall \ (x, 0) \in L \Rightarrow p \circ i = 1_L$(identity map on L). Define $F : Y \times I \to Y$ by the rule $F((x, y), t) = (x, (1 - t)y)$. Then $F((x, y), 0) = (x, y) = 1_Y(hx, y)$, $\forall \ (x, y) \in Y$ and $F((x, y), 1) = (x, 0) = (i \circ p)(x, y), \ \forall \ (x, y) \in Y$ show that $F : 1_Y \simeq i \circ p \Rightarrow p \in C(Y, L)$ is a homotopy equivalence $\Rightarrow Y \simeq L$. Again $L \approx I \Rightarrow L \simeq I$. Moreover, I being a contractible space, I is of the same homotopy type of a one-point space and hence L is of the same homotopy type of one-point space. Consequently, Y is of the same homotopy type of one-point space. In other words, Y is contractible by Theorem 2.6.8.

Second part: Any small neighborhood V of $(0, 1)$ has infinite number of path components. Let D be the open disk around $(0, 1)$ of radius $\frac{1}{2}$. Then the neighborhood $U = D \cap Y$ of $(0, 1)$ in Y cannot have any neighborhood V each of whose points can be joined to $(0, 1)$ by a path entirely lying in $U \Rightarrow Y$ is not contractible relative to $\{(0, 1)\}$ by Theorem 2.6.12, otherwise we would reach a contradiction by the same theorem. ❑

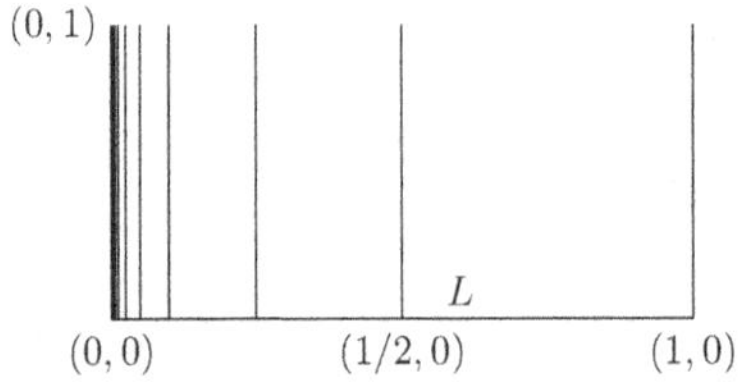

Fig. 2.17 Comb space

Remark 2.6.23 The concept of relative homotopy is stronger than the concept of homotopy.

Let A be a subspace of X and $f, g : X \to Y$ be two continuous maps such that $f \simeq g$ rel A. Then $f \simeq g$. But its converse is not true.

Example 2.6.24 Let Y be the comb space, $1_Y : Y \to Y$ be the identity map and $c : Y \to Y$ be the constant map defined by $c(x, y) = (0, 1)$, $\forall\ (x, y) \in Y$. Then I_Y and c agree on $\{(0, 1)\}$ and hence $1_Y \simeq c$ by Corollary 2.6.5, since Y is contractible. But the comb space Y is not contractible relative to $\{(0, 1)\}$ (see Proposition 2.6.22).

2.7 Retraction and Deformation

This section mainly studies inclusion maps from the viewpoint of homotopy theory. We consider whether an inclusion map $i : A \hookrightarrow X$ has a left inverse or a right inverse or a left homotopy inverse or a right homotopy inverse or two sided inverse or two-sided homotopy inverse. More precisely, the concepts of retraction and weak retraction are introduced and it is proved that these two concepts coincide under suitable homotopy extension property (HEP).

Let A be a subspace of a topological space X and $i : A \hookrightarrow X$ be the inclusion map. Then a continuous map $f : A \to Y$ from A to a subspace Y is said to have a continuous extension over X if $\exists$ a continuous map $F : X \to Y$ such that the diagram in Fig. 2.18 is commutative, i.e., $F \circ i = f$. Thus F is said to be a continuous extension of f over X if $F|_A = f$.

Definition 2.7.1 A subspace A of X is called a retract of X if there exists a continuous map $r : X \to A$ such that $r \circ i = 1_A$, i.e., if i has a left inverse in the category $\mathcal{T}op$ of topological spaces and continuous maps, i.e., if $r(x) = x$, $\forall\ x \in A$. Such a map r is called a retraction of X to A. On the other hand, if $i \circ r \simeq 1_X$, A is called a deformation retract of X and r is called a deformation retraction.

Thus A is a retract of X if $\exists$ a continuous map $r : X \to A$ making the diagram in Fig. 2.19 is commutative.

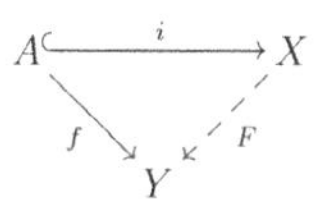

Fig. 2.18 Continuous extension of f

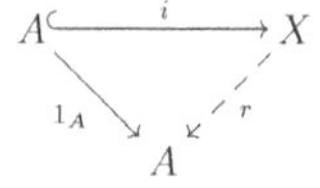

Fig. 2.19 Retraction and retract

Remark 2.7.2 The main property of a retract A of X is that any continuous map $f : A \to Y$ has at least one continuous extension $\tilde{f} : X \to Y$, namely, $\tilde{f} = f \circ r$, where $r : X \to A$ is a retraction.

Example 2.7.3 The circle A as shown in Fig. 2.20 is a retract of the annulus X. The arrows indicate the action of the retraction. The whole of X is mapped onto A keeping points in A fixed.

Example 2.7.4 Consider the inclusion map $i : D^n \hookrightarrow \mathbf{R}^n$. Define a map $r : \mathbf{R}^n \to D^n$ by the rule

$$r(x) = \begin{cases} \frac{x}{||x||}, & \text{if } ||x|| > 1 \\ x, & \text{if } ||x|| \leq 1. \end{cases}$$

Then r is a retraction and hence D^n is a retract of $\mathbf{R}^n$. Geometrically, the map r fixes points in D^n and shifts points x outside of D^n along a straight line from the origin to x onto the boundary S^{n-1} of D^n.

Definition 2.7.5 A subspace A of a topological space X is called a weak retract of X if there exists a continuous map $r : X \to A$ such that $r \circ i \simeq 1_A$, i.e., if i has a left homotopy inverse, i.e., if i has a left inverse in the homotopy category $\mathcal{H}tp$ of topological spaces and continuous maps.

Thus A is a weak retract of X if $\exists$ a continuous map $r : X \to A$ making the diagram in Fig. 2.21 homotopy commutative. Such a map r is called a weak retraction of X to A.

Remark 2.7.6 A is retract of $X \Rightarrow A$ is a weak retract of X, because $r \circ i = 1_A \simeq 1_A$. But its converse is not true.

Example 2.7.7 Consider $X = I^2$ and A =comb space (see Example 2.6.21). Then $A \subsetneqq X$. As A and X are both contractible spaces, the inclusion map $i : A \hookrightarrow X$ is

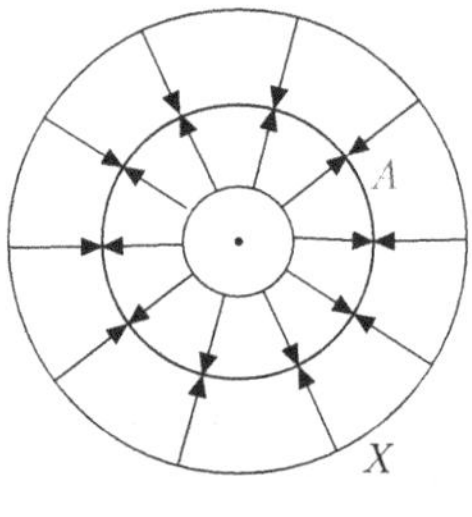

Fig. 2.20 The circle A (*deep black*) is a retract of the annulus X

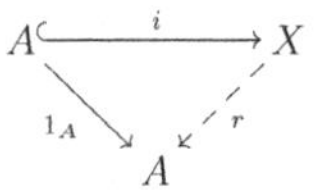

Fig. 2.21 Weak retraction and weak retract

a homotopy equivalence by Corollary 2.6.9. Hence $\exists$ a continuous map $r : X \to A$ such that $r \circ i \simeq 1_A$. This shows that r is a weak homotopy equivalence. Clearly $r \circ i \neq 1_A$. Consequently, A is a weak retract of X but not a retract of X.

We now search conditions under which the concepts of retraction and weak retraction coincide. For this purpose we introduce the concept of Homotopy Extension Property for the pair of spaces (X, A).

Definition 2.7.8 Let (X, A) be pair of topological spaces and Y be an arbitrary topological space. Then the pair (X, A) is said to have the Homotopy Extension Property (HEP) with respect to the space Y if given continuous maps $g : X \to Y$ and $G : A \times I \to Y$ such that $g(x) = G(x, 0)$, $\forall\ x \in A$, there is a continuous map $F : X \times I \to Y$ with the property $F(x, 0) = g(x)$, $\forall\ x \in X$ and $F|_{A \times I} = G$.

If $h_0(x) = (x, 0)$, $\forall\ x \in X$, the existence of F is equivalent to the existence of a continuous map represented by the dotted arrow which makes the diagram in Fig. 2.22 is commutative.

Thus (X, A) has the HEP with respect to Y if $\exists$ a continuous map $F : X \times I \to Y$ such that the square and the two triangles in the diagram in Fig. 2.22 are commutative.

Proposition 2.7.9 *If (X, A) has the HEP with respect to Y and if $f_0, f_1 : A \to Y$ are homotopic, then f_0 has a continuous extension over X iff f_1 has also a continuous extension over X.*

Proof Let $f_0, f_1 : A \to Y$ be two continuous maps such that $f_0 \simeq f_1$. Then $\exists$ a homotopy $G : A \times I \to Y$ such that $G(x, 0) = f_0(x)$ and $G(x, 1) = f_1(x)$, $\forall\ x \in A$. Let $f_0 : A \to Y$ have a continuous extension $g_0 : X \to Y$. Then $G(x, 0) = f_0(x) = g_0(x)$, $\forall\ x \in A$. As (X, A) has the HEP with respect to Y, $\exists$ a map $F : X \times I \to Y$ extending $G : A \times I \to Y$ and therefore the diagram in Fig. 2.23 is commutative. The existence of F follows from the HEP of (X, A) with respect to Y. Define a map $g_1 : X \times Y$ by $g_1(x) = F(x, 1)$, $\forall\ x \in X$. Then g_1 is an extension

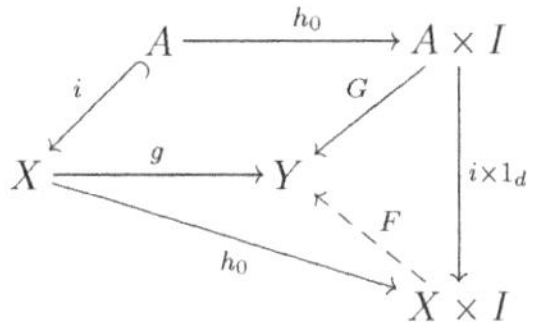

Fig. 2.22 Homotopy extension property (HEP)

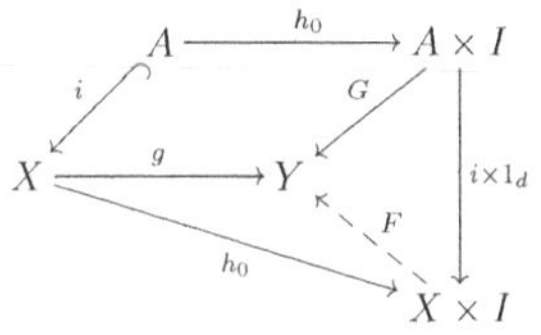

Fig. 2.23 Homotopy extension property of (X, A) w.r.t. Y

of f_1 over X, because, $g_1(a) = F(a, 1) = G(a, 1) = f_1(a)$, $\forall\ a \in A \Rightarrow g_1|_A = f_1$. Moreover g_1 is continuous and hence f_1 has a continuous extension g_1 over X. ❑

Remark 2.7.10 A continuous map $f : A \to Y$ can or cannot be extended over X is a property of the homotopy class of that map. Thus the homotopy extension property implies that the extension problem for continuous maps $A \to Y$ is a problem in the homotopy category.

The map $r : \mathbf{R}^n \to D^n$ defined in Example 2.7.4 is a deformation retraction. To show this define a homotopy $F : \mathbf{R}^n \times I \to \mathbf{R}^n$ by the rule

$$F(x, t) = \begin{cases} (1-t)x + tx/||x||, & \text{if } ||x|| \geq 1 \\ x, & \text{if } ||x|| < 1 \end{cases}$$

Then $F : 1_d \simeq i \circ r$ shows that r is a deformation retraction.

Geometrically, F fixes points in D^n and shifts points x outside of D^n linearly from x to $r(x)$ along the straight line determined by x and the origin $0 = (0, 0, \ldots, 0)$.

Theorem 2.7.11 *If (X, A) has the HEP with respect to A, then A is a weak retract of X iff A is a retract of X.*

Proof Let $A \subset X$ be a retract of X and $r : X \to A$ be a retraction. Then $r \circ i = 1_A \Rightarrow r \circ i \simeq 1_A \Rightarrow A$ is a weak retract of X. Conversely, let $r : X \to A$ be a weak retraction. Then $r \circ i \simeq 1_A$, where $i : A \hookrightarrow X$ is the inclusion map. Then $\exists$ a homotopy $G : A \times I \to Y$ such that $G(x, 0) = r(x)$, $G(x, 1) = 1_A(x) = x$, $\forall\ x \in A$. As (X, A) has the HEP with respect ot A, $\exists$ a continuous map $F : X \times I \to A$ extending $G : A \times I \to A$. Hence $F(x, 0) = r(x)$, $\forall\ x \in X$ and $F|_{A \times I} = G$. Define a map $r' : X \to A$ by the rule $r'(x) = F(x, 1)$, $\forall\ x \in X$. Now, for all $a \in A$, $(r' \circ i)(a) = r'(i(a)) = F(a, 1) = G(a, 1) = a \Rightarrow r' \circ i = 1_A \Rightarrow A$ is a retract of X and r' is a retraction of X into A. ❑

We can as well consider inclusion maps with right homotopy inverses.

Definition 2.7.12 Given a subspace $X' \subset X$, a deformation D of X' in X is a homotopy $D : X' \times I \to X$ such that $D(x', 0) = x'$, $\forall\ x' \in X'$. If moreover, $D(X' \times 1) \subset A \subset X'$, D is said to a deformation of X' into A and X' is said to be deformable in X into A. If $X = X'$, then a space X is said to be deformable into a subspace A of X if it is deformable in itself into A.

Theorem 2.7.13 *A topological space X is deformable into a subspace A of X iff the inclusion map $i : A \hookrightarrow X$ has a right homotopy inverse.*

Proof Let X be deformable into a subspace A of X. Then $\exists$ a continuous map $D : X \times I \to X$ such that $D(x, 0) = x$ and $D(x, 1) \in A \subset X$, $\forall\ x \in X$. Let $f : X \to A$ be defined by the equation $(i \circ f)(x) = D(x, 1)$, $\forall\ x \in X$. Then $D : 1_X \simeq i \circ f \Rightarrow i$ has a right homotopy inverse. Conversely, let $i : A \subset X$ has a right homotopy inverse $f : X \to A$. Then $1_X \simeq i \circ f$. Let $F : X \times I \to X$ be such that $F : 1_X \simeq i \circ f$. Then $F(x, 0) = 1_X(x) = x$, $\forall\ x \in X$ and $F(X \times 1) = (i \circ f)(X) \subset A$(i.e., $F(x, 1) = i(f(x)) = f(x)$, $\forall\ x \in X) \Rightarrow X$ is deformable into A. ❑

Remark 2.7.14 The homotopy D which starts with identity map $1_X : X \to X$, simply moves each point of X continuously, including the points of A and finally, pushes every point into a point of A. In particular, if X is deformable into a point $x_0 \in X$, then X is contractible and vice verse (see Ex. 4 of Sect. 2.11).

Remark 2.7.15 An inclusion map $i : A \hookrightarrow X$ has never a right inverse in the category of topological spaces and continuous maps in the trivial case when $A = X$.

We now consider inclusion maps which are homotopy equivalences.

Definition 2.7.16 A subspace A of a topological space X is called a weak deformation retract of X if the inclusion map $i : A \hookrightarrow X$ is a homotopy equivalence.

Theorem 2.7.17 *A subspace A of a topological space X is a weak deformation retract of X iff A is a weak retract of X and X is deformable into A.*

Proof Let A be a weak deformation retract of X. Then $i : A \hookrightarrow X$ is a homotopy equivalence. Hence $\exists$ a continuous map $r : X \to A$ such that $r \circ i \simeq 1_A$ and $i \circ r \simeq 1_X$. Thus i has a right homotopy inverse and also a left homotopy inverse. Consequently, A is a weak retract of X and X is deformable into A.

Conversely, let A be a weak retract of X and X be deformable into A. Then $i : A \hookrightarrow X$ has a left homotopy inverse f(say) and a right homotopy inverse g(say). Now $f \circ i \simeq 1_A$ and $i \circ g \simeq 1_X \Rightarrow f = f \circ 1_X \simeq f \circ (i \circ g) = (f \circ i) \circ g \simeq 1_A \circ g = g \Rightarrow f \simeq g$. Hence $f \circ i \simeq 1_A$ and $i \circ f \simeq 1_X \Rightarrow i$ is a homotopy equivalence $\Rightarrow A$ is a weak deformation retract of X. ❑

We now consider a deformation D which deforms X into A, but the points of A do not move at all.

This led to the concept of strong deformation retract introduced by Borsuk in 1933.

Definition 2.7.18 A subspace A of a topological space X is called a strong deformation retract of X if there exists a retraction $r : X \to A$ such that $1_X \simeq i \circ r$ rel A. If $F : 1_X \simeq i \circ r$ rel A, then F is called a strong deformation retract of X to A.

There is an intermediate concept between the concepts of weak deformation retraction and strong deformation retraction.

Definition 2.7.19 A subspace A of a topological space X is called a deformation retract of X if $\exists$ a retraction $r : X \to A$ such that $1_X \simeq i \circ r$. If $F : 1_X \simeq i \circ r$, then F is called a deformation retraction of X to A.

Remark 2.7.20 A homotopy $F : X \times I \to X$ is a deformation retraction of X to A iff $F(x, 0) = 1_X(x) = x,\ \forall\ x \in X$, $F(x, 1) = x,\ \forall\ x \in A$ and $F(X \times 1) \subset A$. A map F is called a strong deformation retraction iff it also satisfies the condition $F(x, t) = x,\ \forall\ x \in A$ and $\forall\ t \in I$. Thus F is a strong deformation retraction of X to A iff F satisfies the conditions:

(i) $F(x, 0) = x,\ \forall\ x \in X$;

(ii) $F(x, 1) = x, \ \forall \ x \in A$;
(iii) $F(X \times 1) \subset A \subset X$;
(iv) $F(x, t) = x, \ \forall \ x \in A$ and $\forall \ t \in I$.

Example 2.7.21 Let $X = \mathbf{R}^{n+1} - \{0\}$ and S^n be the n-sphere ($n \geq 1$). Then $S^n \subset X$ is a strong deformation retract of X. Let $i : S^n \hookrightarrow X$ be the inclusion. Define a retraction $r : X \to S^n$ by $r(x) = \frac{x}{||x||}$. Geometrically, this map shifts points $x \in X$ to the boundary S^n along a straight line from the origin. Define a continuous map $F : X \times I \to X$ by $F(x, t) = (1 - t)x + \frac{tx}{||x||}, \ \forall \ x \in X, \ \forall \ t \in I$. Then

(i) $F(x, 0) = x, \ \forall \ x \in X$;
(ii) $F(x, 1) = x, \ \forall \ x \in S^n$;
(iii) $F(X \times 1) \subset S^n \subset X$ and
(iv) $F(x, t) = x, \ \forall \ x \in S^n$ and $\forall \ t \in I$.

Clearly, $r \circ i$ is the identity map on S^n and $i \circ r$ is homotopic to the identity map on X. Hence i is a homotopy equivalence. Geometrically, this homotopy F moves linearly along the straight path defined above from x to $r(x)$. Hence F is a strong deformation retraction of X to S^n. Therefore S^n is a strong deformation retract of $X = \mathbf{R}^{n+1} - \{0\}$.

Remark 2.7.22 We now explain Example 2.7.21 geometrically for $n = 1$ as shown in Fig. 2.24.

Let l be an arbitrary half-line starting from the origin $0 = (0, 0)$. Then it intersects the circle S^1 at exactly one point l_P(say). Since $0 = (0, 0)$ is not a point of $\mathbf{R}^2 - \{0\}$, the lines $l - \{0\}$ are disjoint and their union is $\mathbf{R}^2 - \{0\}$. Define a map $r : \mathbf{R}^2 - \{0\} \to S^1$ by $r(x) = l_P, \ \forall \ x \in l$. Then r is a retraction and S^1 is a retract of $\mathbf{R}^2 - \{0\}$. Define a deformation $D : (\mathbf{R}^2 - \{0\}) \times I \to \mathbf{R}^2 - \{0\}$ by $D(x, t) = (1 - t)x + t\frac{x}{||x||}$. Then as before D is a strong deformation retraction of $\mathbf{R}^2 - \{0\}$ relative to S^1 into S^1.

Example 2.7.23 Consider the product space $X = D^n \times I$ and the subspace $A = (S^{n-1} \times I) \cup (D^n \times \{0\})$. If P is the point $(0, 2)$ in $\mathbf{R}^n \times \mathbf{R}$, a retraction $r : X \to A$ is defined by taking $r(x)$ to be the point where the line joining P and x meets A. Consequently, the map

$$F : X \times I \to X, (x, t) \mapsto F(x, t) = t\,r(x) + (1 - t)x$$

is such that $F(x, 0) = x$ and $F(x, 1) = r(x)$. This shows that F is a strong deformation retraction.

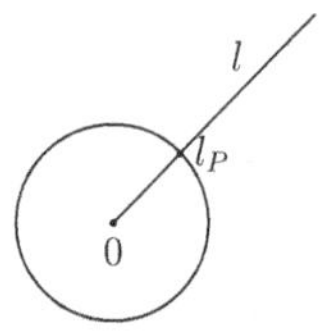

Fig. 2.24 Half line starting from the origin and intersecting the circle

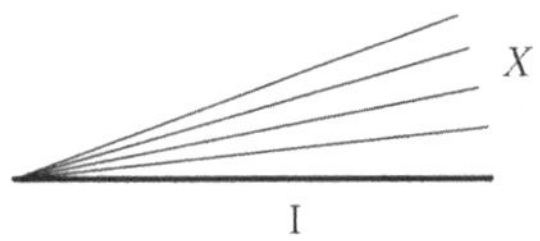

Fig. 2.25 Construction of X

Example 2.7.24 Let X be the topological space given by I together with a family of segments approaching it as shown in Fig. 2.25.

Then I is a deformation retract of X but not a strong deformation retract.

Proposition 2.7.25 *A topological space X is deformable into a retract $A \subset X$ if and only if A is a deformation retract of X.*

Proof Let X be deformable into a retract $A \subset X$ and $i : A \hookrightarrow X$ be the inclusion map. Then there exists a retraction $r : X \to A$ such that $r \circ i = 1_A$. Hence r is a left inverse of i and thus r is a left homotopy inverse of i. Again as X is deformable into A, i has a right homotopy inverse by Theorem 2.7.13, which is also r, i.e., $1_X \simeq i \circ r$. Consequently, A is a deformation retract of X. Conversely, let A be a deformation retract of X. Then $\exists$ a retraction $r : X \to A$ such that $1_X \simeq i \circ r$ and $r \circ i = 1_A$. Consequently, X is deformable into a retract A. ❑

We now show that if (X, A) has the HEP with respect to A, then the concepts of weak deformation retraction and deformation retraction coincide.

Theorem 2.7.26 *If (X, A) has the HEP with respect to A, then A is a weak deformation retract of X if and only if A is a deformation retract of X.*

Proof Suppose (X, A) has the HEP with respect to A. Then A is a weak retract of X and X is deformable into A if and only if A is a retract of X and X is deformable into A by Theorem 2.7.11. Then the theorem follows from Proposition 2.7.25. ❑

We now show that under suitable HEP the concepts of strong deformation retraction and deformation retraction coincide.

Theorem 2.7.27 *If $A \subset X$ and $(X \times I, (X \times \{0\}) \cup (A \times I) \cup (X \times 1))$ has the HEP with respect to X and A is closed in X, then A is a deformation retract of X if and only if A is a strong deformation retract of X.*

Proof Let $X' = X \times I$ and $A' = (X \times \{0\}) \cup (A \times I) \cup (X \times 1)$. Then $A' \subset X'$. Let (X', A') has the HEP with respect to X and A be closed in X. Suppose $i : A \hookrightarrow X$ and A is a deformation retract of X. We claim that A is also a strong deformation retract of X. Since A is a deformation retract of X, $\exists$ a retraction $r : X \to A$ such that $1_X \simeq i \circ r$. Let $F : X \times I \to X$ be a continuous map such that $F : 1_X \simeq i \circ r$. Then $F(x, 0) = 1_X(x) = x$, $\forall\ x \in X$ and $F(x, 1) = (i \circ r)(x) = r(x)$, $\forall\ x \in X$. We now define a map $G : A' \times I \to X$ by the equations

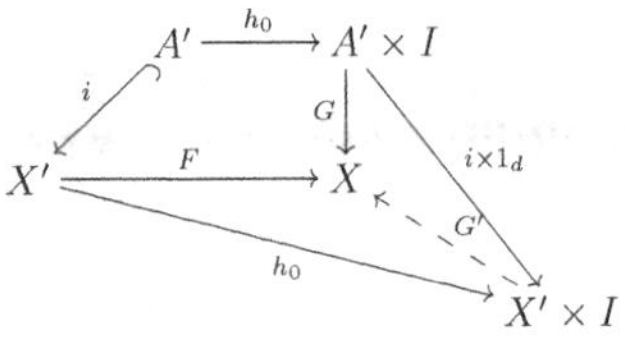

Fig. 2.26 Homotopy extension property of (X', A') w.r.t. X

$$G((x,0),t') = x,\ \forall\ x \in X,\ \forall\ t' \in I \tag{2.11}$$

$$G((x,t),t') = F(x,(1-t')t),\ \forall\ x \in A,\ \forall\ t,t' \in I \tag{2.12}$$

$$G((x,t),t') = F(r(x),1-t'),\ \forall\ x \in X,\ \forall\ t,t' \in I \tag{2.13}$$

Then G is well defined, because for $x \in A$, $G((x,0),t') = x = F(x,0)$ by the first two Eqs. (2.11) and (2.12) and $G((x,1),t') = F(x,1-t') = F(r(x),1-t')$ by the last two Eqs. (2.12) and (2.13). Again G is continuous, because its restriction to each of the closed sets $(X \times \{0\}) \times I$, $(A \times I) \times I$ and $(X \times 1) \times I$ is continuous. For $(x,t) \in A'$, $G((x,t),0) = F(x,t)$, because $F(x,0) = x$ and since $r : X \to A$ is a retraction.

$F(r(x),1) = (i \circ r)(r(x)) = r(x) = F(x,1)$. Therefore G restricted to $A' \times \{0\}$ can be extended to $(X \times I) \times \{0\}$. Then by HEP of (X', A') w.r.t X in the hypothesis, $\exists$ a homotopy $G' : X' \times I \to X$ extending $G : A' \times I \to X$ (see Fig. 2.26). Define $H : X \times I \to X$ by $H(x,t) = G'((x,t),1)$. Then we have the equations
$H(x,0) = G'((x,0),1) = G((x,0),1) = x,\ \forall\ x \in X;$
$H(x,1) = G'((x,1),1) = F(r(x),0) = r(x),\ \forall\ x \in X;$
and $H(x,t) = G'((x,t),1) = G((x,t),1) = F(x,0) = x,\ \forall\ x \in A,\ \forall\ t \in I.$

Therefore $H : 1_X \simeq i \circ r$ rel A. Hence A is a strong deformation retract of X. Conversely, if A is a strong deformation retract of X, then A is automatically a deformation retract of X. ❑

2.8 NDR and DR Pairs

This section defines the concepts of NDR-pair and DR-pair which are closely related to the concepts of retraction and homotopy extension property for compactly generated spaces (see Sect. B.4 of Appendix B). N. Steenrod (1910–1971) proved in 1967 the equivalence between the NDR condition and the homotopy extension property (Steenrod 1967).

We now use the concept of compactly generated space defined in Appendix B.

Definition 2.8.1 Let X be a compactly generated topological space and $A \subset X$ be a subspace. Then (X, A) is said to be an NDR-pair (NDR stands for 'neighborhood deformation retract') if there exist continuous maps $u : X \to I$ and $h : X \times I \to X$ such that

NDR(i) $A = u^{-1}(0)$;
NDR(ii) $h(x, 0) = x, \forall x \in X$;
NDR(iii) $h(a, t) = a, \forall t \in I, a \in A$;
NDR(iv) $h(x, 1) \in A$ for all $x \in X$ such that $u(x) < 1$.

In particular, A is a retract of its neighborhood $U = \{x \in X : u(x) < 1\}$, and hence is a neighborhood retract of X.

Definition 2.8.2 A pair (X, A) is called a DR-pair (DR stands for "deformation retract") if in addition to **NDR(i)–NDR(iii),** another condition **NDR(v)**: $h(x, 1) \in A$ (instead of **NDR(iv)**) holds for all $x \in X$.

Remark 2.8.3 The concepts of DR-pair and NDR-pair are closely related to the concepts of retraction and HEP (see Ex. 32 of Sect. 2.11).

2.9 Homotopy Properties of Infinite Symmetric Product Spaces

This section conveys homotopy properties of infinite symmetric product spaces defined for spaces in $\mathcal{T}op_*$ (see Sect. B.2.5 of Appendix B). These spaces link homotopy theory with homology theory via Elienberg–MacLane spaces (see Chaps. 11 and 17) and form an important class of topological spaces in the study of algebraic topology. So it has become necessary to study such spaces from homotopy viewpoint.

We have constructed in Sect. B.2.5 of Appendix B the finite symmetric product SP^nX and infinite symmetric product $SP^\infty X$ of a pointed topological space X. Both SP^n and SP^∞ are functors from the category $\mathcal{T}op_*$ to itself (see Sect. B.2.5 of Appendix B). A continuous map $f : X \to Y$ in $\mathcal{T}op_*$ induces maps $f^n : X^n \to Y^n, (x_1, x_2, \dots, x_n) \mapsto (f(x_1), f(x_2), \dots, f(x_n))$. These maps are compatible with the action of the symmetric group S_n of the set $\{1, 2, \dots, n\}$ and hence induce maps $SP^n(f) : SP^nX \to SP^nY$ between the corresponding orbit spaces and also induce maps $SP^\infty(f) = f_* : SP^\infty X \to SP^\infty Y$ (see Sect. B.2.5 of Appendix B).

Theorem 2.9.1 *If $f, g : X \to Y$ are in $\mathcal{T}op_*$ and $f \simeq g$, then $SP^\infty(f) \simeq SP^\infty(g)$.*

Proof Let $F : X \times I \to Y$ be a map such that $F : f \simeq g$. For all $n \geq 1$, define $F^n : X^n \times I \to Y^n, (x_1, x_2, \dots, x_n, t) \mapsto (F(x, t), F(x_2, t), \dots, F(x_n, t))$. Then F^n is continuous, because its projection onto each coordinate is continuous. Since S_n acts on $X^n \times I$ by permuting the coordinate of X^n and fixing I, and F^n respects this action, F^n induces maps $SP^n(F) : SP^nX \to SP^nY$, which passing to the limit induces a map $SP^\infty(F) : SP^\infty X \times I \to SP^\infty Y$. Define

$$h_t : X \to Y, x \mapsto F(x, t) \text{ and } SP^\infty(F) : SP^\infty X \times I \to SP^\infty Y, (x, t) \mapsto SP^\infty(h_t)(x).$$

Hence $SP^\infty(F) : SP^\infty(f) \simeq SP^\infty(g)$. ❑

Corollary 2.9.2 *If spaces X and Y in* $\mathcal{T}op_*$ *are homotopy equivalent, then the spaces* $SP^\infty X$ *and* $SP^\infty Y$ *are also homotopy equivalent. In particular, if X is contractible, then* $SP^\infty X$ *is contractible.*

Proof Let $f : X \to Y$ be a homotopy equivalence with homotopy inverse $g : Y \to X$. Then $SP^\infty(g)$ is a homotopy inverse of $SP^\infty(f)$. Consequently, the spaces $SP^\infty X$ and $SP^\infty Y$ are homotopy equivalent. Again $SP^\infty\{*\} = \{*\}$ proves the second part. ❑

Theorem 2.9.3 $SP^\infty : \mathcal{H}tp_* \to \mathcal{H}tp_*$ *is a covariant functor.*

Proof It follows from Theorem 2.9.1 and Proposition B.2.18. ❑

2.10 Applications

This section presents some interesting immediate applications of homotopy. It deals with some extension problems and proves 'Fundamental Theorem of Algebra' by using homotopic concepts.

2.10.1 *Extension Problems*

This subsection solves some extensions problems with the help of homotopy.

Theorem 2.10.1 *A continuous map* $f : S^n \to Y$ *from* S^n *to any space Y can be continuously extended over* D^{n+1} *if and only if f is nullhomotopic, i.e., iff f is homotopic to a constant map.*

Proof Let $c : S^n \to Y$ be a constant map defined by $c(S^n) = y_0 \in Y$ such that $f \simeq c$. Then *exists* a homotopy $H : S^n \times I \to Y$ such that $H(x, 0) = f(x)$ and $H(x, 1) = c(x) = y_0$, $\forall\ x \in S^n$. We now construct a map $F : D^{n+1} \to Y$ by the rule

$$F(x) = \begin{cases} y_0, & 0 \le ||x|| \le 1/2 \\ H\left(\frac{x}{||x||}, 2 - 2||x||\right), & 1/2 \le ||x|| \le 1. \end{cases}$$

Since at $||x|| = \frac{1}{2}$, $H(\frac{x}{||x||}, 1) = y_0$, F is well defined. Again, since its retraction to each of the closed sets $C_1 = \{x \in D^{n+1} : 0 \le ||x|| \le 1/2\}$ and $C_2 = \{x \in D^{n+1} : 1/2 \le ||x|| \le 1\}$ is continuous, F agrees on $C_1 \cap C_2$ and $D^{n+1} = C_1 \cup C_2$, F is continuous by Pasting lemma. Moreover, $\forall\ x \in S^n$, $||x|| = 1$ and hence $F(x) = H(x, 0) = f(x) \Rightarrow F$ is a continuous extension of f over D^{n+1}. Thus $f \simeq c \Rightarrow f$ has a continuous extension over D^{n+1}. Conversely, let $F : D^{n+1} \to Y$ be a continuous extension of $f : S^n \to Y$. Then $\forall\ x \in S^n$, $F(x) = f(x)$. Suppose $p_0 \in S^n$ and $f(p_0) = y_0 \in Y$. We now define a mapping $H : S^n \times I \to Y$ by $H(x, t) =$

$F((1-t)x + tp_0)$. H is well defined, because D^{n+1} is a convex set. Moreover, H is continuous and $H : f \simeq c$. ❑

Theorem 2.10.2 *Any continuous map from S^n to a contractible space has a continuous extension over D^{n+1}.*

Proof Let $c : S^n \to Y$ be a constant map from S^n to a contractible space Y and $f : S^n \to Y$ be an arbitrary continuous map. Then $f \simeq c$ by Corollary 2.6.4. Hence it follows by Theorem 2.10.1 that f has a continuous extension over D^{n+1}. ❑

Theorem 2.10.3 *Let p_0 be an arbitrary point of S^n and let $f : S^n \to Y$ be continuous. Then the following statements are equivalent.*

(a) *f is nullhomotopic.*
(b) *f can be continuously extended over D^{n+1}.*
(c) *f is nullhomotopic relative to $\{p_0\}$.*

Proof **(a)** $\Rightarrow$ **(b)** follows from Theorem 2.10.1.
(b) $\Rightarrow$ **(c)** Let $F : D^{n+1} \to Y$ be a continuous extension of f over D^{n+1}. Suppose $f(p_0) = y_0 \in Y$. Define a map $H : S^n \times I \to Y$ by $H(x,t) = F((1-t)x + tp_0)$ as shown in Fig. 2.27.
Then $\forall\ x \in S^n$, $H(x,0) = F(x) = f(x)$, $H(x,1) = F(p_0) = f(p_0) = y_0 = c(x)$ and $H(p_0,t) = F(p_0) = f(p_0) = y_0 = c(p_0)$, $\forall\ t \in I$. Hence $H : f \simeq c \text{ rel } \{p_0\}$.
(c) $\Rightarrow$ **(a)** It follows trivially.

❑

Proposition 2.10.4 *There exists a continuous map $f : D^n \to S^{n-1}$ with $f \circ i = 1_d$ iff the identity map $1_d : S^{n-1} \to S^{n-1}$ is nullhomotopic.*

Proof Suppose there exists such a map $f : D^n \to S^{n-1}$. Define a homotopy

$$H : S^{n-1} \times I \to S^{n-1}, (x,t) \mapsto f(tx).$$

Then $H(x,1) = x$, $\forall x \in S^{n-1}$ and $H(x,0) = f(0)$, $\forall x \in S^{n-1}$, i.e., $H(x,0)$ is independent of x. Hence 1_d is homotopic to a constant map. Conversely, let there exist $H : S^{n-1} \times I \to S^{n-1}$ such that $H(x,0) = c$ and $H(x,1) = x$. Define

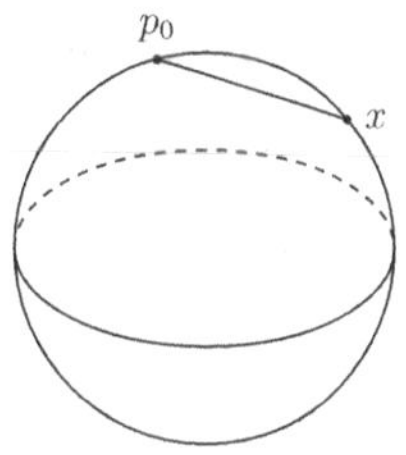

Fig. 2.27 Construction of H

$$f : D^n \to S^{n-1}, x \mapsto \begin{cases} H\left(\frac{x}{||x||}, ||x||\right), & \text{if } x \neq 0 \\ c, & \text{if } x = 0 \end{cases}$$

Since S^{n-1} is compact, H is uniformly continuous. Hence for every $\epsilon > 0$, e a $\delta > 0$ (depends on ϵ but not on x) such that $||H(x,t) - c|| < \delta$ if $t < \epsilon$. This shows that f is continuous at $0 \in D^n$. ❑

Proposition 2.10.5 *Let (X, A) be a normal pair (i.e., X is normal and A is closed in X) such that $X \times I$ is normal and $f : X \to S^n$ be a continuous function. Then every homotopy of $f|_A$ can be extended to a homotopy of f.*

Proof Let $H : A \times I \to S^n$ be a homotopy of $f|_A$. Then H can be extended to a continuous map $\tilde{H} : X \times \{0\} \cup A \times I$ be setting $H(x, 0) = f(x)$ for all $x \in X$. Then by using Ex. 22 of Sect. 1.16 of Chap. 1 it follows that there exists an extension $F : X \times I \to S^n$, which is required homotopy. ❑

Proposition 2.10.6 *If a continuous map $f : X \to S^n$ is essential, then $f(X) = S^n$ (i.e., f is a surjection).*

Proof If $f(X) \neq S^n$, then there exists an element $y \in S^n - f(X)$. Since $S^n - y$ is contractible to a point and $f(X) \subset S^n - y$, it follows that f is inessential. This is a contradiction. ❑

Proposition 2.10.7 *Let (X, A) be a normal pair such that $X \times I$ is normal. Then every inessential map $f : A \to S^n$ admits an inessential extension $\tilde{f} : X \to S^n$.*

Proof Let $g : X \to S^n$ be a map of X into a single point of S^n. Then $g|_A$ is homotopic to f. Hence the existence of $\tilde{f}$ follows from Proposition 2.10.5. ❑

Definition 2.10.8 A topological space X is said to be aspherical if every continuous map $f : S^n \to X$ extends to a continuous map $\tilde{f} : D^{n+1} \to X$.

Example 2.10.9 Every convex subspace of Euclidean space and every contractible space are aspherical.

2.10.2 *Fundamental Theorem of Algebra*

This subsection applies the tools of homotopy to prove the celebrated fundamental theorem of algebra which shows that the field of complex numbers is algebraically closed. There are several methods to prove the fundamental theorem of algebra. We now present a proof by homotopy. For an alternative proof see Theorem 3.8.1 of Chap. 3.

Theorem 2.10.10 *Let $\mathbf{C}$ denote the field of complex numbers, and $C_\rho \subset \mathbf{C} \approx \mathbf{R}^2$ denote the circle at the origin and of radius ρ. Let $f_\rho^n : C_\rho \to \mathbf{C} - \{0\}$ be the restriction to C_ρ to the map $z \mapsto z^n$. If none of the maps f_ρ^n is nullhomotopic ($n \geq 1$ and $\rho > 0$), then every nonconstant polynomial over $\mathbf{C}$ has a root in $\mathbf{C}$.*

Proof of Fundamental Theorem of Algebra by Homotopy:

Proof Without loss of generality we consider the polynomial $g(z) = a_0 + a_1 z + \cdots + a_{n-1}z^{n-1} + z^n$, $a_i \in \mathbf{C}$. We choose

$$\rho > \max\left\{1, \sum_{i=0}^{n-1} |a_i|\right\} \tag{2.14}$$

We define a map $F : C_\rho \times I \to \mathbf{C}$ by $F(z,t) = z^n + \sum_{i=0}^{n-1}(1-t)a_i z^i$. Then $F(z,t) \neq 0$ for any $(z,t) \in C_\rho \times I$. Otherwise, $F(z,t) = 0$ for some $z \in C_\rho$ and $t \in I$ would imply $z^n = -\sum_{i=0}^{n-1}(1-t)a_i z^i$. This implies $\rho^n \leq \sum_{i=0}^{n-1}(1-t)|a_i|\rho^i \leq \sum_{i=0}^{n-1}|a_i|\rho^i \leq \sum_{i=0}^{n-1}|a_i|\rho^{n-1}$ for $\rho > 1$, because $\rho^i \leq \rho^{n-1}$ for $\rho > 1$.

Hence $\rho \leq \sum_{i=0}^{n-1}|a_i|$, by canceling ρ^{n-1}
$\Rightarrow$ a contradiction to the relation (2.14).

In other words, $F(z,t) \neq 0$ for any z with $|z| = 1$ and for any $t \in I$. We now assume that g has a root in $\mathbf{C}$. We define $G : C_\rho \times I \to \mathbf{C} - \{0\}$ by $G(z,t) = g((1-t)z)$. Since g has no root in $\mathbf{C}$, $G(z,t) \neq 0$ and hence the values of G must lie in $\mathbf{C} - \{0\}$. Now $G : g|_{C_\rho} \simeq k$, where k is the constant map $z \mapsto g(0) = a_0$ at a_0. Hence $g|_{C_\rho}$ is nullhomotopic. Again $g|_{C_\rho} \simeq f_\rho^n$. Thus f_ρ^n is nullhomotopic by symmetric and transitive properties of the relation $\simeq$. This contradicts the hypothesis. Consequently, g has a complex root. ❑

2.11 Exercises

1. For all $n \geq 0$, show that the topological spaces $S^1 \wedge S^n$ and S^{n+1} are homeomorphic.
[Hint: Let S^{n+1} be the $(n+1)$-sphere in $\mathbf{R}^{n+2}$, S^n be equator, D^{n+1} be the $n+1$-disk embedded in D^{n+2}, H_+^{n+1} be upper hemisphere, H_-^{n+1} be lower hemisphere and $s_0 = (1, 0, 0, \ldots, 0)$ be base point. Now proceed as in Proposition 2.5.6.]
2. Given a collection of pointed topological spaces $X_\alpha, Y_\alpha (\alpha \in A)$, and maps $f_\alpha \simeq g_\alpha : X_\alpha \to Y_\alpha$, show that $\times f_\alpha \simeq \times g_\alpha$.
[Hint: Let $F_\alpha : X_\alpha \times I \to Y_\alpha$ be a homotopy between f_α and g_α. Then $F : (\times X_\alpha) \times I \to \times Y_\alpha$, defined by $F((x_\alpha), t) = (F_\alpha(x_\alpha, t))$, $\forall\, t \in I$ is continuous and a homotopy between $\times f_\alpha$ and $\times g_\alpha$ (relative to base point).]

3. Consider the homotopy set $[A, X]$, where A is a fixed space. Show that a continuous map $f : X \to Y$ induces a function $f_* : [A, X] \to [A, Y]$ satisfying the following properties:

(i) If $f \simeq g$, then $f_* = g_*$;
(ii) If $1_X : X \to X$ is the identity map, then $1_{X^*} : [A, X] \to [A, X]$ is the identity function;
(iii) If $g : Y \to Z$ is another continuous map, then $(g \circ f)_* = g_* \circ f_*$.
Deduce that if $X \simeq Y$, then $\exists$ a bijection between the sets $[A, X]$ and $[A, Y]$. What are the corresponding results for the sets $[X, A]$ for a fixed space A? [See Theorems 2.3.1 and 2.3.5 and their corollaries.]

4. Show that

(a) $S^1 = \{z \in \mathbf{C} : |z| = 1\}$ is a topological group under usual multiplication of complex numbers.
(b) For any space X, pointwise multiplication endows the set of continuous maps $X \to S^1$ with the structure of an abelian group. It is compatible with homotopy and then the set $[X, S^1]$ acquires the structure of a group.
(c) If $f : Y \to X$ is continuous then $f^* : [X, S^1] \to [Y, S^1]$ is a homomorphism.

5. Show that a space X is contractible iff it is deformable into one of its points.
6. Show that if A is a deformation retract of X, then A and X have the same homotopy type.
7. Show that any one-point subset of a convex subspace Y of $\mathbf{R}^n$ is a strong deformation retract of Y.
8. Let X be the closed unit square and A be the comb space. Show that A is weak deformation retract of X but not a deformation retract of X.
9. Show that the point $(0, 1)$ of the comb space X is a deformation retract of X but not a strong deformation retract of X.
10. Let X be a Hausdorff space and $A \subset X$ be a retract of X. Prove that A is closed in X. Hence show that an open interval $(0, 1)$ cannot be a retract of any closed subset of the real line $\mathbf{R}^1$.
11. Show that

(a) A continuous map $f : X \to Y$ is nullhomotopic iff it has a continuous extension over the cone $CX = (X \times I)/X \times \{1\}$.
(b) Given a continuous map $f : X \to Y$, its mapping cylinder $M_f = (X \times I) \cup Y = (X \times I) \cup Y/\sim$, where for all $x \in X$, $\sim$ identifies $(x, 1)$ with $f(x)$.
(c) $S^1 = \{z \in \mathbf{C} - \{0\} : |z| = 1\}$ is a strong deformation retract of $\mathbf{C} - \{0\}$.
(d) S^1 and $\mathbf{C} - \{0\}$ have the same homotopy type.
(e) For all $f : X \to Y$, the space Y is a deformation retract of the its mapping cylinder M_f.
(f) Any continuous map from a closed subset of $\mathbf{R}^n$ into a sphere is extendable over the whole of $\mathbf{R}^m$ iff f is essential.

(g) Two constant maps $k_i : X \to Y, x \mapsto y_i, i = 0, 1$ are homotopic iff $\exists$ a continuous curve $\gamma : I \to Y$ from y_0 to y_1.

12. Let $[X, Y]$ denote the set of homotopy classes of maps $f : X \to Y$. Show that

(i) for any space X, $[X, I]$ has a single element;
(ii) if X is path-connected, then $[I, X]$ has a single element;
(iii) a contractible space is path-connected;
(iv) if X is contractible, then for any space Y, $[Y, X]$ has a single element;
(v) if X is contractible, and Y is path-connected, then $[X, Y]$ has a single element.

13. Show that a retract of a contractible space is contractible.
14. Show that $\mathbf{R}^{n+1} - \{0\}$ is homotopy equivalent to S^n.
15. Show that the space $X = \{(x, y, z) \in \mathbf{R}^3 : y^2 > xz\}$ is homotopy equivalent to a circle. Interpret this result by considering the roots of the equation $ax^2 + 2hxy + by^2 = 0$.
16. Let $X = \{(p, q) \in S^n \times S^n : p \neq -q\}$. Show that the map $f : S^n \to X$ defined by $f(p) = (p, p)$ is a homotopy equivalence.
17. In $\mathbf{R}^2$, define $A_1 = \{(x_1, x_2) \in \mathbf{R}^2 : (x_1 - 1)^2 + x_2^2 = 1\}$, $A_2 = \{(x_1, x_2) \in \mathbf{R}^2 : (x_1 + 1)^2 + x_2^2 = 1\}$. Suppose $Y = A_1 \cup A_2$, $X = Y \setminus \{(2, 0), (-2, 0)\}$, $A = 0 = \{(0, 0)\}$. Show that A is a strong deformation retract of X.
18. **(a)** Let X and Y be pointed topological spaces. Show that there exists a bijection $\psi : [\Sigma X, Y] \to [X, \Omega Y]$ such that it is natural in X and in Y in the sense that if $f : X' \to X$ and $g : Y \to Y'$ are base point preserving continuous maps, then the diagrams in the Fig. 2.28 and in Fig. 2.29 are commutative, where the horizontal arrows represents the corresponding isomorphism.
(b) Show that for $n \geq 2$ and any pointed Hausdorff space X the iterated loop spaces $\Omega^n X (= \Omega(\Omega^{n-1} X) = (\Omega^{n-1} X)^{S^1})$ are homotopy commutative H-groups. Hence prove that for $n \geq 2$ and pointed spaces X, Y, the groups $[X, \Omega^n Y]$ are abelian.
[Hint: See Theorem 2.5.14.]
19. Let (X, A) have the absolute homotopy extension property (AHEP) (in the sense that A has HEP in X with respect to every space Y) and A be contractible. Show that the identification map $p : X \to X/A$ is a homotopy equivalence.

Fig. 2.28 Naturality of ψ in X

$$\begin{array}{ccc} [\Sigma X, Y] & \xrightarrow{\cong} & [X, \Omega Y] \\ \downarrow{\scriptstyle (\Sigma f)^*} & & \downarrow{\scriptstyle f^*} \\ [\Sigma X', Y]_* & \xrightarrow{\cong} & [X', \Omega Y] \end{array}$$

Fig. 2.29 Naturality of ψ in Y

$$\begin{array}{ccc} [\Sigma X, Y] & \xrightarrow{\cong} & [X, \Omega Y] \\ \downarrow{\scriptstyle g_*} & & \downarrow{\scriptstyle (\Omega g)_*} \\ [\Sigma X, Y']_* & \xrightarrow{\cong} & [X, \Omega Y'] \end{array}$$

20. **(a)** Let G be a fixed H-group with base point e with continuous multiplication $\mu : G \times G \to G$ and homotopy inverse $\phi : G \to G$. Show that there exists a contravariant functor $\pi^G : \mathcal{H}tp_* \to \mathcal{G}rp$.

(b) For each homotopy associative H-space K, show that π^K is a contravariant from $\mathcal{H}tp_*$ to the category of monoids and their homomorphisms.

(c) Show that π^G is homotopy type invariant for each H-group G.

(d) Let G be a pointed topological space such that π^G assumes values in $\mathcal{G}rp$. Show that G is an H-group. Moreover, for any pointed space X, show that the group structure on $\pi^G(X)$ and $[X, G]$ coincide.

(e) Let $\alpha : G \to H$ be a homomorphism of H-groups. Show that α induces a natural transformation $N(\alpha) : \pi^G \to \pi^H$, where $N(\alpha)(X) : [X, G] \to [X, H]$ is defined by $N(\alpha)(X)([f]) = [\alpha \circ f]$, $\forall [f] \in [X, G]$.

21. Given a closed curve C in the plane $\mathbf{R}^2 \times \{0\}$, show that there exists a continuous deformation deforming C into a spherical closed curve $\tilde{C}$ and conversely given a spherical curve $\tilde{C}$, show that there exists a continuous deformation deforming $\tilde{C}$ into a closed curve C in the plane $\mathbf{R}^2 \times \{0\}$ such that total normal twists of C and $\tilde{C}$ remain the same.

22. (M. Fuchs) Prove that two topological spaces X and Y have the same homotopy type iff they are homeomorphic to a strong deformation retract of a space Z.

23. Using the notation of Theorem 2.4.18, show that a pointed space P is an H-space iff there is a continuous map $\mu : P \times P \to P$ such that $\mu \circ i_1 = \mu \circ i_2 = c$. The map μ satisfies the condition $[\mu] = [p_1]\cdot[p_2]$ and if $f, g : X \to P$ are base point preserving continuous maps, then $[f]\cdot[g]$ is the homotopy class of the composite

$$X \xrightarrow{\Delta} X \times X \xrightarrow{f \times g} P \times P \xrightarrow{\mu} P.$$

24. Let X and Y be topological spaces and $f : X \to Y$ be a continuous map. Show that Y is a strong deformation retract of its mapping cylinder M_f.

25. Show that a continuous map $f : X \to Y$ has a left homotopy inverse iff X is a retract of its mapping cylinder M_f.

26. Show that a continuous map $f : X \to Y$ has a right homotopy inverse iff the mapping cylinder M_f deforms into X.

27. Show that

(a) A continuous map $f : X \to Y$ is a homotopy equivalence iff X is a deformation retract of the mapping cylinder M_f;

(b) If D is such a deformation retraction, then $D|_{Y \times \{1\}}$ is a homotopy inverse to f and for any homotopy inverse g, there is a deformation retract of M_f into X which gives g.

(c) Let X be a normal space. If $A \subset X$ is the set of zeros of a continuous map $f : X \to I$, and if A is a strong deformation retract of a neighborhood U of A in X, then $(X \times \{0\} \cup A \times I)$ is a strong deformation retract of $X \times I$.

(d) S^1 is a deformation retract of $\mathbf{R}^2 \setminus \{0\} (= \mathbf{R}^2 - \{0\})$.

(e) Möbius strip is homotopy equivalent to S^1.

28. Show that the suspension

$$\Sigma : \mathcal{H}tp_* \to \mathcal{H}tp_*$$

is an endofunctor (i.e., a functor from $\mathcal{H}tp_*$ to itself).

29. Let X and Y be pointed Hausdorff spaces. Show that

(i) Both $[\Sigma X, Y]$ and $[X, \Omega Y]$ are groups.
(ii) The groups $[\Sigma X, Y]$ and $[X, \Omega Y]$ are isomorphic.
(iii) If X is an H-cogroup and Y is an H-group, then the products available on $[X, Y]$ determine isomorphic groups which are abelian;

30. Let A be a closed (or open) subspace of X in $\mathcal{T}op_*$. Then the inclusion $i : A \hookrightarrow X$ induces closed (or open) inclusions $SP^\infty(i) : SP^\infty A \hookrightarrow SP^\infty X$.

31. Let (X, A) be a pair of topological spaces such that X is a compact Hausdorff space, A is closed in X and A is a strong deformation retract of X. Let $p : X \to X/A$ be the identification map and $p(A) = y \in X/A$. Show that $\{y\}$ is a strong deformation retract of X/A.

32. (Steenrod) Let the space X be compactly generated and A be closed in X. Show that the following statements are equivalent:

(i) (X, A) is an NDR-pair;
(ii) $(X \times I, X \times \{0\} \cup A \times I)$ is a DR-pair;
(iii) $X \times \{0\} \cup A \times I$ is a retract of $X \times I$;
(iv) (X, A) has the homotopy extension property (HEP) with respect to arbitrary topological spaces.

33. **(i)** Let $f, g : (X, A) \to (Y, B)$ be two continuous maps of pair of spaces such that $f \simeq g$. Show that that their induced maps $\tilde{f}, \tilde{g} : X/A \to Y/B$ are also homotopic.
(ii) Let $f : (X, A) \to (Y, B)$ be a homotopy equivalence. Show that the induced map $\tilde{f} : X/A \to Y/B$ is a (based) homotopy equivalence.

34. Let (X, A) be a pair of topological spaces such that A is closed in X and $X \times I$ is a normal space. If there is a neighborhood U such that U is a retract of $(X \times \{0\} \cup (A \times I))$, show that any continuous map $G : (X \times \{0\}) \cup (A \times I) \to Y$ has a continuous extension over $X \times I$.

2.12 Additional Reading

[1] Adams, J.F., *Algebraic Topology: A student's Guide*, Cambridge University Press, Cambridge, 1972.
[2] Adhikari, M.R., and Adhikari, Avishek, *Basic Modern Algebra with Applications*, Springer, New Delhi, New York, Heidelberg, 2014.

[**3**] Adhikari, Avishek and Rana, P.K., *A Study of Functors Associated with Topological Groups*, Studia Universiatis, Babes-Bolyai Mathematica, XLVI(4), 3–14, 2001

[**4**] Arkowitz, Martin, *Introduction to Homotopy Theory*, Springer, New York, 2011.

[**5**] Armstrong, M.A., *Basic Topology*, Springer-Verlag, New York, 1983.

[**6**] Aguilar, Gitler, S., Prieto, C., *Algebraic Topology from a Homotopical View Point*, Springer-Verlag, New York, 2002.

[**7**] Chatterjee, B.C., Ganguly, S., and Adhikari, M.R., *A Textbook of Topology*, Asian Books Pvt.Ltd., New Delhi, 2002.

[**8**] Croom, F.H., *Basic Concepts of Algebraic Topology*, Springer-Verlag, New York, Heidelberg, Berlin, 1978.

[**9**] Dugundji, J., *Topology*, Allyn & Bacon, Newtown, MA, 1966.

[**10**] Dieudonné, J., *A History of Algebraic and Differential Topology*, 1900–1960, Modern Birkhäuser, 1989.

[**11**] Gray, B., *Homotopy Theory, An Introduction to Algebraic Topology*, Academic Press, New York, 1975.

[**12**] Hilton, P.J., *An introduction to Homotopy Theory*, Cambridge University Press, Cambridge, 1983.

[**13**] Hu, S.T., *Homotopy Theory*, Academic Press, New York, 1959.

[**14**] Massey, W.S., *A Basic Course in Algebraic Topology*, Springer-Verlag, New York, Berlin, Heidelberg, 1991.

[**15**] Munkres, J.R., *Topology, A First Course*, Prentice-Hall, New Jersey, 1975.

[**16**] Rotman, J.J., *An Introduction to Algebraic Topology*, Springer-Verlag, New York, 1988.

[**17**] Satya, Deo *Algebraic Topology: A primer*, Hindustan Book Agency, New Delhi, 2003.

[**18**] Switzer, R.M., *Algebraic Topology-Homotopy and Homology*, Springer-Verlag, Berlin, Heidelberg, New York, 1975.

[**19**] Whitehead, G.W., *Elements of Homotopy Theory*, Springer-Verlag, New York, Heidelberg, Berlin, 1978.

References

Adams, J.F.: Vector fields on spheres. Ann. Math. **75**, 603–632 (1962)

Adams, J.F.: Algebraic Topology: A Student's Guide. Cambridge University Press, Cambridge (1972)

Adhikari, A., Rana, P.K.: A study of functors associated with topological groups, Studia Universiatis, Babes-Bolyai Mathematica, vol. XLVI(4), pp. 3–14 (2001)

Adhikari, M.R., Adhikari, A.: Basic Modern Algebra with Applications. Springer, New York (2014)

Arkowitz, M.: Introduction to Homotopy Theory. Springer, New York (2011)

Armstrong, M.A.: Basic Topology. Springer, New York (1983)

Aguilar, M., Gitler, S., Prieto, C.: Algebraic Topology from a Homotopical View Point. Springer, New York (2002)

Barratt, M.G.: Track groups I. Proc. Lond. Math. Soc. **5**(3), 71–106 (1955)

Chatterjee, B.C., Ganguly, S., Adhikari, M.R.: A Textbook of Topology. Asian Books Pvt. Ltd., New Delhi (2002)

Croom, F.H.: Basic Concepts of Algebraic Topology. Springer, New York (1978)

Dugundji, J.: Topology. Allyn & Bacon, Newtown (1966)

Dieudonné, J.: A History of Algebraic and Differential Topology, 1990-1960. Modern Birkhäuser, Basel (1989)

Eilenberg, S., Steenrod, N.: Foundations of Algebraic Topology. Princeton University Press, Princeton (1952)

Gray, B.: Homotopy Theory, An Introduction to Algebraic Topology. Acamedic Press, New York (1975)

Hatcher, A.: Algebraic Topology. Cambridge University Press, Cambridge (2002)

Hilton, P.J.: An Introduction to Homotopy Theory. Cambridge University Press, Cambridge (1983)

Hu, S.T.: Homotopy Theory. Academic Press, New York (1959)

Hurewicz, W.: Beitrage der Topologie der Deformationen. Proc. K. Akad. Wet. Ser. A **38**, 112–119, 521–528 (1935)

Massey, W.S.: A Basic Course in Algebraic Topology. Springer, New York (1991)

Maunder, C.R.F.: Algebraic Topology. Van Nostrand Reinhhold, London (1970)

Munkres, J.R.: Topology, A First Course. Prentice-Hall, New Jersey (1975)

Rotman, J.J.: An Introduction to Algebraic Topology. Springer, New York (1988)

Satya, D.: Algebraic Topology: A Primer. Hindustan Book Agency, New Delhi (2003)

Spanier, E.: Algebraic Topology. McGraw-Hill Book Company, New York (1966)

Steenrod, N.: A convenient category of topological spaces. Mich. Math J. **14**, 133–152 (1967)

Switzer, R.M.: Algebraic Topology-Homotopy and Homology. Springer, Berlin (1975)

Whitehead, G.W.: Elements of Homotopy Theory. Springer, New York (1978)

Chapter 3
The Fundamental Groups

This chapter continues the study of homotopy theory though the concept of fundamental groups invented by H. Poincaré (1854–1912) in 1895, which conveys the first transition from topology to algebra by assigning a group structure on the set of relative homotopy classes of loops in a functorial way. Its group structure is proved in Sects. 3.1 and 3.2 in two different ways. This group earlier called Poincaré group, is now known as fundamental group. It plays an influential role in the study of algebraic topology.

Properties and characteristics which are shared by homeomorphic spaces are called topological properties and topological invariants; on the other hand those by homotopy equivalent spaces are called homotopy properties and homotopy invariants. The Euler characteristic invented by L. Euler (1703–1783) in 1752 is an integral invariant, which distinguishes non-homeomorphic spaces. The search of other invariants has established connections between topology and modern algebra in such a way that homeomorphic spaces have isomorphic algebraic structures. Historically, the concept of fundamental group introduced by Poincaré in 1895 is the first important invariant of homotopy theory which came from such a search. His work explained the difference between curves deformable to one another and curves bounding a larger space. The first one led to the concepts of homotopy and fundamental group. Fundamental group is one of the basic homotopy invariants. It is a very powerful invariant in algebraic topology and is the first of a series of algebraic invariants π_n associated with a topological space with a base point.

The classification of topological spaces up to homeomorphism is the main problem of topology. Given two topological spaces, either we have to find an explicit expression for a homeomorphism between them or we have to show that no such homeomorphism exists. In the latter case, it does not suffice to consider any special mapping, and it is impossible to consider all the mappings. So for proving nonexistence of a homeomorphism we use indirect arguments. In particular, we find a property or a characteristic shared by homeomophic spaces. This is the basic motivation of invention of homotopy and homology groups in algebraic topology.

M.R. Adhikari, *Basic Algebraic Topology and its Applications*,
DOI 10.1007/978-81-322-2843-1_3

Using the tools of the fundamental groups, this chapter introduces the concept of degree function of a continuous map $f : (I, \dot{I}) \to (S^1, 1)$ and develops the necessary tools to compute and study the fundamental group of the circle. It also studies Brouwer fixed point theorem for dimension 2, fundamental theorem of algebra, vector field problems on D^2, and knot groups, and finally computes fundamental groups of some important spaces.

For this chapter the books Armstrong (1983), Bredon (1993), Croom (1978), Massey (1991), Maunder (1970), Munkres (1975), Rotman (1988), Switzer (1975), Whitehead (1978) and some others are referred in the Bibliography.

3.1 Fundamental Groups: Introductory Concepts

This section introduces the concept of fundamental groups and starts studying the basic elementary properties of fundamental groups with an eye to apply them as tools for the study of subsequent chapters. It is one of the several key homotopy invariants which exist associated with topological spaces.

3.1.1 Basic Motivation

The basic motivation of the concept of the fundamental group is given by a geometric approach. Consider the disk X with a hole and another disk Y without a hole as shown in Figs. 3.1 and 3.2.

Any loop in Fig. 3.1 cannot be continuously shrunk to a point; on the other hand any loop in Fig. 3.2 can be continuously shrunk to a point.

For example, the loop α in Fig. 3.1 cannot be continuously shrunk to a point due to existence of a hole in X, but some loops in X such as β may be continuously shrunk to a point but not all loops. This characterizes the difference between the spaces X and Y. This difference leads to the concept of fundamental group of a pointed topological space.

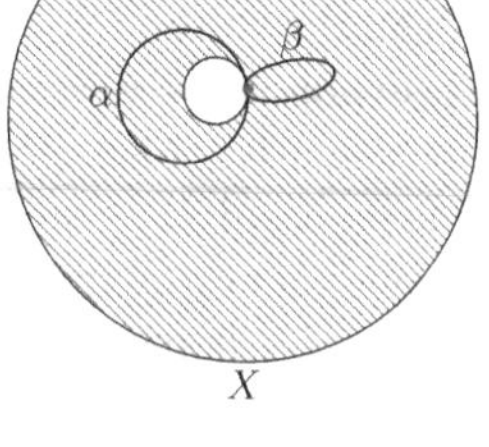

Fig. 3.1 A disk X with a hole

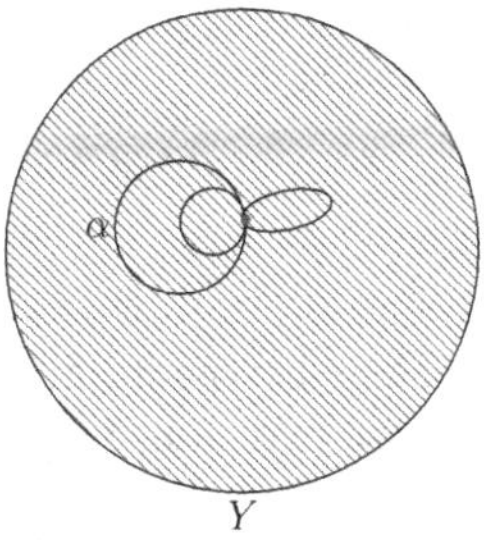

Fig. 3.2 A disk Y without a hole

3.1.2 *Introductory Concepts*

Fundamental group is the first of a sequence of functors π_n (see Chap. 7), called homotopy group functors from the category of pointed topological spaces to the category of groups. Such functors occupy a vast territory in algebraic topology and are still the subject of intensive study. More precisely, given a pointed topological space (X, x_0), the set $\pi_1(X, x_0)$ is defined to be the set of homotopy classes of paths $f : I \to X$ that send 0 and 1 to x_0. Each such path is called a loop in X based at x_0. It is shown that $\pi_1(X, x_0)$ admits a group structure.The group $\pi_1(X, x_0)$ depends on X as well as on $x_0 \in X$ and is called the fundamental group or Poincaré group of the space X based at x_0. It is a homotopy type invariant in the sense that homotopy equivalent spaces (X, x_0) and (Y, y_0) have the isomorphic fundamental groups $\pi_1(X, x_0)$ and $\pi_1(Y, y_0)$.

Definition 3.1.1 Let X be a topological space and $u : I \to X$ be a continuous map. Then u is said to be a path in X, $u(0)$ is called the initial point and $u(1)$ is called the terminal point of the path u.

If u and v are two paths in X such that $u(1) = v(0)$, then we can define a new path, called the product of u and v denoted by $u * v$ as follows:

$$(u * v)(t) = \begin{cases} u(2t), & 0 \leq t \leq 1/2 \\ v(2t-1), & 1/2 \leq t \leq 1 \end{cases} \tag{3.1}$$

$u * v : I \to X$ is continuous by Pasting Lemma. The initial point of $u * v$ is the initial point of u and the terminal point of $u * v$ is the terminal point of v.

If w is a third path in X such that $v(1) = w(0)$, then the paths $u * (v * w)$ and $(u * v) * w$ are defined by

$$(u * (v * w))(t) = \begin{cases} u(2t), & 0 \leq t \leq 1/2 \\ v(4t-2), & 1/2 \leq t \leq 3/4 \\ w(4t-3), & 3/4 \leq t \leq 1 \end{cases}$$

and

$$((u * v) * w)(t) = \begin{cases} u(4t), & 0 \leq t \leq 1/4 \\ v(4t-1), & 1/4 \leq t \leq 1/2 \\ w(2t-1), & 1/2 \leq t \leq 1 \end{cases}$$

These two paths in X are not necessarily the same paths, because at $t = \frac{1}{2}$, images of these two paths may not be the same, since $u(1)$ and $v(1)$ may not be equal. This shows that the product of paths is in general not an associative operation. Even for a fixed $x_0 \in X$, the product of loops in X based at x_0 need not be associative, because their respective images at $t = \frac{1}{4}$ may not be equal. To overcome this difficulty we consider an equivalence relation on the set $\Omega(X, x_0)$ of all loops in X based at $x_0 \in X$.

Definition 3.1.2 A path $u : I \to X$ is called a loop in X based at $x_0 \in X$ if $u(0) = u(1) = x_0$. If $\dot{I} = \{0, 1\}$, then a loop f in X based at x_0 is a continuous map $u : (I, \dot{I}) \to (X, x_0)$. In particular, the constant map $c : I \to X, t \mapsto x_0, \forall\ t \in I$, is called a constant path or a null loop in X at x_0.

Definition 3.1.3 Let $u, v : (I, \dot{I}) \to (X, x_0)$ be two loops in X based at x_0. Then u and v are said to be homotopic relative to the subspace $\dot{I} = \{0, 1\}$ of I denoted by $u \simeq v$ rel $\dot{I}$, if $\exists$ a continuous map
$F : I \times I \to X$ such that
$F(t, 0) = u(t), \ \forall\ t \in I,$
$F(t, 1) = v(t), \ \forall\ t \in I,$
and $F(0, s) = F(1, s) = x_0, \ \forall\ s \in I.$

Let $\Omega(X, x_0)$ be the set of all loops in X based at x_0. Then it follows from Theorem 2.1.37 of Chap. 2 that '$\simeq$' is an equivalence relation on $\Omega(X, x_0)$. This gives the set of homotopy classes of loops relative to $\dot{I} = \{0, 1\}$, denoted by $\pi_1(X, x_0)$. Thus $\pi_1(X, x_0)$ is the quotient set $\Omega(X, x_0)/\simeq$.

We want to define a composition on $\pi_1(X, x_0)$ to make it a group. First, we define composition $*$ on $\Omega(X, x_0)$ and then we carry it to $\Omega(X, x_0)/\simeq\ = \pi_1(X, x_0)$.

Definition 3.1.4 Given $u, v \in \Omega(X, x_0)$ their product $u * v : (I, \dot{I}) \to (X, x_0)$ is defined by

$$(u * v)(t) = \begin{cases} u(2t), & 0 \leq t \leq 1/2 \\ v(2t-1), & 1/2 \leq t \leq 1 \end{cases} \tag{3.2}$$

Then at $t = \frac{1}{2}$, $u(2t) = u(1) = x_0 = v(0) = v(2t-1)$ shows that $u * v$ is well defined and continuous by Pasting Lemma. Moreover $(u * v)(0) = u(0) = x_0 = v(0) = (u * v)(1) \implies u * v$ is a loop in X based at $x_0 \implies u * v \in \Omega(X, x_0)$

We now extend this definition for the product of three loops. Given loops $u, v, w : (I, \dot{I}) \to (X, x_0)$, their product $u * v * w : I \to X$ is defined by

$$(u * v * w)(t) = \begin{cases} u(3t), & 0 \leq t \leq 1/3 \\ v(3t-1), & 1/3 \leq t \leq 2/3 \\ w(3t-2), & 2/3 \leq t \leq 1 \end{cases}$$

Then as before, $u * v * w \in \Omega(X, x_0)$.

Definition 3.1.5 If $u \in \Omega(X, x_0)$, then its inverse $u^{-1} : (I, \dot{I}) \to (X, x_0)$ is defined by $u^{-1}(t) = u(1-t), \forall\, t \in I$.

Clearly, $u^{-1} \in \Omega(X, x_0)$. Thus $u \in \Omega(X, x_0) \implies u^{-1} \in \Omega(X, x_0)$.

Remark 3.1.6 u and u^{-1} give the same set of points of X but their directions are opposite.

Proposition 3.1.7 *If $u_1, u_2, v_1, v_2 \in \Omega(X, x_0)$ and $u_1 \simeq u_2$ rel $\dot{I}$, $v_1 \simeq v_2$ rel $\dot{I}$, then $u_1 * v_1 \simeq u_2 * v_2$ rel $\dot{I}$.*

Proof Let $F : u_1 \simeq u_2$ rel $\dot{I}$ and $G : v_1 \simeq v_2$ rel $\dot{I}$. Then $F(t, 0) = u_1(t)$, $F(t, 1) = u_2(t)$, $\forall\, t \in I$, $F(0, s) = x_0 = F(1, s)$, $\forall\, s \in I$ and $G(t, 0) = v_1(t)$, $G(t, 1) = v_2(t)$, $\forall\, t \in I$, $G(0, s) = x_0 = G(1, s)$, $\forall\, s \in I$.

Define a map $H : I \times I \to X$ by

$$H(t, s) = \begin{cases} F(2t, s), & 0 \leq t \leq 1/2 \\ G(2t-1, s), & 1/2 \leq t \leq 1 \end{cases}$$

Then H is well defined. Moreover, it is continuous by Pasting Lemma, since its restrictions to $I \times [0, \frac{1}{2}]$ and $I \times [\frac{1}{2}, 1]$ are continuous, and both functions agree on $\{\frac{1}{2}\} \times I$ and their restrictions to $[0, \frac{1}{2}] \times I$ and $[\frac{1}{2}, 1] \times I$ are continuous. Again,

$$\begin{aligned} H(t, 0) &= \begin{cases} F(2t, 0), & 0 \leq t \leq 1/2 \\ G(2t-1, 0), & 1/2 \leq t \leq 1 \end{cases} \\ &= \begin{cases} u_1(2t), & 0 \leq t \leq 1/2 \\ v_1(2t-1), & 1/2 \leq t \leq 1 \end{cases} \\ &= (u_1 * v_1)(t), \quad \forall\, t \in I. \end{aligned}$$

Similarly, $H(t, 1) = (u_2 * v_2)(t)$, $\forall\, t \in I$, $H(0, s) = x_0 = F(0, s)$, $\forall\, s \in I$ and $H(1, s) = x_0 = G(1, s)$, $\forall\, s \in I$. Consequently, $H : u_1 * v_1 \simeq u_2 * v_2$ rel $\dot{I}$. ❑

Proposition 3.1.8 *If $u, v \in \Omega(X, x_0)$ and $u \simeq v$ rel $\dot{I}$, then $u^{-1} \simeq v^{-1}$ rel $\dot{I}$.*

Proof Let $F : u \simeq v$ rel $\dot{I}$. Then $F(t, 0) = u(t)$, $F(t, 1) = v(t)$, $\forall\, t \in I$ and $F(0, s) = x_0 = F_1(1, s)$.

Define $G : I \times I \to X$ by $G(t, s) = F(1-t, s)$.

Then G is a continuous function such that

$$G(t,0) = F(1-t,0) = u(1-t) = u^{-1}(t),\ \forall\ t \in I,$$
$$G(t,1) = F(1-t,1) = v(1-t) = v^{-1}(t),\ \forall\ t \in I$$
$$\text{and } G(0,s) = F(1,s) = x_0,\ \ G(1,s) = F(0,s) = x_0.$$

Consequently, $G : u^{-1} \simeq v^{-1} \text{ rel } \dot{I}$. ❑

Proposition 3.1.9 *If $u, v, w \in \Omega(X, x_0)$, then $u * (v * w) \simeq (u * v) * w \text{ rel } \dot{I}$*

Proof $u * (v * w) : (I, \dot{I}) \to (X, x_0)$ is defined by

$$(u * (v * w))(t) = \begin{cases} u(2t), & 0 \le t \le 1/2 \\ (v * w)(2t-1), & 1/2 \le t \le 1 \end{cases} = \begin{cases} u(2t), & 0 \le t \le 1/2 \\ v(4t-2), & 1/2 \le t \le 3/4 \\ w(4t-3), & 3/4 \le t \le 1 \end{cases}$$

Then $u * (v * w)$ is well defined, continuous by Pasting Lemma and a loop in X based at x_0 and therefore $u * (v * w) \in \Omega(X, x_0)$. On the other hand,

$$((u * v) * w)(t) = \begin{cases} (u * v)(2t), & 0 \le t \le 1/2 \\ w(2t-1), & 1/2 \le t \le 1 \end{cases} = \begin{cases} u(4t), & 0 \le t \le 1/4 \\ v(4t-1), & 1/4 \le t \le 1/2 \\ w(2t-1), & 1/2 \le t \le 1 \end{cases}$$

As before, $(u * v) * w \in \Omega(X, x_0)$.
Define a map $H : I \times I \to X$ by the rule

$$H(t,s) = \begin{cases} u(4t/(1+s)), & 0 \le t \le (1+s)/4 \\ v(4t-1-s), & (1+s)/4 \le t \le (2+s)/4 \\ w(1-(4(1-t)/(2-s))), & (2+s)/4 \le t \le 1 \end{cases}$$

Then H is well defined. Moreover it is continuous by Pasting Lemma. Now

$$H(t,0) = \begin{cases} u(4t), & 0 \le t \le 1/4 \\ v(4t-1), & 1/4 \le t \le 1/2 \\ w(2t-1), & 1/2 \le t \le 1 \end{cases}$$
$$= ((u * v) * w)(t), \quad \forall\ t \in I,$$

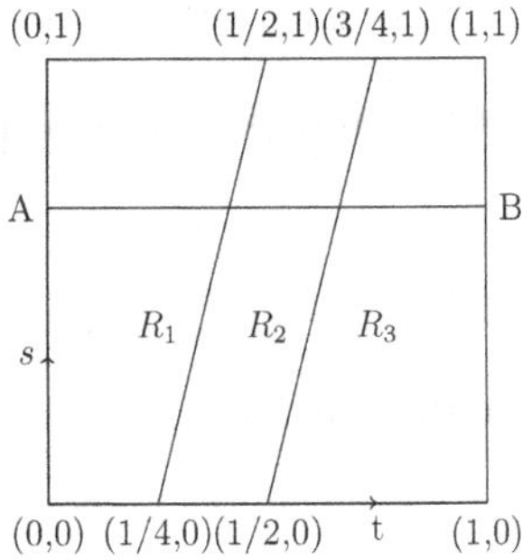

Fig. 3.3 Construction of homotopy H

$$H(t,1) = \begin{cases} u(2t), & 0 \leq t \leq 1/2 \\ v(4t-2), & 1/2 \leq t \leq 3/4 \\ w(4t-3), & 3/4 \leq t \leq 1 \end{cases}$$
$$= (u * (v * w))(t), \quad \forall\ t \in I,$$

$H(0,s) = u(0) = x_0$, and $H(1,s) = w(1) = x_0$.
Hence $(u * v) * w \simeq u * (v * w)$ rel $\dot{I}$. ❑

The motivation for writing H comes from the diagram in Fig. 3.3.
Divide the square $I \times I$ into the three regions R_1, R_2, and R_3 given by

$$R_1 : 0 \leq t \leq (s+1)/4, 0 \leq s \leq 1;$$
$$R_2 : (s+1)/4 \leq t \leq (s+2)/4, 0 \leq s \leq 1;$$
$$R_3 : (s+2)/4 \leq t \leq 1, 0 \leq s \leq 1.$$

The two slanted lines are given by the equations: $s = 4t - 1$ and $s = 4t - 2$.

For a fixed $s \in I$, the horizontal line AB has three pieces. When s moves from 0 to 1, these pieces also change their positions. For $s = 0$, we obtain a partition defining $(u * v) * w$ and for $s = 1$, we obtain a partition defining $u * (v * w)$. The map H defined by u on R_1, v on R_2 and w on R_3, each of which is continuous. On their common boundary, each pair of maps agree. Then by Pasting Lemma H is continuous and yields the required homotopy.

Proposition 3.1.10 *If $u \in \Omega(X, x_0)$ and $c : I \to X$ is the constant loop at x_0 defined by $c(t) = x_0$, $\forall\ t \in I$, then $u * c \simeq u$ rel $\dot{I}$ and $c * u \simeq u$ rel $\dot{I}$*

Proof $u * c : I \to X$ is defined by

$$(u * c)(t) = \begin{cases} u(2t), & 0 \leq t \leq 1/2 \\ c(2t-1), & 1/2 \leq t \leq 1 \end{cases}$$
$$= \begin{cases} u(2t), & 0 \leq t \leq 1/2 \\ x_0, & 1/2 \leq t \leq 1 \end{cases}$$

Then $u * c \in \Omega(X, x_0)$.

Define a map $H : I \times I \to X$ by

$$H(t,s) = \begin{cases} u(2t/(1+s)), & 0 \le t \le (1+s)/2 \\ x_0, & (1+s)/2 \le t \le 1 \end{cases}$$

Then $H : u * c \simeq u \text{ rel } \dot{I}$. Similarly, $c * c \simeq u \; rel \; \dot{I}$. ❑

Proposition 3.1.11 *If $u \in \Omega(X, x_0)$, then $u * u^{-1} \simeq c \text{ rel } \dot{I}$ and $u^{-1} * u \simeq c \text{ rel } \dot{I}$*

Proof $u * u^{-1} : I \to X$ is given by

$$\begin{aligned}(u * u^{-1})(t) &= \begin{cases} u(2t), & 0 \le t \le 1/2 \\ u^{-1}(2t-1), & 1/2 \le t \le 1 \end{cases} \\ &= \begin{cases} u(2t), & 0 \le t \le 1/2 \\ u(1-\overline{2t-1}), & 1/2 \le t \le 1 \end{cases} \\ &= \begin{cases} u(2t), & 0 \le t \le 1/2 \\ u(2-2t), & 1/2 \le t \le 1 \end{cases}\end{aligned}$$

Then $u * u^{-1} \in \Omega(X, x_0)$. Define $H : I \times I \to X$ by

$$H(t,s) = \begin{cases} u(2t(1-s)), & 0 \le t \le 1/2 \\ u((2-2t)(1-s)), & 1/2 \le t \le 1 \end{cases}$$

Then $H : u * u^{-1} \simeq c \text{ rel } \dot{I}$.
Similarly, $u^{-1} * u \simeq c \text{ rel } \dot{I}$. ❑

Theorem 3.1.12 *$\pi_1(X, x_0)$ is a group.*

Proof Let $[u], [v] \in \pi_1(X, x_0)$. Then $u, v \in \Omega(X, x_0)$ and $u * v (\in \Omega(X, x_0))$ is defined by (3.1). This law of composition '$*$' is carried over to $\pi_1(X, x_0)$ to give the composition '$\circ$' by the rule $[u] \circ [v] = [u * v]$. The composition '$\circ$' is well defined by Proposition 3.1.7, because it is independent of the choice of the representatives of the classes. This composition is associative by Proposition 3.1.9, $[c]$ is the identity element by Proposition 3.1.10 and any element $[u] \in \pi_1(X, x_0)$ has an inverse $[u^{-1}] \in \pi_1(X, x_0)$ by Proposition 3.1.11. Consequently, $\pi_1(X, x_0)$ is a group under the composition '$\circ$.' ❑

Definition 3.1.13 $\pi_1(X, x_0)$ is called the Fundamental group or Poincar*é* group of X based at x_0.

Remark 3.1.14 For an equivalent definition of $\pi_1(X, x_0)$, and an alternative proof of its group structure see Sect. 3.2.

Remark 3.1.15 The index '1' in the notation $\pi_1(X, x_0)$ appeared later than the notation $\pi(X, x_0)$ used by Poincaré in 1895. It is sometimes called the first or one-dimensional homotopy group. There is an infinite sequence of groups $\pi_n(X, x_0)$

with $n = 1, 2, 3, \ldots$, the first of them is the fundamental group. The higher dimensional homotopy groups (see Chap. 7) were introduced by W. Hurewicz in 1935. For $n = 0$, $\pi_0(X, x_0)$, which is the set of path-connected components of X, is not a group as a rule.

Example 3.1.16 **(a)** If X is a contractible space and $x_0 \in X$, then $\pi_1(X, x_0) = 0$.
(b) $\pi_1(\mathbf{R}^n, x) = 0$ for any $x \in \mathbf{R}^n$.
(c) $\pi_1(D^n, d_0) = 0$ for any $d_0 \in D^n$

[Hint: **(a)** Let X be a contractible space. Then any continuous map $f : (I, \dot{I}) \to (X, x_0)$ is homotopic to the constant map c at x_0 relative to $\dot{I} \implies f \simeq c \text{ rel } \dot{I} \implies [f] = [c] \implies \pi_1(X, x_0) = [c] = 0$. **(b)** and **(c)** follow from **(a)**.]

Example 3.1.17 If X is any convex set in $\mathbf{R}^n$, then $\pi_1(X, x_0) = 0$.

We shall compute the fundamental group of the circle a little later.

It is natural to ask: does $\pi_1(X, x_0)$ depend on the choice of the base point x_0? How are $\pi_1(X, x_0)$ and $\pi_1(X, x_1)$ related for two different points $x_0, x_1 \in X$? If X is an arbitrary topological space, then a loop in X at x_0 being itself path-connected, lies completely in the path-component of x_0. On the other hand, if x_0 and x_1 are points in distinct path components of X, then $\pi_1(X, x_0)$ and $\pi_1(X, x_1)$ are not at all related. If x_0 and x_1 lie in the same path component of X, we shall show that the groups $\pi_1(X, x_0)$ and $\pi_1(X, x_1)$ are isomorphic.

Theorem 3.1.18 *If X is a path-connected space and x_0, x_1 are two distinct points of X, then the groups $\pi_1(X, x_0)$ and $\pi_1(X, x_1)$ are isomorphic.*

Proof As X is path-connected and $x_0, x_1 \in X$, then $\exists$ a path $u : I \to X$ in X from x_0 to x_1 with inverse path $\bar{u} : I \to X$ defined by $\bar{u}(t) = u(1 - t)$ from x_1 to x_0 as shown in Fig. 3.4.

We now define a map

$$\beta_u : \pi_1(X, x_0) \to \pi_1(X, x_1), [f] \mapsto [\bar{u} * f * u].$$

Thus $\beta_u([f]) = [\bar{u} * f * u]$, $\forall\ [f] \in \pi_1(X, x_0)$.

Let $g \in [f]$. Then $f \simeq g \text{ rel } \dot{I} \implies \bar{u} * f * u \simeq \bar{u} * g * u \text{ rel } \dot{I} \implies [\bar{u} * f * u] = [\bar{u} * g * u] \implies \beta_u([f]) = \beta_u([g]) \implies \beta_u$ is well defined. Again β_u is homomorphism. Because, $\beta_u([f] \circ [g]) = \beta_u([f * g]) = [\bar{u} * (f * g) * u] = [\bar{u} * f * u * \bar{u} * g * u]$, since $u * \bar{u} \simeq c \text{ rel } \dot{I} = [(\bar{u} * f * u) * (\bar{u} * g * u)] = [\bar{u} * f * u] \circ [\bar{u} * g * u] = \beta_u[f] \circ \beta_u[g]$, $\forall\ [f], [g] \in \pi_1(X, x_0)$.

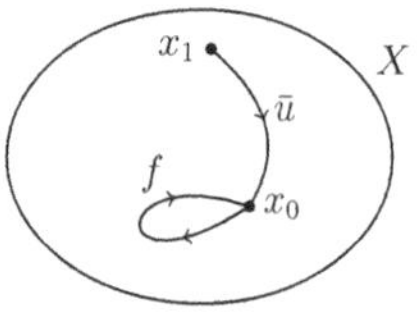

Fig. 3.4 Isomorphism of fundamental groups in a path-connected space

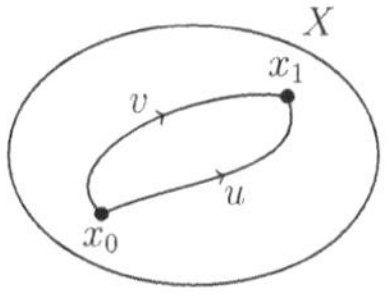

Fig. 3.5 Path homotopic maps

Finally, β_u is an isomorphism with inverse $\beta_{\bar{u}} : \pi_1(X, x_1) \to \pi_1(X, x_0)$, because $(\beta_{\bar{u}} \circ \beta_u)([f]) = \beta_{\bar{u}}([\bar{u} * f * u]) = [\bar{\bar{u}} * \bar{u} * f * u * \bar{u}] = [u * \bar{u} * f * u * \bar{u}] = [f]$, since $u * \bar{u} \simeq c$ rel $\dot{I} \implies \beta_{\bar{u}} \circ \beta_u =$ identity homomorphism. Similarly $\beta_u * \beta_{\bar{u}}=$ identity homomorphism. Hence β_u is an isomorphism of groups. ❑

Corollary 3.1.19 *If u is a path in X from x_0 to x_1, then u induces an isomorphism*

$$\beta_u : \pi_1(X, x_0) \to \pi_1(X, x_1).$$

Remark 3.1.20 If X is path-connected, the group $\pi_1(X, x_0)$ is, up to isomorphism, independent of the choice of the base point x_0. In this case, the notation $\pi_1(X, x_0)$ is abbreviated to $\pi_1(X)$.

We now consider the following situation.

Proposition 3.1.21 *If u and v are two paths in X joining x_0 to x_1 which are path homotopic, then their induced isomorphisms β_u and β_v are identical.*

Proof If u and v are path homotopic as shown in Fig. 3.5, then $\bar{u}$ and $\bar{v}$ are also path homotopic. Therefore, it follows that for any loop f in X based at x_0, $\bar{u} * f * u$ is path homotopic to $\bar{v} * f * v$. Consequently, $\beta_u([f]) = \beta_v([f])$, $\forall\ [f] \in \pi_1(X, x_0) \implies \beta_u = \beta_v$. ❑

We now characterize the commutativity of $\pi_1(X, x_0)$ for a path-connected space.

Theorem 3.1.22 *Let X be a path-connected space and $x_0, x_1 \in X$. Then the group $\pi_1(X, x_0)$ is abelian if and only if for each pair of paths u, v from x_0 to x_1, $\beta_u = \beta_v$.*

Proof Let the group $\pi_1(X, x_0)$ be abelian. Then $[u * \bar{v}] \circ [f] = [f] \circ [u * \bar{v}]$ for each $[f] \in \pi_1(X, x_0)$, since $u * \bar{v}$ is a loop in X based at x_0 as shown in Fig. 3.6.
$\implies u * \bar{v} * f \simeq f * u * \bar{v}$ rel $\dot{I}$

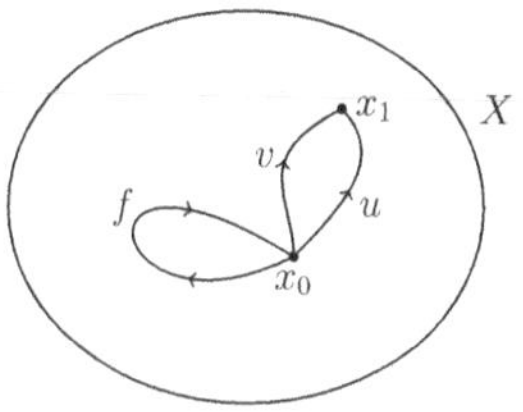

Fig. 3.6 Loop in a path-connected space

$\Longrightarrow \bar{u} * u * \bar{v} * f * v \simeq \bar{u} * f * u * \bar{v} * v \text{ rel } \dot{I}$
$\Longrightarrow [\bar{v} * f * v] = [\bar{u} * f * u]$
$\Longrightarrow \beta_v([f]) = \beta_u([f]), \ \forall \ [f] \in \pi_1(X, x_0)$
$\Longrightarrow \beta_v = \beta_u.$

Conversely, let $[f], [g] \in \pi_1(X, x_0)$. Let u be a path in X joining x_0 to x_1. Then $g * u$ is also a path in X joining x_0 to x_1. By hypothesis, $\beta_{g*u}([f]) = \beta_u([f])$

$\Longrightarrow [\overline{g * u} * f * (g * u)] = [\bar{u} * f * u]$
$\Longrightarrow \bar{u} * \bar{g} * f * g * u \simeq \bar{u} * f * u \text{ rel } \dot{I}$
$\Longrightarrow u * \bar{u} * \bar{g} * f * g * u * \bar{u} \simeq u * \bar{u} * f * u * \bar{u} \text{ rel } \dot{I}$
$\Longrightarrow \bar{g} * f * g \simeq f \text{ rel } \dot{I}$
$\Longrightarrow f * g \simeq g * f \text{ rel } \dot{I}$
$\Longrightarrow [f * g] = [g * f]$
$\Longrightarrow [f] \circ [g] = [g] \circ [f]$
$\Longrightarrow \pi_1(X, x_0)$ is abelian. □

3.1.3 *Functorial Property of π_1*

This subsection studies π_1 from the view point of category theory (see Appendix B). Recall that for every pointed topological space (X, x_0) there exists the fundamental group $\pi_1(X, x_0)$. We now show that for every continuous map $f : (X, x_0) \to (Y, y_0)$, there exists a homomorphism $f_* : \pi_1(X, x_0) \to \pi_1(Y, y_0)$ satisfying some interesting properties. The construction of the fundamental group shows that π_1 is functorial.

Theorem 3.1.23 *Every continuous map $f : (X, x_0) \to (Y, y_0)$ induces a group homomorphism $f_* : \pi_1(X, x_0) \to \pi_1(Y, y_0)$ with the following properties:*

- **(i)** *If $1_X : (X, x_0) \to (X, x_0)$ is the identity map, then $1_{X*} : \pi_1(X, x_0) \to \pi_1(X, x_0)$ is the identity automorphism;*
- **(ii)** *If $f : (X, x_0) \to (Y, x_0)$ and $g : (Y, y_0) \to (Z, x_0)$ are two continuous maps, then $(g \circ f)_* = g_* \circ f_* : \pi_1(X, x_0) \to \pi_1(Z, z_0)$.(These properties are called functorial properties);*
- **(iii)** *If $f, g : (X, x_0) \to (Y, y_0)$ are two continuous maps such that $f \simeq g$ rel $\{x_0\}$, then $f_* = g_*$;*
- **(iv)** *If $f : (X, x_0) \to (Y, y_0)$ has a left (resp. right) homotopy inverse, then $f_* : \pi_1(X, x_0) \to \pi_1(Y, y_0)$ is a monomorphism (resp. an epimorphism);*
- **(v)** *If (X, x_0) and (Y, y_0) are homotopy equivalent spaces, then the groups $\pi_1(X, x_0)$ and $\pi_1(Y, y_0)$ are isomorphic, i.e., if $(X, x_0) \simeq (Y, y_0)$, then $\pi_1(X, x_0) \cong \pi_1(Y, y_0)$;*

(vi) *If $f : (X, x_0) \to (Y, y_0)$ is a homotopy equivalence, then $f_* : \pi_1(X, x_0) \to \pi_1(Y, y_0)$ is an isomorphism.*

Proof Define $f_* : \pi_1(X, x_0) \to \pi_1(Y, y_0)$ by the rule $f_*([u]) = [f \circ u]$. Then f_* is well defined. Because, $f \circ u$ is a loop in Y based at y_0 and if $u \simeq v$ rel $\dot{I}$ by a homotopy F, then $f \circ u \simeq f \circ v$ rel $\dot{I}$ by a homotopy $f \circ F$.

We now show that f_* is a homomorphism.

Let $[u], [v] \in \pi_1(X, x_0)$. Then

$$f_*([u] \circ [v]) = f_*([u * v]) = [f \circ (u * v)]. \tag{3.3}$$

$$\text{Now} \quad (f \circ (u * v))(t) = f((u * v)(t))$$
$$= \begin{cases} f(u(2t)), & 0 \le t \le 1/2 \\ f(v(2t-1)), & 1/2 \le t \le 1 \end{cases}$$
$$= \begin{cases} (f \circ u)(2t), & 0 \le t \le 1/2 \\ (f \circ v)(2t-1)), \text{`` `` ``} & 1/2 \le t \le 1 \end{cases}$$
$$= ((f \circ u) * (f \circ v))(t), \quad \forall\ t \in I$$

$$\implies f \circ (u * v) = (f \circ u) * (f \circ v) \tag{3.4}$$

Hence from (3.3) and (3.4) it follows that $f_*([u] \circ [v]) = [(f \circ u) * (f \circ v)] = f_*([u]) \circ f_*([v])$, $\forall\ [u], [v] \in \pi_1(X, x_0) \implies f_*$ is a group homomorphism.

(i) Let $1_X : (X, x_0) \to (X, x_0)$ be the identity map. Then $1_X \circ u = u$ for each loop u in X based at $x_0 \implies 1_{X^*}([u]) = [1_X \circ u] = [u]$, $\forall\ [u] \in \pi_1(X, x_0) \implies 1_{X^*}$ is the identity automorphism on $\pi_1(X, x_0)$.

(ii) For any $[u] \in \pi_1(X, x_0)$, $(g \circ f)_*([u]) = [(g \circ f) \circ u] = [g \circ (f \circ u)] = g_*([f \circ u]) = g_*(f_*([u])) = (g_* \circ f_*)[u]$, $\forall\ [u] \in \pi_1(X, x_0) \implies (g \circ f)_* = g_* \circ f_*$.

(iii) Let $f \simeq g$ rel $\{x_0\}$. Then $\forall\ [u] \in \pi_1(X, x_0)$, $f \circ u \simeq g \circ u$ rel $\{y_0\} \implies f_*([u]) = [f \circ u] = [g \circ u] = g * ([u]) \implies f_* = g_*$.

(iv) It follows from **(iii)**.

(v) Suppose $(X, x_0) \simeq (Y, y_0)$. Hence $\exists$ two continuous functions $f : (X, x_0) \to (Y, y_0)$ and $g : (Y, y_0) \to (X, x_0)$ such that $g \circ f \simeq 1_X$ and $f \circ g \simeq 1_Y$. Hence $(g \circ f)_* = 1_{X^*}$ and $(f \circ g)_* = 1_{Y^*}$ by (iii). Thus it follows from (i) and (ii) that $g_* \circ f_* =$ identity automorphism and $f_* \circ g_* =$ identity automorphism. Consequently, f_* is an isomorphism with g_* its inverse. For the second part, proceed as in first part.

(vi) It follows from **(v)**. ❑

Corollary 3.1.24 **(i)** *The fundamental group of a pointed topological space is invariant under homeomorphisms, and hence it is a topological invariant.*

(ii) *The fundamental group of a pointed topological space is invariant under homotopy equivalences, and hence it is a homotopy invariant.*

We now express the results of Theorem 3.1.23 in the language of category theory.

Theorem 3.1.25 **(a)** π_1 *is a covariant functor from the category Top*$_*$ *of pointed topological spaces and their base point preserving continuous maps to the category Grp of groups and their homomorphisms. Moreover, if* $f, g : (X, x_0) \to (Y, y_0)$ *are continuous maps and* $f \simeq g$ rel $\{x_0\}$, *then* $\pi_1(f) = f_* = g_* = \pi_1(g)$.

(b) π_1 *is a covariant functor from homotopy category Htp*$_*$ *of pointed topological spaces and their homotopy classes of maps to the category Grp.*

Proof **(a)** The object function is defined by $(X, x_0) \mapsto \pi_1(X, x_0)$ and the morphism function is defined by $f \mapsto \pi_1(f) = f_*$. Then **(a)** follows from Theorem 3.1.23.

(b) follows from **(a)** and Theorem 3.1.23 (iii). ❑

3.1.4 Some Other Properties of π_1

This subsection conveys the behavior of fundamental groups of some special spaces such as H-spaces, simply connected spaces, product spaces, and some other spaces.

Proposition 3.1.26 *Let X be a topological space and A be a subspace of X. If* $i : A \hookrightarrow X$ *is the inclusion map and* $r : X \to A$ *is a retraction, then* $r_* : \pi_1(X, a) \to \pi_1(A, a)$ *is an epimorphism for each* $a \in A$ *and* $i_* : \pi_1(A, a) \to \pi_1(X, a)$ *is a monomorphism for each* $a \in A$.

Proof For each $a \in A$, the composite map $(A, a) \xrightarrow{i} (X, a) \xrightarrow{r} (A, a)$ is the identity map on (A, a) (see Definition 2.7.1 of Chap. 2). Consequently, the composite homomorphism

$$\pi_1(A, a) \xrightarrow{i_*} \pi_1(X, a) \xrightarrow{r_*} \pi_1(A, a)$$

is the identity automorphism on $\pi_1(A, a)$.

Hence r_* is an epimorphism and i_* is a monomorphism. ❑

Proposition 3.1.27 *Let A be a strong deformation retract of a space X. Then for each* $a \in A$, *the groups* $\pi_1(A, a)$ *and* $\pi_1(X, a)$ *are isomorphic.*

Proof Let A be a strong deformation retract of a topological X. Then $\exists$ a retraction $r : (X, a) \to (A, a)$ for each $a \in A$ such that $1_X \simeq i \circ r$ rel $A \implies i_* \circ r_* = \text{id} \implies i_*$ is an epimorphism. Again $r \circ i = \text{id} \implies r_* \circ i_* = \text{id} \implies i_*$ is a monomorphism. Consequently, $i_* : \pi_1(A, a) \to \pi_1(X, a)$ is an isomorphism. ❑

Let X be a topological space. Let C be the path component of X containing x_0. Then $\pi_1(C, x_0) = \pi_1(X, x_0)$, because all loops and homotopies in X based at x_0 lie entirely in the subspace C. Therefore, $\pi_1(X, x_0)$ depends only on the path component of X containing x_0 and provides no information about the set $X - C$. So it is usual to deal only with path-connected spaces while studying the fundamental groups.

Definition 3.1.28 A topological space X is called simply connected if it is path-connected and $\pi_1(X, x_0) = 0$ for some $x_0 \in X$ (hence for every $x_0 \in X$).

Theorem 3.1.29 *If X is a simply connected space, then any two paths in X having the same initial and final points are homotopic.*

Proof Let X be simply connected and u and v be two paths in X from x_0 to x_1 as shown in Fig. 3.7. Then $u * \bar{v}$ is a loop in X based at x_0.

Since X is simply connected, $u*\bar{v} \simeq c_{x_0} \implies u*\bar{v}*v \simeq c_{x_0}*v$. Hence it follows that $[(u*\bar{v})*v] = [c_{x_0}*v] = [v]$. But $[(u*\bar{v})*v] = [u*(\bar{v}*v)] = [u*c_{x_1}] = [u]$. Consequently, $[u] = [v] \implies u \simeq v$. ❑

Theorem 3.1.30 *Every contractible space is simply connected.*

Proof Let X be a contractible space. Then there is a point $x_0 \in X$ and homotopy $H : X \times I \to X$ such that $H(x, 0) = x$ and $H(x, 1) = x_0$, $\forall\ x \in X$.

We claim that X is path-connected. If $x \in X$, the function

$$\sigma_x = H(x, -) : I \to X, t \mapsto \sigma_x(t) = H(x, t)$$

is a path in X from $\sigma_x(0) = H(x, 0) = x$ to $\sigma_x(1) = H(x, 1) = x_0$. Similarly, for any $y \in X, \sigma_y$ is a path from y to x_0 and hence $\bar{\sigma}_y$(the inverse path of σ_y) is a path in X from x_0 to y. Thus any two paths x and y can be joined by the path $\sigma_x * \bar{\sigma}_y$ in X. Hence X is a path-connected space. Moreover, $\pi_1(X, x_0) = 0$ (see Example 3.1.16). Consequently, X is simply connected. ❑

Corollary 3.1.31 $\mathbf{R}^n$, *D^n and any convex subset of* $\mathbf{R}^n$ *is simply connected.*

Definition 3.1.32 Let $X \subset \mathbf{R}^n$ be a subspace of $\mathbf{R}^n$. Then X is said to be star convex if for some $x_0 \in X$, all the line segments joining x_0 to any other point x of X lie entirely in X, i.e., $(1-t)x + tx_0 \in X$, $\forall t \in I$.

Proposition 3.1.33 *Let $X \subset \mathbf{R}^n$ be a star convex space. Then X is simply connected.*

Proof As X is star convex, there is a continuous map $H = X \times I \to X, (x, t) \mapsto (1-t)x + tx_0$. Hence X is a contractible space. Consequently, X is simply connected by Theorem 3.1.30. ❑

We now characterize simply connected spaces as follows:

Theorem 3.1.34 *A path-connected space X is simply connected if and only if any two paths in X having the same initial point and same final point are homotopic.*

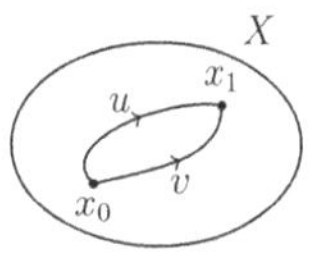

Fig. 3.7 Paths in a simply connected space

Proof Suppose X is simply connected. If u and v be two paths in X from x_0 to x_1, then $u \simeq v$ by Theorem 3.1.29. For the converse let X be path-connected and $[f] \in \pi_1(X, x_0)$. Then by hypothesis, $f \simeq c_{x_0} \implies [f] = [c_{x_0}] \implies \pi_1(X, x_0) = 0$. As X is path-connected and $\pi_1(X, x_0) = 0$, X is simply connected. □

Let X_1 and X_2 be two simply connected spaces. Is $X = X_1 \cup X_2$ simply connected? The answer is negative. For an example, we shall show that the unit circle S^1 in the complex plane can be expressed as the union of two contractible arcs but S^1 is not simply connected (see Corollary 3.3.12). Note that contractible spaces are simply connected by Theorem 3.1.30.

We now present the following interesting theorem.

Theorem 3.1.35 *Let $\{U_i : i \in \Lambda\}$ be an open covering of a space X, where each U_i is simply connected. Then X is itself simply connected if*

(a) $\cap U_i \neq \emptyset$
(b) $i \neq j \in \Lambda$, *$U_i \cap U_j$ is path-connected.*

Proof The space X is path-connected, because by hypothesis each of the open sets U_i is path-connected and their intersection is nonempty. It is now sufficient to prove that $\pi_1(X, x) = 0$ for some $x \in X$. Suppose $x_0 \in \bigcap_i U_i$. Let $u : (I, \dot{I}) \to (X, x_0)$ be a loop at x_0. Then $\{u^{-1}(U_i)\}$ is an open covering of I. Since I is compact, this covering will have a Lebesgue number μ(say)> 0. This implies that $\exists$ a partition. $0 = t_0 < t_1 < t_2 < \cdots < t_n = 1$ of I such that for $0 \leq j \leq n-1$, $u[t_j, t_{j+1}]$ is contained in some U_i. Without loss of generality, we assume that $u[t_j, t_{j+1}] \subset U_j$, $0 \leq j \leq n-1$. For each j, we define a path u_j in X by $u_j(s) = u((1-s)t_j + st_{j+1})$. Then $U_j(I)$ is contained in the simply connected open set U_j for each j and $[u] = [u_0 * u_1 * \cdots * u_{n-1}]$. Clearly, $u(t_1) \in U_0 \cap U_1$ and $U_0 \cap U_1$ is path-connected containing the base point x_0. Hence, we obtain a path $v_1 : I \to X$ from x_0 to $u(t_1)$ such that $v_1(I) \subset U_0 \cap U_1$. Similarly, we can find a path v_j from x_0 to $u(t_j)$ lying entirely in $U_{j-1} \cap U_j$, for $j = 1, 2, \ldots, n-1$. If $\bar{v}_j$ denotes the reverse path of v_j, then we have

$$\begin{aligned} [u] &= [u_0 * \bar{v}_1 * v_1 * u_1 * \bar{v}_2 * v_2 * u_2 * \cdots * \bar{v}_{n-1} * v_{n-1} * u_{n-1}] \\ &= [u_0 * \bar{v}_1] \circ [v_1 * u_1 * \bar{v}_2] \circ \cdots \circ [v_{n-1} * u_{n-1}] \end{aligned} \quad (3.5)$$

The first term in the right hand side of (3.5) is a loop based at x_0 and lying entirely in the simply connected space U_0. Similarly the second term is a loop lying entirely in the simply connected space U_1, and so on. Hence, each term is null homotopic in U_j for some j and so is in X. Consequently, $[u]$ is the zero element of $\pi_1(X, x_0)$. Since $[u]$ is an arbitrary element of $\pi_1(X, x_0)$, it follows that $\pi_1(X, x_0) = 0$. Hence X is itself simply connected. □

Corollary 3.1.36 *The n-sphere $S^n (n \geq 2)$ is simply connected.*

Proof Let $p = (0, 0, \ldots, 0, 1) \in \mathbf{R}^{n+1}$ be the north pole of S^n and $q = (0, 0, \ldots, 0, -1) \in \mathbf{R}^{n+1}$ be the south pole of S^n. Then $U = S^n - \{p\}$ and $V = S^n - \{q\}$ are both open sets homeomorphic to $\mathbf{R}^n$ by stereographic projection and hence they are simply connected. We now prove that $U \cap V$ (to apply Theorem 3.1.35) is path-connected. Clearly, $U \cap V = S^n - \{p\} - \{q\} \approx \mathbf{R}^n - \{0\}$(under stereographic projection). We now show that $\mathbf{R}^n - \{0\}$ is path-connected.

Any point $x \in \mathbf{R}^n - \{0\}$ can be joined to the point $x_0 = (1, 0, \ldots, 0)$ by the straight line path in $\mathbf{R}^n - \{0\}$, except for point x of the form $(a, 0, \ldots, 0)$, where $a < 0$. For the case $a < 0$, we can take the straight line path from x to $x_1 = (0, 1, 0, \ldots, 0)$, followed by the straight line path from x_1 to x_0 (This is possible for $n \geq 2$). Consequently, S^n is simply connected by Theorem 3.1.35. ❑

We now give a relation between the fundamental group of a product space and the fundamental groups of its factors. We recall that if A and B are groups with operation '$\cdot$' then the cartesian product $A \times B$ can be endowed with a group structure by the composition $(a, b) \cdot (a', b') = (a \cdot a', b \cdot b')$. Moreover, if $\alpha : G \to A$ and $\beta : G \to B$ are group homomorphisms, then the map $\psi : G \to A \times B$ defined by $\psi(g) = (\alpha(g), \beta(g))$ is a group homomorphism.

Again we recall a basic property of the product topology. Let $p_1 : (X \times Y) \to X$, $p_2 : X \times Y \to Y$ be the canonical projections. Given a pair of continuous maps $f : I \to X, g : I \to Y$, there is a continuous map $(f, g) : I \to X \times Y$ defined by $(f, g)(t) = (f(t), g(t))$. Conversely, any continuous map $h : I \to X \times Y$ defines a pair of continuous maps $p_1 \circ h : I \to X,\ p_2 \circ h : I \to Y$.

We are now equipped to prove the following Theorem.

Theorem 3.1.37 *Let X and Y be two topological spaces with base points $x_0 \in X$ and $y_0 \in Y$, respectively. Then the fundamental groups $\pi_1(X \times Y, (x_0, y_0))$ and $\pi_1(X, x_0) \times \pi_1(Y, y_0)$ are isomorphic.*

Proof Let $f : (I, \dot{I}) \to (X \times Y, (x_0, y_0))$ be a loop in $X \times Y$ at (x_0, y_0). Then the canonical projections $p_1 : (X \times Y) \to X,\ p_2 : (X \times Y) \to Y$ are continuous maps of product spaces and hence they induce homomorphisms $p_{1*} : \pi_1(X \times Y, (x_0, y_0)) \to \pi_1(X, x_0)$ and $p_{2*} : \pi_1(x \times Y, (x_0, y_0)) \to \pi_1(Y, y_0)$ defined by $p_{1*}([f]) = [p_1 \circ f]$ and $p_{2*}([f]) = [p_2 \circ f]$. Then the map $\psi = (p_{1*}, p_{2*}) : \pi_1(X \times Y, (x_0, y_0)) \to \pi_1(X, x_0) \times \pi_1(Y, y_0)$ defined by $\psi([f]) = (p_*([f]), q_*([f])) = ([p \circ f], [q \circ f])$ is a group homomorphism.

We claim that ψ is an isomorphism.

ψ **is a monomorphism**:

$$\begin{aligned}\ker \psi &= \{[f] \in \pi_1(X \times Y, (x_0, y_0)) : \psi([f]) = \text{identity element of} \\ &\qquad \pi_1(X, x_0) \times \pi_1(Y, y_0)\} \\ &= \{[f] \in \pi_1(X \times Y, (x_0, y_0)) : p_1 \circ f \simeq c_{x_0} \text{ and } p_2 \circ f = c_{y_0}\}.\end{aligned}$$

Let $f : (I.\dot{I}) \to (X \times Y, (x_0, y_0)$ be defined by $f(t) = (g(t), h(t))$, where $g : (I, \dot{I}) \to (X, x_0)$ and $h : (I, \dot{I}) \to (Y, y_0)$ be the corresponding loops. Let M :

$p_1 \circ f \simeq c_{x_0}$ and $H : p_2 \circ f \simeq c_{y_0}$. Then $M(t, 0) = (p_1 \circ f)(t) = p_1(f(t)) = g(t)$, $\forall\ t \in I$ and $M(t, 1) = c_{x_0}(t) = x_0$, $\forall\ t \in I$.

Similarly, $H(t, 0) = h(t)$ and $H(t, 1) = y_0$, $\forall\ t \in I$. We now define $F : I \times I \to Y \times Y$ by $F(t, s) = (M(t, s), H(t, s))$. Then $F(t, 0) = (M(t, 0), H(t, 0)) = (g(t), h(t)) = f(t)$, $\forall\ t \in I$ $F(t, 1) = (M(t, 1), H(t, 1)) = (x_0, y_0) = c_{(x_0, y_0)}$, $\forall\ t \in I$,

Thus $F : f \simeq c_{(x_0, y_0)} \implies [f]$ is the identity element $[c_{(x_0, y_0)}]$. Consequently, $\ker \psi = \{0\} \implies \psi$ is a monomorphism.

ψ **is an epimorphism**: Let $g : (I, \dot{I}) \to (X, x_0)$ and $h : (I, \dot{I}) \to (Y, y_0)$ be two loops. Then $\exists$ a continuous map $(g, h) : (I, \dot{I}) \to (X \times Y, (x_0, y_0))$ defined by $(g, h)(t) = (g(t), h(t))$. Let $f : (I, \dot{I}) \to (X \times Y, (x_0, y_0))$ be the continuous map defined by the rule $f(t) = (g(t), h(t))$. Then $\psi([f]) = ([p_1 \circ f], [p_2 \circ f]) = ([g], [h]) \implies \psi$ is an epimorphism.

Consequently, ψ is an isomorphism. ❑

Corollary 3.1.38 *The maps*

$$\theta : \pi_1(X, x_0) \times \pi_1(Y, y_0) \to \pi_1(X \times Y, (x_0, y_0))$$

defined by $\theta([g], [h]) = [g, h]$ *is well defined and is the inverse of* ψ.

Proof It follows from the definition of ψ. ❑

We will see that the fundamental groups of arbitrary spaces may be abelian or nonabelian. But the next theorem shows that the fundamental group of an H-space is always abelian. We recall that if G and H are groups, $x \in G$ and $y \in H$, then in the direct product

$$G \times H, (x, e) \cdot (e', y) = (x, y) = (e', y) \cdot (x, e').$$

holds.

Theorem 3.1.39 *If* (X, x_0) *is an* H*-space with multiplication* μ *and* x_0 *a homotopy identity, then* $\pi_1(X, x_0)$ *is abelian.*

Proof The map

$$\theta : \pi_1(X, x_0) \times \pi_1(Y, y_0) \to \pi_1(X \times Y, (x_0, y_0)), ([g], [h]) \mapsto [(g, h)],$$

where $(g, h) : I \to X \times Y$ is defined by $t \mapsto (g(t), h(t))$ is the inverse of the isomorphism ψ and hence θ is an isomorphism (see Corollary 3.1.38). Let $[f], [g] \in \pi_1(X, x_0)$. Then

$$\begin{aligned}[g] &= (\mu \circ (c, 1_X))_*[g] = \mu_*([(c, 1_X) \circ g]) = \mu_*([(c \circ g, g)]) \\ &= (\mu_* \circ \theta)([c \circ g], [g]) = (\mu_* \circ \theta)(e, [g]),\end{aligned}$$

where e is the identity element of $\pi_1(X, x_0)$. Similarly, $[f] = (\mu_* \circ \theta)([f], e)$. Since

$$\mu_* \circ \theta : \pi_1(X, x_0) \times \pi_1(X, x_0) \to \pi_1(X, x_0))$$

is a homomorphism, we have

$$\begin{aligned}(\mu_* \circ \theta)(([f], [g])) &= (\mu_* \circ \theta)((e, [g]), ([f], e)) \\ &= (\mu_* \circ \theta)(e, [g]) \cdot (\mu_* \circ \theta)([f], e) = [g] \circ [f].\end{aligned}$$

On the other hand, $(\mu_* \circ \theta)(([f], [g])) = (\mu_* \circ \theta))(([f], e), (e, [g])) = (\mu_* \circ \theta)(([f], e)) \cdot (\mu_* \circ \theta)(e, [g]) = [f] \circ [g]$. Hence

$$[f] \circ [g] = [g] \circ [f], \ \forall \ [f], [g] \in \pi_1(X, x_0).$$

Consequently, $\pi_1(X, x_0)$ is abelian. ❑

Corollary 3.1.40 *If G is a topological group with identity e then the $\pi_1(G, e)$ is abelian.*

Proof Since every topological group is an H-group (hence it is an H-space), the Corollary follows from Theorem 3.1.39. ❑

Remark 3.1.41 If for any pointed space (X, x_0), the fundamental group $\pi_1(X, x_0)$ is not abelian (such X exists namely, figure-eight, double torus), then there is no way to define a multiplication on X making it a topological group. Even we cannot equip such X with the structure of an H-space. Otherwise, we would have a contradiction to Theorem 3.1.39 and Corollary 3.1.40.

3.2 Alternative Definition of Fundamental Groups

This section presents an alternative definition of the fundamental group given by Hurewicz (1904–1956) equivalent to its definition given in Sect. 3.1, which is convenient at many situations. The unit circle $S^1 = \{e^{2\pi it} : 0 \leq t \leq 1\}$ in the complex plane is the prototype of a loop. The elements of $\pi_1(X, x_0)$ may be equally well considered as homotopy classes of maps $f : (S^1, 1) \to (X, x_0)$.

The basic aim of fundamental groups is to classify all loops up to homotopy equivalences. This leads to give an alternative definition of fundamental groups. The fundamental group $\pi_1(X, x_0)$ is now defined as the homotopy classes of maps $f : (S^1, 1) \to (X, x_0)$, instead of path classes, which is equivalent to the former definition.

If $f : I \to X$ is a loop in X based at x_0, then f determines a pointed map

$$\tilde{f} : (S^1, 1) \to (X, x_0), e^{2\pi it} \mapsto f(t).$$

Conversely a pointed map $\alpha : (S^1, 1) \to (X, x_0)$ determines a loop f_α in X based at x_0, defined by $f_\alpha : I \to X, t \mapsto \alpha(e^{2\pi i t})$. Hence it follows that

Proposition 3.2.1 *The function*

$$\psi : \pi_1(X, x_0) \to [(S^1, 1), (X, x_0)], [f] \mapsto [\tilde{f}]$$

is a bijection.

Remark 3.2.2 For our subsequence study, we use this identification implicitly and make no difference between f and $\tilde{f}$.

Using this identification map ψ we again prove the group structure of $\pi_1(X, x_0)$.

Theorem 3.2.3 $\pi_1(X, x_0)$ *is group.*

Proof Given two two loops $f, g : I \to X$ in X based at x_0, there exist two pointed maps

$$[\tilde{f}], [\tilde{g}] : (S^1, 1) \to (X, x_0), e^{2\pi i t} \mapsto f(t), g(t).$$

Define a multiplication in the set $[(S^1, 1) \to (X, x_0)]$ by the rule

$$[\tilde{f}] \circ [\tilde{g}] = [\widetilde{(f * g)}].$$

Hence it follows that $\pi_1(X, x_0)$ is a group with the constant map $c : S^1 \to \{x_0\}$ representing the identity element and the inverse of the map $f : (S^1, 1) \to (X, x_0)$ is represented by the map $g : (S^1, 1) \to (X, x_0), e^{i\theta} \mapsto f(e^{-\theta}), 0 \leq \theta \leq 2\pi$ and the bijection

$$\psi : \pi_1(X, x_0) \to [(S^1, 1), (X, x_0)], [f] \mapsto [\tilde{f}]$$

gives an isomorphism. ❑

Remark 3.2.4 By using the H-cogroup structure of S^1, the group structure of $\pi(X, x_0)$ also follows.

Theorem 3.2.5 *Let S^0 consist of two points -1 and 1 and let 1 be its based point. Then the continuous map*

$$\lambda : I \to \Sigma(S^0), t \mapsto [-1, t]$$

induces an isomorphism

$$\tilde{\lambda} : [\Sigma(S^0), 1] \to \pi_1(X, x_0), [g] \mapsto [g \circ \lambda].$$

Proof Since $\Sigma(S^0) \approx S^1$, and S^1 is an H-cogroup, it follows that $[(S^1, 1), (X, x_0)]$ is a group. Moreover, the products in $[(S^1, 1), (X, x_0)]$ and $\pi_1(X, x_0)$ show that $\tilde{\lambda}$ is a group homomorphism. Again as $\tilde{\lambda}$ is a bijection, it follows that $\tilde{\lambda}$ is an isomorphism of groups. ❑

Remark 3.2.6 It is sometimes convenient to consider the elements of $\pi_1(X, x_0)$ as homotopy classes of continuous maps $f : (S^1, 1) \to (X, x_0)$. The circle S^1 is a group under usual multiplication of complex numbers. The composition law in $\pi_1(S^1, 1)$ can also be defined by using this multiplication.

3.3 Degree Function and the Fundamental Group of the Circle

This section introduces the concept of degree function of a continuous map $f : (I, \dot{I}) \to (S^1, 1)$ and develops the necessary tools to compute and study the fundamental group of the circle and utilizes degree function to exhibit a space which is not simply connected, i.e., a space X such that $\pi_1(X, x_0) \neq 0$. Since $\pi_1(X, x_0)$ consists of relative homotopy class of maps $f : S^1 \to X$, the space $X = S^1$ demands its natural consideration. We consider the particular case of $[X, Y]$ when $X = Y = S^1$. Let $f : S^1 \to S^1$ be a continuous map. As X moves around S^1, $f(x)$ will move around S^1 some integer number of times. This integer is called the degree of f. We now formalize this concept of degree. For example, the homotopy classes of loops on S^1 based at 1 is completely characterized by their degrees, which are integers.

Recall that $S^1 = \{z \in \mathbf{C} : |z| = 1\}$ is a topological group with 1 as its identity element under the usual multiplication of complex numbers. The circle S^1 is studied through the real line $\mathbf{R}$. The homotopy class of a loop is determined by the number of times it winds around. This leads to the concept of degree function of a continuous $f : (I, \dot{I}) \to (S^1, 1)$.

We compute $\pi_1(S^1)$ with the concept of degree function. We use a map $p : \mathbf{R} \to S^1$, called the exponential map, defined by $p(t) = e^{2\pi i t}$, $\forall\ t \in \mathbf{R}$ as shown in Fig. 3.8. Then p is a continuous onto map which wraps the real line $\mathbf{R}$ onto the circle infinite number of times. p is a group homomorphism from $(\mathbf{R}, +)$ to $(S^1, 1)$ with $\ker p = \{t \in \mathbf{R} : p(t) = 1\} = \{t \in \mathbf{R} : e^{2\pi i t} = 1\} = \mathbf{Z}$.

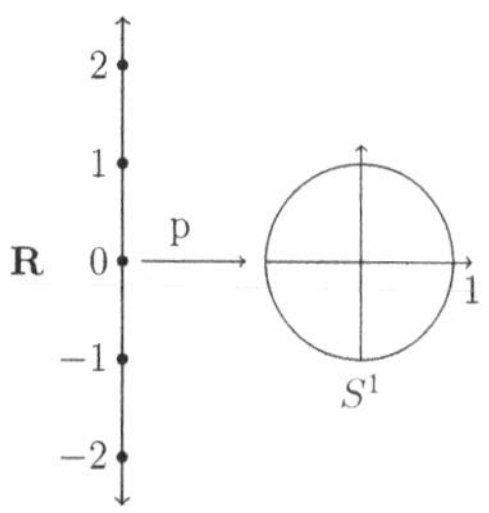

Fig. 3.8 Projection of the real line on a circle in a complex plane

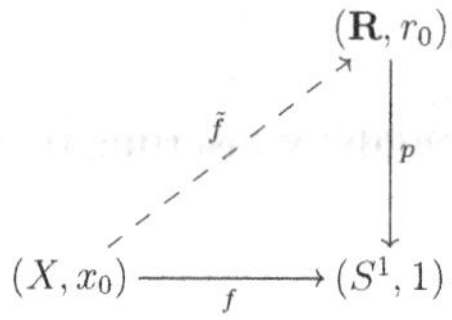

Fig. 3.9 Lifting of a map

Definition 3.3.1 A continuous map $f : (X, x_0) \to (S^1, 1)$ is said to have a lifting $\tilde{f} : (X, x_0) \to (\mathbf{R}, r_0)$, where $r_0 \in \ker p = \mathbf{Z}$ if there exists a continuous map $\tilde{f} : (X, x_0) \to (\mathbf{R}, r_0)$ such that $p \circ \tilde{f} = f$, i.e., the triangle in Fig. 3.9 is commutative.

We now show that under suitable conditions, $\tilde{f}$ is unique.

Proposition 3.3.2 *Let X be a connected space and $\tilde{f}, \tilde{g}$ are two liftings of $f : (X, x_0) \to (S^1, 1)$. If $\tilde{f}(x_0) = \tilde{g}(x_0)$, then $\tilde{f} = \tilde{g}$.*

Proof Using the group structure of $(\mathbf{R}, +)$, we define a map $h : (X, x_0) \to (\mathbf{R}, r_0)$ by $h(x) = \tilde{f}(x) - \tilde{g}(x)$. Now, for any $x \in X$, $(p \circ h)(x) = p(h(x)) = e^{2\pi i h(x)} = e^{2\pi i(\tilde{f}(x) - \tilde{g}(x))} = e^{2\pi i \tilde{f}(x)}/e^{2\pi i \tilde{g}(x)} = (p \circ \tilde{f})(x)/(p \circ \tilde{g})(x) = f(x)/f(x) = 1 \implies h(x) \in \ker p = \mathbf{Z}$. Therefore $h : X \to \mathbf{Z}$ is integral valued. Since X is connected and h is continuous, then it follows from discreteness of $\mathbf{Z}$ that h is constant $\implies$ image set $h(X)$ must be singleton. But by hypothesis, $h(x_0) = \tilde{f}(x_0) - \tilde{g}(x) = 0 \implies h(x) = 0, \ \forall\ x \in X \implies \tilde{f}(x) = \tilde{g}(x), \ \forall\ x \in X \implies \tilde{f} = \tilde{g}$. ❑

We now show that any path in S^1 starting at 1 can be lifted to a unique path in $\mathbf{R}$ starting at the origin 0 of $\mathbf{R}$ and any homotopy between two given paths in S^1 starting at 1, can be lifted to a unique homotopy between the two lifted paths starting at the origin 0 of $\mathbf{R}$. These two results follow as corollaries of the following theorem.

Theorem 3.3.3 *Let X be a compact convex subset of the Euclidean space $\mathbf{R}^n$ for some n. Let $f(X, x_0) \to (S^1, 1)$ be continuous, and $z_0 \in \mathbf{Z}$. Then $\exists$ a unique continuous map $\tilde{f} : (X, x_0) \to (\mathbf{R}, z_0)$ with $p \circ \tilde{f} = f$ i.e., making the diagram in Fig. 3.10 commutative.*

Proof Since X is a compact metric space, f must be uniformly continuous. Hence there is an $\epsilon > 0$ such that whenever $||x - x'|| < \epsilon$, then $|f(x) - f(x')| < 2$. Here $||;||$ is Euclidean norm of $\mathbf{R}^n$ and $|f(x)|$ denotes the modulus of the complex number $f(x)$. We choose $2 = diam\ S^1$ to ensure that $f(x)$ and $f(x')$ are not antipodal points i.e., $f(x)/f(x') \neq -1$. Since X is bounded, $\exists$ a positive integer n such that

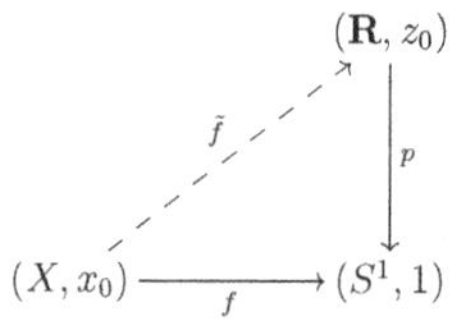

Fig. 3.10 Lifting of f in $\mathbf{R}$

$||x - x_0||/n < \epsilon, \ \forall\ x \in X$. For each $x \in X$, we subdivide the line segment having end points x_0 and x (which entirely lies in X by convexity) into n intervals of equal length by inserting (uniquely determined) points $x_0, x_1, \ldots, x_n = x$. Then

$$||x_{j+1} - x_j|| = ||x - x_0||/n < \epsilon \implies f(x_{j+1})/f(x_j) \neq -1.$$

Now for each j with $0 \leq j \leq n-1$, the function $g_j : X \to S^1 - \{-1\}$ defined by $g_j(x) = f(x_{j+1})/f(x_j)$ is continuous, because multiplication map $S^1 \times S^1 \to S^1$ and inversion map $S^1 \to S^1$ are continuous. Then for all $x \in X$, we can write (by taking $j = 0, 1, 2, \ldots, n-1$ successively)

$$f(x_0)g_0(x) = f(x_1),\ f(x_1)g_1(x) = f(x_2), \ldots, f(x_{n-1})g_{n-1}(x) = f(x)$$

and hence $f(x) = f(x_0)g_0(x)g_1(x)\ldots g_{n-1}(x)$ (called telescoping product in S^1). Define a map $\tilde{f} : X \to \mathbf{R}$ by

$$\tilde{f}(x) = z_0 + \log g_0(x) + \log g_1(x) + \cdots + \log g_{n-1}(x).$$

Then $\tilde{f}$ is the sum of n continuous functions and hence it is continuous. Moreover, $g_j(x_0) = 1, \ \forall\ j$. Consequently, $\tilde{f}(x_0) = z_0$ and $p \circ \tilde{f} = f$. ❑

Corollary 3.3.4 *Let* $f : (I, \dot{I}) \to (S^1, 1)$ *be continuous.*

(i) *Then there exists a unique continuous map* $\tilde{f} : I \to \mathbf{R}$ *with* $p \circ \tilde{f} = f$ *and* $\tilde{f}(0) = 0$ *(Path lifting Property) (Fig. 3.11).*
(ii) *If* $g : (I, \dot{I}) \to (S^1, 1)$ *is continuous and* $f \simeq g$ rel $\dot{I}$, *then* $\tilde{f} \simeq \tilde{g}$ rel $\dot{I}$ *(where* $p \circ \tilde{g} = g$ *and* $\tilde{g}(0) = 0$*). Moreover,* $\tilde{f}(1) = \tilde{g}(1)$. *(Homotopy Lifting Property).*

Proof **(i)** follows from the Theorem 3.3.3 by taking in particular $X = I = [0, 1] \subset \mathbf{R}$, and Proposition 3.3.2.
(ii) $I \times I$ is compact convex. We choose $(0, 0)$ as a base point of $I \times I$. Let $F : f \simeq g$ rel $\dot{I}$. Then Theorem 3.3.3 gives a continuous map $\tilde{F} : I \times I \to \mathbf{R}$ such that $p \circ \tilde{F} = F$ and $\tilde{F}(0, 0) = 0$. We show that $\tilde{F} : \tilde{f} \simeq \tilde{g}$ rel $\dot{I}$ i.e., F can be lifted. Let $\psi_0 : I \to \mathbf{R}$ be defined by $\psi_0(t) = \tilde{F}(t, 0)$. Then $p \circ \psi_0(t) = p \circ \tilde{F}(t, 0) = F(t, 0) = f(t)$. Since $\psi_0(0) = \tilde{F}(0, 0) = 0$, uniqueness of lifting shows that $\psi_0 = \tilde{f}$. Again define $\phi_0 : I \to \mathbf{R}$ by $\phi_0(t) = \tilde{F}(0, t)$. Then proceeding as above, we show that ϕ_0 is the constant function $\phi_0(t) = 0$. Hence it follows that $\tilde{F}(0, 1) = 0$. Again define $\psi_1 : I \to \mathbf{R}$ by $\psi_1(t) = \tilde{F}(t, 1)$. Then

Fig. 3.11 Path lifting of f

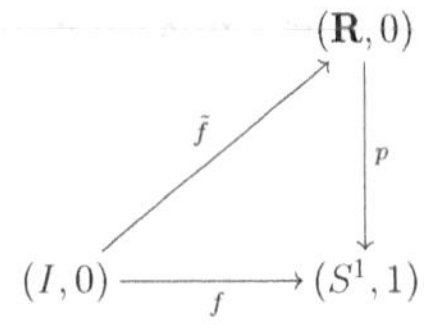

$p \circ \psi_1(t) = F(t, 1) = g(t)$ and $\psi_1(0) = \tilde{F}(0, 1) = 0$. Hence $\psi_1 = \tilde{g}$. Finally, define $\phi_1 : I \to \mathbf{R}$ by $\phi_1(t) = \tilde{F}(1, t)$. Now $p \circ \phi_1$ is the constant function c with value $f(1)$, and $\phi_1(0) = \tilde{f}(1)$. Hence the constant function at $\tilde{f}(1)$ is a lifting of c, and uniqueness of lifting shows that $\phi_1(t) \equiv \tilde{f}(1)$, $\forall\ t \in I$. Consequently, $\tilde{g}(1) = \tilde{f}(1)$ and $\tilde{F} : \tilde{f} \simeq \tilde{g}$ rel $\dot{I}$. ❑

Corollary 3.3.5 *Let $f, g : (I, \dot{I}) \to (S^1, 1)$ be two continuous functions such that $f \simeq g$ rel $\dot{I}$. If $w(f)$ denotes the winding number of f, then $w(f) = w(g)$.*

Proof $w(f) = \tilde{f}(1) - \tilde{f}(0) = \tilde{g}(1) - \tilde{g}(0)$ (by Corollary 3.3.4) $= w(g)$. ❑

Definition 3.3.6 Let $f : (I, \dot{I}) \to (S^1, 1)$ be a continuous map. The degree of f denoted $\deg f$ is defined by $\deg f = \tilde{f}(1)$, where $\tilde{f}$ is the unique lifting of f with $\tilde{f}(0) = 0$.

Remark 3.3.7 It shows by Corollary 3.3.4 that degree of circular maps is an example of a homotopy invariant.

Remark 3.3.8 $\deg f$ is an integer. Because

$$p \circ \tilde{f} = f \implies (p \circ \tilde{f})(1) = f(1) = 1 \implies \tilde{f}(1)$$

lies in the $\ker p = \mathbf{Z} \implies \deg f \in \mathbf{Z}$ for every $f : (I, \dot{I}) \to (S^1, 1)$. If $f(z) = z^m$ i.e., if $f(t) = p(mt) = e^{2\pi imt}$, then $\tilde{f}(1) = m$. This explains the term degree.

Theorem 3.3.9 *The degree function $d : \pi_1(S^1, 1) \to \mathbf{Z}, [f] \mapsto \deg f$ is an isomorphism of groups.*

Proof By Corollary 3.3.4(ii), d is independent of the choice of the representatives of the classes $[f] \in \pi_1(S^1, 1)$ and hence the function d is well defined.
d is a homomorphism: Let $[f], [g]$ be any two elements of $\pi_1(S^1, 1)$ such that $\deg f = m$ and $\deg g = n$. To compute the $\deg(f * g)$ we define a path $\tilde{h} : I \to \mathbf{R}$ such that $p \circ \tilde{h} = f * g$ and $\tilde{h}(0) = 0$. Then $\deg(f * g) = \tilde{h}(1)$.

Let $\tilde{g}$ be the unique lifting of g such that $\tilde{g}(0) = 0$. Define $\tilde{\gamma} : I \to \mathbf{R}$ by $\tilde{\gamma}(t) = m + \tilde{g}(t)$. Then $\tilde{\gamma}$ is a path in $\mathbf{R}$ from m to $m + n$. Again let $\tilde{f}$ be the lifting of f such that $\tilde{f}(0) = 0$ and $\tilde{f}(1) = m$. Then $\tilde{f} * \tilde{\gamma}$ is a path in $\mathbf{R}$ such that $(\tilde{f} * \tilde{\gamma})(0) = 0$ and $(\tilde{f} * \tilde{\gamma})(1) = m + n$, because $\tilde{f} * \tilde{\gamma} : I \to \mathbf{R}$ is defined by

$$(\tilde{f} * \tilde{\gamma})(t) = \begin{cases} \tilde{f}(2t), & 0 \leq t \leq 1/2 \\ \tilde{\gamma}(2t - 1), & 1/2 \leq t \leq 1 \end{cases}$$

We claim that $\tilde{f} * \tilde{\gamma}$ is the lifting of $f * g$:

$$(p \circ (\tilde{f} * \tilde{g}))(t) = \begin{cases} (p \circ \tilde{f})(2t), & 0 \leq t \leq 1/2 \\ (p \circ \tilde{\gamma})(2t - 1), & 1/2 \leq t \leq 1 \end{cases}$$

.

Now $(p\circ\tilde{f})(t) = f(t)$, $\forall\ t \in I$ and $(p\circ\tilde{\gamma})(t) = p(m+\tilde{g}(t)) = e^{2\pi i m}.p(\tilde{g}(t)) = g(t)$, $\forall\ t \in I \implies p\circ(\tilde{f}*\tilde{\gamma}) = f*g \implies \deg(f*g) = (\tilde{f}*\tilde{\gamma})(1) = m+n \implies d$ is a homomorphism.

d is a monomorphism: Let $[f] \in \ker d$. Then $\deg f = 0 \implies \tilde{f}(1) = 0 \implies \tilde{f}$ is a closed path in $\mathbf{R}$ at 0, since $\tilde{f}(0) = 0$. The continuous map $p : (\mathbf{R}, 0) \to (S^1, 1)$ induces a homomorphism $p_* : \pi_1(\mathbf{R}, 0) \to \pi_1(S^1, 1)$, defined by $p_*([\tilde{f}]) = [p \circ \tilde{f}] = [f]$. But $\mathbf{R}$ is contractible and hence $\pi_1(\mathbf{R}, 0) = 0 \implies [\tilde{f}] = [c] \implies [f]$ is the identity element of $\pi_1(S^1, 1) \implies \ker d$ is trivial $\implies d$ is a monomorphism.

d is an epimorphism: Let $n \in \mathbf{Z}$. Define a loop $f : I \to S^1$ by $f(t) = e^{2\pi i n t}$. Then the path $\tilde{f} : I \to \mathbf{R}$ defined by $\tilde{f}(t) = nt$, starts at the origin O of $\mathbf{R}$ and lifts the path f. Then $\deg f = \tilde{f}(1) = n \implies d$ is an epimorphism. Consequently, d is an isomorphism. ❑

Corollary 3.3.10 *$\pi_1(S^1, s)$ is isomorphic to $\mathbf{Z}$ for any $s \in S^1$.*

Proof For any $s \in S^1, \pi_1(S^1, s)$ is isomorphic to $\pi_1(S^1, 1)$, since S^1 is path-connected. Hence the Corollary follows. ❑

Definition 3.3.11 A topological space X is called simply connected if it is path-connected and for every $x_0 \in X, \pi(X, x_0) = 0$.

Corollary 3.3.12 *S^1 is not simply connected.*

Proof S^1 is path-connected but $\pi_1(S^1, s) \neq 0$. Hence S^1 is not simply connected.❑

Remark 3.3.13 The homotopy classes of loops on S^1 based at 1 can be completely characterized with the help of their degrees.

Theorem 3.3.14 *Let $f, g : (I, \dot{I}) \to (S^1, 1)$ be continuous. Then $f \simeq g$ rel $\dot{I}$ if and only if $\deg f = \deg g$.*

Proof $f \simeq g$ rel $\dot{I} \implies \tilde{f} \simeq \tilde{g}$ *rel* $\dot{I}$ with $\tilde{f}(1) = \tilde{g}(1) \implies \deg f = \deg g$. Conversely, $\deg f = \deg g \implies [f] = [g]$, because the degree function d is injective $\implies f \simeq g$ rel $\dot{I}$. ❑

3.4 The Fundamental Group of the Punctured Plane

This section studies punctured Euclidean plane from homotopy view-point and computes its the fundamental group. It is an important space in geometry and topology.

We now proceed to calculate $\pi_1(\mathbf{R}^2 - \{0\})$, the fundamental group of the punctured plane.

Theorem 3.4.1 *Let $s_0 \in S^1$. The inclusion map $i : (S^1, s_0) \to (\mathbf{R}^2 - \{0\}, s_0)$ induces an isomorphism. $i_* : \pi_1(S^1, s_0) \to \pi_1(\mathbf{R}^2 - \{0\}, s_0)$.*

Proof $r : \mathbf{R}^2 - \{0\} \to S^1$ be the continuous map defined by $r(x) = \frac{x}{||x||}$. The map r can be depicted as collapsing each radial ray in $\mathbf{R}^2 - \{0\}$ onto the point where the ray intersects S^1. In particular, it maps each point $x \in S^1$ to itself.

We claim that the induced homomorphism r_* is the inverse of the induced homomorphism i_*. We consider the composite map

$$(S^1, s_0) \xrightarrow{i} (\mathbf{R}^2 - \{0\}, s_0) \xrightarrow{r} (S^1, s_0).$$

Then $r \circ i = 1_{S^1} \implies r_* \circ i_* =$ identity automorphism of $\pi_1(S^1, s_0)$. On the other hand, we also show that $i_* \circ r_*$ is the identity automorphism of $\pi_1(\mathbf{R}^2 - \{0\}, s_0)$. Let f be a loop in $\mathbf{R}^2 - \{0\}$ based at s_0. Then $(i_* \circ r_*)([f]) = [i \circ r \circ f]$ where $i \circ r \circ f = g : I \to \mathbf{R}^2 - \{0\}$ is a loop given by the equation

$$g(t) = (i \circ r)(f(t)) = i(\frac{f(t)}{||f(t)||}) = \frac{f(t)}{||f(t)||}.$$

Define $F : I \times I \to \mathbf{R}^2 - \{0\}$ by the equation $F(t, s) = s\frac{f(t)}{||f(t)||} + (1-s)f(t)$. Then $F(t, s) \neq 0$, $\forall\ (t, s) \in I \times I$. Otherwise, $s\frac{f(t)}{||f(t)||} + (1-s)f(t) = 0$ would imply

$$f(t)[\frac{s}{||f(t)||} + (1-s)] = 0 \implies s + (1-s)||f(t)|| = 0,$$

since $f(t) \neq 0$, which is not true, since $0 \leq s \leq 1$.

Then $F : f \simeq g$ rel $\dot{I}$.

Consequently, $(i_* \circ r_*)([f]) = [i \circ r \circ f] = [g] = [f]$, $\forall\ [f] \in \pi_1(\mathbf{R}^2 - \{0\}, s_0) \implies i_* \circ r_*$ is the identity automorphism of $\pi_1(\mathbf{R}^2 - \{0\}, s_0)$.
Thus we conclude that i_* is an isomorphism. ❑

Corollary 3.4.2 *S^1 is a strong deformation retract of $\mathbf{R}^2 - \{0\}$ and hence $\pi_1(S^1) \cong \pi_1(\mathbf{R}^2 - \{0\}) \cong \mathbf{Z}$.*

Proof Let $i : S^1 \hookrightarrow \mathbf{R}^2 - \{0\} = X$(say). Define a map $F : X \times I \to X$ by the rule $F(x, t) = (1-t)x + t\frac{x}{||x||}$.

Then $F(x, t) \neq 0$ for any $(x, t) \in X \times I$. Clearly, F is a continuous map such that $F(x, 0) = x$, $\forall\ x \in X$, $F(x, 1) = \frac{x}{||x||} \in S^1$, $\forall\ x \in X$, $F(a, t) = a$, $\forall\ a \in S^1$.

Consequently, S^1 is a strong deformation retract of $\mathbf{R}^2 - \{0\}$. Thus $\pi_1(S^1) \cong \pi_1(\mathbf{R}^2 - \{0\}) \cong \mathbf{Z}$. ❑

3.5 Fundamental Groups of the Torus

This section considers the torus and computes its fundamental group. A torus is a connected 2-manifold homeomorphic to the product of two circles $S^1 \times S^1$. Surfaces are very important in geometry, topology, and complex analysis. Recall that a surface is a Hausdorff space with a countable basis, every point of which has a neighborhood which is homeomorphic to an open disk in $\mathbf{R}^2$. We shall consider other familiar surfaces: the sphere S^2, the projective plane $\mathbf{R}P^2$, double torus etc. in the next chapter. The simplest example of a compact surface is the 2-sphere S^2. We now consider another example which is the torus. A torus may be considered as any surface homeomorphic to the surface of a doughnut or a solid ring.

Theorem 3.5.1 *The fundamental group of the torus $T = S^1 \times S^1$ is isomorphic to the group $\mathbf{Z} \times \mathbf{Z}$.*

Proof It follows from Theorems 3.1.37 and 3.3.9. ❑

3.6 Vector Fields and Fixed Points

This section studies vector fields on D^2 and applies the fundamental group to the following 2 problems:

(a) The existence of vector fields tangent to given surfaces.
(b) Given a topological space X does every continuous map $f : X \to X$ necessarily has a fixed point?
Moreover this section presents a proof of Brouwer fixed point theorem for D^2 using the concept of vector field.

Definition 3.6.1 A vector field on D^2 is an ordered pair $(x, v(x))$, where $x \in D^2$ and $v : D^2 \to \mathbf{R}^2$ is a continuous map.

We say that a vector field is nonvanishing if $v(x) \neq 0$ for every $x \in D^2$; in such a case, we may consider $v : D^2 \to \mathbf{R}^2 - \{0\}$ mapping D^2 into $\mathbf{R}^2 - \{0\}$.

Theorem 3.6.2 *Given a nonvanishing vector field on D^2, there exists a point of S^1 where the vector field points directly inward and a point of S^1 where it points directly outward.*

Proof First we show that given a vector field v, $\exists$ a point of S^1 where v points directly inward. We consider the map $u : S^1 \to \mathbf{R}^1 - \{0\}$ obtained by restriction of v to S^1. If the assertion is not true, $\exists$ no point $x \in S^1$ at which v points directly inward. In other words, $\exists$ no $x \in S^1$ such that $u(x)$ is a negative multiple of x. It follows that u is homotopic to the inclusion map $i : S^1 \hookrightarrow \mathbf{R}^2 - \{0\}$, under the homotopy $F : S^1 \times I \to \mathbf{R}^2 - \{0\}$ given by the equation $F(x, t) = tx + (1-t)u(x)$. Clearly, $F(x, t) \neq 0$, otherwise $(1 - t)u(x) = -tx$. This is not true for $t = 0$ or $t = 1$, since $x \in S^1$ and $u(x) \neq 0$. Again for $0 < t < 1$, we have from above, $u(x) = -tx/(1 - t) \implies u(x)$ is a negative multiple of x, which is not true by assumption. Since u is homotopic to the inclusion map $i : S^1 \hookrightarrow \mathbf{R}^2 - \{0\}$, it follows that u is extendable to the continuous map $v : D^2 \to \mathbf{R}^2 - \{0\}$. Hence we reach a contradiction. Thus we conclude that v points directly inward at some point of S^1.

For the second part, we consider the non-vanishing vector field $(x, -v(x))$. By the first part, it points directly inward at some point of S^1. Then v points directly outward at that point. ❑

Remark 3.6.3 For nonvanishing continuous vector fields on S^n see Chap. 14.

We present a proof of Brouwer Fixed Point Theorem for D^2 by using the concept of vector field. For an alternative proof see Theorem 3.8.8.

Theorem 3.6.4 (Brouwer Fixed Point Theorem for the disk D^2) *If $f : D^2 \to D^2$ is continuous, then there exists a point $x \in D^2$ such that $f(x) = x$.*

Proof Suppose $f(x) \neq x$ for every $x \in D^2$. Then the map v defined by $v(x) = f(x) - x$ gives a non-vanishing vector field $(x, v(x))$ on D^2. But the vector field v cannot point directly outward at any point $x \in S^1$, otherwise $f(x) - x = ax$ for some real $a > 0$. Then $f(x) = (1 + a)x$ lies outside the unit ball D^2. Thus we reach a contradiction. ❑

Corollary 3.6.5 *Let M be a 3×3 matrix of positive real numbers. Then M has a positive real eigenvalue.*

Proof Let $T : \mathbf{R}^3 \to \mathbf{R}^3$ be the linear transformation whose matrix representation is M relative to the standard basis for $\mathbf{R}^3$. Let $O_1 = \{(x_1, x_2, x_3) \in \mathbf{R}^3 : x_1 \geq 0, x_2 \geq 0, x_3 \geq 0\}$, be the first octant of $\mathbf{R}^3$ and $B = S^2 \cap O_1$. Then B is homeomorphic to the ball D^2. Therefore the Brouwer fixed point theorem holds for continuous maps of B into itself. Now if $x = (x_1, x_2, x_3) \in B$, then all the components of x are non-negative and at least one is positive. Since all the entries of M are positive, the vector $T(x)$ is a vector all of whose components are positive. Consequently, the map $B \to B, x \mapsto \frac{T(x)}{||T(x)||}$ is continuous and hence it has a fixed point x_0(say). Therefore, $x_0 = \frac{T(x_0)}{||T(x_0)||}$ shows that $T(x_0) = ||T(x_0)||x_0$. This implies that T has the positive real eigenvalue $||T(x_0)||$ and hence the matrix M has a positive real eigenvalue. ❑

3.7 Knot and Knot Groups

This section conveys a study of knots, which returns to geometry and considers various ways of embeddings the circle as a subspace of $\mathbf{R}^3$ or $\mathbf{S}^3$. The scientists working in physics and biochemistry find interesting applications of knot theory. H. Tietze (1880–1964) contributed to the the foundations of knot theory. Fundamental groups play an important role in the study of knot theory.

Definition 3.7.1 A knot K is a subspace of Euclidean 3-space $\mathbf{R}^3$ which is homeomorphic to the circle and knot group of K is the fundamental group $\pi_1(\mathbf{R}^3 - K)$ of the complement of K in $\mathbf{R}^3$.

Remark 3.7.2 Properties of complement of the knot K in $\mathbf{R}^3$ are significant, because it is how the knot is embedded in $\mathbf{R}^3$ is crucial.

Definition 3.7.3 The standard embedding $K = S^1 \subset \mathbf{R}^2 \subset \mathbf{R}^3 \subset S^3$ is called the trivial knot (or unknot).

Remark 3.7.4 A knot is represented by its projection in the plane of the paper. Hence 'trivial knot' or 'unknot' consists of the unit circle in the xy-plane. Two knots are said to be same if there exists a homeomorphism of $\mathbf{R}^3$ which sends one knot onto the other knot.

Definition 3.7.5 Two knots K_1, K_2 are said to be equivalent if there is a homeomorphism $h : \mathbf{R}^3 \to \mathbf{R}^3$ such that $h(K_1) = K_2$.

Remark 3.7.6 Two knots K_1 and K_2 are equivalent if one can be continuously deformed into other.

Definition 3.7.7 A knot K is said to be polygonal if it is made up of a finite number of line segments. A tame knot is a knot equivalent to a polygonal knot.

For picturing knots and work with them effectively, it needs projecting them into the plane in a 'nice' way in the sense that the projection only crosses itself at a finite number of points, at most two pieces of the knot meet at such crossing, and they do so at 'right angles' as shown in Figs. 3.12, 3.13 and 3.14. These knots are called 'trefoil knot', 'figure-eight knot', and 'square knot' respectively.

Another important knot called 'torus knot' is now described. For any relatively prime pair of positive integers (p, q), the image of the torus of the line with the equation $px = qy$ in $\mathbf{R}^3$ is a knot that winds p times around the torus one way and it winds q times around the other way. This is called a torus knot of type (p, q).

Definition 3.7.8 For relatively prime integers p and q, the torus knot $K = K_{p,q} \subset \mathbf{R}^3$ is the image of the embedding $f : S^1 \to S^1 \times S^1 \subset \mathbf{R}^3$, $z \mapsto (z^p, z^q)$, where the torus $S^1 \times S^1$ is embedded in $\mathbf{R}^3$ in the usual way.

Fig. 3.12 Trefoil knot

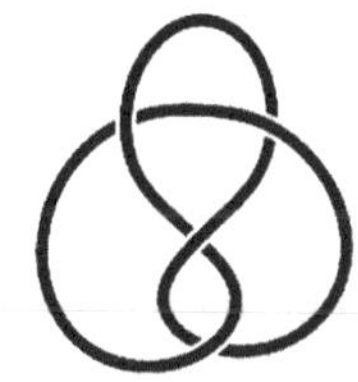

Fig. 3.13 Figure-eight knot

Fig. 3.14 Square knot

To make the above map f injective, the integers p and q are assumed to be relatively prime.

Remark 3.7.9 Geometrically, the torus knot $K = K_{p,q}$ winds the torus a total of p times in the longitudinal direction and q times in the meridian direction.

S^3 is considered as the one-point compactification of $\mathbf{R}^3$.

Proposition 3.7.10 *Let K be any knot. If $\mathbf{R}^3 - K$ and $S^3 - K$ are the complements of K in $\mathbf{R}^3$ and 3-sphere S^3 respectively, then the inclusion map $i : (\mathbf{R}^3 - K) \to (S^3 - K)$ induces an isomorphism*

$$i_* : \pi_1(\mathbf{R}^3 - K) \to \pi_1(S^3 - K).$$

Proof Let K be any knot. Then K is a compact subset of $\mathbf{R}^3$. Moreover, $S^3 - K$ is the union of $\mathbf{R}^3 - K$ and an open ball B formed by the compactification point together with the complement of a large closed ball in $\mathbf{R}^3$ containing K. Both B and $B \cap (\mathbf{R}^3 - K)$ are simply connected. As $B \cap (\mathbf{R}^3 - K)$ is homeomorphic to $S^2 \times \mathbf{R}$, van Kampen theorem (see Chap. 6) implies that the inclusion $\mathbf{R}^3 - K \hookrightarrow S^3 - K$ induces an isomorphism. Hence if $i : (\mathbf{R}^3 - K) \to (S^3 - K)$ is the inclusion map, then the induced homomorphism

$$i_* : \pi_1(\mathbf{R}^3 - K) \to \pi_1(S^3 - K).$$

is an isomorphism. ❑

Remark 3.7.11 The simplest knot is a circle which we may think of as the unit circle in the x_1x_2-plane. Its knot group is $\mathbf{Z}$.

Definition 3.7.12 A homeomorphism $h : \mathbf{R}^3 \to \mathbf{R}^3$ is said to be isotopic to the identity if there is a homotopy $H : \mathbf{R}^3 \times I \to \mathbf{R}^3$ such that the map

$$H_t : \mathbf{R}^3 \to \mathbf{R}^3, x \mapsto H(x, t)$$

is a homeomorphism with $H_0 = 1_d$ and $H_1 = h$.

Remark 3.7.13 If there is a homeomorphism $h : \mathbf{R}^3 \to \mathbf{R}^3$ which is isotopic to the identity such that $h(K_1) = K_2$ for two knots K_1 and K_2, then the knots $H_t(K_1)$ give a continuous family of maps which move gradually from K_1 to K_2 as t increases from 0 to 1. Since S^3 is the one-point compactification of $\mathbf{R}^3$, a homeomorphism $h : \mathbf{R}^3 \to \mathbf{R}^3$ has a unique extension to a homeomorphism $\tilde{h} : S^3 \to S^3$.

Definition 3.7.14 A homeomorphism $h : \mathbf{R}^3 \to \mathbf{R}^3$ is said to be orientation preserving (or orientation reversing) if its extension homeomorphism $\tilde{h} : S^3 \to S^3$ preserves (or reverses) the orientation of S^3.

Example 3.7.15 A homeomorphism which is isotopic to the identity is orientation preserving, because we can extend each homeomorphism $H_t : \mathbf{R}^3 \to \mathbf{R}^3$ to the homeomorphism $\widetilde{H}_t : S^3 \to S^3$, since homotopic maps have the same degree.

Example 3.7.16 Reflection in a plane is a homeomorphism of $\mathbf{R}^3$ and transforms a knot to its mirror image. It is orientation reversing and cannot be isotopic to the identity.

Remark 3.7.17 Any orientaion preserving homeomorphism of $\mathbf{R}^3$ is isotopic to the identity.

Proposition 3.7.18 *Equivalent knots have homeomorphic complements in* $\mathbf{R}^3$.

Proof Let K_1 and K_2 be equivalent knots. Then there is a homeomorphism $h : \mathbf{R}^3 \to \mathbf{R}^3$ such that $h(K_1) = h(K_2)$. Restricting h to $\mathbf{R}^3 - K_1$ defines a homeomorphism $\overline{h} : \mathbf{R}^3 - K_1 \to \mathbf{R}^3 - K_2$. This shows that equivalent knots have homeomorphic complements in $\mathbf{R}^3$. ❑

Remark 3.7.19 As the knot group $\pi_1(\mathbf{R}^3 - K)$ of K is the complement of K in $\mathbf{R}^3$, it can be used to distinguish various knots.

Definition 3.7.20 A knot K is said to be untied if there is an isotopy of $\mathbf{R}^3$ that would take K to the standard circle $S^1 \subset \mathbf{R}^3$.

Remark 3.7.21 Circle knot is a trivial knot. If a knot K is trivial, then the fundamental group of its complement (which is homeomorphic to the solid torus) is the infinite cyclic group. Hence the knot group of K is abelian. This shows that if the knot group of a knot K is not abelian, then K can not be a trivial knot which means that K can not be untied.

Remark 3.7.22 For some sort of reasonable presentation for a knot group in terms of generators and relations the book (Armstrong 1983) is referred.

Remark 3.7.23 There is an interesting link between a knot and an Eilenberg–Maclane space: see Chap. 11.

3.8 Applications

This section applies fundamental group and degree function to prove some important theorems such as fundamental theorem of algebra, Brouwer fixed point theorem for dimension 2, and Borsuk–Ulam theorem. Finally this section applies winding numbers and exponential map $p : \mathbf{R} \to S^1, t \mapsto e^{2\pi i t}$ to prove Cauchy integral theorem of complex analysis.

3.8.1 *Fundamental Theorem of Algebra*

This subsection provides an alternative proof of fundamental theorem of algebra given in Theorem 2.10.10. Although algebraic topology is usually 'algebra serving topology,' the role is reversed in the following proof of fundamental theorem of algebra. The concept of degree function is now utilized to prove this theorem of algebra.

Theorem 3.8.1 (Fundamental Theorem of Algebra) *Every nonconstant polynomial with coefficients in* **C** *has a root in* **C**.

Proof It is sufficient to prove the theorem for a polynomial of the form: $p(z) = a_0 + a_1 z + \cdots + a_n z^{n-1} + z^n$, $a_0 \neq 0$, $n \geq 1$ over **C**. Suppose the theorem is not true. Then $z \mapsto p(z)$ is a mapping from the complex plane **C** to $\mathbf{C} - \{0\}$. The restriction of this mapping to the different circles: $|z| = r$ for different values of $r \geq 0$ are loops in $\mathbf{C} - \{0\}$.

Consider the mappings $G : I \times [0, \infty) \to S^1 \subset \mathbf{C}$ defined by

$$G(t, r) = \frac{p(re^{2\pi it})/p(r)}{|p(re^{2\pi it})/p(r)|}$$

and $F : I \times I \to S^1 \subset \mathbf{C}$ defined by

$$F(t, s) = \begin{cases} G(t, s/(1-s)), & 0 \leq t \leq 1, 0 \leq s < 1 \\ e^{2\pi int}, & 0 \leq t \leq 1, s = 1. \end{cases}$$

Since G is continuous, $\lim\limits_{s \to 1} F(t, s) = \lim\limits_{s \to 1} G(t, s/(1-s)) = \lim\limits_{r \to \infty} G(t, r) = e^{2\pi int}$ and hence F is continuous. If $F(t, 0) = f_0(t)$ and $F(t, 1) = f_1(t)$, then $F : f_0 \simeq f_1$ rel $\dot{I}$. Consequently, deg f_0 = deg f_1. But deg $f_0 = 0$ and deg $f_1 = n$. This implies a contradiction, since $n \geq 1$. Thus we conclude that $p(z)$ has a root in **C**. ❑

Definition 3.8.2 A field F is said to be algebraically closed (or complete) if every polynomial $f(x)$ over F with degree ≥ 1, has a root in F.

Corollary 3.8.3 *The field* **C** *of complex numbers is algebraically closed.*

Proof It follows from Theorem 3.8.1. ❑

Corollary 3.8.4 *The field* **R** *of real numbers is embedded in the algebraically closed field* **C**.

Proof It follows from Corollary 3.8.3. ❑

Remark 3.8.5 The Corollary 3.8.3 proves the algebraic completeness of the field of complex numbers.

3.8.2 An Alternative Proof of Brouwer Fixed Point Theorem

We now prescribe an **alternative proof** of Brouwer Fixed Point Theorem given in Theorem 3.6.4 for dimension 2.

Theorem 3.8.6 *The circle S^1 is not a retract of the disk D^2.*

Proof Suppose there is a retraction $r : D^2 \to S^1$. Then $r \circ i = 1_{S^1}$, where $i : S^1 \hookrightarrow D^2$ is the inclusion map. This implies that the composite homomorphism $\pi_1(S^1, 1) \xrightarrow{i_*} \pi_1(D^2, 1) \xrightarrow{r_*} \pi_1(S^1, 1)$ is the identity automorphism by Theorem 3.1.23. But this is not possible, since $\pi_1(D^2, 1) = 0$ and $\pi_1(S^1, 1) = \mathbf{Z}$. ❑

Corollary 3.8.7 *The identity map $1_{S^1} : S^1 \to S^1$ cannot be continuously extended over D^2.*

Proof Suppose 1_{S^1} is continuously extendable over D^n. Then there must exist a continuous map $r : D^2 \to S^1$ such that $r|_{S^1} = 1_{S^1}$. In other words, $r \circ i = 1_{S^1}$. Consequently, S^1 is a retract of D^2. This contradicts Theorem 3.8.6. ❑

Theorem 3.8.8 (Brouwer Fixed Point Theorem for dimension 2) *Any continuous map $f : D^2 \to D^2$ has a fixed point, that is, there exists a point $x \in D^2$ such that $f(x) = x$.*

Proof Suppose to the contrary that $f(x) \neq x$ for any $x \in D^2$. We can define a map $r : D^2 \to S^1$ by letting $r(x)$ to the point of S^1, where the ray in $\mathbf{R}^2$ starting from $f(x)$ and passing through x meets S^1. This is well defined, since the ray meets S^1 at exactly one point which we call $r(x)$. This means there is a $t > 0$ such that $x = (1-t)f(x) + tr(x) \implies r(x) = \frac{(x-(1-t)f(x))}{t} \implies r$ is a continuous function of x. Clearly, $\forall\ x \in S^1, r(x) = x \implies r$ is a retraction from D^2 to $S^1 \implies S^1$ is a retract of D^2. This contradicts Theorem 3.8.6. Hence we conclude that f has a fixed point. ❑

Remark 3.8.9 Brouwer fixed point theorem for D^n was first proved and studied by L.E.J. Brouwer (1881–1967) during 1910–2012. Now this Theorem is proved by using the homology or homotopy groups. But Brouwer used neither of them, which had not been invented at that time. Instead, he used the notion of degree of spherical maps $f : S^n \to S^n$.

Remark 3.8.10 A generalization of the Corollary 3.6.5 is given in the Perron–Frobenius theorem in $\mathbf{R}^n$. It asserts that any square matrix with positive entries has a unique eigenvector with positive entries (up to a multiplication by a positive constant) and the corresponding eigenvalue has multiplicity one and is strictly greater than the absolute value of any other eigenvalue.

Definition 3.8.11 If x is a point of S^n, then its antipode is the point $-x \in S^n$. A continuous map $f : S^n \to S^m$ is said to be antipode-preserving or antipodal if $f(-x) = -f(x), \forall x \in S^n$.

Proposition 3.8.12 *Let $f_n : S^n \to S^n$ be the antipodal map. If there exists a vector field on S^n, then $f_n \simeq 1_d$.*

Proof Let $f : S^n \to \mathbf{R}^{n+1}$ be a map such that $f(x) \neq 0$ and $\langle f(x), x\rangle = 0$. Construct a path from x to $f_n(x)$ in the plane determined by x and $f(x)$ and on the sphere:

$$H(x,t) = \alpha(t)x + \beta(t)f(x), ||H(x,t)||^2 = 1.$$

This gives the equation

$$\alpha(t)^2 + \beta(t)^2\langle f(x), f(x)\rangle = 1.$$

Choose $\alpha(t) = 1 - 2t$. Then $\beta(t)^2 = 4(t - t^2)/||f(x)||^2$, since $f(x) \neq 0$, gives $\beta(t) = 2\sqrt{t - t^2}/||f(x)||$. Consequently, $H(x,t) = (1-2t)x + 2\sqrt{t - t^2}f(x)/||f(x)||$ shows that $H : 1_d \simeq f_n$. ❑

Proposition 3.8.13 *If $f : S^1 \to S^1$ is nullhomotopic, then*

(i) *f has a fixed point;*
(ii) *f maps some point $x \in S^1$ to its antipode $-x$.*

Proof **(i)** $f : S^1 \to S^1$ is nullhomotopic $\implies$ f has a continuous extension $\tilde{f}$ over D^2. Consider the map $\tilde{f} : D^2 \to S^1 \subset D^2$. Then by Brouwer Fixed Point Theorem, $\tilde{f}$ has a fixed point x_0 (say). Now Im $\tilde{f} \subset S^1$ and $\tilde{f}(x_0) = x_0 \implies x_0 \in S^1$, which is clearly a fixed point of f.
(ii) Define a map $g : S^1 \to S^1, x \mapsto -x$. Then $g \circ f : S^1 \to S^1$ is also null-homotopic. Hence $g \circ f$ has a fixed point x_0 (say) by **(i)**. This implies that $x_0 = (g \circ f)(x_0) = g(f(x_0)) = -f(x_0) \implies f(x_0) = -x_0$. ❑

Corollary 3.8.14 *There is no antipode-preserving continuous map $f : S^2 \to S^1$.*

3.8.3 Borsuk–Ulam Theorem

We now prove the Borsuk–Ulam theorem for dimension 2. For general case see Chap. 14.

Theorem 3.8.15 (Borsuk–Ulam Theorem) *For every continuous map $f : S^2 \to \mathbf{R}^2$, there exists a pair of antipodal points x and $-x$ in S^2 such that $f(x) = f(-x)$.*

Proof Suppose $f(x) \neq f(-x)$ for all $x \in S^2$. Define

$$g : S^2 \to S^1, x \mapsto \frac{f(x) - f(-x)}{||f(x) - f(-x)||}.$$

Then g is a continuous map such that $g(-x) = -g(x)$, $\forall x \in S^2$. This contradicts the Corollary 3.8.14. ❑

Corollary 3.8.16 *The 2-sphere S^2 can not be put in the plane $\mathbf{R}^2$.*

Proof If possible, let $f : S^2 \to \mathbf{R}^2$ be an embedding. Then it is continuous and hence f must map at least one pair of points to a single point of $\mathbf{R}^2$ by Theorem 3.8.15. This implies a contradiction, because f cannot be injective. ❑

Example 3.8.17 (Physical interpretation) Consider the earth as 2-sphere S^2. Let T, P are functions on the earth defining temperature and barometric pressure at any point of time and at a place on the earth. These are continuous functions and define a map

$$f : S^2 \to \mathbf{R}^2, x \mapsto (T(x), P(x)).$$

The the Borsuk–Ulam Theorem 3.8.15 says that at any point of time, there exists a pair of points x and $-x$ on the earth S^2 such that the temperature and barometric pressure both are identical at x and $-x$.

3.8.4 Cauchy's Integral Theorem of Complex Analysis

This subsection proves Cauchy's integral theorem (homotopy version) of complex analysis. This theorem is one of the central theorems in the study of functions of a complex variable for analytic functions. We now utilize winding number $w(f; a)$ of a differentiable loop parametrized by f in the complex plane with respect to a point a not in Im f (see Corollary 3.3.5) and the exponential map $p : \mathbf{R} \to S^1$ to study complex line integral.

Lemma 3.8.18 *Let f be a piecewise differentiable loop in the complex plane and a be a point in $\mathbf{C}$ but not in* Im f. *Then*

$$w(f; a) = \frac{1}{2\pi i} \int_f \frac{dz}{z - a}.$$

Proof Let $p : \mathbf{R} \to S^1$ be the usual covering map. Define

$$g : I \to S^1, t \mapsto \frac{f(t) - a}{||f(t) - a||}.$$

Then g is a loop in S^1. Let $\tilde{g} : I \to \mathbf{R}$ be the unique lifting of g. Define $\theta(t)$ by $\theta(t) = 2\pi\tilde{g}(t)$. Then $f(t) - a = ||f(t) - a||e^{i\theta(t)}$. Then $\int_f \frac{dz}{z-a} = 2\pi i[\tilde{g}(1) - \tilde{g}(0)] = 2\pi i.w(f; a)$. ❑

Theorem 3.8.19 (Cauchy's Integral Theorem) *Let X be an open subset of the complex plane $\mathbf{C}$ and $f : X \to \mathbf{C}$ be an analytic function. If α is a simple closed piecewise differentiable curve in $X \subset \mathbf{C}$ such that α is nullhomotopic, then $\int_\alpha f = 0$.*

Proof Let $\alpha_1 = \alpha$ and α_0 be a constant curve such that $\alpha_1 \simeq \alpha_0$. Let $F : \alpha_1 \simeq \alpha_0$. Define $h(t) = w(\alpha_t; \beta)$, where $\alpha_t(s) = F(s, t)$ for $0 \leq s, t \leq 1$ and β is fixed in $\mathbf{C} - X$. We claim that h is continuous on I and since h is an integral valued function and $h(0) = 0$, it follows that $h(t) \equiv 0$. In particular, $w(\alpha; \beta) = 0$ for all β in $\mathbf{C} - X$. Hence the theorem follows from Lemma 3.8.18. ❑

Remark 3.8.20 Theorem 3.8.19 follows immediately from Corollary 3.3.5 and Lemma 3.8.18, since α is nullhomotopic.

3.9 Exercises

1. Let $\{X_i : i \in \alpha\}$ be a family of spaces and let for each $i \in \alpha$, $x_i \in X_i$ be a base point. Then generalize the Theorem 3.1.37 to prove that $\pi_1(\prod X_i, (x_i)) \cong \prod \pi_1(X_i, x_i)$.
2. Show that $\mathbf{R}^n - \{0\}$ is simply connected for $n > 2$.
   [Hint: $\mathbf{R}^n - \{0\} \approx S^{n-1} \implies \pi_1(\mathbf{R}^n - \{0\}) \cong \pi_1(S^{n-1})$ for $n \geq 3 \implies \pi_1(\mathbf{R}^n - \{0\}) = 0$, $\forall\ n > 2$.]
3. Prove that $\mathbf{R}^n$ and $\mathbf{R}^2$ are not homeomorphic for $n > 2$.
   [Hint: Suppose $n > 2$. Then deleting a point p from $\mathbf{R}^n$ leave $\mathbf{R}^n - p$ simply connected. On the other hand deleting a point from $\mathbf{R}^2$ does not so.]
4. **(a)** If $s_0 \in S^{n-1}$, show that the inclusion map $i : (S^{n-1}, s_0) \hookrightarrow (\mathbf{R}^n - \{0\}, s_0)$ induces an isomorphism of fundamental groups.
   [Hint: Proceed as in Theorem 3.4.1.]
   (b) Show that the map $H : (\mathbf{R}^n - \{0\}) \to \mathbf{R}^n - \{0\}$ defined by $H(x, t) = t\frac{x}{||x||} + (1 - t)x$ is a strong deformation of $\mathbf{R}^n - \{0\}$ onto S^{n-1}.
5. Let A denote the z-axis in $\mathbf{R}^3$. Show that $\pi_1(\mathbf{R}^3 - A) \cong \mathbf{Z}$.
6. Show that none of the following subspaces A of X are retract of X:
   (i) $X = S^1 \times D^2$ and A is its boundary torus $S^1 \times S^1$.
   (ii) $X = \mathbf{R}^3$ and A is any subspace of X homeomorphic to S^1.
   [Hint: **(i)** Using the results that $\pi_1(S^1) \cong \mathbf{Z}, \pi_1(D^2) \cong 0$, we get $\pi_1(S^1 \times D^2) \cong \mathbf{Z}$ and $\pi_1(S^1 \times S^1) \cong \mathbf{Z} \times \mathbf{Z}$, it follows that for any homomorphism $\psi : \mathbf{Z} \times \mathbf{Z} \to \mathbf{Z}$, $\psi((1, 0)) = n$ and $\psi((0, 1)) = m$ for some integer m and n. Clearly $\psi((m, 0)) = nm$ and $\psi((0, n)) = nm$ imply that ψ is not injective. Consequently, there exists no retraction of X to A by Proposition 3.1.26.
   (ii) Use the results that $\pi_1(\mathbf{R}^3) \cong 0$ and $\pi_1(A) \cong \pi_1(S^1) \cong \mathbf{Z}$ to show that there does not exists any retraction of $\mathbf{R}^3$ to any subspace A homeomorphic to S^1.]
7. Show that there is no antipodal map $f : S^n \to S^1$ for $n > 1$.
8. Show that the n-sphere S^n has a nonvanishing tangent vector field if and only if n is odd.(The Brouwer- Poincaré Theorem).
   [Hint: If n is odd, then $n = 2m + 1$. A vector field f on S^n can be defined

by $f(x_1, x_2, \ldots, x_{2m+2}) = (x_2, -x_1, x_4, -x_3, \ldots, x_{2m+2}, -x_{2m+1})$, $\forall$ $(x_1, x_2, \ldots, x_{2m+2}) \in S^n$. Then f is a continuous function from S^n to S^n. For each $x \in S^n$, $\langle x, f(x)\rangle = (x_1x_2 - x_1x_2) + (x_3x_4 - x_3x_4) + \cdots + (x_nx_{n+1} - x_nx_{n+1}) = 0$. If n is even, no such vector field exists. See Chap. 11.]

9. Show that the sphere S^2 has no nonvanishing tangent vector field.

10. **(a)** Let X be a path-connected space. Show that the fundamental group of $\pi_1(X)$ based at any point $x \in X$ is abelian iff all base point change homomorphisms β_u depend only on the end points of the path u in X.
[Hint: Proceed as in Theorem 3.1.22.]

(b) Let (X, x_0) be a pointed space and $P(X)$ denote the space of paths $(I, 0) \to (X, x_0)$, i.e., the space of paths in X starting at x_0. If α_0 is not the constant path in X based at x_0, show that $\pi_1(P(X), \alpha_0) = 0$.

11. Let X be an annulus surrounded by two concentric circles C_1 and C_2 and C be any concentric circle lying completely in X. Show that the circle C is a strong deformation retract of X and $\pi_1(X) \cong \mathbf{Z}$.
[Hint: From the centre O draw a half line in each direction through a point x as shown in Fig. 3.15. Each such line meets the circle C in a unique point, $r(x)$(say). Let $r : X \to C$ be the mapping which maps all points x of the half line to the corresponding point $r(x)$. Define $H : X \times I \to X$ by $H(x, t) = (1-t)x + tr(x)$. Then H is a strong deformation retraction.]

12. **(a)** Show that there is a bijection

$$\psi : \pi_1(X, x_0) \to [(S^1, 1), (X, x_0)].$$

[Hint: Let $\alpha : I \to X$ be a loop based at x_0. Then α determines a base point preserving map

$$\tilde{\alpha} : (S^1, 1) \to (X, x_0), e^{2\pi i t} \mapsto \alpha(t).$$

Conversely, a base point preserving continuous map $f : (S^1, 1) \to (X, x_0)$ determines a loop α_f based at x_0 given by $\alpha_f(t) = f(e^{2\pi i t})$.]

(b) Show that every group is the fundamental of some path-connected topological space.

(c) Let X be a path-connected space and let

$$\psi : \pi_1(X, x_0) \to [S^1, 1], [\alpha] \mapsto [\tilde{\alpha}] \text{ (ignoring base points).}$$

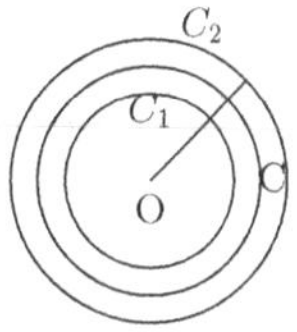

Fig. 3.15 Annulus surrounded by concentric circles

Show that ψ is a surjective map such that if $\beta, \delta \in \pi_1(X, x_0)$, then $\psi(\beta) = \psi(\delta)$ iff β and δ are conjugate elements of $\pi_1(X, x_0)$.

13. Let M be the Möbius band. Show that $\pi_1(M) \cong \mathbf{Z}$.
[Hint: The equatorial loop $\alpha : I \to M$ such that $\alpha(t) = q(t, \frac{1}{2})$, where $q : I \times I \to M$ is the canonical identification, represents a generator of $\pi_1(M)$.]

14. Let $f : S^1 \to S^1$ be a continuous map such that $\deg f = n$. Show that the homomorphism $f_* : \pi_1(S^1) \to \pi_1(S^1)$ is given by $f_*(z) = z^n$.

15. Let $f, g : S^1 \to S^2$ be the two continuous maps defined by $f(z) = z^n$ and $g(z) = z^{-n}$. Compute their induced homomorphisms f_* and g_* of the infinite cyclic group $\pi_1(S^1, s_0)$ into itself.

16. Prove that m-torus T^m and n-torus T^n are homeomorphic iff $m = n$.

17. Let A be a deformation retract of a space X. Show that for each $a \in A$, the groups $\pi_1(X, a)$ and $\pi_1(A, a)$ are isomorphic.

18. Show that $\pi_1(\mathbf{R}^2 - \mathbf{Q}^2)$ is uncountable.

19. Let X be a finite graph. Define Euler characteristic $\chi(X)$ to be the number of vertices minus the number of edges. Show that $\chi(X) = 1$ if X is a tree and the rank (number of elements in a basis) of $\pi_1(X)$ is $1 - \chi(X)$ if X is connected.

20. Show that $\pi_1(\mathrm{SO}(3)) \cong \mathbf{Z}_2$, where SO(3) is the Lie group of rotations in $\mathbf{R}^3$.

21. Prove that the fundamental groups of the following spaces are commutative.

(i) $\mathbf{R}P^n$;
(ii) $V_{m,n} = \mathrm{Hom}\,(\mathbf{R}^m, \mathbf{R}^n)$;
(iii) the space of non-degenerate real $n \times n$ matrices $\mathrm{GL}(n, R) = \{A : \det A \neq 0\}$;
(iv) the space of orthogonal real $n \times n$ matrices $\mathrm{O}(n, \mathbf{R}) = \{A : AA^t = 1_d\}$;
(v) the space of special unitary complex $n \times n$ matrices $\mathrm{SU}(n) = \{A : A\bar{A}^t = 1_d \text{ and } \det A = 1\}$.

22. The bouquet of the topological spaces $X_1, X_2, \ldots, X_n$ each with a base point, is the space obtained from the union $X_1 \cup X_2 \cup \cdots \cup X_n$ by identifying together all the base points.
Show that the fundamental group of the bouquet of n circles is the free group on n generators.

23. Show that the trefoil knot

(i) is not equivalent to the trivial knot;
(ii) cannot be untied.

24. Show that abelianizing a knot group gives the infinite cyclic group.

25. Show that the homomorphism $\pi_1(X^{(1)}) \to \pi_1(X)$ induced by the inclusion of the one-skeleton of a polyhedron X is surjective.

26. Show that the fundamental groups of a sphere with n handles has the following presentation:

$$\langle a_1, b_1, \ldots, a_n, b_n : a_1 b_1 a_1^{-1} b_1^{-1} \cdots a_n b_n a_n^{-1} b_n^{-1} \rangle.$$

27. Let X be a path-connected space. Show that there is a bijective correspondence between the conjugacy classes of elements of $\pi_1(X, x_0) = [(S^1, 1) \to (X, x_0)]$ and the set $[S^1, X]$, the free homotopy classes of maps $f : S^1 \to X$ (i.e., having no base points).

28. Let G be a free group and k-elements and let H be a subgroup of G of index n. Using fundamental group, show that H is a free group on $(k-1)n+1$ elements.

29. Given $[f] \in \pi_1(S^1, 1)$, let γ be the contour $\{f(t) : t \in I\} \subset \mathbf{C}$. Define

$$w(f) = \frac{1}{2\pi} \int_{\gamma} \frac{dz}{z}.$$

Using fundamental group, show that

(i) $w(f)$ is an integer;
(ii) $w(f)$ is independent of the choice of representative $f \in [f]$;
(iii) $w(f) = \deg f$.

30. Show that

(i) two continuous maps $f, g : S^1 \to S^1$ are homotopic iff they have have the same degree;
(ii) the degree function of maps $f : S^1 \to S^1$ sets up a one to one correspondence between the set of homotopy classes $[S^1, S^1]$ and the set $\mathbf{Z}$.

3.10 Additional Reading

[1] Adhikari, M.R., and Adhikari, Avishek, *Basic Modern Algebra with Applications*, Springer, New Delhi, New York, Heidelberg, 2014.
[2] Aguilar, M., Gitler, S., Prieto, C., *Algebraic Topology from a Homotopical View Point*, Springer-Verlag, New York, 2002.
[3] Arkowitz, Martin, *Introduction to Homotopy Theory*, Springer, New York, 2011.
[4] Chatterjee, B.C., Ganguly, S., and Adhikari, M.R., *A Textbook of Topology*, Asian Books Pvt.Ltd., New Delhi, 2002.
[5] Crowell, R.H., and Fox, R.H., *Introduction to Knot Theory*, Ginn and Company, Boston, 1963.
[6] Dieudonné, J., *A History of Algebraic and Differential Topology*, 1900–1960, Modern Birkhäuser, 1989.
[7] Dugundji, J., *Topology*, Allyn & Bacon, Newtown, MA, 1966.
[8] Eilenberg, S., and Steenrod, N., *Foundations of Algebraic Topology*, Princeton University Press, Princeton, 1952.
[9] Gray, B., *Homotopy Theory, An Introduction to Algebraic Topology*, Academic Press, New York, 1975.
[10] Hatcher, Allen, *Algebraic Topology*, Cambridge University Press, 2002.

[11] Hilton, P.J., *An introduction to Homotopy Theory*, Cambridge University Press, Cambridge, 1983.
[12] Hu, S.T., *Homotopy Theory*, Academic Press, New York, 1959.
[13] Spanier, E., *Algebraic Topology*, McGraw-Hill Book Company, New York, 1966.

References

Adhikari, M.R., Adhikari, A.: Basic Modern Algebra with Applications. Springer, New Delhi (2014)

Aguilar, M., Gitler, S., Prieto, C.: Algebraic Topology from a Homotopical View Point. Springer, New York (2002)

Arkowitz, M.: Introduction to Homotopy Theory. Springer, New York (2011)

Armstrong, M.A.: Basic Topology. Springer, New York (1983)

Bredon, G.: Topology and Geometry. Springer, GTM 139 (1993)

Chatterjee, B.C., Ganguly, S., Adhikari, M.R.: A Textbook of Topology. Asian Books Pvt. Ltd., New Delhi (2002)

Croom, F.H.: Basic Concepts of Algebraic Topology. Springer, New York (1978)

Crowell, R.H., Fox, R.H.: Introduction to Knot Theory. Ginn and Company, Boston (1963)

Dieudonné, J.: A History of Algebraic and Differential Topology, 1900–1960. Modern Birkhäuser, Basel (1989)

Dugundji, J.: Topology. Allyn & Bacon, Newtown (1966)

Eilenberg, S., Steenrod, N.: Foundations of Algebraic Topology. Princeton University Press, Princeton (1952)

Gray, B.: Homotopy Theory, An Introduction to Algebraic Topology. Acamedic Press, New York (1975)

Hatcher, A.: Algebraic Topology. Cambridge University Press, Cambridge (2002)

Hilton, P.J.: An Introduction to Homotopy Theory. Cambridge University Press, Cambridge (1983)

Hu, S.T.: Homotopy Theory. Academic Press, New York (1959)

Massey, W.S.: A Basic Course in Algebraic Topology. Springer, New York (1991)

Maunder, C.R.F.: Algebraic Topology. Van Nostrand Reinhhold, London (1970)

Munkres, J.R., Topology, A First Course. Prentice-Hall, New Jersey (1975)

Rotman, J.J.: An Introduction to Algebraic Topology. Springer, New York (1988)

Spanier, E.: Algebraic Topology. McGraw-Hill Book Company, New York (1966)

Switzer, R.M.: Algebraic Topology-Homotopy and Homology. Springer, Berlin (1975)

Whitehead, G.W.: Elements of Homotopy Theory. Springer, New York (1978)

Chapter 4
Covering Spaces

This chapter continues the study of the fundamental groups and is designed to utilize the power of the fundamental groups through a study of covering spaces. The fundamental groups are deeply connected with covering spaces. Algebraic features of the fundamental groups are expressed by the geometric language of covering spaces. Main interest in the study of this chapter is to establish an exact correspondence between the various connected covering spaces of a given base space B and subgroups of its fundamental group $\pi_1(B)$, like Galois theory, with its correspondence between field extensions and subgroups of Galois groups, which is an amazing result. Historically, the systemic study of covering spaces appeared during the late 19th century and early 20th century through the theory of Riemann surfaces. But its origin was found before the invention of the fundamental groups by H. Poincaré in 1895. Poincaré introduced the concept of universal covering spaces in 1883 to prove a theorem on analytic functions.

The theory of covering spaces is of great importance not only in topology but also in other branches of mathematics such as complex analysis, geometry, Lie groups and also in some areas beyond mathematics. A covering space is a locally trivial map with discrete fibers. The objects of this nature can be classified by algebraic objects related to fundamental groups. The exponential map $p : \mathbf{R} \to S^1$ defined by $p(x) = e^{2\pi i x}$, $x \in \mathbf{R}$ is a powerful covering projection and $(\mathbf{R}, p)$ is the universal covering space of S^1. Chapter 3 has utilized this map as a tool for computing $\pi_1(S^1)$. Covering spaces likewise provide useful general tools for computation of fundamental groups. The fundamental group is instrumental for classifying the topological spaces which can be covering spaces of a given base space B. For a large class of spaces, the possible covering spaces of B are determined by the subgroups of $\pi_1(B)$. Moreover, the theory of covering spaces facilitates to determine the fundamental groups of several spaces.

More precisely, this chapter considers a class of mappings $p : X \to B$, called the 'covering projections' from a space X, called a covering space, to a space B, called base space, to which the properties of the exponential map p are extended. Moreover,

M.R. Adhikari, *Basic Algebraic Topology and its Applications*,
DOI 10.1007/978-81-322-2843-1_4

this chapter introduces the concepts of fibrations and cofibrations born in geometry and topology and proves some classical results such as Borsuk–Ulum theorem and Hurewicz theorem for a fibration.

For this chapter the books Croom (1978), Hatcher (2002), Rotman (1988), Spanier (1966), Steenrod (1951), and some others are referred in the Bibliography.

4.1 Covering Spaces: Introductory Concepts and Examples

This section introduces the concept of covering spaces. Covering spaces displays the first example of the power of the fundamental groups in classifying topological spaces. Algebraic features of the fundamental groups $\pi_1(B)$ of the base space B are expressed in the geometric language of covering spaces of B.

4.1.1 Introductory Concepts

This subsection introduces the concept of a covering space with illustrative examples. Recall that a topological space X is path-connected if each pair of points in X can be joined by a path in X. A space that satisfies this property locally is called 'locally path-connected.' If X is a disconnected space, a maximal path-connected subset of the space X is called a path component and is not a proper subset of any path-connected subset of X. The path components of a subset B of X are the path components of B in its subspace topology. For example, each interval and each ray in the real line are both path-connected and locally path-connected. On the other hand, the subspace $[-1, 0) \cup (0, 1]$ of $\mathbf{R}$ is not path-connected but it is locally path-connected. The deleted comb space is path-connected but not locally path-connected. The space of rationals $\mathbf{Q}$ is neither connected nor locally connected.

Definition 4.1.1 Let X and B be topological spaces and let $p : X \to B$ be a continuous surjective map. An open set U of B is said to be evenly covered by p if $p^{-1}(U)$ is a union of disjoint open sets S_i, called sheets such that $p|_{S_i} : S_i \to U$ is a homeomorphism for each i and U is called an admissible open set in B.

Example 4.1.2 Consider the exponential map $p : \mathbf{R} \to S^1$ defined by

$$p(x) = e^{2\pi ix} = \cos 2\pi x + i \sin 2\pi x, x \in \mathbf{R}.$$

Then the open set $U = S^1 - \{1\}$ is evenly covered by p, since $p^{-1}(U) = \bigcup\limits_{n \in \mathbf{Z}} (n - \frac{1}{2}, n + \frac{1}{2})$. Clearly, the sheets are open intervals.

Definition 4.1.3 Let B be a topological space. The pair (X, p) is called a covering space of B if

(i) X is a path-connected topological space;
(ii) the map $p : X \to B$ is continuous;
(iii) each $b \in B$ has an open neighborhood which is evenly covered by p.

The map p is called the covering projection and an open set in B which is evenly covered by p is called p-admissible or simply admissible.

Remark 4.1.4 Some authors do not assume X to be path-connected but assume p to be surjective while defining a covering space.

Example 4.1.5 The exponential map $p : \mathbf{R} \to S^1$ defined in Example 4.1.2 is a covering projection and hence $(\mathbf{R}, p)$ is a covering space of S^1. Because the open sets $U_1 = S^1 - \{-1\}$ and $U_2 = S^1 - \{1\}$ are evenly covered by p. Thus each point of S^1 has an admissible open neighborhood in S^1. In fact, any proper connected arc of S^1 is evenly covered by p. The same argument shows that the map $p : \mathbf{R} \to S^1$ defined by $p(t) = e^{i\alpha t}$, where $\alpha \in \mathbf{R}$ is a fixed nonzero real number, is also a covering projection.

Example 4.1.6 For any positive integer n, let $p_n : S^1 \to S^1$ be the map defined by $p_n(z) = z^n, z \in S^1$. Then (S^1, p_n) is a covering space of S^1. Because, in polar coordinates, p_n is given by $p_n(1, \theta) = (1, n\theta)$. The map p_n wraps the circle around itself n times. Let U be an open arc on S^1 subtended by an angle θ, $0 \leq \theta \leq 2\pi$, and containing a point x. Then $p^{-1}(U)$ consists of n open arcs each determining an angle θ/n and each containing one nth root of x. Each of these n open arcs is mapped homeomorphically onto U. Thus any proper arc in S^1 is an admissible neighborhood. Consequently, (S^1, p_n) is a covering space of S^1.

Example 4.1.7 Consider the map $f : \mathbf{R}^2 \to S^1 \times S^1$ from the plane to the torus defined by $f(t_1, t_2) = (e^{2\pi i t_1}, e^{2\pi i t_2})$, $(t_1, t_2) \in \mathbf{R}^2$. Then $(\mathbf{R}^2, f)$ is a covering space of $S^1 \times S^1$.

For any point $(z_1, z_2) \in S^1 \times S^1$, let U be a small rectangle formed by the product of two open arcs in S^1 containing z_1 and z_2, respectively. Then U is an admissible neighborhood whose inverse image consists of a countably infinite family of open rectangles in the plane $\mathbf{R}^2$. This example is essentially a generalizaton of the covering projection $p : \mathbf{R} \to S^1$.

Theorem 4.1.8 *Let (X_1, p_1) be a covering space of B_1, (X_2, p_2) be a covering space of B_2, then $(X_1 \times X_2, p_1 \times p_2)$ is a covering space of $B_1 \times B_2$, where $p_1 \times p_2 : X_1 \times X_2 \to B_1 \times B_2$ is defined by $(p_1 \times p_2)(x, y) = (p_1(x), p_2(y))$.*

Proof Let $(b_1, b_2) \in B_1 \times B_2$ and U_1 be an open neighborhood of b_1 and U_2 be an open neighborhood of b_2 which are evenly covered by p_1 and p_2, respectively. Then $U_1 \times U_2$ is a neighborhood of (b_1, b_2) in $B_1 \times B_2$ which is evenly covered by $p_1 \times p_2$. ❑

Example 4.1.9 Consider the exponential map $p : \mathbf{R} \to S^1$ defined by

$$p(x) = e^{2\pi i x} = \cos 2\pi x + i \sin 2\pi x, \; x \in \mathbf{R}.$$

Then the map $(p, p) : \mathbf{R} \times \mathbf{R} \to S^1 \times S^1$ is a covering projection. In fact for every positive integer n, the product map $p \circ p \circ \cdots \circ p = p^n : \mathbf{R}^n \to T^n$ is a covering projection, where $T^n = \prod_1^n S^1$ is the n-dimensional torus.

Theorem 4.1.10 *Let $p : (X, x_0) \to (B, b_0)$ be a covering projection. If X is path-connected, then there is a surjection $\psi : \pi_1(B, b_0) \to p^{-1}(b_0)$. If X is simply connected, then ψ is a bijection.*

Proof Following the technique for computation of $\pi_1(S^1, 1)$ (see Theorem 3.3.9 of Chap. 3) the theorem can be proved. ❑

Remark 4.1.11 Everytopological space is not necessarily a covering space. The following is an example of a topological space X which is not a covering space of Y.

Example 4.1.12 Let X be a rectangle which is mapped by the projection p onto the first coordinate to an interval Y. Let U be an interval in Y. Then $p^{-1}(U)$ is a strip in X consisting of all points above U (as shown in Fig. 4.1).

This strip cannot be mapped by p homeomorphically onto U. Hence U is not evenly covered by p. Consequently, (X, p) is not a covering space of Y.

Example 4.1.13 (*Infinite and finite spirals*) Let X be an infinite spiral, and $p : X \to S^1$ be the projection described in Fig. 4.2.

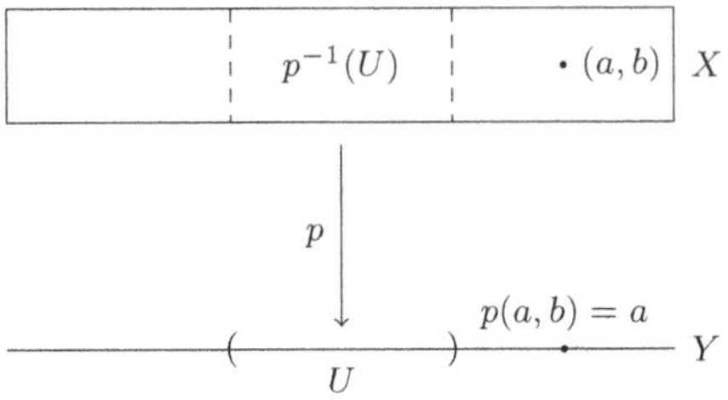

Fig. 4.1 Example of a neighborhood which is not evenly covered

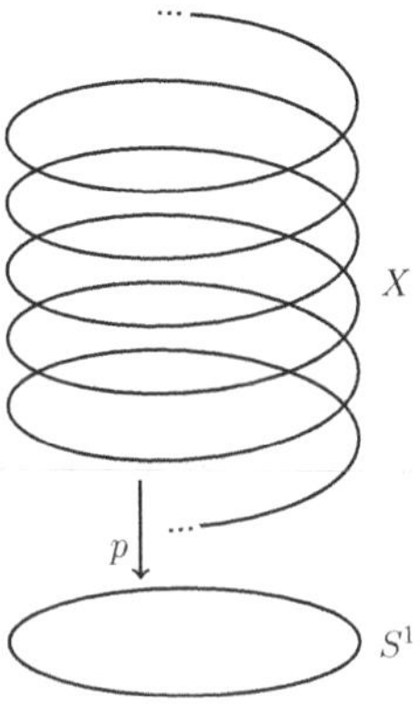

Fig. 4.2 Infinite spiral with projection p

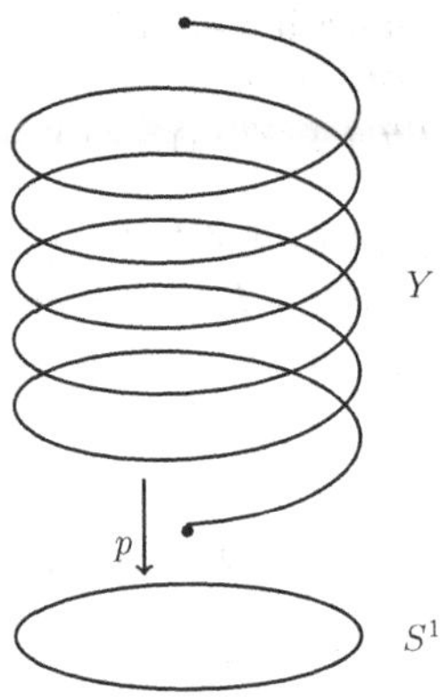

Fig. 4.3 Finite spiral with projection p

Each point of X is projected by p to the point on the circle directly below it. Then (X, p) is a covering space of S^1. On the other hand, if $p : Y \to S^1$ is a finite spiral projection as shown in Fig. 4.2, then (Y, p) is not covering space of S^1, because if x_0 and x_1 are the end points of the spiral Y, then the points $p(x_0)$ and $p(x_1)$ as shown in Fig. 4.3 have no admissible neighborhoods.

4.1.2 Some Interesting Properties of Covering Spaces

This subsection presents some properties of covering spaces.

Proposition 4.1.14 *Let $p : (X, x_0) \to (B, b_0)$ be a covering space. Then the induced homomorphism $p_* : \pi_1(X, x_0) \to \pi_1(B, b_0)$ is a monomorphism and the subgroup $p_*(\pi_1(X, x_0))$ in $\pi_1(B, b_0)$ consists of homotopy class of loops in B based at b_0 which lifts to X starting at x_0 are loops.*

Proof Let an element $\alpha \in \ker p_*$ be represented by a loop $\tilde{f}_0 : I \to X$ with a homotopy $F_t : I \to B$ of $f_0 = p \circ \tilde{f}_0$ to the trivial loop $\tilde{f}_1$ (Fig. 4.4).

Hence there exists a lifted homotopy of loop $\tilde{F}_t : I \to X$ started at $\tilde{f}_0$ and ending with a constant loop (because the lifted homotopy $\tilde{F}_t$ is a homotopy of paths fixing the end points, since t varies each point of $\tilde{F}_t$ gives a path lifting a constant path, which is therefore constant). Hence $[\tilde{f}_0] = 0$ in $\pi_1(X, x_0)$ shows that p_* is injective. ❑

We now state the following two other properties of covering spaces whose proofs are given in Sect. 4.5.2.

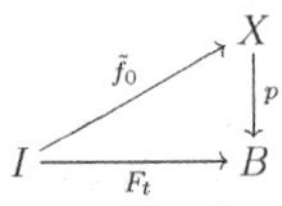

Fig. 4.4 Homotopy diagram corresponding to a lifting of f_0 to $\tilde{f}_0$

Proposition 4.1.15 *Let $p : (X, x_0) \to (B, b_0)$ be a covering space and $f : (Y, y_0) \to (B, b_0)$ be a map, where Y is path-connected and locally path-connected. Then a lift $\tilde{f} : (Y, y_0) \to (X, x_0)$ of f exists iff $f_*(\pi_1(Y, y_0)) \subset p_*(\pi_1(X, x_0))$.*

Proposition 4.1.16 *Given a covering space $p : X \to B$ and a map $f : Y \to X$ with two lifting $\tilde{f}_1, \tilde{f}_2 : Y \to X$ that agree at some point of Y, then if Y is connected, $\tilde{f}_1 = \tilde{f}_2$, i.e., $\tilde{f}_1(y) = \tilde{f}_2(y)$, $\forall\, y \in Y$.*

4.1.3 Covering Spaces of $\mathbf{R}P^n$

This subsection studies covering spaces of real projective spaces $\mathbf{R}P^n$ and computes fundamental group of $\mathbf{R}P^2$.

Definition 4.1.17 (*Real projective plane*) Let $\mathbf{R}P^2$ be the real projective plane defined as a quotient space of the 2-sphere S^2 obtained by identifying each point x of S^2 with its antipodal point $-x$ and $p : S^2 \to \mathbf{R}P^2$ be the natural map which identifies each pair of antipodal points i.e., p maps each x to its equivalence class. We topologize $\mathbf{R}P^2$ by defining V to be open in $\mathbf{R}P^2$ if and only if $p^{-1}(V)$ is open in S^2. With this topology $\mathbf{R}P^2$ becomes a topological space.

Theorem 4.1.18 *The projective space $\mathbf{R}P^2$ is a surface and (S^2, p) is a covering space of $\mathbf{R}P^2$.*

Proof First we show that $p : S^2 \to \mathbf{R}P^2$ is a covering map. Given $y \in \mathbf{R}P^2$, we choose $x \in p^{-1}(y)$. We then choose an ϵ-neighborhood U of x in S^2 for some $\epsilon < 1$, using the Euclidean metric d of $\mathbf{R}^3$. If $A : S^2 \to S^2$ is the antipodal map sending z to its antipodal point $-z$, then U contains no pair $\{z, A(z)\}$ of antipodal points of S^2, since $d(z, a(z)) = 2$. Consequently, the map $p : U \to p(U)$ is bijective. The antipodal map $A : S^2 \to S^2$, given by $A(z) = -z$ is a homeomorphism of S^2 and hence $A(U)$ is open in S^2. Since $p^{-1}(p(U)) = U \cup A(U)$, this set is also open in S^2. Consequently, $p(U)$ is open in $\mathbf{R}P^2$ and hence p is an open map. Thus the bijective map $p : U \to p(U)$ is continuous and open. Hence it is a homeomorphism. Similarly, $p : A(U) \to p(A(U)) = p(U)$ is a homeomorphism. The set $p^{-1}(p(U))$ is thus the union of two open sets U and $A(U)$, each of which is mapped homeomorphically by p onto $p(U)$. Hence $p(U)$ is a neighborhood of $p(x) = y$, which is evenly covered by p. Consequently, (S^2, p) is a covering space of $\mathbf{R}P^2$. For the first part, let $\{U_n\}$ be countable basis of S^2. Then $\{p(U)\}$ is a countable basis of $\mathbf{R}P^2$. Clearly, $\mathbf{R}P^2$ is a Hausdorff space. Let y_1 and y_2 be two points of $\mathbf{R}P^2$. The set $p^{-1}(y_1) \cup p^{-1}(y_2)$ consists of four points. Let 2ϵ be the minimum distance between them. Let U_1 be the ϵ-neighborhood of one of the points $p^{-1}(y_1)$ and U_2 be the ϵ-neighborhood of one of the points $p^{-1}(y_2)$. Then the sets $U_1 \cup A(U_1)$ and $U_2 \cup A(U_2)$ are disjoint. Consequently, $p(U_1)$ and $p(U_2)$ are disjoint neighborhoods of y_1 and y_2, respectively, in $\mathbf{R}P^2$. Since S^2 is a surface and every point of $\mathbf{R}P^2$ has a neighborhood homeomorphic to an open subset of S^2, the space $\mathbf{R}P^2$ is also a surface. ❑

A generalization of Theorem 4.1.18 for $n > 1$ is now given.

Theorem 4.1.19 *(S^n, p) is a covering space of $\mathbf{R}P^n$, where p is the map identifying antipodal points of S^n for $n > 1$.*

Proof The sets $E_i^+ = \{(x_1, x_2, \ldots, x_{n+1}) \in S^n : x_i > 0\}$ and $E_i^- = \{(x_1, x_2, \ldots, x_{n+1}) \in S^n : x_i < 0\}$ are open sets and cover S^n. The map $p|_{E_i^+}$ is 1-1, continuous and open. Hence if $U_i = p(E_i^+) = p(E_i^-)$, then $p^{-1}(U_i) = E_i^+ \cup E_i^-$. The sets E_i^+ and E_i^- are disjoint open sets, and homeomorphic to U_i. This shows that $p : S^n \to \mathbf{R}P^n$ is a covering space. This asserts that (S^n, p) is a covering space of $\mathbf{R}P^n$. ❑

Definition 4.1.20 The multiplicity of a covering space (X, p) of B is the cardinal number of a fiber. If the multiplicity is n, we say that (X, p) is an n-sheeted covering space of B or that (X, p) is an n-fold cover of B.

Example 4.1.21 **(i)** (S^2, p) is a double covering of $\mathbf{R}P^2$.

(ii) The number of sheets of $(\mathbf{R}, p)$ of S^1 is countably infinite.
Because, p identifies pairs of antipodal points, the number of sheets of this covering in (i) is 2. On the other hand, for the (ii) covering projection $p : \mathbf{R} \to S^1$ (see Example 4.1.2) maps each integer and only the integers to $1 \in S^1$. Thus $p^{-1}(1) = \mathbf{Z}$ and hence the number of sheets of this covering is countably infinite.

Theorem 4.1.22 $\pi_1(\mathbf{R}P^2, y) \cong \mathbf{Z}_2$.

Proof The projection $p : S^2 \to \mathbf{R}P^2$ is covering map by Theorem 4.1.18. Since S^2 is simply connected, we apply Theorem 4.1.10, which gives a bijective correspondence between $\pi_1(\mathbf{R}P^2, y)$ and the set $p^{-1}(y)$. Since $p^{-1}(y)$ is a two-element set, $\pi_1(\mathbf{R}P^2, y)$ is a group of order 2. Since any group of order 2 is isomorphic to $\mathbf{Z}_2$, it follows that $\pi_1(\mathbf{R}P^2, y) \cong \mathbf{Z}_2$. ❑

Remark 4.1.23 For computing $\pi_1(\mathbf{R}P^n, y)$ by using the universal covering space (S^n, q) of $\mathbf{R}P^n$, where q identifies the antipodal points of S^n, (see Sect. 4.6.2), use topological group action see Corollary 4.10.4.

4.2 Computing Fundamental Groups of Figure-Eight and Double Torus

We now consider some topological spaces whose fundamental groups are nonabelian. This section constructs covering spaces for computation of fundamental groups of some spaces such as figure-eight and double torus whose fundamental groups are not abelian. For computing the fundamental group of figure-eight by graph-theoretic method see Sect. 4.10.6.

Example 4.2.1 (*figure-eight*) The figure-eight F is the union of two circles A and B with a point x_0 in common. We now describe a certain covering space X for F.

Let X be the subspace of the plane consisting of the x-axis and the y-axis, along with the small circles tangent to these axes, one circle tangent to the x-axis at each nonzero integer point and one circle tangent to the y-axis at each nonzero integer point as shown in Fig. 4.5.

The projection map p wraps the x-axis around the circle A and wraps the y-axis around the other circle B; in each case the integer points are mapped by p into the base point x_0 of F. Then each circle tangent to an integer point on the x-axis is mapped homeomorphically by p onto B; on the other hand, each circle tangent to an integer point on the y-axis is mapped homeomorphically onto A; in each case the point of tangency is mapped onto the point x_0. Then p is a covering map.

Theorem 4.2.2 *The fundamental group of the figure-eight is not abelian.*

Proof Let $\tilde{f} : I \to X$ be the path $\tilde{f}(t) = (t, 0)$, going along the x-axis from the origin $(0, 0)$ to the point $(1, 0)$. Let $\tilde{g} : I \to X$ be the path $\tilde{g}(t) = (0, t)$, going along the y-axis from the origin $(0, 0)$ to the point $(0, 1)$. Let $f = p \circ \tilde{f}$ and $g = p \circ \tilde{g}$. Then f and g are loops, in the figure-eight F based at x_0, going around the circles A and B, respectively. We claim that $f * g$ and $g * f$ are not path homotopic. We lift each of these paths to a path in X beginning at the origin. The path $f * g$ lifts to a path that goes along the x-axis from the origin to $(1, 0)$, and then goes once around the circle tangent to the x-axis at $(1, 0)$. But the path $g * f$ lifts to a path in X that goes along the y-axis from the origin to $(0, 1)$, and then goes once around the circle tangent to the y-axis at $(0, 1)$. Since the lifted paths do not end at the same point, $f * g$ and $g * f$ cannot be path homotopic. Therefore we conclude that $[f * g] \neq [g * f]$ and hence $[f] \cdot [g] \neq [g] \cdot [f]$ proves that the fundamental group of the figure eight is not abelian. ❑

Remark 4.2.3 For computing the fundamental group of figure-eight by graph-theoretic method see Sect. 4.10.6.

Corollary 4.2.4 *The fundamental group of the double torus T_2 is not abelian.*

Fig. 4.5 Figure-eight

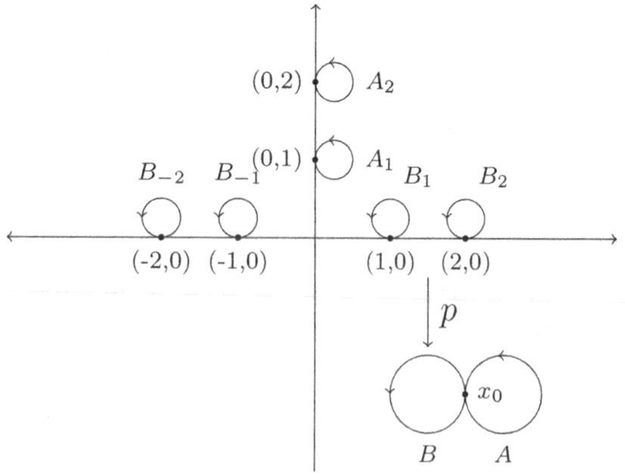

Proof Figure-eight F is a retract of $T_2 \Rightarrow$ the inclusion map $i : (F, x_0) \hookrightarrow (T_2, x_0)$ induces a monomorphism $i_* : \pi_1(F, x_0) \to \pi_1(T_2, x_0) \Rightarrow \pi_1(T_2, x_0)$ is not abelian, since $\pi_1(F, x_0)$ is not abelian. ❑

Remark 4.2.5 For computing fundamental groups of some orbit spaces see Sect. 4.10.2.

4.3 Path Lifting and Homotopy Lifting Properties

This section continues the study of covering spaces and displays basic properties of covering spaces such as path lifting and homotopy lifting properties (PLP and HLP). We begin with characterization of locally path-connected spaces.

Recall the following definitions.

Definition 4.3.1 A topological space X is said to be locally path-connected if for each point $x \in X$ and every neighborhood U_x of x, there is an open set V with $x \in V \subset U_x$ such that any two points in V can be joined by a path in U_x.

Definition 4.3.2 A topological space X is said to be semilocally path-connected if for every point $x \in X$, there is an open neighborhood U_x of x such that every closed path in U_x at x is nullhomotopic in X.

Proposition 4.3.3 *A topological space X is locally path-connected if and only if each path component of each open subset of X is open.*

Proof Left as an exercise. ❑

Theorem 4.3.4 *Every covering projection $p : X \to B$ is an open mapping for any locally path-connected space X.*

Proof Let X be a locally path-connected space such that $p : X \to B$ be a covering projection and V be an open set in X. We claim that $p(V)$ is open in B. Let $b \in p(V)$ and $x \in p^{-1}(b)$ and U be an admissible neighborhood for b. Then x is a point of V such that $p(x) = b$. Let W be the component of $p^{-1}(U)$ which contains x. Since X is locally path connected, W is open in X by Proposition 4.3.3. Since p maps W homeomorphically onto U, p maps the open set $W \cap V$ to the open subset $p(W \cap V)$ in B. Then $b \in p(W \cap V) \subseteq p(V)$. Since b is an arbitrary point of $p(V)$, it follows that $p(V)$ is a union of open sets and hence $p(V)$ is an open set. Consequently p is an open mapping. ❑

Theorem 4.3.5 *Let (X, p) be a covering space of B and Y be a space. If f and g are continuous maps from Y to X for which $p \circ f = p \circ g$, as shown in Fig. 4.6, then the set $A = \{y \in Y : f(y) = g(y)\}$ (i.e., the set of points of Y at which f and g agree) is both open and closed in Y. (Y is not assumed to be path-connected or locally path-connected).*

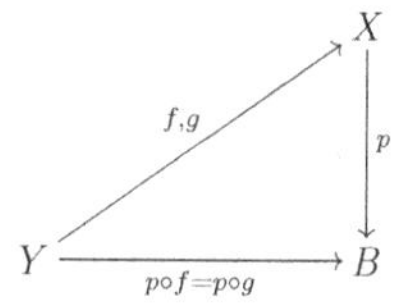

Fig. 4.6 Triangular diagram involving f, g and p

Proof To prove that A is open, let $y \in A$ and U be an admissible neighborhood of $(p \circ f)(y)$. Then the path component V of $p^{-1}(U)$ to which $f(y)$ belongs is an open set in X and hence $f^{-1}(V)$ and $g^{-1}(V)$ are open in Y. Since $f(y) \in V$ and $f(y) = g(y)$, then $y \in f^{-1}(V) \cap g^{-1}(V)$. We claim that $f^{-1}(V) \cap g^{-1}(V)$ is a subset of A and conclude that A is open, since it contains a neighborhood of each of its points. Let $t \in f^{-1}(V) \cap g^{-1}(U)$. Then $f(t), g(t) \in V$ and $(p \circ f)(t) = (p \circ g)(t)$. Since p maps V homeomorphically onto U, it follows that $f(t) = g(t)$ and hence $t \in A$. Thus it follows that A is an open set.

Next we prove that A is closed. Suppose A is not closed and let t be a limit point of A not in A. Then $f(t) \neq g(t)$. The point $(p \circ f)(t) = (p \circ g)(t)$ has an elementary neighborhood U such that the points $f(t)$ and $g(t)$ must be in distinct path components V_1 and V_2 of $p^{-1}(U)$. Since $t \in f^{-1}(V_1) \cap g^{-1}(V_2)$ which is an open set in Y, $f^{-1}(V_1) \cap g^{-1}(V_2)$ must contain a point $y \in A$. But this implies a contradiction, since $V_1 \cap V_2 = \emptyset$ and $f(y) = g(y) \in V_1 \cap V_2$. Hence all limit points of A must lie in A and therefore A is closed. ❑

Corollary 4.3.6 *Let (X, p) be a covering space of B, and let f, g be continuous maps from a connected space Y into X such that $p \circ f = p \circ g$. If f and g agree at a point of Y, then $f = g$.*

Proof Let Y be a connected space. Then the only sets that are both open and closed in Y are Y and $\emptyset$. Hence by Theorem 4.3.5 it follows that either $A = Y$ or $A = \emptyset$. This implies that either $f(y) = g(y)$ at every $y \in Y$ or $f(y) \neq g(y)$ at every $y \in Y$. By hypothesis $f(y) = g(y)$ at some $y \in Y$. Thus $A \neq \emptyset$ and hence $A = Y$. Consequently, $f(y) = g(y)$, $\forall\, y \in Y$ shows that $f = g$. ❑

Remark 4.3.7 The Corollary 4.3.6 gives the uniqueness of the lifting of a map and generalizes Proposition 3.3.2 of Chap. 3.

We now consider lifting problems. What is lifting problem?

Let $p : X \to B$ be a continuous surjective map (not necessarily a covering projection). Given a subspace A of X and a continuous map $f : A \to B$, does there exist a continuous map $\tilde{f} : A \to X$ such that $p \circ \tilde{f} = f$?

In other words, can we find a continuous map $\tilde{f} : A \to X$ making the diagram in Fig. 4.7 commutative? If such $\tilde{f}$ exists, then $\tilde{f}$ is called a lift of f. The satisfactory answer is available if p is covering projection.

Fig. 4.7 Lifting of a map f

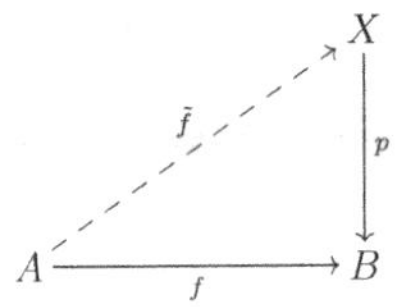

Fig. 4.8 Lifting of a path f

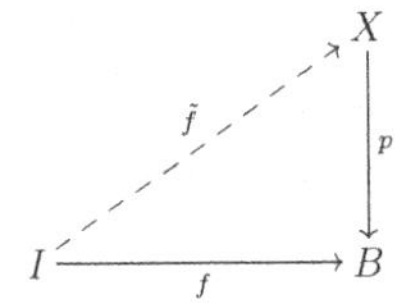

Definition 4.3.8 Let (X, p) be a covering space of B and let $f : I \to B$ be a path in B. A path $\tilde{f} : I \to X$ in X such that $p \circ \tilde{f} = f$, is called a lifting or covering path of f,i.e., if it makes the diagram as shown in Fig. 4.8 commutative.

If $F : I \times I \to B$ be a homotopy, then a homotopy $\tilde{F} : I \times I \to X$ for which $p \circ \tilde{F} = F$, is called a lifting or covering homotopy of F.

We now generalize Theorem 3.3.3 and its Corollary 3.3.4. of Chap. 3.

Theorem 4.3.9 (The Path Lifting Property) *Let (X, p) be a covering space of B and $f : I \to B$ be a path in B beginning at a point $b_0 \in B$. If $x_0 \in p^{-1}(b_0)$, then there is a unique covering path $\tilde{f} : I \to X$ as shown in Fig. 4.9 of f beginning at x_0 such that $p \circ \tilde{f} = f$.*

Proof **Existence of $\tilde{f}$**: Suppose $[a, b] \subset I$ is such that $f([a, b]) \subset U$, where U is an admissible neighborhood of $y = f(a)$ in B. Let $x \in f^{-1}(y)$. Then x lies in a unique sheet S (say). Define

$$\tilde{g} : ([a, b], a) \to (X, x), \text{ by } \tilde{g} = (p|_S)^{-1} \circ (f|_{[a,b]})$$

such that $p \circ \tilde{g} = f|_{[a,b]}$. Let U_t be an admissible neighborhood of $f(t)$ for each $t \in I$. Then $\{f^{-1}(U_t), t \in I\}$, being an open cover of the compact metric space I has a Lebesgue number λ. This shows that if $0 < \delta < \lambda$ and Y is a subset of I of diameter less than δ, then $Y \subset f^{-1}(U_t)$ for some $t \in I$. Thus $f(Y) \subset U_t$ partitions I with points $t_1 = 0, t_2, \ldots, t_k = 1$, where $t_{i+1} - t_i < \delta$ for $1 \leq i \leq k - 1$. Then there is a continuous map $\tilde{g}_1 : [0, t_2] \to X$ satisfying $p \circ \tilde{g}_1 = f|_{[0,t_2]}$ and $\tilde{g}_1(o) = x_0$.

Fig. 4.9 Path lifting property (PLP)

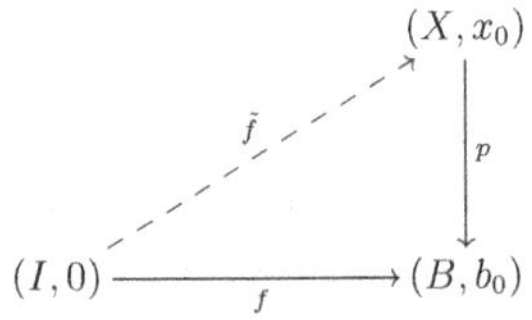

Similarly, there is a continuous map $\tilde{g}_2 : [t_2, t_3] \to X$ satisfying $p \circ \tilde{g}_2 = f|_{[t_2,t_3]}$ and $\tilde{g}_2(t_2) = \tilde{g}_1(t_2)$. In this way, for $1 \leq i \leq k-2$, there is a continuous map

$$\tilde{g}_{i+1} : [t_{i+1}, t_{i+2}] \to X$$

satisfying $p \circ \tilde{g}_{i+1} = f|_{[t_{i+1},t_{i+2}]}$ and $\tilde{g}_{i+1}(t_{i+1}) = \tilde{g}_i(t_{i+1})$. Using gluing lemma, and assembling the functions g_i, we obtain a continuous function $\tilde{f} : I \to X$, where $\tilde{f}(t) = \tilde{g}_i(t)$ if $t \in [t_i, t_{i+1}]$.
The uniqueness of $\tilde{f}$: It follows from Corollary 4.3.6, because I is connected, and by assumption any two lifts of f agree at the point $0 \in I$. ❑

Corollary 4.3.10 (Homotopy Lifting Property) *Let (X, p) be a covering space of B and $F : I \times I \to B$ be a homotopy such that $F(0,0) = b_0$. If $x_0 \in p^{-1}(b_0)$, then there exists a unique homotopy $\tilde{F} : I \times I \to X$ such that $\tilde{F}(0,0) = x_0$.*

Proof Proceed as in Theorem 4.3.9 by subdividing $I \times I$ into rectangles (in place of I). ❑

We can prove in a similar way the general form of the Homotopy Lifting Property.

Theorem 4.3.11 (The Generalized Homotopy Lifting Property) *Let (X, p) be a covering space of B and A be a compact space. If $f : A \to X$ is continuous and $F : A \times I \to B$ is a homotopy starting from $p \circ f$, then there is a homotopy $\tilde{F} : A \times I \to X$ starting from f and lifts F. Furthermore, if F is a homotopy relative to a subset A' of A, then $\tilde{F}$ is also so.*

4.4 Lifting Problems of Arbitrary Continuous Maps

This section gives a necessary and sufficient condition for lifting of an arbitrary continuous map $f : A \to X$ by applying the tools of fundamental groups. More precisely, given a covering space $(\tilde{X}, p)$ of B and a continuous map $f : A \to X$, can we find a continuous map $\tilde{f} : A \to X$ such that $p \circ \tilde{f} = f$? The answer is positive if f is a path or a homotopy between paths by the Path Lifting Property (Theorem 4.3.9), and the Homotopy Lifting Property (Corollary 4.3.10), respectively. To the contrary the answer is negative for an arbitrary continuous map f. For more results see Chap. 16.

Example 4.4.1 The exponential map $p : \mathbf{R} \to S^1$ defined by $p(t) = e^{2\pi i t}$ is a covering projection. The identity map $1_{S^1} : S^1 \to S^1$ cannot be lifted to a continuous map $\psi : S^1 \to \mathbf{R}$ making the triangle in Fig. 4.10 commutative. Otherwise, $p \circ \psi = 1_{S^1} \Rightarrow \psi$ is injective $\Rightarrow \psi$ is an embedding of S^1 into $\mathbf{R}$, since S^1 is compact $\Rightarrow \psi(S^1)$ is a closed interval homeomorphic to S^1, since any compact connected subset of $\mathbf{R}$ must be a closed interval. This is impossible, since a closed interval cannot be homeomorphic to S^1.

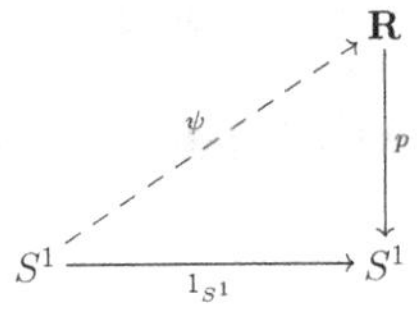

Fig. 4.10 Covering projection for exponential map p

Remark 4.4.2 We now give a necessary and sufficient condition under which an arbitrary continuous map $f : A \to X$ can be lifted. The methods of algebraic topology are now applied to solve such problems.

Theorem 4.4.3 (Lifting Theorem) *Let (X, p) be a covering space of B. Given a connected and locally path-connected space A, let $f : A \to B$ be any continuous map. Then given any three points $a_0 \in A$, $b_0 \in B$ and $x_0 \in X$ such that $f(a_0) = b_0$ and $p(x_0) = b_0$, there exists a unique continuous map $\tilde{f} : A \to X$ satisfying $\tilde{f}(a_0) = x_0$ such that $p \circ \tilde{f} = f$ if and only if $f_*(\pi_1(A, a_0)) \subset p_*(\pi_1(X, x_0))$.*

Proof Suppose that $\exists$ a continuous map $\tilde{f} : A \to X$ satisfying the given conditions. Then the diagram in Fig. 4.11 is commutative. Hence the diagram in Fig. 4.12 is also commutative (by the functorial property of π_1). Consequently, $f_*(\pi_1(A, a_0)) = p_*(\tilde{f}_*(\pi_1(A, a_0))) \subseteq p_*(\pi_1(X, x_0))$. Conversely, let the algebraic condition $f_*(\pi_1(A, a_0)) \subset p_*(\pi_1(X, x_0))$ holds.

Since A is connected, A has only one component. Again since A is locally path-connected, this component is a path component. Hence A is path-connected.

Let $a \in A$. We take a path $u : I \to A$ such that $u(0) = a_0$ and $u(1) = a$. Then $f \circ u : I \to B$ is a path such that $(f \circ u)(0) = f(u(0)) = f(a_0) = b_0$. By path Lifting Property, Theorem 4.3.9, $\exists$ a unique path $\tilde{u} : I \to X$ that lifts $f \circ u$ in X with $\tilde{u}(0) = x_0$ as shown in Fig. 4.13. Define a map

$$\tilde{f} : A \to X, a \mapsto \tilde{u}(1).$$

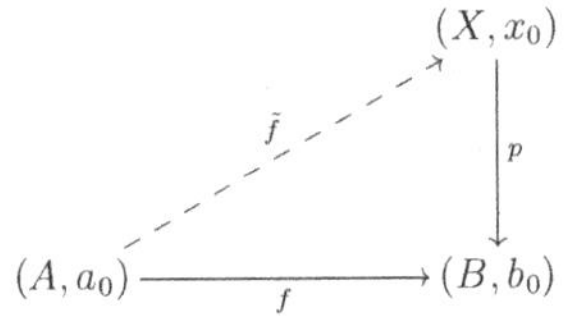

Fig. 4.11 Lifting of f to $\tilde{f}$

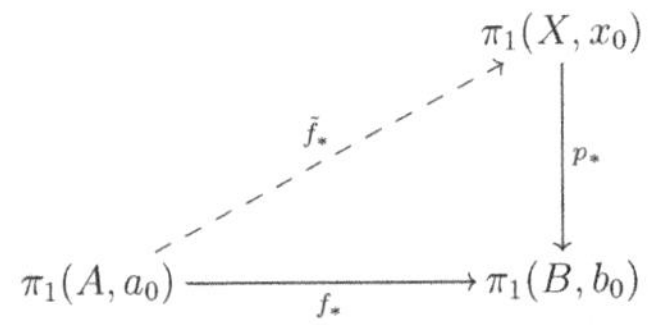

Fig. 4.12 Induced homomorphisms of f and $\tilde{f}$

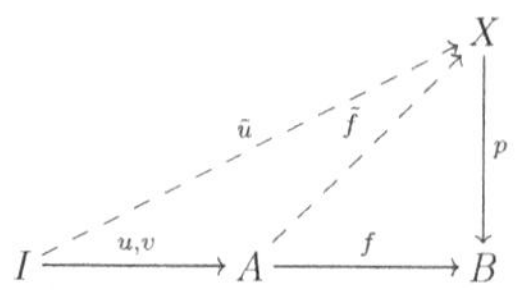

Fig. 4.13 Diagram for lifting theorem

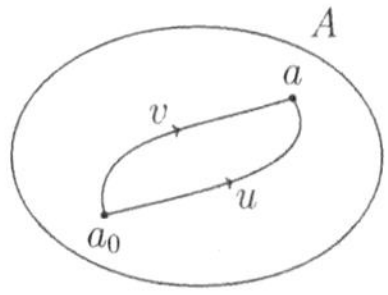

Fig. 4.14 Two paths u and v in A

To show that $\tilde{f}$ is well defined, choose another path v from a_0 to a as shown in Fig. 4.14. Let $\tilde{v}$ be the unique path in X lifting $f \circ v$ for which $\tilde{v}(0) = x_0$, i.e., $p \circ \tilde{v} = f \circ v$ and $\tilde{v}(0) = x_0$.

Now $u * v^{-1}$ is a closed path in A at a_0. Then $f \circ (u * v^{-1}) = (f \circ u) * (f \circ v^{-1})$ is a closed path in B at b_0. Again since

$$[(f \circ u) * (f \circ v^{-1})] = f_*[u * v^{-1}] \in f_*\pi_1(A, a_0) \subseteq p_*\pi_1(X, x_0) \text{(by hypothesis)},$$

there exists a closed path α in X at x_0 such that

$$(f \circ u) * (f \circ v^{-1}) \simeq p \circ \alpha \text{ rel } \dot{I}.$$

Hence

$$(f \circ u) * (f \circ v^{-1}) * (p \circ \tilde{v}) \simeq (p \circ \alpha) * (p \circ \tilde{v}) \text{ rel } \dot{I};$$

$$f \circ u \simeq p \circ (\alpha * \tilde{v}) \text{ rel } \dot{I}, \text{ since } p \circ \tilde{v} = f \circ v.$$

Again by homotopy lifting property (see Corollary 4.3.10)

$$\tilde{u} \simeq \alpha * \tilde{v} \text{ rel } \dot{I} \text{ with } \tilde{u}(1) = (\alpha * \tilde{v})(1) = \tilde{v}(1).$$

This shows that $\tilde{f}$ is well defined.

$\tilde{f}$ **is continuous**: Let $a \in A$ and U be an open neighborhood of $\tilde{f}(a)$. To show the continuity of $\tilde{f}$, we have to find an open neighborhood V_a of a with $\tilde{f}(V_a) \subset U$. We take an open admissible neighborhood V of $p\tilde{f}(a) = f(a)$ such that $V \subset p(U)$. Let W be the path component of $p^{-1}(V)$ which contains the point $\tilde{f}(a)$, and let V' be an open admissible neighborhood of $f(a)$ such that $V' \subseteq p(U \cap W)$. Then the path component of $p^{-1}(V')$ containing $\tilde{f}(a)$ must be contained in U. Since f is continuous and path-connected A is locally connected, $\exists$ a path-connected neighborhood V_a of a such that $f(V_a) \subset V$. Then $\tilde{f}(V_a) \subset U$. ❑

Corollary 4.4.4 *Let A be simply connected and locally path-connected and $f : (A, a_0) \to (B, b_0)$ be continuous. If (X, p) is a covering space of B and if $x_0 \in p^{-1}(b_0)$, then $\exists$ a unique lifting $\tilde{f} : (A, a_0) \to (X, x_0)$ of f.*

Proof A is simply connected $\Rightarrow \pi_1(A, a_0) = 0 \Rightarrow p_*\pi_1(A, a_0) = \{0\} \subset p_*\pi_1 (X, x_0)$. Then $\exists$ a unique lifting $\tilde{f} : (A, a_0) \to (X, x_0)$ of f. ❑

Corollary 4.4.5 *Let B be a connected and locally path-connected space, and (X, p) and (Y, q) be covering spaces of B. Let $b_0 \in B$ and $x_0 \in X$, $y_0 \in Y$ be base points with $p(x_0) = b_0 = q(y_0)$. If $p_*\pi_1(X, x_0) = q_*\pi_1(Y, y_0)$, then there exists a unique continuous map $f : (Y, y_0) \to (X, x_0)$ such that $p \circ f = q$.*

Example 4.4.6 (S^n, p) is a covering space of $\mathbf{R}P^n$ of multiplicity 2. Since S^n is simply connected for $n \geq 2$, it follows that if $x_0 \in p^{-1}(b_0)$, $b_0 \in \mathbf{R}P^n$, then for any continuous map $f : (S^n, s_0) \to (\mathbf{R}P^n, b_0)$, there exists a unique lifting $\tilde{f} : (S^n, s_0) \to (S^n, x_0)$.

4.5 Covering Homomorphisms: Their Classifications and Galois Correspondence

This section defines covering homomorphisms between covering spaces of the base space B and classify the covering spaces with the help of conjugacy classes of the fundamental group $\pi_1(B)$. This classification establishes an exact correspondence between the various connected covering spaces of a given space B and subgroups of its fundamental group $\pi_1(B)$, like Galois theory, with its correspondence between field extensions and subgroups of Galois groups. There is a natural question: given a space B, how many distinct covering spaces of B, we can find? Before answering this question, we explain what is meant by distinct covering spaces of B.

4.5.1 Covering Homomorphisms and Deck Transformations

This subsection introduces the concepts of covering homomorphisms and deck transformations.

Definition 4.5.1 Let (X, p) and (Y, q) be covering spaces of the same space B. A covering homomorphism h from (X, p) to (Y, q) is a continuous map $h : X \to Y$ such that the diagram in Fig. 4.15 is commutative. If in addition, h is a homeomorphism, then h is called an isomorphism. If there is an isomorphism from (X, p) to (Y, q), then they are called isomorphic or equivalent covering spaces, otherwise, they are said to be distinct covering spaces. An isomorphism of a covering space onto itself is called an automorphism or a deck transformation.

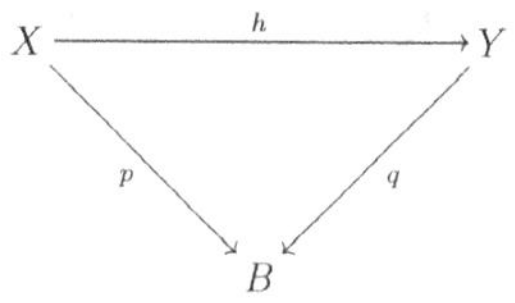

Fig. 4.15 Covering homomorphism

Remark 4.5.2 A homomorphism of covering spaces is a covering projection i.e.,if $h : X \to Y$ is a homomorphism of covering spaces, then (X, h) is a covering space of Y.

Proposition 4.5.3 *Covering spaces of a space B and their homomorphisms form a category.*

Proof We take covering spaces of B as the class of objects and their homomorphisms as the class of morphisms. Let (X, p) be a covering space of B. Then $1_X : X \to X$ is a covering homomorphism. If (X, p), (Y, q) and (Z, r) are covering spaces of B and $h : X \to Y, g : Y \to Z$ are covering homomorphisms, then $g \circ h : X \to Z$ is also a covering homomorphism from (X, p) to (Z, r). ❑

Isomorphisms in this category are just the isomorphisms of covering spaces as defined above.

Let $\mathcal{A}ut(X/B)$ be the set of all automorphisms of covering space (X, p) of B.

Proposition 4.5.4 *$(\mathcal{A}ut(X/B), \circ)$ is a group under usual composition of maps.*

Proof The identity map $1_X : X \to X$ is itself an automorphism and the inverse of an automorphism is again an automorphism. Consequently, $\mathcal{A}ut(X/B)$ is a group under usual composition of maps. ❑

Definition 4.5.5 $\mathcal{A}ut(X/B)$ is called the automorphism group of the covering space (X, p) of B. These automorphisms are also known as the covering transformations or deck transformations of the covering space (X, p) of B.

Let $p : X \to B, q : Y \to B$ be covering projections. Then (X, p) and (Y, q) are covering spaces of B. Suppose $g, h : X \to Y$ are two covering homomorphisms. We now consider each of g and h as liftings of the map $p : X \to B$ with respect to the covering projection $q : Y \to B$ (Fig. 4.16).

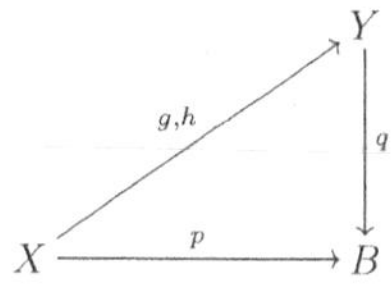

Fig. 4.16 Uniqueness of lifting

Consequently, if X is connected and g and h both agree at a single point of X, then $g = h$. This proves the following proposition.

Proposition 4.5.6 *Let $g, h : X \to Y$ be two covering homomorphisms from the covering space (X, p) to the covering space (Y, q) of B. If X is connected and $g(x_0) = h(x_0)$ for some $x_0 \in X$, then $g = h$.*

4.5.2 *Classification of Covering Spaces by Using Group Theory*

This subsection characterizes and classifies covering spaces of a space B with the help of conjugacy classes of the group $\pi_1(B)$. The following two results of algebra are used in this subsection.

(i) If H and K are subgroups of a group G, then they are conjugate subgroups iff $H = g^{-1}Kg$ for some $g \in G$.
(ii) If H and K are subgroups of a group G, then the G-sets G/H and G/K are G-isomorphic iff H and K are conjugate subgroups in G.

Theorem 4.5.7 *Let (X, p) be a covering space of B, where X and B are path-connected. If $b_0 \in B$, then the groups $p_*\pi_1(X, y)$, as y runs over $Y = p^{-1}(b_0)$, form a conjugacy class of subgroups of $\pi_1(B, b_0)$.*

Proof To prove the theorem we have to prove:

(a) for any $y_0, y_1 \in Y$, the subgroups $p_*\pi_1(X, y_0)$ and $p_*\pi_1(X, y_1)$ are conjugate;
(b) any subgroup of $\pi_1(B, b_0)$ conjugate to $p_*\pi_1(X, y_0)$ is equal to $p_*\pi_1(X, y)$ for some $y \in Y$.

(a) Let $u : I \to X$ be a path from y_0 to y_1. Then the function $\beta_u : \pi_1(X, y_0) \to \pi_1(X, y_1)$ defined by $\beta_u([f]) = [\bar{u} * f * u]$, $\forall [f] \in \pi_1(X, y_0)$, is an isomorphism (by Theorem 3.1.18). In particular, $\beta_u\pi_1(X, y_0) = \pi_1(X, y_1) \Rightarrow (p_* \circ \beta_u)\pi_1(X, y_0) = p_*\pi_1(X, y_1)$. It follows from the definition of β_u that $(p_* \circ \beta_u)\pi_1(X, y_0) = [p \circ u]^{-1}p_*\pi_1(X, y_0)[p \circ u] \Rightarrow p_*\pi_1(X, y_1)$ and $p_*\pi_1(X, y_0)$ are conjugate subgroups of $\pi_1(B, b_0)$.
(b) Let H be a subgroup of $\pi_1(B, b_0)$ such that H is conjugate to $p_*\pi_1(X, y_0)$ for some $[g] \in \pi_1(B, b_0)$. Then $H = [g]^{-1}p_*\pi_1(X, y_0)[g]$. Let $\tilde{g}$ be the unique lifting of g in X starting at y_0. Then $\tilde{g}(1) = y(\text{say}) \in Y$. Now proceeding as in (a), we have

$$\begin{aligned} p_*\pi_1(X, y) &= [p \circ \tilde{g}]^{-1}p_*\pi_1(X, y_0)[p \circ \tilde{g}] \\ &= [g]^{-1}p_*\pi_1(X, y_0)[g] = H \\ \Rightarrow p_*\pi_1(X, y) &= H. \end{aligned}$$

We conclude that the set $\{p_*\pi_1(X, y) : y \in Y\}$ forms a complete conjugate class of subgroups of the group $\pi_1(B, b_0)$.

❑

Definition 4.5.8 The conjugacy class of subgroups $\{p_*\pi_1(X, y) : y \in Y = p^{-1}(b_0)\}$ described above is called the conjugate class determined by the covering space (X, p) of B.

We now characterize covering spaces of a base space B with the help of conjugacy classes of subgroups of $\pi_1(B)$.

Theorem 4.5.9 *Let B be path-connected and locally path-connected. Let (X, p) and (Y, q) be path-connected covering spaces of B; let $p(x_0) = q(y_0) = b_0$. Then the covering spaces (X, p) and (Y, q) are isomorphic if and only if $p_*\pi_1(X, x_0)$ and $q_*\pi_1(Y, y_0)$ are conjugate subgroups of $\pi_1(B, b_0)$ (i.e., iff they determine the same conjugacy class of subgroups of $\pi_1(B, b_0)$).*

Proof Suppose that the covering spaces (X, p) and (Y, q) are isomorphic. Then there exists a homeomorphism $h : Y \to X$ such that $p \circ h = q$ i.e., making the diagram in Fig. 4.17 commutative.

Let $h(y_0) = x_1$. Then h induces an isomorphism $h_* : \pi_1(Y, y_0) \to \pi_1(X, x_1) \Rightarrow h_*(\pi_1(Y, y_0)) = \pi_1(X, x_1) \Rightarrow (p_* \circ h_*)(\pi_1(Y, y_0)) = p_*(\pi_1(X, x_1))$. Hence $q_*(\pi_1(Y, y_0)) = p_*\pi_1(X, x_1)$. By Theorem 4.5.7, $p_*\pi_1(X, x_1)$ is a subgroup of $\pi_1(B, b_0)$ and conjugate to the subgroup $p_*\pi_1(X.x_0)$. Consequently, $p_*\pi_1(X, x_0)$ and $q_*\pi_1(Y, y_0)$ are conjugate subgroups of $\pi_1(B, b_0)$. For the converse, let the two subgroups of $\pi_1(B, b_0)$ be conjugate. By Theorem 4.5.7 we can choose a different base point y_0 in Y such that the two groups are equal. We now consider the diagram in Fig. 4.18 where q is a covering map. The space X is path-connected; it is also locally path-connected, being locally homeomorphic to B. Moreover, $p_*\pi_1(X, x_0) \subseteq q_*\pi_1(Y, y_0)$. In fact, these two groups are equal. By Theorem 4.5.7, we can lift the map p to $\tilde{p} : X \to Y$ such that $\tilde{p}(x_0) = y_0$. Then $q \circ \tilde{p} = p$.

Reversing the role of X and Y in this discussion, we see that $q : Y \to B$ can also be lifting to $\tilde{q} : Y \to X$ such that $\tilde{q}(y_0) = x_0$ as shown in Fig. 4.19.

We claim that $\tilde{p}$ and $\tilde{q}$ as shown in Fig. 4.20 are inverses of each other. Consider the diagram in Fig. 4.21.

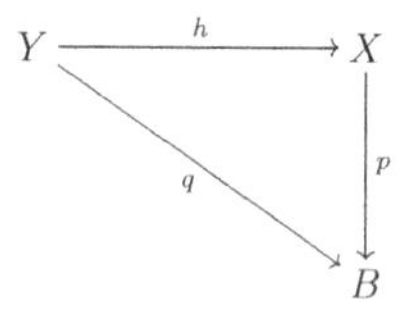

Fig. 4.17 Isomorphisms of covering spaces

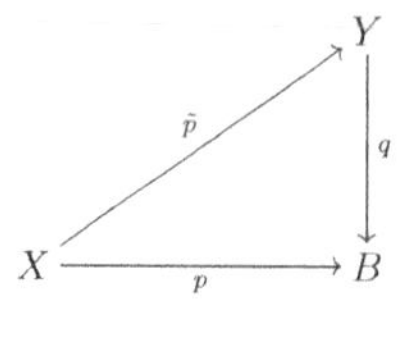

Fig. 4.18 Lifting of p

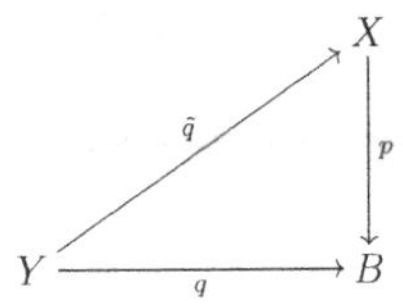

Fig. 4.19 Lifting of q

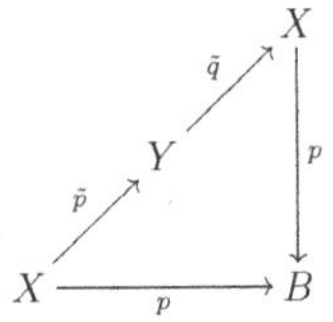

Fig. 4.20 Liftings of p and q

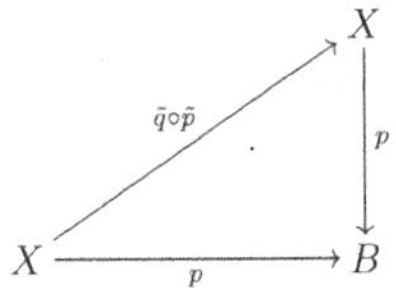

Fig. 4.21 Lifting of p involving $\tilde{p}$ and $\tilde{q}$

Now $\tilde{q} \circ \tilde{p} : X \to X$ is a lifting of the map $p : X \to B$ satisfying the condition $(\tilde{q} \circ \tilde{p})(x_0) = x_0$. The identity map $1_X : X \to X$ is another such lifting of p. Hence by uniqueness of lifting it follows that $\tilde{q} \circ \tilde{p} = 1_X$. Similarly, $\tilde{p} \circ \tilde{q} = 1_Y$. Consequently, $\tilde{p} : X \to Y$ is a homeomorphism and hence the covering spaces (X, p) and (Y, q) are isomorphic. ❑

Remark 4.5.10 For any covering space (X, p) of B, the subgroups $\{p_*(\pi_1(X, x)) : x \in p^{-1}(b)\}$ form a conjugacy class of subgroups of $\pi_1(B, b)$. The above Theorem 4.5.9 shows that a conjugacy class of a subgroup of $\pi_1(B, b)$ determines completely the covering spaces upto isomorphisms.

Recall that

Definition 4.5.11 A topological space X is said to be simply connected if it is path-connected and $\pi_1(X, x_0) = 0$ for some $x_0 \in X$ (hence for every $x_0 \in X$).

Example 4.5.12 Consider the covering spaces of S^1. $\pi_1(S^1, 1)$ is abelian $\Rightarrow$ two subgroups of $\pi_1(B, b_0)$ are conjugate if and only if they are equal. Consequently, two covering spaces of S^1 are isomorphic if and only if they correspond to the same subgroup of $\pi_1(S^1) \cong \mathbf{Z}$. The subgroups of $\mathbf{Z}$ are given by $< n >$, consisting of all multiples of n, for $n = 0, 1, 2, \ldots$. The covering space $(\mathbf{R}, p)$ of S^1 corresponds to the trivial subgroup of $\mathbf{Z}$, because $\mathbf{R}$ is simply connected. On the other hand, the covering space (S^1, p) of S^1 defined by $p(z) = z^n$ corresponds to the subgroup $< n >$ of $\mathbf{Z}$. We conclude that every path-connected covering space of S^1 is isomorphic to one of these coverings i.e., any covering space of S^1 must be isomorphic either to $(\mathbf{R}, p)$ or to one of the coverings (S^1, q_n), where $q_n(z) = z^n$, $z \in S^1$ wraps S^1 around itself n times.

Fig. 4.22 Lifting of f to X

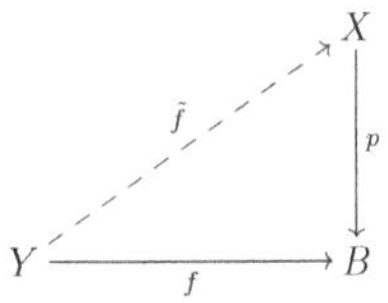

Example 4.5.13 Consider the double covering (S^1, p) over $\mathbf{R}P^2$. Since S^2 is simply connected, $\pi_1(S^2, s) = 0$ and hence the conjugacy class contains only the trivial subgroup.

Example 4.5.14 The plane $\mathbf{R}^2$ is simply connected. Consequently, the conjugacy class of $(\mathbf{R}^2, r)$ over the torus also contains only the trivial subgroup.

Example 4.5.15 Let X denote an infinite spiral and let $q : X \to S^1$ denote the projection map projecting each point on X to the point on the circle directly beneath it. Then (X, q) is a covering space of S^1. Since X is contractible, it has trivial fundamental group. Consequently, (X, q) determines the conjugacy class of $\pi_1(S^1)$ consisting of only the trivial subgroup. $(\mathbf{R}, p)$ also determines the conjugacy class of $\pi_1(S^1)$ consisting of only the trivial subgroup. We conclude that (X, q) and $(\mathbf{R}, p)$ are isomorphic covering spaces of S^1 by Theorem 4.5.9.

We recall the following proposition (see Sect. 4.1.2).

Proposition 4.5.16 *Let $p : (X, x_0) \to (B, b_0)$ be a covering space. Then the induced homomorphism $p_* : \pi_1(X, x_0) \to \pi_1(B, b_0)$ is a monomorphism and the subgroup $p_*(\pi_1(X, x_0))$ in $\pi_1(B, b_0)$ consists of homotopy classes of loops in B based at b_0 which lift to X starting at x_0 are loops.*

Remark 4.5.17 If $p : X \to B$ is a covering map, then p is also onto. But its induced homomorphism

$$p_* : \pi_1(X, x_0) \to \pi_1(B, b_0)$$

need not be an epimorphism. However, p_* is a monomorphism.

Proposition 4.5.18 *Let $p : (X, x_0) \to (B, b_0)$ be a covering space and $f : (Y, y_0) \to (B, b_0)$ be a map, where Y is path-connected and locally path-connected. Then a lift $\tilde{f} : (Y, y_0) \to (X, x_0)$ of f (as shown in Fig. 4.22) exists iff $f_*(\pi_1(Y, y_0)) \subset p_*(\pi_1(X, x_0))$.*

Proof Since $f_*([\alpha]) = (p * \circ \tilde{f}_*)[\alpha] \in p_*(\pi_1(X, x_0)),\ \forall\, [\alpha] \in \pi_1(Y, y_0)$, it follows that $f_*(\pi_1(Y, y_0)) \subset p_*(\pi_1(X, x_0))$.

Conversely, let $y \in Y$ and β be a path in Y from y_0 to y. Then the path $f \circ \beta$ in B starting at b_0 has a unique lifting $\widetilde{(f \circ \beta)}$ starting at x_0. Define

$$\tilde{f} : (Y, y_0) \to (X, x_0),\ y \mapsto \widetilde{(f \circ \beta)}(1).$$

Clearly, $\tilde{f}$ is well defined and continuous. ❑

Proposition 4.5.19 *Given a covering space $p : X \to B$ and a map $f : Y \to X$ with two liftings $\tilde{f}_1, \tilde{f}_2 : Y \to X$ that agree at some point of Y, if Y is connected, then $\tilde{f}_1 = \tilde{f}_2$, i.e., $\tilde{f}_1(y) = \tilde{f}_2(y), \forall\, y \in Y$.*

Proof Let $y \in Y$ and U be an open neighborhood of $f(y)$ in B such that $p^{-1}(U)$ is a disjoint union of open sets $\widetilde{U}_i$ each of which is mapped homeomorphically onto U by p. Suppose $\widetilde{U}_1$ and $\widetilde{U}_2$ are the $\widetilde{U}_i$'s containing $\tilde{f}_1(y)$ and $\tilde{f}_2(y)$, respectively. By continuity of $\tilde{f}_1$ and $\tilde{f}_2$ there is neighborhood N_y of y mapped into $\widetilde{U}_1$ by $\tilde{f}_1$ and $\widetilde{U}_2$ by $\tilde{f}_2$. If $\tilde{f}_1(y) \neq \tilde{f}_2(y)$, then $\widetilde{U}_1 \neq \widetilde{U}_2$. Hence $\widetilde{U}_1$ and $\widetilde{U}_2$ are disjoint open sets and $\tilde{f}_1 \neq \tilde{f}_2$ throughout the neighborhood N_y. Again if $\tilde{f}_1(y) = \tilde{f}_2(y)$, then $\widetilde{U}_1 = \widetilde{U}_2$ and hence $\tilde{f}_1 = \tilde{f}_2$ on N_y, because $p \circ \tilde{f}_1 = p \circ \tilde{f}_2$ and p is injective on $\widetilde{U}_1 = \widetilde{U}_2$. This shows that the set of points where $\tilde{f}_1$ and $\tilde{f}_2$ agree is a both open and closed set in Y. ❑

4.5.3 *Classification of Covering Spaces and Galois Correspondence*

This subsection considers the problem of classifying all different covering spaces of a fixed base space B. The main thrust of this classification is given in the Galois correspondence between connected covering spaces of B and subgroups of $\pi_1(B)$. The Galois correspondence ψ arises from the function that assigns to each covering space $p : (X, x_0) \to (B, b_0)$ the subgroup $p_*(\pi_1(X, x_0))$ of $\pi_1(B, b_0)$. By Proposition 4.5.16, this correspondence ψ is injective. To show that ψ is surjective, we have to show that corresponding to each subgroup G of $\pi_1(B, b_0)$, there is a covering space $p : (X, x_0) \to (B, b_0)$ such that $p_*\pi_1(X, x_0) = G$.

Definition 4.5.20 A topological space X is said to be semilocally simply connected if each point $x \in X$ has a neighborhood U_x such that the map induced by inclusion $i : U_x \hookrightarrow X$ is trivial, i.e., $i_* : \pi_1(U_x, x) \to \pi_1(X, x)$ is trivial (equivalently, every closed path in U_x at x is nullhomotopic in X).

Definition 4.5.21 A topological space X is said to be semilocally path-connected if for every point $x \in X$, there is an open neighborhood U_x of x such that every closed path in U_x at x is nullhomotopic in X.

Theorem 4.5.22 *Let B be a path-connected, locally path-connected and semilocally path space. Then for each subgroup G of $\pi_1(B, b_0)$ there is a covering space $p : X_G \to B$ such that $p_*(\pi_1(X_G, x_0)) = G$ for some suitable chosen base point $x_0 \in X_G$.*

Proof Let $b \in B$. Since B is semilocally path-connected, there is an open neighborhood W_b of b such that every closed path in W_b at b is nullhomotopic in B. Again since X is locally path-connected, $\exists$ an open connected neighborhood U_b of b such that $b \in U_b \subset W_b$. Clearly, every closed path in U_b at b is null homotopic in B and U_b is evenly covered by p.

Construction of X_G: Let $P(B, b_0)$ be the family of all paths f in B with $f(0) = b_0$, topologized by the compact open topology. Define a binary relation $f_1 \sim f_2$ mod G iff $f_1(1) = f_2(1)$ and $[f_1 * f_2^{-1}] \in G$. Then '$\sim$' is an equivalence relation. The equivalence class of $f \in P(B, b_0)$ is denoted by $[f]$. Let X_G denote the set of all such equivalence classes, topologized by the quotient topology. If c_0 is the constant path at b_0, define $x_0 = \langle c_0 \rangle_G \in X_G$ and $p : X_G \to B, [f]_G \mapsto f(1)$. Then $p(x_0) = b_0$. Since any two paths in the basic neighborhoods $U_{[f_1]_G}$ and $U_{[f_2]_G}$ are identified in X_G, the whole neighborhoods are identified. Consequently, the natural projection $p : X_G \to B$ is a covering space with $p(x_0) = b_0$. Then the image of $p_* : \pi_1(X_G, x_0) \to \pi_1(B, b_0)$ is precisely G. Because, for any loop β in B based at b_0, its lifting to X_G starting at $x_0 = \langle c_0 \rangle_G$ ends at $[\beta]_G$ and hence the image of this lifted path in X_G is a loop iff $[\beta]_G \sim [c_0]_G$ (equivalently, $[\beta] \in G$). ❑

Remark 4.5.23 Every group G can be realized as the fundamental group of the topological space X_G.

Corollary 4.5.24 *Let B be a connected, locally path-connected, semilocally simply connected space. Then every covering space $q : Y \to B$ is isomorphic (equivalent) to a covering spaces of the form $p : X_G \to B$.*

Proof Let $b_0 \in B$ be a base point of B and $y_0 \in Y$ lie in the fiber over b_0. If $G = q_*\pi_1(Y, y_0)$, then $p_*\pi_1(X_G, x_0) = G$. Hence Theorem 4.5.9 shows that the covering spaces $p : X_G \to B$ and $q : Y \to B$ are isomorphic. ❑

Corollary 4.5.25 *Let B be a connected, locally path-connected, semilocally simply connected space. If $p : X \to B$ is a covering space of B, then every open contractible set V in B is evenly covered by p.*

Proof Since if V is an open path-connected set in B for which every closed path in V is nullhomotopic in B, then V is evenly covered by p. In particular, if $b \in V$, then $p^{-1}(V) = \bigcup_{x \in p^{-1}(b)} (V, x)$ and contractible open sets are evenly covered in every covering space of the form $p : X_G \to B$. Then the corollary follows from Corollary 4.5.24. ❑

Corollary 4.5.26 *Let B be a connected and locally path-connected space. Then B has a universal covering space X (i.e., X is simply connected) iff X is semilocally simply connected.*

Proof Theorem 4.5.22 proves sufficiency of the condition. Definition 4.5.20 gives the necessity of the condition. ❑

Theorem 4.5.27 (Classification theorem) *Let B be a path-connected and locally path-connected space. Then the two coverings $p : X \to B$ and $q : Y \to B$ are isomorphic via a homeomorphism $f : X \to Y$ taking a base point $x_0 \in p^{-1}(b_0)$ to a base point $y_0 \in q^{-1}(b_0)$ iff $p_*(\pi_1(X, x_0)) = q_*(\pi_1(Y, y_0))$.*

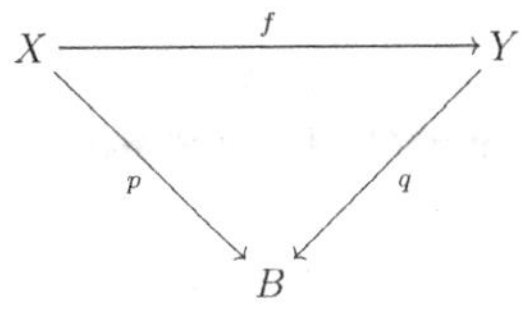

Fig. 4.23 Diagram for two isomorphic coverings of B

Proof Suppose there is a homeomorphism $f : (X, x_0) \to (Y, y_0)$ as shown in Fig. 4.23. Then the two relations $p = q \circ f$ and $q = p \circ f^{-1}$ show that $p_*(\pi_1(X, x_0)) = q_*(\pi_1(Y, y_0))$.

For the converse, let $p_*(\pi_1(X, x_0)) = q_*(\pi_1(Y, y_0))$. Then by the lifting criterion, we may lift p to $\tilde{p} : (X, x_0) \to (Y, y_0)$ with $q \circ \tilde{p} = p$. Similarly, we obtain $\tilde{q} : (Y, y_0) \to (X, x_0)$ with $p \circ \tilde{q} = q$. Then by unique lifting property, it follows that $\tilde{p} \circ \tilde{q} = 1_d$ and $\tilde{q} \circ \tilde{p} = 1_d$, since these composed lifts fix the base points. Consequently, p_* and q_* are inverse isomorphisms. ❑

Remark 4.5.28 We now present a generalization of the above classification theorem in the following form.

Theorem 4.5.29 (Classification theorem in general form) *Let B be a path-connected, locally path-connected and semilocally simply connected space. Then there exists a bijection between the set of base point preserving isomorphism classes of path-connected covering spaces $p : (X, x_0) \to (B, b_0)$ and the set of subgroups of $\pi_1(B, b_0)$, obtained by assigning the subgroups $p_*(\pi_1(X, x_0))$ to the covering spaces (X, x_0). If the base points are ignored, this correspondence gives a bijection between isomorphism classes of path-connected covering spaces $p : X \to B$ and conjugacy classes of subgroups of $\pi_1(B, b_0)$.*

Proof The first part follows from Theorem 4.5.27. For the proof of the second part, we claim that covering space $p : X \to B$, changing the base point x_0 within $\pi^{-1}(b_0)$ corresponds exactly to changing $p_*(\pi_1(X, x_0))$ to a conjugate subgroup of $\pi_1(B, b_0)$. Suppose x_1 is another base point $p^{-1}(b_0)$. Let $\tilde{\alpha}$ is a path from x_0 to x_1. Then $\tilde{\alpha}$ projects to a loop α in B, which represents some element $g \in \pi_1(B, b_0)$. Define G_i by $G_i = p_*(\pi_1(X, x_i))$ for $i = 0, 1$. Then we have an inclusion $g^{-1}G_0 g \subset G_1$, since for $\tilde{f}$ a loop at x_0, $\tilde{\gamma}^{-1} * f * \tilde{\gamma}^{-1}$ is a loop at x_1. Similarly, $gG_1g^{-1} \subset G_0$. Using conjugation the latter relation by g^{-1} we have $G_1 \subset g^{-1}G_0 g$ and hence $g^{-1}G_0 g = G_1$. Consequently, changing the base point from x_0 to x_1 changes G_0 to the conjugate subgroup $G_1 = g^{-1}G_0 g$. Conversely, to change G_0 to a conjugate subgroup $G_1 = g^{-1}G_0 g$, choose a loop β represents g, that lifts to a path $\tilde{\beta}$ starting at x_0 and let $x_1 = \tilde{\beta}(1)$. The earlier argument proves that $G_1 = g^{-1}G_0 g$. ❑

Theorem 4.5.30 (Galois correspondence) *Let B be path-connected and locally path-connected space. The Galois correspondence ψ arising from the function that assigns to each covering space $p : (X, x_0) \to (B, b_0)$ the subgroup $p_*(\pi_1(X, x_0))$ of $\pi_1(B, b_0)$ is a bijection.*

Proof ψ **is injective**: it follows from Proposition 4.5.16.

ψ **is surjective**: it follows from classification Theorem 4.5.27, since to each subgroup G of $\pi_1(B, b_0)$, there is a covering space $p : (X, x_0) \to (B, b_0)$ such that $p_*\pi_1(X, x_0) = G$.

Hence this correspondence ψ is a bijection. ❑

Definition 4.5.31 ψ defined in Theorem 4.5.30 is called a Galois correspondence.

4.6 Universal Covering Spaces and Computing $\pi_1(\mathbf{R}P^n)$

This section introduces the concept of a special class of covering spaces, called universal covering spaces and studies them with the help of fundamental groups of their base spaces and computes $\pi_1(\mathbf{R}P^n)$.

4.6.1 Universal Covering Spaces

This subsection opens with the concept of universal covering spaces. For a topological space B, $(B, 1_B)$ is a covering space over B. This covering space does not create in general much interest because it corresponds to the conjugacy class of the entire fundamental group $\pi_1(B, b)$. On the other hand, the covering space corresponding to the conjugacy class of the trivial subgroup $\{0\}$ of $\pi_1(B, b)$ is interesting. This covering space, if it exists for some B, is called the 'universal covering space'.

We now examine the relation between a base space B and its universal covering space.

Definition 4.6.1 Let B be a topological space. A covering space (X, p) of B for which X is simply connected (i.e., X is path-connected and $\pi_1(X, x_0) = 0$ for every $x_0 \in X$) is called the universal covering space of B.

Remark 4.6.2 We now explain the name of the term "universal covering space".

Theorem 4.6.3 **(i)** *Any two universal covering spaces of the same base space B are isomorphic.*

(ii) *If (X, p) is the universal covering space of B and (Y, q) is a covering space of B, then there is a continuous map*

$$\tilde{p} : X \to Y$$

such that $(X, \tilde{p})$ is a covering space of Y.

Proof **(i)** Any universal covering space of B determines the conjugacy class of the trivial subgroup $\Rightarrow$ any two universal covering spaces of B are isomorphic by Theorem 4.5.9.

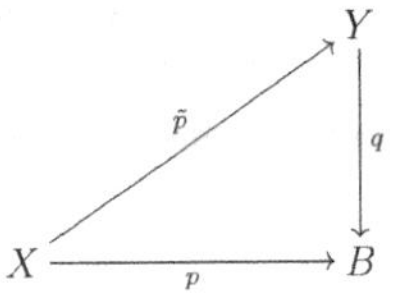

Fig. 4.24 Lifting of p to $\tilde{p}$

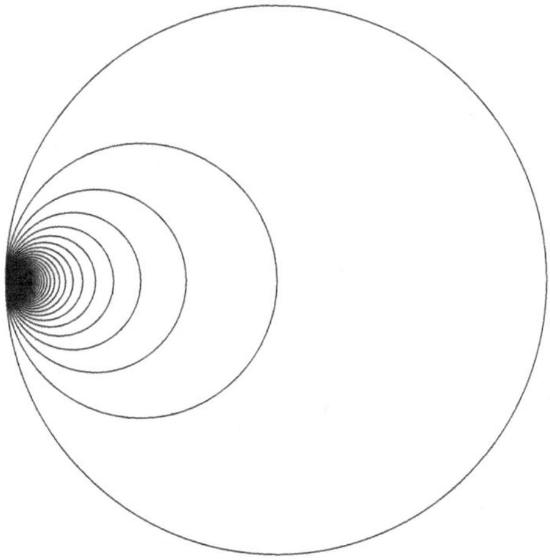

Fig. 4.25 Infinite earring

(ii) We consider the commutative diagram in Fig. 4.24 and choose base points $x_0 \in X$, $b_0 \in B$ and $y_0 \in Y$ such that $p(x_0) = q(y_0) = b_0$. Since $\pi_1(X, x_0) = 0$, $p_*\pi_1(X, x_0) \subset q_*\pi_1(Y, y_0)$. Hence Lifting Theorem 4.4.3 shows the existence of a continuous map $\tilde{p} : (X, x_0) \to (Y, y_0)$ such that $q \circ \tilde{p} = p$ and therefore $\tilde{p}$ is a covering projection. In other words, $(X, \tilde{p})$ is a covering space of Y. ❑

Example 4.6.4 **(i)** $(\mathbf{R}, p)$ is the universal covering space of S^1, where $p(t) = e^{2\pi i t}$, since the space of real numbers $\mathbf{R}$ is simply connected.

(ii) $(\mathbf{R}^2, r)$ (in Example 4.5.14) is a universal covering space over the torus, since $\mathbf{R}^2$ is simply connected.

(iii) (S^2, p) is the universal covering space of $\mathbf{R}P^2$.

(iv) (S^n, p_n) is a universal covering space of $\mathbf{R}P^n$, where $p_n : S^n \to S^n$ is the map identifying antipodal points of S^n for $n > 1$ (see Theorem 4.1.19).

Remark 4.6.5 A space may not have a universal covering. We now present an example of a space which has no universal covering.

Example 4.6.6 (Infinite earring or shrinking wedge of circles) Let C_n be the circle of radius $1/n$ in $\mathbf{R}^2$ with center at $(1/n, 0)$, for each $n \geq 1$. Let X be the subspace of $\mathbf{R}^2$ that is the union of these circles as shown in Fig. 4.25.

Then X is the union of a countably infinite collection of circles. The space X is called the 'infinite earring' or 'shrinking wedge of circles' in the plane $\mathbf{R}^2$. Let b_0 the origin. We claim that if U is a neighborhood of b_0 in X, then the homomorphism of

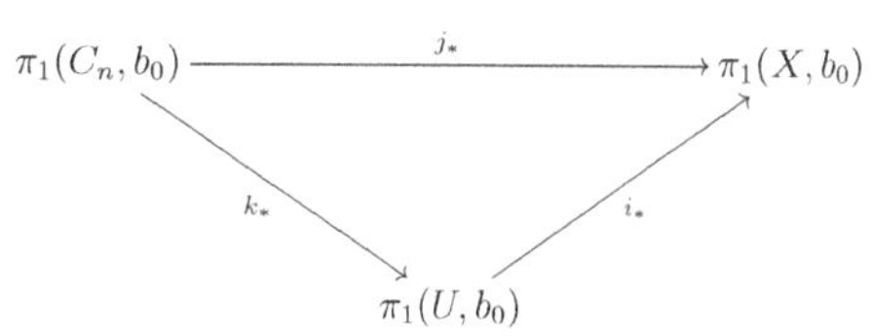

Fig. 4.26 Homomorphisms induced by inclusion maps

fundamental groups induced by the inclusion $i : U \hookrightarrow X$ is not trivial. To show this, let n be a given integer, there is a retraction $r : X \to C_n$ defined by letting r maps each circle C_i for $i \neq n$ to the point b_0. We can choose n sufficiently large such that inclusion $j : U \hookrightarrow X$ and inclusion $k : U \hookrightarrow U$ and thus for sufficiently large n, C_n lies in U. Then in the commutative diagram of groups and homomorphisms induced by inclusions k_* and j_* as shown in Fig. 4.26, j_* is injective.

Hence j_* can not be trivial. This asserts that X has no universal covering.

4.6.2 *Computing* $\pi_1(\mathbf{R}P^n)$

We now present an interesting result of the universal covering space and utilize this result to compute $\pi_1(\mathbf{R}P^n)$. For an alternative method see Corollary 4.10.4.

Theorem 4.6.7 *Let (X, p) be the universal covering space of B and $\mathcal{A}ut(X/B)$ be the group of all automorphisms of (X,B). Then the automorphism group $\mathcal{A}ut(X/B)$ is isomorphic to the fundamental group $\pi_1(B)$ of B. Moreover, if $|\pi_1(B)|$ is the order of the group $\pi_1(B)$, then $|\pi_1(B)|$=number of sheets of the universal covering space.*

Proof To prove the first part, let $x_0 \in X$ and $p(x_0) = b_0$. We define a map $\psi : \mathcal{A}ut(X/B) \to \pi_1(B, b_0)$ as follows:
$f \in \mathcal{A}ut(X/B) \Rightarrow f$ permutes the points of the fiber $p^{-1}(b_0)$. The point $f(x_0) \in p^{-1}(b_0)$, since $(p \circ f)(x_0) = b_0$. Let u be the path in X joining x_0 and $f(x_0)$. Then $p \circ u$ is a loop in B based at b_0. We define a mapping $\psi : \mathcal{A}ut(X/B) \to \pi_1(B)$ given by $\psi(f) = [p \circ u]$.

ψ **is well defined**: Let v be any other path joining x_0 and $f(x_0)$. Since X is simply connected, u is equivalent to v and hence $[p \circ u] = [p \circ u] \Rightarrow \psi$ is well defined.

ψ **is a homomorphism**: Let $f, g \in \mathcal{A}ut(X/B)$ and u, v be two paths in X joining x_0 to $f(x_0)$ and to $g(x_0)$, respectively. Then $\psi(f) = [p \circ u]$ and $\psi(g) = [p \circ v]$. Clearly, $f \circ v$ is a path joining $f(x_0)$ to $f(g(x_0))$ and hence $u * (f \circ v)$ is a path in X joining x_0 to $f(g(x_0))$. Again $\psi(fg) = [p \circ (u * (f \circ v))] = [(p \circ u) * (p \circ f \circ v)] = [p \circ u][p \circ f \circ v]$. Since $p \circ f = p$, we have $\psi(fg) = [p \circ u * p \circ v] = [p \circ u][p \circ v] = \psi(f)\psi(g)$.

ψ **is a monomorphism**: Let $\psi(f) = \psi(g)$. Then $[p \circ u] = [p \circ v]$, where u, v are paths in X starting at x_0 and ending at $f(x_0)$ and $g(x_0)$, respectively. Consequently,

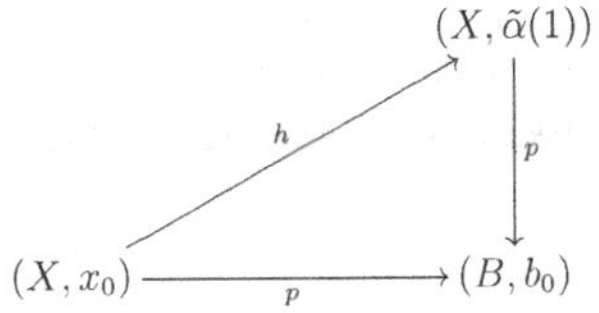

Fig. 4.27 Commutativity of the triangle for lifting of p to h

$p_*[u] = p_*[v] \Rightarrow u$ and v must have the same terminal point by Monodromy Theorem 4.9.3 i.e., $f(x_0) = g(x_0)$ and hence $f = g$ by Proposition 4.5.6, since X is connected.

ψ **is an epimorphism**: Let $\alpha \in \pi_1(B, b_0)$ and $\tilde{\alpha}$ be the unique lifting of the path α in X such that $\tilde{\alpha}(0) = x_0 \in X$. Consider the commutative diagram in Fig. 4.27 obtained by applying Lifting Theorem 4.4.3 to define a continuous lifting h of p such that $h(x_0) = \tilde{\alpha}(1)$.

Since X is a simply connected covering space of B, there exists a homeomorphism $h : X \to X$ such that $h(x_0) = \tilde{\alpha}(1)$. By the same argument, there is also a homeomorphism $k : X \to X$ such that $k(\tilde{\alpha}(1)) = x_0$. Since the homeomorphism $k \circ h : X \to X$ maps x_0 to itself and hence by Proposition 4.5.6 it follows that $h \circ k = 1_X$. This implies that $h \in \mathcal{A}ut(X/B)$ and by definition, $\psi(h) = [p \circ \tilde{\alpha}] = [\alpha]$. This shows that ψ is an isomorphism.

Proof of the last part: Since ψ is one-to-one, it establishes a one-to-one correspondence between $p^{-1}(b_0)$ and a subset of $\pi_1(B, b_0)$. While proving ψ is onto, we showed that every homotopy class $[\alpha]$ in $\pi_1(B, b_0)$ corresponds to a point $\tilde{\alpha}(1)$ in $p^{-1}(b_0)$. Hence it follows that $|p^{-1}(b_0)|$ = number of sheets of (X, p), is the order of $\pi_1(B, b_0)$. ❑

Remark 4.6.8 Last part of the Theorem 4.6.7 also follows from Theorem 4.10.1(iii), since $\pi_1(X) = 0$.

Theorem 4.6.9 $\pi_1(\mathbf{R}P^n) \simeq \mathbf{Z}_2$ *for* $n \geq 2$.

Proof Consider the universal covering space (S^n, q) of $\mathbf{R}P^n$ where q identifies the antipodal points of S^n. Then $|\pi_1(\mathbf{R}P^n)| = 2 \Rightarrow \pi_1(\mathbf{R}P^n) \cong \mathbf{Z}_2$. ❑

Theorem 4.6.10 *The automorphism group* $G = \mathcal{A}ut(X/B)$ *of a universal covering space* (X, p) *of* B *acts on* X *freely.*

Proof It is sufficient to prove that if $g \in G$ and $g(x) = x$ for some $x \in X$, then $g = 1_X$. The group *homeo*(X) of all homeomorphisms of a space X acts on the set X by the action defined by $g \cdot x = g(x)$, where $g \in$ Homeo (X) and $x \in X$. Since the group $\mathcal{A}(X/B)$ is a subgroup of Homeo (X), $\mathcal{A}ut(X/B)$ also acts on the space X by the above action. Thus $g, 1_X : X \to X$ are two covering homomorphisms of corresponding covering projections such that $g(x) = x = 1_X(x)$ for some $x \in X$. This shows that $g = 1_X$ by Proposition 4.5.6, since every path-connected space is connected. ❑

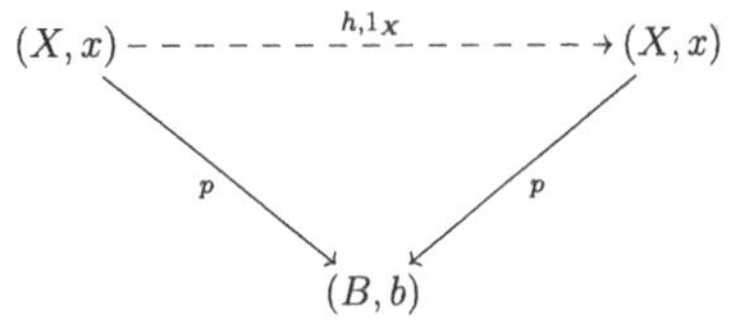

Fig. 4.28 Commutativity of the triangle for the covering space (X, p)

Corollary 4.6.11 *Let (X, p) be covering space of B.*

(i) *If $h \in \text{Cov}\,(X/B) = \mathcal{A}ut(X/B)$ and $h \neq 1_X$, then h has no fixed point.*
(ii) *If $h, g \in \mathcal{A}ut(X/B)$ and $\exists\, x \in X$ with $h(x) = g(x)$, then $h = g$.*

Proof **(i)** Let $\exists\, x \in X$ with $h(x) = x$; let $b = p(x)$. Consider the commutative diagram in Fig. 4.28.
Since both h and 1_X complete the diagram in Fig. 4.28, it follows that $h = 1_X$, a contradiction.
(ii) The map $h^{-1}g \in \mathcal{A}ut(X/B)$ has a fixed point, namely x and so by (i) $h^{-1}g = 1_X \Rightarrow h = g$. ❑

4.7 Fibrations and Cofibrations

This section gives a systematic approach to the lifting and extension problems through representation of maps as fibrations or cofibrations which are dual concepts of each other in some sense and form two important classes of maps in algebraic topology. They are central concepts in homotopy theory. Every continuous map is equivalently expressed up to homotopy as a fibration and also as a cofibration. The concept of fibration first appeared in 1937 implicitly in the work of K. Borsuk (1905–1982). This concept born in geometry and topology provides important strong mathematical tools to invade many other branches of mathematics. More precisely, this section introduces the concepts of fibrations and cofibrations and establishes a connection between a fibration and a covering projection.

The concept of homotopy lifting property (HLP) is very important in algebraic topology, specially in homotopy theory. It is the dual concept of the homotopy extension property (HEP). The concept of HLP leads to the concept of fibration. There is a dual theory to fibration leading to the concept of cofibration. This is a very nice duality principle in homotopy theory.

4.7.1 Homotopy Lifting Problems

This subsection discusses homotopy lifting problems of a map. It is an important problem of algebraic topology and dual to the extension problem. Let $p : X \to B$

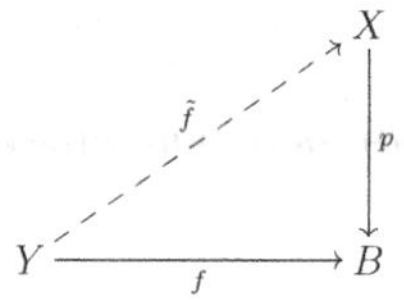

Fig. 4.29 Lifting of f

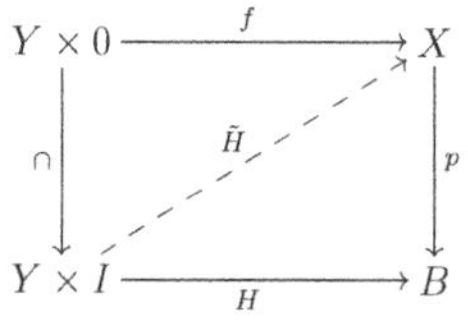

Fig. 4.30 Homotopy Lifting of H

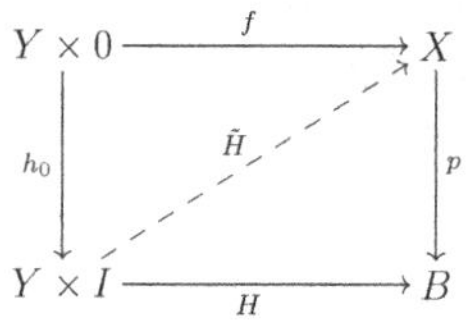

Fig. 4.31 Homotopy Lifting Problem

be a map and Y be a space. If $f : Y \to B$ is a map, then the lifting problem for f is to determine whether there is a continuous map $\tilde{f} : Y \to X$ such that the diagram in Fig. 4.29 is commutative, i.e., $f = p \circ \tilde{f}$. If there exists such a map $\tilde{f} : Y \to X$, we say that f can be lifted to X, and $\tilde{f}$ is called a lifting or a lifting of f. To show that the lifting problem is a problem in the homotopy category, we need the concept of homotopy lifting property(HLP) which is similar to the concept of HEP.

Definition 4.7.1 A continuous map $p : X \to B$ is said to have the HLP with respect to a space Y, if given maps $f : Y \to X$ and $H : Y \times I \to B$ such that $H(y, 0) = pf(y)$ for all $y \in Y$, there is a continuous map $\tilde{H} : Y \times I \to X$ such that $\tilde{H}(y, 0) = f(y)$ for all $y \in Y$ and $H = p \circ \tilde{H}$. If f is regarded as a map of $Y \times 0$ to X, the existence of $\tilde{H}$ is equivalent to the existence of a map represented by the dotted arrow that makes the diagram in Fig. 4.30 commutative.

Let $p : X \to B$ be a map and Y be a space. A homotopy lifting problem is sometimes symbolized by the commutative diagram in Fig. 4.31 where $h_0(y) = (y, 0)$ for all $y \in Y$ and the maps $f : Y \to X$, $H : Y \times I \to B$ are said to constitute the data for the problem in question. The map H is a homotopy of $p \circ f$ and a solution to the problem is a homotopy $\tilde{H} : Y \times I \to X$ of f such that $p \circ \tilde{H} = H$. Thus $\tilde{H}$ lifts the homotopy of H of $p \circ f$ to a homotopy of f.

Proposition 4.7.2 *Let $p : X \to B$ has the HLP with respect to a space Y. If $f \simeq g : Y \to B$, then f can be lifted to X iff g can be lifted to X.*

Proof Similar to the proof of Corollary 3.3.4 of Chap. 3. ❑

Remark 4.7.3 Let $p : X \to B$ and $f : Y \to B$ be two continuous maps. Then f can or cannot be lifted to X is a property of the homotopy class. This implies that the lifting problem for maps $f : Y \to B$ to X is a problem of homotopy category.

4.7.2 *Fibration: Introductory Concepts*

This subsection introduces the concept of a fibration first implicitly appeared in the work of K. Borsuk in 1937 but explicitly in the work of Whiteney during 1935–1940, first on sphere bundles. Fibrations form an important class of maps in algebraic topology. Covering map is a fibration. The homotopy lifting property leads to the concept of fibration (or Hurewicz fiber space) (Hurewicz 1955). More precisely, a continuous map $p : X \to B$ has the HLP with respect to a space Y if and only if every problem symbolized by the commutative diagram in Fig. 4.31 has a solution.

Definition 4.7.4 A pointed continuous map $p : X \to B$ is called a fibration (or fiber map or Hurewicz fiber space) if p has the HLP with respect to every space. X is called the total space and B is called the base space of the fibration. For $b \in B$, $p^{-1}(b) = F$ is called the fiber over b. A Serre fibration is map $X \to B$ satisfying HLP with respect to disk D^n, $\forall n$. It is sometimes called a weak fibration.

We use the notation "$F \hookrightarrow X \xrightarrow{p} B$ is a fibration" to mean that $p : X \to B$ is a fibration, F is the fiber space over some specific point of B, and $i : F \hookrightarrow X$ is the inclusion map.

Example 4.7.5 The projection

$$p : B \times F \to B, (b, f) \mapsto b$$

is a fibration.

Definition 4.7.6 A fibration $p : X \to B$ is called principal fibration if there is also a space C and a map $g : B \to C$ and a homotopy equivalence (over B, i.e., commuting through B) of X with mapping path space of g defined by

$$P_g = \{(b, \sigma) \in B \times C^I : \sigma_g(0) = *, \sigma_g(1) = g(b)\},$$

$$p_1 : P_g \to B, (b, \sigma) \mapsto b.$$

as shown in Fig. 4.32; C is called the classifying space and g is called the classifying map for the principal fibration.

Theorem 4.7.7 *Given a principal fibration $p : X \to B$, a lifting $\tilde{f}$ of f exists if and only if $g \circ f$ is homotopic to a constant map, where $g : B \to C$ is the classifying map.*

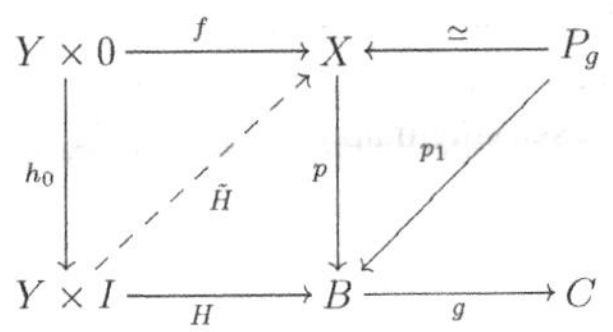

Fig. 4.32 Diagram of the classifying space and classifying map

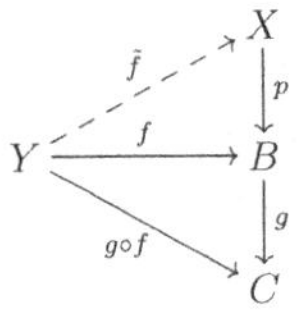

Fig. 4.33 Existence of lifting for a principal fibration

Proof Since p is a principal fibration, there is a homotopy equivalence (over B) $X \simeq P_g$ and hence there exist maps

$$h : X \to P_g \text{ and } k : P_g \to X$$

such that

$$k \circ h \simeq 1_X \text{ and } h \circ k \simeq 1_{P_g} \text{ and } p_1 \circ h = p, \, p \circ k = p_1,$$

where $p_1 : P_g \to B, (b, \sigma) \mapsto b$.

Given $\tilde{f} : Y \to X$, we obtain a homotopy $g \circ f \simeq c$, where $c : Y \to C$ is the constant map $y \mapsto * \in C$ as the composite

$$H : Y \times I \to P_g \to C, (y, t) \mapsto (h \circ \tilde{f})(y) = (f(y), \sigma_y) \mapsto \sigma_y(t) = H_t(y)$$

as shown in Fig. 4.33.

Conversely, let $G : g \circ f \simeq c$. Define

$$G_y : I \to C, t \mapsto G(y, t);$$

$$\tilde{f} : Y \to P_g \to X, y \mapsto (f(y), G_y) \mapsto k(f(y), G_y).$$

Hence

$$(p \circ \tilde{f})(y) = (pok)(f(y), G_y) = p_1(f(y), G_y) = f(y), \; \forall \, y \in Y \Rightarrow p \circ \tilde{f} = f$$

□

Theorem 4.7.8 *A lifting $\tilde{f}$ of a principal fibration $p : X \to B$ exists iff there exists a map $\tilde{g} : C_f \to C$ extending the classifying map g in the diagram as shown in Fig. 4.34.*

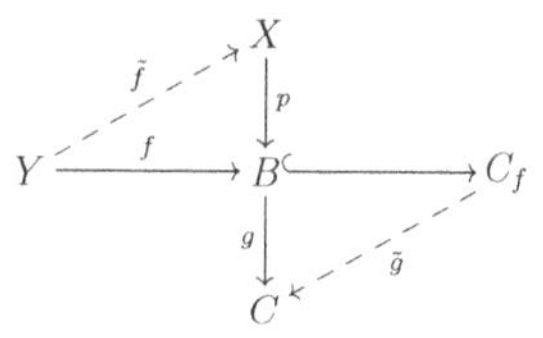

Fig. 4.34 Existence of a map extending the classifying map g

Proof In the category $\mathcal{T}op_*$ of pointed topological spaces the mapping cone C_f is obtained from the mapping cylinder M_f by identifying $Y \times \{0\} \cup \{*\} \times I$ with $*$ in B.

Suppose there is a homotopy

$$H : c \simeq g \circ f : Y \to C,$$

where $c : Y \to * \in C$ is the given constant map. Define

$$\tilde{g} : C_f \to C, \begin{cases} (y, t) \mapsto H(y, t), \\ b \mapsto g(b) \text{ for } b \notin f(Y). \end{cases}$$

Then

$$\tilde{g}(y, 0) = * \text{ and } \tilde{g}(y, 1) = gf(y), \ \forall\, y \in Y.$$

This shows that $\tilde{g}$ is the required extension of g.

Conversely, let $\tilde{g}$ be an extension of g. Then there is a homotopy

$$G : Y \times I \to C, (y, t) \mapsto \tilde{g}(y, t.)$$

Consequently,

$$G(y, 0) = \tilde{g}(y, 0) = *$$

and

$$G(y, 1) = \tilde{g}(y, 1) = (g \circ f)(y), \ \forall\, y \in Y.$$

Hence $g \circ f \simeq c$. ❑

Proposition 4.7.9 *Let $p : X \to B$ be a fibration and α be any path in B such that $\alpha(0) \in p(X)$. Then α can be lifted to a path $\tilde{\alpha}$ in X.*

Proof α can be regarded as a homotopy $\alpha : \{p_0\} \times I \to B$, where $\{p_0\}$ is a one-point space. Let x_0 be a point in X such that $p(x_0) = \alpha(0)$. Then there exists a map $f : \{p_0\} \to X$ such that $pf(p_0) = \alpha(p_0, 0)$. Hence it follows from the HLP of p that there exists a path $\tilde{\alpha}$ in X such that $\tilde{\alpha}(0) = x_0$ and $p \circ \tilde{\alpha} = \alpha$. This shows that $\tilde{\alpha}$ is a lifting of α. ❑

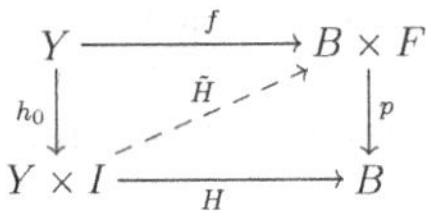

Fig. 4.35 Trivial fibration

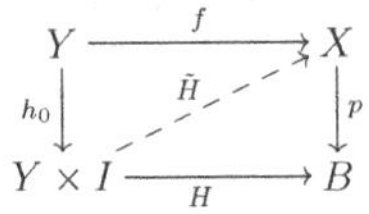

Fig. 4.36 Homotopy lifting property

Example 4.7.10 Let F be any space and $p : B \times F \to B$ be the projection to the first factor. Then p is a trivial fibration and for any $b \in B$, the fiber $p^{-1}(b)$ over b is homeomorphic to F. Because, if the diagram in Fig. 4.35 symbolizes homotopy lifting problem, then the map $\tilde{H} : Y \times I \to B \times F$ defined by $\tilde{H}(y, t) = (H(y, t), pf(y))$ is a solution of the lifting problem.

The projection $p : B \times I \to B$ is said to be a trivial fibration.

Example 4.7.11 For any space X, let $P(X) = M(I, X)$ be the space of all paths in X. Then the map $p : P(X) \to X \times X$, defined by $p(\alpha) = (\alpha(0), \alpha(1))$ is a fibration. Again

$$p_i : P(X) \to X, \alpha \mapsto \alpha(0), \alpha(1)$$

for $i = 1, 2$, respectively, are also fibrations.

Example 4.7.12 Let $p : X \to Y$ be a fibration and $q : Y \to B$ be also a fibration, then their composite $q \circ p : X \to B$ is also a fibration.

Theorem 4.7.13 *Every covering projection is a fibration.*

Proof Let $p : X \to B$ be a covering projection and the diagram in Fig. 4.36 symbolizes a homotopy lifting problem. Then for each $y \in Y$, there exists a unique path $\alpha_y : I \to X$ such that $\alpha_y(0) = f(y)$ and $p\alpha_y(t) = H(y, t)$. Then the map $\tilde{H} : Y \times I \to X, (y, t) \mapsto \alpha_y(t)$ is a continuous map and p is a fibration. ❑

Remark 4.7.14 For a covering projection the lifting is unique but it is not true for an arbitrary fibration.

4.7.3 Cofibration: Introductory Concepts

This subsection conveys the concept of cofibration and studies it in the category $\mathcal{T}op_*$ of pointed topological spaces and pointed maps. Cofibrations form an important class of maps in topology. Geometrically, the concept of cofibrations is less complicated than that of fibrations. There is a very nice duality principle in homotopy theory.

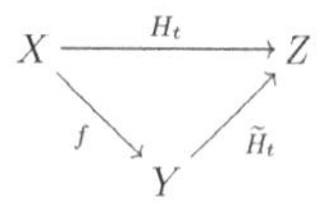

Fig. 4.37 Commutative triangle for cofibration

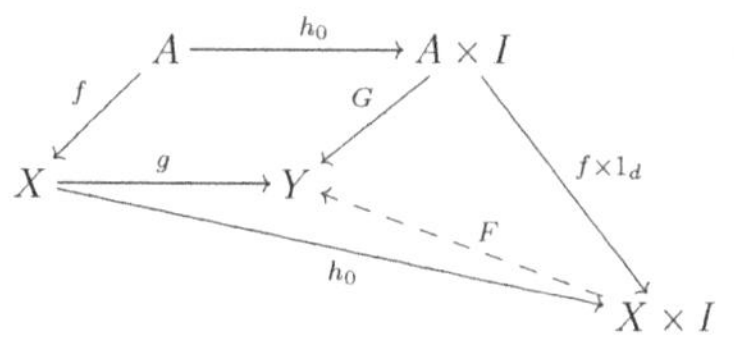

Fig. 4.38 Cofibration of f

For example, if in the definition of a fibration as a map satisfying homotopy lifting property we reverse the directions of all rows, we obtain the dual notion, called a cofibration. This is a continuous map $f : X \to Y$ satisfying the property: given $\tilde{g} : Y \to Z$ in $\mathcal{T}op_*$ and a homotopy $\widetilde{H}_t : Y \to Z$ such that there is a continuous map $H_t : X \to Z$ with the property $\widetilde{H}_t \circ f = H_t$, i.e., making the triangle in Fig. 4.37 commutative.

In this subsection, we work in $\mathcal{T}op_*$ unless specified otherwise.

Definition 4.7.15 A continuous map $f : A \to X$ is said to be a cofibration if for every topological space Y and given a continuous map $g : X \to Y$ and a homotopy $G : A \times I \to Y$ starting with $g \circ f$, there exists a homotopy $F : X \times I \to Y$ starting with g, such that $G = F \circ (f \times 1_d)$ as shown in Fig. 4.38.

Remark 4.7.16 It follows from Definition 4.7.15 that if A is a subspace of X, the inclusion map $i : A \hookrightarrow X$ is a cofibration if the pair (X, A) has the homotopy extension property (HEP) with respect to the given space Y.

Example 4.7.17 For any space A in $\mathcal{T}op_*$, let $CA = A \times I / A \times \{1\} \cup \{*\} \times I$ be the cone of A and $i : A \to CA, a \mapsto [a, 0]$ be the inclusion. Then i is a cofibration.

Proposition 4.7.18 *Given a map $f : (X, x_0) \to (Y, y_0)$ in $\mathcal{T}op_*$, the inclusion $i : Y \hookrightarrow Y \bigcup_f CX$ is a cofibration.*

Proof Let $r : I \times I \to I \times \{0\} \cup \dot{I} \times I$ be a retraction. Then given maps $f : Y \bigcup_g CX \to Z$ in $\mathcal{T}op_*$ and $G : Y \times I \to Z$ with $G(y, 0) = f(y)$, $\forall\, y \in Y$, define $H : CX \times \{0\} \cup X \times I \to Z$ by the rule $H|_{CX \times \{0\}} = f|_{CX}$ and $H|_{X \times I} = G \circ (g \times 1_d)$. Again define $F : (Y \bigcup_g CX) \times I$ by the rule $F|_{Y \times I} = G$ and $F([s, x], t) = H([p_1 \circ r(s, t), x], p_2 \circ r(s, t))$, $\forall\, [s, t] \in CX, t \in I$, where p_1, p_2 are the restrictions to $I \times \{0\} \cup \dot{I} \times I$ of the projections $p_1, p_2 : I \times I \to X$. Then F is well defined and is a continuous map such that $F(y, 0) = f(y)$. ❑

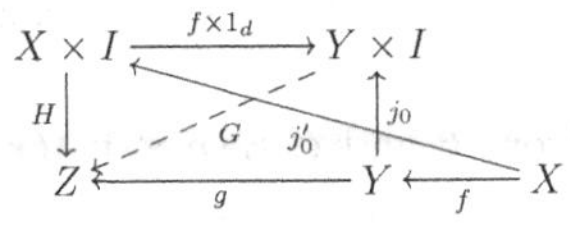

Fig. 4.39 Diagram for cofibration

Proposition 4.7.19 *If $f : X \to Y$ is a cofibration, then f is injective, and in fact it is a homeomorphism onto its image.*

Proof Consider the mapping cylinder M_f of f, the quotient space of $X \times I \cup Y$ in which $(x, 1)$ is identified with $f(x)$. Let $H_t : X \to M_f$ be the homotopy, mapping $x \in X$ to the image $(x, 1-t) \in X \times I$ in M_f, and let $\widetilde{H}_t : Y \hookrightarrow M_f$ be the inclusion. Then the cofibration property of f shows that $\widetilde{H}_t : Y \to M_f$ is such that $\widetilde{H}_t \circ f = H_t$. Restriction to a fixed $t > 0$, shows that f is injective, since H_t is so. Moreover, as H_t is a homeomorphism onto its image $X \times \{1-t\}$, the relation $\widetilde{H}_t \circ f = H_t$ holds. ❑

There is an equivalent definition of cofibration in $\mathcal{T}op_*$.

Definition 4.7.20 A continuous map $f : X \to Y$ in $\mathcal{T}op_*$ is said to be a cofibration if given a topological space Z, a continuous map $g : Y \to Z$ and a homotopy $H : X \times I \to Z$ starting from $g \circ f$, there exists a homotopy $G : Y \times I \to Z$, starting from g such that $H = G \circ (f \times 1_d)$ i.e., making the three triangles as shown in Fig. 4.39 commutative, where $j_0(y) = (y, 0)$, $\forall\, y \in Y$ and $j_0'(x) = (x, 0)$, $\forall \in X$.

Remark 4.7.21 Let A be a subspace of a topological space X. Then the inclusion map $i : A \hookrightarrow X$ is a cofibration if the pair (X, A) has the absolute homotopy extension property (see Chap. 2). The converse is not true in general. Because the definition of a cofibration refers to $\mathcal{T}op_*$ but the absolute homotopy extension property refers to maps and homotopies that are not necessarily based.

Theorem 4.7.22 *Every continuous map $f : X \to Y$ in $\mathcal{T}op_*$ is the composite of a cofibration and a homotopy equivalence.*

Proof Let M_f be the mapping cylinder in $\mathcal{T}op_*$ obtained from Y and $(X \times I)/x_0 \times I$ by identifying, for each $x \in X$, the points $(x, 1)$ and $f(x)$. Suppose $g : X \to M_f, x \mapsto [(x, 0)]$ is the inclusion map. Let $h : M_f \to Y$ be the map induced by the identity map 1_Y of Y and the map from $X \times I$ to Y that sends each $[(x, t)]$ to $f(x)$. Then $f = h \circ g$. We claim that h is a homotopy equivalence and g is a cofibration. We first show that g is a cofibration. Given a map $k : M_f \to Z$ in $\mathcal{T}op_*$, and a homotopy $H : X \times I \to Z$ starting from $k \circ g$, define maps

$$G_Y : Y \times I \to Z, (y, s) \mapsto k(y), \ \forall s \in I$$

and $G_X : (X \times I) \times I \to Z, (x, t, s) \mapsto \begin{cases} k(x, (2t - s)/(2 - s)), & 0 \le s \le 2t \\ H(x, s - 2t), & 2t \le s \le 1. \end{cases}$

Clearly, G_X is continuous and $G_X(x, 1, s) = k(x, 1) = (k \circ f)(x) = G_Y(f(x), s)$. Hence G_X and G_Y give together a homotopy $G : M_f \times I \to Z$ such that G starts from k and $G \circ (g \times 1_d)(x, s) = G(x, 0, s) = H(x, s)$. Hence $G \circ (g \times 1_d) = H$ shows that g is a cofibration.

Finally we show that h is a homotopy equivalence. Define $j : Y \to M_f$ to be (the restriction) of the identification map onto M_f. Then $h \circ j = 1_Y$ and

$$j \circ h : M_f \to M_f, y \mapsto y \text{ and } (x, t) \mapsto f(x).$$

Define a homotopy

$$H : M_f \times I \to M_f, (y, s) \mapsto y \text{ and } (x, t, s) \mapsto (x, t + s(1 - t)).$$

Clearly, H is continuous and $H : 1_d \simeq j \circ h$. This shows that h is a homotopy equivalence. ❑

Remark 4.7.23 The dual of the Theorem 4.7.22 is true in the sense that every continuous map $f : X \to Y$ in $\mathcal{T}op_*$ is also the composite of a homotopy equivalence and a fiber map.

Theorem 4.7.24 *Let A be a closed subset of a topological space X. Then the inclusion $i : A \hookrightarrow X$ is a cofibration iff $X \times \{0\} \cup A \times I$ is a retract of $X \times I$.*

Proof If $i : A \hookrightarrow X$ is a cofibration, then the given continuous maps $f : X \to X \times \{0\} \cup A \times I, x \mapsto (x, 0)$ and $G : A \times I \to X \times \{0\} \cup A \times I, (a, t) \mapsto (a, t)$ determine a map $H : X \times I \to X \times \{0\} \cup A \times I$ which is a retraction. Conversely, suppose there is a retraction $r : X \times I \to X \times \{0\} \cup A \times I$. Then given a space Y, a map $f : X \to Y$, and a homotopy $G : A \times I \to Y$ with the property that $H(a, 0) = f(i(a)), \forall a \in A$, define a map

$$H : X \times I \to Y, (x, t) \mapsto \begin{cases} (f \circ p_X \circ r)(x, t), & \text{if } (x, t) \in r^{-1}(X \times \{0\}) \\ (G \circ r)(x, t), & \text{if } (x, t) \in r^{-1}(A \times I). \end{cases}$$

Since $X \times \{0\}$ and $A \times I$ are closed in $X \times I$, it follows that H is continuous. Consequently, i is a cofibration. ❑

4.8 Hurewicz Theorem for Fibration and Characterization of Fibrations

This section continues the study of fibrations, characterizes path liftings of fibrations with the help of their fibers and studies Hurewicz theorem. This theorem is due to W. Hurewicz (1904–1956). It gives a sufficient condition for a map $p : X \to B$ to be a fibration (Hurewicz 1955).

Theorem 4.8.1 (Hurewicz) *Let $p : X \to B$ be a continuous map. Suppose B is paracompact and there is an open covering $\{V_i\}$ of B such that, for each V_i, $p|p^{-1}(V_i) : p^{-1}(V_i) \to V_i$ is a fibration. Then p is a fibration.*

Proof The proof is long and complicated. (Dugundji 1966) is referred. ❑

An immediate important consequence of Hurewicz Theorem 4.8.1 gives a sufficient condition for a projection map $p : X \to B$ of a fiber bundle (Chap. 5) to be a fibration.

Corollary 4.8.2 *Let $p : X \to B$ be the projection of a fiber bundle, and suppose B is paracompact, then p is a fibration.*

Theorem 4.8.3 *Let $p : X \to B$ be a covering projection and let $f, g : Y \to X$ be liftings of the same map (i.e., $p \circ f = p \circ g$). If Y is connected and $f(y_0) = g(y_0)$ for some point y_0 of Y, then $f = g$.*

Proof Let $A = \{y \in Y : f(y) = g(y)\}$. Then $A \neq \emptyset$ and A is an open set in Y. To show this, let $y \in A$ and U be an open neighborhood of $pf(y)$ evenly covered by p and let $\tilde{U}$ be an open subset of X containing $f(y)$ such that p maps $\tilde{U}$ homeomorphically onto U. Then $f^{-1}(\tilde{U}) \cap g^{-1}(\tilde{U})$ is an open subset of Y containing y and contained in A. Again let $B = \{y \in Y : f(y) \neq g(y)\}$. If X is assumed to be Hausdorff, then B is open in Y. Otherwise, let $y \in B$ and U be an open neighborhood of $pf(y)$ evenly covered by p. Since $f(y) \neq g(y)$, there are disjoint open sets $\tilde{V}_1$ and $\tilde{V}_2$ of X such that $f(y) \in \tilde{V}_1$ and $g(y) \in \tilde{V}_2$ and p maps each of the sets $\tilde{V}_1$ and $\tilde{V}_2$ homeomorphically onto U. Then $f^{-1}(\tilde{V}_1) \cap g^{-1}(\tilde{V}_2)$ is an open subset of Y containing y and contained in B. Finally, $Y = A \cup B$ and A and B are disjoint open sets imply from the connectedness of Y that either $A = \emptyset$ or $A = Y$. By hypothesis $A \neq \emptyset$ and hence $A = Y$ shows that $f = g$. ❑

Definition 4.8.4 A continuous map $p : X \to B$ is said to have a unique path lifting if, given paths α and β in X such that $p \circ \alpha = p \circ \beta$ and $\alpha(0) = \beta(0)$, then $\alpha = \beta$.

Proposition 4.8.5 *If a continuous map $p : X \to B$ has unique path lifting property, then it has path lifting property for path-connected spaces.*

Proof Let $p : X \to B$ has unique path lifting property and Y be path connected space. If $f, g : Y \to X$ and maps are such that $p \circ f = p \circ g$ and $f(y_0) = g(y_0)$ for some $y_0 \in Y$, we claim that $f = g$. Let $y \in Y$ and α be a path in Y from y_0 to y. Then $f \circ \alpha$ and $g \circ \alpha$ are paths in X that are liftings of some path in B and have the same initial point. Since p has unique path lifting, $f \circ \alpha = g \circ \alpha$ and hence $f(y) = (f \circ \alpha)(1) = (g \circ \alpha)(1) = g(y)$ implies $f = g$, since $\alpha(1) = y$. ❑

We now characterize path liftings of fibrations with the help of their fibers.

Theorem 4.8.6 *Let $p : X \to B$ be a fibration. Then the fibration has unique path lifting iff every fiber has no nonconstant paths.*

Proof Let $p : X \to B$ be a fibration with unique path lifting. Let α be a path in the fiber $p^{-1}(b)$ and β be the constant path in $p^{-1}(b)$ such that $\beta(0) = \alpha(0)$. Then $p \circ \alpha = p \circ \beta \Rightarrow \alpha = \beta \Rightarrow \alpha$ is a constant path. Conversely, let $p : X \to B$ be a fibration such that every fiber has no nontrivial path. If α and β are paths in X such that $p \circ \alpha = p \circ \beta$ and $\alpha(0) = \beta(0)$, then for $t \in I$ define a path γ_t in X by

$$\gamma_t(t') = \begin{cases} \alpha((1-2t')t), \ 0 \le t' \le 1/2 \\ \beta((2t'-1)t), \ 1/2 \le t' \le 1. \end{cases}$$

Then for each $t \in I$, $\gamma_t : I \to X$ is a path in X from $\alpha(t)$ to $\beta(t)$ and $p \circ \gamma_t$ is a closed path in B, which is homotopic rel $\dot{I}$ to the constant path at $p\alpha(t)$. By HLP of p, there is a map $H : I \times I \to X$ such that $H(t', 0) = \gamma_t(t')$ and H maps $0 \times I \cup I \times 1 \cup 1 \times I$ to the fiber $p^{-1}(p\alpha(t))$. Every Since $p^{-1}(p\alpha(t))$ has no nonconstant paths, F maps $0 \times I$, $I \times 1$ and $1 \times I$ to a single point. Hence it follows that $F(0,0) = F(1,0)$. Consequently, $\gamma_t(0) = \gamma_t(1)$ and $\alpha(t) = \beta(t)$. ❑

Proposition 4.8.7 *Let X be pointed topological space with base point x_0 and $P(X)$ be the space of paths in X starting at x_0, then the map*

$$p : P(X) \to X, \alpha \mapsto \alpha(1)$$

is fibration with fiber $\Omega(X)$.

Proof Let Y be an arbitrary space. Given maps $f : Y \to P(X)$ and $G : Y \times I \to X$ with $G_0 = p \circ f : Y \to X$, define a function

$$H : Y \times I \times I \to X,$$

$$(y, t, s) \mapsto \begin{cases} (f(y))(s(t+1)), & 0 \le s \le \frac{1}{t+1} \\ G(y, s(t+1) - 1), & \frac{1}{t+1} \le s \le 1 \end{cases}$$

Then H is continuous and defines a map $F : Y \times I \to X^I$ such that

$$F(y, t)(0) = f(y)(0) = x_0, \ \forall\, y \in Y, t \in I.$$

Hence $F \in P(X)$ and $F(y, 0)(s) = f(y)s$, $\forall\, y \in Y$, $\forall\, s \in I$. Consequently, $F_0 = f$ and $p \circ F = G$. This implies that F is the required lifting of G. Moreover, $p^{-1}(x_0) = \{\alpha \in P(x) : \alpha(1) = x_0\} = \Omega(X)$. ❑

4.9 Homotopy Liftings and Monodromy Theorem

This section continues the study of covering spaces by presenting some interesting applications of the path lifting property and homotopy lifting property of covering projections and proves Monodromy Theorem which provides a necessary and sufficient condition for two liftings of a covering projection to be equivalent.

4.9.1 *Path Liftings and Homotopy Liftings*

This subsection discusses path lifting property of a covering projection by using the homotopy lifting property.

Theorem 4.9.1 *Let (X, p) be a covering space of B and $b_0 \in B$. If $x_0 \in p^{-1}(b_0)$, then for any path $f : I \to B$, with $f(0) = b_0$, there exists a unique path $\tilde{f} : I \to X$ such that $\tilde{f}(0) = x_0$ and $p \circ \tilde{f} = f$.*

Proof Let $A = \{a\}$ be a singleton space. We consider the map $f : A \to B$ defined by $f(a) = b_0$. The path f defines a homotopy $F : A \times I \to B$ on A given by $F(a, t) = f(t)$. Then by the Homotopy Lifting Property, $\exists$ a map $\tilde{F} : A \times I \to X$ such that $\tilde{F}(a, 0) = x_0$ and $p \circ \tilde{F} = F$. Consequently, $\tilde{f} : I \to X$ defined by $\tilde{f}(t) = \tilde{F}(a, t), t \in I$, is a path in X starting from x_0 and having the property:

$$(p \circ \tilde{f})(t) = (p \circ \tilde{F}(a, t)) = F(a, t) = f(t),\ \forall t \in I \text{ i.e., } p \circ \tilde{f} = f.$$

Clearly the path $\tilde{f} : I \to X$ is unique. ❑

Remark 4.9.2 If $p : X \to B$ is a covering map, then p is also onto. But its induced homomorphism

$$p_* : \pi_1(X, x_0) \to \pi_1(B, b_0)$$

need not be a epimorphism. However, p_* is a monomorphism.

4.9.2 *Monodromy Theorem*

This subsection gives a criterion for two path liftings in X to be equivalent through a result known as 'Monodromy theorem'.

Theorem 4.9.3 (The Monodromy Theorem) *Let (X, p) be a covering space of B and $\tilde{f}, \tilde{g}$ are paths in X with same initial point x_0. Then $\tilde{f}$ and $\tilde{g}$ are equivalent (i.e., $\tilde{f} \simeq \tilde{g}$ rel $\dot{I}$) if and only if $p \circ \tilde{f}$ and $p \circ \tilde{g}$ are equivalent paths in B.*

Proof Let $\tilde{f}, \tilde{g}$ be equivalent paths in X. Then $\exists$ a homotopy $\tilde{F} : \tilde{f} \simeq \tilde{g}$ rel $\dot{I} \Rightarrow p \circ \tilde{F} : I \times I \to B$ is a continuous map such that $p \circ \tilde{F} : p \circ \tilde{f} \simeq p \circ \tilde{g}$ rel $\dot{I}$ and hence $p \circ \tilde{f}$ and $p \circ \tilde{g}$ are equivalent paths in B. Conversely, let $p \circ \tilde{f}$ and $p \circ \tilde{g}$ be equivalent paths in B. Then $\exists$ a continuous map $G : I \times I \to B$ such that $G : p \circ \tilde{f} \simeq p \circ \tilde{g}$ rel $\dot{I}$. By Homotopy Lifting Property, $\exists$ a unique homotopy $\tilde{G} : I \times I \to X$ such that $\tilde{G}(0, 0) = x_0$ and $p \circ \tilde{G} = G$. Restricting $\tilde{G}$ on $(t, 0), t \in I$, we have a path $t \mapsto \tilde{G}(t, 0)$, starting from x_0 and lifting $p \circ \tilde{f}$. Then $t \mapsto \tilde{f}(t)$ is also a path in X starting from x_0 and lifting $p \circ \tilde{f}$. Hence the uniqueness property of the covering paths, $\tilde{G}(t, 0) = \tilde{f}(t)$, $\forall t \in I$. Similarly, $\tilde{G}(t, 1) = \tilde{g}(t)$. Again by restricting $\tilde{G}$ on $(0, s)$, $s \in I$, we have a path $s \mapsto \tilde{G}(0, s)$, which projects under p to the constant path at $b_0 = p(x_0)$. A constant path $s \mapsto x_0$ in X also projects under p to the constant path $s \mapsto x_0$ in X. Hence by uniqueness theorem $s \mapsto \tilde{G}(0, s)$ is a constant path based at x_0. Similarly, the path $s \mapsto \tilde{G}(s, t)$ is a constant path based at some point $x_1 \in p^{-1}(b_0)$. This shows that $\tilde{G}$ is a homotopy between $\tilde{f}$ and $\tilde{g}$ rel $\dot{I}$. Consequently $\tilde{f}$ and $\tilde{g}$ are equivalent paths in X. ❑

Corollary 4.9.4 *Let (X, p) be a covering space of B and $b_0 \in B$, $x_0 \in p^{-1}(b_0)$. Then the induced homomorphism $p_* : \pi_1(X, x_0) \to \pi_1(B, b_0)$ is a monomorphism.*

Proof Let $[\tilde{f}], [\tilde{g}] \in \pi_1(X, x_0)$ and $[\tilde{f}] \neq [\tilde{g}]$. Then $p_*([\tilde{f}]) = [p \circ \tilde{f}]$ and $p_*([\tilde{g}]) = [p \circ \tilde{g}]$. Now $p \circ \tilde{f} \simeq p \circ \tilde{g}$ rel $\dot{I} \Leftrightarrow \tilde{f} \simeq \tilde{g}$ rel $\dot{I}$. But $\tilde{f} \not\simeq \tilde{g}$ rel $\dot{I} \Leftrightarrow p \circ \tilde{f} \not\simeq p \circ \tilde{g}$ rel I, otherwise we arrive at a contradiction by Theorem 4.9.3. This shows that p_* is well defined and injective; hence p_* is a monomorphism. ❑

4.10 Applications and Computations

This section presents applications of covering spaces and computes fundamental groups of some interesting spaces. Finally it presents an application of Galois correspondence arising from the function that assigns to each covering space $p : (X, x_0) \to (B, b_0)$ the subgroup $p_*(\pi_1(X, x_0))$ of $\pi_1(B, b_0)$.

4.10.1 Actions of Fundamental Groups

This subsection considers action of the fundamental group of the base space of a covering space on a fiber. This action plays an important role in the study of the covering space.

Let (X, p) be a covering space of B and $b_0 \in B$. We now consider the action of the fundamental group $\pi_1(B, b_0)$ on the fiber $p^{-1}(b_0) = Y$.

Theorem 4.10.1 *Let (X, p) be a covering space of B and $b_0 \in B$. Let $Y = p^{-1}(b_0)$ be the fiber over b_0. Then*

(i) *$\pi_1(B, b_0)$ acts transitively on Y;*
(ii) *If $x_0 \in Y$, then the isotropy group $G_{x_0} = p_*\pi_1(X, x_0)$; and*
(iii) *$|Y| = [\pi_1(B, b_0) : p_*\pi_1(X, x_0)]$.*

Proof First we show that Y is a (right) $\pi_1(B, b_0)$-set. We define $\sigma : Y \times \pi_1(B, b_0) \to Y$ by the rule $\sigma(x, [f]) = x \cdot [f] = \tilde{f}(1)$, where $\tilde{f}$ is the unique lifting of $f : (I, 0) \to (B, b_0)$ such that $\tilde{f}(0) = x$. This definition does not depend on the choice of the representative of the class $[f]$ by the Monodromy Theorem 4.9.3. If f is a constant path at b_0, then $\tilde{f}$ is also a constant path at $x \in Y$. Hence $x \cdot [f] = \tilde{f}(1) = x$. Next suppose $[f], [g] \in \pi_1(B, b_0)$. Let $\tilde{f}$ be the lifting of f with $\tilde{f}(0) = x$ and $\tilde{g}$ be the lifting of g with $\tilde{g}(0) = \tilde{f}(1)$. Then $\tilde{f} * \tilde{g}$ is a lifting of $f * g$ that begins at x and ends at $\tilde{g}(1)$. Consequently, $x \cdot [f * g] = (x \cdot [f])[g]$. As a result σ is an action of $\pi_1(B, b_0)$ on Y.

(i) σ **is transitive**: Let $x_0 \in Y$ and x be any point in Y. Since X is path-connected, $\exists$ a path $\tilde{\lambda}$ in X from x_0 to x. Then $p \circ \tilde{\lambda}$ is a closed path in B at b_0 whose lifting with initial point x_0 is $\tilde{\lambda}$. Thus $[p \circ \tilde{\lambda}] \in \pi_1(B, b_))$ and $x_0 \cdot [p \circ \tilde{\lambda}] = \tilde{\lambda}(1) = x$. Hence it follows that $\pi_1(B, b_0)$ acts transitively on Y.

(ii) Let f be a closed path in B at b_0 and $\tilde{f}$ be the lifting of f with $\tilde{f}(0) = x_0$. Let $G = \pi_1(B, b_0)$. Then $G_{x_0} = \{[f] \in \pi_1(B, b_0) : x_0 \cdot [f] = x_0\}$. Hence $[f] \in G_{x_0} \Rightarrow x_0 \cdot [f] = x_0 \Rightarrow \tilde{f}(1) = x_0 = \tilde{f}(0) \Rightarrow \tilde{f} \in \pi_1(X, x_0)$ and $[f] = [p \circ \tilde{f}] \in p_*\pi_1(X, x_0) \Rightarrow \pi_1(B, b_0) = G \subseteq p_*\pi_1(X, x_0)$. For the reverse inclusion, assume $[f] = [p \circ \tilde{g}]$ for some $[\tilde{g}] \in \pi_1(X, x_0)$. Then $\tilde{f} = \tilde{g}$, since both are liftings of f and both have initial point $x_0 \Rightarrow \tilde{f}(1) = \tilde{g}(1) \Rightarrow x_0 \cdot [f] = \tilde{f}(1) = x_0 \Rightarrow [f] \in G_{x_0} \Rightarrow p_*\pi_1(X, x_0) \subseteq G_{x_0}$. Consequently, $G_{x_0} = p_*\pi_1(X, x_0)$.

(iii) Recall that if a group G acts on a set Y and $x_0 \in Y$, then $|\text{orbit of } x_0| = [G : G_{x_0}]$. In particular, if G acts transitively, $|Y| = [G : G_{x_0}]$. Hence in this case, $G = \pi_1(B, b_0)$ and $G_{x_0} = p_*\pi_1(X, x_0)$ by (ii). Consequently, $|Y| = [\pi_1(B, b_0) : p_*\pi_1(X, x_0)]$. ❑

Corollary 4.10.2 *Let (X,p) be the universal covering space of B, then* $|Y| = |\pi_1(B, b_0)|$.

Proof The corollary follows from Theorem 4.10.1(iii), since $\pi_1(X, x_0) = 0$. ❑

Corollary 4.10.3 *If* $n \geq 2$, *then* $\pi_1(\mathbf{R}P^n) \cong \mathbf{Z}_2$.

Proof Since (S^n, p) is a covering space of $\mathbf{R}P^n$ of multiplicity 2, it follows that $[\pi_1(\mathbf{R}P^n, x_0) : p_*\pi_1(S^n, y_0)] = 2$. Again, S^n is simply connected for $n \geq 2 \Rightarrow p_*\pi_1(S^n, y_0) = 0 \Rightarrow |\pi_1(\mathbf{R}P^n, x_0)| = 2 \Rightarrow \pi_1(\mathbf{R}P^n, x_0) \cong \mathbf{Z}_2$. ❑

Corollary 4.10.4 *Let* (X, p) *be a covering space of* $b_0 \in B$, $x_0 \in p^{-1}(b_0)$. *If* $p_* : \pi_1(X, x_0) \to \pi_1(B, b_0)$ *is onto, then the map* $p : X \to B$ *induces an isomorphism*

$$p_* : \pi_1(X, x_0) \to \pi_1(B, b_0).$$

Proof p_* is a monomorphism by Corollary 4.9.4 and hence the corollary follows from the given condition. ❑

4.10.2 Fundamental Groups of Orbit Spaces

This subsection computes the fundamental groups of some important spaces which are obtained as orbit spaces. For example, projective spaces, lens spaces, figure-eight and Klein's bottles are interesting spaces. We represent them as orbit spaces and compute their fundamental groups. A topological group G with identity e acting on a topological space X is said to satisfy the condition **(A)**: if for each $x \in X, \exists$ a neighborhood U_x such that, $\Phi_g(U_x) \cap U_x \neq \emptyset \Rightarrow g = e$, where

$$\Phi_g : X \to X, x \mapsto gx$$

is a homeomorphism. This special kind of group action of the group G of homeomorphisms of X is said to act on X properly discontinuously. For example, any action of a finite group on a Hausdorff space is properly discontinuous.

Example 4.10.5 The automorphism group $\mathcal{A}ut(X/B)$ of (X, p) of B satisfies the condition **(A)**.

Definition 4.10.6 A covering space (X, p) of B is said to be regular if $p_*\pi_1(B, b_0)$ is a normal subgroup of $\pi_1(B, b_0)$.

Example 4.10.7 Let B be a connected, locally path-connected space and G satisfies the condition **(A)** on X then (X, p) is a regular covering space of

$$X \bmod G, \text{ where } p : X \to X \bmod G, x \mapsto Gx$$

is the natural projection.

Theorem 4.10.8 *If an action of a topological group G on a topological space X satisfies the condition* **(A)**, *then*

(i) *if X is path-connected, then G is the group of deck transformations of the covering space*

$$p : X \to X \bmod G, \ x \mapsto Gx$$

(ii) *if X is path-connected and locally path-connected, then G is isomorphic to the quotient group $\pi_1(X \bmod G)/p_*\pi_1(X)$.*

(iii) *for any simply connected space X, the groups $\pi_1(X \bmod G)$ and G are isomorphic.*

Proof **(i)** Let X be path-connected. The deck transformation group contains G as a subgroup and equals this group, since if f is any deck transformation, then given any point $x \in X$, x and $f(x)$ are in the same orbit and hence there is some $g \in G$ such that $g(x) = f(x)$. Consequently, $f = g$, since deck transformation of a connected covering space are uniquely determined under this situation.

(ii) It follows from Ex. 28 of Sect. 4.11.

(iii) Let $y \in X \bmod G$. Since X is simply connected, $\pi_1(X, x_0) = \{e\}$, $\forall x_0 \in p^{-1}(y)$ and hence $p_*\pi_1(X, x_0) = \{e\}$. Consequently, Theorem 4.10.8(iii) follows from Theorem 4.10.8(ii). ❑

Remark 4.10.9 We first make geometrical constructions of some orbit spaces and then compute their fundamental groups.

4.10.3 Fundamental Group of the Real Projective Space $\mathbf{R}P^n$

This subsection computes the fundamental group of $\mathbf{R}P^n$ by using group action. We have computed $\pi_1(\mathbf{R}^n) \cong \mathbf{Z}_2$ for $n \geq 2$ in Theorem 4.6.9. Here we give an alternative approach.

Definition 4.10.10 Let $S^n = \{x \in \mathbf{R}^{n+1} : ||x|| = 1\}$ be the n-sphere and $\mathbf{R}P^n$ be the n-dimensional real projective n-space. The antipodal map $A : S^n \to S^n, x \mapsto -x$, generates an action of the two element group $G = \{+1, -1\}$ given by the relation $(+1)x = x$ and $(-1)x = -x$. Then its orbit space S^n mod G is $\mathbf{R}P^n$, the real projective n-space.

Theorem 4.10.11 $\pi_1(\mathbf{R}^n) \cong \mathbf{Z}_2$ *for* $n \geq 2$.

Proof As S^n is simply connected for $n \geq 2$, so from the covering space $p : S^n \to \mathbf{R}P^n$ it follows by the Theorem 4.10.8 that the fundamental group of orbit space is G. Thus $\pi_1(S^n \text{ mod } G) = G \Rightarrow \pi(\mathbf{R}^n) = G \cong \mathbf{Z}_2$ for $n \geq 2$. ❑

Remark 4.10.12 The above action is free in the sense that $gx = x \Rightarrow g = e$. Does there exist any other finite group G acting freely on S^n and defining covering space $S^n \to S^n$ mod G? The answer is Z_2 is the only non-trivial group that can act freely on S^n if n is even (see Chap. 14).

Remark 4.10.13 A generator for $\pi_1(\mathbf{R}P^n)$ is any loop obtained by projecting a path in S^n connecting two antipodal points.

4.10.4 The Fundamental Group of Klein's Bottle

This subsection computes the fundamental group of Klein's bottle. Let G be the group of transformations of the plane generated by a and b. Consider the action of G on $\mathbf{R}^2$ by $a(x, y) = (x + 1, y)$ and $b(x, y) = (1 - x, y + 1)$, $\forall\, (x, y) \in \mathbf{R}^2$. Then $a^{-1}(x, y) = (x - 1, y)$ and $b^{-1}(x, y) = (1 - x, y - 1)$. Hence $\mathbf{R}^2$ is simply connected and the action satisfies condition **(A)**, then by Theorem 4.10.8, $\pi_1(\mathbf{R}^2 \text{ mod } G) \simeq G$. Now

$$\begin{aligned} b^{-1}ab(x, y) &= b^{-1}a(1 - x, y + 1) \\ &= b^{-1}(2 - x, y + 1) = (1 - 2 + x, y) \\ &= (x - 1, y) = a^{-1}(x, y),\ \forall\, (x, y) \in \mathbf{R}^2 \Rightarrow b^{-1}ab = a^{-1}. \end{aligned}$$

Therefore $\mathbf{R}^2$ mod G is the Klein's bottle. This gives a representation of Klein's bottle as an orbit space whose fundamental group is generated by a and b.

4.10.5 The Fundamental Groups of Lens Spaces

This subsection computes the fundamental group of lens spaces defined by H. Tietze (1888–1971) in 1908, which are are 3-manifolds. Such spaces constitute an important class of objects in the study of algebraic topology.

Let $m > 1$ be an integer space and p be an integer relatively prime to m and $S^3 = \{(z_1, z_2) \in \mathbf{C}^2 : |z_1|^2 + |z_2|^2 = 1\} \subset \mathbf{C}^2$. Let $\rho = e^{\frac{2\pi i}{m}}$ be a primitive m-th root of unity.

Define a map

$$h : S^3 \to S^3, (z_1, z_2) \mapsto (\rho z_1, \rho^p z_2) = (e^{\frac{2\pi i}{m}} z_1, e^{\frac{2\pi i p}{m}} z_2).$$

Then h is a homeomorphism of S^3 onto itself of period m, i.e., $h^m = 1_d$. Thus h induces an action of $\mathbf{Z}_m$ on S^n by the rule $\mathbf{Z}_m \times S^3 \to S^3, k(z_1, z_2) = h^k(z_1, z_2)$, where k denotes the residue class of the integer k modulo m, i.e., the action is generated by the rotation $z \mapsto e^{\frac{2\pi i}{m}} z$ of the unit sphere $S^3 \subset \mathbf{C}^2 = \mathbf{R}^4$. This action has no fixed point, because the equation $z = e^{\frac{2\pi i r}{m}} z$, where r is an integer such that $0 < r < m$ has a solution $z = 0$ but $z = 0$ is not a point of S^3.

The orbit spaces S^3 mod $\mathbf{Z}_m$ is called a lens space and is denoted by $L(m, p)$. Then the lens space is the quotient space $S^3/\sim$ given by an equivalence relation $\sim$ on S^3, defined by $(z_1, z_2) \sim (z_1', z_2')$ if there exists an integer k such that $k(z_1, z_2) = (z_1', z_2')$, i.e., $(z_1', z_2') = h^k(z_1, z_2)$.

As S^3 is Hausdorff, $\mathbf{Z}_m$ is finite and $\mathbf{Z}_m$ acts on S^3 without fixed point. The above action of $\mathbf{Z}_m$ on S^3 satisfies condition (**A**). Hence $\mathbf{Z}_m \simeq \pi_1(S^3 \bmod \mathbf{Z}_m) = \pi_1(L(m, p))$. We now extend the method of construction to construct generalized lens spaces. Let $m > 1$ be an integer and $p_1, p_2, \ldots, p_{n-1}$ be integers relatively prime to m and $S^{2n-1} = \{(z_1, z_2, \ldots, z_n) \in \mathbf{C}^n : |z_1|^2 + |z_2|^2 + \cdots + |z_n|^2 = 1\} \subset \mathbf{C}^n$. Let $\rho = e^{\frac{2\pi i}{m}}$ be a primitive mth root of unity. Define

$$h : S^{2n-1} \to S^{2n-1}, (z_1, z_2, \ldots, z_n) \mapsto (\rho z_1, \rho^{p_1} z_2, \ldots, \rho^{p_{n-1}} z_n) = (e^{\frac{2\pi i}{m}} z_1, e^{\frac{2\pi i p_1}{m}} z_2, \ldots, e^{\frac{2\pi i p_{n-1}}{m}} z_n)$$

Then h is a homeomorphism of S^3 with period m, i.e., $h^m = 1_d$. Thus as before h induces an action of $\mathbf{Z}_m$ on S^{2n-1} without fixed point by the rule $\mathbf{Z}_m \times S^{2n-1} \to S^{2n-1}, k(z_1, z_2, \ldots, z_n) \mapsto h^k(z_1, z_2, \ldots, z_n), \forall h \in \mathbf{Z}_m$.

The orbit spaces S^{2n-1} mod Z_m is called a generalized lens space and is denoted by $L(m, p_1, \ldots, p_{n-1})$. As before $\pi_1(L(m, p_1, \ldots, p_{n-1})) \cong \mathbf{Z}_m$. As a particular case, for $\mathbf{Z}_2 = \{1_d, a\}$, S^2 mod $\mathbf{Z}_2 = L(2, 1)$, where $1_d : S^2 \to S^2$ is the identity map and $a : S^2 \to S^2$ is the antipodal map and hence $a^2 = 1_d$.

As the action of $\mathbf{Z}_2$ on S^2 yields S^2 mod $\mathbf{Z}_2 = \mathbf{R}P^2$, and its fundamental group $\pi_1(S^2 \bmod \mathbf{Z}_2) \cong \mathbf{Z}_2$.

4.10.6 *Computing Fundamental Group of Figure-Eight by Graph-theoretic Method*

This subsection computes the fundamental group of figure-eight by graph-theoretic method. We have shown in Theorem 4.2.2 that fundamental group of figure-eight is not abelian.

For an alternative proof let G be a free group on two letters a and b. Define a graph $X = \text{Graph}\,(G, a, b)$ as follows:

The vertices of X are the elements of G. Hence the vertices are the reduced words a and b. The edges of X are of the two types: (g, ga) and (g, gb), $g \in G$. Again (g, ga), (g, gb), (ga^{-1}, g) and are the only four edges corresponding to the vertex g. Now define a map $G \times X \to X$, given by $h \cdot g = hg$, for every $g, h \in G$. Then $h \cdot (g, ga) = (hg, hga)$ and $h \cdot (g, gb) = (hg, hgb)$, for edges (g, ga) and (g, gb).

Let 1_G be the identity element in G. Then $1_G \cdot g = g$ and $1_G \cdot (g, ga) = (g, ga)$ and $1_G \cdot (g, gb) = (g, gb)$. Again $(h_1h_2) \cdot g = h_1h_2 \cdot g$ and $h_1 \cdot (h_2 \cdot g) = h_1 \cdot (h_2 \cdot g) = h_1h_2 \cdot g$, $\forall\, h_1, h_2 \in G$ (Since h_2g be a vertex in X).

$$\begin{aligned}\text{Also } h_1 \cdot (h_2 \cdot (g, ga)) &= h_1 \cdot (h_2g, h_2ga) = h_1(h_2g, h_1(h_2ga)) = (h_1(h_2)) \cdot (g, ga)\\ &= h_1(h_2 \cdot (g, gb)) = h_1(h_2g, h_2gb)\\ &= h_1(h_2g, h_1(h_2gb)) = h_1(h_2) \cdot (g, gb).\end{aligned}$$

Clearly, G acts on X. The orbit space $X \bmod G$ is the Figure-Eight space whose two loops are the images of the edges (g, ga) and (g, gb). As X is simply connected, by using Theorem 4.10.8 it follows that $\pi_1(X \bmod G, *) \cong G$, which is the free group on two generators.

4.10.7 *Application of Galois Correspondence*

This subsection presents an interesting application of Galois correspondence.

Theorem 4.10.14 *Let (X, p) be a covering space of B. If B is connected, locally path-connected, and semilocally simply connected, then*

(a) *The components of X are in one-to-one correspondence with orbits of the action of $\pi_1(B, b_0)$ on the fiber $p^{-1}(b_0)$;*

(b) *Under the Galois correspondence between connected covering spaces of B and subgroups of $\pi_1(X, x_0)$, the subgroup corresponding to the component of X containing a given lift $\tilde{b}_0 = x_0$ of b_0 is the stabilizer group G_{x_0} of $x_0 \in X$, whose action on the fiber leaves x_0 fixed.*

Proof **(a)** Let $x_0, x_1 \in p^{-1}(b_0)$. If they are in different components of X, $\pi_1(X, x_0)$ cannot map one to the other, since there exists no path-connecting them. Claim that $\pi_1(X, x_0)$ acts transitively on each of the components of X to obtain a bijection. By hypothesis B is locally path-connected, hence X is locally path connected. Clearly, the notions of connected components and path-connected components are the same. If x_0 and x_1 lie in the same component, there exists a path $\alpha : I \to X$ such that $\alpha(0) = x_0, \alpha(1) = x_1$. Then $[p \circ \alpha]$ is an element of $\pi_1(B, b_0)$ whose action on $p^{-1}(b_0)$ maps x_0 to x_1. Hence this action is transitive. Then the set of elements of $p^{-1}(b_0)$ in a given component constitutes an orbit, and this produces a bijection.

(b) Choose a given lift x_0 of b_0 in some component X' of X. Under the Galois correspondence, the subgroup of $\pi_1(B, b_0)$ corresponding to X' is the image of $G = \pi_1(X', x_0)$ in the inclusion $p_* : G \to \pi_1(B, b_0)$. Any loop $\alpha \in p_*(G)$ lifts back to a loop in X' by the unique lifting property. Hence α sends x_0 to itself and is an element of the stabilizer group G_{x_0} of x_0.

Conversely, if $\beta \in \pi_1(B, b_0)$ is in the stabilizer group of x_0, then the lift $\tilde{\beta}$ of β is a loop from x_0 to itself and hence $\tilde{\beta} \in G$, which implies $\beta \in p_*(G)$. This shows that $p_*(G)$ is the stabilizer group of x_0. ❑

4.11 Exercises

1. Assume that $f : S^n \to \mathbf{R}^n$ is a continuous map such that $f(-x) = -f(x)$ for any $x \in S^n$. Show that there exists a point $x \in S^n$ such that $f(x) = 0$.
2. Assume that $f : S^n \to \mathbf{R}^n$ is a continuous map. Show that there exists a point $x \in S^n$ such that $f(x) = f(-x)$.
3. Prove that no subspace of $\mathbf{R}^n$ is homeomorphic to S^n.
4. Show that there is no continuous antipode-preserving map $f : S^2 \to S^1$. Use this result to prove Borsuk–Ulum theorem for dimension 2.
5. Let X and B be path-connected spaces and (X, p) be a covering space of B. Let $b_0 \in B$ and $Y = p^{-1}(b_0)$ be the fiber over b_0. Prove the following:
   - **(i)** If $x_0, x_1 \in Y$, then $p_*\pi_1(X, x_0)$ and $p_*\pi_1(X, x_1)$ are conjugate subgroups of $\pi_1(B, b_0)$;
   - **(ii)** If H is a subgroup of $\pi_1(X, x_0)$ which is conjugate to $p_*\pi_1(X, x_0)$ for some $x_0 \in Y$, then there exists a point $x_1 \in Y$ such that $H = p_*\pi_1(X, x_1)$. [Hint: Use Theorem 4.5.7.]
   - **(iii)** A covering space (X, p) of B is said to be regular if $p_*\pi_1(X, x_0)$ is a normal subgroup of $\pi_1(B, b_0)$ for every $b_0 \in B$. If (X, p) is regular covering space of B, show that $p_*\pi_1(X, x_0) = p_*\pi_1(X, x_1)$ for every pair of point x_0, x_1 in the same fiber.
   - **(iv)** If X is simply connected, prove that every covering space (X, p) of B is regular.
   - **(v)** If $\pi_1(B, b_0)$ is abelian, then every covering space of B is regular.
6. Let B be a connected and locally path-connected space and let $b_0 \in B$. Then show that a covering space (X, p) of B is regular if and only if the group $\mathcal{A}ut(X/B)$ acts transitively on the fiber over b_0.

7. Let B be locally path-connected and $b_0 \in B$. Show that two covering spaces (X, p) and (Y, q) of B are isomorphic if and only if the fibers $p^{-1}(b_0)$ and $q^{-1}(b_0)$ are isomorphic $G = \pi_1(B, b_0)$-sets.
8. Let a group G act transitively as a set Y, and let $x, y \in Y$. Prove that $G_x = G_y$ if and only if there exists $f \in \mathcal{A}ut(Y)$ with $f(x) = y$.
9. Show that the graph X described in Sect. 4.10.6 has no cycles.
10. Let (X, p) be a covering space of B, where X is locally path-connected. Let $b_0 \in B$. Given $x_0, x_1 \in Y = p^{-1}(b_0)$, show that there exists an $h \in \text{Cov}\,(X/B)$ with $h(x_0) = x_1$ if and only if there exists $f \in \mathcal{A}ut(Y)$ with $f(x_0) = x_1$.
11. Let (X, p) be a covering space of B, where B is locally path-connected. Let $b_0 \in B$ and let the fiber $p^{-1}(b_0) = y$ be viewed as a $G = \pi_1(B, b_0)$-set. Then show that $\psi : \text{Cov}\,(X/B) \to \mathcal{A}ut(Y)$ defined by $\psi(h) = h|Y$ is isomorphism.
12. Let G be a group acting transitively on a set Y and let $y_0 \in Y$. Let $N_G(G_0)$ denote the normalizer of the isotropy group G_0 of y_0. Show that $\mathcal{A}ut(Y) \cong N_G(G_0)/G_0$.
13. Let (X, p) be a covering space of B, where B is locally path-connected. Show that for $b_0 \in B$ and $x_0 \in p^{-1(b_0)}$, $\mathcal{A}ut(X/B) \cong N_G(p_*\pi(X, x_0))/p_*\pi_1(X, x_0)$. Hence show that $\pi_1(S^1, 1) \cong \mathbf{Z}$.
14. Let (X, p) be a regular covering space of B, where B is locally path-connected. For $b_0 \in B$ and $x_0 \in p^{-1}(b_0)$, show that $\mathcal{A}ut(X/B) \cong \pi_1(B, b_0)/p_*\pi_1(X, x_0)$ by the monodromy group of the regular covering space.
15. Let (X, p) be a universal covering space of B, where B is locally path connected. Show that for any $b_0 \in B$, $\mathcal{A}ut(X/B) \cong \pi_1(B, b_0)$.
16. If B is an H-space, prove that every covering space of B is regular. [Hint: $\pi_1(B, b_0)$ is abelian for $b_0 \in B \Rightarrow$ every covering space of B is regular.]
17. Let G be a path-connected topological group and H be a discrete normal subgroup of G. If $p : G \to G/H$ is the natural homomorphism, show that (G, p) is covering space of G/H.
18. Let (X, p) be a covering space of B and $b_0, b_1 \in B$. If F_0 and F_1 are the fibers over b_0 and b_1, respectively, show that $|F_0| = |F_1|$ and any two fibers of (X, p) a [Hint: Use Theorem 4.10.1(iii). Since each fiber is discrete and any two fibers have the same cardinal numbers, it follows that any two fibers are homeomorphic.]
19. Show that the map $p : S^1 \to S^1, z \mapsto z^2$ is a covering map. Generalize to the map $p : S^1 \to S^2, z \mapsto z^n$.
20. If $S^1 \to S^1$ is continuous and antipode preserving, show that f is not nullhomotopic.
21. Let B be a path-connected and locally path-connected space. Suppose (X, p) and (Y, q) are covering spaces of B. Let $b_0 \in B, x_0 \in X$ and $y_0 \in Y$ be base points with $p(x_0) = b_0 = q(y_0)$. If $q_*\pi_1(Y, y_0) \subset p_*\pi_1(X, x_0)$, show that

    (i) there exists a unique continuous map $f : (Y, y_0) \to (X, x_0)$ such that $p \circ f = q$;
    (ii) (Y, f) is a covering space of X and so X is a quotient space of Y.

22. Let X, B, Y be path-connected and locally path-connected spaces such that (X, p) is a covering space of B. If $x_0 \in X$, $y_0 \in Y$ and $b_0 \in B$ with $p(x_0) = b_0$, show that for every continuous map $f : (Y, y_0) \to (B, b_0)$ with $f_*\pi_1(Y, y_0) \subset$

$p_*\pi_1(B, b_0)$, there exists a continuous map $\tilde{f} : (Y, y_0) \to (X, x_0)$ such that $p \circ \tilde{f} = f$.

23. Let (X, p) be a covering space of B and $x_0 \in X$, $b_0 \in b$ such that $p(x_0) = b_0$. If X is simply connected, show that b_0 has a neighborhood U such that the inclusion map $i : U \hookrightarrow B$ induces the trivial homomorphism

$$i_* : \pi_1(U, b_0) \to \pi_1(B, b_0).$$

24. Let X be a normal space. Show that the inclusion $i : A \hookrightarrow X$ is a cofibration iff the inclusion $j : A \hookrightarrow U$ is a cofibration for some open neighborhood U of A in X.

25. Let $p : X \to B$ be a fibration and $f : A \to B$ be a continuous map. Show that there exists a bijection between the homotopy sets $C = [g : A \to X : p \circ g = f]$ and $D = [\tilde{g} : A \to X : p \circ \tilde{g} \simeq f]$.

26. Is the map $p : (0, 3) \to S^1,\ x \mapsto e^{2\pi i x}$ a covering map? Justify your answer.

27. Find nontrivial coverings of Möbius strip by itself.

28. Let B be path-connected, locally path-connected and $p : (X, x_0) \to (B, b_0)$ be a covering space. If H is the subgroup $p_*(\pi_1(X, x_0))$ of $\pi_1(B, b_0)$, show that

- **(i)** The automorphism group $\mathcal{A}ut(X/B)$ is isomorphic to the quotient group $N(H)/H$, where $N(H) = \{g \in \pi_1(B, b_0) : gHg^{-1} = H\}$ is the normalizer of H in $\pi_1(B, b_0)$.
- **(ii)** The group $\mathcal{A}ut(X/B)$ is isomorphic to the group $\pi_1(B, b_0)/H$ if X is a regular covering.
- **(iii)** If $p : (X, x_0) \to (B, b_0)$ is universal covering, then $\mathcal{A}ut(X/B) \cong \pi_1(B, b_0)$.

29. Let (X, p) be a universal covering space of a connected topological space B. If $b_0 \in B$ and $x_0 \in X$ are base points such that $x_0 \in p^{-1}(b_0)$, show that the induced homomorphism $p_* : \pi_1(X, x_0) \to \pi_n(B, b_0)$ is an isomorphism for $n \geq 2$. Hence show that $\pi_n(\mathbf{R}P^m) \cong \pi_n(S^m)$ for $n \geq 2$.

30. Let B be a path-connected space and X be a connected covering space of B. Let $p : X \to B$ be a covering projection. Let $b_0 \in B$ and $x_0 \in p^{-1}(b_0)$. Show that for every $n \geq 2$, $p_* : \pi_1(X, x_0) \to \pi_1(B, b_0)$ is an isomorphism. Hence show that for every $n \geq 2$, $\pi_n(S^1, 1) = 0$.

31. Let $f : A \to X$ be a continuous map and $i : A \to M_f$ be the inclusion $i(a) = [a, 0]$. Show that the inclusion $i : A \to M_f$ is cofibration.

[Hint. Use Steenrod theorem, Chap. 2.]

32. Let $p : X \to B$ be a fibration with fiber $F = p^{-1}(b_0)$ and B be path-connected. Let Y be any space. Show that the sequence of sets

$$[Y, F] \xrightarrow{i_*} [Y, X] \xrightarrow{p_*} [Y, B]$$

is exact.

33. Let $i : A \hookrightarrow X$ be a cofibration, with cofiber X/A and $q : X \to X/A$ denote the quotient map. If Y is any path-connected space, then show that the sequence of sets

$$[X/A, Y] \xrightarrow{q_*} [X, Y] \xrightarrow{i_*} [A, Y]$$

is exact.

4.12 Additional Reading

[1] Arkowitz, Martin, *Introduction to Homotopy Theory*, Springer, New York, 2011.
[2] Armstrong, M.A., *Basic Topology*, Springer-Verlag, New York, 1983.
[3] Aguilar, Gitler, S., Prieto, C., *Algebraic Topology from a Homotopical View Point*, Springer-Verlag, New York, 2002.
[4] Bredon, G.E., *Topology and Geometry*, Springer-Verlag, New York, Inc.1993.
[5] Davis, J.F. and Kirk, P. *Lecture Notes in Algebraic Topology*, Indiana University, Bloomington, IN (http://www.ams.org/bookstore-getitem/item=GSM-35).
[6] Dieudonné, J., *A History of Algebraic and Differential Topology*, 1900–1960, Modern Birkhäuser, 1989.
[7] Dodson, C.T.J., and Parker, P.E., *A User's Guide to Algebraic Topology*, Kluwer, Dordrecht 1997.
[8] Gray, B., *Homotopy Theory, An Introduction to Algebraic Topology*, Acamedic Press, New York, 1975.
[9] Hilton, P.J., *An introduction to Homotopy Theory*, Cambridge University Press, Cambridge, 1983.
[10] Hu, S.T., *Homotopy Theory*, Academic Press, New York, 1959.
[11] Mayer, J. *Algebraic Topology*, Prentice-Hall, New Jersy, 1972.
[12] Massey, W.S., *A Basic Course in Algebraic Topology*, Springer-Verlag, New York, Berlin, Heidelberg, 1991.
[13] Maunder, C.R.F., *Algebraic Topology*, Van Nostrand Reinhhold, London, 1970.
[14] Munkres, J.R., *Topology, A First Course*, Prentice-Hall, New Jersey, 1975.
[15] Munkres, J.R., *Elements of Algebraic Topology*, Addition-Wesley-Publishing Company, 1984.
[16] Switzer, R.M., *Algebraic Topology-Homotopy and Homology*, Springer-Verlag, Berlin, Heidelberg, New York, 1975.
[17] Wallace , A.H., Algebraic Topology, Benjamin, New York, 1980.
[18] Whitehead, G.W., *Elements of Homotopy Theory*, Springer-Verlag, New York, Heidelberg, Berlin, 1978.

References

Aguilar, M., Gliter, S., Prieto, C.: Algebraic Topology from a Homotopical View Point. Springer, New York (2002)

Arkowitz, M.: Introduction to Homotopy Theory. Springer, New York (2011)

Armstrong, M. A.: Basic Topology. Springer, New York (1983)

Bredon, G.E.: Topology and Geometry. Springer, New York, Inc (1993)

Croom, F.H.: Basic Concepts of Algebraic Topology. Springer, New York (1978)

Davis, J.F., Kirk, P.: Lecture Notes in Algebraic Topology, Indiana University, Bloomington, IN. http://www.ams.org/bookstore-getitem/item=GSM-35

Dieudonné, J.: A History of Algebraic and Differential Topology, 1900–1960. Modern Birkhäuser, Boston (1989)

Dodson, C.T.J., Parker, P.E.: A User's Guide to Algebraic Topology. Kluwer, Dordrecht (1997)

Dugundji, J.: Topology. Allyn & Bacon, Newtown (1966)

Gray, B.: Homotopy Theory, An Introduction to Algebraic Topology. Acamedic Press, New York (1975)

Hatcher, A.: Algebraic Topology. Cambridge University Press, Cambridge (2002)

Hilton, P.J.: An introduction to Homotopy Theory. Cambridge University Press, Cambridge (1983)

Hu, S.T.: Homotopy Theory. Academic Press, New York (1959)

Hurewicz, W.: On the concept of fibre space. Proc. Natl. Acad. Sci., USA **41**, 956–961 (1955)

Mayer, J.: Algebraic Topology. Prentice-Hall, New Jersy (1972)

Massey, W.S.: A Basic Course in Algebraic Topology. Springer, New York (1991)

Maunder, C.R.F.: Algebraic Topology. Van Nostrand Reinhhold, London (1970)

Munkres, J.R.: Topology, A First Course. Prentice-Hall, New Jersey (1975)

Munkres, J.R.: Elements of Algebraic Topology. Addison-Wesley-Publishing Company, Menlo Park (1984)

Rotman, J.J.: An Introduction to Algebraic Topology. Springer, New York (1988)

Spanier, E.: Algebraic Topology. McGraw-Hill Book Company, New York (1966)

Steenrod, N.: The Topology of Fibre Bundles. Prentice University Press, Prentice (1951)

Switzer, R.M.: Algebraic Topology-Homotopy and Homology. Springer, Berlin (1975)

Wallace, A.H.: Algebraic Topology. Benjamin, New York (1980)

Whitehead, G.W.: Elements of Homotopy Theory. Springer, New York (1978)

Chapter 5
Fiber Bundles, Vector Bundles and K-Theory

This chapter continues the study of homotopy theory through fiber bundles, vector bundles, and K-theory. Fiber bundles and vector bundles form special classes of bundles with additional structures. They are closely related to the homotopy theory and are important objects in the study of algebraic topology. A fiber bundle is a bundle with an additional structure derived from the action of a topological group on the fibers. On the other hand, a vector bundle is a bundle with an additional vector space structure on each fiber. Covering spaces provide tools to study the fundamental groups. Fiber bundles provide likewise tools to study higher homotopy groups (which are generalizations of fundamental groups). The notion of fiber spaces is the most fruitful generalization of covering spaces. The importance of fiber spaces was realized during 1935–1950 to solve several problems relating to homotopy and homology. The motivation of the study of fiber bundles and vector bundles came from the distribution of signs of the derivatives of the plane curves at each point.

A fiber bundle is a locally trivial fibration and has covering homotopy property. J. Feldbau reduced in 1939 the classification problem of principal fiber bundles with a given base S^n for $n \geq 2$ to a problem in homotopy theory (Feldbau 1939). Fiber bundles carry nice homotopy properties and play a key role in geometry and physics. (see Chaps. 7, 14 and 17). This subject also marks a return of algebraic topology to its origin. If we consider the tangent plane at each point of a surface, to get global information about the surface, we investigate how the planes change as we move the point on the surface. Again to investigate a higher dimensional smooth geometrical object such as differential manifold, we consider the linear space tangent at each point of the manifold. This leads to the concept of tangent bundles of manifolds, general vector bundles, and fiber bundles.

The concept of fiber bundles arose through the study of some problems in topology and geometry of manifolds around 1930. Its first general definition was given by H. Whitney (1907–1989). His work and that of H. Hopf (1894–1971), E. Stiefel (1909– 1978), J. Feldbau (1914–1945) and many others displayed the importance of the subject for the application of topology to different areas of mathematics during

M.R. Adhikari, *Basic Algebraic Topology and its Applications*,
DOI 10.1007/978-81-322-2843-1_5

1935–1940. Since then the subject has attracted general interest because of some of the finest applications of topology to other fields, and promising many more applications. On the other hand, the concept of a vector bundle arose through the study of tangent vector fields to smooth manifolds such as spheres, projective spaces, and manifolds in general. Although this notion had appeared in the literature before 1955, the definition introduced by W. Hurewicz (1904–1956) in 1955 is much more general and useful.

K-theory born in connecting the rich structure of vector bundles over a paracompact space B with the set of homotopy classes of continuous maps from B into the Grassmann manifold $G_n(F^\infty)$ of n-dimensional subspaces in infinite-dimensional space (F^∞) plays a vital role in applications of algebraic topology to analysis, algebraic geometry, topology, ring theory, and number theory. The two most surprising applications of topological K-theory are: J.F. Adams (1930 –1989) solved the Hopf invariant one problem in 1962 by doing a computation with his Adams operations. Then he proved an upper bound for the number of linearly independent vector fields on spheres (see Chap. 17).

More precisely, this chapter studies the theory of fiber bundles with a special attention to vector bundles with fibers of different dimensions, homotopy classification of vector bundles, and K-theory (which is a generalized cohomology theory) and interlinks vector bundles with homotopy theory. This chapter also studies Hopf maps, Hopf bundles, and Hurewicz fibering.

Milnor's construction of a universal fiber bundle for any topological group G and homotopy classification of numerable principal G-bundles are given and hence the classification of numerable principal G-bundles has been reduced to homotopy theory. Finally, it has been shown that for every topological group G, there exists a topological space B_G, called classifying space having the property that for every pointed topological space B there is a bijective correspondence between isomorphism classes of numerable principal G-bundles over B and $[B, B_G]$, the homotopy classes of base point preserving maps from B to B_G. There also exists a bijective correspondence between the set of isomorphism classes of F-vector bundles over a paracompact space B and the set $[B, G_n(F^\infty)]$ of homotopy classes of continuous maps from B to Grassmann manifold $G_n(F^\infty)$, which leads to define a group $K_F(B)$, called the K-theory introduced by M. Atiyah and F. Hirzebruch in 1961.

For this chapter the books Gray (1975), Husemöller (1966), Luke and Mischenko (1984), Nakahara (2003), Spanier (1966), Steenrod (1951), Switzer (1975) and some others are referred in the Bibliography.

5.1 Bundles, Cross Sections, and Examples

This section introduces the concept of bundles and their cross sections. Fiber bundles and vector bundles form special classes of bundles with additional structure which are important in the study of algebraic topology and they are closely related to the homotopy theory. The recognition of bundles in mathematics was realized during

1935–1940 through the work of Whitney, H. Hopf and E. Stiefel and some others. Since then the subject has created a general interest.

5.1.1 Bundles

This subsection studies the concept of bundles which plays an important role in the theory of fiber bundles and vector bundles. So, it is natural to introduce the concept of bundles at the beginning. A bundle is the basic underlying structure leading to the concepts of fiber bundles and vector bundles. It is a triple consisting of two topological spaces, one is called total space and the other is called base space connected by a continuous map from the total space to the base space, called the projection of the bundle. Roughly speaking, a bundle is a union of fibers parametrized by its base space and glued together by the topology of the total space.

Let E and B be two topological spaces and $p : E \to B$ be a continuous map.

Definition 5.1.1 A bundle $\xi = (E, p, B)$ is an ordered triple consisting of a topological space E, called the total space of ξ, a topological space B, called the base space of ξ and a continuous map $p : E \to B$, called the projection of the bundle ξ.

For each $b \in B$, $E_b = p^{-1}(b)$ (it is nonempty as p is onto by assumption), is called the fiber of ξ over b, which has the topology induced by the inclusion in E. Clearly, $E = \bigcup\limits_{b \in B} E_b = \bigcup\limits_{b \in B} p^{-1}(b)$ and every two fibers E_b and $E_{b'}$ are disjoint if $b \neq b'$ and hence every point of E lies in exactly one fiber. Sometimes we write $E(\xi)$ for the total space and $B(\xi)$ for the base space of the bundle ξ to avoid any confusion.

Example 5.1.2 For the bundle $\xi = (E, p, B)$ as shown in Fig. 5.1, the total space E is decomposed into fibers of four types: a point, a point together with a segment, two segments, and a segment.

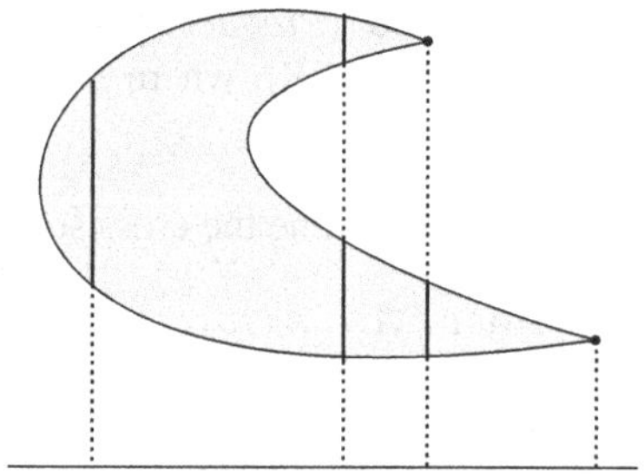

Fig. 5.1 fibers of bundle

Definition 5.1.3 Let $\xi = (X, p, B)$ be a bundle. If Y be a subspace of X and $q = p|Y : Y \to B$, then (Y, q, B) is said to be a subbundle of ξ. In particular, if A is a subspace of B, then the bundle (Y, q, A) is a subbundle of ξ.

Example 5.1.4 Let (X, p, B) be a bundle. If there are inclusion maps $f : Y \hookrightarrow X$ and $g : A \hookrightarrow B$, then the bundle (Y, q, A) is a subbundle of (X, p, B), where q is the restriction of p over Y. For interesting examples see Sect. 5.3.

Definition 5.1.5 (*Induced bundle*) Let $\xi = (X, p, B)$ be a bundle and $f : A \to B$ be a continuous map from a topological space A to the topological space B. The induced bundle $f^*(\xi) = (Y, q, A)$ of ξ under f is the bundle (Y, q, A), where $Y = \{(a, x) \in A \times X : f(a) = p(x)\}$, and $q : Y \to A, (a, x) \mapsto a$.

Proposition 5.1.6 *Let $\xi = (X, p, B)$ be a bundle and $f : A \to B$ be a continuous map from a topological space A to the topological space B. If $f^*(\xi) = (Y, q, A)$ is the induced bundle of ξ under f and p is an open map, then q is also so.*

Proof Left as an exercise. □

5.1.2 *Cross Sections*

We now introduce the concept of cross sections of a bundle. Its importance lies in the fact that the cross sections of certain bundles are identified with familiar geometric objects.

Definition 5.1.7 Let $\xi = (E, p, B)$ be a bundle. A cross section (or in brief section) s of ξ is a continuous map $s : B \to E$ such that $p \circ s = 1_B$ (identity map on B).

For every $b \in B$, $(p \circ s)(b) = 1_B(b) = b$ shows that $s(b) \in p^{-1}(b)$, and hence $s(b)$ lies in the fiber $p^{-1}(b)$ for each b of the base space B. The condition $p \circ s = 1_B$ shows that the map $s : B \to E$ is injective. Otherwise, if for some $b_1 \neq b_2 \in B$, $s(b_1) = s(b_2)$, then $p(s(b_1)) = b_1$ and $p(s(b_2)) = b_2$ implies $b_1 = b_2$.

Definition 5.1.8 (*Product bundle*) The product bundle over B with fiber F is the bundle $(B \times F, p, B)$, where $p : B \times F \to B, (b, x) \mapsto b$ is the projection on the first factor.

We now determine the cross sections of a product bundle.

Proposition 5.1.9 *A cross section s of the product bundle $(B \times F, p, B)$ is precisely of the form $s(b) = (b, f_s(b))$, where $f_s : B \to F$ is a continuous map uniquely determined by s.*

Proof Let $s : B \to B \times F$ be a cross section of the product bundle $(B \times F, p, B)$. Then s takes the form $s : B \to B \times F, b \mapsto (g_s(b), f_s(b))$ such that $p \circ s = 1_B$, where $g_s : B \to B$ and $f_s : B \to F$ are maps uniquely determined by s.

Hence $p(s(b)) = 1_B(b) \Rightarrow p(g_s(b), f_s(b)) = b \Rightarrow g_s(b) = b, \forall b \in B \Rightarrow s(b) = (b, f_s(b))$ for every $b \in B$. Conversely, let $s : B \to B \times F$ be a continuous map such that $s(b) = (b, f_s(b))$ for every $b \in B$. Then $(p \circ s)(b) = b, \forall\ b \in B \Rightarrow p \circ s = 1_B \Rightarrow s$ is a cross section of the bundle $(B \times F, p, B)$. ❑

The map $f_s : B \to F$ determined in Proposition 5.1.9 gives the following Corollary.

Corollary 5.1.10 *Given a product bundle $\xi = (B \times F, p, B)$, let $S(\xi)$ be the set of all cross sections of ξ and $C(\xi)$ be the set of all continuous maps $B \to F$. Then the map $\psi : S(\xi) \to C(\xi), s \mapsto f_s$ is a bijection.*

Remark 5.1.11 Let $\eta = (X', p', B)$ be a subbundle of a bundle $\xi = (X, p, B)$. Then s is a cross section of η iff $s(b) \in X'$ for every $b \in B$.

5.1.3 *Morphisms of Bundles*

This subsection introduces the concept of bundle morphisms with an aim to utilize this concept in the study of fiber and vector bundles. While comparing vector bundles or fiber bundles over the same or different base spaces the concept of morphisms becomes necessary like group homomorphisms. A bundle morphism is intuitively a fiber preserving map and is similar to the concept of a group homomorphism.

Definition 5.1.12 Let $\xi = (X, p, B)$ and $\eta = (Y, q, A)$ be bundles. A bundle morphism or a fiber map $(f, g) : \xi \to \eta$ is a pair of continuous maps $f : X \to Y$ and $g : B \to A$ such that the diagram in Fig. 5.2 is commutative, i.e., $q \circ f = g \circ p$.

Notation: For the bundle $\xi = (X, p, B)$, we use sometimes the notation $E(\xi)$ for its total space X.

Remark 5.1.13 The map f in Fig. 5.2 is fiber preserving. Since for every $x \in X$, $(q \circ f)(x) = (g \circ p)(x)$ holds, and hence the pair $(x, p(x))$ is mapped into the pair $(f(x), g(p(x)))$ by (f, g). Consequently, for every $b \in B$ we have $f(p^{-1}(b)) \subset q^{-1}(f(b))$. This implies that f carries fibers of ξ over b into the fibers of η over $f(b)$ for each $b \in B$.

Remark 5.1.14 The particular case, when ξ and η are both bundles over the same base space B, is interesting.

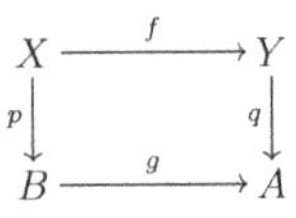

Fig. 5.2 Morphism of bundles

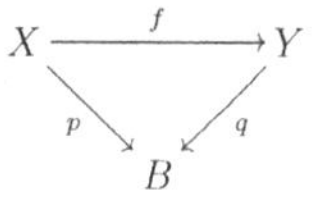

Fig. 5.3 B-morphism of bundles

Definition 5.1.15 Given two bundles $\xi = (X, p, B)$ and $\eta = (Y, q, B)$ over the same base space B, a bundle B-morphism $f : \xi \to \eta$ is a continuous map $f : X \xrightarrow{f} Y$ such that the triangle in the Fig. 5.3 is commutative i.e., $p = q \circ f$. Clearly, $q \circ f = p$ implies that $f(p^{-1}(b)) \subset q^{-1}(b)$ for every $b \in B$. Hence f is a fiber preserving map.

We now show that a cross section of a bundle carries the general property of a bundle morphism.

Proposition 5.1.16 *The cross sections of a bundle $\xi = (X, p, B)$ over B are precisely the B-morphisms $s : (B, 1_B, B) \to (X, p, B)$.*

Proof Let s be cross section of ξ. Then $p \circ s = 1_d$. This implies the triangle in Fig. 5.4 is commutative. Hence s is a B-morphism. Conversely, let $s : (B, 1_d, B) \to (X, p, B)$ be a B-morphism. Then clearly, s is a cross section of ξ. ❑

Remark 5.1.17 Every general property of bundle morphisms is equally valid for cross sections also.

Definition 5.1.18 A bundle (Y, q, A) is a subbundle of (X, p, B) if there are inclusion maps $f : Y \hookrightarrow X$ and $g : A \hookrightarrow B$. Then the pair $(f, g) : (Y, q, A) \to (X, p, B)$ is a bundle morphism.

Definition 5.1.19 Let $\xi = (X, p, B)$ and $\eta = (Y, q, A)$ be two bundles. A bundle map $(f, g) : \xi \to \eta$ is said to be a bundle isomorphism if both the maps $f : X \to Y$ and $g : B \to A$ are homeomorphisms.

Remark 5.1.20 If $(f, g) : \xi \to \eta$ is a bundle isomorphism, then the pair $(f^{-1}, g^{-1}) : \eta \to \xi$ is also a bundle isomorphism such that $(f \circ f^{-1}, g \circ g^{-1})$ and $(f^{-1} \circ f, g^{-1} \circ g)$ are both identity bundle morphisms.

Definition 5.1.21 Let $\xi = (X, p, B)$ be a bundle and A be a nonempty subset of B. The restricted bundle of ξ to A, denoted by $\xi|A$, is the bundle (Y, q, A), where $Y = p^{-1}(A)$ and $q = p|Y$.

Example 5.1.22 Let $\xi = (X, p, B)$ be a bundle. If X' is a nonempty subspace of X and $p' = p|X' : X' \to B$, then (X', p', B) is a restricted bundle of ξ.

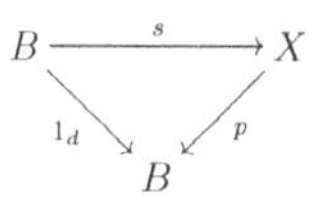

Fig. 5.4 Cross section of bundle

Definition 5.1.23 Let $\xi = (X, p, B)$ and $\eta = (Y, q, B)$ be two bundles over the same base space B. If $f : X \to Y$ is a homeomorphism, then f is called a B-isomorphism. It is said to be locally isomorphic if to every point $b \in B$, there is an open neighborhood U_b of b and an U_b-isomorphism between the restricted bundles $\xi|U_b$ and $\eta|U_b$.

Remark 5.1.24 The relation of being locally isomorphic is an equivalence relation on the set of all bundles over B and hence if ξ is locally isomorphic to a locally trivial bundle, then ξ is locally trivial.

Definition 5.1.25 (*Canonical morphism*) Let $\xi = (X, p, B)$ be a bundle and $f : A \to B$ be a continuous map. If $f^*(\xi) = (Y, q, A)$ is the induced bundle under f, then the pair of maps $\psi : Y \to X, (a, x) \mapsto x$ and $f : A \to B$ form a bundle morphism $(\psi, f) : f^*(\xi) \to \xi$ called the canonical morphism of the induced bundle.

It has some interesting properties:

Proposition 5.1.26 *Let $\xi = (X, p, B)$ be a bundle and $f : A \to B$ be a continuous map. If $f^*(\xi) = (Y, q, A)$ is the induced bundle under f and $(\psi, f) : f^*(\xi) \to \xi$ is the canonical bundle morphism, then for each $a \in A$, the restriction $\psi|q^{-1}(a) : q^{-1}(a) \to p^{-1}(f(a))$ is a homeomorphism.*

Proof Left as an exercise. ❑

Proposition 5.1.27 *Let $h : C \to A$ and $g : A \to B$ be two continuous maps and ξ be a bundle over B. Then the induced bundle $1_B^*(\xi)$ and the bundle ξ are B-isomorphic and the induced bundles $h^*(g^*(\xi))$ and $(g \circ h)^*(\xi)$ are C-isomorphic.*

Proof Left as an exercise. ❑

Proposition 5.1.28 *Let $(f, g) : (X, p, B) \to (Y, q, A)$ be a bundle morphism and $s : B \to X$ be a cross section of (X, p, B). If the map $g : B \to A$ is a homeomorphism, then $s' = f \circ s \circ g^{-1} : A \to Y$ is a cross section of (Y, q, A).*

Proof Left as an exercise. ❑

Corollary 5.1.29 *Let (X, p, B) and (Y, q, B) be two bundles over the same base space B and $f : (X, p, B) \to (Y, q, B)$ is a bundle B-morphism. Then to every cross section s of (X, p, B), there exists an induced cross section s' of (Y, q, B) given by $s' = f \circ s$ (Fig. 5.5).*

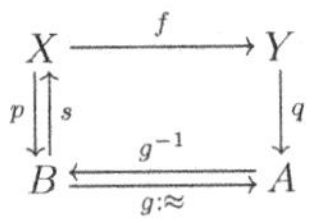

Fig. 5.5 Construction of cross section of (Y, q, A)

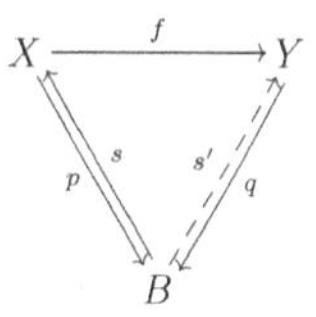

Fig. 5.6 Induced cross section of (Y, q, A)

Proof It follows from Proposition 5.1.28 by taking $g = 1_B$ (see diagram in Fig. 5.6). ❑

Definition 5.1.30 (*Trivial bundle*) Let $\xi = (X, p, B)$ be a bundle. A space F is said to be the fiber space of the bundle ξ if for every $b \in B$, the fiber $p^{-1}(b)$ (topologized by the induced topology from X) is homeomorphic to F. The bundle ξ is said to be a trivial bundle with fiber F if ξ is B-isomorphic to the product bundle $(B \times F, q, B)$, with the projection map

$$q : B \times F \to B, (b, f) \mapsto b.$$

Remark 5.1.31 A bundle is locally trivial if locally it is a direct product.
All fibers of a locally trivial bundle with connected base are homeomorphic.

Remark 5.1.32 Let X and B be connected spaces. A bundle $p : E \to B$ is a covering space if its fiber is discrete (i.e., if the fiber consists of isolated points which means that each point is an open set).

5.1.4 Examples

This subsection presents some interesting examples illustrating the concepts discussed earlier. Here $\langle , \rangle$ denotes the inner product in $\mathbf{R}^{n+1}$.

Example 5.1.33 (*Tangent bundle and normal bundle*) Let $S^n = \{x \in \mathbf{R}^{n+1} : \|x\| = \sqrt{\langle x, x \rangle} = 1\}$ be the n-sphere. The tangent bundle ξ_T over S^n in $\mathbf{R}^{n+1}$ is the subbundle $(T(S^n), p, S^n)$ of the product bundle $(S^n \times \mathbf{R}^{n+1}, p, S^n)$, whose total space is

$$T(S^n) = \{(b, x) \in S^n \times \mathbf{R}^{n+1} : \langle b, x \rangle = 0\},$$

and projection is

$$p : T(S^n) \to S^n, (b, x) \mapsto b.$$

An element of $T(S^n)$ is called a tangent vector to S^n at the point $b \in S^n$. Clearly, the fiber $p^{-1}(b) \subset T(S^n)$ is a vector space of dimension n. A cross section of the tangent bundle ξ_T over S^n is called a tangent vector field (or simply vector field) over S^n.

Similarly, the normal bundle ξ_N over S^n is the subbundle $(N(S^n), q, S^n)$ of the product bundle $(S^n \times \mathbf{R}^{n+1}, p, S^n)$ whose total space is

$$N(S^n) = \{(b, x) \in S^n \times \mathbf{R}^{n+1} : x = tb \text{ for some } t \in \mathbf{R}\}$$

and projection is

$$q : N(S^n) \to S^n, q(b, x) = b.$$

An element of $N(S^n)$ is called a normal vector to S^n at the point $b \in S^n$. Clearly, the fiber $q^{-1}(b) \subset T(S^n)$ is a vector space of dimension 1. A cross section of the normal bundle ξ_N over S^n is called a normal vector field on S^n.

Example 5.1.34 (*Orthonormal r-frames*) The bundle ξ_r of orthonormal r-frames over S^n for $r \leq n$ is the subbundle (X, p, B) of the product bundle $(S^n \times (S^n)^r, p, S^n)$, where total space X is defined by

$$\begin{aligned} X = \{(b, y_1, y_2, \dots, y_r) \in S^n \times (S^n)^r : \langle b, y_i \rangle = 0 \text{ and } \langle y_i, y_j \rangle \\ = \delta_{ij} (\text{Kronecker delta}), 1 \leq i, j \leq r\}, \end{aligned}$$

where δ_{ij} means

$$\delta_{ij} = \begin{cases} 1, & \text{if } i = j \\ 0, & \text{otherwise} \end{cases}$$

An element $(b, y_1, y_2, \dots, y_r)$ of X is an orthonormal system of r-tangent vectors to S^n at $b \in S^n$. A cross section of ξ_r over S^n is called a field of r-frames.

Example 5.1.35 (*Canonical vector bundle* γ^n) Let $V_r(\mathbf{R}^n)$ be the Stiefel manifold (variety) of orthonormal r-frames in $\mathbf{R}^n$ defined by $V_r(\mathbf{R}^n) = \{(y_1, y_2, \dots, y_r) \in (S^{n-1})^r : \langle y_i, y_j \rangle = \delta_{ij}\}$ and $G_r(\mathbf{R}^n)$ be the Grassmann manifold (variety) of r-frames in $\mathbf{R}^n (r \leq n)$. Then $G_r(\mathbf{R}^n)$ is the set of r-dimensional subspaces of $\mathbf{R}^n$ with the quotient topology defined by the identification map

$$\pi : V_r(\mathbf{R}^n) \to G_r(\mathbf{R}^n), (y_1, y_2, \dots, y_r) \mapsto \langle y_1, y_2, \dots, y_r \rangle,$$

where $\langle y_1, y_2, \dots, y_r \rangle$ is an r-dimensional subspace in $\mathbf{R}^n$ with a basis $\{y_1, y_2, \dots, y_r\}$. The canonical r-dimensional vector bundle $\gamma_r^n = (X, p, G_r(\mathbf{R}^n))$ on $G_r(\mathbf{R}^n)$ is the subbundle of the product bundle $(G_r(\mathbf{R}^n) \times \mathbf{R}^n, p, G_r(\mathbf{R}^n))$ with the total space consisting of the subspace of pairs $(V, x) \in G_r(\mathbf{R}^n) \times \mathbf{R}^n$ with $x \in V$ and the orthogonal complement vector bundle of γ_r^n, denoted by γ_r^{*n} is the subbundle of $(G_r(\mathbf{R}^n) \times \mathbf{R}^n, p, G_r(\mathbf{R}^n))$ defined by $\gamma_r^{*n} = (Y, p, G_r(\mathbf{R}^n))$, where $Y = \{(V, x) \in G_r(\mathbf{R}^n) \times \mathbf{R}^n : \langle V, x \rangle = 0$ (i.e., x is orthogonal to V)$\}$. In particular, if $r = 1$, then γ_1^n on $\mathbf{R}P^{n-1} = G_1(\mathbf{R}^n)$, is one-dimensional and is called the canonical line bundle.

By natural inclusion $G_r(\mathbf{R}^n) \subset G_r(\mathbf{R}^{n+1})$ and $G_r(\mathbf{R}^\infty)$ is defined by $G_r(\mathbf{R}^\infty) = \bigcup_{r \leq n} G_r(\mathbf{R}^n)$ with induced (weak) topology. Similarly, $G_r(\mathbf{C}^n)$, $G_r(\mathbf{H}^n)$ and $G_r(\mathbf{C}^\infty)$ and $G_r(\mathbf{H}^\infty)$ are defined, where $\mathbf{H}$ is the division ring of quaternions.

If $F = \mathbf{R}$, $\mathbf{C}$ or $\mathbf{H}$, then the canonical vector bundle over $G_r(\mathbf{F}^\infty)$ is denoted by γ_r. As $G_r(\mathbf{R}^n) \subset G_r(\mathbf{R}^{n+t})$ for integers $t \geq 1$, we may view γ_r^n as $\gamma_r^n = \gamma_r^{n+t}|G_r(\mathbf{R}^n)$, which is a restriction of the bundle γ_r^{n+t} over $G_r(\mathbf{R}^n)$. Similarly, the restricted bundle over $G_r(\mathbf{C}^n)$ (or $G_r(\mathbf{H}^n)$) is defined.

Example 5.1.36 (*Bundle of groups*) We now consider a special type of covering spaces with additional structure on fibers. A bundle of groups $\xi = (X, p, B)$ is a covering space such that all the fibers $p^{-1}(b)$ are isomorphic to a fixed group G in the following sense:

Every point b of B has a neighborhood U for which there exists a homeomorphism $f_U : p^{-1}(U) \to U \times G$ taking each $p^{-1}(b)$ to $\{b\} \times G$ by a group isomorphism. Since G is endowed with discrete topology, the projection p is a covering map. The bundle ξ is called a bundle of groups with fiber G.

5.2 Fiber Bundles: Introductory Concepts

This section considers a class of fibrations, called fiber bundles which are frequently used in geometry, topology, and theoretical physics. Fiber bundles over paracompact spaces are always fibrations. Fiber bundles form a special family of topological spaces in the study of algebraic topology. The concept of fiber bundles arose through the study of some problems in topology and geometry of manifolds around 1930. There exists an infinite exact sequence corresponding to any fiber space (see Chap. 7). A manifold of dimension n is a topological space which looks locally like $\mathbf{R}^n$, but not necessarily globally so. By introducing a chart, a local Euclidean structure to a manifold is provided, which facilitates to use the conventional calculus of several variables. A fiber bundle is a topological space which likewise looks locally a direct product of two topological spaces.

A fiber bundle with a discrete fiber space is a covering space. Conversely, a covering space whose all fibers have the same cardinality is a fiber bundle with discrete fiber. For example, a covering space over a connected space is a fiber bundle with a discrete fiber. Covering spaces provide tools to study fundamental groups. Likewise fiber bundles provide tools to study higher homotopy groups. The first general definition of fiber bundles was given by H. Whitney. The work of H. Whitney, H. Hopf and E. Stienfel established the importance of fiber bundles for applications of topology to geometry around 1940. Since then, this topic has created general interest for its finest applications to other fields such as general relativity and gauge theories and has promised many more. It also makes a return of algebraic topology to its origin and revitalized this topic from its origin in the study of classical manifolds.

A covering space is locally the product of its base space and a discrete space. For the covering space $p : \mathbf{R} \to S^1, t \mapsto e^{2\pi i t}$, we say that $\mathbf{R}$ is a fiber space over S^1. This example introduces the concept of fiber bundles in this section. Roughly speaking, a fiber bundle looks locally a trivial fibration, because, the total space of a fiber bundle is locally the product of its base space and its fiber. For the general theory of fiber bundles see Steenrod (1951).

Definition 5.2.1 A fiber bundle is an ordered quadruple $\xi = (X, p, B, F)$, consisting of a total space X, a base space B, a projection $p : X \to B$ such that B has an open covering $\{U_j\}_{j\in J}$ and for each $j \in J$, there is a homeomorphism $\phi_j : U_j \times F \to p^{-1}(U_j)$, with the property that the composite $p \circ \phi_j$ is the projection

$$p_{U_j} : U_j \times F \to U_j$$

to the first factor (i.e., $p \circ \phi_j = p_{U_j}$).
$p^{-1}(b)$ is said to be the fiber of ξ over b and p is called a fiber bundle projection.

For a fiber bundle with fiber F, the fiber bundle projection $p : X \to B$ and the projection $B \times F \to B$ are locally isomorphic (See Definition 5.1.23). A fiber bundle $\xi = (X, p, B, F)$ is sometimes symbolized like $F \hookrightarrow X \xrightarrow{p} B$. It is called a covering of B if F is discrete and X is called a covering space over B and p is called a covering projection. Then $p^{-1}(b_0)$ is discrete for $b_0 \in B$.

Remark 5.2.2 The space F is homeomorphic to $p^{-1}(b)$ for each $b \in B$. Usually, there is a structure group G for ξ consisting of homeomorphisms of F leading to the concept of G-bundles. If B is a paracompact space, the map $p : X \to B$ is a fibration.

Example 5.2.3 Consider the circle S^1 in the complex plane, i.e., $S^1 = \{z \in \mathbf{C} : |z| = 1\}$. Define $p : \mathbf{R} \to S^1, t \mapsto e^{2\pi i t}$. Then $(\mathbf{R}, p, S^1, \mathbf{Z})$ is a fiber bundle. To show it, let $U_1 = S^1 - \{1\}$ and $U_2 = S^1 - \{-1\}$. Then $p^{-1}(U_1) = \mathbf{R} - \mathbf{Z}$ and there is a homeomorphism $\phi_1 : U_1 \times \mathbf{Z} \to p^{-1}(U_1)$ making the triangle in Fig. 5.7 commutative, where ϕ_1 is defined by $\phi_1(z, n) = n + (1/2\pi i)\log z$, and $\log z$ is the principal value of the logarithm function on $\mathbf{C} - \{t \in \mathbf{R} : t \geq 0\}$. Its inverse $\phi_1^{-1} : p^{-1}(U) \to U_1 \times \mathbf{Z}$ is defined by $\phi_1^{-1}(t) = (e^{2\pi i t}, [t])$, where $[t]$ denotes the greatest integer $< t$ for $t \in \mathbf{R} - \mathbf{Z} = p^{-1}(U_1)$. Similarly we can define $\phi_2 : U_2 \times \mathbf{Z} \to p^{-1}(U_2)$ by $\phi_2(z, n) = n + (1/2\pi i)\log z$, where $\log z$ is the principal value of the logarithm function on $\mathbf{C} - \{t \in \mathbf{R} : t \leq 0\}$. Its inverse ϕ_2^{-1} is defined by $\phi_2^{-1} = (e^{2\pi i t}, [t + \frac{1}{2}])$, for $t \in p^{-1}(U_2)$.

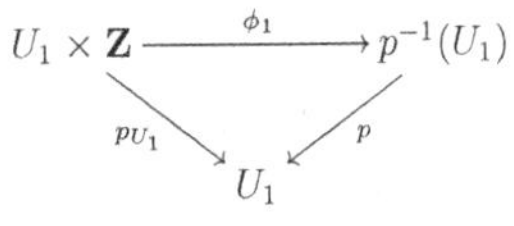

Fig. 5.7 fiber bundle over S^1

Example 5.2.4 Let B be a connected space and $b_0 \in B$. If $p : X \to B$ is a covering projection, then $(X, p, B, p^{-1}(b_0))$ is a fiber bundle.

Example 5.2.5 The n-dimensional torus T^n is a fiber bundle. Consider T^n defined by the n-fold cartesian product $T^n = S^1 \times S^1 \times \ldots \times S^1$ and the map $p : \mathbf{R}^n \to T^n$ defined by

$$(t_1, t_2, \ldots, t_n) \mapsto (e^{2\pi i t_1}, e^{2\pi i t_2}, \ldots, e^{2\pi i t_n}).$$

Then $(\mathbf{R}^n, p, T^n, F)$ forms a fiber bundle with fiber F which is the set of integer lattice points in $\mathbf{R}^n$.

We now consider a continuous action

$$X \times G \to X, (x, g) \mapsto xg$$

of a discrete topological group G on a topological space X with an aim to construct a fiber bundle over the orbit space X mod G.

Definition 5.2.6 A discrete topological group G with identity e is said to act (from the right) properly discontinuously on a topological X if

PD(i) for every $x \in X$, there is a neighborhood U_x of x such that $U_x g \cap U_x \neq \emptyset$ implies $g = e$; and

PD(ii) for every pair of elements $x, y \in X$, $y \notin xG$, there are neighborhoods V_x and V_y of x and y, respectively, such that $V_x g \cap V_y = \emptyset$, for all $g \in G$.

Remark 5.2.7 The condition **PD(ii)** shows that the orbit space $X/G = X$ mod G (topologized by the identification map $p : X \to X$ mod $G, x \mapsto xG$) is Hausdorff and hence the condition **PD(i)** shows that the projection $p : X \to X$ mod $G, x \mapsto xG$ is a covering map.

Proposition 5.2.8 *If a discrete topological group G acts on a topological space X properly discontinuously, then $(X, p, X$ mod $G, G)$ is a fiber bundle.*

Proof Let $y \in X$ mod G. Then there is some $x \in X$ such that $p(x) = y$ and a neighborhood U_x of x in X by **PD(i)**. Let $V_y = p(U_x)$. Then $p^{-1}(V_y) = \bigcup_{g \in G} U_x g$ implies that V_y is open and $y \in V_y$. Moreover, $p|U_x : U_x \to V_y$ is a homeomorphism. Define a homeomorphism

$$\phi_y : V_y \times G \to p^{-1}(V_y), (z, g) \mapsto ((p|U_x)^{-1}(z))g,$$

$g \in G$ and $z \in V_y$. Then ϕ_y^{-1} is defined by

$$\phi_y^{-1}(z'g) = (p(z'), g),$$

for $z'g \in gU_x$. Consequently, ϕ_y is a homeomorphism such that $p \circ \phi_y = p_{V_y}$. Then open sets $\{V_y\}$ from a covering of the space X mod G and hence $(X, p, X$ mod $G, G)$ is a fiber bundle. ❑

Definition 5.2.6 is now redefined for convenience of future discussion.

Definition 5.2.9 A group G of homeomorphisms of a topological space X is said to be discontinuous if the orbits in orbit space $X/G = X \bmod G$ are discrete subsets and G is said to be properly discontinuous if for $x \in X$, there is a neighborhood, U_x of x in X such that for $g, g' \in G$ if $U_x g$ meets $U_x g'$, then $g = g'$. The group is said to act without fixed points if the only element of G having fixed point is the identity element.

Remark 5.2.10 The Proposition 5.2.8 prescribes an important way in which fiber bundles arise.

We now study the local cross sections of fiber bundles.

Definition 5.2.11 A map $p : X \to B$ is said to have a local cross section s at a point $b \in B$ if there is a neighborhood U_b in B and a map $s : U_b \to X$ is such that $p \circ s = 1_{U_B}$.

Remark 5.2.12 Every bundle $\xi = (E, p, B)$ may not have a cross section.

Example 5.2.13 The bundle $\xi = (D^2, p, S^1)$ has no cross section. If possible let ξ have a cross section $s : S^1 \to D^2$. Then $p \circ s = 1_d$ implies that s is injective. Otherwise there exists at least one pair of elements b_1, b_2 such that $b_1 \neq b_2$ but $s(b_1) = s(b_2)$. Then $p(s(b_1)) = p(s(b_2))$ implies $b_1 = b_2$, which is a contradiction. Again since $s : S^1 \to D^2$ is continuous and injective, it contradicts the result, that every continuous map $f : S^1 \to D^2$ sends at least one pair of antipodal points of S^1 to the same point (see Borsuk–Ulam Theorem).

Remark 5.2.14 Since every fiber bundle has a cross sections, we assume the existence of cross section in the following Propositions and Theorem (If G is Lie group and H is a closed subgroup of G, then the cross section of the natural projection $p : G \to G/H$ exists).

Proposition 5.2.15 *Let G be a topological group and H be a closed subgroup of G. Then the projection $p : G \to G/H, g \mapsto Hg$ has a local cross section at any point of G/H.*

Proof Consider the action σ of G on the space G/H, given by $G \times G/H \to G/H, (g, g'H) \mapsto (gg')H$. To prove the proposition it suffices to prove that p has a local cross section at the coset H. Suppose (U, σ) is a local cross section for p at H. Then for any other point gH of G/H, the set Ug is a neighborhood of gH and the function $\sigma_g : g \cdot U \to G$, given by

$$g'H \mapsto g \cdot (\sigma(g^{-1}g'H)), \text{ for } g'H \in g' \cdot U,$$

is continuous and such that $p \circ \sigma_g = 1_{g \cdot U}$. This proves the proposition. ❑

Theorem 5.2.16 *Let G be a topological group and H be a closed subgroup of G. If $p : G \to G/H$, $g \mapsto gH$ has a local cross section at H, then for any closed subgroup $A \subset H$, the natural projection $q : G/A \to G/H$, $Ag \mapsto Hg$, is a fiber bundle with fiber H/A.*

Proof By Proposition 5.2.15 the projection $p : G \to G/H$ has a local cross section at every point of G/H. Let $x \in G/H$ and (U, σ) be a local cross section of p at x. Define $\psi : U \times H/A \to G/A$, given by $(y, hA) \mapsto \sigma(y).hA$ for $y \in U, h \in H$. Then ψ is a continuous function such that $(q \circ \psi)(y, hA) = q(\sigma(y) \cdot hA) = \sigma(y) \cdot hA = \sigma(y) \cdot hA = \sigma(y) \cdot H = p(\sigma(y)) = y = 1_U(y)$ for all $y \in U$ and $h \in H$. Again Define $\phi : (q)^{-1}(U) \to U \times HA$, given by, $gA \to (gH, \sigma(gH)^{-1} \cdot gA)$ for all $gA \in q^{-1}(U)$. Then ϕ is a continuous map such that $\phi \circ \psi = 1d$ and $\psi \circ \phi = 1_d$. ❑

5.3 Hopf and Hurewicz Fiberings

This section studies the various fiberings of spheres discovered by H.Hopf and W. Hurewicz.

5.3.1 *Hopf Fibering of Spheres*

This subsection discusses Hopf fiberings given by Hopf (1931, 1935) and considers the early examples of bundles spaces: three fiberings of spheres: $p : S^{2n-1} \to S^n, n = 2, 4, 8$. The simplest of them is the map $p : S^3 \to S^2$ of the 3-sphere on the 2-sphere given by Hopf in 1935, known as a Hopf map.

Theorem 5.3.1 *The 3-sphere is decomposed into a family of great circles, called fibers, with the 2-sphere as a decomposition space.*

Proof Let S^3 be represented in $\mathbf{C}^2$ as $S^3 = \{(z_1, z_2) \in \mathbf{C}^2 : |z_1|^2 + |z_2|^2 = 1\}$ and S^2 be represented as the complex projective line (i.e., as pairs $[z_1, z_2]$ of complex numbers, not both zero, with the equivalence relation $[z_1, z_2] \sim [\lambda z_1, \lambda z_2]$, where $\lambda(\neq 0)$). Define

$$p : S^3 \to S^2, (z_1, z_2) \mapsto [z_1, z_2]/(|z_1|^2 + |z_2|^2)^{1/2}.$$

Then S^3 is a bundle space over S^2 relative to p. If $(z_1, z_2) \in S^3$ and $|\lambda| = (|z_1|^2 + |z_2|^2)^{1/2} = 1$, then $(\lambda z_1, \lambda z_2)$ is also in S^3 and they have the same image point in S^2. Conversely, if $p(z_1, z_2) = p(z_1', z_2')$, then $(z_1', z_2') = (\lambda z_1, \lambda z_2)$ for some λ having $|\lambda| = 1$. Hence the inverse image of a point of S^2 is obtained by any point of the inverse image by multiplying it by $e^{i\theta} (0 \leq \theta \leq 2\pi)$. Hence the inverse image is just a great circle of S^3. This shows that the 3-sphere is decomposed into a family of great circles, called fibers, with the 2-sphere as a decomposition space. ❑

Remark 5.3.2 We now generalize the Hopf map $p : S^3 \to S^2$ through discussion of some spaces that arise in projective geometry. Let F denote one of the fields $\mathbf{R}$ of real numbers, $\mathbf{C}$ of complex numbers or division ring $\mathbf{H}$ of quaternions and F^n be the right vector space whose elements are ordered sets of n elements of F. If $x = (x_1, x_2, \ldots, x_n) \in F^n$ and $\lambda \in F$, then $x\lambda = (x_1\lambda, \ldots, x_n\lambda)$. Define the inner product x and y in F^n by $\langle x, y\rangle = \sum_1^n \overline{x}_i y_i$, where $\overline{x}_i$ is the conjugate of x_i. Then

$$\langle y, x\rangle = \overline{\langle x, y\rangle}, \langle x\lambda, y\rangle = \overline{\lambda}\langle x, y\rangle, \langle x, (y\lambda)\rangle = \langle x, y\rangle\lambda.$$

In particular, $\langle x, y\rangle = 0$ iff $\langle y, x\rangle = 0$. This shows that the relation of orthogonality is symmetric. Let S be the unit sphere in F^n which is the locus $\langle x, x\rangle = 1$. If G_n is the orthogonal, unitary, or sympletic group according as $F = \mathbf{R}, \mathbf{C}$ or $\mathbf{H}$, then each G_n is a compact Lie group (see Appendix A). Let FP^n be the projective space associated with F, then it can be thought of the set of all lines through the origin in $F^{n+1} = \overbrace{F \oplus F \oplus \cdots \oplus F}^{n+1}$. Then $\mathbf{R}P^n$, $\mathbf{C}P^n$, and $\mathbf{H}P^n$ are called n-dimensional real, complex, and quaternionic projective spaces. We topologize FP^n be considering it as a quotient space of $F^{n+1} - \{0\}$. Then each point of $F^{n+1} - \{0\}$ determines a line through the origin 0. Thus if x and y are nonzero elements of F^{n+1}, we say that $x \sim y$ iff there is an element $\lambda (\neq 0) \in F$ such that $y = x\lambda$. This is an equivalence relation and define FP^n to be the set of equivalence classes with the quotient topology.

There is a natural map $F^{n+1} - \{0\} \to FP^n$, which is continuous, and gives, on restriction to the unit sphere of F^n, maps

$$p_n : S^n \to \mathbf{R}P^n,$$

$$q_n : S^{2n+1} \to \mathbf{C}P^n,$$

and

$$r_n : S^{4n+3} \to \mathbf{H}P^n.$$

Generally we write p in place of p_n, q_n or r_n, where there is no confusion.

Example 5.3.3 **(i)** (Complex Hopf bundle) $\xi = (S^{2n+1}, p, \mathbf{C}P^n, S^1)$ is a trivial fiber bundle with fiber S^1.

(ii) (Quaternionic Hopf bundle) $\gamma = (S^{4n+3}, p, \mathbf{H}P^n, S^3)$ is a locally trivial fiber bundle with fiber S^3.

(iii) (Real Hopf bundle) $\eta = (S^n, p, \mathbf{R}P^n, \mathbf{Z}_2)$ is a locally trivial fiber bundle with fiber $\mathbf{Z}_2$.

Remark 5.3.4 The Example 5.3.3 indicates the importance of bundle theory to compute the homotopy groups of sphere where results are only partly known (see Chap. 7).

5.3.2 *Hurewicz Fibering*

This subsection studies Hurewicz fibering. This fibering is due to W. Hurewicz and named after him.

Definition 5.3.5 A Hurewicz fibering is a map $p : X \to B$ that has the homotopy lifting property (HLP) with respect to any space Y.

Let $f : Y \to Z$ be continuous. Define $E_f = \{(y, w) \in Y \times Z^1 : w(0) = f(x)\} \subset Y \times Z^1$ with the induced topology. Then there are maps

$$\lambda : E_f \to Z, \mu : E_f \to Y, \alpha : Y \to E_f,$$

defined by $\lambda(y, w) = w(1)$, $\mu(y, w) = y$, $\alpha(y) = (y, c_{f(y)})$, where , $c_{f(y)}$ is the constant path at $f(y)$. Hence the diagram in Fig. 5.8 commutes, i.e., $\lambda \circ \alpha = f$ and $\mu \circ \alpha = 1_d$.

The existence of maps

$$\lambda : E_f \to Z, \mu : E_f \to Y, \alpha : Y \to E_f,$$

proves the following proposition:

Proposition 5.3.6 *Let $f : Y \to Z$ be continuous. Define $E_f = \{(y, w) \in Y \times Z^1 : w(0) = f(x)\} \subset Y \times Z^1$ with the induced topology. Then*

(i) $Y \simeq E_f$.
(ii) $\lambda : E_f \to Z$ *is a Hurewicz fibering with fiber* $F_f = \{(y, w) \in Y \times Z^1 : w(0) = f(x), w(1) = *\}$.

Proof **(i)** Define a homotopy $H : E_f \times I \to E_f, (y, w, t) \mapsto (y, w_t)$, where $w_t(s) = w(st)$. Then $\alpha \circ \mu \simeq 1_d$. Again $\mu \circ \alpha = 1_d$ and hence $Y \simeq E_f$.

(ii) Consider the commutative diagram in Fig. 5.9. Let $h(a) = (h_1(a), h_2(a))$. Define $\beta : A \times I \to E_f \subset Y \times Z^1, (a, t) \mapsto (\beta_1(a, t), \beta_2(a, t))$, where $\beta_1 (a, t) = h_1(a)$ and

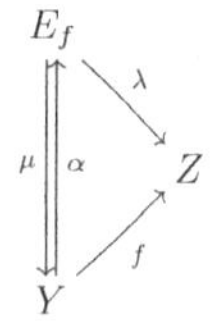

Fig. 5.8 Commutative triangle for Hurewicz fibering

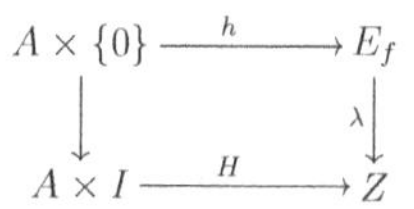

Fig. 5.9 Commutative rectangular diagram for Hurewicz fibering

$$\beta_2(a,t)(s) = \begin{cases} h_2(a)(s(1+t)), & 0 \leq s \leq 1/(1+t) \\ H(a,(1+t)s-1), & 1/(1+t) \leq s \leq 1. \end{cases}$$

Then $\beta(a,t) \in E_f$, because, the equality $(f \circ \beta_1)(a,t) = \beta_2(a,t)(0)$ follows from $(f \circ h_1)(a) = h_2(a)(0)$. The map β is clearly continuous, since β_1 and β_2 are also. Moreover, $\beta(a,0)(s) = h(s)$ and $(\lambda \circ \beta)(a,t) = H(a,t)$. Clearly, F_f is its fiber.

❑

Theorem 5.3.7 (Hurewicz) *Let $p : X \rightarrow B$ be a covering map. Suppose B is paracompact, and there is an open covering $\{U_j\}$ of B such that for each $U_i \in \{U_j\}$, $p|_{p^{-1}(U_i)} : p^{-1}(U_i) \rightarrow U_i$ is a fibration. Then p is a fibration.*

Remark 5.3.8 The proof is long and complicated. We omit the proof referring the reader to [Dugundji, pp. 400].

An important consequence of Hurewicz theorem.

Proposition 5.3.9 *Let $p : X \rightarrow B$ be the projection of a fiber bundle (X,p,B,F) such that the base space B is paracompact. Then p is a fibration.*

Proof It follows from Theorem 5.3.7. ❑

Remark 5.3.10 The decompositions of compact Lie groups modulo their closed subgroups produce fiber bundles, which serve as a valuable source of examples. We recommend the survey articles by [H. Samelson, Topology of Lie groups, Bull. Amer. Math. Soc.58 (1952), 2–37] and Borel (1955).

5.4 *G*-Bundles and Principal *G*-Bundles

This section studies G-bundles and principal G-bundles defined by transformation groups G and shows that if X is a simply connected space and G is a properly discontinuous group of homeomorphisms of X, then the fundamental group $\pi_1(X \bmod G)$ of the orbit space $X \bmod G$ is isomorphic to G.

A G-bundle is a bundle with an additional structure derived from the action of the topological group G on a topological space. For a Lie group G named after Sophus Lie (1842–1899), the principal G-bundles are studied in Sect. 5.11. Transformation groups obtained by actions of topological groups on topological spaces (see Appendix A), are now used to study G-bundles defined by transformations groups G.

Definition 5.4.1 Let X be a right G-space. Then the set $X \bmod G = \{xG : x \in X\}$ is the set of all orbits of X under the action of G on X, equipped with the quotient topology, which is the largest topology such that the projection map $p : X \rightarrow X \bmod G$, given by $x \mapsto xG$, is continuous. The quotient space $X \bmod G$ is called the orbit space of X.

Clearly, p is an identification map.

Proposition 5.4.2 *Let X be a right G-space. Then for each $g \in G$, the map $\phi_g : X \to X, x \mapsto xg$ is a homeomorphism and the projection $p : X \to X \bmod G, x \mapsto xG$ is an open map.*

Proof Clearly, ϕ_g is a homeomorphism with its inverse $\phi_{g^{-1}}$ for each $g \in G$. For the second part, let U be an open subset of X. Then $p^{-1}(p(U)) = \bigcup_{g \in G} Ug$ is a union of open sets in X and hence $\bigcup_{g \in G} Ug$ is an open set in $X \bmod G$. Consequently, $p(U)$ is an open set of $X \bmod G$ for each open set U of X. ❑

Definition 5.4.3 Let X and Y be two right G-spaces. A map $f : X \to Y$ is said to be a G-morphism if $f(x \cdot g) = f(x) \cdot g$ holds for all $x \in X$ and for all $g \in G$.

Clearly, $f(xG) \subset f(x)G$ for each $x \in X$.

Definition 5.4.4 A bundle (X, p, B) is said to be G-bundle if the bundles (X, p, B) and $(X, p_X, X \bmod G)$ are isomorphic for some G-space structure on X by an isomorphism $(1_d, f) : (X, p_X, X \bmod G) \to (X, p, B)$ i.e., there exists a homeomorphism $f : X \bmod G \to B$ making the diagram in Fig. 5.10 commutative.

Proposition 5.4.5 *Given G-spaces X and Y and a G-morphism $f : X \to Y$, there exists a bundle morphism $(f, \tilde{f}) : (X, p_X, X \bmod G) \to (Y, p_Y, Y \bmod G)$.*

Proof Let $\xi(X) = (X, p_X, X \bmod G)$ be the bundle corresponding to a G-space X. Then $\xi(Y) = (Y, p_Y, Y \bmod G)$. Clearly, the map $f : X \to Y$ induces a quotient map $\tilde{f} : X \bmod G \to Y \bmod G$, given by $\tilde{f}(xG) \mapsto f(x)G$. Since the diagram in the Fig. 5.11 is commutative, the pair $(f, \tilde{f})$ is a bundle morphism. Clearly, every G-space X determines a bundle $(X, p_X, X \bmod G)$, where $p_X : X \to X \bmod G, x \mapsto xG$. ❑

Let G be the group of covering transformations(see Chap. 4) of a covering projection $p : X \to B$. Then the action of G on X is properly discontinuous. Is its converse true ?

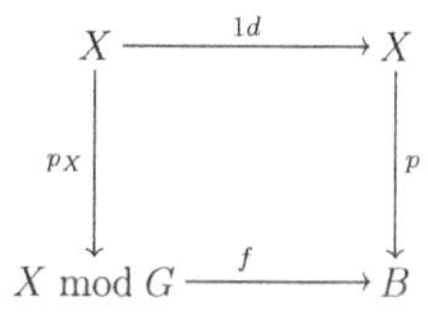

Fig. 5.10 G-bundle (X, p, B)

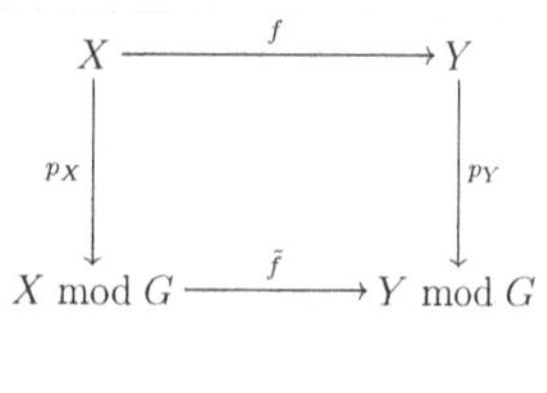

Fig. 5.11 G-bundle morphism

Theorem 5.4.6 *If G is a properly discontinuous group of homeomorphisms of a space X, then the projection $p : X \to X \bmod G$ is a covering projection. Moreover, if X is connected, then this covering projection is regular and G is its group of covering transformations.*

Proof Let $p : X \to X \bmod G$ be the projection. Then p is continuous and open by Proposition 5.4.2. Let U be an open subset of X satisfying the property that whenever Ug meets Ug', then $g = g'$. We claim that $p(U)$ is evenly covered by p. By assumption, U shows that $\{Ug\}_{g \in G}$ is a disjoint collection of open sets whose union is $p^{-1}(p(U))$. It is sufficient to prove that $p|Ug : Ug \to p(U)$ is a bijection. If $x \in U$, then $p(xg) = p(x)$ and hence $p(Ug) = p(U)$. If $p(xg) = p(x'g)$, with $x, x' \in U$, then for some $s \in G, xg = xsg$. Consequently, Ug meets $x'sg$ and $g = sg$. Hence $s = e$ and $xg = x'g$. Thus $p : Ug \to p(U)$. Since G is properly discontinuous, the sets $p(U)$ are evenly covered by p and form an open covering of $X \bmod G$. Since $p(xg) = p(x)$, it follows that G is contained in the group of covering transformations of p. Hence the group G and the group of covering transformations are same. Finally, as the group of covering transformations is transitive on each fiber, it follows that the covering projection is regular. ❑

Corollary 5.4.7 *Let X be a simply connected space and G be a properly discontinuous group of homeomorphisms of X. Then the fundamental group $\pi_1(X \bmod G)$ is isomorphic to G.*

Proof It follows by using Theorem 5.4.6 that G is the group of covering transformations of the regular covering projection $p \to X \bmod G$. Then the Corollary follows from Theorem 5.4.6. ❑

We now study from the homotopical view point special G-bundles, known as, principal G-bundles, which come with an action of some topological G.

Definition 5.4.8 Let G be a topological group with identity e. A (locally trivial) principal G-bundle is a fiber bundle $\xi = (X, p, B, G)$ with a continuous right action of G on X, i.e., a continuous map $X \times G \to X, (x, g) \mapsto x \cdot g$ such that there is an open covering $\{U_j : j \in J\}$ of B and for each $j \in J$, there is a homeomorphism $\phi_j : U_j \times G \to p^{-1}(U_j)$ with the conditions:

(i) $(p \circ \phi_j)(b, y) = b$ and
(ii) $\phi_j(b, g) = \phi_j(b, e) \cdot g$ for $b \in U_j$ and $g \in G$.

Example 5.4.9 (*Product principal G-bundle*) The product G-space $B \times G$ is principal under the action of G given by $(b, t)s = (b, ts)$.

We now define the morphisms of G-bundles.

Definition 5.4.10 Let $\xi = (X, p, B, G)$ and $\xi' = (X', p', B', G)$ be two principal G-bundles. A morphism $\phi : \xi \to \xi'$ of principal G-bundles is a pair of maps (f, h), where $f : X \to X', h : B \to B'$ are such that the diagram in the Fig. 5.12 is commutative and $f(x \cdot g) = f(x) \cdot g$ for all $g \in G$ and $x \in X$.

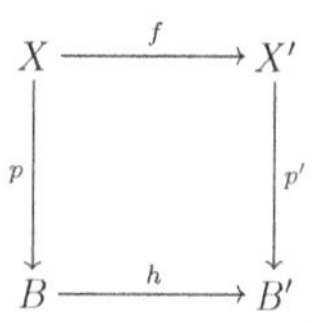

Fig. 5.12 Morphism of G-bundles

Remark 5.4.11 Let $\xi = (X, p, B, G)$ be a principal G-bundle. Then the action of G on X is free and transitive on each $b \in B$ and $p(x_1) = p(x_2)$ iff there exists some g such that $x_2 = x_1 g$.

Definition 5.4.12 Let $\xi = (X, p, B, G)$ and $\eta = (Y, q, B, G)$ be two principal G-bundles over B. Then they are said to be isomorphic or equivalent, denoted by $\xi \cong \eta$ if there exists a homeomorphism $f : X \to Y$ such that

(i) f is equivariant, i.e., $f(xg) = f(x)g$ for all $x \in X$ and $g \in G$;
(ii) $q \circ f = p$.

Definition 5.4.13 Let $\xi = (X, p, B, G)$ be a G-bundle. Then the family $\{(U_j, \phi_j)\}_{j \in J}$ for ξ given in Definition 5.4.8 is called trivializing cover for ξ and the bundle ξ is called numerable if there is a partition of unity subordinate to a trivializing cover for ξ.

Example 5.4.14 Any principal G-bundle over a paracompact space is numerable.

Let B be a space and $[\xi]$ denote the isomorphism class of numerable principal G-bundle ξ over B. The set of isomorphism classes of numerable principal G-bundles over B is denoted by $K_G(B)$.

Definition 5.4.15 Let $\xi = (X, p, B)$ be a principal G-bundle. If $f : A \to B$ is a map, then f induces a principal G-bundle $f^*(\xi) = (Y, q, A, G)$, where $Y = \{(a, x) \in A \times X\}$ such that $f(a) = p(x)$ and as projection $q : Y \to A$, is given by $(a, x) \mapsto a$ and the action of G on Y is given by $(a, x) \cdot g = (a, x \cdot g)$ for all $a \in A, x \in X$ and $g \in G$. If $\tilde{f} : Y \to X, (a, x) \mapsto x$, then the diagram in the Fig. 5.13 is commutative.

Definition 5.4.16 Let $\xi = (X, p, B)$ be a principal G-bundle and $f : A \to B$ be a given map. Then the induced G-bundle $f^*(\xi) = (Y, q, A, G)$ over A is called the pull-back of ξ or G-bundle induced by f.

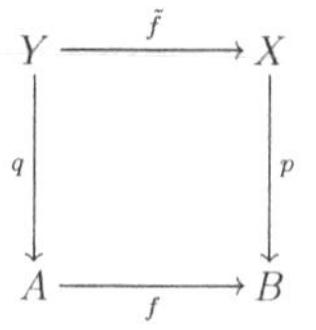

Fig. 5.13 Induced G-bundle

We recall the following definition of an action of a topological group.

Definition 5.4.17 An action of a group G on X is said to be free if for all $g(\neq e) \in G$, $g \cdot x \neq x$ for every $x \in X$. It is called effective if for all $g(\neq e) \in G$ there exists an element $x \in X$ such that $g \cdot x \neq x$.

Definition 5.4.18 (*Principal G-bundle*) Let G be a topological group and B be a topological space. A principal G-bundle over B consists of a fiber bundle $p : X \to B$ together with an action $X \times G \to X$ such that

PG(i) the shearing map

$$T : X \times G \to X \times X, (x, g) \mapsto (x, x \cdot g)$$

maps $X \times G$ homeomorphically to its image;

PG(ii) $B = X \bmod G$ and $p : X \to X \bmod G$ is the quotient map;

PG(iii) for all $b \in B$, $\exists$ an open neighborhood U of b such that $p : p^{-1}(U) \to U$ is G-bundle isomorphic to the trivial bundle $q : U \times G \to U$ (i.e., there exists a homeomorphism $\phi : p^{-1}(U) \to U \times G$ satisfying $q \circ \phi = p$, where $(x, g') \cdot g = (x, gg')$).

Remark 5.4.19 The shearing map T is injective iff the action of G on X is free and hence by **PG(i)**, the action of G on the total space X of a principal bundle is always free. If G and X are compact, then a free action satisfies **PG(i)**. Moreover, a free action produces a translation function $\rho : Y \to G$, where $Y = \{(x, x \cdot g) \in X \times X\}$ is the image of the shearing map T. Condition **PG(i)** is equivalent to a free action with a continuous translation function.

Let $X_1, X_2, \ldots, X_n, \ldots$ topological spaces such that $X_1 \subset X_2 \subset \cdots \subset X_n \subset \cdots$ are inclusions. Let $X_\infty = \lim\limits_{n\to\infty} \bigcup\limits_n X_n$, with weak topology (i.e., $A \subset X$ is closed iff $A \cap X_n$ is closed in X_n for each n).

Proposition 5.4.20 *Let G be a topological group. If $X_1 \subset X_2 \subset \cdots \subset X_n \cdots$ be inclusions such that $X_\infty = \lim\limits_{n\to\infty} \cup X_n$. Let $\sigma : X_\infty \times G \to X_\infty$ be a free action which restricts to an action of G on X_n for each n (i.e., $\sigma(X_n \times G) \subset X_n$ for each n). Then this action of G on X_n is free and if the translation function for this action is continuous for all n, then the translation function for σ is continuous*

Proof Since the action of G on X_∞ is free, the action of G on each X_n is free for all n. Let

$$T : X_n \times G \to X_n \times X_n, (x, g) \mapsto (x, x \cdot g)$$

be the shearing function for X_n and $Y_n = \text{Im } T \subset X_n \times X_n$. Let $\rho_n : Y_n \to G$ be the translation function for X_n. Then as a topological space $Y_\infty = \bigcup\limits_n Y_n = \lim\limits_{n\to\infty} Y_n$ and $\rho_\infty|_{Y_n} = \rho_n$. This shows that ρ_∞ is continuous. ❑

Remark 5.4.21 Let G be a topological group and $p : X \to B$ be a principal G-bundle. Let Y be a G-space on which the action of G is effective. The fiber bundle with structure group G obtained from p and Y is given by $q : (Y \times X) \bmod G \to B$, where $(y, x) \cdot g = (y \cdot g, x \cdot g)$ and $q(y, x) = p(x)$. This is sometimes called "G-bundle for a fiber bundle with structure group G". In particular, when $Y = \mathbf{R}^n (\mathbf{C}^n)$, and G is the orthogonal group $O(n, \mathbf{R})$ (unitary group $U(n.\mathbf{C})$) acting in the standard way, and the restrictions of the trivialization maps to each fiber are linear transformations, such a fiber bundle is called an n-dimensional real (complex) vector bundle. A one-dimensional vector bundle is called a line bundle.

5.5 Homotopy Properties of Numerable Principal G-Bundles

This section studies numerable principal G-bundles over B from the view point of homotopy theory and constructs a contravariant functor $K_G : \mathcal{H}tp \to \mathcal{S}et$ corresponding to a given topological group G. This functor plays an important role in the study of homotopy theory.

Theorem 5.5.1 *Let ξ be a numerable principal G-bundle over $B \times I$. Then the bundles ξ, $(\xi|(B \times \{1\})) \times I$ and $(\xi|(B \times \{0\})) \times I$ are G-isomorphic. If $h_i : B \to B \times I$ is the map $h_i(b) = (b, i)$ for $i = 0, 1$, then the bundles $h_0^*(\xi)$ and $h_1^*(\xi)$ are B-isomorphic.*

Proof Let $f(b, t) = (b, 1)$. Then by Ex. 11(b) of Sect. 5.13, $f^*(\xi)$ and ξ are isomorphic principal G-bundles over $B \times I$. Again $f^*(\xi)$ and $(\xi|(B \times \{1\}) \times I)$ are isomorphic principal G-bundles over $B \times I$. Hence it follows that ξ and $(\xi|(B \times \{1\}) \times I)$ are also isomorphic principal G-bundles. In a similar way, it follows that ξ and $(\xi|(B \times \{0\}) \times I)$ are also isomorphic. Again since $f \circ h_0 = h_1$ and the bundles $f^*(\xi)|(B \times \{0\})$ and $\xi|(B \times \{0\})$ are G-isomorphic, it follows that $h_1^*(\xi) = h_0^* \circ f^*$ and $h_0^*(\xi)$ are isomorphic principal G-bundles. ❑

Corollary 5.5.2 *Let $\xi = (X, p, B)$ be a numerable principal G-bundle over B and $f, g : A \to B$ are homotopic maps. Then the principal bundles $f^*(\xi)$ and $g^*(\xi)$ are isomorphic over A.*

Proof Let $H_t : f \simeq g : A \to B$ and $h_i : A \to A \times I$ be the map defined by $h_i(a) = (a, i)$ for $i = 0, 1$. Then $H_0 \circ h_0 = f$ and $H_1 \circ h_1 = g$. Hence the principal bundles $f^*(\xi)$ and $h_0^* \circ f^*(\xi)$ are isomorphic by Theorem 5.5.1. ❑

Theorem 5.5.3 *For each space B, let $K_G(B)$ be the set of isomorphism classes of numerable principal G-bundles over B. Then $K_G : \mathcal{H}tp \to \mathcal{S}et$ is a contravariant functor from the homotopy category $\mathcal{H}tp$ of topological spaces and their homotopy classes of maps to the category $\mathcal{S}et$ of sets and their functions.*

Proof Define the object function and morphism function as follows:
For each object B in $\mathcal{H}tp$, $K_G(B)$ is assigned to be the set of isomorphism classes of numerable principal G-bundle over B. For a homotopy class $[f]$ of $f : A \to B$, define the morphism function $K_G([f] : K_G(B) \to K_G(A)$, by $K_G([f]) = \{f^*(\xi)\}$. By Corollary 5.5.2, the function $K_G([f])$ is well defined. Consequently, $K_G([f]) : K_G(B) \to K_G(A)$ is a function. Moreover, if $f : A \to B$ and $g : B \to C$ are two maps and η is a numerable principal G-bundle over C, then the bundle $(g \circ f)^*(\eta)$ and $f^*(g^*(\eta))$ are isomorphic for every numerable principal G-bundles η over C. This shows that $K_G([g] \circ [f]) = K_G([f]) \circ K_G([g])$. Similarly, η and $1_{C^*}(\eta)$ are isomorphic and hence the function $K_G([1_C])$ is the same as the identity function on $K_G([C])$. Consequently, K_G is a contravariant functor. ❑

Corollary 5.5.4 *Let ξ be a numerable principal G-bundle over B. If $f : A \to B$ is a homotopy equivalence, then $K_G([f]) : K_G(B) \to K_G(A)$ is a bijection.*

Proof Let $f : A \to B$ be a homotopy equivalence. Then there exists a map $g : B \to A$ such that $g \circ f \simeq 1_A$ and $f \circ g \simeq 1_B$. Hence by using the property of the contravariant functor K_G(see Theorem 5.5.3), it follows that $K_G([g \circ f]) = K_G([f]) \circ K_G([g])$ is the identity function. This shows that $K_G([f])$ is a surjective function. Similarly, $K_G([f \circ g]) = K_G([g]) \circ K_G([f])$ is the identity function implying that $K_G([f])$ is a monomorphism. Hence $K_G([f])$ is a bijection. ❑

Corollary 5.5.5 *Let $\xi = (X, p, B)$ be a numerable principal G-bundles over B. If B is contractible, then ξ is trivial.*

Proof Let the space B be contractible. Then the space B is homotopy equivalent to a point $\{*\}$. Again $K_G(\{*\})$ has only one point, the isomorphism class of trivial bundle. Hence the corollary follows from Corollary 5.5.4. ❑

Let $\xi_0 = (X, p_0, B_0, G)$ be a fixed numerable principal G-bundle. For each space B, define a function $\phi_{\xi_0}(B) : [B, B_0] \to K_G(B)$ by $\phi_{\xi_0}(B)([f]) = [f^*(\xi_0)], [f] \in [B, B_0]$. This function is well defined by Corollary 5.5.2.

Hence it follows that

Theorem 5.5.6 *Given a fixed numerable principal G-bundle $\xi_0 = (X_0, p_0, B_0, G)$, $\phi_{\xi_0} : [-, B_0] \to K_G$ is a contravariant functor from the category $\mathcal{H}tp$ to the category $\mathcal{S}et$.*

We now define a class of principal G-bundles with the help of ϕ_{ξ_0}.

Definition 5.5.7 Let $\xi_0 = (X, p_0, B_0, G)$ be a principal G-bundle. Then ξ_0 is said to be universal if

(i) ξ_0 is numerable; and
(ii) $\phi_{\xi_0} : [-, B_0] \to K_G$ is a natural equivalence.

We now characterize universal G-bundles.

Proposition 5.5.8 *A numerable principal G-bundle $\xi_0 = (X_0, p_0, B_0, G)$ is universal iff*

(i) *for each numerable principal G-bundle* $\xi = (X, p, B, G)$, *there exists a map* $f : B \to B_0$ *such that* ξ *and* $f^*(\xi_0)$ *are B-isomorphic; and*
(ii) *if* $f, g : B \to B_0$ *are two maps such that* $f^*(\xi_0)$ *and* $g^*(\xi_0)$ *are isomorphic, then* $f \simeq g$.

Proof The condition **(i)** shows that the function $\phi_{\xi_0}(B) : [B, B_0] \to K_G(B)$ is surjective and condition **(ii)** shows that ϕ_{ξ_0} (B) is injective. Hence the proposition follows. ❑

5.6 Classifying Spaces: The Milnor Construction

This section presents a simple elegant method of construction of a classifying space and a universal principal fiber space associated with a principal fiber space (X, p, B, G), the method invented by John Willard Milnor (1931-) in 1956, where G is any topological group. The classification of numerable principal G- bundles up to homotopy classes which is a very important problem. This classification asserts that for every topological group G, there exists a topological space B_G such that for every pointed topological space B, there is a bijective correspondence between the isomorphism classes of numerable principal G-bundles over B and the homotopy classes$[B, B_G]$ of base point preserving continuous maps from B to B_G. It is an important example of a numerable principal G-bundle. Milnor was awarded Fields Medal in 1962 and the Abel Prize in 2011.

Example 5.6.1 (*Milnor construction*) Let G be a topological group. The universal fiber space is defined as an infinite join $X_G = G * G * \cdots * G \cdots$. An element of X_G denoted by $\langle x, t\rangle$ is written as

$$\langle x, t\rangle = (t_0x_0, t_1x_1, \ldots, t_rx_r, \ldots),$$

where each $x_i \in G$ and $t_i \in [0, 1]$ such that only a finite number of $t_i \neq 0$ and $\sum_{0\leq} t_i = 1$. We say that $\langle x, t\rangle = \langle x', t'\rangle$ in the set X_G iff $t_i = t_i'$ for each i and $x_i = x_i'$ for all i with $t_i = t_i' > 0$. We note that if $t_i = t_i' = 0$, then x_i and x_i' may be different but $\langle x, t\rangle = \langle x', t'\rangle$ in the set X_G. We define an action of G from the right $X_G \times G \to X_G$ by the relation

$$\langle x, t\rangle g = \langle xg, t\rangle \text{ or } (t_0x_0, t_1x_1, \ldots)g = (t_0x_0g, t_1x_1g, \ldots) \tag{5.1}$$

for all $g \in G$. We define a topology on X_G in such a way that X_G becomes a G-space. We consider two families of functions $f_i : X_G \to [0, 1]$ for $0 \leq i$, which assigns to the element $(t_0x_0, t_1x_1, \ldots)$ the component $t_i \in [0, 1]$ and $g_i : f_i^{-1}(0, 1] \to G$ for $0 \leq i$, which assigns to the element $(t_0x_0, t_1x_1, \ldots)$ the component $x_i \in G$. We observe that x_i cannot be uniquely defined outside $f_i^{-1}(0, 1]$ in a natural way. For

$\alpha \in X_G$ and $g \in G$, there are the following relations between the action of G and the functions f_i and g_i : $g_i(\alpha g) = g_i(\alpha)g$ and $f_i(\alpha g) = f_i(\alpha)$. The set X_G is made into a topological space by endowing X_G the smallest topology such that each of the functions $f_i : X_G \to [0, 1]$ and $g_i : f^{-1}(0, 1] \to G$ is continuous, where $f_i^{-1}(0, 1]$ has the subspace topology. From the relations in (5.1), it follows that X_G is a G-space where the G-set structure map $X_G \times G \to X_G$ is continuous. We denote the orbit space X_G mod G by B_G, the quotient map $p : X_G \to B_G$ and the resulting bundle $\omega_G = (X_G, p, B_G)$. This is known as Milnor construction and ω_G is an example of principal G-bundle.

We now illustrate this principal G-bundle ω_G in some concrete situations.

Example 5.6.2 Let S^n be the n-sphere in $\mathbf{R}^{n+1} (n \geq 1)$. If $a : S^n \to S^n, x \mapsto -x$ is the antipodal map, then $a^2 = a \circ a$ is identity 1_{S^n}. The group $G = \{1, a\} = \mathbf{Z}_2$ is a group of homeomorphisms of S^n and the space $X_G = S^n$. The action of G on X_G is given by $S^n \times G \to S^n$,

$$\begin{aligned} x \cdot g &= x \text{ if } g = 1 \\ &= -x \text{ if } g = a. \end{aligned}$$

Hence the quotient space X_G mod $G = B_G = \mathbf{R}P^n$. Hence $\omega_G = (S^n, p, \mathbf{R}P^n)$ is a numerable principal G-bundle for dimensions $\leq (n-1)$.

Example 5.6.3 If $G = S^1$ and S^{2n+1} is the $(2n+1)$-sphere in $\mathbf{R}^{2n+2}$, then the action of S^1 on S^{2n+1} is given by $(z_0, z_1, \ldots, z_n)e^{i\theta} = (e^{i\theta}z_0, e^{i\theta}z_1, \ldots, e^{i\theta}z_n)$. If $X_G = G * G * \cdots * G * \cdots$ (infinite join), then X_G mod $G = B_G = \mathbf{C}P^n$. Hence $w_G = (S^{2n+1}, p, \mathbf{C}P^n)$, is a principal numerable G-bundle of dimensions $\leq 2n$.

Definition 5.6.4 (Milnor 1956) Let G be a topological group. The functor $K_G : \mathcal{H}tp \to \mathcal{S}et$ is said to be representation if there exists a space B_G, called the classifying space of K_G and an element $\xi_G = (X_G, p_G, B_G)$, called universal bundle in $K_G(B_G)$, such that there is a natural equivalence between the functor K_G and $[-, B_G]$, defined from the category $\mathcal{H}tp$ to the category $\mathcal{S}et$.

Remark 5.6.5 The Definition 5.6.4 implies that for any space, the function $\psi(B) : [B, B_G] \to K_G(B), [f] \mapsto [f^*(\xi_G)])$ is a bijection. Moreover, any space homotopy equivalent to B_G is also a classifying space for K_G.

Remark 5.6.6 J. Feldbau reduced in 1939 the classification of principal fiber bundles (X, p, B, G) with a given base S^n for $n \geq 2$ to a problem in homotopy theory (Feldbau 1939).

We now consider the effect of continuous homomorphism on the corresponding isomorphism classes of principal bundles.

Let G and H be topological groups and $\alpha : H \to G$ be a continuous homomorphism. Define an action of H on G, $H \times G \to G, h \cdot g = \alpha(h)g$ for all $h \in H$ and $g \in G$. Let $\xi = (X, p, B, H)$ be a principal H-bundle. Define an action:

$$(X \times G) \times H \to X \times G, (x, g) \cdot h \mapsto (x \cdot h, h^{-1} \cdot g).$$

Let $(X \times G)$ mod H be the resulting orbit space and $q : (X \times G)$ mod $G \to B$, $[x, g] \mapsto p(x)$ be the projection map. Define another action

$$((X \times G) \text{ mod } G) \times G \to (X \times G) \text{ mod } G, ([x, g], g') \mapsto [x, gg'].$$

Let $\{\phi_j : U_j \times H \to p^{-1}(U_j)\}$ be a trivial covering for ξ. Define

$$\psi_j : U_j \times G \to q^{-1}(U_j), (b, g) \mapsto [\phi_j(b, e), e].$$

Then the family $\{U_j, \psi_j\}$ is a trivializing family for q. Hence we have a principal G-bundle $(X \times G)$ mod (G, q, B, G) and a natural transformation $\alpha_* : K_H \to K_G$ under the function

$$\alpha_*(B) : K_H(B) \to K_G(B), \{\xi\} \mapsto \{\alpha^*(\xi)\}.$$

Definition 5.6.7 Let $\xi = (X, p, B, G)$ be a principal G-bundle. ξ is said to have an H-structure if there exists a principal H-bundle $\eta = (Y, q, B, H)$ and a continuous homomorphism $\alpha : H \to G$ such that $\{\alpha^*(\eta)\} = \{\xi\}$.

We now study the H-structures on ξ from the view point homotopy theory

Proposition 5.6.8 *Let H and G be topological groups and $\alpha : H \to G$ be a continuous homomorphism. Then the natural transformation $\alpha_* : K_H \to K_G$ induces a unique map(upto homotopy) $B_H \to B_G$*

Proof Clearly, $K_H(Y) \cong [Y, B_H]$ and $K_G(Y) = [Y, B_G]$. Let $B\alpha : B_H \to B_G$ be the classifying map for the principal G-bundle $(XH \times G)$ mod $H \to B_H$. Then we have the commutative diagram in Fig. 5.14 where $B\alpha_*[f] = [B\alpha \circ f]$ for any map $f : Y \to B_H$. ❑

We now use the notation of Milnor construction (see Example 5.6.1)

Theorem 5.6.9 (Milnor) *Let G be a topological group. Then the G-bundle $\omega_g = (X_G, p, B_G)$ is a numerable principal G-bundle and this bundle is a universal G-bundle.*

Proof See Milnor construction (see Example 5.6.1) and Husemöller (1966). ❑

The above discussion showing the classification of numerable principal G-bundles reduced to homotopy theory is summarized in the basic and important theorem:

$$\begin{array}{ccc} K_H(Y) & \xrightarrow{\alpha} & K_G(Y) \\ \downarrow \cong & & \downarrow \cong \\ [Y, B_H] & \xrightarrow{B\alpha_*} & [Y, B_G] \end{array}$$

Fig. 5.14 A natural transformation

Theorem 5.6.10 *Given any topological group G, there exists a topological space called classifying space B_G with the property that for any space B, there exists a bijection between the set $[B, B_G]$ of homotopy classes of base point preserving maps from B to B_G and the set of isomorphism classes of numerable principal G-bundle over B.*

Example 5.6.11 Let $G = S^3$ be the multiplicative group nonzero quaternions and S^{4n+1} be the $(4n+1)$-sphere in $\mathbf{R}^{4n+2}$. If $X_G = G * G * \cdots * G$ (infinite join), then $X_G \bmod G = B_G = \mathbf{H}P^n$. Hence $w_G = (S^{4n+1}, p, \mathbf{H}P^n)$ is a principal numerable G-bundle.

5.7 Vector Bundles: Introductory Concepts

This section conveys introductory concepts of vector bundles. Vector bundles and their homotopy classifications play a very important role in mathematics and physics. Vector bundles form a special class of fiber bundles for which every fiber has the structure of a vector space compatible on neighbouring fibers and the structure group is a group of linear automorphisms of the vector space. The concept of vector bundles arose through the study of tangent vector fields to smooth geometric objects such as spheres, projective spaces, and more generally, manifolds. This section introduces the concepts of vector bundles, Gauss maps named after C.F. Gauss (1777–1855) and also gives homotopy classifications of vector bundles. For more homotopy classifications of vector bundles see Sect. 5.9.

A vector bundle over a topological B is a family of vector spaces continuously parametrized by B. If the fiber of a vector bundle ξ is $\mathbf{R}^n$, then ξ is said to be finite-dimensional with $\dim\xi = n$. On the other hand if the fiber of ξ is an infinite-dimensional Banach space and the structure group is the group of invertible bounded operators of the Banach space, the bundle ξ is said to be infinite-dimensional.

Let F denote one of the fields $\mathbf{R}$, $\mathbf{C}$ or division ring $\mathbf{H}$ of quaternionic numbers.

Definition 5.7.1 An n-dimensional F-vector bundle over B is a fiber bundle $\xi = (X, p, B, F^n)$ together with the structure of an n-dimensional vector space over F on each fiber $E_b = p^{-1}(b)$ such that there is an open covering $\{U_j : j \in J\}$ of B and for each $j \in J$, a homeomorphism $\phi_j : U_j \times F^n \to p^{-1}(U_j)$ with $p \circ \phi_j = p_{U_j}$ and $(\phi_j|\{b\} \times F^n) : \{b\} \times F^n \to p^{-1}(b)$ is an isomorphism of vector spaces over F for each $b \in U_j$. Each ϕ_j is called a coordinate transformation.

Remark 5.7.2 An n-dimensional vector bundle $\xi = (X, p, B, F^n)$ satisfies the following local triviality condition:

Each point b of B has an open neighborhood U and an U-isomorphism $\psi : U \times F^n \to p^{-1}(U)$ with the property that $\psi|_{\{b\}\times F^n} : \{b\} \times F^n \to p^{-1}(b)$ is a vector space isomorphism for each $b \in U$. The F-vector bundle ξ is said to be a real, complex or quaternionic vector bundle according as $F = \mathbf{R}$, $\mathbf{C}$ or $\mathbf{H}$. Sometimes we say that $\xi = (X, p, B)$ is an n-dimensional F-vector bundle over B.

Example 5.7.3 **(i)** For any space B, the trivial bundle $(B \times F^n, p, B, F^n)$ is an n-dimensional F-vector bundle.

(ii) For $n \geq 1$, the tangent bundle $T(S^n)$ of the n-sphere S^n is the fiber bundle $T(S^n) = (X, p, S^n, \mathbf{R}^n)$, where $X = \{(x, y) \in \mathbf{R}^{n+1} \times \mathbf{R}^{n+1} : ||x|| = 1$ and $\langle x, y\rangle = 0\}$, $p : X \to S^n$ is the projection given by $(x, y) \mapsto x$, is an n-dimensional real vector bundle. To see this we take $U_i \subset S^n$ to be the open set $U_i = \{x \in \mathbf{R}^{n+1} : ||x|| = 1, x_i \neq 0\}$, $1 \leq i \leq n+1$ and define

$$\phi_i : U_i \times \mathbf{R}^n \to p^{-1}(U_i), (x, y) \mapsto (x, f_i(y) - \langle x, f_i(y)\rangle x),$$

where $f_i : \mathbf{R}^n \to \mathbf{R}^{n+1}$ is given by

$$(y_1, y_2, \ldots, y_n) \mapsto (y_1, y_2, \ldots, y_{i-1}, 0, y_i, \ldots, y_n).$$

Then ϕ_i are linear maps on each fiber such that ϕ_i are homeomorphisms and satisfy the relation $p \circ \phi_i = p_{U_i}$.

(iii) For $n \geq 1$, the normal bundle $N(S^n)$ over S^n is the fiber bundle $\xi = (X, q, S^n, \mathbf{R}^1)$, where

$$X = \{(x, y) \in \mathbf{R}^{n+1} \times \mathbf{R}^{n+1} : ||x|| = 1, y = rx, r \in \mathbf{R}\}$$

and $q : X \to S^n$, $(x, y) \mapsto x$. Define

$$\phi : S^n \times \mathbf{R}^1 \to X, (x, r) \mapsto (x, rx)$$

and

$$\psi : X \to S^n \times \mathbf{R}^1, (x, y) \mapsto (x, \langle x, y\rangle).$$

Then ϕ is a homeomorphism with inverse ψ. Consequently, $\xi = N(S^n)$ is an 1-dimensional real trivial bundle.

(iv) The bundle $\gamma_r^n = (X, p, G_r(F^n), F^n)$, where

$$X = \{(V, y) \in G_r(F^n) \times F^n : p(V, y) \text{ is the orthogonal projection of } y \text{ into } V\},$$

is an n-dimensional F-vector bundle, where $G_r(F^n)$ is the Grassmann manifold of r-dimensional subspaces of F^n.

We now show that the local triviality condition of a vector bundle provides the following continuity proposition.

Proposition 5.7.4 *Let $\xi = (X, p, B, F^n)$ be an n-dimensional vectors bundle. Then $p : X \to B$ is an open map. Moreover, the fiber preserving function $f : X \oplus X \to X$, $(x, x') \mapsto x + x'$ and $g : F \times X \to X$, $(\alpha, x) \mapsto \alpha x$, $\alpha \in F$, are continuous.*

Proof Let $\phi : U \times F^n \to p^{-1}(U)$ be a local coordinate of ξ. Then for $f|(p^{-1}(U) \oplus p^{-1}(U))$, the above statement is valid. Similarly, for the function restricted to

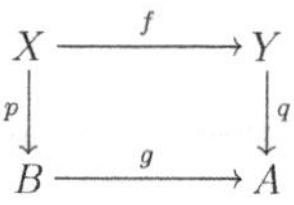

Fig. 5.15 Morphism of vector bundles

$p^{-1}(U)$, the above statement is also valid. Since the family of $p^{-1}(U)$ is an open covering of X, the functions f and g are continuous. ❑

We now show that the set of cross sections of an F-vector bundle $\xi = (X, p, B)$ form a module over the ring R of all F-vector functions continuous on B.

Proposition 5.7.5 *Let $\xi = (X, p, B)$ be an n-dimensional F-vector bundle over B. The cross sections of ξ of forms a module over the ring of continuous F-valued functions on B.*

Proof Let s, s' be two cross sections of ξ and $f : B \to F$ be a map. Then the function $s + s' : B \to X$, given by

$$(s + s')(b) = s(b) + s'(b)$$

is a cross section of ξ and the function $fs : B \to X$, given by $(fs)(b) = f(b)s(b)$ is also a cross section of ξ for all functions $f : B \to F$. Finally, the map $b \mapsto 0 \in p^{-1}(b)$ is a cross section (zero cross section). Let $\phi : U \times F^n \to p^{-1}(U)$ be a local coordinate of ξ over U. Suppose $\phi^{-1}(s(b)) = (b, g(b))$ and $\phi^{-1}(s'(b)) = (b, g'(b))$ for all $b \in B$, where $g : U \to F^n$ and $g' : U \to F^n$ are maps. Then

$$\phi^{-1}((s + s')(b)) = (b, g(b) + g'(b)), \phi^{-1}((fs)(b)) = (b, f(b)g(b))$$

and

$$\phi^{-1}(0)(b) = (b, 0)$$

for all $b \in$ U. Consequently, $s + s'$, fs and 0 are continuous and hence they are all cross sections. Consequently, the proposition follows. ❑

Definition 5.7.6 Let $\xi = (X, p, B)$ and $\eta = (Y, q, A)$ be two vector bundles. A vector bundle morphism $(f, g) : \xi \to \eta$ consists of a pair of maps $f : X \to Y$ and $g : B \to A$ such that the diagram in the Fig. 5.15 is commutative (i.e., $q \circ f = g \circ p$) and $f|_{p^{-1}(b)} : p^{-1}(b) \to q^{-1}(g(b))$ is a linear transformation for each $b \in B$. For the particular case, when $B = A$, a B-morphism of vector bundles $f : \xi \to \eta$ is defined by a morphism of the form $(f, 1_B) : \xi \to \eta$.

Remark 5.7.7 If $f : \xi \to \eta$ is a B-morphism of vector bundles, then $q \circ f = p$ and $f|_{p^{-1}(b)} : p^{-1}(b) \to q^{-1}(b)$ is a linear transformation for each $b \in B$ (Fig. 5.16).

Definition 5.7.8 Let $\xi = (X, p, B, F^n)$ and $\eta = (Y, q, A, F^n)$ be two n-dimensional F-vector bundles. A vector bundle morphism $(f, g) : \xi \to \eta$ is said

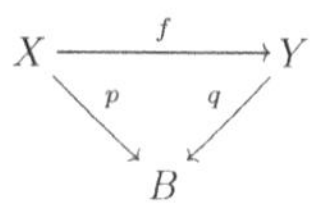

Fig. 5.16 B-morphism of vector bundles

to be an isomorphism or an equivalence if $f : X \to Y$ is a homeomorphism such that for each $b \in B$, $f|_{p^{-1}(b)} : p^{-1}(b) \to q^{-1}(g(b))$ is a linear isomorphism. In particular, $f : \xi \to \eta$ is called an equivalence over B, if $B = A$ and $g = 1_B$. In this case we write $\xi \cong \eta$.

Remark 5.7.9 Equivalent bundles have the same dimension.

Proposition 5.7.10 *Relation of being equivalent of bundles over B is an equivalence relation.*

Proof It follows from Definition 5.7.8 reflexivity and transitivity of the given relation. To show symmetry of the relation, let $\xi = (X, p, B, F^n)$ and $\eta = (Y, q, B, F^n)$ be two bundles; $f : \xi \to \eta$ be an equivalence. Then $f : X \to Y$ is a continuous 1-1 map. We claim that f is open. To prove this it is sufficient to show that $f|_{p^{-1}(U_\alpha)}$ is open, where $\{U_\alpha\}$ is an open covering of B. In terms of local coordinates, this is given by $(x, v) \to (x, A_x v)$, where A_x is a nonsingular linear transformation depending continuously on x. This map has a continuous inverse, because matrix inversion is continuous. This shows that $f|_{p^{-1}(U_\alpha)}$ is a homeomorphism and hence f is open. ❑

Definition 5.7.11 A vector bundle $\xi = (X, p, B, F^n)$ is said to be trivial if it is isomorphic to the product bundle $B \times F^n \to B$.

Example 5.7.12 Let $\xi = (B \times F^n, p, B, F^n)$ be an n-dimensional product bundle and $\eta = (B \times F^m, p, B, F^m)$ be an m-dimensional product bundle. Then the B-morphisms $f : \xi \to \eta$ are of the form $f(b, x) = (b, g(b, x))$, where $g : B \times F^n \to F^m$ is a map such that $g(b, x)$ is linear in x. If $L(F^n, F^m)$ denotes the vector space of all linear transformations $T : F^n \to F^m$, then the vector spaces $L(F^n, F^m)$ and F^{mn} are isomorphic as vector spaces. Thus $g : B \times F^n \to F^m$ is continuous iff the $\psi : B \to L(F^n, F^m), b \mapsto g(b, -)$ is a continuous function.

Theorem 5.7.13 *Let $\xi = (X, p, B)$ be a vector bundle over B and $1_B : B \to B$ be the identity map. Then the vector bundles 1_B^* and ξ are B-isomorphic. Moreover, if given a pair of maps (continuous)*

$$B_2 \xrightarrow{f} B_1 \xrightarrow{g} B,$$

the induced vector bundles $f^*(g^*(\xi))$ and $(g \circ f)^*(\xi)$ are B_2-isomorphic.

Proof We now recall that if $g^*(\xi) = (E_1, p_1, B_1)$, then

$$E_1 = \{(b_1, x) \in B_1 \times X : g(b_1) = p(x)\}, p_1 : E_1 \to B_1, (b_1, x) \mapsto b_1.$$

Hence if in particular, $g = 1_B : B \to B$, then $1_B^*(\xi)$ and ξ are B-isomorphic. Again the B_2-vector bundle induced by the map $g \circ f : B_2 \to B$ is given by

$$E((g \circ f)^*(\xi)) = \{(b_2, x) \in B_2 \times X : (g \circ f)(b_2) = p(x)\}.$$

On the otherhand, the total space of $f^*(g^*(\xi))$ is

$$\begin{aligned} E(f^*(g^*(\xi))) &= \{(b_2, y) \in B_2 \times E_1 : f(b_2) = p_1(y)\} \\ &= \{(b_2, (b_1, x)) : g(b_1) = p(x) \text{ and } f(b_2) = p_1(y) = b_1\} \\ &= \{(b_2, (f(b_2), x)) : (g \circ f)(b_2) = p(x)\}. \end{aligned}$$

Since the map $\psi : (g \circ f)^*(\xi) \to f^*(g^*(\xi))$, given by $(b_2, x) \mapsto (b_2, (f(b_2), x))$ is a bundle isomorphism from $(g \circ f)^*(\xi)$ to $f^*(g^*(\xi))$, it follows that the induced vector bundles $f^*(g^*(\xi))$ and $(g \circ f)^*(\xi)$ are B_2-isomorphic. ❑

Corollary 5.7.14 *There is a contravariant functor from the category of topological spaces $\mathcal{T}$**op** to the category of sets $\mathcal{S}et$ which assigns to a topological space B (which is an object in $\mathcal{T}$**op**), the vector bundle* $\mathrm{Vect_n}(B)$ *equals to the set of isomorphism classes of n-dimensional vector bundles over B and to any continuous map $f : B_1 \to B$ (which is a morphism in category $\mathcal{T}$**op**), there exists a function m $f^* : Vect_n(B) \to Vect_n(B_1)$ defined by $\{\xi\} \mapsto \{f^*(\xi)\}$, where $\{\xi\}$ denotes the isomorphism class of a vector bundle ξ over B.*

Definition 5.7.15 A Gauss map for an F-vector bundle $\xi = (X, p, B)$ is a continuous map $G : X \to F^m$ for some m, $1 \leq m \leq \infty$ such that for each $b \in B$, $G|_{p^{-1}(b)}$ is a linear monomorphism.

Proposition 5.7.16 *Let $\xi = (X, p, B, F^n)$ be an F-vector bundle of dimension n. If B is paracompact, then there exists a Gauss map for ξ.*

Proof Let $U_{j1}, U_{j2}, \ldots, U_{jm}$ be a finite collection of coordinate neighborhoods that cover B. Then we can choose an associated partition of unity f_{ji}.

Define

$$F_{ji} : X \to F^n$$

by the rule

$$F_{ji}(x) = \begin{cases} f_{ji}(p(x)) \cdot q\phi_{ji}^{-1}(y), & \text{if } p(x) \in U_{ji} \\ 0, & \text{if } p(x) \notin U_{ji}, \end{cases}$$

where $q : U_{ji} \times F^n \to F^n$ is the projection and ϕ_{ji} are associated coordinate transformations. This shows that F_{ji} is a linear monomorphism on $p^{-1}(b)$ if $f_{ji}(b) \neq 0$. A Gauss map $G : X \to F^{mn}$ is then given by $x \mapsto (F_{ji}(x), \ldots, F_{jm}(x))$. ❑

Proposition 5.7.17 *Let ξ be a vector bundle over a paracompact space such that $\xi|_{U_i}, i \in I$ (indexing set) is trivial, where $\{U_i\}, i \in I$, is an open covering of B.*

Then there is a countable open covering $\{V_j\}$, $1 \leq j$, *of* B *such that* $\xi|_{V_j}$ *is trivial. Moreover, if each* $b \in B$ *is an element of almost n sets* U_i, *there exists a finite open covering* $\{V_j\}$, $1 \leq j \leq n$, *of* B *such that* $\xi|_{V_j}$ *is trivial.*

Proof Since B is paracompact, there exists a partition of unity. Let $\{\lambda_i\}, i \in I$ be such that $w_i = \lambda_i^{-1}(0, 1] \subset U_i$. Then for each $b \in B$, let $T(b)$ be the finite set of $i \in I$ such that $\lambda_i(b) > 0$. For each finite subset $T \subset I$, let $V(T)$ be the open subset of all $b \in B$ such that $\lambda_i(b) > \lambda_j(b)$ for each $i \in T$ and $j \notin T$. If T and T' are two distinct subsets of I each with m elements, then $V(T) \cap V(T') = \emptyset$. Let V_m be the union of all $V(T(b))$ such that $T(b)$ has m elements. Since $i \in T(b)$ gives the relation $V(T(b)) \subset W_i$, then bundle $\xi|_{V(T(b))}$ is trivial. Since V_m is a disjoint union, $\xi|_{V_m}$ is trivial. Moreover, under the last hypothesis, $V_j = \emptyset$ for $j > n$. ❑

Theorem 5.7.18 *For each n-dimensional vector space* $\xi = (X, p, B, F^n)$ *over a paracompact space* B, *there is a Gauss map* $g : E(\xi) \to F^\infty$, *where* $E(\xi) = X$. *If* B *has an open covering of sets* $\{U_i\}$, $1 \leq i \leq m$, *such that* $\xi|_{U_i}$ *is trivial, then* ξ *has a Gauss map* $G : E(\xi) \to F^{nm}$.

Proof By Proposition 5.7.17, we may assume that $\{U_i\}$ is the countable or finite open covering of B such that $\xi|_{U_i}$ is trivial. Let $\psi_i : U_i \times F^n \to \xi|_{U_i}$ be U_i-isomorphisms, and let $\{\lambda_i\}$ be a partition of unity with closure of $\lambda_i^{-1}((0, 1]) \subset U_i$. Define a map

$$g : E(\xi) \to \sum_i F^n, g \mapsto \Sigma g_i,$$

where $g_i|_{E(\xi|_{U_i})}$ is $(\lambda_i p)(p_2\psi_i^{-1})$ and $p_2 : U \times F^n \to F^n$ is the projection to the second factor. The map g_i is zero outside $E(\xi|_{U_i})$. Since $g_i : E(\xi) \to F^n$ is a monomorphism on the fibers of $E(\xi)$ over b with $\lambda_i(b) > 0$ and the images of g_i are in complementary subspaces, the map g is a Gauss map. In general, $\sum_i F^n$ is F^∞, but if there are only m sets U_i, then ΣF^n is F^{mn}. ❑

Corollary 5.7.19 *Every n-dimensional vector bundle* $\xi = (X, p, B, F^n)$ *over a paracompact* B *is* B*-isomorphic to* $f^*(\gamma_n^\infty)$ *for some* $f : B \to G_n(F^\infty)$.

Remark 5.7.20 Theorem 5.7.18 and Corollary 5.7.19 form a homotopy classification theorem for vector bundles. For other homotopy classification theorems see Sect. 5.9

We now use the following notations for Bott Periodicity Theorem:

Notations: If $F = \mathbf{R}$, $G_n(F^\infty)$ is written BO;
If $F = \mathbf{C}$, $G_n(F^\infty)$ is written BU;
If $F = \mathbf{H}$, $G_n(F^\infty)$ is written BS_p

Theorem 5.7.21 (Bott Periodicity Theorem)

(i) $\Omega^2 BU \simeq BU \times \mathbf{Z}$
(ii) $\Omega^4 BS_p \simeq BO \times \mathbf{Z}$
(iii) $\Omega^4 BO \cong BS_p \times \mathbf{Z}$

Proof For proof see (Bott 1957, 1959) and Remark 5.7.22 ❑

Remark 5.7.22 All of the basic three methods of proving 'Bott Periodicity Theorem' are complicated. The first method prescribed by R. Bott in 1957 and 1959 appeared in his papers [The stable homotopy groups of classical groups, Proc. Nat. Acad. Sci., USA, 43(1957), 933-935 and Ann. of Maths, **70**(1959), 313-337] by using Morse Theory to analyze ΩX for a Lie group X. The other references in this respect are Milnor (1963), Toda (1959), Dyer and Lashof (1961) and for the general case Atiyah and Bott (1964).

Remark 5.7.23 An important relation between Bott Periodicity Theorem and stable homotopy groups of spheres π_n^S came through the so-called stable J-homomorphism from the (unstable) homotopy groups of the (stable) classical groups to these stable homotopy groups π_n^S. Following the original description given by G. W. Whitehead, it became the subject of the famous Adams conjecture posed by J.F. Adams in 1963. This conjecture is true which was proved by Daniel Quillen in 1971.

5.8 Charts and Transition Functions of Bundles

This section studies charts and transition functions of bundles and establishes a one-one correspondence between the set of equivalence classes of principal G-bundles ξ over a space B and the set of equivalence classes of transition functions associated with an atlas of ξ. Let G be a topological group with identity e and F be a G-space. We assume in this section that all principal bundles are G-bundles and all fibers have fiber F(or F^n). Given a space B, let $(B \times F, p, B)$ denote the product fiber bundle.

Definition 5.8.1 Let $\xi = (X, p, B, G)$ be a principal G-bundle. A chart (U, ϕ) of ξ is a pair consisting of an open set $U \subset B$ and a homeomorphism $\phi : U \times G \to p^{-1}(U)$ such that $p \circ \phi = p_U$ and $\phi(b, g) = \phi(b, e) \cdot g$, $\forall\, b \in U$ and $\forall\, g \in G$. An atlas is a family of charts $\{(U_j, \phi_j) : j \in J\}$ such that $\{U_j : j \in J\}$ is an open covering of B and each homeomorphism $\phi_j : U_j \times G \to p^{-1}(U_j)$ is such that $p \circ \phi_j = p_{U_j}$ and $\phi_j(b, g) = \phi_j(b, e) \cdot g$, $\forall\, b \in U_j$ and $\forall\, g \in G$. An atlas is said to be complete if it includes all the charts of ξ.

Example 5.8.2 Every principal G-bundle has at least one atlas.

Definition 5.8.3 Given a space B and a topological group G, a set of transition functions $\tilde{\xi}$ for B and G consists of an open covering $\{U_j : j \in J\}$ of B and a family of maps $g_{jk} : U_j \cap U_k \to G$ such that $U_j \cap U_k \neq \emptyset$ and $g_{jl}(b) = g_{jk}(b) g_{kl}(b)$, $\forall\, b \in U_j \cap U_k \cap U_l (\neq \emptyset)$. Each function g_{jk} is called a transition function defined on $U_j \cap U_k$.

For $j = k = l$, $g_{jj}(b) = g_{jj}(b)g_{jj}(b) \in G$ for all $b \in U_j$ implies that $g_{jj}(b) = e$ for all $b \in B$. Again for $j = l$, $e = g_{jj}(b) = g_{jk}(b)g_{kj}(b) \in G$ for all $b \in U_j \cap U_k$ implies that $g_{jk}(b) = g_{kj}^{-1}(b)$ for all $b \in U_j \cap U_k$.

Definition 5.8.4 Let $\xi = (X, p, B, G)$ and $\xi' = (X', p', B', G)$ be two principal G-bundles and $\tilde{\xi} = \{(U_j, g_{jk}) : j, k \in J\}$ and $\tilde{\xi}' = \{(U'_a, g'_{ab}) : a, b \in A\}$ be the sets of transition functions of ξ and ξ', respectively. A morphism $f : \tilde{\xi} \to \tilde{\xi}'$ between two sets of transition functions is a pair consisting of a map $\tilde{f} : B \to B'$ and a collection of maps $f_{aj} : U_j \cap \tilde{f}^{-1}(U'_a) \to G$ such that

$$f_{aj(b)}g_{jk}(b) = g'_{ab}(\tilde{f}(b))f_{bk} \text{ for all } b \in U_j \cap U_k \cap \tilde{f}^{-1}(U'_a) \cap \tilde{f}^{-1}(U'_b).$$

We now consider the particular case, when $\tilde{f} : B \to B$ is 1_B.

Definition 5.8.5 Two sets of transition functions $\{(U_j, g_{jk}) : j, k \in J\}$ and $\{(U'_a, g'_{ab}) : a, b \in A\}$ for the same space B, are said to be equivalent if there exist maps $f_{aj} : U_j \cap U'_a \to G$, such that

$$g'_{ab}(b) = f_{aj}(b)g_{jk}(b)f_{bk}^{-1}(b) \text{ for all } b \in U_i \cap U_j \cap U'_a \cap U'_b, i, j \in J \text{ and } a, b \in A.$$

We now recall the following definition.

Definition 5.8.6 Let $\xi = (X, p, B, F^n)$ and $\xi' = (X', p', B', F^n)$ be two n-dimensional F- vector bundles. A morphism $\phi : \xi \to \xi'$ is a pair of maps $\phi : X \to X'$ and $\tilde{\phi} : B \to B'$ such that $p' \circ \phi = \tilde{\phi} \circ p$ and $\phi|p^{-1}(b) : p^{-1}(b) \to p'^{-1}(\tilde{\phi}(b))$ is a linear map for each $b \in B$. The identity maps $1_X : X \to X$ and $1_B : B \to B$ define an identity morphism $1_d : \xi \to \xi$.

We now establish a relation between the concepts of a vector bundle and a principal G-bundle. To do this we first show that given an atlas of a principal G-bundle, there exists a unique set of transition functions.

Proposition 5.8.7 **(a)** *Corresponding to a principal G-bundle over B, there exists a unique set of transition functions $\tilde{\xi} = \{(U_j, g_{jk}) : j, k \in J\}$ such that the map $\phi_k : U_k \times G \to p^{-1}(U_k)$ satisfies the relation*

$$\phi_k(b, g) = \phi_j(b, g_{jk}(b)g), \forall\, b \in U_j \cap U_k, g \in G.$$

(b) *Let $\xi = (X, p, B, G)$ and $\xi' = (X', p', B', G)$ be two G-bundles and $\tilde{\xi}$ and $\tilde{\xi}'$ be two sets of transition functions as in (a). If $\phi : \xi \to \xi'$ is a bundle morphism, then ϕ induces a unique morphism of sets of transition functions $f : \tilde{\xi} \to \tilde{\xi}'$ such that*

$$\begin{aligned} \tilde{f} = \tilde{\phi} : B \to B' \text{ and } \phi \circ \phi_j(b, g) \\ = \phi'_a(\tilde{\phi}(b), f_{aj}(b)g), \forall\, b \in U_j \cap \tilde{\phi}^{-1}(U'_a), g \in G. \end{aligned}$$

Proof **(a)** Let $j, k \in J$. Then the map

$$\psi_{jk} = \phi_j^{-1} \circ (\phi_k|(U_j \cap U_k) \times G) : (U_j \times U_k) \times G \to (U_j \times U_k) \times G$$

is such that $p_{U_j \cap U_k} \circ \psi_{jk} = p_{U_j \cap U_k}$. Now we can write ψ_{jk} in the form $\psi_{jk}(b, f_{jk}(b, g))$ for some $f_{jk} : (U_i \cap U_k) \times G \to G$. Hence

$$\phi_k(b, g) = \phi_j(b, f_{jk}(b, g)), \ \forall\, b \in U_j \cap U_k, g \in G.$$

Consequently,

$$\begin{aligned}\phi_j(b, f_{jk}(b, g)) &= \phi_k(b, g) = \phi_k(b, e) \cdot g = \phi_j(b, f_{jk}(b, e)) \cdot g \\ &= \phi_j(b, f_{jk}(b, e) \cdot g), \ \forall\, b \in U_j \cap U_k\end{aligned}$$

and $g \in G$. This shows that $f_{jk}(b, g) = f_{jk}(b, e) \cdot g$. Hence if we take $g_{jk}(b) = f_{jk}(b, e)$, $\forall\, b \in U_j \cap U_k$, $j, k \in J$, then the requisite condition for ϕ_k is satisfied. Next let, $b \in U_j \cap U_k \cap U'_a$. Then

$$\phi_j(b, g_j(b)g) = \phi_a(b, g) = \phi_k(b, g_{ka}(b)g) = \phi_j(b, g_{jk}(b)g_{ka}(b)g).$$

This shows that $g_{ja}(b) = g_{jk}(b)g_{ka}(b)$. Hence $\tilde{\xi} = \{(U_j, g_{jk}) : j, k \in J\}$ is a set of transition functions.

(b) Let $j, k \in J$ and $a, b \in A$. Then the maps

$$\begin{aligned}\phi_{aj} = {\phi'_a}^{-1} \circ \phi(\phi_j|(U_j \cap \tilde{\phi}^{-1}(U'_a) \times G) &: U_j \cap \tilde{\phi}^{-1}(U_j \cap \tilde{\phi}^{-1}(U'_a) \\ &\times G \to U'_a \times G)\end{aligned}$$

are such that $p_{U'_a} \circ \theta_{aj} = \tilde{\phi} \circ p_{U_j} \cap \tilde{\phi}^{-1}(U'_a)$, where θ_{aj} is of the form $\theta_{aj}(b, g) = (\tilde{\phi}(b), h_{aj}(b, g))$ for some $h_{aj} : (U_j \cap \tilde{\phi}^{-1}(U'_a)) \times G \to G$. Hence

$$\phi \circ \phi_j(b, g) = \phi'_a(\tilde{\phi}(b), h_{aj}(b, g)), \ \forall\, b \in U_j \cap \tilde{\phi}^{-1}(U'_a)$$

and $\forall\, g \in G$. Consequently,

$$\begin{aligned}\phi'_a(\tilde{\phi}(b), h_{aj}(b, g)) &= (\phi \circ \phi_j(b, e)) \cdot g = \phi'_a(\tilde{\phi}(b), h_{aj}(b, e) \cdot g) \\ &= \phi'_a(\tilde{\phi}(b), h_{aj}(b, e)g).\end{aligned}$$

Hence it follows that $h_{aj}(b, g) = h_{aj}(b, e)g$. If we take $f_{aj}(b) = h_{aj}(b, e)$, $\forall\, b \in U_j \cap \tilde{\phi}^{-1}(U'_a)$, $j \in J$ and $a \in A$, then the requisite condition for (b) is satisfied. Next suppose that $\tilde{\xi} = \{(U_j, g_{jk}) : j, k \in J\}$ and $\tilde{\xi}' = \{(U'_a : g'_{ab}) : a, b \in A\}$. Hence it follows that

$$\phi'_a(\tilde{\phi}(b), f_{aj}(b)g_{jk}(b)g) = \phi \circ \phi_j(b, g_{jk}(b)g) = \phi \circ \phi_k(b, g)$$
$$= \phi'_b(\tilde{\phi}(b), f_{bk}(b)g) = \phi'_a(\tilde{\phi}(b), g'_{ak}(\tilde{\phi}(b)) f_{bk}(b)g).$$

Consequently, it follows that $f_{aj}(b)g_{jk}(b) = g'_{al}(\tilde{\phi}(b)) f_{lk}(b)$. Hence $\{f_{aj}\}$ is a morphism of sets of transition functions. ❑

Theorem 5.8.8 *Given a topological group G, there exists a (1-1)- correspondence between the equivalence classes of principal G-bundles ξ over a fixed base space B and the equivalence classes of sets of transition functions associated with an atlas of ξ.*

Proof Let ξ and ξ' be two principal G-bundles over B and $\{\tilde{\xi}\}$ be the set of equivalence classes of the sets of transition functions associated with an atlas of ξ given in Proposition 5.8.7(a). If $\phi : \xi \to \xi'$ is an equivalence of G-bundles, then by Proposition 5.8.7(b), there exists a morphism $f(\phi) : \tilde{\xi} \to \tilde{\xi}'$. Since here $\tilde{\phi} : B \to B$ is 1_B, this correspondence is well defined. Next suppose that $\tilde{\xi}$ is any set of transition functions. Then by Ex. 5 of Sect. 5.13, we have a principal G-bundle ξ and an atlas $\{(U_j : \phi_j) : j \in J\}$ of ξ such that $\tilde{\xi}$ is the corresponding set of transition functions. Hence $f(\phi)$ is surjective. To show that $f(\phi)$ is injective, let ξ and ξ' be two G-bundles such that the corresponding sets $\tilde{\xi}$ and $\tilde{\xi}'$ of transition functions are equivalent, by an equivalence $f : \tilde{\xi} \to \tilde{\xi}'$. Then by Ex. 5 of Sect. 5.13, there is a morphism $\phi : \xi \to \xi'$ inducing f. In particular, $\tilde{\phi} : B \to B$ is 1_B. Moreover, we have the morphism $f^{-1} : \tilde{\xi}' \to \tilde{\xi}$ defined by

$$f^{-1} = \{f_{aj}^{-1} : a \in A, j \in J\}.$$

Then the corresponding morphism $\psi : \xi' \to \xi$ of G-bundles is the inverse of ϕ. Because,

$$(\psi \circ \phi \circ \phi_j)(b, g) = (\psi \circ \phi)(\phi_j(b, g)) = \psi \circ \phi'_a(b, f_{aj}(b)g)$$
$$= \phi_j(b, f_{aj}^{-1}(b) f_{aj}(b)g) = \phi_j(b, g),$$

$\forall\, b \in U_j \cap U'_a$ and $\forall\, g \in G, j \in J, a \in A$. This shows that $\psi \circ \phi = 1_d$. Similarly, $\phi \circ \psi = 1_d$. Hence ξ and ξ' are equivalent principal G-bundles. Consequently the correspondence $f(\phi)$ is injective. Hence f is a bijection. ❑

Let GL $(n, F) = G$ denote the group of all nonsingular $n \times n$ matrices over F. This is a topological group. We show that there exists a one-one correspondence between the equivalence classes of n-dimensional F-vector bundles over B and the equivalence classes of sets of transition functions for B and $G =$ GL (n, F) and also a one-one correspondence with principal GL (n, F)-bundles.

We now define chart and atlas of a vector bundle in a way analogous to Definition 5.8.1.

Definition 5.8.9 Let $\xi = (X, p, B, F^n)$ be an n-dimensional F-vector bundle over B. A chart (U, ϕ) of ξ is a pair consisting of an open set $U \subset B$ and a homeomorphism

$\phi : U \times F^n \to p^{-1}(U)$ such that $p \circ \phi = p_U$ and ϕ is linear on all fibers p^{-1}, $b \in B$. An atlas is a family $\{(U_j, \phi_j) : j \in J\}$ of charts such that $\{U_j : j \in J\}$ is an open covering of B.
ξ has at least one atlas.

Construction 5.8.10 *Given an atlas of a vector bundle, a method of construction of a set of transition functions is prescribed:*
Let $\xi = (X, p, B, F^n)$ be an n-dimensional F-vector bundle over B and $\{(U_j, \phi_j) : j \in J\}$ be a given atlas of ξ over B. We now construct a set of transition functions $\{(U_j, g_{jk}) : j, k \in J\}$ for B and the group GL (n, F) *as follows:*
for $j, k \in J$, the maps

$$\psi_{jk} = {\phi_j}^{-1} \circ (\phi_k|(U_j \cap U_k) \times F^n) : (U_j \cap U_k) \times F^n \to (U_j \cap U_k) \times F^n$$

are of the form $\psi_{jk}(b, u) = (b, f_{jk}(b, u))$ for some $f_{jk} : (U_j \cap U_k) \times F^n \to F^n$. Clearly, given a fixed $b \in U_j \cap U_k$, the map $f_{jk}(b, -) : F^n \to F^n$ is a linear isomorphism and hence $f_{jk}(b, -) \in$ GL (n, F). *Taking $g_{jk}(b) = f_{jk}(b, -)$, we have $f_{jk}(b, b) = g_{jk}(b) \cdot u$. Hence*

$$\phi_k(b, u) = \phi_j(b, g_{jk}(b) \cdot u), \; \forall\, b \in U_j \cap U_k, u \in F^n.$$

Following the above construction, a basic and important theorem is obtained.

Theorem 5.8.11 *There exists a one-one correspondence between equivalence classes of n-dimensional F-vector bundles over a space B and the equivalence classes of the set of transition functions for B and the group* GL (n, F).

5.9 Homotopy Classification of Vector Bundles

This section presents two main theorems on the homotopy classification of vector bundles. The problems of homotopy classification of vector bundles are very interesting in algebraic topology. This section studies the homotopy classification (see Theorem 5.9.5 and Corollary 5.9.7) of vector bundles which leads to define K-theory. The reader is referred to the book Husemöller (1966).

Theorem 5.9.1 *Let B be a paracompact space and $f, g : B \to A$ be two homotopic maps. If ξ is a vector bundle over the space A, then the induced bundles $f^*(\xi)$ and $g^*(\xi)$ over B are B-isomorphic.*

Proof As $f \simeq g : B \to A$, there exists a map $H : B \times I \to A$ such that

$$H(x, 0) = f(x), H(x, 1) = g(x), \; \forall\, x \in B.$$

Hence $f^*(\xi)$ and $H^*(\xi)$ are both vector bundles(see Ex. 4 of Sect. 5.13) such that $f^*(\xi)$ and $H^*(\xi)|(B \times \{0\})$ are B-isomorphic. Similarly, $g^*(\xi)$ and $H^*(\xi)|(B \times \{1\})$ are B-isomorphic. Since there exists an isomorphism

$$(\alpha, \beta) : H^*(\xi) : (B \times \{0\}) \to H^*(\xi)|(B \times \{1\})(\text{see Ex. 4 of Sect. 5.13}),$$

it follows that $f^*(\xi)$ and $g^*(\xi)$ are B-isomorphic. ❑

Corollary 5.9.2 *Every vector bundle over a contractible paracompact space is trivial.*

Proof Let B be a contractible paracompact space. As B is contractible, the identity map $1_B : B \to B$ and the constant map $f : B \to B$ are homotopic. Let ξ be an n-dimensional vector bundle over B. Then $1_B^*(\xi)$ and ξ are B-isomorphic and $f^*(\xi)$ is B-isomorphic to the product bundle $(B \times F^n, p, B)$. Since $1_B \simeq f$, it follows by Theorem 5.9.1 that ξ is B-isomorphic to the product bundle $(B \times F^n, p, B)$ which is a trivial bundle. ❑

We now prove a classification theorem of vector bundles by proving that there is a (1-1) correspondence between isomorphism classes of n-dimensional vector bundles over a paracompact space B and the homotopy classes of maps from B to Grassmann manifold $G_n(F^\infty)$. Let $\mathcal{H}$ denote the category of paracompact spaces and their homotopy classes, and $\mathcal{S}et$ denote the category of sets and their functions. Let $\text{Vect}_\text{n}(B)$ denote the set of B-isomorphic classes of n-dimensional vector bundles over B. Given an n-dimensional vector bundle ξ over B, let $\{\xi\}$ denote the B-isomorphism class in $\text{Vect}_\text{n}(B)$ of ξ and $[f]$ denote the homotopy class of $f : A \to B$ between two paracompact spaces A and B. We now study $\text{Vect}_n(-)$.

Theorem 5.9.3 $\text{Vect}_\text{n} : \mathcal{H} \to \mathcal{S}et$ *is a contravariant functor.*

Proof The object function is defined by assigning to each object B in $\mathcal{H}$, the set $\text{Vect}_\text{n}(B)$ of B-isomorphic classes of n-dimensional vector bundles over B which is an object in the category $\mathcal{S}et$. Define the morphism function

$$\text{Vect}_\text{n}([f]) : \text{Vect}_\text{n}(B) \to \text{Vect}_\text{n}(A)$$

given by

$$\text{Vect}_\text{n}([f])(\{\xi\}) = \{f^*(\xi)\}$$

for every morphism $[f]$ in $\mathcal{H}$, where $f : A \to B$ is a continuous map between paracompact spaces. This function is well defined by Theorem 5.9.1. For the identity map $1_B : B \to B$, $1_B^*(\xi)$ and ξ are B-isomorphic. This shows that $\text{Vect}_\text{n}([1_d])$ is the identity function. Finally, let $[g]$ denote the homotopy class of $g : C \to A$ between the paracompact spaces C and A. Then $f \circ g : C \to B$. Hence f induces a vector bundle over A and g, $f \circ g$ induce vector bundles over C such that $g^*(f^*(\xi))$ and $(f \circ g)^*(\xi)$ are C-isomorphic. This shows that $\text{Vect}_n([f][g]) = \text{Vect}_n[g]\text{Vect}_n[f]$. Consequently, Vect_n is a contravariant functor from $\mathcal{H}$ to $\mathcal{S}et$. ❑

Let $F = \mathbf{R}, \mathbf{C}$ or $\mathbf{H}$. Then the natural inclusion $G_n(F^m) \subset G_n(F^{m+1})$ defines a space $G_n(F^\infty) = \bigcup_{n \leq m} G_n(F^m)$ with induced topology, called Grassmann manifold $G_n(F^\infty)$. Let γ_n^∞ denote the n-dimensional vector bundle over Grassmann manifold $G_n(F^\infty)$. Clearly, $\phi_n = [-, G_n(F^\infty])$ and $\mathrm{Vect_n}$ are both contravariant functors from $\mathcal{H}$ to $\mathcal{S}et$.

Theorem 5.9.4 *The natural transformation* $\psi : [-, G_n(F^\infty)] \rightarrow \mathrm{Vect_n}$ *is a natural equivalence.*

Proof For each object B in the category $\mathcal{H}$, define the function $\psi(B) : [B, G_n(F^\infty)] \rightarrow \mathrm{Vect_n}(B)$ given by $\psi(B)([f]) = \{f^*(\gamma_n^\infty)\}$. Clearly, $\psi(B)$ is well defined. We claim that ψ is a natural transformation of contravariant functors. Let $[f]$ be the homotopy class of the map $f : A \rightarrow B$ between paracompact spaces A and B. Then the diagram in Fig. 5.17 is commutative. To show this, let $[g] \in [B, G_n(F^\infty)]$. Then

$$(\mathrm{Vect}_n([f]) \circ \psi(B))([g]) = \mathrm{Vect}_n([f])\{g^*(\gamma_n^\infty)\} = \{(f^*g^*(\gamma_n)).$$

On the otherhand,

$$\psi(A)\phi([f][g]) = \psi(A)([g \circ f]) = \{(g \circ f)^*(\gamma_n^\infty)\}.$$

For each B in $\mathcal{H}$, $\psi(B)$ is injective by Ex. **9(b)** of Exercises 5.13 and surjective by Corollary 5.7.19. Hence for each B, $\psi(B)$ is a bijection. Consequently, ψ is an equivalence. ❑

Theorem 5.9.4 gives the following classification of vector bundles:

Theorem 5.9.5 (Homotopy classification of vector bundles) *There exists a one-one correspondence between isomorphism classes of n-dimensional F-vector bundles on a paracompact space and the homotopy classes of maps from B to Grassmann manifold* $G_n(F^\infty)$.

Proof For each paracompact space B, the function $\psi(B) : [B, G_n(F^\infty)] \rightarrow Vect_n(B)$ defined by $\psi(B)([f]) = \{f^*(\gamma_n^\infty)\}$ is a bijection. Hence the theorem follows. ❑

Definition 5.9.6 The natural equivalence $\psi : [-, G_n(F^\infty)] \rightarrow Vect_n$ is called a representation of the contravariant functor $\mathrm{Vect_n}$.

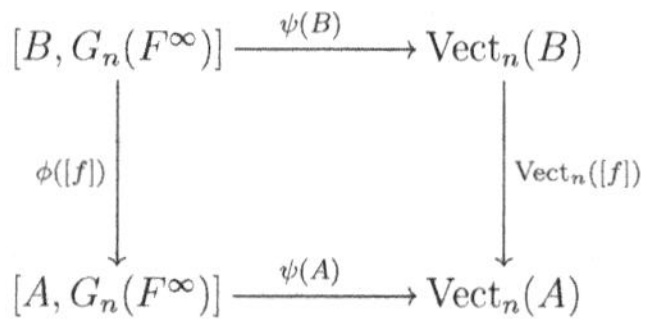

Fig. 5.17 Natural equivalence

Corollary 5.9.7 (Classification of vector bundles) *Every n-dimensional F-vector bundle over a paracompact space B is isomorphic to the vector bundle induced by a map from the base space B to the Grassmann manifold* $G_n(F^\infty)$.

Proof Let $\xi = (E, p, B)$ be an n-dimensional F-vector bundle over a paracompact space B. Since $\psi(B)$ is a bijection, there exists a continuous map $f : B \to G_n(F^\infty)$ such that $f^*(\{\gamma_n^\infty\}) = \{\xi\}$. This implies that $f^*(\gamma_n^\infty)$ and ξ are B-isomorphic. ❑

5.10 *K*-Theory: Introductory Concepts

This section conveys introductory concept of K-theory (topological) which is a branch of algebraic topology. This branch created around 1960 by Alexander Grothendieck (1928–2014) in his study of intersection theory on algebraic varieties is a surprising theory. Topological K-theory is a branch of algebraic topology. It is the first example of generalized cohomology theories (see Chap. 15). These are groups in the sense of abstract algebra. They contain detailed information about the original object but are very difficult to compute; for example, an important outstanding problem is to compute the K-theory of the integers. This theory connects algebraic topology with algebraic geometry, analysis, ring theory, and number theory.

The concept of K-theory arose through the study of vector bundles. The rich structure of vector bundles establishes that the set of isomorphism classes of n-dimensional vector bundles over a paracompact space B has a natural bijective correspondence with the set of homotopy classes of maps from B into a Grassmann manifold of n-dimensional subspaces in an infinite-dimensional space. This result motivated M. F. Atiyah (1929 -) and F. E. Peter Hirzebruch (1927–2012) to introduce K-theory' in 1961 by using the Grothendick construction. The early work on topological K-theory is due them. Given a compact Hausdorff space X and $F = \mathbf{R}$ or $\mathbf{C}$ or $\mathbf{H}$, let $K_F(X)$ be the Grothendieck group of the abelian monoid of isomorphism classes of finite-dimensional F-vector bundles over X under Whitney sum. Tensor product of bundles gives K_F-theory a commutative ring structure. $K(X)$ usually denotes complex K-theory; on the other hand, real K-theory is sometimes denoted by $KO(X)$.

Whitney sum of two vector bundles, which is a generalization of the concept of direct sum of vector spaces to vector bundles is now defined.

Definition 5.10.1 Let $F = \mathbf{R}$ or $\mathbf{C}$ or $\mathbf{H}$ and ξ and η be F- vector bundles over B. Then their Whitney sum $\xi \oplus \eta$ is the vector bundle over B such that the fibers of $\xi \oplus \eta$ is the direct sum of the fibers in ξ and η.

Construction 5.10.2 (of $\xi \oplus \eta$) *Consider the F- vector bundles $\xi = (X, p, B)$ and $\eta = (X', p', B)$. Then $\xi \times \eta = (X \times X', p \times p', B \times B)$. Let*

$$d : B \to B \times B, b \mapsto (b, b).$$

Define

$$\xi \oplus \eta = d^*(\xi \times \eta).$$

Then its total space is

$$E(\xi \oplus \eta) = \{(x, x') \in (X \times X') : p((x) = p'(x')\}.$$

If ξ_1, ξ_2 *and* ξ_3 *are vector bundles over* B, *then*

$$\xi_1 \oplus \xi_2 \cong \xi_2 \oplus \xi_1;$$
$$\xi_1 \oplus (\xi_2 \oplus \xi_3) \cong (\xi_1 \oplus \xi_2) \oplus \xi_3;$$
$$0 \oplus \xi = \xi;$$

where 0 *is the* 0*-plane bundle. Moreover, if* $\xi_1 \cong \xi_2$ *and* $\xi_3 \cong \xi_4$, *then* $\xi_1 \oplus \xi_3 \cong \xi_2 \oplus \cong \xi_4$. *Consequently, the set of* B*-isomorphism classes of vector bundles over* B *is an abelian monoid under the operation* $\oplus$.

There is a standard construction of group (ring) completion of an abelian monoid (semiring), called Grothendieck Construction given by Grothendieck.

Grothendieck Construction

(i) *(Grothendieck group) Let* M *be an abelian monoid. Take the quotient of the free abelian group generated by the elements of* M *bu the subgroup generated by the set of elements of the form* $x + y - (x \oplus y)$, *where* $\oplus$ *is the sum on* M. *The morphism* $i : M \to G(M)$ *of abelian monoids is universal for any homomorphism of monoids* $f : M \to G$, *where* G *is an abelian group. Then there is a unique homomorphism of groups* $\tilde{f} : G(M) \to G$ *such that* $\tilde{f} \circ i = f$.

(ii) *(Grothendieck ring) Let* M *be a semiring. Then its multiplication induces a multiplication on* $G(M)$ *such that* $G(M)$ *is a ring, called the Grothendieck ring of* M. *If the semiring* M *is commutative, then the ring* $G(M)$ *is also commutative.*

We now use the Grothendieck construction to its group completion as follows: Let $G(B)$ *be the free group generated by isomorphism classes of vector bundles over* B. *Let* $[\xi]$ *be the isomorphism class in* $G(B)$ *corresponding to the vector bundle* ξ. *Let* S *be the subgroup generated by all elements of the form* $[\xi] \oplus [\eta] - [\xi \oplus \eta]$ *and* $K_F(B)$ *be the quotient group.*

Definition 5.10.3 The K-theory on B, denoted by $K_F(B)$, is the Grothendieck ring of the semiring $\mathcal{V}ect_F(B)$. We write $[\xi]$ for the element of $K_F(B)$ determined by a vector bundle ξ. Grothendieck group $K_F(B)$ is also called K-theory on the category of all F-vector bundles over the base space B.

Example 5.10.4 The K-theory of a point are the integers, because vector bundles over a point are trivial and hence classified by their rank and the Grothendieck group of the natural numbers are the integers.

Remark 5.10.5 The relation $[\xi] \oplus [\eta] - [\xi \oplus \eta] = 0$ holds. This relation shows that in the group $K_F(B)$ the direct sum corresponds to the group operation.

Remark 5.10.6 For $F = \mathbf{R}$ or $\mathbf{C}$ or $\mathbf{H}$, $K_F(B)$ is the group generated by all real vector bundles, complex vector bundles or quaternionic vector bundles over B, respectively.

Construction 5.10.7 *[of $\xi \otimes_F \eta$] Let ξ be an F-vector bundle over B of dimension m and η be an F-vector bundle over B' of dimension n. We construct an F-vector bundle $\xi \otimes_F \eta$ over $B \times B'$ of dimension mn with fibers $\xi_b \otimes_F \eta_{b'}$ over (b, b') as follows:*
take the largest atlases

$$\{(U_a, h_a) : a \in A\}, \{(V_c, h'_c) : c \in C\}$$

for ξ, η with corresponding system

$$\{U_a, g_{ak} : a, k \in A\}, \{V_c, g'_{cd} : c, d \in C\}$$

of transition functions. Then

$$\{U_a \times V_c, f_{nm} \circ (g_{ak} \times g'_{cd}) : a, b \in A, c, d \in C\}$$

is a system of transition functions for GL (nm, F) *on $B \times B'$, where for the standard basis $e_1, e_2 \cdots, e_n$ of F^n,*

$$f_{nm} : \mathrm{GL}\,(n, F) \times \mathrm{GL}\,(m, F) \to \mathrm{GL}\,(nm, F)$$

is the tensor product homomorphism obtained by identifying $F^n \otimes F^m$ with F^{nm} by the unique isomorphism which sends $e_i \times e_j$ to $e_{(i-1)m+j}$, for $1 \leq i \leq n$, $1 \leq j \leq m$ with $f_{nm}(M, N)$ to be the usual automorphism

$$M \otimes N : F^n \otimes F^m \to F^n \otimes F^m.$$

This gives an nm-dimensional vector bundle over $B \times B'$, denoted by $\xi \otimes_F \eta$.

We now describe the isomorphism class of $\xi \otimes_F \eta$ as follows:
If we write $\xi = \xi'(F^n)$, $\eta = \eta'(F^m)$, where ξ' is a principal GL (n, F)-bundle and η' is a principal GL (m, F)-bundle, then $(f_{nm})_*(\xi' \times \eta')$ is a GL (nm, F)-bundle with

$$\xi \otimes_F \eta \cong ((f_{nm})_*(\xi' \times \eta'))[F^n \otimes_F F^m] \tag{5.2}$$

The relation (5.2) gives the following properties of $\otimes_F$.

Proposition 5.10.8 **(i)** *$\otimes_F$ is a functor of two variables.*
(ii) *$(\xi \otimes_F \eta) \otimes_F \zeta \cong \xi \otimes_F (\eta \otimes_F \zeta)$.*
(iii) *If $T : B \times C \to C \times B$ is the switch map, then $T^*(\eta \otimes_F \xi) \cong \xi \otimes_F \eta$.*

(iv) $\xi \otimes_F (\eta \times \zeta) \cong (\xi \otimes_F \eta) \times (\xi \otimes_F \zeta)$.
(v) $(g, h)^*(\xi \otimes_F \eta) \cong g^*(\xi) \otimes_F h^*(\eta)$ *for maps* $g : B' \to B$ *and* $h : C' \to C$.

Proof Proposition follows from the formula (5.2). ❑

Remark 5.10.9 Given two F-vector bundles ξ, η over the same base space B, we apply a diagonal map Δ to obtain an internal tensor product $\xi \otimes_F \eta$ such that $(\xi \otimes_F \eta)_b = \xi_b \otimes_F \eta_b$ for all $b \in B$.

Remark 5.10.10 Tensor products distribute over the Whitney sum, the group $K_F(B)$ admits also the natural ring structures and the exterior power operations define natural transformations

$$\lambda^i : K_F(B) \to K_F(B) \text{ such that}$$
$$\lambda^0(b) = 1, \lambda^1(b) = b$$
$$\lambda^k(b + s) = \sum_{i+j=k} \lambda^i(b)\lambda^j(s)$$

Details of these constructions can be available in Husemöller (1966).

The above discussion can be summarized in the basic and important result:

Theorem 5.10.11 *For $F = \mathbf{R}$ or $\mathbf{C}$, $K_F(B)$ is the Grothendieck ring whose sum is induced by $[\xi] + [\eta] = [\xi \oplus \eta]$ and whose product is given by $[\xi] \cdot [\eta] = [\xi \otimes \eta]$. Moreover, given a continuous map $f : B' \to B$, there ia a ring homomorphism $f^* : K_F(B) \to K_F(B')$ such that $f^*([\xi]) = [f^*(\xi)]$.*

Proposition 5.10.12 *Let B be the paracompact space. If $f \simeq g : B' \to B$, then $f^*(\xi) = g^*(\xi) : K_F(B) \to K_F(B')$.*

Proof Let $\xi = (X, p, B)$ be a vector bundle, B be a paracompact space and $f \simeq g : B' \to B$. Then the induced bundle $f^*(\xi)$ and $g^*(\xi)$ over B are B-isomorphic (see Theorem 5.9.1). Hence it follows that $f^*(\xi) = g^*(\xi) : K_F(B) \to K_F(B')$. ❑

Corollary 5.10.13 *The correspondence $B \mapsto K_F(B)$ is a functor from the homotopy category of paracompact spaces to the category of rings.*

Proof The proof follows from Theorem 5.10.11 and Proposition 5.10.12. ❑

Definition 5.10.14 The functor $K_F(B)$ is called K-theory on the category of all F-vector bundles over the base space B. The element $[\xi]$ of $K_F(B)$ is determined by a vector bundle ξ.

Remark 5.10.15 The K-theory introduced by Atiyah and Hirzebruch in 1961 is the first example of generalized cohomology theories (see Chap. 15).

Remark 5.10.16 K-theory over contractible spaces is always $\mathbf{Z}$.

5.11 Principal G-Bundles for Lie Groups G

This section continues the study of principal G-bundles over differentiable manifolds when G is a Lie group. Throughout this section G denotes an arbitrary Lie group.

Definition 5.11.1 A principal (differentiable) G-bundle is a triple (E, p, M) such that $p : E \to M$ is a differentiable mapping of differentiable manifolds. Furthermore, E is given a differentiable right G-action $E \times G \to E$ such that the following conditions hold:

(i) $E_x = p^{-1}(x), x \in M$ are the orbits for the G-action.
(ii) (Local trivialization). Every point in M has an open neighborhood U and a diffeomorphism $\psi : p^{-1}(U) \to U \times G$ such that the diagram in Fig. 5.18 commutes, i.e., $\psi_x = \psi|E_x$ maps E_x to $\{x\} \times G$; and ψ is equivariant i.e.,

$$\psi(xg) = \psi(x)g, \ \forall x \in p^{-1}(U), g \in G,$$

where G acts on $U \times G$ by $(x, g')g = (x, g'g)$.
E is called the total space, M the base space and $E_x = p^{-1}(x)$ the fiber at $x \in M$. Sometimes we use the notation E to denote the G-bundle (E, p, M).

Remark 5.11.2 **(i)** Let (E, p, M) be a principal G-bundle. Then p is surjective and open.
(ii) The orbit space E mod G is homeomorphic to M.
(iii) The G-action is free i.e., $x \cdot g = x \Rightarrow g = e, \ \forall x \in E, g \in G$.
(iv) For each $x \in E$, the mapping $G \to E_x$ given by $g \mapsto x \cdot g$, is a diffeomorphism.
(v) If $N \subset M$ is a submanifold (e.g., if N is an open subset), then the restriction to N
$E|N = (p^{-1}(N), p, N)$
is again a principal G-bundle with base space N.

Example 5.11.3 **(i)** For an n-dimensional real vector bundle (V, p, M) the associated frame bundle $(F(V), \tilde{p}, M)$ is a principal $G = \mathrm{GL}\,(n, \mathbf{R})$-bundle.
(ii) For an n-dimensional real vector bundle V equipped with Riemannian metric, $(F_0(V), \tilde{p}, M)$ is a principal $O(n, \mathbf{R})$-bundle.
(iii) Let G be any Lie group and M be a differential manifold. Then $(M \times G, p, M)$ with p the projection onto the first factor, is a principal G-bundle called the product bundle.

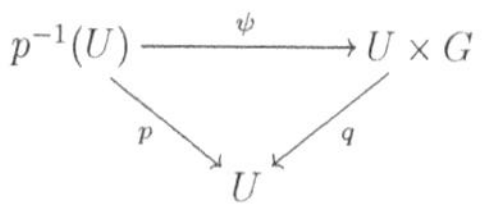

Fig. 5.18 Local trivialization

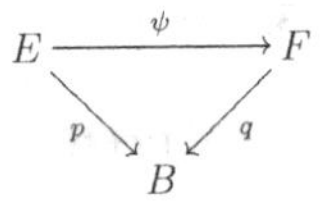

Fig. 5.19 Isomorphism of principal G-bundles

Definition 5.11.4 Let G be a Lie group and (E, p, B) and (F, q, B) be two principal G-bundles over the same base space B. An isomorphism $\psi : E \to F$ is a diffeomorphism of the total spaces such that

(i) The diagram in Fig. 5.19 commutes i.e., $\psi_b = \psi|E_b$ maps $E_b = p^{-1}(b)$ to $F_b = q^{-1}(b)$ and
(ii) ψ is equivariant i.e., $\psi(xg) = \psi(x)g,\ \forall\, x \in E,\ \forall\, g \in G$.

Remark 5.11.5 The map $\psi_b : E_b \to F_b$ is also a diffeomorphism for each $b \in B$.

5.12 Applications

For applications of fiber bundles to higher homotopy groups, see Chap. 7. For some other important applications of fiber bundles in determining the existence or nonexistence of cross sections of a particular tensor bundle see Steenrod (1951) and to theoretical physics see Nakahara (2003) and also Chap. 14 of this present book. Atiyah and Hirzebruch defined in 1961 the K-theory by using stability class of vector bundles to study manifolds. Since then K-theory is applied for investigation of manifolds by constructing powerful new topological invariants. For example, J.F. Adams solved the vector field problems for spheres S^n (if n is odd, the problem is to determine the maximum number of linearly independent nowhere vanishing vector fields on S^n, see Chap. 14), using K-theory see Adams (1962). For further applications of K-theory, see Atiyah and Singer (1963).

5.13 Exercises

For this section the books Husemöller (1966), Spanier (1966), and Switzer (1975) are referred.

1. Let $\xi = (X, p, B, F^n)$ and $\eta = (Y, q, B, F^n)$ be two n-dimensional vector bundles over the space B and $f : \xi \to \eta$ be a B-morphism. Show that f is a vector bundle isomorphism iff the map $f : p^{-1}(b) \to q^{-1}(b)$ is a linear isomorphism for each $b \in B$.
2. Let $\xi = (X, p, B, F^n)$ be an n-dimensional vector bundle over B and $f : B_1 \to B$ be a map. Show that the induced bundle $f^*(\xi) = (Y, q, B, F^n)$ is a vector bundle over B_1 such that $(f_\xi, f) : f^*(\xi) \to \xi$ is a morphism of vector bundles, where $f_\xi : Y \to X, (b, x) \mapsto x$.

3. Let $\xi = (X, p, B, F^n)$ be an n-dimensional vector bundle over $B = B_1 \cup B_2$, where $B_1 = A \times [a, c]$ and $B_2 = A \times [c, b]$, $a < c < b$. If $\xi|B_1 = (X_1, p_1, B_1)$ and $\xi|B_2 = (X_2, p_2, B_2)$ are trivial bundles, show that ξ is also trivial.

4. Prove the following:

(a) Let $p : X \to B$ be the projection of a fiber bundle, and suppose B is paracompact. Then p is a fibration.

(b) If ξ is a vector bundle over $B \times I$, then there exists an open covering $\{U_j\}_{j \in J}$ of B such that $\xi|(U_j \times I)$ is trivial.

(c) If $f : B \times I \to B \times I, (b, t) \mapsto (b, 1)$ is continuous and $\xi = (X, p, B \times I, F^n)$ is a vector bundle over $B \times I$, where B is a paracompact space, then there exists a map $g : X \to X$ such that $(g, f) : \xi \to \xi$ is a morphism of vector bundles and g is a linear isomorphism on each fiber.

(d) The vector bundles ξ over $B \times I$ and the vector bundle $f^*(\xi|(B \times \{1\}))$ are isomorphic.

(e) The vector bundles ξ and $\xi|((B \times 1) \times I)$ over $B \times I$ are isomorphic.

(f) There exists an isomorphism $(f, g) : \xi|(B \times \{0\}) \to \xi|(B \times \{1\})$ of vector bundles.

5. **(a)** Let $\tilde{\xi} = \{(U_j, g_{ij}) : i, j \in J\}$ be a set of transition functions for the space B and the topological group G. Show that there exists a principal G-bundle $\xi = (X, p, B, G)$ and an atlas $\{(U_j, \phi_j) : j \in J\}$ for ξ such that $\tilde{\xi}$ is the set of transition functions for the atlas.

(b) Let $\xi = (X, p, B, G)$ and $\xi' = (X', p', B', G)$ be two principal G-bundles with atlases $\{(U_j, \phi_j) : j \in J\}$ and $\{(U'_a, \phi'_a) : a \in A\}$ and the corresponding sets of transition functions of $\tilde{\xi}$ and $\tilde{\xi}'$. If $f : \tilde{\xi} \to \tilde{\xi}'$ is a morphism of sets of transition functions, show that there is a morphism $\phi : \xi \to \xi'$ of principal G-bundles inducing f.

(c) Let ξ and ξ' be two principal G-bundles over the same space B and $\psi = (\phi, \tilde{\phi}) : \xi \to \xi'$ is a morphism of principal G-bundles such that $\tilde{\phi} : B \to B$ is 1_B. Show that ψ is an equivalence.

6. A Gauss map of an n-dimensional vector bundle $\xi = (X, p, B, F^n)$ in $F^m (n \leq m \leq \infty)$ is a map $f : X \to F^m$ such that $f|p^{-1}(b) : p^{-1}(b) \to F^n$ is a linear monomorphism. Show that for each n-dimensional vector bundle ξ over a paracompact space B, there is Gauss map $f : X \to F^\infty$ and moreover, if B has an open covering $\{U_j\}$, $1 \leq j \leq m$, such that $\xi|U_j$ is trivial, then ξ has a Gauss map $f : X \to F^{mn}$.

7. Show that

(a) A properly discontinuous group of homeomorphisms is discontinuous and acts without fixed points.

(b) A finite group of homeomorphisms acting without fixed points on a Hausdorff space is properly discontinuous.

8. Let $\xi = (X, p, B, F^n)$ and $\eta = (Y, q, B, F^n)$ be vector bundles and $f : X \to Y$ is a map such that $f|p^{-1}(b) : p^{-1}(b) \to q^{-1}(b)$ is a linear isomorphism for each $b \in B$. Show that f is an isomorphism of vector bundles.

9. **(a)** Show that every n-dimensional vector bundle over a paracompact space is B-isomorphic to $f^*(\gamma_n^\infty)$ for some $f : B \to G_n(F^\infty)$;

(b) Let $f, g : B \to G_n(F^m)$ be two maps such that $f^*(\gamma_n{}^m)$ and $g^*(\gamma_n{}^m)$ are B-isomorphic. If

$$i : G_n(F^m) \hookrightarrow G_n(F^{2m})$$

is the natural inclusion, show that the maps $i \circ f$ and $i \circ g$ are homotopic for $1 \leq m \leq \infty$.

10. Let $\xi = (X, p, A, G)$ be a principal G-bundle. Show that ξ has an H-structure iff there exists a map $\tilde{f} : A \to B_H$ such that $B\alpha \circ \tilde{f} \simeq f_\xi$, where $f_\xi : A \to B_G$ is the unique map(upto homotopy) such that $f_\xi^*(\xi_G) \cong \xi$.

11. **(a)** Let $f : B \times I \to B \times I$ be the map defined by $f(b, t) = (b, 1)$ and ξ be a numberable principal G-bundle over $B \times I$. Show that there exists a G-morphism $(g, f) : \xi \to \xi$.

(b) Using notation of **(a)**, show that the principal G-bundles ξ and $f^*(\xi)$ are isomorphic over $B \times I$.

12. Let B be a paracompact space. Show that a fiber bundle $p : X \to B$ is a fibration.

13. Let H be a closed subgroup of a Lie group G. Show that every natural subgroup $N \subset H$ determines an H/N-fiber bundle with bundle map

$$p : G/N \to G/H, gN \mapsto gH.$$

14. If $\xi = (X, p.B)$ is a principal G-bundle, show that ξ is a bundle with fiber G.

15. Let $G = \{+1, -1\}$ be the two-element group and the n- sphere S^n be the G-space with action given by the relation $x(+1) = x, x(-1) = -x$. Show that this principal $\mathbf{Z}_2$ space gives a principal $\mathbf{Z}_2$- bundle with the real n-dimensional projective space $\mathbf{R}P^n$ as its base space.

16. Let $p : X \to B$ be a covering projection. Show that it is a principal G-bundle, where G is the group of covering transformations with the discrete topology.

17. If $p : X \to B$ be a fiber bundle, show that p is an open map.

18. For any G-space X, the automorphisms of the trivial G-bundle $p_2 : X \times B \to B$ are in (1-1)- correspondence with continuous functions $f : B \to G$.
[Hint: Any bundle automorphism of p comes from a morphism of the underlying trivial G-bundle $p_2 : G \times B \to B$. If $f : B \to G$ is a map, define

$$\phi_f : B \times G \to B \times G, (b, g) \mapsto (b, f(b)g).$$

Since $(b, g) = (e, b) \cdot g$, the map is completely determined by $\phi_f(e, b)$. Conversely given a G-bundle map $\phi : B \times G \to B \times G$, define $f(b)$ to be the second component of $\phi(e, b)$.]

19. Given a space B, let η^n denote the trivial n-dimensional vector bundle $p_2 : F^n \times B \to B$, where $F = \mathbf{R}$ or $\mathbf{C}$.

(a) Let $\xi : X \to B$ be a vector bundle which has a cross section $s : B \to X$ such that $s(b) \neq 0$ for all b. Show that ξ has a subbundle isomorphic to the trivial bundle and this $\xi \cong \eta^1 \oplus \xi'$ for some bundle ξ'.
[Hint. Define

$$\psi : F \times B \to X, (\lambda, b) \mapsto \lambda s(b),$$

for $(\lambda, b) \in F \times B$ under the right hand product on F_b of the scalar λ. Since s is nowhere zero, Im ψ is a subbundle of ξ and $\psi : \eta^1 \to \text{Im}\,\psi$ is an isomorphism. Consequently, $\xi = \eta^1 \oplus \xi'$, where $\xi' = (\text{Im}\,\psi)^{\perp}$.]

(b) Let $\xi : X \to B$ be a vector bundle which has an m linearly independent cross sections $s_i : B \to X$ for $i = 1, 2, \ldots, m$. Show that $\xi \cong \eta^m \oplus \xi'$ for some bundle ξ'. In particular, if an n-dimensional vector bundle has n-linearly independent cross sections, show that it is isomorphic to the trivial bundle η^n.

20. Show that the tangent bundle $T(S^n)$ to S^n is trivial only for $n = 1, 3$ or 7.

21. Let H be a subgroup of a topological map G. Show that the quotient map X_G mod $H \to B_G$ is locally trivial with fiber homeomorphic to the space G/H of cosets.
Let G be a topological group.

(a) Show that

(i) the space X_G is contractible.

(ii) If $\xi' = (X'_G, p', B'_G)$ is any universal G-bundle, then the space X'_G is contractible.

(b) Let ϕ_g be an inner automorphism of the topological group G. Show that $B\phi_g \simeq 1_{B_G}$.

[Hint: Use the Milnor's universal bundle $\xi : X_G \to B_G$].

22. Show that $F_0(V) \subset F(V)$ is a submanifold and $\tilde{p}|F_0(V) : F_0(V) \to M$ is differentiable.

23. Show that there exists a differentiable right $O(n)$-action $F_0(V) \times O(n) \to F_0(V)$ such that for $b \in M$, the orbits are the sets $F_0(V_b)$ for $b \in M$.

24. Represent S^3 as $S^3 = \{(z_1, z_2) \in \mathbf{C}^2 : z\bar{z_1} + z_2\bar{z}_2 = 1\}$, S^2 as the complex projective line. Define $p : S^3 \to S^2, (z_1, z_2) \mapsto [z_1, z_2]/(|z_1|^2 + |z_2|^2)^{1/2}$, called the Hopf map. Show that (S^3, p, S^2, S^1) is a fiber bundle with fiber S^1.

25. Generalize the Hopf map $p : S^3 \to S^2$ to the Hopf map of the form

$$p : S^7 \to S^4$$

with the fiber as the unit quaternion $S^3 = SU(2, \mathbf{H})$.

26. Let L be the real line bundle over S^1, such that L is either the cylinder $S^1 \times \mathbf{R}$ or the Möbius strip. Show that the Whitney sum $L \oplus L$ is a trivial bundle.

27. Give two fiber bundles over S^1.

28. Let $\xi : X \xrightarrow{p} B$ be a fiber bundle with fiber F and $f, g : X \to B$ be homotopic maps. Show that $f^*(\xi) = g^*(\xi)$.
[Hint. See Steenrod (1951)].

29. Let $\xi : X \xrightarrow{p} B$ be a fiber bundle. Show that ξ is trivial if B is contractible to a point.

30. Consider a fiber bundle (X, B, p, F) with total space X, base space B, fiber F and projection $p : X \to B$. Show that the following statements are equivalent:

(i) p is a weak fibration.
(ii) If the base space B is paracompact, then $p : X \to B$ is a fibration.

31. Let $\xi : X \xrightarrow{p} B$ be a fiber bundle with base space B a paracompact space. Show that if p is a weak fibration, then it is a fibration.

32. Let $G_{n,k} = G_k(\mathbf{R}^n)$ be the Grassmann manifold of k-planes through the origin in $\mathbf{R}^n$. Show that

(i) The orthogonal group $O(n, \mathbf{R})$ acts transitively on $G_{n,k}$;
(ii) The isotropy group of the standard $\mathbf{R}^k \subset \mathbf{R}^n$ is $O(k, \mathbf{R}) \times O(n-k, \mathbf{R})$;
(iii) $G_{n,k} \cong O(n, \mathbf{R})/O(k, \mathbf{R}) \times O(n-k, \mathbf{R})$. See Bredon p 464.

33. Let $K(B)$ be the Grothendieck group of all vector bundles over the base space B. Show that

(i) each element $x \in K(B)$ can be represented as a difference of two vector bundles:
$$x = [\xi] - [\eta].$$
(ii) two vector bundles ξ and η define the same element in the group $K(B)$ iff there is a trivial bundle θ such that $\xi \oplus \theta = \eta \oplus \theta$.

34. Show that given a vector bundle ξ over a finite CW-complex X, there is a vector bundle η over X such that $\xi \oplus \eta = \theta$ for some trivial bundle θ.

35. Show that tensor product of vector bundles induces a ring structures in the additive Grothendieck group $K(B)$. Moreover, show that

(i) if $B = \{b_0\}$ is a one point space, then the ring $K(B)$ is isomorphic to the ring $\mathbf{Z}$;
(ii) if $B = \{b_0, b_1\}$, then $K(B) \cong \mathbf{Z} \oplus \mathbf{Z}$.

5.14 Additional Reading

[1] Adams, J.F., *Algebraic Topology: A student's Guide*, Cambridge University Press, Cambridge, 1972.
[2] Atiyah, M. F., *K-Theory*, Benjamin, New York, 1967.
[3] Arkowitz, Martin, *Introduction to Homotopy Theory*, Springer, New York, 2011.
[4] Bredon, G.E., *Topology and Geometry*,Springer-Verlag, New York, Inc.1993.

[5] Dugundji, J., *Topology*, Allyn & Bacon, Newtown, MA, 1966.
[6] Dieudonné, J., *A History of Algebraic and Differential Topology*, 1900-1960, Modern Birkhäuser, 1989.
[7] Dupont, J., *Fiber Bundles and Chern-Weil Theory*, Aarhus Universitet, 2003.
[8] Eilenberg, S., and Steenrod, N., *Foundations of Algebraic Topology*, Princeton University Press, Princeton, 1952.
[9] Hatcher, Allen, *Algebraic Topology*, Cambridge University Press, 2002.
[10] Maunder, C.R.F., *Algebraic Topology*, Van Nostrand Reinhhold, London, 1970.
[11] Whitehead, G.W., *Elements of Homotopy Theory*, Springer-Verlag, New York, Heidelberg, Berlin, 1978.

References

Adams, J.F.: Vector fields on spheres. Ann. Math. **75**, 603–632 (1962)
Adams, J.F.: Algebraic Topology: A student's Guide. Cambridge University Press, Cambridge (1972)
Atiyah, M.F.: K-Theory. Benjamin, New York (1967)
Atiyah, M.F., Bott, R.: On the periodicity theorem for complex vector bundles. Acta Math. **112**, 229–247 (1964)
Atiyah, M.F., Hizebruch, F.: Vector bundles and homogeneous spaces. Proc. Symp. Pure Maths. Amer. Math. Soc. 7–38 (1961)
Atiyah, M.F., Singer, I.M.: The index of elliptic operators on compact manifolds. Bull. Amer. Math. Soc. **69**, 422–433 (1963)
Arkowitz, Martin: Introduction to Homotopy Theory. Springer, New York (2011)
Borel, A.: Topology of Lie groups and characteristic classes. Bull. Amer. Math. Soc. **61**, 397–432 (1955)
Bott, R.: The stable homotopy groups of classical groups. Proc. Nat. Acad. Sci., USA, **43**, 933–935 (1957)
Bott, R.: The stable homotopy of the classical groups. Ann. Math. **70**, 313–337 (1959)
Bredon, G.E.: Topology and Geometry. Springer, New York (1993)
Dugundji, J.: Topology. Allyn & Bacon, Newtown, MA (1966)
Dyer, E., Lashof, R.: A topological proof of the Bott. Periodicity Theorems. Ann. Math. Pure Appl. **54(4)**, 231–254 (1961)
Dieudonné, J.: A History of Algebraic and Differential Topology. pp. 1900–1960. Modern Birkhäuser (1989)
Dupont, J.: Fiber Bundles and Chern-Weil Theory, Aarhus Universitet (2003)
Eilenberg, S., Steenrod, N.: Foundations of Algebraic Topology. Princeton University Press, Princeton (1952)
Feldbau, J.: Sur la classification des espaces fibrés. C. R. Acad. Sci. Paris **208**, 1621–1623 (1939)
Gray, B.: Homotopy Theory, An Introduction to Algebraic Topology. Academic Press, New York (1975)
Hatcher, A.: Algebraic Topology. Cambridge University Press, Cambridge (2002)
Hopf, H.: Über die Abbildungen der dreidimensionalen Sphre auf die Kugelfiche. Math. Ann. **104**(1), 637–665 (1931)
Hopf, H.: Über die Abbildungen von Sphren auf Sphren niedrigerer Dimension. Fund. Math. **25**, 427–440 (1935)
Husemöller, D.: Fibre Bundles, McGraw-Hill, New York (1966)
Luke, G., Mischenko, A.: Vector Bundles and their Applications. Kluwer Academic Publisher, Boston (1984)

Maunder, C.R.F.: Algebraic Topology. Van Nostrand Reinhhold, London (1970)

Milnor, J.: Construction of Universal Bundles I, II. Ann. Math. **63**(2), 272–284 (1956), **63**(2), 430–436 (1956)

Milnor, J.: 'Morse Theory', Notes by M. Spivak and R. Wells. Princeton University Press, New Jersey (1963)

Nakahara, M.: Geometry, Topology and Physics, Institute of Physics Publishing, Bristol (2003)

Spanier, E.: Algebraic Topology. McGraw-Hill Book Company, New York (1966)

Steenrod. N.: The Topology of Fibre Bundles. Prentice University Press, Prentice (1951)

Switzer, R.M.: Algebraic Topology-Homotopy and Homology. Springer, Berlin (1975)

Whitehead, G.W.: Elements of Homotopy Theory. Springer, New York (1978)

Toda, H.: A topological proof of theorems of Bott and Borel-Hirzeburch for homotopy groups of unitary groups, Memoirs of the College of Science. Univ. Kyoto. Series A: Math. **32**(1), 103–119 (1959)

Chapter 6
Geometry of Simplicial Complexes and Fundamental Groups of Polyhedra

This chapter conveys the geometry of finite simplicial complexes which provides a convenient way to study manifolds and builds up interesting topological spaces called polyhedra from these complexes followed by a study of their homotopy properties with computing their fundamental groups and develops certain analytical geometric tools for subsequent chapters. These are: simplex, complex, subcomplex, simplicial map, triangulation, polyhedron, and simplicial approximation. Simplicial complexes provide useful tools in computing fundamental groups of simple compact spaces. The combinatorial device, now called abstract complex was systematically used by W. Mayer (1887–1947) in 1923.

The geometrical objects such as points, edges, triangles, and tetrahedra are examples of low dimensional simplexes. Many important spaces are constructed from certain familiar subsets of Euclidean spaces $\mathbf{R}^n$. One of them is a simplex. A simplex S is just a generalization to n dimensions of a triangle or a tetrahedron and these are fitted together to form a geometric complex K, known as simplicial complex in such a way that two simplexes are either disjoint or they meet in a common edge or face and every proper face of S is also in K. Simplicial complexes provide a convenient tool for the study of manifolds. For example, Poincaré duality theorem given by H. Poincaré (1854–1912) in 1895 is one of the earliest theorems in topology. Simplicial complexes form building blocks for homology theory which begins in Chap. 10. For example, simplicial homology invented and studied by Henry Poincaré during 1895–1904, is one of the most fundamental influential inventions in mathematics. He started with a geometric object (a space) which is given by combinatorial data (a simplicial complex). Then the linear algebra and boundary relations by these data are used to construct simplicial homology groups.

It is easy to define algebraic invariants such as fundamental groups, higher homotopy groups and homology groups, etc., of different classes of topological spaces but difficult to compute them as the supply of useful tools provided by topological invariants corresponding to arbitrary spaces are quickly exhausted. To facilitate such computation, this chapter works with topological spaces that can be broken up into pieces which fit together in a nice way. Such spaces are called triangulable spaces.

M.R. Adhikari, *Basic Algebraic Topology and its Applications*,
DOI 10.1007/978-81-322-2843-1_6

A polyhedron is a topological space which admits a triangulation by a simplicial complex.

Simplicial approximation is an important concept in algebraic topology. It is sometimes convenient to utilize a good feature of simplicial complexes: arbitrary continuous maps between their polyhedra can always be deformed to maps that are linear on the simplexes of some subdivision of the domain complex. This leads to the concept of simplicial approximation theorem first proved by J.W. Alexander (1888–1971) in 1915 and later by O. Veblen (1880–1960) in 1922. Its more revised version was given by E.C. Zeeman (1925–2016) in 1964. This theorem is used to calculate fundamental groups, and to examine the topological invariance of the homology groups of a space.

This chapter utilizes the concept of triangulation to solve extension problems and that of edge group $E(K, v)$ (which is isomorphic to the fundamental group $\pi_1(|K|, v)$ for any simplicial complex K) is applied to graph theory and proves van Kampen theorem by using graph-theoretic results. It also proves simplicial approximation theorem given by Brouwer and Alexander around 1920 by utilizing a certain good feature of simplicial complexes introduced by J.W. Alexander in (1915). This theorem plays a key role in the study of homotopy and homology theories.

For this chapter the books and papers (Armstrong 1983), (Eilenberg and Steenrod 1952), (Gray 1975), (Hilton and Wylie 1960), (Maunder 1970), (Singer and Thrope 1967), (Veblen 1922), (Zeeman 1964) and some others are referred in Bibliography.

6.1 Geometry of Finite Simplicial Complexes

This section studies geometry of finite simplicial complexes to facilitate the construction of simplicial complexes which provide a convenient way to study manifolds. Such study is important in the study of algebraic topology. The term "simplicial complex" is derived from the term "simplex". Such complexes are also called geometrical complexes. Historically, the simplicial techniques were gradually modified until introduction of singular homology by S. Eilenberg (1915–1998) in a topological invariant manner (see Chap. 10). The concepts of 1-dimensional and 2-dimensional simplicial complexes date back at least to L. Euler (1707–1783) and their higher dimensional analog was first studied by J.B. Listing (1808–1882) in 1862.

We start with the concept of a simplex. Let $\mathbf{R}^n$ be the Euclidean n-space. It is an n-dimensional vector space over $\mathbf{R}$. The standard n-dimensional simplex

$$\Delta^n = \{(x_1, x_2, \ldots, x_{n+1}) \in \mathbf{R}^{n+1} : 0 \leq x_i \leq 1, \sum x_i = 1\}.$$

More generally, given $p + 1$ points $v_0, v_1, \ldots, v_p$ of $\mathbf{R}^n$, they are said to be affinely independent if the equations

$$\sum_{i=1}^{p} a_i v_i = 0,\ a_i \in \mathbf{R} \text{ and } \sum_{i=1}^{p} a_i = 0 \text{ imply } a_i = 0 \text{ for all } i.$$

We now give the formal definition of "independent points" in $\mathbf{R}^n$.

Definition 6.1.1 A set S of $(p+1)$ of distinct points $v_0, v_1, \ldots, v_p$ in $\mathbf{R}^n$ is said to be(geometrically) independent if the vectors $v_1 - v_0, v_2 - v_0, \ldots, v_p - v_0$ are linearly independent.

This is equivalent to the statement that the equations

$$\sum_{i=0}^{p} a_i v_i = 0,\ a_i \in \mathbf{R} \text{ and } \sum_{i=0}^{p} a_i = 0 \text{ imply that } a_0 = a_1 = \cdots = a_n = 0.$$

It shows that the Definition 6.1.1 does not depend on the order of the points $v_0, v_1, \ldots, v_p$. Thus if the vectors in S are independent, then no three of them lie on a line, no four of them lie on a plane, and no $(m+1)$ of them lie in a hyperplane of dimension $m-1$ or less.

Example 6.1.2 The points v_0, v_1, v_2 in Fig. 6.1 are geometrically independent and the points v_0, v_1, v_2 in Fig. 6.2 are geometrically dependent.

Definition 6.1.3 Let $\{v_0, v_1, \ldots, v_p\}$ be a set of geometrically independent points in some Euclidean space $\mathbf{R}^n$. Then a geometric p-simplex s_p is the set of points

$$\sum_{i=0}^{p} a_i v_i,\ a_i \in \mathbf{R} \text{ such that } a_i \geq 0 \text{ for all } i \text{ and } \sum_{i=0}^{p} a_i = 1.$$

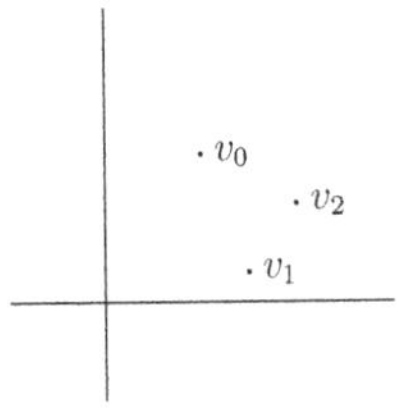

Fig. 6.1 Geometrically independent points in $\mathbf{R}^2$

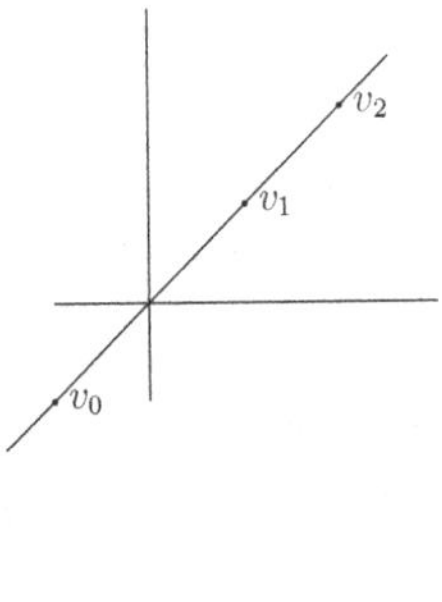

Fig. 6.2 Geometrically dependent points in $\mathbf{R}^2$

Clearly, $s_p = \{v = \sum_{i=0}^{p} a_i v_i, a_i \geq 0 \text{ and } \sum_{i=0}^{p} a_i = 1\} \subset \mathbf{R}^n$, and s_p is given the subspace topology and is a compact metric space. s_p is denoted by $s_p = \langle v_0 v_1 v_2 \ldots v_p \rangle$. The points $v_0, v_1, \ldots, v_p$ are called vertices of s_p and the set $\text{Vert}(s_p) = \{v_0, v_1, \ldots, v_p\}$ is called the vertex set of s_p and p is called the dimension of s_p.

The subspace of s_p consisting of all those points $\sum_{i=0}^{p} a_i v_i$ with $a_i > 0$ of s_p for all i is called the interior of s_p, denoted by $\mathring{S}$. The particular point in $\mathring{S}$ defined by

$$\hat{S} = \frac{1}{p+1}(v_0 + v_1 + \cdots + v_p)$$

is called the barycenter of s_p.

If $\{v_{i_0}, v_{i_1}, \ldots, v_{i_r}\}$ is any subset of the set of vertices $\{v_0, v_1, \ldots, v_p\}$ of s_p, the subspace of s_p consisting of those points linearly dependent of $v_{i_0}, v_{i_1}, \ldots, v_{i_r}$ is called an r-face of s_p. We write $s_p \prec \sigma$ if s_p is a face of σ. Let s_p be a p-dimensional simplex. A face of s_p may be empty or s_p itself.

Definition 6.1.4 Let s_p be a p-simplex. A face of s_p is said to be proper if it is neither empty nor the whole of s_p. The number p is called the dimension of s_p.

Example 6.1.5 The faces of a 2-simplex $s_2 = \langle v_0 v_1 v_2 \rangle$ are the 2-simplex s_2 itself, three 1- simplexes $\langle v_0 v_1 \rangle$, $\langle v_1 v_2 \rangle$, and $\langle v_2 v_0 \rangle$; and three 0-simplexes $\langle v_0 \rangle$, $\langle v_1 \rangle$ and $\langle v_2 \rangle$.

Definition 6.1.6 Let S be a p-simplex. Define an open p-simplex S° by

$$S^\circ = \begin{cases} s_p & \text{if } p = 0 \\ S - \dot{S} & \text{if } p > 0 \end{cases}$$

where $\dot{S}$ is the boundary of S, which is the set of all faces of S other than S itself. Clearly, a simplicial complex is the disjoint union of its open simplexes.

Remark 6.1.7 An open p-simplex $\mathring{s}_p$ is the interior of s_p. For example, an open 1-simplex $\mathring{s}_1$ is an open interval.

To build up a geometric complex from a collection of simplexes in a nice way, we need the following concept.

Definition 6.1.8 Two simplexes s_t and s_p are said to be properly joined if their intersection is either empty or is a common face.

A geometric complex or a simplicial complex is a finite family K of simplexes which are properly joined and have the property that each face of a member of K is also a member of K. Its formal definition is now given.

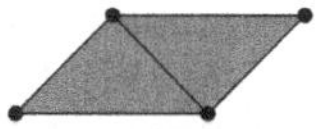

Fig. 6.3 Example of a geometrical simplicial complex

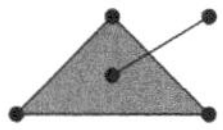

Fig. 6.4 Example which is not geometrical simplicial complex

Definition 6.1.9 A geometric finite simplicial complex K is a finite set of simplexes, all contained in some Euclidean space $\mathbf{R}^n$ such that

(i) if s_p is a simplex of K, and s_q is a face of s_p, then s_q is in K;
(ii) if s_p and s_q are simplexes of K, then $s_p \cap s_q$ is either empty, or is a common face of s_p and s_q.

The example in Fig. 6.3 is a geomertical simplicial complex and example in Fig. 6.4 is not so.

6.2 Triangulations and Polyhedra

This section conveys the concepts of triangulations and polyhedra which are very important in computing fundamental groups and the homology groups of a certain class topological spaces. It is easy to define the algebraic invariant such as fundamental groups but difficult to compute them in general. So it has become necessary to compute them for at least a reasonably large class of spaces. This problem can be solved effectively with topological spaces that can be broken up into pieces which fit together in a nice way. Such spaces are called triangulable spaces obtained by triangulations. A polyhedron is a topological space which admits a triangulation by a simplicial complex.

Definition 6.2.1 Given a simplicial complex K, the set of points of $\mathbf{R}^n$ that lie in at least one of the simplexes of K, topologized as a subspace of $\mathbf{R}^n$, is a topological space, called the polyhedron of K, denoted by $|K|$.

Remark 6.2.2 A geometrical simplicial complex K is not a topological space. It is a set whose elements are geometric simplexes. But $|K|$ is a topological space. It denotes the point set union of the simplexes of K with the Euclidean subspace topology and is sometimes called a carrier of K or the polyhedron associated with K or a realization of K

Definition 6.2.3 Let X be a topological space. If there is a geometric complex K whose carrier $|K|$ is homeomorphic to X, then X is said to be a triangulable space or a polyhedron and the complex K is called a triangulation of X. The space $|K|$ is called a realization of K. More precisely, if $h : |K| \to X$ is a homeomorphism, the ordered pair (K, h) is called a triangulation of X and X is said to be a polyhedron.

Definition 6.2.4 Let K be a finite simplicial complex. The dimension of K denoted by dim K is defined by $\dim K = \sup_{S \in K}\{\dim S\}$.

Thus dim K is the largest positive integer m such that K has an m-simplex; in particular, an n-simplex has dimension n.

Theorem 6.2.5 (Invariance of dimensions) *Let K and M be two finite simplicial complexes. If there exists a homeomorphism $f : |K| \to |M|$, then* $\dim K = \dim M$.

Proof Suppose $\dim K = n$ and $\dim M = m$. If possible let $n > m$. Let $S = s_n$ be an n-simplex in K and $\mathring{S} = S - \dot{S}$ be its interior. Then $\mathring{S}$ is an open set in $|K|$. Again, since f is a homeomorphism, $f(\mathring{S})$ is open in $|M|$. Consequently, there exists some p-simplex σ_p in M (where $p \leq m < n$) such that $f(\mathring{S}) \cap \sigma_p^0 = W$(say), a nonempty open set in $|M|$. Take a homeomorphism $\psi : \Delta^n \to S$ such that $\psi(\dot{\Delta}^n) = \dot{S}$. Define U by $U = (\psi^{-1} \circ f^{-1})(W)$. Then U is an open subset of $(\Delta^n)^{\circ}$. Since $p < n$, there exists an embedding $j : \Delta^p \to (\Delta^n)^{\circ}$ such that image of j contains nonempty open subset of $(\Delta^n)^{\circ}$. As U is open and $j(W)$ is not open but both U and $j(W)$ are homeomorphic subset $(\Delta^n)^{\circ}$. Hence we reach a contradiction. Thus $n \not> m$. Similarly, $m \not> n$. Consequently, $m = n$ ❑

Theorem 6.2.5 defines the dimensions of a polyhedron.

Definition 6.2.6 The dimension of a polyhedron X is defined to be the common dimension of the simplicial complexes associated with triangulations of X.

Definition 6.2.7 Let X be a polyhedron. Then the dimension of X is the common dimension of the associated simplicial complexes involved in triangulations of X.

Example 6.2.8 For the standard 2-simplex $\Delta^2 \subset \mathbf{R}^2$, define K to be the family of all vertices and 1-simplexes of Δ^2 (i.e., which is the family of all proper faces of Δ^2). Then K is a simplicial complex such that $|K|$ is the perimeter of the triangle Δ^2 in $\mathbf{R}^2$. If $X = S^1$, given distinct points $v_0, v_1, v_2 \in S^1$, define a homeomorphism $h : |K| \to S^1$ by $h(e_i) = v_i$, for $i = 0, 1, 2$ and h taking each 1-simplex $[e_i, e_{i+1}]$ onto the arc joining v_i to v_{i+1}. Then (K, h) is a triangulation of S^1 and hence S^1 is a polyhedron.

Example 6.2.9 Let X be the 2-sphere S^2 defined by $S^2 = \{(x_1, x_2, x_3) \in \mathbf{R}^3 : \sum_{i=1}^{3} x_i^2 = 1\}$. Consider a closed 3-simplex $S_3 = \langle v_0 v_1 v_2 v_3 \rangle$. Then, the complex K whose simplexes are the proper faces of S_3 is a geometrical simplicial complex. Clearly, $|S_3|$ = geometric carrier of S_3 is the boundary of a tetrahedron, and hence, it is homeomorphic to S^2. This shows that S^2 is triangulable with K as one triangulation.

Example 6.2.10 Let K be a family of all proper faces of an n-simplex S. If there is a triangulation (K, h) of S^{n-1}, then this simplicial complex K is denoted by $\dot{S}$(this notation is borrowed, since $|K|$ is the boundary $\dot{S}$, which is homeomorphic to S^{n-1}).

Example 6.2.11 Every p-simplex s_p determines a simplicial complex K, consisting of the family of all (not necessarily proper) faces of s_p.

Definition 6.2.12 Let K be a simplicial complex in $\mathbf{R}^n$. A subcomplex L of K is a subset of K such that L is itself a simplicial complex (i.e., L satisfies the conditions of Definition 6.1.9. If L is a subcomplex of the complex K, then $|L|$ is called a subpolyhedron of $|K|$.

Definition 6.2.13 The n-skeleton K^n of a simplicial complex K is that subcomplex consisting of m-faces of simplexes of K for $m \leq n$. By convention, the empty set is the (-1)-skeleton.

Example 6.2.14 Let s_p be a p-simplex in $\mathbf{R}^n$. The simplex s_p alone does not from a simplicial complex. But s_p and all faces of s_p taken together form a simplicial complex denoted by $K(s_p)$ such that $|K(s_p)| = s_p$. On the otherhand, the set of all faces of s_p other than s_p forms the boundary $\dot{s_p}$ of s_p; and $s_p - \dot{s}_p = \mathring{s}_p$ is called the interior of s_p, sometimes it is denoted by $\text{Int}(s_p)$.

Definition 6.2.15 A simplicial pair (K, L) consists of a simplicial complex K and a subcomplex L of K.

Definition 6.2.16 Let K be a simplicial complex and v is a vertex in its vertex set Vert (K). Then, the star of v, denoted by st (v), is the subset

$$\text{st}\,(v) = \{s : s \text{ is a simplex of } K \text{ and } v \text{ is a vertex of } s\}.$$

On the other hand, the open star of v denoted by $\text{ost}(v)$ is a subset of $|K|$ defined by

$$\text{ost}(v) = \bigcup_{\substack{s \in K \\ v \in \text{Vert}\,(K)}} \mathring{s} \subset K$$

Example 6.2.17 The open shaded region in Fig. 6.5 consisting of all the open simplexes of which v is a vertex, is the set $\text{st}(v)$.

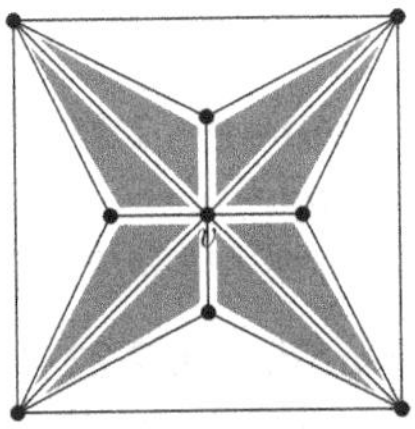

Fig. 6.5 Star of v

Proposition 6.2.18 *Let K be a simplicial complex in $\mathbf{R}^n$. Then*

(i) *$|K|$ is a closed compact subspace of $\mathbf{R}^n$;*
(ii) *every point of $|K|$ lies in the interior of exactly one simplex of K;*
(iii) *if the subspaces of K are taken separately and their union is endowed with the identification topology, then $|K|$ is exactly obtained;*
(iv) *if $|K|$ is a connected space, then it is path-connected.*

Proof **(i)** Since the simplicial complex K is finite and each simplex of K is both closed and bounded, it follows that $|K|$ is compact.

(ii) Let A and B be two simplexes of K such that their interiors overlap. Since K is a simplicial complex, A and B must meet in a common face. But the only face of a simplex which contains interior points is the whole simplex itself. This shows that $A = B$.

(iii) Since simplexes of K are closed in $\mathbf{R}^n$, they are closed subsets of $|K|$. Hence if A is a subset of $|K|$, and if $A \cap B$ is closed in A for each simplex B of K, then $A \cap B$ is closed in $|K|$. Consequently, the finite union $\cup\{A \cap B : B \in K\}$ is closed in $|K|$. Thus, the closed subsets of $|K|$ are precisely those which intersect each simplex of K is a closed set. Hence $|K|$ has the identification topology.

(iv) Let $|K|$ be a connected space. Given an arbitrary point $x \in |K|$, if L denotes the subcomplex of K defined by $L = \{s \in K : x \notin s\}$ and ϵ is the distance from x to $|L|$, then for a positive number $\eta < \epsilon$, the set $B(x, \eta) \cap |K|$ is path- connected. This is because any point in the set can be joined to x by a straight line segment in some simplex of K. This shows that $|K|$ is locally path-connected. Hence it follows that $|K|$ is path-connected.

❑

Remark 6.2.19 Given a simplicial complex K, the space $|K|$ is topologized as a subspace of the Euclidean n-space $\mathbf{R}^n$. An alternative description of the topology in $|K|$ is given.

Proposition 6.2.20 *Let K be a finite simplicial complex. Then a subset A of $|K|$ is closed iff $A \cap s$ is closed in s, for each simplex s in K.*

Proof Since each s is closed in $\mathbf{R}^n$, s is also closed in $|K|$. Hence $A \cap s$ is closed in s. Consequently, it is also closed in $|K|$. Thus $A = \bigcup\limits_{s \in K} A \cap s$ is closed, since K is a finite set of simplexes. The converse part follows trivially. ❑

Corollary 6.2.21 *Let K be a finite simplicial complex. Then the topology of $|K|$ as a subspace of the Euclidean space $\mathbf{R}^n$ coincides with the topology of $|K|$, considered as the space obtained from simplexes by identifying together the various intersections.*

Definition 6.2.22 Let K be a simplicial complex. For each point $x \in |K|$, the simplicial neighborhood of x, $N_K(x)$ is the set of all simplexes of K that contain x, together with all their faces. The link of $x \in |K|$, $L_K(x)$ is the subset of simplexes of $N_K(x)$ that donot contain x.

Definition 6.2.23 Let K be a simplicial complex. For each simplex s of K, the star of s, denoted by $st_K(s)$, is defined to be the union of the interiors of the simplexes of K that have s as a face.

Remark 6.2.24 $N_K(x)$ and $L_K(x)$ are subcomplexes of K; $st_K(s)$ is an open set for each simplex s of K.

6.3 Simplicial Maps

This section considers simplicial maps which are maps from one simplicial complex to another simplicial complex preserving in some sense the simplicial structures. The concept of a simplicial map is an analogous concept of a group homomorphism.

Definition 6.3.1 Let K and L be simplicial complexes. A simplicial map $f : K \to L$ is such that

(i) if v is a vertex of a simplex of K, then $f(v)$ is a vertex of a simplex of L;
(ii) if $\{v_0, v_1, \ldots, v_n\}$ spans a simplex of K, then $\{f(v_0), f(v_1), \ldots, f(v_n)\}$ spans a simplex of L (repetitions among $f(v_0), f(v_1), \ldots, f(v_n)$ are allowed);
(iii) if $x = \sum \lambda_i v_i$ is in a simplex $\langle v_0.v_1 \ldots v_n \rangle$ of K, then $f(x) = \sum \lambda_i f(v_i)$; (i.e., f is a linear on each simplex).

Remark 6.3.2 Given simplicial complexes K and L. Some authors define a simplial map as a function $f : |K| \to |L|$ between their corresponding polyhedra which takes simplexes of K linearly onto simplexes of L

We now use this definition to prove the following proposition which gives the continuity of a simplicial map.

Proposition 6.3.3 *Given simplicial complexes K and L, a simplicial map $f : |K| \to |L|$ between their corresponding polyhedra is continuous.*

Proof Let A be a closed subset of $|L|$. Then $A \cap s$ is closed in s for each simplex s of L. Since the restriction of f to any simplex of K is linear, it is continuous. Hence $f^{-1}(A) \cap s'$ is closed in s' for each simplex s' in K. Consequently, $f^{-1}(A)$ is closed in $|K|$ by Proposition 6.2.20. Hence f is continuous. ❑

The definition of simplicial map is now extended for simplicial pairs.

Definition 6.3.4 A simplicial pair (K, L) consists of a simplicial complex K and a subcomplex L of K.

Definition 6.3.5 A simplicial map $f : (K, L) \to (A, B)$ for simplicial pairs is just a simplicial map $f : K \to A$ such that $f(L) \subset B$.

Remark 6.3.6 The composite of two simplicial maps is another simplicial map.

6.4 Barycentric Subdivisions

This section introduces the concept of barycentric subdivisions of a simplicial complex. This concept is very significant in providing useful method of changing the structure of a simplicial complex K without changing the underlying set $|K|$ or its topology. The aim of barycentric subdivision is to prescribe a process of repeating it to make the simplexes of a simplicial complex as small as we please according to our need.

Definition 6.4.1 Let $s_p = \langle v_0 v_1 \dots v_p \rangle$ be a p-simplex in $\mathbf{R}^n$. Then the barycenter of s_p, denoted by $B(s_p)$, is the point in the open simplex (s_p) defined by

$$B(s_p) = \frac{1}{p+1} \sum_{i=0}^{p} v_i.$$

This is the center of gravity of the vertices in the usual sense. Barycenter comes from the Greek word 'barys' meaning heavy.

Definition 6.4.2 A partial ordering is defined in a simplicial complex by $s_i \preceq s_{i+1}$ if s_i is a face of s_{i+1}. The notation $s_i \prec s_{i+1}$ means $s_i \preceq s_{i+1}$ but $s_i \neq s_{i+1}$.

Definition 6.4.3 Let K be a simplicial complex. A barycentric subdivision of K is a simplicial complex K' such that

(i) the vertices of K' are the barycenters of simplexes of K;
(ii) the simplexes of K' are the simplexes $\langle B(s_0) \dots B(s_m) \rangle$, where $s_i \prec s_{i+1}$.

If a barycentric subdivision of a complex exists, it is unique. K' is called the first barycentric subdivision of K, denoted by $K^{(1)}$. For $n > 1$, the nth barycentric subdivision $K^{(n)}$ of K is defined inductively by taking $K^{(n)}$ the first barycentric subdivision of $K^{(n-1)}$.

Example 6.4.4 A barycentric subdivision $K^{(2)}$ of a 2-simplex is given in Fig. 6.6.

Example 6.4.5 **(i)** $\langle v_0 \rangle$ is a 0-simplex and consists of one point, which is its barycenter.

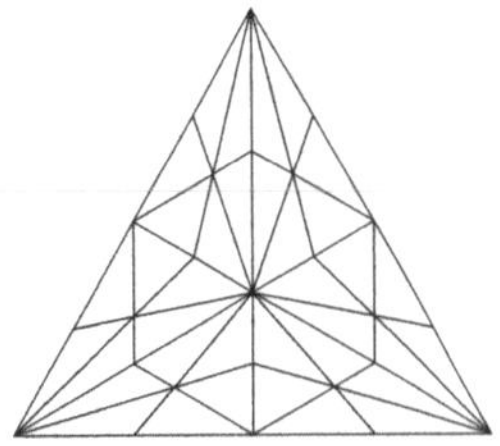

Fig. 6.6 Barycentric subdivision $K^{(2)}$

(ii) The 1-simplex $\langle v_0 v_1 \rangle = \{t v_0 + (1-t) v_1 : t \in I\}$ is the closed line segment with end points v_0, v_1. The barycenter $\frac{1}{2}(v_0 + v_1)$ is the midpoint of the line segment.

(iii) The 2-simplex $\langle v_0 v_1 v_2 \rangle$ is a triangle (with interior) having barycentric subdivision $K^{(2)}$ as shown in Fig. 6.6. This is the center of gravity of the vertices in the usual sense.

Lemma 6.4.6 *Let L be a subcomplex of a simplicial complex K. If K has a barycentric subdivision K', then L has a barycentric subdivision L' and it consists of all simplexes of K' which lie in $|L|$.*

Proof The simplexes of K' contained in $|L|$ form a subcomplex of K'. If s_i are simplexes of L and $s_0 \prec \cdots \prec s_m$, then $\langle B(s_0)B(s_1)\ldots B(s_m)\rangle \subset s_m \subset |L|$ satisfying the conditions **(i)** and **(ii)** of Definition 6.4.3. ❑

Proposition 6.4.7 *Let K be a simplicial complex. If a barycentric subdivision K' of K exists, then $|K'| = |K|$.*

Proof The simplexes of K' form a subcomplex of K'. Let s_i be simplexes of K and barycentric division of K' of K exist. If $s' = \langle B(s_0)\ldots B(s_n)\rangle$ is a simplex of K', then $s' \subset s_n \subset |K|$. This shows that $|K'| \subset |K|$. Again let $x \in s \subset |K|$. We order the vertices of $s = \langle v_0 v_1 \ldots v_n \rangle$ so that if $x = \sum r_i v_i, r_0 \geq r_1 \geq \cdots \geq r_n$. Let $s_i = \langle v_0 v_1 \ldots v_i \rangle$. Then $x = \sum t_i B(s_i)$, where $t_i = (i+1)(r_i - r_{i+1})$. Hence $t_i \geq 0$ and $\sum t_i = \sum r_i = 1$. Consequently, $x \in \langle B(s_0)\ldots B(s_n)\rangle \in K'$. Thus $|K| \subseteq |K'|$. Hence $|K'| = |K|$. ❑

Suppose K and M be simplicial complexes in $\mathbf{R}^n$. Let $K \cap M$ denote the set of all simplexes which are in both K and M and $K \cup M$ denote the set of all simplexes which are in either K or L. Clearly, $K \cap M$ is a subcomplex of K and M but in general $K \cup M$ is not a simplicial complex. We now give a sufficient condition under which $K \cup M$ is a simplicial complex.

Proposition 6.4.8 *Let K and M be simplicial complexes in $\mathbf{R}^n$. If $|K \cap M| = |K| \cap |M|$, then $K \cup M$ is a simplicial complex.*

Proof Let $s \in K$ and $t \in M$. We claim that $s \cap t$ is a face of both s and t. Let L and N be subcomplexes of K and M respectively. Then $|L \cap N| \subset |L| \cap |N| \cap |K \cap M|$. Suppose $x \in s \in L$, $x \in t \in N$, and $x \in w$ for some simplex w in $K \cap M$. Since t and w are both simplexes in M, $t \cap w \prec w$. Similarly, $s \cap w \prec s$. Hence $s \cap t \cap w = (s \cap w) \cap (t \cap w) \prec s \cap w \prec s$. Similarly, $s \cap t \cap w \prec t$. Consequently, $x \in s \cap t \cap w \in L \cap N$ and $x \in |L \cap N|$. Again let L be the simplicial complex consisting of s and all of its faces; N the simplicial complex consisting of t and all of its faces. Then

$$\begin{aligned} |L \cap N| = s \cap t \cap |K \cap M| &= s \cap t \cap |K| \cap |M| (\text{ by hypothesis } |K \cap M| = |K| \cap |M|) \\ &= (s \cap |K|) \cap (t \cap |M|) \\ &= s \cap t \end{aligned}$$

As $L \cap N$ is a subcomplex of L and N, $s \cap t$ is a subcomplex of s and t. Again as $s \cap t$ is convex, it is a face of s and t. Consequently, $K \cup M$ is a simplicial complex. ❑

If A is a subset of $\mathbf{R}^n$, then the diameter of A, denoted by diam (A) is defined by

$$\text{diam}\,(A) = \sup\{d(x, y) : x, y \in A\},$$

where $d(x, y) = ||x - y||$.

Proposition 6.4.9 *Let s be a simplex of positive dimension. Then* diam $(s) = ||u - v||$ *for some pair of vertices u and v of s, i.e., the diameter of s is the length of its largest edge.*

Proof Let $s = \langle v_0 v_1 \dots v_n \rangle$ and $x, y \in s$. If $y = \sum_{i=0}^{p} \lambda_i v_i$, where λ_i are barycentric coordinates of y, then

$$\begin{aligned} ||x - y|| = ||\Big(\sum_{i=0}^{p} \lambda_i\Big)x - \sum_{i=0}^{p} \lambda_i v_i|| &= ||\sum_{i=0}^{p} \lambda_i (x - v_i)|| \\ &\leq \sum_{i=0}^{p} \lambda_i ||x - v_i|| \\ &\leq \max(||x - v_i|| : 0 \leq i \leq p) \end{aligned} \tag{6.1}$$

If we replace y by v_i, then we have

$$||x - v_i|| \leq \max\{||v_j - v_i|| : 0 \leq j \leq p\} \tag{6.2}$$

Hence it follows from (6.1) and (6.2) that

$$||x - y|| \leq \max\{||v_j - v_i|| : 0 \leq i, j \leq p\} \tag{6.3}$$

Consequently, it follows from (6.3) that $\text{diam}(s) = ||u - v||$ for some pair of over-prices u and v of s. ❑

Definition 6.4.10 Let $K \subset \mathbf{R}^n$ be a simplicial complex and dim (s) denote the dimension of a simplex s of K. The mesh of K, written $\mu(K)$, is the maximum of the diameters of its simplexes i.e., mesh $(K) = \max\{\dim(s) : s \text{ is a simplex of } K\}$.

Proposition 6.4.11 *If the dimension of a simplicial complex $K \subset \mathbf{R}^m$ is n, then* $\mu(K') \leq \frac{n}{n+1}\mu(K)$.

Proof For proof it needs only measure the length of the 1-simplexes of K'. Let $\langle b_0 b_1 \rangle$ be a 1-simplex with $b_0 < b_1$. Then b_1 is the barycenter of a k-simplex $s = \langle v_0 v_1 \dots v_k \rangle$ in K. Given vectors $u_0 \dots, u_n$, u and numbers t_i with $\sum t_i = 1$, we have

$$\|u - \sum t_i u_i\| = \|\sum t_i(u - u_i)\| \leq \sum t_i \|u - u_i\|.$$

As $b_0 \in \langle v_0 v_1 \dots v_k \rangle$, $\|b_1 - b_0\| = \|b_1 - \sum t_i v_i\| \leq \sum t_i \|b_1 - v_i\|$. Again,

$$\|v_i - b_1\| = \|v_i - \frac{v_0 + \cdots + v_k}{k+1}\| \leq \frac{1}{k+1} \sum_j \|v_i - v_j\| \leq \frac{k}{k+1}\mu(s).$$

Consequently, $\|b_1 - b_0\| \leq \frac{k}{k+1}\mu(s)$. As $k \leq n$, it follows that $\frac{k}{k+1} \leq \frac{n}{n+1}$.
Hence, we have $\|b_1 - b_0\| \leq \frac{n}{n+1}\mu(K)$. This proves that $\mu(K') \leq \frac{n}{n+1}\mu(K)$. ❑

Corollary 6.4.12 $\lim_{r\to\infty} \mu(K^{(r)}) = 0$.

Proof By induction, it follows that $\mu(K^{(r)}) \leq \left(\frac{n}{n+1}\right)^r \mu(K)$. The corollary follows, since $\lim_{r\to\infty} \left(\frac{n}{n+1}\right)^r = 0$. ❑

6.5 Simplicial Approximation

This section introduces the concept of simplicial approximation and proves simplicial approximation theorem, first given by J.W. Alexander in (1915). This theorem is utilized in calculating fundamental groups, and in the study of topological invariance of the homology groups of a topological space.

Given topological spaces X and Y with triangulations $h : |K| \to X$ and $k : |L| \to Y$, any continuous map $f; X \to Y$ induces a continuous map $k^{-1} \circ f \circ h : |K| \to |L|$. Moreover, any continuous map between polyhedra may be approximated by a simplicial map in the sense that a continuous map g is considered as an 'approximation' to a continuous map f if $f \simeq g$.

Definition 6.5.1 Let $f : |K| \to |L|$ be a map between polyhedra. Given a point $x \in |K|$, the point $f(x)$ lies in the interior of a unique simplex of L. This unique simplex is called the carrier of $f(x)$.

We now consider simplicial maps which take simplexes to simplexes and are linear on the corresponding simplexes to define 'simplicial approximation' of a simplicial map.

Definition 6.5.2 Let K and L be simplicial complexes and $f : |K| \to |L|$ be a continuous map between polyhedra. A simplicial map $g : K \to L$ is said to be a simplicial approximation of $f : |K| \to |L|$ if $g(x)$ lies in the carrier of $f(x)$ for each $x \in |K|$ (i.e., for every $x \in |K|$ and for every simplex $t \in L$, $g(x) \in t$ implies $f(x) \in t$).

In many problems, such as for computation of the fundamental group of a triangulable space, it needs approximate a given map by a simplicial map. We claim that a simplicial approximation is homotopic to the original map.

Theorem 6.5.3 *Let K and L be simplicial complexes and $f : |K| \to |L|$ be a continuous map. Then*

(a) *any simplicial approximation g to f is homotopic to f;*
(b) *the homotopy in (a) is relative to the subspace $A = \{x \in |K| : f(x) = g(x)\}$ of $|K|$.*

Proof Let $g : |K| \to |L|$ be a simplicial approximation to $f : |K| \to |L|$. Suppose $|L|$ lies in $\mathbf{R}^n$, and $F : |K| \times I \to \mathbf{R}^n$ is the straight line homotopy given by $F(x, t) = (1 - t)g(x) + tf(x)$. Given $x \in |K|$, there is some simplex of L which contains $g(x)$ and $f(x)$; and as every simplex is convex, all points $(1 - t)g(x) + tf(x), t \in I$, must lie in this complex. Consequently, image of F is contained in $|L|$, and F is a homotopy from f to g. By construction this homotopy is relative to this subspace $A = \{x \in |K| : f(x) = g(x)\}$ of $|K|$. ❑

Corollary 6.5.4 *Let (K, L) and (B, C) be simplicial pairs and $f : (|K|, |L|) \to (|B|, |C|)$ be a map of pairs. If g is any simplicial approximation to $f : |K| \to |B|$, then $g(|L|) \subset |C|$, and $f \simeq g$ as maps of pairs.*

Proof If $x \in |L|$, then $f(x)$ is in the interior of the unique simplex of C. This simplex also contains $g(x)$ and hence $g(x) \in |C|$. Moreover, the line segment joining $f(x)$ and $g(x)$ lies entirely in $|C|$. Hence the corollary follows. ❑

Remark 6.5.5 Simplicial approximations do not always exist.

Example 6.5.6 Let $|K| = |L| = [0, 1]$ with K having vertices at the points $0, \frac{1}{3}, 1$ and L having vertices at $0, \frac{2}{3}, 1$ as shown in Fig. 6.7.

We shall show that the map $f : |K| \to |L|, x \mapsto x^2$ has no simplicial approximation. Because if $g : |K| \to |L|$ is a simplicial approximation to $f : |K| \to |L|$, then g must agree with f on inverse image of every vertex of L, giving $g(0) = 0$ and $g(1) = 1$. But as g is simplicial, $g(\frac{1}{3})$ must be $\frac{2}{3}$. Hence, g takes the closed segment $[0, \frac{1}{3}]$ linearly onto $[0, \frac{2}{3}]$ and $[\frac{1}{3}, 1]$ linearly onto $[\frac{2}{3}, 1]$. Since carrier of $f(\frac{1}{2})$ is $[0, \frac{2}{3}]$ and this does not contain $g(\frac{1}{2})$, we have a contradiction. Similarly, there is no simplicial approximation to $f : |K'| \to |L|$. But simplicial approximation to $f : |K^{(2)}| \to |L|$ exists for its second barycentric subdivision.

Remark 6.5.7 The above example shows that it may be possible to have a simplicial approximation by replacement of K by a suitable barrycentric subdivision $K^{(m)}$ of K. Moreover, simplicial approximations are not unique.

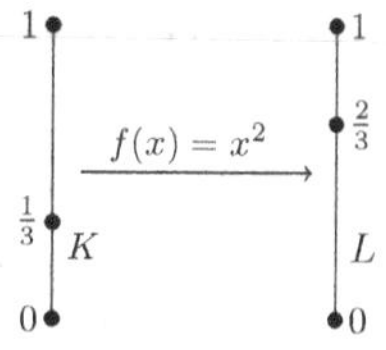

Fig. 6.7 Nonexistence of a simplicial approximation

We now characterize simplicial complexes with the help of open stars. We recall that if v is a vertex of K, the open star of v in K is the union of the interiors of those simplexes of K which have v as a vertex.

Proposition 6.5.8 *The vertices $v_0, v_1, \ldots, v_r$ of a simplicial complex K span (i.e., are the vertices of) a simplex of K iff the intersection of their open stars is nonempty.*

Proof Let $v_0, v_1, \ldots, v_r$ be the vertices of the simplex s of K. Then $\mathring{s} = \text{Int}(s) \subset \text{ost}(v_i)$, i.e., the whole of the interior of s lies in $\text{ost}(v_i)$ for each $i = 0, 1, \ldots, r$. Hence, $\emptyset \neq \text{Int}(s) \subset \bigcap_{i=0}^{r} \text{ost}(v_i)$. Conversely, let $\bigcap_{i=0}^{r} \text{ost}(v_i) \neq \emptyset$ and $x \in \bigcap_{i=0}^{r} \text{ost}(v_i)$. Then for each $i = 0, 1, 2, \ldots, r$, there is a simplex s_i in K such that $x \in \text{Int}(s_i)$ and v_i is a vertex of s_i. Since the set of all interiors of all simplexes of K constitutes a partition of $|K|$, there is a unique simplex of K whose interior contains x, which is the carrier of x. This shows that $s_0 = s_1 = s_2 = \cdots = s_r$. Hence $v_0, v_1, \ldots, v_r$ are vertices of the simplex s and hence these vertices span some face of s of K. ❑

The following theorem is the most basic form of "Simplicial approximation theorem."

Theorem 6.5.9 (Simplicial approximation) *Let $f : |K| \to |L|$ be a continuous map between polyhedra. If r is chosen sufficiently large, then there is a simplicial approximation $g : |K^{(r)}| \to |L|$ to $f : |K^{(r)}| \to |L|$.*

Proof Case I. We first suppose that for each vertex v of K there exists a vertex u of L such that

$$f(\text{ost}(v)) \subset \text{ost}(u) \tag{6.4}$$

Define a function $h : \text{Vert}(K) \to \text{Vert}(L)$ by choosing a u for each v and assigning $h(v) = u$. Then by using Proposition 6.5.8 and the inclusion (6.4) it follows that if $\{v_0, v_1, \ldots, v_r\}$ spans a simplex of K, then $\{h(v_0), \ldots, h(v_r)\}$ spans a simplex of L. Now extend h linearly over each simplex of K to obtain a simplicial map $h : |K| \to |L|$. We claim that h is a simplicial approximation to f. For $x \in |K|$, let $v_0, v_1, \ldots, v_r$ be the vertices of its carrier. Then $x \in \bigcap_{0}^{r} \text{ost}(v_i)$. Hence by the inclusion (6.4), $f(x) \in \bigcap_{0}^{r} \text{ost}(f(v_i))$ shows that carrier of $f(x)$ in L has the simplex spanned by $h(v_0), h(v_1), \ldots h(v_r)$ as a face. Hence it must contain the point $h(x)$.

Case II. For the general case, we replace K by a suitable barycentric subdivision $K^{(m)}$ and proceed as follows:

Since the open stars of the vertices of L form an open cover of $|L|$ and $f : |K| \to |L|$ is continuous, the inverse images of these open sets under f form an open cover of $|K|$. Again as $|K|$ is a compact metric space, there is a Lebesgue number η of this open cover by Lebesgue Lemma 1.11.5. Choose m sufficiently large such that $\mu(K^{(m)}) < \eta/2$. Given a vertex v of $K^{(m)}$, the diameter of its open star in $K^{(m)}$ is less than η. Hence $\text{ost}(v)$ in $K^{(m)}$ and $\text{ost}(u)$ in $|L|$ are such that $\text{ost}(v) \subset f^{-1}(\text{ost}(u))$ for some vertex u of L. Now proceed as in the first part to prove the general case. ❑

6.6 Computing Fundamental Groups of Polyhedra

This section presents a method of computing fundamental groups of a special class of topological spaces. More precisely, this section computes fundamental groups of polyhedra through a study of loops which are closed paths along 1-simplexes called edge loops.

Let K be a simplicial complex. We now define a group called edge group of K based at a vertex of K. An edge path in K is a sequence $v_0v_1 \ldots v_k$ of vertices such that each consecutive pairs v_i, v_{i+1} spans a simplex of K (the possibility $v_i = v_{i+1}$ is permissible for technical reason). If $v_0 = v_k = v$ for the edge path $v_0v_1 \ldots v_k$, then the edge path is called an edge loop of K based at v. To define the edge group of K, we define a simplicial version of the concept of homotopy. Two edge paths are said to be equivalent if we can obtain one from the other by a finite number of operations of types:

(i) if $v_{i-1} = v_i$, replace $\ldots v_{i-1}v_i \ldots$ by $\ldots v_i, \ldots$, or conversely, replace $\ldots v_i \ldots$ by $\ldots v_iv_i \ldots$; or

(ii) if $\{v_{i-1}, v_i, v_{i+1}\}$ spans a simplex of K (not necessarily 2-simplex), replace $\ldots v_{i-1}v_iv_{i+1} \ldots$ by $\ldots v_{i-1}v_{i+1} \ldots$, or conversely (geometrically this condition implies that two sides of a triangle can be replaced by the third side and vice-versa).

This equivalent relation sets up an equivalence relation between edge paths.
We denote the equivalence class of the edge path $v_0v_1v_2 \ldots v_k$ by $[v_0v_1 \ldots v_k]$. The set of equivalence classes of edge loops of K based at a vertex v of K forms a group under the binary operations of juxtaposition

$$[vv_1 \ldots v_{k-1}v]\cdot[vu_1 \cdots u_{r-1}v] = [vv_1 \cdots v_{k-1}vu_1 \cdots u_{r-1}v] \tag{6.5}$$

The identity element is the equivalence class $[v]$ and the inverse of $[vv_1 \cdots v_{k-1}v]$ is the class $[vv_{k-1} \cdots v_1v]$.

Definition 6.6.1 The set of equivalence classes of edge loops of K based at a vertex v of a simplicial complex K forms a group under the binary operations given by (6.5), called the edge group of K based at v, denoted by $E(K, v)$.

This group is closely related to the fundamental group $\pi_1(|K|, v)$ as given by the Theorem 6.6.2. It is convenient to present the edge group as a set of generators and relations.

Theorem 6.6.2 *The group $E(K, v)$ is isomorphic to the fundamental group $\pi_1(|K|, v)$ for any simplicial complex K.*

Proof Define a function $\psi : E(K, v) \to \pi_1(|K|, v)$ as follows: consider each edge loop in K as a loop in $|K|$. Then given an edge loop $vv_1 \dots v_{r-1}v$, divide I into r equal segments and let $f : I \to |K|$ be the linear extension of

$$f(0) = f(1) = v,\ f(i/r) = v_i,\ 1 \le i \le r - 1.$$

Then f is a loop in $|K|$ based at v. Since equivalent edge paths give homotopic loops, define

$$\psi : E(K, v) \to \pi_1(|K|, v),\ [vv_1 \dots v_{r-1}v] \mapsto [f].$$

Use simplicial approximation to show that ψ is onto. Then verify that ψ is an isomorphism. ❑

6.7 Applications

This section presents some interesting applications. We apply triangulation to prove an extension problem. Moreover, we apply the concepts of edge path and edge group $E(K, v)$ (which is isomorphic to the fundamental group $\pi_1(|K|, v)$) for any simplicial complex K to graph theory and prove van Kampen Theorem with the help of graph-theoretic results obtained.

6.7.1 Application to Extension Problem

One of the basic aims of algebraic topology is to solve extension problems of continuous maps. This subsection studies the extendability of the continuous map $f : S^m \to S^n$ to D^{n+1} for positive integers m and n with $m < n$.

Theorem 6.7.1 *Let $T = (K, h)$ be a triangulation of a space X such that dimension of $K < n$. Then every map $f : X \to S^n$ is inessential, i.e., f is homotopic to a map of X into a single point of S^n.*

Proof Choose a triangulation $T' = (K', h')$ of S^n with dim $K' = n$. Using simplicial approximation theorem there exists a map $g : X \to S^n$ homotopic to f and which is related to triangulations $T^{(r)}, T'$, where $T^{(r)} = (K^{(r)}, h^{(r)})$. Since the dimension of a simplicial complex is not changed under barycentric subdivision, it shows that dim $K^{(r)} < n$. Again since g is simplicial, it follows that $g(X) \neq S^n$. If possible, let g be essential. Then by Proposition 2.10.6 of Chap. 2, $g(X) = S^n$, which is a contradiction. This shows that g is inessential and hence f is inessential. ❑

Proposition 6.7.2 *If m and n are two positive integers such that $m < n$. Then every map $f : S^m \to S^n$ is inessential and admits an extension $\tilde{f} : D^{n+1} \to S^n$.*

Proof Since dimension of S^m is $m < n$, it follows from Theorem 6.7.1 that the map $f : S^m \to S^n$ is inessential. Moreover, since (D^{n+1}, S^n) forms a normal pair it follows from Proposition 2.10.7 of Chap. 2 that f admits an extension $\tilde{f} : D^{n+1} \to S^n$. ❑

Proposition 6.7.3 *If (X, A) is a finite triangulable pair, then A has the homotopy extension property (HEP) with respect to every space.*

Proof Let $f : X \to Y$ be a given continuous map and $H_t : A \to Y$ be a partial homotopy of f in the sense that $f|A = H_0$. Consider, the product space $P = X \times I$ and its closed subspace $C = (X \times \{0\} \cup (A \times I)$ ❑

Define a map

$$H : C \to Y, (x, t) \mapsto \begin{cases} f(x), & \text{if } x \in X, t = 0 \\ H_t(x), & \text{if } x \in A, t \in I \end{cases}$$

As (X, A) is a finite triangulable pair, C is a retract of P. Hence there is a retraction $r : P \to C$. Define a homotopy

$$G_t : X \to Y, x \mapsto H(r(x, t)).$$

Then G_t is an extension of H_t such that $G_0 = f$.

6.7.2 Application to Graph Theory

This subsection presents some graph-theoretic results which are used as tools to prove van Kampen Theorem. A graph is a one-dimensional simplicial complex.

Definition 6.7.4 A one-dimensional subcomplex of a complex K whose polyhedron is both path-connected and simply connected is called a tree. A tree T is said to be maximal if T' is a tree such that T' contains T, then $T' = T$.

Theorem 6.7.5 *A maximal tree T in a complex K contains all the vertices of K.*

Proof If possible, T does not contain all the vertices of K. Then there exists some vertex v which is in $K - T$. Choose a vertex u of T. Since $|K|$ is path-connected, there is a path joining u and v in $|K|$. By simplicial approximation Theorem 6.5.9, this path is replaced by an edge path $uv_1v_2 \ldots v_kv$. If v_i is the last vertex of this edge path which lies in T, a new subcomplex T' is formed by adding the vertex v_{i+1} and the edge spanned by v_iv_{i+1} to T. The space $|T'|$ is the same as $|T|$ with a 'spike' attached. Clearly, $|T'|$ is a deformation retract of $|T|$. Hence T' is a tree, which contradicts the assumption that T is a maximal tree. ❑

6.7.3 van Kampen Theorem

This subsection proves van Kampen theorem given by van Kampen (1908–1942), which prescribes a method for computing the fundamental groups of topological spaces that can be decomposed into simpler spaces whose fundamental groups are already known. It is convenient to present the edge group as a set of generators and relations. Some authors call van Kampen theorem Seifert–van Kampen theorem. This theorem is proved by using the algebraic concept of free product of two groups (see Chap. 1).

Let L be a subcomplex of K such that $|L|$ is simply connected. Then edge loops in L has no contribution to $E(K, v)$. This shows that the simplexes of L may be effectively ignored in calculation of $E(K, v)$. If we list the vertices of K as $v = v_0, v_1, v_2, \ldots, v_t$ and denote $G(K, L)$ for the group which is determined by generators g_{ij}, one for each ordered pair of vertices v_i, v_j that span a simplex of K with the relations $g_{ij} = 1$ if v_i, v_j span a simplex of L, and $g_{ij}g_{jk} = g_{ik}$ if v_i, v_j, v_k span a simplex of K. If $i = j$, then $g_{ii} = 1$, and for $i = k$, $g_{ji} = g_{ij}^{-1}$. This implies that we introduce a generator g_{ij} for each pair of vertices v_i, v_j which span an edge of $K - L$ and for which $i < j$. So we consider only the relations $g_{ij}g_{jk} = g_{ik}$, whenever v_i, v_j, v_k span a 2-simplex of $K - L$ and $i < j < k$. In particular, if any two of these vertices v_i, v_j span a simplex of L, we take $g_{ij} = 1$.

Theorem 6.7.6 *The group $G(K, L)$ is isomorphic to the group $E(K, v)$ when $|L|$ is simply connected.*

Proof We construct homomorphisms

$$\psi : G(K, L) \to E(K, v) \quad \text{and} \quad \theta : E(K, v) \to G(K, L)$$

as follows: we join v to each vertex v_i of K by an edge path E_i in L, by taking $E_0 = v$ and define ψ on the generators of $G(K, L)$ by $\psi(g_{ij}) = [E_i v_i v_j E_j^{-1}]$. If v_i, v_j span a simplex of L, then $E_i, v_i, v_j E_j^{-1}$ is a loop which lies entirely in L. Hence, it represents the identity element of $E(K, v)$, since $|L|$ is simply connected. Moreover, if v_i, v_j, v_k span a simplex of K, then

$$\begin{aligned}\psi(g_{ij})\psi(g_{jk}) &= [E_i v_i v_j E_j^{-1}][E_j v_j v_k E_k^{-1}] \\ &= [E_i v_i v_k E_k^{-1}] \\ &= \psi(g_{ik})\end{aligned}$$

This proves that ψ is a homomorphism.

Define $\theta : E(K, v) \to G(K, L), [v v_k v_l v_m \ldots v_n v] \mapsto g_{ok} g_{kl} g_{lm} g_{no}$. Then θ defines also a homomorphism such that $\theta \circ \psi$ is the identity. Moreover, $\psi \circ \theta$ is the identity. Hence ψ is an isomorphism with θ as its inverse. ❑

Remark 6.7.7 Let L, K be two simplicial complexes in the same Euclidean space which intersect in a common subcomplex. If $|L|, |K|, |L \cap K|$ are all path-connected spaces and their fundamental groups are known, then we can calculate the fundamental group $\pi_1(|L \cup K|)$.

Case 1: If L and K intersect in a single vertex, then any edge loop in $L \cup K$ based at this vertex is a product of loops, each of which lies in either L or K. We now obtain the free product $\pi_1(|L|) * \pi_1(|K|)$ for the fundamental group of $|J \cup K|$.

Case 2: In the general case, similar arguments hold, except that the free product $\pi_1(|L|) * \pi_1(|K|)$ effectively counts the homotopy classes of these loops which lie in $|L \cap K|$ twice (one in each of $\pi_1(|L|), \pi_1(|K|)$). So we need in this case some extra relations.

The above discussion leads to prove "van Kampen Theorem."

Theorem 6.7.8 (van Kampen Theorem) *Let L, K be two simplicial complexes in the same Euclidean space such that $|L|$, $|K|$ and $|L \cap K|$ are all path-connected spaces. If $i : |L \cap K| \hookrightarrow |L|$ and $j : |L \cap K| \hookrightarrow |K|$ are inclusion maps and a vertex v of $L \cap K$ is taken as a base point of $L \cap K$, then the fundamental group of $|L \cup K|$ based at v is the free product $\pi_1(|L|, v) * \pi_1(|K|, v)$ with the relations $i_*(x) = j_*(x)$ for all $x \in \pi_1(|L \cap K|, v)$.*

Proof Let T_M be a maximal tree in $L \cap K$. We extend it to a maximal tree T_{M_1} in L and a maximal tree T_{M_2} in K. Then $T_{M_1} \cup T_{M_2}$ is a maximal tree in $L \cup K$. Using Theorems 6.7.5 and 6.7.6, the group $\pi_1(|L \cup K|)$ is generated by elements g_{ij} corresponding to edges of $L \cup K - T_{M_1} \cup T_{M_2}$, with relations $g_{ij}g_{jk} = g_{ik}$ given by the triangles of $L \cup K$. But this is precisely the group obtained by taking a generator b_{ij} for each edge of $L - T_{M_1}$, a generator c_{ij} for each edge of $K - T_{M_2}$, with relations of the form $b_{ij}b_{jk} = b_{ik}$, $c_{ij}c_{jk} = c_{ik}$ corresponding to the triangles of L, K with additional relations $b_{ij} = c_{ij}$, whenever b_{ij} and c_{ij} correspond to the same edge of $L \cap K$. Since the edges of $L \cap K - T_M$, when considered as edges of L, give a set of generators for $i_*(\pi_1(|L \cap K|))$. Similarly the same edges, when considered as edges of K give a set of generators for $j_*(\pi_1(|L \cap M|))$. ❑

Remark 6.7.9 For another form of van Kampen Theorem see Theorem 14.7.1 of Chap. 14.

6.8 Exercises

1. Let X be a polyhedron and ϵ be an arbitrary small positive real number. Show that there is a simplicial complex K such that $|K| = X$ and $\mu(K) < \epsilon$.
[Hint: Let $K = X^{(r)}$. If $n = \dim X$, then $\mu(K) \leq (\frac{n}{n+1})^r \mu(X)$. Choose r sufficiently large to obtain $\mu(K) < \epsilon$.]
2. Let $L \subset K$ be a subcomplex and $f : K \to M$ be a map such that $f|_L$ is simplicial. If g is a simplicial approximation to f, show that $g \simeq f$ rel L.

[Hint: Let $L \subset \mathbf{R}^n$. Define $F_t : K \to \mathbf{R}^n$ by $F_t(x) = tg(x) + (1-t)f(x)$. Since both $g(x)$ and $f(x)$ lie in some simplex s, F_t is well defined. Hence $F_t(K) \subset L$. Consequently, $F_t : f \simeq g$ rel L.]

3. **(a)** Show that every simplicial complex has a barycentric subdivision.

(b) Let K be a simplicial complex and K' be its first barycentric subdivision of K. Show that each simplex of K' is contained in a simplex of K.

[Hint: **(a)** Use double induction: first on the dimension of the complex and second on the number of simplexes.

(b) Let s be a simplex of K' and barycenters $B(s_0), B(s_1), \ldots, B(s_k)$, are the vertices of s, where $v_i \in K$ and $v_0 < v_1 < \cdots < v_r$. Hence all vertices s of s lie in v_r. Hence s is contained in s. $s_i \in K$ and $s_0 < s_1 < \cdots < s_k$. Hence all these vertices of s lie in s_k. This implies that $s \subset s_k$.]

4. Let s_p be a geometric p-simplex in $\mathbf{R}^n$. Show that

(a) s_p is a closed convex connected subspace of $\mathbf{R}^n$, is the closure of its interior;

(b) faces of s_p is a closed subspace of s_n;

(c) s_p determines its vertices;

(d) any two simplexes are identical iff they have the same set of vertices. [Hint: Use the result that a point of s_p is a vertex iff it is not a point of an open line segment lying inside s_p].

5. Let $s_p = \langle v_0 v_1 \ldots v_p \rangle$ be a geometric p-simplex in $\mathbf{R}^n$ and $s'_p = \langle v'_0, v'_1, \ldots, v'_p \rangle$ be a geometric p-simplex in $\mathbf{R}^m$. Show that s_p and s'_p are linearly homeomorphic.

[Hint: Define $f : s_p \to s'_p$, $\sum_{i=0}^{p} a_i v_i \mapsto \sum_{i=0}^{p} a_i v'_i$ for all points of s_p. Then f is a linear homeomorphism].

6. Show that a geometric p-simplex s_p is completely characterized by its dimension. [Hint: Use Ex 5].

7. Let K be a simplicial complex in $\mathbf{R}^n$ and A be a subset of $|K|$. Show that

(a) A is closed in $|K|$ iff $A \cap S$ is closed in S for every simplex S of K.

(b) if S is a simplex in K of largest dimension, then $\mathring{S} = S - \dot{S}$ is an open subset of $|K|$.

(c) The topology of $|K|$ as a subspace of $\mathbf{R}^n$, coincides with the topology of $|K|$, considered as the space obtained from its simplexes by identifying together with the various intersections.

8. Let (K, L) be a simplicial pair in $\mathbf{R}^n$. Show that

(i) $|K|$ is a closed and compact subspace of $\mathbf{R}^n$;

(ii) every point of $|K|$ is in the interior of exactly one simplex of K;

(iii) $|L|$ is a closed subspace of $|K|$;

(iv) if C is another subcomplex of K, then $L \cup C$ and $L \cap C$ are both subcomplexes of K;

(v) If $|K|$ is a connected space, then it is path-connected.

9. Let K be a simplicial simplex. Prove that for each vertex v of Vert (K),

(i) st (v) is an open subset of $|K|$;
(ii) the family of all such stars is an cover of $|K|$,
(iii) if $x \in$ st (v), then the line segment with end points x and v is contained in st (v).

10. Given a simplex s of K and any point x in the interior of s, show that $st_K(s) = |N_K(x)| - |L_K(x)|$.

11. If $v_0, v_1, \ldots, v_m$ are in the vertex set Vert (K) of a simplicial complex K. Show that $\{v_0, v_1, \ldots, v_n\}$ spans a simplex of K iff $\bigcap_{i=0}^{m} \text{st}\,(v_i) \neq \emptyset$.

[Hint. See Proposition 6.5.8]

12. Using van Kampen Theorem, show that

(i) $\pi_1(S^n) = 0$ for $n > 1$;
(ii) If X is the union of two circles in $\mathbf{R}^2$ with one point in common, then $\pi_1(X)$ is the free group on two generators.

(Compare the fundamental group of 'figure-eight' described in Chap. 4).

13. Show that a group G is finitely presented iff there exists a polyhedron X such that $G \cong \pi_1(X, x_0)$.

14. (Tietze) If X is a connected polyhedron, show that fundamental group $\pi_(X, x_0)$ is finitely presented in the sense that $\pi_1(X, x_0)$ has a presentation with only finitely many generators and finitely many relations.

6.9 Additional Reading

[1] Croom, F.H., *Basic Concepts of Algebraic Topology*, Springer-Verlag, New York, Heidelberg, Berlin, 1978.
[2] Hatcher, Allen, *Algebraic Topology*, Cambridge University Press, 2002.
[3] Lahiri B. K., *First Course In Algebraic Topology*, Narosa Publishing House, New Delhi, 2005.
[4] Rotman, J.J., *An Introduction to Algebraic Topology*,Springer-Verlag, New York, 1988.
[5] Spanier, E., *Algebraic Topology*, McGraw-Hill Book Company, New York, 1966.
[6] Switzer, R.M., *Algebraic Topology-Homotopy and Homology*, Springer-Verlag, Berlin, Heidelberg, New York, 1975.
[7] Wallace, A.H., Algebraic Topology, Benjamin, New York, 1980,
[8] Whitehead, G.W., *Elements of Homotopy Theory*, Springer-Verlag, New York, Heidelberg, Berlin, 1978.

References

Alexander, J.W.: A proof of the invariance of certain constants of analysis situs. Trans. Amer. Math. Soc. **16**, 148–154 (1915)

Armstrong, M.A.: Basic Topology. Springer, New York (1983)

Croom, F.H.: Basic Concepts of Algebraic Topology. Springer, New York (1978)

Eilenberg, S., Steenrod, N.: Foundations of Algebraic Topology. Princeton University Press, Princeton (1952)

Gray, B.: Homotopy Theory: An Introduction to Algebraic Topology. Academic, New York (1975)

Hatcher, A.: Algebraic Topology. Cambridge University Press, Cambridge (2002)

Hilton, P.J., Wylie, S.: Homology Theory. Cambridge University Press, Cambridge (1960)

Lahiri, B.K.: First Course In Algebraic Topology. Narosa Publishing House, New Delhi (2005)

Maunder, C.R.F.: Algebraic Topology. Van Nostrand Reinhold, London (1970)

Rotman, J.J.: An Introduction to Algebraic Topology. Springer, New York (1988)

Singer, I.M., Thrope, J.A.: Lecture Notes on Elementary Topology and Geometry. Springer, New York (1967)

Spanier, E.: Algebraic Topology. McGraw-Hill Book Company, New York (1966)

Switzer, R.M.: Algebraic Topology-Homotopy and Homology. Springer, Berlin (1975)

Veblen, O.: Analysis Situs, vol. 2. Part II. American Mathematical Society Colloquium Publications, New York (1922)

Wallace, A.H.: Algebraic Topology. Benjamin, New York (1980)

Whitehead, G.W.: Elements of Homotopy Theory. Springer, New York (1978)

Zeeman, E.C.: Relative simplicial approximation. Proc. Camb. Phil. Soc. **60**, 39–43 (1964)

Chapter 7
Higher Homotopy Groups

This chapter continues to study homotopy theory displaying construction of a sequence of covariant functors π_n given by W. Hurewicz (1904–1956) in 1935 from topology to algebra by extending the concept of fundamental group, which is the first influential functor of homotopy theory invented by H. Poincaré (1854–1912) in 1895. It also studies Hopf map and Freudenthal suspension theorem. Prior to Hurewicz, the idea of higher homotopy groups was originated by E.Čech (1893–1960) in 1932 but the notation used by Hurewicz has become standard and it is followed. The higher dimensional homotopy groups provide fundamental tools of classical homotopy theory and are the most powerful basic invariants in algebraic topology. There is an infinite exact sequence of homotopy groups associated with a fiber space which is utilized to study Hopf fibering and to compute higher homotopy groups of certain spaces. Weak fibration has a key role in the study of higher homotopy groups.

The basic problem of n-dimensional homotopy groups is to classify all continuous maps from S^n to pointed topological spaces X up to homotopy equivalence. For the study of pointed topological spaces X of low dimension, the fundamental group $\pi_1(X)$ is very useful. But it needs refined tools for the study of higher dimensional spaces. For example, fundamental group can not distinguish spheres S^n with $n \geq 2$. Such a limitation of low dimension can be removed by considering the natural higher dimensional analogs of $\pi_1(X)$. The complete determination of higher homotopy groups of spheres is still one of the major unsolved problems in topology. The classification problem of continuous maps of an n-sphere S^n to a given pointed topological space (X, x_0) up to homotopy equivalence led to the discovery of 'homotopy groups.' For $n = 1$, recall that given a pointed topological space (X, x_0), $\pi_1(X, x_0)$ defined by $[(I, \dot{I}), (X, x_0)]$ as the set of homotopy classes of loops $f : (I, \dot{I}) \to (X, x_0)$ admits a group structure, called the fundamental group of (X, x_0). For each integer $n > 1$, the definition of the nth (absolute) homotopy group $\pi_n(X, x_0)$ is strictly analogous to that of the fundamental group. This means that $\pi_n(X, x_0) = [(I^n, \partial I^n), (X, x_0)]$, where ∂I^n is the boundary of the n-cube I^n. An element of $\pi_n(X, x_0)$ can be also well defined as a homotopy class relative to

M.R. Adhikari, *Basic Algebraic Topology and its Applications*,
DOI 10.1007/978-81-322-2843-1_7

s_n of the continuous maps $f : (S^n, s_n) \to (X, x_0)$, where S^n is the unit sphere and $s_n = (1, 0, \ldots, 0) \in \mathbf{R}^{n+1}$ is regarded as the base point of S^n. For $n > 1$, there is a rotation of the n-sphere S^n providing a homotopy interchanging its two hemispheres. It implies the interesting property that $\pi_n(X, x_0)$ is abelian for $n > 1$.

More precisely, this chapter defines nth (absolute) homotopy group and generalizes it to (relative) homotopy group of a triplet and studies algebraic, functorial and fibering properties with exactness of homotopy sequence of fibering along with Hopf maps introduced by H. Hopf (1894–1971) in 1935 for investigation of certain homotopy groups of S^n. This chapter also presents action of π_1 on π_n, Freudenthal suspension theorem given by H. Freudenthal (1905–1990) in 1937 for investigation of the homotopy groups $\pi_m(S^n)$ for $0 < m < n$ and the nth cohomotopy set $\pi^n(X, A)$ on which K. Borsuk (1905–1982) endowed in 1936 an abelian group structure under certain conditions on (X, A). This chapter also discusses some interesting applications of higher homotopy groups. Computing the homotopy group $\pi_n(X, x_0)$ is very difficult even in the cases of some simple spaces, such as certain subspaces of Euclidean spaces. These computations were the object of investigations of many prominent topologists, like E. Cartan (1869–1951), S. Eilenberg (1915–1998), J.P. Serre (1926–) and many others.

For this chapter, the books Croom (1978), Gray (1975), Hatcher (2002), Hu (1959), Switzer (1975) and some others are referred in Bibliography.

7.1 Absolute Homotopy Groups: Introductory Concept

This section defines and studies absolute homotopy groups $\pi_n(X, x_0)(n > 1)$, which are generally called higher homotopy groups of a pointed topological space (X, x_0). For an alternative approach given by W. Hurewicz see Sect. 7.2.

Let I^n be the topological product of n-copies of I for $n > 1$. The nth absolute homotopy group $\pi_n(X, x_0)$ is defined in a way analogous to the construction of the fundamental group $\pi_1(X, x_0)$ of a pointed topological space (X, x_0) by replacing I by n-cube I^n and $\dot{I}$ by the boundary ∂I^n of I^n. Every point $t \in I^n$ is represented by $t = (t_1, t_2, \ldots, t_n)$, $t_i \in I$. The real number t_i is called the i-th coordinate of t. Thus,

$$I^n = \{(t_1, t_2, \ldots, t_n) : 0 \leq t_i \leq 1\}.$$

An $(n-1)$ - face of I^n is given by setting some coordinates t_i to be 0 or 1. The union of all $(n-1)$ - faces of I^n is called the boundary of I^n, denoted by $\dot{I}^n$ or ∂I^n. Thus $\partial I^n = \dot{I}^n = \{(t_1, t_2, \ldots, t_n) \in I^n : \text{ some } t_i \in \dot{I}\}$, which is topologically equivalent to $(n-1)$- sphere S^{n-1}. Let Int (I^n) denote the interior of I^n and it is considered as a subspace of I^n.

Proposition 7.1.1 Int $(I^n) = \{(t_1, t_2, \ldots, t_n) : 0 < t_i < 1\}$.

Proof Let $0 < t_i < 1$ for each i and $\epsilon = \min_i(1 - t_i, t_i)$. Then a disk of radius ϵ with center $t = (t_1, t_2, \ldots, t_n)$ is contained in I^n. This shows that $t \in$ Int (I^n). Again if for some i, $t_i = 1$ or 0, then a disk of radius r (however small it may be) with center t contains points with $t_i > 1$. This implies that such points are not in Int (I^n). ❑

Remark 7.1.2 We write $\partial I^n = I^n - \text{Int}\,(I^n)$.

We now consider the set

$$F_n(X, x_0) = \{f : (I^n, \partial I^n) \to (X, x_0) : f \text{ is continuous }\}.$$

If $F_n(X, x_0)$ is topologized by compact open topology (see Chap. 1), then the homotopy set $[(I^n, \partial I^n), (X, x_0)]$ relative to ∂I^n, denoted by $\pi_n(X, x_0)$ is the set of all path- components of the space $F_n(X, x_0)$.

We recall the definition of relative homotopy.

Definition 7.1.3 Let f, g be two continuous maps in $F_n(X, x_0)$. Then f, g are said to be homotopic related to ∂I^n denoted by $f \simeq g\ rel\ \partial I^n$ if there exists a continuous map $F : I^n \times I \to X$ such that

$$F(t_1, t_2, \ldots, t_n, 0) = f(t_1, t_2, \ldots, t_n),\quad F(t_1, t_2, \ldots, t_n, 1) = g(t_1, t_2, \ldots, t_n),\ \forall\,(t_1, t_2, \ldots, t_n) \in I^n$$

and $F(t_1, t_2, \ldots, t_n, s) = x_0,\ \forall\,(t_1, t_2, \ldots, t_n) \in \partial I^n$ and $s \in I$.

We have shown in Chap. 2 that the relative homotopy relation is an equivalence relation. If $[f]$ denotes the equivalence class of $f \in F_n(X, x_0)$, then $\pi_n(X, x_0)$ is the set defined by $\pi_n(X, x_0) = \{[f] : f \in F_n(X, x_0)\}$. We are now in a position to endow the set $\pi_n(X, x_0)$ with a group operation.

Define a composition $'*'$ in $F_n(X, x_0)$ as follows : For $f, g \in F_n(X, x_0)$, $f * g$ is defined by

$$(f * g)(t) = \begin{cases} f(2t_1, t_2, \ldots, t_n), & \text{if } 0 \le t_1 \le 1/2 \\ g(2t_1 - 1, t_2, \ldots, t_n), & \text{if } 1/2 \le t_1 \le 1 \end{cases}$$

for every $t = (t_1, t_2, \ldots, t_n) \in I^n$.

At $t_1 = \frac{1}{2}$, $f(1, t_1, \ldots, t_n) = x_0$ and $g(0, t_2, \ldots, t_n) = x_0$, since $(1, t_1, \ldots, t_n)$, $(0, t_2, \ldots, t_n) \in \partial I^n$. Again for the point $(t_1, t_2, \ldots, t_n) \in \partial I^n$, the points $(2t_1, t_2, \ldots, t_n)$ and $(2t_1 - 2, t_2, \ldots, t_n) \in \partial I^n$ and hence $(f * g)((t_1, t_2, \ldots, t_n) = x_0$ for any $(t_1, t_2, \ldots, t_n) \in \partial I^n$. By pasting lemma (see Chap. 1), $f * g$ is continuous. Clearly, $f * g \in F_n(X, x_0)$.

Lemma 7.1.4 *Given $f_1, f_2 \in [f]$ and $g_1, g_2 \in [g]$, there is a homotopy $F : I^n \times I \to X$ such that*

$$F : f_1 * g_1 \simeq f_2 * g_2 \text{ rel } \partial I^n.$$

Proof As $f_1 \simeq f_2$ rel ∂I^n and $g_1 \simeq g_2$ rel ∂I^n, $\exists$ homotopies G and H such that

$$G : f_1 \simeq f_2 \text{ rel } \partial I^n \text{ and } H : g_1 \simeq g_2 \text{ rel } \partial I^n.$$

Define $F : I^n \times I \to X$ by the rule

$$F(t_1, t_2, \ldots, t_n, s) = \begin{cases} G(2t_1, t_2, \ldots, t_n, s), & \text{if } 0 \leq t_1 \leq 1/2 \\ H(2t_1 - 1, t_2, \ldots, t_n, s), & \text{if } 1/2 \leq t_1 \leq 1. \end{cases}.$$

Then $F : f_1 * g_1 \simeq f_2 * g_2$ rel ∂I^n. ❑

Define a composition '$\circ$' in $\pi_n(X, x_0)$ by the rule

$$[f] \circ [g] = [f * g] \tag{7.1}$$

This composition '$\circ$' is clearly independent of the choice of the representatives of the classes and hence it is well defined.

Theorem 7.1.5 *$\pi_n(X, x_0)$ is a group under the composition '$\circ$' for $n \geq 1$.*

Proof For $n = 1$, $\pi_1(X, x_0)$ has been shown (in Theorem 3.1.12 Chap. 3) to be a group, called the fundamental group of (X, x_0) based at x_0. For $n > 1$, proceed in a similar way to show that $\pi_n(X, x_0)$ is a group under the composition defined in (7.1). The zero element of the group in the homotopy class of the unique constant map $c : I^n \to x_0$. The inverse element of $[f] \in \pi_n(X, x_0)$ is the homotopy class $[f^{-1}]$, of the composite map f and ψ, where $\psi : I^n \to I^n, t \mapsto (1 - t_1, t_2, \ldots, t_n)$, $\forall t \in t = (t_1, t_2, \ldots, t_n) \in I^n$, i.e., where $f^{-1} : I^n \to X$ is the map defined by

$$f^{-1}(t_1, t_2, \ldots, t_n) = f(1 - t_1, t_2, \ldots, t_n), \ \forall\, (t_1, t_2, \ldots, t_n) \in I^n.$$

To show this it is sufficient to prove that $f * f^{-1} \simeq c$ rel ∂I^n. Consider the continuous map $F : I^n \times \partial I^n$ defined by

$$F(t_1, t_2, \ldots, t_n, s) = \begin{cases} x_0, & \text{if } 0 \leq t_1 \leq s/2 \\ f(2t_1 - s, t_2, \ldots, t_n, s), & \text{if } s/2 \leq t_1 \leq 1/2 \\ f^{-1}(2t_1 + s - 1, t_2, \ldots, t_n, s), & \text{if } 1/2 \leq t_1 \leq (1-s)/2 \\ x_0, & \text{if } (1-s)/2 \leq t_1 \leq 1 \end{cases}$$

Clearly, $f * f^{-1} \simeq c$ rel ∂I^n. ❑

Definition 7.1.6 The group $\pi_n(X, x_0)$ is called the (absolute) homotopy group of (X, x_0) for $n \geq 1$. For $n = 1$, this group is called the fundamental group.

Remark 7.1.7 The homotopies defined in the proof of the set $\pi_n(X, x_0)$ to be a group for $n \geq 1$ are precisely the same homotopies defined in the proof for the fundamental group $\pi_1(X, x_0)$. While defining the homotopies for $\pi_n(X, x_0)$ all the actions are in t_1 by keeping the other coordinates unchanged.

Proposition 7.1.8 *Let (X, x_0) be a pointed topological space and X_0 be the path component of X containing x_0, then $\pi_1(X_0, x_0) \cong \pi_n(X, x_0)$ for $n \geq 1$.*

Proof Since I^n is path- connected, the proof is immediate. □

7.2 Absolute Homotopy Groups Defined by Hurewicz

This section conveys an alternative method of construction of higher homotopy groups introduced by Hurewicz (Hurewicz 1935). Given a pointed topological space X, he defined in 1935 higher homotopy groups by endowing a group structure on the set $\pi_n(X) = [S^n, X]$ to classify continuous maps $f : S^n \to X$ up to homotopy equivalence. This classification made the discovery of homotopy groups. His approach of construction was: if the boundary ∂I^n of the n-cube I^n is identified to a point, a quotient space which is homeomorphic to an n-sphere S^n with a base point $s_n \in S^n$ is obtained and hence an element of $\pi_n(X, x_0)$ can be equally well defined as a homotopy class relative to $s_n \in S^n$ of the maps $f : (S^n, s_n) \to (X, x_0)$. This motivated to define the absolute homotopy groups $\pi_n(X, x_0)$. The group $\pi_1(X, x_0)$ is in general nonabelian. On the other hand the groups $\pi_n(X, x_0)$ are all abelian for all $n \geq 2$.

Definition 7.2.1 For every pointed space (X, x_0) and every integer $n \geq 0$, the n-th homotopy set is denoted by the set $\pi_n(X, x_0) = [(S^n, s_n); (X, x_0)]$. For $n \geq 2$, $\pi_n(X, x_0)$, it is a group called a (higher) homotopy group and is sometimes called an absolute homotopy group.

Remark 7.2.2 The higher homotopy group has some interesting properties different from fundamental group. If $n > 1$, there exists a rotation of S^n which keeps its base point s_n fixed and interchanges the two hemispheres of S^n. This implies intuitively that for $n > 1$, the group $\pi_n(X, x_0)$ is abelian. Its analytical proof is given in Theorem 7.2.5. Moreover, we show as a consequence that every homotopy group of a pointed space can be expressed as a fundamental group of some other space (see Corollary 7.2.4). Let (X, x_0) be a pointed space. The fundamental group $\pi_1(X, x_0)$ is not in general commutative (but commutative for some spaces).

We now use the concepts of loop spaces and suspension spaces defined in the Chap. 2.

Theorem 7.2.3 *For every integer $n > 1$, the groups $\pi_n(X)$ and $\pi_{n-r}(\Omega^r X)$ are isomorphic for all $1 \leq r \leq n-1$ for any pointed space X, where ΩX is the loop space of X and $\Omega^r = \Omega \times \Omega \times \ldots \times \Omega$ (r-times).*

Proof If Σ is the suspension functor, then Ω and Σ are adjoint functors (see Proposition 2.5.1). Now

$$\pi_n(X) = [S^n, X] \cong [S^{n-r}, \Omega^r X] = \pi_{n-r}(\Omega^r X).$$

□

Corollary 7.2.4 *If X is a pointed space, then for $n \geq 2$, the groups $\pi_n(X)$ and $\pi_1(\Omega^{n-1}X)$ are isomorphic.*

Proof It follows from Theorem 7.2.3 as a particular case. This shows that for $n \geq 2$, $\pi_n(X)$ can be equivalently defined by $\pi_n(X) = \pi_1(\Omega^{n-1}X) = [S^1, \Omega^{n-1}X]$. ❑

Theorem 7.2.5 *If X is a pointed space, then the group $\pi_n(X)$ is abelian for all $n \geq 2$.*

Proof For $n \geq 2$, $\Omega^{n-1}X$ is a loop space of X and hence $\Omega^{n-1}X$ is an Hopf group (H-group). Hence by Theorem 3.1.39 of Chap. 3, $\pi_1(\Omega^{n-1}X)$ is abelian. ❑

The following theorem plays an important role in algebraic topology. This is sometimes referred as 'Dimension Axiom.'

Theorem 7.2.6 *If X is a singleton space, then $\pi_n(X) = 0$ for all $n \geq 0$.*

Proof As X is a singleton space, there exists only one map $c : S^n \to X$, which is the constant function. Hence $[S^n, X]$ has only one element $[c]$, denoted by 0, $\forall n \geq 0$. ❑

Given pointed spaces (X, x_0) and (Y, y_0), like fundamental groups, the groups $\pi_n(X, x_0)$, $\pi_n(Y, y_0)$ and $\pi_n(X \times Y, (x_0, y_0))$ have similar relations.

Theorem 7.2.7 *The groups $\pi_n(X, x_0) \times \pi_n(Y, y_0)$ and $\pi_n(X \times Y, (x_0, y_0))$ are isomorphic for all $n \geq 1$.*

Proof Define a function $\psi : \pi_n(X, x_0) \times \pi_n(Y, y_0) \to \pi_n(X \times Y, (x_0, y_0))$, $([f], [g]) \mapsto [(f, g)]$. It can be shown that ψ is a homomorphism and is also a bijection. ❑

7.3 Functorial Properties of Absolute Homotopy Groups

This section presents functorial properties of higher homotopy groups by defining homomorphisms induced by a base point preserving continuous map between two pointed topological spaces to the corresponding higher homotopy groups.

Theorem 7.3.1 *If $f : (X, x_0) \to (Y, y_0)$ is continuous, then f induces a homomorphism $f_* : \pi_n(X, x_0) \to \pi_n(Y, y_0)$ for $n \geq 1$.*

Proof For any $\alpha \in F_n(X, x_0)$, the composition $f \circ \alpha \in F_n(Y, y_0)$ and the assignment $\alpha \mapsto f \circ \alpha$ defines a map $f_\square : F_n(X, x_0) \to F_n(Y, y_0)$. The continuity of $f_\square$ shows that $f_\square$ carries the path- components of $F_n(X, x_0)$ into the path components of $F_n(Y, y_0)$. Hence $f_\square$ determines an induced transformation

$$f_* : \pi_n(X, x_0) \to \pi_n(Y, y_0), [\alpha] \mapsto [f \circ \alpha], \forall [\alpha] \in \pi_n(X, x_0).$$

Clearly, f_* sends the zero element of $\pi_n(X, x_0)$ into the zero the element of $\pi_n(Y, y_0)$. Moreover, for any two maps $\alpha, \beta \in F_n(X, x_0)$, it can be shown that $f(\alpha * \beta) = (f \circ \alpha) * (f \circ \beta$ in $F_n(Y, y_0)$. Hence $f_* : \pi_n(X, x_0) \to \pi_n(Y, y_0)$ defined by $f_*([\alpha \circ \beta]) = (f * [\alpha]) \circ [f_*[\beta]]$ is a group homomorphism. ❑

Definition 7.3.2 The homomorphism $f_* : \pi_n(X, x_0) \to \pi_n(Y, y_0)$ defined in Theorem 7.3.1 is called the homomorphism induced by $f : (X, x_0) \to (Y, y_0)$.

Corollary 7.3.3 *Let $f, g : (X, x_0) \to (Y, y_0)$ be two homotopic maps, then their induced transformations $f_*, g_* : \pi_n(X, x_0) \to \pi_n(Y, y_0)$ are the same for every n.*

Corollary 7.3.4 *If $f : (X, x_0) \to (X, x_0)$ is the identity map, then $f_* : \pi_n(X, x_0) \to \pi_n(X, x_0)$ is the identity transformation on $\pi_n(X, x_0)$ for every n.*

Corollary 7.3.5 *Let $f : (X, x_0) \to (Y, y_0)$ and $g : (Y, y_0) \to (Z, z_0)$ be two base point preserving maps, then $(g \circ f)_* = g_* \circ f_* : \pi_n(X, x_0) \to \pi_n(Z, z_0)$ for every n.*

Corollary 7.3.6 *If $f : (X, x_0) \to (Y, y_0)$ is a homotopy equivalence, then f_* induces an isomorphism $f_* : \pi_n(X, x_0) \to \pi_n(Y, y_0)$ of groups for every $n \geq 1$.*

Proof It follows from Corollaries 7.3.3–7.3.5. ❑

Corollary 7.3.7 (Homotopy Invariance) *If two pointed spaces (X, x_0) and (Y, y_0) have the same homotopy type, then for each $n \geq 1$, there is a group isomorphism $\psi : \pi_n(X, x_0) \to \pi_n(Y, y_0)$.*

Proof It follows from Corollaries 7.3.3–7.3.6. ❑

Proposition 7.3.8 *If X is a path-connected space, then for any two points x_0 and x_1, there is an isomorphism $\psi : \pi_n(X, x_0) \to \pi_n(X, x_1)$ for every $n \geq 1$.*

Proof Consider a continuous curve x_t in X $(t \in I)$, which connects x_0 and x_1. Define a family of maps from I^n to X, which takes the boundary of I^n along the curve. This induces a family of homomorphisms

$$h_t : \pi_n(X, x_0) \to \pi_n(X, x_t)$$

for every $n \geq 1$ and $t \in I$.

Similarly, we have another family of homomorphisms

$$g_t : \pi_n(X, x_t) \to \pi_n(X, x_0).$$

Hence $h_1 \circ g_1 = 1_d$ and $g_1 \circ h_1 = 1_d$. Consequently, there is an isomorphism

$$\psi = h_1 : \pi_n(X, x_0) \to \pi_n(X, x_1)$$

for every $n \geq 1$. ❑

Corollary 7.3.9 *If the topological spaces X and Y are path- connected and homotopy equivalent, then groups $\pi_n(X)$ and $\pi_n(Y)$ are isomorphic for every $n \geq 1$.*

Proof Let $f : X \to Y$ be a homotopy equivalence and $x_0 \in X, y_0 \in Y$. Then f induces an isomorphism $f_* : \pi_n(X, x_0) \to \pi_n(Y, y_0)$ for every $n \geq 1$. As X is path-connected, the groups $\pi_n(X) \cong \pi_n(Y)$ for every $n \geq 1$. ❑

We summarize the above discussion in the basic and important functorial property of π_n like π_1.

Theorem 7.3.10 *Let $\mathcal{H}tp_*$ be the homotopy category of pointed topological spaces and $\mathcal{A}b$ be the category of abelian groups and homomorphisms, then $\pi_n : \mathcal{H}tp_* \to \mathcal{A}b$, is a covariant functor for each $n > 2$. For $n = 1$, π_1 is a covariant functor from $\mathcal{H}tp_*$ to the category $\mathcal{G}rp$ of groups and homomorphisms.*

Proof It follows from Corollaries 7.3.3–7.3.5. ❑

Like fundamental groups, higher homotopy groups of a pointed topological space and higher homotopy groups of its covering spaces have a close relation.

Theorem 7.3.11 *Let $(\tilde{X}, p)$ be a covering space of X and $x_0 \in X$. If $\tilde{x}_0 \in p^{-1}(x_0)$, then p induces an isomorphism $p_* : \pi_n(\tilde{X}, \tilde{x}_0) \to \pi_n(X, x_0)$ for every $n \geq 2$.*

Proof Let $p : \tilde{X} \to X$ be a covering map such that $p(\tilde{x}_0) = x_0$ and $[\tilde{f}] \in \pi_n(\tilde{X}, \tilde{x}_0) = [S^n, \tilde{X}]$. Then $p_* : \pi_n(\tilde{X}, \tilde{x}_0) \to \pi_n(X, x_0)$ given by $p_*[\tilde{f}] = [p \circ \tilde{f}] \in \pi_n(X, x_0)$ is a homomorphism. We claim that p_* is an isomorphism.

Monomorphism: Let $[p \circ \tilde{f}] = [p \circ \tilde{g}]$, where $\tilde{f}, \tilde{g} : S^n \to \tilde{X}$ are pointed maps. Then $p \circ \tilde{f} \simeq p \circ \tilde{g}$. By using covering homotopy theorem, $\tilde{f} \simeq \tilde{g}$ and hence $[\tilde{f}] = [\tilde{g}]$.

Epimorphism: Let $[f] \in \pi_n(X, x_0)$. We now consider the diagram in Fig. 7.1

As S^n is simply connected for $n \geq 2$, there exists a unique lifting $\tilde{f} : S^n \to \tilde{X}$ of $f : S^n \to X$ such that $p \circ \tilde{f} = f$. Hence $p_*[\tilde{f}] = [f]$. ❑

Theorem 7.3.12 $\pi_n(S^1) = 0$ *for all $n \geq 2$.*

Proof Recall that $p : \mathbf{R} \to S^1$, where $p(t) = e^{2\pi i t}$ is covering map. Thus by using Theorem 7.3.11, p induces an isomorphism $p_* : \pi_n(\mathbf{R}) \to \pi_n(S^1)$ for every $n \geq 2$. Since the space $\mathbf{R}$ is contractible, $\pi_n(\mathbf{R}) = 0$ for every $n \geq 2$. This shows that $\pi_n(S^1) = 0$ for every $n \geq 2$. ❑

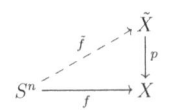

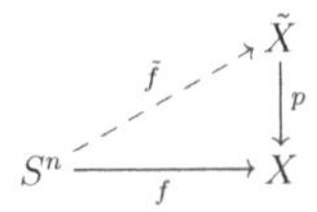

Fig. 7.1 Lifting of f

7.4 The Relative Homotopy Groups: Introductory Concepts

This section generalizes the concept of absolute homotopy groups $\pi_n(X, x_0)$ introduced in Sect. 7.1 by defining the relative homotopy groups $\pi_n(X, A, x_0)$ (with $n \geq 2$) for any triplet (X, A, x_0) which is an important concept in homotopy theory. This generalization was given by Hurewicz (1904–1956) and studied by him during 1935–1936.

Definition 7.4.1 A triplet (X, A, x_0) consists of a topological space X, a nonempty subspace A of X and a point $x_0 \in A$. If x_0 is the only point of A, then the triplet is simply (X, x_0).

Construction of $\pi_n(X, A, x_0)$

Let I^n be the cartesian product of the unit interval $I = [0, 1]$ for $n > 0$, called the n-cube. Thus $I^n = \{(t_1, t_2, \ldots, t_n) : t_i \in I, i = 1, \ldots, n\}$. The initial $(n-1)$-face of I^n defined by $t_n = 0$, is identified with I^{n-1} hereafter. The union of all remaining $(n-1)$-faces of I^n is denoted by J^{n-1}. Hence $\partial I^n = I^{n-1} \cup J^{n-1}$ and $\partial I^{n-1} = I^{n-1} \cap J^{n-1}$.

A map $f : (I^n, I^{n-1}, J^{n-1}) \to (X, A, x_0)$ is a continuous function $f : I^n \to X$ such that $f(I^{n-1}) \subset A$ and $f(J^{n-1}) = x_0$. In particular, $f : (\partial I^n, \partial I^{n-1}) \mapsto (A, x_0)$ is a continuous function. Let $F_n = F_n(X, A, x_0)$ be the set of all such maps. It is topologized by the compact open topology. Two maps $f, g \in F_n(X, A, x_0)$ are said to be homotopic relative to the system $\{I^{n-1}, A; J^{n-1}, x_0\}$ if there exists a homotopy $H_t : I^n \to X, t \in I$ such that $H_0 = f$, $H_1 = g$ and $H_t \in F_n(X, A, x_0)$. In notation, $f \simeq g$ rel $\{I^{n-1}, A; J^{n-1}, x_1\}$. Let $\pi_n(X, A, x_0)$ be the set of homotopy classes of all these maps relative to the system $\{I^{n-1}, A; J^{n-1}, x_0\}$). Let $[f]$ denote the homotopy class of $f \in F_n$ and 0 the homotopy class of the constant map $c(I^n) = x_0$. Then the set $\pi_n(X, A, x_0)$ is the set of all path components of the space F_n. For $n \geq 1$, define a composition '$*$' in $F_n(X, A, x_0)$ as follows: For $f, g \in F_n(X, A, x_0)$, $f * g$ is defined by

$$(f * g)(t) = \begin{cases} f(2t_1, t_2, \ldots, t_i, \ldots, t_n), & \text{if } 0 \leq t_1 \leq 1/2 \\ g(2t_1 - 1, t_2, \ldots, t_i, \ldots, t_n), & \text{if } 1/2 \leq t_1 \leq 1. \end{cases}$$

for every $t = (t_1, t_2, \ldots, t_n) \in I^n$.

Then $f * g$ is continuous and $f * g \in F_n(X, A, x_0)$.

Define a composition in $\pi_n(X, A, x_0)$ by the rule $[f] \circ [g] = [f * g]$ for all $[f], [g] \in \pi_n(X, A, x_0)$. This composition is independent of the choice of the representatives of the classes and hence it is well defined.

Remark 7.4.2 For $f, g \in F_n(X, A, x_0)$, $f * g$ may be equally well-defined by

$$(f * g)(t) = \begin{cases} f(t_1, t_2, \ldots, 2t_i, \ldots, t_n), & \text{if } 0 \leq t_i \leq 1/2 \\ g(t_1, t_2, \ldots, 2t_i - 1, \ldots, t_n), & \text{if } 1/2 \leq t_i \leq 1. \end{cases}$$

So, it is immaterial which coordinate we use to define $f * g$.

Theorem 7.4.3 *$\pi_n(X, A, x_0)$ is a group under the composition '$\circ$' for $n \geq 1$.*

Proof Proceed as in the case of $\pi_n(X, x_0)$ (see Theorem 7.1.5). ❑

Remark 7.4.4 If J^{n-1} is identified to a point s_0, then the triplet (I^n, I^{n-1}, J^{n-1}) admits a configuration equivalent to the triplet (D^n, S^{n-1}, s_0) consisting of the unit n-cell D^n, its boundary S^{n-1}, and a base point $s_0 \in S^{n-1}$. This shows that an element $\pi_n(X, A, x_0)$ can be equally well defined as a homotopy class (relative to the system $\{S^{n-1}, A; s_0, x_0\}$) of the maps $f : (D^n, S^{n-1}, s_0) \to (X, A, x_0)$.

Proposition 7.4.5 *For $n > 2$, the group $\pi_n(X, A, x_0)$ is abelian.*

Proof An element of $\pi_n(X, A, x_0)$ is a homotopy class of a map $f : (D^n, S^{n-1}, s_0) \to (X, A, x_0)$. If $n > 2$, there exists a rotation of D^n which leaves the point s_0 fixed and interchanges two halves of D^n. Hence the group $\pi_n(X, A, x_0)$ is abelian for $n > 2$. ❑

We now search the maps $f : (D^n, S^{n-1}, s_0) \to (X, A, x_0)$ which define the zero element of $\pi_n(X, A, x_0)$.

Proposition 7.4.6 *If an element $[f] \in \pi_n(X, A, x_0)$ is represented by a map $f \in F_n(X, A, x_0)$ such that $f(I^n) \subset A$, then $[f] = 0$.*

Proof Since $f \in F_n(X, A, x_0)$ and $f(I^n) \subset A$, there exists a homotopy $H_t \in F_n(X, A, x_0)$, $t \in I$ defined by

$$H_t(t_1, t_2, \ldots, t_{n-1}, t_n) = f(t_1, \ldots, t_{n-1}, t + t_n - tt_n).$$

Then $H_0 = f$, and $H_1(I^n) = x_0$. Hence $[f] = 0$. ❑

7.5 The Boundary Operator and Induced Transformation

This section conveys the algebraic properties of boundary operator and induced transformation by defining boundary operator $\partial : \pi_n(X, A, x_0) \to \pi_{n-1}(A, x_0)$ for $n > 0$ and the transformation $f_* : \pi_n(X, A, x_0) \to \pi_n(Y, B, y_0)$ induced by a continuous map $f : (X, A, x_0) \to (Y, B, x_0)$. They play a central role in the study of homotopy sequence.

7.5.1 Boundary Operator

This subsection studies the boundary operator ∂ which is an important concept in higher homotopy groups.

Definition 7.5.1 Let (X, A, x_0) be a triplet and $n > 0$ be an integer. If $[f] \in \pi_n(X, A, x_0)$, then it is represented by a continuous map $f : (I^n, I^{n-1}, J^{n-1}) \to (X, A, x_0)$. If $n = 1$, $f(I^{n-1})$ is a point of A which gives a path component $[g] \in \pi_{n-1}(A, x_0)$ of A. If $n > 1$, then the map $f|_{I^{n-1}} : (I^{n-1}, \partial I^{n-1}) \to (A, x_0)$ represents an element $[g] \in \pi_{n-1}(A, x_0)$, independent of the choice of the representative map f. The boundary operator $\partial : \pi_n(X, A, x_0) \to \pi_{n-1}(A, x_0)$ for $n > 0$ is defined by setting $\partial([f]) = [g]$.

We now give some properties of ∂.

Proposition 7.5.2 *The boundary operator ∂ sends the zero element of $\pi_n(X, A, x_0)$ into the zero element of $\pi_{n-1}(A, x_0)$.*

Proof It follows from definition of $\partial : \pi_n(X, A, x_0) \to \pi_{n-1}(A, x_0)$. ❑

Proposition 7.5.3 *For $n > 1$, the boundary operator $\partial : \pi_n(X, A, x_0) \to \pi_{n-1}(A, x_0)$ is a homomorphism.*

Proof It follows from the definition of ∂. ❑

7.5.2 *Induced Transformations*

This subsection generalizes the concept of induced transformations given in Theorem 7.3.1 and relates it to the boundary operator ∂ for triplets. Let $f : (X, A, x_0) \to (Y, B, y_0)$ be a continuous map which means that f is a continuous map from X to Y such that $f(A) \subset B$ and $f(x_0) = y_0$. Then f sends the path components of X into the path components of Y. Hence f induces a transformation $f_* : \pi_0(X, x_0) \to \pi_0(Y, y_0)$, which sends the zero element of $\pi_0(X, x_0)$ into the zero element of $\pi_0(Y, y_0)$.

Theorem 7.5.4 *If $f : (X, A, x_0) \to (Y, B, y_0)$ is continuous, then f induces a homomorphism $f_* : \pi_n(X, A, x_0) \to \pi_n(Y, B, y_0)$ for $n \geq 1$.*

Proof If $n \geq 1$, then for any map $g \in F_n(X, A, x_0)$, the composite map $f \circ g \in F_n(Y, B, y_0)$ and hence the corresponding $g \mapsto f \circ g$ defines a continuous function $f_\square : F_n(X, A, x_0) \to F_n(Y, B.y_0)$ such that $f_\square$ sends the path components of $F_n(X, A, x_0)$ to the path components of $F_n(Y, B, y_0)$. Hence $f_\square$ induces a transformation $f_* : \pi_n(X, A, x_0) \to \pi_n(Y, B, y_0)$, which carries the zero element of $\pi_n(X, A, x_0)$ to the zero element of $\pi_n(Y, B, y_0)$. For $n = 1$, $A = \{x_0\}$, $B = \{y_0\}$ or for $n > 1$, $\pi_n(X, A, x_0)$ and $\pi_n(Y, B, y_0)$ are groups and f_* is a homomorphism, called the homomorphism induced by f. ❑

Corollary 7.5.5 *The identity map $1_d : (X, A, x_0) \to (X, A, x_0)$ induces the identity transformation $1_{d*} : \pi_n(X, A, x_0) \to \pi_n(X, A, x_0)$ for all $n \geq 0$.*

Proof It follows from the induced transformation 1_{d*}. ❑

Fig. 7.2 Commutativity of the rectangle for ∂ and induced transformations

$$\begin{array}{ccc} \pi_n(X, A, x_0) & \xrightarrow{f_*} & \pi_n(Y, B, y_0) \\ \downarrow \partial & & \downarrow \partial \\ \pi_{n-1}(A, x_0) & \xrightarrow{g_*} & \pi_{n-1}(B, y_0) \end{array}$$

Corollary 7.5.6 *If $f : (A, X, x_0) \to (Y, B, y_0)$ and $g : (Y, B, y_0) \to (Z, C, z_0)$ are maps of triples, then for every $n \geq 0$, $(g \circ f)_* = g_* \circ f_* : \pi_n(X, A, x_0) \to \pi_n(Z, C, z_0)$.*

Proof It follows from the definition of induced transformations f_* and g_*. ❑

We now give a relation between the boundary operator and the induced transformation.

Proposition 7.5.7 *If $f : (X, A, x_0) \to (Y, B, y_0)$ is a continuous function and if $f|_A = g : (A, x_0) \to (B, y_0)$ is the restriction of f, then the diagram in Fig. 7.2 is commutative for every $n \geq 1$, i.e., $\partial \circ f_* = g_* \circ \partial$ holds.*

Proof It follows from definitions of ∂, induced transformations f_*, and g_* that the diagram in Fig. 7.2 is commutative. ❑

7.6 Functorial Property of the Relative Homotopy Groups

This section continues to study homotopy theory by displaying the functorial properties of the relative homotopy groups $\pi_n(X, A, x_0)$ $(n \geq 2)$ and homotopy properties of maps $f \in F_n(X, A, x_0)$ and considers the homotopy equivalence of a map $f \in F_n(X, A, x_0)$.

Definition 7.6.1 Let $f, g : (X, A, x_0) \to (Y, B, b_0)$ be two continuous maps. They are said to be homotopic relative to the system $\{A, B, x_0, y_0\}$ (or simply homotopic) if there exists a map $H_t : (X, A, x_0) \to (Y, B, b_0), t \in I$ such that $H_0 = f$ and $H_1 = g$.

Proposition 7.6.2 *Let $f, g : (X, A, x_0) \to (Y, B, b_0)$ be two homotopic maps. Then their induced transformation are equal, i.e., $f_* = g_* : \pi_n(X, A, x_0) \to (Y, B, b_0)$ for every n.*

Proof To prove it, we have to show that $f_*(\alpha) = g_*(\alpha)$, $\forall\, \alpha \in \pi_n(X, A, x_0)$.
Case 1: If $n = 0$, $A = x_0$ and $B = y_0$, then α is a path component of X. If $x \in \alpha$, then $f_*(\alpha)$ and $g_*(\alpha)$ are path components of Y containing the points $f(x)$ and $g(x)$, respectively. Let $H_t : f \simeq g$. Define a path $\beta : I \to Y, t \mapsto H_t(x)$. Then the path β joins $f(x)$ to $g(x)$ and hence $f_*(\alpha) = g_*(\alpha)$, $\forall\, \alpha \in \pi_n(X, A, x_0)$. This shows that $f_* = g_*$.

Case 2: If $n > 0$, choose a map $m \in F_n(X, A, x_0)$ such that $[m] = \alpha$. Then $f_*(\alpha)$ and $g_*(\alpha)$ are represented by the maps $f \circ m$ and $g \circ m$, respectively. Hence $H_t \circ m : f \circ m \simeq g \circ m$ shows that $f_*([m]) = g_*([m])$. Consequently, $f_*(\alpha) = g_*(\alpha)$, $\forall \alpha \in \pi_n(X, A, x_0)$ implies $f_* = g_*$. ❑

Definition 7.6.3 A map $f : (X, A, x_0) \to (Y, B, b_0)$ is said to be a homotopy equivalence if there exists a map $g : (Y, B, b_0) \to (X, A, x_0)$ such that $g \circ f$ is homotopic to the identity map on (X, A, x_0) and $f \circ g$ is homotopic to the identity map on (Y, B, b_0).

Proposition 7.6.4 *If $f : (X, A, x_0) \to (Y, B, b_0)$ is a homotopy equivalence, then the induced transformation $f_* : \pi_n(X, A, x_0) \to \pi_n(Y, B, b_0)$ is an isomorphism for every $n > 1$.*

Proof It follows from Definition 7.6.3 and Proposition 7.6.2. ❑

Corollary 7.6.5 *If the topological spaces X and Y are path - connected and homotopy equivalent, then the groups $\pi_n(X, A, x_0)$ and $\pi_n(Y, B, y_0)$ are isomorphic for every $n > 1$.*

Proof Proceed as in Corollary 7.3.9. ❑

Corollary 7.6.6 *$\pi_n(X, A, x_0)$ depends on the homotopy type of (X, A, x_0).*

Corollary 7.6.7 *If $n \geq 1$ and A is a strong deformation retract of X, then $i_* : \pi_n(A, x_0) \to \pi_n(X, x_0)$ is an isomorphism.*

Proof The corollary follows from Corollary 7.6.6. ❑

Theorem 7.6.8 *Let $\mathcal{H}tp^2$ be the homotopy category of triplets and their cotinuous maps, and $\mathcal{A}b$ be the category of abelian groups and homomorphisms. Then $\pi_n : \mathcal{H}tp^2 \to \mathcal{A}b$ is a covariant functor for each $n > 2$.*

Proof It follows Theorem 7.5.4, Proposition 7.6.2 and Corollaries 7.5.5–7.5.6. ❑

7.7 Homotopy Sequence and Its Exactness

This section defines homotopy sequence and proves its exactness, which provides powerful tools for the study of homotopy theory, specially for computing homotopy groups of certain spaces and proves also some immediate consequences of exactness of homotopy sequences.

7.7.1 Homotopy Sequence and Its Exactness

Given a triplet (X, A, x_0), the inclusion maps $i : (A, x_0) \hookrightarrow (X, x_0)$, and $j : (X, x_0) \hookrightarrow (X, A, x_0)$ induce transformations $i_* : \pi_n(A, x_0) \to \pi_n(X, x_0)$, $j_* : \pi_n(X, x_0) \to \pi_n(X, A, x_0)$. The transformations

i_*, j_* and ∂ give rise to a beginningless sequence:

$$\cdots \to \pi_{n+1}(X, x_0) \xrightarrow{j_*} \pi_{n+1}(X, A, x_0) \xrightarrow{\partial} \pi_n(A, x_0) \xrightarrow{i_*} \pi_n(X, x_0) \xrightarrow{j_*} \pi_n(X, A, x_0)$$
$$\xrightarrow{\partial} \cdots \pi_1(X, A, x_0) \xrightarrow{\partial} \pi_0(A, x_0) \xrightarrow{i_*} \pi_0(X, x_0) \quad (7.2)$$

called the homotopy sequence of the triplet (X, A, x_0), denoted by $\pi(X, A, x_0)$. Every set in (7.2) has its zero element and transformation in (7.2) sends the zero element into the zero element.

Definition 7.7.1 The sequence (7.2) of any triplet (X, A, x_0) is said to be exact if the kernel of each transformation is the same as the image of the preceding transformation

Theorem 7.7.2 (Exactness of homotopy sequence) *The homotopy sequence (7.2) of any triplet (X, A, x_0) is exact.*

Proof To prove the exactness, we show that

(i) Im $j_* = \ker \partial$, i.e., Im $j_* \subseteq \ker \partial \subseteq$ Im j_*;
(ii) Im $\partial = \ker i_*$, i.e., Im $\partial \subseteq \ker i_* \subseteq$ Im ∂;
(iii) Im $i_* = \ker j_*$, i.e., Im $i_* \subseteq \ker j_* \subseteq$ Im i_*.

(i) Im $j_* = \ker \partial$: For each $n > 0$, let $[f] \in \pi_n(X, x_0)$. Then for each representative $f \in F_n(X, x_0)$, $(\partial \circ j_*)([f])$ is determined by the restriction $j \circ f|_{I^{n-1}} = f|_{I^{n-1}}$. Since $f(I^{n-1}) = x_0$, it follows that $\partial \circ j_* = 0$. Hence Im $j_* \subseteq \ker \partial$. For the reverse inclusion, let $n > 1$ and $f \in F_n(X, A, x_0)$ represent $[f] \in \pi_n(X, A, x_0)$. Then $\partial[f] = 0$ shows that there exists a homotopy $H_t : I^{n-1} \to A$ such that $H_0 = f|_{I^{n-1}}$, $H_1(I^{n-1}) = x_0$ and $H_t(\partial I^{n-1}) = x_0$, $\forall t \in I$. Define a map

$$F_t : \partial I^n \to A, s \mapsto \begin{cases} H_t(s), & \text{if } s \in I^{n-1} \\ x_0, & \text{if } s \in J^{n-1} \end{cases}$$

for all $t \in I$. Then $F_0 = f|_{\partial I^n}$. Hence by HEP, the homotopy F_t has a extension $\widetilde{F}_t : I^n \to X$ such that $\widetilde{F}_0 = f$. Again since $\widetilde{F}_1(\partial I^n) = F_1(\partial I^n) = x_0$, $\widetilde{F}_1$ represents an element $[g] \in \pi_n(X, x_0)$. Since $\widetilde{F}_t \in F_n(X, A, x_0)$, it follows that $j_*([g]) = [f]$. If $n = 1$, $[f]$ is represented by a path $f : I \to X$ such that $f(0) \in A$ and $f(1) = x_0$. The given condition $\partial([f]) = 0$ shows that $f(0)$ is contained in the same path-component of A as x_0. Hence there exists a homotopy $F_t : I \to X$ such that $F_0 = f$, $F_t(0) \in A$, $F_t(1) = x_0$ and $F_1(0) = x_0$. Consequently, F_1 represents an element $[g] \in \pi_1(X, x_0)$ and hence the homotopy F_t shows that $j_*([g]) = [f]$.Thus $\ker \partial \subseteq$ Im j_*. Consequently, Im $j_* = \ker \partial$.

(ii) Im $\partial = \ker i_*$: For each $n > 0$, let $[f] \in \pi_n(X, A, x_0)$ and $f \in F_n(X, A, x_0)$ represent $[f]$. Then the element $(i_* \circ \partial)([f])$ is given by $g = f|_{I^{n-1}}$. Define a homotopy

$$G_t : I^{n-1} \to X, (t_1, \ldots, t_{n-1}) \mapsto f(t_1, t_2, \ldots, t_{n-1}, t).$$

Then $G_0 = g, G_1(I^{n-1}) = x_0$ and $G_t \in F_{n-1}(X, x_0)$ for $n > 1$. Hence $(i_* \circ \partial)([f]) = 0$ implies $i_* \circ \partial = 0$. This shows that $\text{Im}\,\partial \subseteq \ker i_*$. For the converse inclusion, first let $n > 1$ and $[f] \in \pi_n(A, x)$ be represented by $f \in F_{n-1}(A, x_0)$ such that $i_*[f] = 0$. Then there exists a homotopy $F_t : I^{n-1} \to X$ such that $F_0 = f, F_1(I^{n-1}) = x_0$ and $F_t(\partial I^{n-1}) = x_0$. Define a map

$$g : I^n \to X, (t_1, t_2, \ldots, t_{n-1}, t_n) \mapsto F_{t_n}(t_1, t_2, \ldots, t_{n-1}).$$

Then $g \in F_n(X, A, x_0)$ represents an element $[g] \in \pi_n(X, A, x_0)$. Since $g|_{I^{n_1}} = f$, it follows that $\partial([g]) = [f]$. The remaining part for $n = 1$ is left as an exercise. Hence $\ker i_* \subseteq \text{Im}\,\partial$. Consequently, $\text{Im}\,\partial = \ker i_*$.

(iii) $\text{Im}\, i_* = \ker j_*$: For $n \geq 1$, we claim that $j_* \circ i_* = 0$. Let $[f] \in \pi_n(A, x_0)$ and $f \in F_n(A, x_0)$ represent $[f]$. Then the element $(j_* \circ i_*)([f]) \in \pi_n(X, A, x_0)$ is represented by $j \circ i \circ f \in F_n(X, A, x_0)$. Since $(j \circ i \circ f)(I^n) \subset A$, it follows by Proposition 7.4.6 that $(j_* \circ i_*)([f]) = 0, \forall [f] \in \pi_n(A, x_0)$. This shows that $j_* \circ i_* = 0$. Hence $\text{Im}\, i_* \subseteq \ker j_*$. To show the reverse inclusion, let $[f] \in \pi_n(X, x_0)$ be such that $j_*([f]) = 0$. Choose $f \in F_n(X, x_0)$ a representative of $[f]$. Then $j_*([f]) = 0$ shows that there is a homotopy $F_t : I^n \to X$ such that $F_0 = f, F_1(I^n) = x_0$, and $F_t \in F_n(X, A, x_0)$. Define a homotopy

$$G_t : I^n \to X, (t_1, t_2, \ldots, t_{n-1}, t_n) \mapsto \begin{cases} F_{2t_n}(t_1, t_2, \ldots, t_{n-1}, 0), & \text{if } 0 \leq 2t_n \leq t \\ F_t(t_1, t_2, \ldots, t_{n-1}, \frac{2t_n - t}{2-t}), & \text{if } t \leq 2t_n \leq 2. \end{cases}$$

Then $G_0 = f, G_1(I^n) \subset A$ and $G_t(\partial I^n) = x_0$ for all $t \in I$. Hence G_1 represents an element $[g] \in \pi_n(A, x_0)$ and the homotopy G_t shows that $i_*([g]) = [f]$. Hence $\ker j_* \subseteq \text{Im}\, i_*$. Consequently, $\text{Im}\, i_* = \ker j_*$. ❑

7.7.2 *Some Consequences of the Exactness of the Homotopy Sequence*

This subsection presents some immediate consequences of the exactness of the homotopy sequence (7.2).

Proposition 7.7.3 *Let (X, A, x_0) be a triplet, A be a retract of X and $x_0 \in A$. Then $\pi_n(X, x_0) \cong \pi_n(A, x_0) \oplus \pi_n(X, A, x_0)$ for any $n \geq 2$ and the inclusion map $i : A \hookrightarrow X$ induces a monomorphism $i_* : \pi_n(A, x_0) \to \pi_n(X, x_0)$ for any $n \geq 1$.*

Proof Let $r : X \to A$ be a retraction. Then $r \circ i = 1_A$ shows that $r_* \circ i_*$ is the identity automorphism on $\pi_n(A, x_0)$ for every $n \geq 1$. Consequently, i_* is a monomorphism and r_* is an epimorphism for every $n \geq 1$. Again for $n \geq 2$, since the group $\pi_n(X, x_0)$ is abelian, it follows from $r_* \circ i_* = 1_d$ that the group $\pi_n(X, x_0)$ decomposes into the direct sum $\pi_n(X, x_0) = B \oplus C$, where $B = \operatorname{Im} i_*$ and $C = \ker r_*$. Since i_* is a monomorphism, $B \cong \pi_n(A, x_0)$. Again it follows from the exactness of the homotopy sequence of (X, A, x_0) that $j_* : \pi_n(X, x_0) \to \pi_n(X, A, x_0)$ is an epimorphism for any $n \geq 2$, and $\ker j_* = \operatorname{Im} i_* = B$ and j_* maps C isomorphically onto $\pi_n(X, A, x_0)$. Consequently, $C \cong \pi_n(X, A, x_0)$. ❑

Remark 7.7.4 For a given triplet (X, A, x_0), if A is a retract of X, then the group $\pi_2(X, A, x_0)$ is abelian.

Definition 7.7.5 Let (X, A, x_0) be a triplet. Then X is said to be deformable into A relative to a point $x_0 \in A$, if there exists a homotopy $H_t : X \to X$ such that $H_0(x) = x$, $H_1(x) \in A$ and $H_t(x_0) = x_0$, $\forall x \in X, t \in I$.

Proposition 7.7.6 *Let (X, A, x_0) be a triplet and X be deformable into A relative to a point $x_0 \in A$. If $i : A \hookrightarrow X$ is the inclusion map, then $\pi_n(A, x_0) \cong \pi_n(X, x_0) \oplus \pi_{n+1}(X, A, x_0)$ for every $n \geq 2$ and $i_* : \pi_n(A, x_0) \to \pi_n(X, x_0)$ is an epimorphism for every $n \geq 1$.*

Proof Since X is deformable into A relative to the point $x_0 \in A$, there exists a homotopy $H_t : X \to X$ such that $H_0(x) = x$, $H_1(x) \in A$ and $H_t(x_0) = x_0$, $\forall x \in X$ and $t \in I$. Define a map

$$f : (X, x_0) \to (A, x_0), x \mapsto H_1(x).$$

Clearly, $i \circ f = H_1 \simeq H_0 \operatorname{rel} x_0$. Hence $i_* \circ f_*$ is the identity automorphism on $\pi_n(X, x_0)$. Hence f_* is a monomorphism and i_* is an epimorphism for every $n \geq 1$. If $n \geq 2$, then the group $\pi_n(A, x_0)$ is abelian. Hence $i_* \circ f_* = 1_d$ implies that $\pi_n(A, x_0) = \operatorname{Im} f_* \oplus \ker i_*$, where i_* is an epimorphism and f_* is a monomorphism for any $n \geq 1$. This shows that $\operatorname{Im} f_* \cong \pi_n(X, x_0)$. Again from exactness of the homotopy sequence of (X, A, x_0) it follows that $\partial : \pi_{n+1}(X, A, x_0) \to \pi_n(A, x_0)$ is a monomorphism. Consequently, $\ker i_* = \operatorname{Im} \partial \cong \pi_{n+1}(X, A, x_0)$. ❑

Definition 7.7.7 Let (X, A) be a pair of topological spaces. It is called 0-connected if every path component of X intersects A. The pair (X, A) is called n-connected if (X, A) is 0-connected and $\pi_m(X, A, a) = 0$ for $1 \leq m \leq n$ and for all $a \in A$. A topological space X is called n-connected if $\pi_m(X, x) = 0$ for $0 \leq m \leq n$ and for all $x \in X$.

We now characterize n-connected spaces with the help of inclusion maps.

Proposition 7.7.8 *Let (X, A) be a pair of topological spaces. Then (X, A) is n-connected ($n \geq 0$) iff for the inclusion map $i : (A, x_0) \to (X, x_0)$, the induced map $i_* : \pi_m(A, x_0) \to \pi_m(X, x_0)$ is a bijection for $m < n$ and a surjection for $m = n$, and for all $x_0 \in A$.*

Proof It follows from the homotopy sequence (7.2) of the triplet (X, A, x_0). ❑

7.8 Homotopy Sequence of Fibering and Hopf Fibering

This section studies the homotopy sequence of fibering which exists corresponding to a fiber space. and describes Hopf fibering: $p : S^{2n-1} \to S^n$, for $n = 2, 4, 8$. They provide tools in computing higher homotopy groups of certain topological spaces. H. Hopf (1894–1975) described various fiberings of spheres by spheres in his paper (Hopf 1935).

7.8.1 *Homotopy Sequence of Fibering*

This subsection discusses the homotopy sequence of a fibering. Let $p : X \to B$ be a projection, $b_0 \in B$ and $F = p^{-1}(b_0) \neq \emptyset$ be the fiber space of p. If $x_0 \in F$, then (X, F, x_0) forms a triplet. Since $p(F) = b_0$, the map $p : X \to B$ defines a map $q : (X, F, x_0) \to (B, b_0)$ such that $p = q \circ j$, where $j : (X, x_0) \hookrightarrow (X, F, x_0)$ is the inclusion map.Then by fibering property of p (see Ex.9 of Sect. 7.14), $q_* : \pi_n(X, F, x_0) \to \pi_n(B, b_0)$ is a bijection for $n \geq 1$. Then q_*^{-1} exists. Define $d_* = \partial \circ q_*^{-1} : \pi_n(B, b_0) \to \pi_{n-1}(F, x_0)$ for $n \geq 1$. Hence the homotopy exact sequence (7.2) produces the following exact sequence

$$\begin{aligned} \cdots \xrightarrow{p_*} \pi_{n+1}(B, b_0) \xrightarrow{d_*} \pi_n(F, x_0) \xrightarrow{i_*} \pi_n(X, x_0) \xrightarrow{p_*} \pi_n(B, x_0) \\ \xrightarrow{d_*} \cdots \xrightarrow{p_*} \pi_1(B, b_0) \xrightarrow{d_*} \pi_0(F, x_0) \xrightarrow{i_*} \pi_0(X, x_0) \end{aligned} \tag{7.3}$$

called the homotopy sequence of the fibering $p : X \to B$ based at x_0.

Proposition 7.8.1 *If the fiber F in sequence (7.3) is totally disconnected, then $p_* : \pi_n(X, x_0) \to \pi_n(B, b_0)$ is an isomorphism for $n \geq 2$ and p_* is a monomorphism for $n = 1$.*

Proof By hypothesis, $\pi_n(F, x_0) = 0$ for $n \geq 1$. Hence the proposition follows from the exactness of the homotopy sequence (7.3). □

Proposition 7.8.2 *Let $p : X \to B$ be a fibering such that it admits a cross section $s : B \to X$. Then for every $b_0 \in B$ and $x_0 = s(b_0) \in F = p^{-1}(b_0)$, $\pi_n(X, x_0) \cong \pi_n(B, b_0) \oplus \pi_n(F, x_0)$ for $n \geq 2$ and $p_* : \pi_n(X, x_0) \to \pi_n(B, b_0)$ is an epimorphism for $n \geq 1$.*

Proof $p \circ s = 1_B$ implies that $p_* \circ s_* : \pi_n(B, b_0) \to \pi_n(B, b_0)$ is the identity automorphism. Hence $s_* : \pi_n(B, b_0) \to \pi_n(X, x_0)$ is a monomorphism and $p_* : \pi_n(X, x_0) \to \pi_n(B, b_0)$ is an epimorphism for $n \geq 1$. For $n \geq 2$, $\pi_n(X, x_0)$ is abelian and hence $p_* \circ s_* = 1_d$ shows that $\pi_n(X, x_0)$ is the direct sum

$$\pi_n(X, x_0) = \text{Im}\, s_* \oplus \ker p_*.$$

Since s_* is a monomorphism, $\text{Im}\, s_* \cong \pi_n(B, b_0)$. Again as p_* is an epimorphism for every $n \geq 1$, it follows from (7.3) that i_* is a monomorphism for every $n \geq 1$. Hence $\ker p_* = \text{Im}\, i_* \cong \pi_n(F, x_0)$. □

7.8.2 *Hopf Fiberings of Spheres*

This subsection discusses the problem whether a continuous map $p : S^m \to S^n$ for $m > n > 1$ is necessarily nullhomotopic which was not known until 1935. The problem was solved in 1935 by H. Hopf with the discovery of his famous map $p : S^3 \to S^2$, now called Hopf map. This section studies Hopf fibering: $p : S^{2n-1} \to S^n$, for $n = 2, 4, 8$. Then in each case, the fiber F is an $(n-1)$-sphere S^{n-1} and is contractible in S^{2n-1}. Hence $\pi_m(S^n) \cong \pi_m(S^{2n-1}) \oplus \pi_{m-1}(S^{n-1})$ for every $n = 2, 4, 8$ and every $m \geq 2$.

(i) In particular, if $n = 2$, then $\pi_m(S^2) \cong \pi_m(S^3)$ for $m \geq 3$.
(ii) If $n = 4$ or 8, then

$$\begin{aligned}
&\pi_m(S^4) \cong \pi_{m-1}(S^3), \ \text{for } 2 \leq m \leq 6,\\
&\pi_m(S^8) \cong \pi_{m-1}(S^7), \ \text{for } 2 \leq m \leq 14,\\
&\pi_7(S^4) \cong \mathbf{Z} \oplus \pi_6(S^3),\\
&\pi_{15}(S^8) \cong \mathbf{Z} \oplus \pi_{14}(S^7),
\end{aligned}$$

7.8.3 *Problems of Computing $\pi_m(S^n)$*

This subsection studies the problems of computing the homotopy groups $\pi_m(S^n)$. The spheres S^n are perhaps the simplest noncontractable spaces (see Chap. 10). The homotopy groups $\pi_m(S^n)$ are not completely determined. The homotopy group $\pi_m(S^n)$ for $m \leq n$ are known. Computing the homotopy groups completely is one of the major unsolved problems. It has been shown that

(i) $\pi_m(S^n) = 0$, for $m < n$, by Corollary 7.10.4.
(ii) $\pi_m(S^1) = 0$, for $m > 1$, by Theorem 7.3.12.
(iii) $\pi_1(S^1) \cong \mathbf{Z}$ (see Chap. 3).
(iv) $\pi_n(S^n) \cong \mathbf{Z}$ (see Corollary 7.10.5).

Remark 7.8.3 One may conject that $\pi_m(S^n) = 0$ for $m > n$. But this is not true. H. Hopf first disproves this conjecture in 1931 by showing that $\pi_3(S^2)$ is not trivial

Table 7.1 Table of $\pi_i(S^n)$ for $1 \leq i, n \leq 8$

$i \rightarrow$	1	2	3	4	5	6	7	8
$\downarrow n$								
1	$\mathbf{Z}$	0	0	0	0	0	0	0
2	0	$\mathbf{Z}$	$\mathbf{Z}$	$\mathbf{Z}_2$	$\mathbf{Z}_2$	$\mathbf{Z}_{12}$	$\mathbf{Z}_2$	$\mathbf{Z}_2$
3	0	0	$\mathbf{Z}$	$\mathbf{Z}_2$	$\mathbf{Z}_2$	$\mathbf{Z}_{12}$	$\mathbf{Z}_2$	$\mathbf{Z}_2$
4	0	0	0	$\mathbf{Z}$	$\mathbf{Z}_2$	$\mathbf{Z}_2$	$\mathbf{Z} \times \mathbf{Z}_{12}$	$\mathbf{Z}_2 \times \mathbf{Z}_2$
5	0	0	0	0	$\mathbf{Z}$	$\mathbf{Z}_2$	$\mathbf{Z}_2$	$\mathbf{Z}_{24}$
6	0	0	0	0	0	$\mathbf{Z}$	$\mathbf{Z}_2$	$\mathbf{Z}_2$
7	0	0	0	0	0	0	$\mathbf{Z}$	$\mathbf{Z}_2$
8	0	0	0	0	0	0	0	$\mathbf{Z}$

(see Theorem 7.9.1). It is isomorphic to $\mathbf{Z}$. Many other examples for $\pi_m(S^n)$ (for $m > n$) are known for particular pair of integers m and n (see a sample Table 7.1) but not known in all possible cases.

7.9 More on Hopf Maps

This section studies in general, continuous maps $p : S^{2n-1} \rightarrow S^n$ for $n = 2, 4, 8$ introduced by H. Hopf in 1935 (Hopf 1935), now called Hopf maps while investigating certain homotopy groups of spheres. These three are the early examples of bundle spaces such that p is not homotopic to a constant map which was not known until 1930 whether a given continuous map $p : S^m \rightarrow S^n$ for $m > n > 1$ is not homotopic to a constant map. Hopf presented in 1930 the first example of a continuous map $p : S^3 \rightarrow S^2$ which is not homotopic to a constant map by showing that $\pi_3(S^2) \neq 0$.

Theorem 7.9.1 $\pi_3(S^2) \neq 0$.

Proof To prove this, we consider Hopf map $p : S^3 \rightarrow S^2$. Define $S^3 = \{(z, \omega) \in \mathbf{C} \times \mathbf{C} : |z|^2 + |\omega|^2 = 1\}$. Let ρ be an equivalence relation on S^3 defined by $(z, \omega)\rho(z', \omega') \Leftrightarrow (z, \omega) = (\lambda z', \lambda\omega')$ for some $\lambda \in \mathbf{C}$ such that $|\lambda| = 1$. Let $M = S^3/\rho$ be the quotient space topologized by the quotient topology and $p : S^3 \rightarrow M$ be the projection map, $(z, \omega) \mapsto \langle z, \omega\rangle$. For $\langle z, \omega\rangle \in M$, $p^{-1}(\langle z, \omega\rangle)$, called the fiber over $\langle z, \omega\rangle$, is a great circle of S^3. Clearly, M is homeomorphic to S^2. Hence replacing M by S^2, we obtain the Hopf map $p : S^3 \rightarrow S^2$. In this way, S^3 is decomposed into a family of great circles with S^2 as a quotient space.

We claim that p is not homotopic to a constant map. If possible, let $H : S^3 \times I \rightarrow S^2$ be a homotopy between p and a constant map c. Then it allows a homotopy $\widetilde{H} : S^3 \times I \rightarrow S^3$ such that the triangle in Fig. 7.3 is commutative. The map $\widetilde{H}$ is a homotopy between the identity map on S^3 and a constant map. This shows that S^3 is contractible, which is a contradiction. Hence $\pi_3(S^2) \neq 0$. ❑

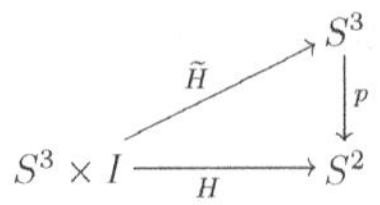

Fig. 7.3 Commutative diagram for Hopf map p

Theorem 7.9.2 $\pi_7(S^4) \neq 0$.

Proof To prove this theorem, we consider the Hopf map $p : S^7 \to S^4$, where S^7 is represented by the unit sphere S^7 given by

$$S^7 = \{(z, \omega) \in \mathbf{H} \times \mathbf{H} : ||z||^2 = 1\},$$

where $\mathbf{H}$ is the division ring of quaternions. Let D be the unit disk in $\mathbf{H}$ defined by

$$D = \{z \in \mathbf{H} : ||z|| \leq 1\}.$$

Identifying the boundary of D to a single point we obtain the quotient space M of the unit disk D. Since the real dimension of D is 4, the quotient space M is homeomorphic to S^4. Proceeding as in Theorem 7.9.1 the Hopf map $p : S^7 \to S^4$ proves that $\pi_7(S^4) \neq 0$. ❑

Theorem 7.9.3 $\pi_{15}(S^8) \neq 0$.

Proof To prove this theorem we consider the Hopf map $p : S^{15} \to S^8$. In $\mathbf{R}^{16}$, we can perform a similar construction as in Theorems 7.9.1 and 7.9.2 which shows that $\pi_{15}(S^8) \neq 0$. ❑

Theorem 7.9.4 *If* $m > 1$, *then* $\pi_m(S^1, s_0) = 0$.

Proof Let $p : \mathbf{R} \to S^1$ be a covering map. Any continuous map $f : S^m \to S^1$ lifts to a map $\tilde{f} : S^m \to \mathbf{R}$ by the lifting property, since S^m is simply connected. Again since S^m is simply connected, $\tilde{f}$ is nullhomotopic. Projecting this homotopy (to a constant map) to S^1 we can define a homotopy of f to a constant map.This implies that $\pi_m(S^1, s_0) = 0$. ❑

7.10 Freudenthal Suspension Theorem and Table of $\pi_i(S^n)$ for $1 \leq I, n \leq 8$

This section studies Freudenthal suspension theorem with its immediate consequences and displays a table showing the values of $\pi_i(S^n)$ for $1 \leq i, n \leq 8$. One of the deepest problems in homotopy theory is computing homotopy groups $\pi_{n+m}(S^n)$. Hans Freudenthal was partially successful in 1937 in solving such problems. Freudenthal suspension theorem is a fundamental theorem in algebraic topology.

It demonstrates the behavior of simultaneously taking suspensions and increasing the index of the homotopy groups of the space in question. The impact of suspension functor comes from the classical theorem of Freudenthal which facilitates to study the deepest problems of homotopy theory to compute the homotopy groups of spheres. Moreover, this theorem implies the concept of stabilization of homotopy groups and ultimately leads to stable homotopy theory which is a generalized cohomology theory (see Chap. 15). Stable homotopy groups of spheres are one of the most important objects in algebraic topology. Moreover, this section gives Table 7.1 displaying a small sample of $\pi_i(S^n)$ extracted from the paper (Toda 1962).

7.10.1 Freudenthal Suspension Theorem

This subsection studies Freudenthal suspension theorem proved by Hans Freudenthal in 1937 which establishes the stable range for homotopy groups. Freudenthal observed that the suspension operation on topological spaces shifts by one their low-dimensional homotopy groups. This observation was important in understanding the special behavior of homotopy groups of spheres, because every sphere can be formed topologically as a suspension of a lower dimensional sphere and this subsequently forms the basis of stable homotopy theory (see Chap. 15.) More precisely, for each pair m and n of positive integers, there is a natural homomorphism $E : \pi_m(S^n) \to \pi_{m+1}(S^{n+1})$. This homomorphism is called the Freudenthal suspension homomorphism defined by H. Freudenthal in 1937 (Freudenthal 1937).

Definition 7.10.1 Consider $\pi_m(S^n)$ as homotopy classes of maps $f : (S^m, 1) \to (S^n, 1)$ and S^n as the subspace of S^{n+1} consisting of all points of S^{n+1} with last coordinate 0. This means S^n is the equator of S^{n+1}. Define the point $(0, 0, \ldots, 1) \in S^{n+1}$ as the north pole and the point $(0, 0, \ldots, -1) \in S^{n+1}$ as their south pole.

Let $[f] \in \pi_m(S^n)$. Then $f : S^m \to S^n$ is a continuous map. Extend f to a continuous function $\tilde{f} : S^{n+1} \to S^{n+1}$ as follows: $\tilde{f}|_{S^m} = f$ and it maps the equator of S^{m+1} to the equator of S^{n+1}. The map is then extended radially as shown in Fig. 7.4.

The arc from the north pole to a point $x \in S^m$ is mapped linearly onto the arc from the north pole of S^{n+1} to $f(x)$. This defines the map $\tilde{f}$ on the northern hemisphere. For the southern hemisphere it is similarly defined. The extended map $\tilde{f}$ is called the suspension of f.

We are now in a position to define 'Suspension homomorphism' E.

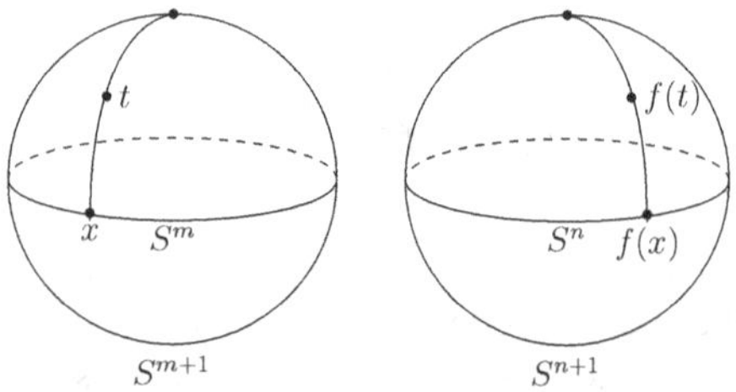

Fig. 7.4 Radially extended map

Definition 7.10.2 (*Suspension homomorphism*) The natural homomorphism

$$E : \pi_n(S^n) \to \pi_{n+1}(S^{n+1}), [f] \mapsto [\tilde{f}]$$

is called the suspension homomorphism.

H. Freudenthal proved the following suspension theorem in 1937 which is very important to study the homotopy theory. In his honor this suspension theorem is known as 'The Freudenthal suspension theorem.'

Theorem 7.10.3 (The Freudenthal suspension theorem) *The suspension homorphism*

$$E : \pi_m(S^n) \to \pi_{m+1}(S^{n+1})$$

is an isomorphism for $m < 2n - 1$ and is onto for $m \leq 2n - 1$.

Proof See [Freudenthal, 1937]. □

An immediate consequence of the Theorem 7.10.3 is the following.

Corollary 7.10.4 (Hurewicz) *The homotopy groups $\pi_m(S^n) = 0$ for $0 < m < n$.*

Proof For any positive integer $p < m$, the integer $m + p + 1 < 2n$. Hence $m - p < 2(n - p) - 1$ proves by Freudenthal Suspension Theorem that $\pi_m(S^n) \cong \pi_{m-1}(S^{n-1}) \cong \ldots \cong \pi_1(S^{n-m+1})$. Since $n - m + 1 > 1$ for $m < n$, it follows that $\pi_1(S^{n-m+1}) = 0$ and its isomorphic image $\pi_m(S^n)$ is also 0. □

Corollary 7.10.5 (Hopf) *For every integer $n \geq 1$, $\pi_n(S^n) \cong$ **Z**. (This result is known as 'Hopf degree theorem').*

Proof It follows from previous argument that $\pi_1(S^1) \cong \pi_2(S^2)$. Since $\pi_1(S^1) \cong \mathbf{Z}$, $\pi_2(S^2) \cong \mathbf{Z}$. If $n \geq 2$, then $n < 2n - 1$ and hence by Frudenthal Suspension Theorem $\pi_2(S^2) \cong \pi_3(S^3) \cong \pi_4(S^4) \cong \ldots \cong \pi_n(S^n)$. □

7.10.2 *Table of $\pi_i(S^n)$ for $1 \leq I, n \leq 8$*

Remark 7.10.6 Table 7.1 displays a small sample of the values of the groups $\pi_i(S^n)$ extracted from the paper (Toda 1962).

7.11 Action of π_1 on π_n

This section considers an important action of π_1 on π_n which provides tools to make the abelian group $\pi_n(X, x_0)$ a module over the group ring $\mathbf{Z}[\pi_1(X, x_0)]$ for $n > 1$, and is used to prove the homological version of Whitehead theorem. The fundamental group $\pi_1(x, x_0)$ acts on $\pi_n(X, x_0)$ as a group of automorphisms for $n \geq 1$. Let X be a path- connected space and $x_0, x_1 \in X$. Given a path $\gamma : I \to X$ from $x_0 = \gamma(0)$ and $x_1 = \gamma(1)$, define a base point changing homomorphism

$$\beta_\gamma : \pi_n(X, x_1) \to \pi_n(X, x_0),\ [f] \mapsto [\gamma \circ f],$$

which is an isomorphism with inverse $\beta_{\overline{\gamma}}$, where $\overline{\gamma}$ is the inverse path of γ defined by $\overline{\gamma}(s) = \gamma(1 - s)$. Hence it follows that if X is a path-connected space, different choices of base point x_0 give isomorphic homotopy groups, written simply as $\pi_n(X)$. So without loss of generality we assume that γ is a loop in X based at x_0. Since $\beta_{\gamma_\eta} = \beta_\gamma \circ \beta_\eta$, there is a homomrphism

$$\psi : \pi_1(X, x_0) \to \mathcal{A}ut\,(\pi_n(X, x_0)),\ [\gamma] \mapsto \beta_\gamma.$$

This is called the action of π_1 on π_n and each element of π_1 acts as an automorphism $[f] \mapsto [\gamma \circ f]$ of π_n. In particular, for $n = 1$, this is the action of π_1 on itself by inverse automorphisms. If $n > 1$, the action makes the abelian group $\pi_n(X, x_0)$ a module over the group ring $\mathbf{Z}[\pi_1(X, x_0)]$, whose elements are finite sums $\sum\limits_i \eta_i\gamma_i$, where $n_i \in \mathbf{Z}$ and $\gamma_i \in \pi_1$, multiplication being defined by distributivity and the multiplication in π_1. The module structure of $\pi_n(X, x_0)$ is given by $(\sum\limits_i \eta_i\gamma_i)\cdot\alpha = \sum\limits_i \eta_i(\gamma\cdot\alpha)$ for $\alpha \in \pi_n(X, x_0)$. Sometimes we say that π_n is a π_1-module instead of $\mathbf{Z}[\pi_1(X, x_0)]$-module.

Definition 7.11.1 A topological space with trivial π_1 action on π_n is called 'n-simple' and is called 'simple' if it is n-simple for all n.

Definition 7.11.2 A topological space X is said to be abelian if it has trivial action on all π_n, since when $n = 1$, this is the condition that π_1 to be abelian.

Remark 7.11.3 The concept of action of π_1 on π_n is used to prove the homological version of Whitehead theorem.

7.12 The *n*th Cohomotopy Sets and Groups

This section studies the concept of cohomotopy sets of pointed topological spaces and pairs of spaces, which form groups under suitable situations. There is a duality between homotopy groups $\pi_m(X, A, x_0)$ and cohomotopy groups $\pi^m(X, A.x_0)$. More precisely, given a pair (n, m) of integers such that $\pi_m(X, A, x_0)$ and $\pi^n(X, A)$ are abelian groups, there is a homomorphism $\psi : \pi_m(X, A, x_0) \otimes \pi^n(X, A) \to \pi_m(S^n, s_0)$. In particular, if $m = n$, there exists a homomorphism $\psi : \pi_n(X, A, x_0) \otimes \pi^n(X, A) \to \mathbf{Z}$. K. Borsuk endowed the abelian group structure in 1936 on the set $\pi^n(X, A)$ under certain conditions on (X, A) (Borsuk 1936).

Let $\mathcal{T}op_*$ and $\mathcal{T}op^2$ denote the category of pointed topological spaces and the category of pairs of topological spaces, respectively.

Definition 7.12.1 For any pointed space (X, x_0) in $\mathcal{T}op_*$, the nth-cohomotopy set $\pi^n(X, x_0)$ is defined to be the set of all homotopy classes $[(X, x_0), (S^n, s_0)]$ of continuous maps $f : (X, x_0) \to (S^n, s_0) \in \mathcal{T}op_*$.

This definition is generalized for (X, A) in $\mathcal{T}op^2$.

Definition 7.12.2 For any pair of topological spaces (X, A) in $\mathcal{T}op^2$, the nth-cohomotopy set $\pi^n(X, A)$ is defined to be the set of all homotopy classes $[(X, A), (S^n, s_0)]$ relative to A of continuous maps $f : (X, A) \to (S^n, s_0) \in \mathcal{T}op^2$. In particular, if $A = \emptyset$, $\pi^n(X, A)$ is defined by $\pi^n(X)$ and called the nth-cohomotopy set of the topological space X.

Remark 7.12.3 $\pi^n(X, A)$ has a distinguished element namely, the homotopy class of the constant map $c : X \to s_0$. The element $[c]$ is denoted by the symbol 0 and called the zero element of $\pi^n(X, A)$.

Definition 7.12.4 $\pi^0(X, A)$ is defined to be the set of all open and closed subspaces of X not intersecting A. The zero element of $\pi^0(X, A)$ is the empty subspace of X.

Definition 7.12.5 Every map $f : (X, A) \to (Y, B)$ in $\mathcal{T}op^2$ induces a transformation $f^* : \pi^n(Y, B) \to \pi^n(X, A), [\alpha] \to [\alpha \circ f]$, called the transformation induced by f.

Remark 7.12.6 Clearly, f^* is well defined and sends the zero element of $\pi^n(Y, B)$ to the zero element of $\pi^n(X, A)$.

Proposition 7.12.7 *$\pi^n : \mathcal{T}op^2 \to \mathcal{S}et$ is a contravariant functor.*

Proof Consider the object function: $(X, A) \mapsto \pi^n(X, A)$ and every $f : (X, A) \to (Y, B)$ in $\mathcal{T}op^2$. Then f^* satisfies the functorial properties. ❑

Let $(X, A) \in \mathcal{T}op^2$ be such that $A \neq \emptyset$. If we identify A to a point $*$, we obtain the quotient space X/A with base point $*$. If $q : (X, A) \to (X/A, *)$ is the natural projection, then q maps $X - A$ homeomorphically onto $X/A - \{*\}$.

Proposition 7.12.8 *The induced map $q^* : \pi^n(X/A, *) \to \pi^n(X, A)$ is a bijection.*

Proof q^* is surjective: Let $\alpha \in \pi^n(X, A)$ be represented by a map $f : (X, A) \to (S^n, s_0)$. Define a function

$$\theta : (X/A, *) \to (S^n, s_0), z \mapsto \begin{cases} f(q^{-1}(z)), & \text{if } z \in X/A - \{*\} \\ s_0, & \text{if } z = *. \end{cases}$$

Clearly, θ is continuous as f is so. Hence $\beta = [\theta] \in \pi^n(X/A, *)$. Again $\theta \circ q = f$ shows that $q^*([\beta]) = \alpha$. Hence q^* is onto.

q^* is injective: Let $\gamma, \delta : (X/A, *) \to (S^n, s_0)$ be two maps such that $q^*([\gamma]) = q^*([\delta])$. Then $\gamma \circ q \cong \delta \circ q$ rel A. Hence there exists a continuous map $H : (X \times$

$I, A \times I) \to (S^n, s_0)$ such that $H(x,0) = (\gamma \circ q)(x)$, $H(x,1) = (\delta \circ q)(x)$, $\forall x \in X$ and $H(a,t) \in A$, $\forall a \in A, t \in I$.

Define a map

$$F : (X/A \times I, \{*\} \times I) \to (S^n, s_0), (z,t) \mapsto \begin{cases} H(q^{-1}(z), t), & \text{if } z \in X/A - \{*\} \\ s_0, & \text{if } z = *. \end{cases}$$

Then $F : \gamma \simeq \delta \text{ rel } \{*\}$ shows that $[\gamma] = [\delta]$. Hence q^* is injective.

Consequently, q^* is a bijection. ❑

Remark 7.12.9 The Proposition 7.12.8 shows that while computing cohomotopy sets of a given pair $(X, A) \in \mathcal{T}op^2$, without loss of generality, we may assume that if $A \neq \emptyset$, the subset A consists of a single point $\{a\}$ say. Then for $n > 1$, since S^n is simply connected, the inclusion map $j : X \hookrightarrow (X, a)$ induces a one-to-one transformation $j^* : \pi^n(X, a) \to \pi^n(X)$, provided that X satisfies certain homotopy extension properties (which hold in particular, if X is a paracompact Hausdorff space and a is any given point of X).

7.13 Applications

This section gives some interesting applications of higher homotopy groups. Homotopy groups play a key role in algebraic topology. For more applications see Exercises 7.14 of this chapter, Chap. 14 and also Chap. 17.

Proposition 7.13.1 *Let (X, A) be a pair of path- connected spaces. Then $\pi_1(X, A, x_0)$ can be identified with the set of cosets aH of the subgroup H of $\pi_1(X, x_0)$ represented by loops in A at x_0.*

Proof $\pi_1(X, A, x_0)$ is the set of homotopy classes of paths in X from a varying point in A. Define a map $\psi : \pi_1(X, x_0) \to \pi_1(X, A, x_0)$ by considering a loop at x_0 as an element of $\pi_1(X, A, x_0)$. Since A is path- connected, every element of $\pi_1(X, A, x_0)$ is homotopic to a loop at x_0. Hence ψ is surjective. Again two loops $\alpha, \beta \in \pi_1(X, x_0)$ are homotopic rel A iff $[\alpha^{-1} * \beta]$ is represented by a loop in A. Hence, we can identify $\pi_1(X, A, x_0)$ with the set of cosets aH. ❑

Proposition 7.13.2 *Let (X, A, x_0) be a triplet such that A is a strong deformation retract of X. If $i : (A, x_0) \to (X, x_0)$ is the inclusion, then $i_* : \pi_n(A, x_0) \to \pi_n(X, x_0)$ is an isomorphism for all $n > 1$.*

Proof Let $A \subset X$ be a retract with retraction $r : X \to A$. Since A is a strong deformation retract of X, $i \circ r \simeq 1_X$ rel A. Then the inclusion i a homotopy equivalence and hence $i_* : \pi_n(A, x_0) \to \pi_n(X, x_0)$ is an isomorphism for all $n > 1$. ❑

Corollary 7.13.3 *If (X, A, x_0) is a triplet such that A is a strong deformation retract of X. Then $\pi_n(X, A, x_0) = 0$ for any integer $n > 0$.*

Proof Since $i_* : \pi_n(A, x_0) \to \pi_n(A, x_0)$ is an isomorphism for $n > 0$,the corollary follows from the exact sequence (7.2) of the triplet (X, A, x_0). ❑

Definition 7.13.4 A map $p : E \to B$ is said to have polyhedra covering homotopy property (PCHP) if it has the covering homotopy property for every triangular space X. If p has PCHP, then p is said to be a fibering.

Proposition 7.13.5 *Let a map $f : (X, A, x_0) \to (Y, B, y_0)$ be given. If $f : X \to Y$ is a fibering and $A = f^{-1}(B)$, then the induced transformations $f_* : \pi_n(X, A, x_0) \to \pi_n(Y, B, y_0)$ is an isomorphism for every $n > 1$.*

Proof f_* is a monomorphism: Clearly, f_* is a homomorphism for $n > 1$. Let $[g], [h] \in \pi_n(X, A, x_0)$ be such that $f_*([g]) = f_*([h])$. Then their representatives $g, h \in \pi_n(X, A, x_0)$ are such that $f \circ g$ and $f \circ h$ represent the same element of $\pi_n(Y, B, y_0)$. Consequently, there exists a map $H : (I^n \times I, I^{n-1} \times I, J^{n-1} \times I) \to (Y, B, y_0)$ such that $H(\omega, 0) = (f \circ g)(\omega)$ and $H(\omega, 1) = (f \circ h)(\omega)$, $\forall \omega \in I^n$. Let $C = (I^n \times 0) \cup (J^{n-1} \times I) \cup (I^n \times 1)$. Then, C is a closed subspace of $I^n \times I$.

Define a map

$$F : C \to X, (\omega, t) \mapsto \begin{cases} g(\omega), & \text{if } \omega \in I^n, t = 0 \\ x_0, & \text{if } \omega \in J^{n-1}, t \in I \\ h(\omega), & \text{if } \omega \in I^n, t = 1. \end{cases}$$

Then $f \circ F = H|_C$. Since C is a strong deformation structure of $I^n \times I$, F has an extension $\widetilde{F}^n : I^n \times I \to X$ such that $f \circ \widetilde{F} = H$. Again since H maps $I^{n-1} \times I$ into B and $A = f^{-1}(B)$, then condition $f \circ \widetilde{F} = H$ shows that $\widetilde{H}(I^{n-1} \times I) \subset A$. Consequently, the map $\widetilde{H} : (I^n \times I, J^{n-1} \times I, J^{n-1} \times I) \to (X, A, x_0)$ is such that $\widetilde{H}(\omega, 0) = g(\omega)$ and $\widetilde{H}(\omega, 1) = h(\omega)$ for all $\omega \in I^n$. This implies that g and h represents the same element of $\pi_n(X, A, x_0)$. In other words, $[g] = [h]$. This shows that f_* is a monomorphism.

f_* **is an epimorphism**: Let $[g] \in \pi_n(Y, B, y_0)$ be an arbitrary element. Then $[g]$ is represented by a map $g : (I^n, I^{n-1}, J^{n-1}) \to (Y, B, y_0)$. Since J^{n-1} is a strong deformation retract of I^n, it follows that $\exists$ a map $h : I^n \to X$ such that $f \circ h = g$ and $h(J^{n-1}) = x_0$. Again since $A = f^{-1}(B)$, $f \circ h = g$, $h(I^{n-1}) \subset A$, it gives a map $h : (I^n, I^{n-1}, J^{n-1}) \to (X, A, x_0)$. Clearly, $f_*([h]) = [g]$ shows that f_* is an epimorphism. ❑

Proposition 7.13.6 *If $X = \{x_0\}$ is a topological space consisting of a single point x_0, then $\pi_n(X, x_0) = 0$ for every $n \geq 0$.*

Proof If $X = \{x_0\}$, then for each n, the map $f : I^n \to X$ is the only map of I^n onto X, which is a constant map. Hence $\pi_n(X, x_0) = 0$. ❑

Proposition 7.13.7 *For $0 < m < n$, the m-th homotopy group $\pi_m(S^n) = 0$.*

Proof Let $[f] \in \pi_m(S^n)$ be represented by a map $f : (S^m, 1) \to (S^n, 1)$. If we represent S^m and S^n as the boundary complex of the simplexes of dimension $m+1$ and $n+1$, f has a simplicial approximation cannot map a simplex onto a simplex of higher dimension, g cannot be onto. Let s_0 be a point of S^n which is not the range of g. Then, $S^n - \{s_0\}$ is homeomorphic to $\mathbf{R}^n$ and hence $S^n - \{s_0\}$ is contractible and g is a map such that $g(S^m)$ is contained in a contractible. Consequently, g is a nullhomotopic, i.e., homotoic to a constant map c. Hence, $[f] = [g] = [c] = 0$. ❑

Theorem 7.13.8 *Let (X, A, x_0) be a triplet and A be contractible relative to a point $x_0 \in A$. Then $\pi_n(X, A, x_0) \cong \pi_n(X, x_0) \oplus \pi_{n-1}(A, x_0)$ for any $n \geq 3$ and i_* maps $\pi_n(A, x_0)$ into the zero element of $\pi_n(X, x_0)$ for any $n \geq 1$.*

Proof By the given condition, $\exists$ a homotopy $H_t : A \to X$ such that $H_0 = i : A \hookrightarrow X$, $H_1(A) = x_0$, and $H_t(x_0) = x_0$. Hence $i_* = 0$ for any $n \geq 1$. For $n \geq 2$, use the exactness of the homotopy sequence (7.2) of (X, A, x_0). Then $i_* = 0$ shows that j_* is a monomorphism and ∂ is an epimorphism. Consequently, $\pi_n(X, A, x_0)$ may be considered as an extension of $\pi_n(X, x_0)$ by $\pi_{n-1}(X, x_0)$. Define a homomorphism $f_* : \pi_{n-1}(A, x_0) \to \pi_n(X, A, x_0)$ for each $n \geq 2$. Let $[g] \in \pi_{n-1}(A, x_0)$ be represented by a map $g : (I^{n-1}, \partial I^{n-1}) \to (A, x_0)$. Define a map

$$h : (I^n, I^{n-1}, J^{n-1}) \to (X, A, x_0), (t_1, t_2, \ldots, t_n) \mapsto (H_{t_n} \circ g)(t_1, t_2, \ldots, t_{n-1}).$$

Set $f_*([g]) = [h]$. Since $h|_{I^{n-1}} = g$, it follows that $\partial \circ f_*$ is the identity automorphism on $\pi_{n-1}(A, x_0)$. This implies that f_* is a monomorphism for every $n \geq 2$. For $n \geq 3$, the group $\pi_n(X, A, x_0)$ is abelian. Hence $\partial \circ f_* = 1_d$ implies that

$$\begin{aligned} \pi_n(X, A, x_0) &= \text{Im } f_* \oplus \ker \partial \\ &\cong \pi_{n-1}(A, x_0) \oplus \pi_n(X, x_0), \end{aligned}$$

because f_* and j_* are monomorphisms and $\ker \partial = \text{Im } j_*$. ❑

Theorem 7.13.9 *Let $F \hookrightarrow E \to B$ be a fiber bundle such that the inclusion $F \hookrightarrow E$ is homotopic to a constant map. Show that the long exact sequence of homotopy groups breaks up into split short exact sequences producing isomorphisms*

$$\pi_n(B) \cong \pi_n(E) \oplus \pi_{n-1}(F).$$

In particular, for the Hopf bundles $S^3 \to S^7 \to S^4$ and $S^7 \to S^{15} \to S^8$ this gives isomorphisms

$$\pi_n(S^4) \cong \pi_n(S^7) \oplus \pi_{n-1}(S^3); \pi_n(S^8) \cong \pi_n(S^{15}) \oplus \pi_{n-1}(S^7).$$

Proof The maps $i_* : \pi_1(F, x_0) \to \pi_n(E, x_0)$ in the long exact sequence (7.3) for the Serre fibration are induced by the inclusion $i : F \hookrightarrow E$. Hence if this is homotopic to a constant map, then $i_* = 0$. Thus for all $n > 0$, we have a short exact sequence

$$0 \longrightarrow \pi_n(E, x_0) \xrightarrow{p_*} \pi_n(B, b_0) \xrightarrow{d_*} \pi_{n-1}(F, x_0) \longrightarrow 0.$$

Again since $p : E \to B$ has the homotopy lifting property with respect to all disks, it follows that there is a splitting map $\pi_n(B, b_0) \to \pi_n(E, x_0)$ such that the above short exact sequence splits. ❑

Corollary 7.13.10 *The groups $\pi_7(S^4)$ and $\pi_{15}(S^8)$ contain $\mathbf{Z}$ summands.*

Proof The corollary follows from Theorem 7.13.9. The corollary also follows from Hopf fiberings of spheres (see section 7.8.2). ❑

Theorem 7.13.11 *If $S^k \to S^m \to S^n$ is a fiber bundle, then $k = n - 1$ and $m = 2n - 1$.*

Proof Consider the relations $n \leq m$ and $k \leq m$ and $k + n = m$. If $k = m$, then $n = 0$, and S^0 is not connected. This contradicts that $S^m \to S^n$ is a surjection. So $k < m$, and hence $S^k \to S^m$ is homotopic to a constant map. Then it follows that $\pi_i(S^n) \cong \pi_i(S^m) \oplus \pi_{i-1}(S^k)$, $\forall i > 0$. This implies $k > 0$ and hence $m > n$. In particular, taking values of $i = 1, 2, \ldots, n$, we have $\pi_i(S^k) = 0$ if $i < n - 1$ and $\pi_{n-1}(S^k) \cong \mathbf{Z}$. Consequently, $k = n - 1$. Hence $m = 2n - 1$. ❑

7.14 Exercises

1. Show that

   (i) $\pi_n(\mathbf{R}^m) = 0$ for every positive integer n and m. Because, $\mathbf{R}^m$ is homotopy equivalent to a point.
   (ii) $\pi_n(D^m) = 0$ for m-disk D^m for every positive integers n and m.

2. Let X be a path- connected space. Show that $\pi_n(X, x_0) \cong \pi_n(X, x_1)$ for all $x_0, x_1 \in X$ for $n \geq 1$.
   [Hint. There is a homeomorphism $h : (X, x_0) \to (X, x_1)$. Hence there exists a homotopy equivalence between (X, x_0) and (X, x_1).]
3. If X is a contractible space, show that $\pi_n(X, x_0) = 0$ for all $n \geq 0$.
   [Hint: As X is contractible, it is homotopy equivalent to a singleton space. Use Dimension Theorem 7.2.6.]
4. **(a)** Let X be a connected covering space of a path- connected space B with covering projection $p : (X, x_0) \to (B, b_0)$ such that $p(x_0) = b_0$. Show that the induced homomorphism $p_* : \pi_n(X, x_0) \to \pi_n(B, b_0)$ is an isomorphism for any $n \geq 2$.
   (b) Show that $\pi_n(S^1) = 0$ for any $n \geq 2$.
   [Hint: $p_* : \pi_n(\mathbf{R}) \to \pi_1(S^1)$ is an isomorphism for any $n \geq 2$ by **(a)**. Since all the groups of the contractible space $\mathbf{R}$ is 0, **(b)** follows.]
5. Let $(X, x_0), (Y, y_0)$ be two topological spaces in $\mathcal{T}op_*$. Show that $\pi_n((X \times Y), (x_0, y_0)) \cong \pi_n(X, x_0) \oplus \pi_n(Y, y_0)$, for $n \geq 1$.

6. Show that the 4-manifold $S^2 \times S^2$ is simply connected, but it is not homeomorphic to S^4.
[Hint: Use the results : $\pi_2(S^2 \times S^2) \cong \mathbf{Z} \oplus \mathbf{Z}$ and $\pi_2(S^4) = 0$.]

7. Let (X, x_0) and (B, b_0) be pointed topological spaces. Consider the homotopy exact sequence (7.3) of the fibering $p : X \to B$. Show that

(a) If F is a retract of B, then $\pi_n(X, x_0) \cong \pi_n(B, b_0) \oplus \pi_n(F, x_0)$ for every $n \geq 2$ and p_* is an epimorphism for every $n \geq 1$.
(b) If X is deformable into F, then $\pi_n(F, x_0) \cong \pi_n(X, x_0) \oplus \pi_{n+1}(B, b_0)$ for every $n \geq 2$ and $p_* = 0$ for every $n \geq 1$.
(c) If F is contractible in X, then $\pi_n(B, b_0) \cong \pi_n(X, x_0) \oplus \pi_{n-1}(F, x_0)$ for every $n \geq 2$ and p_* is a monomorphism for every $n \geq 1$.

[Hint: As $n \geq 2$, use Propositions 7.7.3 and 7.7.6 and exactness property of the homotopy sequence of a fibering $p : X \to B$.]

8. **(a)** Show that for any triplet (X, A, x_0) the formula $a + b - a = (\partial a)b$ holds for all $a, b \in \pi_2(X, A, x_0)$, where $\partial : \pi_2(X, A, x_0) \to \pi_1(A, x_0)$ is the usual boundary operator and $(\partial a)b$ denotes the action of ∂a on b.
(b) Deduce from **(a)** that the image of the map $j_* : \pi_2(X, x_0) \to \pi_2(X, A, x_0)$ lies in the entire of $\pi_2(X, A, x_0)$.

9. Let $p : (X, A, x_0) \to (Y, B, b_0)$ be a given continuous map. If $p : X \to Y$ is a fibering and $A = p^{-1}(B)$, show that the induced transformation $p_* : \pi_n(X, A, x_0) \to \pi_n(Y, B, y_0)$ is a bijection for every $n > 0$.

10. Show that a continuous map $f : (D^n, S^{n-1}, s_0) \to (X, A, x_0)$ defines the zero element in $\pi_n(X, A, x_0)$ iff $f \simeq g$ rel S^{n-1} for some $g : (D^n, S^{n-1}, s_0) \to (X, A, x_0)$ such that $g(D^n) \subset A$.

11. Let $p : X \to B$ be a weak fibration with $p(x_0) = b_0$ If $b_0 \in A \subset B, x_0 \in p^{-1}(b_0), Y = p^{-1}(A)$, show that the induced transformation $p_* : \pi_n(X, Y, y_0) \to \pi_n(B, A, b_0)$ is a bijection for every $n \geq 1$.
[Hint. Use mathematical induction on n starting from $n = 1$.]

12. Let $p : E \to B$ be a locally trivial fiber bundle and $b_0 \in B$. If $F = p^{-1}(b_0)$ and $f_0 \in F$, show that for every $n > 1$, $p_* : \pi_n(E, F, f_0) \to \pi_n(B, b_0, b_0)$ is an isomorphism.

13. Let $p : X \to B$ be a covering of X with discrete fiber F. Suppose $b_0 \in B$ and $x_0 \in p^{-1}(b_0)$, show that

(i) $p_* : \pi_n(X, x_0) \to \pi_n(B, b_0)$ is an isomorphism for all $n > 1$ and a monomorphism for $n = 1$;
(ii) if X is 0-connected, then the points of F are in 1-1 correspondence with the cosets of $p_*(\pi_n(X, x_0))$ in $\pi_1(B, b_0)$.
[Hint: Since F is discrete, $\pi_n(F, x_0) = \pi_n(\{x_0\}, x_0) = 0$ for all $n \geq 1$.]

14. Let $\mathrm{O}(n, \mathbf{R})$ be the topological group of real orthogonal $n \times n$ matrices and $\mathrm{SO}(n, \mathbf{R})$ be the subspace of $\mathrm{O}(n, \mathbf{R})$ of real orthogonal matrices of determinant 1. Show that the inclusion map $i : \mathrm{SO}(n, \mathbf{R}) \hookrightarrow \mathrm{O}(n, \mathbf{R})$ induces an isomorphism

$$i_* : \pi_n(\mathrm{SO}\,(n, \mathbf{R}), 1) \to \pi_n(\mathrm{O}\,(n, \mathbf{R}), 1) \text{ for } n \geq 1.$$

[Hint: Consider the exact homotopy sequence

$$\cdots \to \pi_{n+1}(\mathbf{Z}_2, 1) \to \pi_n(\mathrm{SO}\,(n, \mathbf{R}), 1) \xrightarrow{i_*} \pi_n(\mathrm{O}\,(n, \mathbf{R}), 1) \to \pi_n(\mathbf{Z}_2, 1),$$

where $\pi_n(\mathbf{Z}_2, 1) = 0$ for $n \geq 1$.]

15. Suppose there exist fiber bundles $S^{n-1} \to S^{2n-1} \to S^n$, for all n. Show that the groups $\pi_i(S^n)$ would be finitely generated free abelian groups computable by induction, and nonzero for $i \geq n \geq 2$.

16. Let $p : S^3 \to S^2$ be the Hopf bundle and $q : T^3 \to S^3$ be the quotient map collapsing the complement of a ball in the 3-dimensional torus $T^3 = S^1 \times S^1 \times S^1$ to a point. Show that $p \circ q : T^3 \to S^2$ induces the trivial map $(p \circ q)_* : \pi_n(T^3) \to \pi_n(S^2)$, but not homotopic to a constant map.

17. Let X be a path-connected space with a base point $x_0 \in X$ and $f : S^n \to X$ be a continuous map such that $f(s_0) = x_0$, where s_0 is a base point of S^n. If $Y = X \bigcup_f D^{n+1}$, and $i : X \hookrightarrow Y$ is inclusion, show that induced homomorphism $i_*; \pi_m(X, x_0) \to \pi_m(Y, y_0)$

(i) is an isomorphism if $m < n$;
(ii) is an epimorphism if $m = n$ and
(iii) ker i_* is generated by $\alpha^{-1}[f]\alpha \in \pi_n(X, x_0)$, where $\alpha \in \pi_1(X, x_0)$.

7.15 Additional Reading

[1] Adhikari, M.R., and Adhikari, Avishek, *Basic Modern Algebra with Applications*, Springer, New Delhi, New York, Heidelberg, 2014.
[2] Armstrong, M.A., *Basic Topology*, Springer-Verlag, New York, 1983.
[3] Barratt, M.G., *Track groups I*, Proc. Lond. Math. Soc. **5**(3), 71–106, 1955.
[4] Borel, A., *Topology of Lie groups and characteristic classes*, Bull. Amer. Math. Soc. **61**(1955), 397–432.
[5] Bott, R., *The stable homotopy groups of classical groups*, Proc. Nat. Acad. Sci., USA, **43**(1957), 933–935.
[6] Bott, R., *The stable homotopy of the classical groups*, Annals of Mathematics. Ann. of Math, **70**(1959), 313-337.
[7] Chatterjee, B.C., Ganguly, S., and Adhikari, M.R., *A Textbook of Topology*, Asian Books Pvt.Ltd., New Delhi, 2002.
[8] Dugundji, J., *Topology*, Allyn & Bacon, Newtown, MA, 1966.
[9] Dieudonné, J., *A History of Algebraic and Differential Topology*, 1900-1960, Modern Birkhäuser, 1989.
[10] Eilenberg, S., and Steenrod, N., *Foundations of Algebraic Topology*, Princeton University Press, Princeton, 1952.

[**11**] Hilton, P.J., *An introduction to Homotopy Theory*, Cambridge University Press, Cambridge, 1983.
[**12**] Hu, S.T., *Homotopy Theory*, Academic Press, New York, 1959.
[**13**] Husemöller, D., *fiber Bundles*, McGraw-Hill, Inc, 1966.
[**14**] Mayer, J. *Algebraic Topology*, Prentice-Hall, New Jersy, 1972.
[**15**] Massey, W.S., *A Basic Course in Algebraic Topology*, Springer-Verlag, New York, Berlin, Heidelberg, 1991.
[**16**] Maunder, C.R.F., *Algebraic Topology*, Van Nostrand Reinhhold, London, 1970.
[**17**] Munkres, J.R., *Elements of Algebraic Topology*, Addition-Wesley-Publishing Company, 1984.
[**18**] Rotman, J.J., *An Introduction to Algebraic Topology*,Springer-Verlag, New York, 1988.
[**19**] Spanier, E., *Algebraic Topology*, McGraw-Hill Book Company, New York, 1966.
[**20**] Steenrod. N., *The Topology of fiber Bundles*, Prentice University Press, Prentice, 1951.
[**21**] Wallace, A.H., Algebraic Topology, Benjamin, New York, 1980.
[**22**] Whitehead, G.W., *Elements of Homotopy Theory*, Springer-Verlag, New York, Heidelberg, Berlin, 1978.

References

Adhikari, M.R., Adhikari, A.: Basic Modern Algebra with Applications. Springer, Heidelberg (2014)
Armstrong, M.A.: Basic Topology. Springer, New York (1983)
Barratt, M.G.: Track groups I. Proc. Lond. Math. Soc. **5**(3), 71–106 (1955)
Borel, A.: Topology of lie groups and characteristic classes. Bull. Am. Math. Soc. **61**, 397–432 (1955)
Bott, R.: The stable homotopy groups of classical groups. Proc. Nat. Acad. Sci. USA **43**, 933–935 (1957)
Borsuk, K.: Sur les groupes des classes de transformations continues, pp. 1400–1403. C.R. Acad. Sci, Paris (1936)
Bott, R.: The stable homotopy of the classical groups. Ann. Math **70**, 313–337 (1959)
Chatterjee, B.C., Ganguly, S., Adhikari, M.R.: A Textbook of Topology. Asian Books Pvt. Ltd., New Delhi (2002)
Croom, F.H.: Basic Concepts of Algebraic Topology. Springer, Heidelberg (1978)
Dugundji, J.: Topology. Allyn & Bacon, Boston (1966)
Dieudonné, J.: A History of Algebraic and Differential Topology. Modern Birkhäuser, Boston (1989). 1900–1960
Eilenberg, S., Steenrod, N.: Foundations of Algebraic Topology. Princeton University Press, Princeton (1952)
Freudenthal, H.: Über die Klassen von Sphärenabbildungen. Compositio Mathematica **5**, 299–314 (1937)
Gray, B.: Homotopy Theory, An Introduction to Algebraic Topology. Acamedic Press, New York (1975)
Hatcher, A.: Algebraic Topology. Cambridge University Press, Cambridge (2002)
Hilton, P.J.: An introduction to Homotopy Theory. Cambridge University Press, Cambridge (1983)

Hopf, H.: Üeber die Abbildungen von Sphären niedriger dimension. Fund. Math. **25**, 427–440 (1935)

Hurewicz, W.: Beitrage der Topologie der Deformationen, Proc. K.Akad.Wet., Ser.A **38**, 112–119, 521–528 (1935)

Husemöller, D.: Fibre Bundles. McGraw-Hill, Inc, New York (1966)

Hu, S.T.: Homotopy Theory. Academic Press, New York (1959)

Mayer, J.: Algebraic Topology. Prentice-Hall, New Jersy (1972)

Massey, W.S.: A Basic Course in Algebraic Topology. Springer, Heidelberg (1991)

Maunder, C.R.F.: Algebraic Topology. Van Nostrand Reinhhold, London (1970)

Munkres, J.R.: Elements of Algebraic Topology. Addison-Wesley Publishing Company, Boston (1984)

Rotman, J.J.: An Introduction to Algebraic Topology. Springer, New York (1988)

Spanier, E.: Algebraic Topology. McGraw-Hill Book Company, New York (1966)

Steenrod, N.: The Topology of Fibre Bundles. Prentice University Press, Prentice (1951)

Toda, H.: Composition methods in homotopy groups of spheres. Ann. of Math. Stud. **49**, (1962)

Switzer, R.M.: Algebraic Topology-Homotopy and Homology. Springer, Heidelberg (1975)

Wallace, A.H.: Algebraic Topology. Benjamin, New York (1980)

Whitehead, G.W.: Elements of Homotopy Theory. Springer, Heidelberg (1978)

Chapter 8
CW-Complexes and Homotopy

This chapter conveys a study of a special class of topological spaces, called CW-complexes introduced by J.H.C. Whitehead (1904–1960) in 1949 with their homotopy properties to meet the need for development of algebraic topology. This class of spaces is broader and has some better categorical properties than simplicial complexes, but still retains a combinatorial nature that allows for computation (often with a much smaller complex). Algebraic topologists now feel that the category of CW-complexes is a good category for homotopy and homology theories. So a study of CW-complexes should enter in a basic course of algebraic topology and this study should move up to the theorem that every continuous map between CW-complexes is homotopic to a cellular map. More precisely, this chapter studies the basic aspects of CW-complexes and relative CW-complexes with their homotopy properties and proves Whitehead theorem, Freudenthal suspension theorem (general form) and cellular approximation theorem with their applications.

The concept of CW-complexes is introduced as a natural generalization of the concept of polyhedra by relaxing all 'linearity conditions' in simplicial complexes, instead cells are attached by arbitrary continuous maps starting with a discrete set, whose each point is regarded as a 0-cell. Simplicial structure does not behave well with respect to the usual topological operations such as products and quotients of spaces. There is a natural question: what is the good category of topological spaces in which homotopy theory works well? So one of the problems for systematic study of algebraic topology is to decide a suitable category of topological spaces. J.H.C. Whitehead constructed a new category which is now called the category of CW-complexes and studied it in his two papers (Whitehead 1949a, b). A CW-complex is a Hausdorff space built up by successive adjunctions of cells of dimensions $1, 2, 3, \ldots$; such spaces form an extensive class of topological spaces suitable for the study of algebraic topology, where a weak homotopy equivalence is necessarily a homotopy equivalence.

M.R. Adhikari, *Basic Algebraic Topology and its Applications*,
DOI 10.1007/978-81-322-2843-1_8

There are many advantages of CW-complexes over polyhedra: one is that a polyhedron can be regarded as a CW-complex with fewer cells than there were simplexes originally and another advantage is the permissibility of many constructions such as the product of two polyhedra is a CW-complex in a natural way, since the product of two simplexes is a cell, but not a simplex, in general. Since all CW-complexes are paracompact and all open coverings of a paracompact space are numerable, the results on the homotopy classification of principle G-bundles discussed in Chap. 5 apply to all locally trivial principal G-bundles over a CW-complex. Moreover, CW-complexes readily lend themselves to study homotopy, homology and cohomology theories in a relatively convenient way.

One of the main features of CW-complexes is that it is possible to define a continuous map $f : K \to X$ from a CW-complex K into a topological space X step by step by defining them in succession on the n-skeletons $K^{(n)}$ of K. The construction of a CW-complex is made by stages by successive attachments of cells. Despite the fact that every topological space is not a CW-complex, it is sufficient for many important purposes to consider only CW-complexes (instead of arbitrary topological spaces) by a theorem of Whitehead which says that given any topological space X, there exists a CW-complex K and a weak homotopy equivalence $f : K \to X$.

The following terminology and notations for any integer $n \geq 1$ are used in this chapter.

$\mathbf{R}^n = \{(x_1, x_2, \cdots, x_n) \in \mathbf{R}^n : x_i \in \mathbf{R}\}$ (n-dimensional Euclidean space with norm $\|x\|$).

$D^n = \{x \in \mathbf{R}^n : ||x|| \leq 1\}$ (closed n-dimensional disk or ball).

$e^n = \{x \in \mathbf{R}^n : ||x|| < 1\}$ (n-dimensional cell or open n-dimensional disk or ball).

$S^{n-1} = \{x \in \mathbf{R}^n : ||x|| = 1\}$ (($n-1$)-dimensional sphere).

$\Delta^{n-1} = \{(x_1, x_2, \cdots, x_n) \in \mathbf{R}^n : 0 \leq x_i \leq 1, \Sigma x_i = 1\}$ (n-dimensional simplex).

For this chapter the books and papers Gray (1975), Hatcher (2002), Maunder (1970), Rotman (1988), Spanier (1966), Switzer (1975), Whitehead (1978), and the papers Blakers, A.L. and Massey (1952), Whitehead (1949b), and some others are referred in the Bibliography.

8.1 Cell-Complexes and CW-Complexes: Introductory Concepts

This section introduces the concepts of cell-complexes and CW-complexes. A CW-complex X is a cell complex X satisfying two additional conditions: **CW(4)** and **CW(5)** (i.e., having weak topology and satisfying closure finite property). CW-complexes form an important class of topological spaces that contains all simplicial complexes. A simplicial complex is built up successively by attaching simplexes along their boundaries. A simplex and its boundary form a triangulation (D^n, S^{n-1}) for some n. Thus, a polyhedron is built up successively by attaching simplexes by maps of their boundaries. On the other hand, a CW-complex is built up successively by attaching a family of 1-cells to a discrete space; attaching a family of 2-cells to the result; attaching a family of 3-cells to the result and so on (even allowing more than a finite number of cells). A CW-complex generalizes the concept of polyhedra, because the cells are attached by arbitrary continuous maps starting with a discrete set, whose points are regarded as 0-cells.

8.1.1 Cell-Complexes

This subsection introduces the concept of cell-complexes, given by J.H.C. Whitehead in 1949, which is easier to handle than simplicial complexes at many situations. This concept generalizes the notion of simplicial complexes.

Definition 8.1.1 An n-cell is a pair (X, A) of topological spaces homeomorphic to the pair (D^n, S^{n-1}).

Example 8.1.2 Let $\text{Int}\,\Delta^{n-1} = \{x = (x_1, x_2, \ldots, x_n) \in \Delta^{n-1} : 0 < x_i < 1\}$. Then $(\Delta^{n-1}, \partial\Delta^{n-1})$ is an $(n-1)$-cell, where $\partial\Delta^{n-1} = \Delta^{n-1} - \text{Int}\,\Delta^{n-1}$.

Definition 8.1.3 A cell complex X is a Hausdorff space which is the union of disjoint subspaces $e_\alpha(\alpha \in \mathbf{A})$ called cells satisfying

(i) to each cell, an integer $n \geq 0$ is assigned. This integer is called its dimension. If the cell e_α has dimension n, we use the notation e^n_α for this cell.
The union of all cells e^k_α with $k \leq n$, denoted by $X^{(n)}$ is called the n-skeleton of X.
(ii) If e^n_α is an n-cell, there is a characteristic map $\psi_\alpha : (D^n, S^{n-1}) \to (X, X^{(n-1)})$ such that its restriction $\psi_\alpha|_{D^n - S^{n-1}}$ is a homeomorphism from $D^n - S^{n-1}$ onto e^n_α.

Remark 8.1.4 Some authors prefer to call a cell complex X as a cell complex or simply a complex K on X.

Definition 8.1.5 A continuous map $f : X \to Y$ between two cell-complexes is said to be cellular if $f(X^{(n)}) \subset Y^{(n)}$ for all $n \geq 0$.

Example 8.1.6 The polyhedron of any finite geometric simplicial complex is a cell complex. Each open n-simplex is an n-cell, and in this case, the maps ψ_α are all homeomorphisms.

Example 8.1.7 The n-sphere S^n is a cell complex with two cells e^0, e^n, where $e^0 = \{(1, 0, 0, \ldots, 0)\}$ and $e^n = S^n - e^0$. The cell e^n is called the standard n-cell and it is thought of as n-sphere minus its 'east point' $e^0 = \{(1, 0, 0, \ldots, 0)\}$.

Example 8.1.8 **(i)** $\mathbf{R}P^n$ is a cell complex with one cell of dimension k for each $k \leq n$. It is represented symbolically as $\mathbf{R}P^n = e^0 \cup e^1 \cup \cdots \cup e^n$.

(ii) $\mathbf{C}P^n$ is a cell complex with one cell of dimension $2k$ for each $k \leq n$. It is represented symbolically as $\mathbf{C}P^n = e^0 \cup e^2 \cup \cdots \cup e^{2n}$;

(ii) $\mathbf{H}P^n$ is a cell complex with one cell of dimension $4k$ for each $k \leq n$. It is represented symbolically as $\mathbf{H}P^n = e^0 \cup e^4 \cup \cdots \cup e^{4n}$.

Definition 8.1.9 Let X be a cell complex with characteristic ψ_α and $A \subset X$. Then A is said to be subcomplex of X if A is a union of cells e_α and $\bar{e}_\alpha \subset A$, where $\psi_\alpha(D^n) = \bar{e}_\alpha$.

Example 8.1.10 $X^{(n)}$ is a subcomplex of a cell complex X for every $n \geq 0$. Because, $\bar{e}^n_\alpha = \psi_\alpha(D^n)$.

Example 8.1.11 Let X be a cell complex and $K(A)$ be the intersection of all subcomplexes of X containing A. If $A \subset B$, then $K(A) \subset K(B)$ and $K(A)$ is a subcomplex of the cell complex X.

Example 8.1.12 An n-skeleton $X^{(n)}$ is a subcomplex of a cell complex X.

Definition 8.1.13 A pair (X, A) is called a relative cell complex if X is a Hausdorff space and $X - A$ is a union of disjoint subspaces $e^n_\alpha(\alpha \in \mathbf{A})$ called cells satisfy (i) and (ii) of Definition 8.1.3.

Remark 8.1.14 A relative cell complex generalizes the concept of a cell complex.

Definition 8.1.15 A polytope is the union of finitely many simplices, with the additional property that, for any two simplices that have a nonempty intersection, their intersection is a vertex, edge, or higher dimensional face of the two.

Remark 8.1.16 A polytope is an important concept and this term was generally used before the creation of CW-complex by J.H.C. Whitehead in 1949.

8.1.2 CW-Complexes

This subsection studies an important class of topological spaces, called CW-complexes, which describe cell complexes X which are closure finite and have the weak topology. Such topological spaces constitute a most useful class of spaces in which homotopy theory works well. CW complexes meet the need of homotopy theory and develop both homotopy and homology theories. This class of spaces is broader and has some better categorical properties than simplicial complexes, but still retains a combinatorial nature that allows for computation (often with a much smaller complex). Roughly speaking, a CW complex is made of basic building blocks called cells. An n-dimensional closed cell is the image of an n-dimensional closed ball under an attaching map. For example, a simplex is a closed cell, and more generally, a convex polytope is a closed cell.

One of the problems for systematic study of algebraic topology is to decide a suitable category of topological spaces. J.H.C. Whitehead introduced in 1949 (Whitehead

1949b) a suitable category, which is now called the category of CW-complexes. For a long time the term CW-complexes was not in regular use. Many later authors continued to refer to 'polyhedra' which are now called finite CW-complexes. Computing homotopy and homology groups is in general a difficult problem. One of the difficulties is that given the arbitrary spaces X, Y it is not easy to construct continuous maps $f : X \to Y$. If we pay our attention to a class of spaces obtained step by step out of simple building blocks such as simplicial complexes, then we have a better chance for constructing maps step by step, extending them over the building blocks one at a time. This motivated Whitehead to define CW-complexes in 1949.

A CW-complex is built up by successive adjunctions of cells of dimensions $1, 2, 3, \ldots$. The precise definition asserts how the cells may be topologically glued together. A CW complex is a Hausdorff space X together with a partition of X into open cells (of varying dimensions) which satisfy some additional conditions prescribed below.

Definition 8.1.17 A CW-complex is a Hausdorff space X, together with an indexing set A_n for each integer $n \geq 0$, and maps $\psi_i^n : D^n \to X$ for all $n \geq 0, i \in A_n$, are such that the following conditions are satisfied: If $e^n = \{x \in \mathbf{R}^n : d(x, 0) < 1\}(n \geq 1)$, then

CW(1) $X = \cup \psi_i^n(e_i^n)$, for all $n \geq 0, i \in A_n$ (e^0 and D^0 are each considered as a single point);

CW(2) $\psi_i^n(e^n) \cap \psi_j^m(e^m) = \emptyset$, unless $n = m$ and $i = j$; and $\psi_i|_{e^n}$ is (1-1) for all $n \geq 0, i \in A_n$;

CW(3) if $X^{(n)} = \cup \psi_i^m(e^m)$, for all $m, 0 \leq m \leq n$ and $i \in A_n$, then $\psi_i(S^{n-1}) \subset X^{(n-1)}$ for each $n \geq 1$ and $i \in A_n$;

CW(4) A subset Y of X is closed if and only if $(\psi_i^n)^{-1}(Y)$ is closed in D^n, for each integer $n \geq 0$ and $i \in A_n$;

CW(5) for each integer $n \geq 0$ and $i \in A_n$, the subspace $\psi_i^n(D^n)$ is contained in the union of a finite number of sets of the form $\psi_j^m(e^m)$.

Definition 8.1.18 The maps ψ_i^n are called characteristic maps for the CW-complex X, and the subspaces $\psi_i^n(D^n)$ are called n-cells of X; $X^{(n)}$ is called the n-skeleton of X. If n is the smallest integer such that $X^{(n)} = X$, then X is said to be finite dimensional with dimension n or simply n-dimensional. Otherwise, i.e., if there exists no such n, X is said to be infinite dimensional.

Remark 8.1.19 The condition **CW(4)** says that X has the union topology; frequently called the weak topology and the condition **CW(5)** says that X is closure finite in the sense that each closed cell is covered by a finite union of open cells.

Remark 8.1.20 The original reason for the term 'CW-complex': the symbol C stands for 'closure-finite' and the symbol W stands for 'weak topology'. Hence 'CW-complex' stands for 'closure-finite and weak topology.'

Remark 8.1.21 A CW-complex X is a cell complex X satisfying two additional conditions **CW(4)** and **CW(5)**.

Remark 8.1.22 A simplicial complex is a set of simplexes but not a topological space. On the other hand, a CW-complex is itself a topological space Thus a simplicial complex K is not a CW-complex but the polyhedron $|K|$ is a CW-complex.

Given a simplicial K, the spaces $|K|$ provide an extensive class of CW-complexes.

Proposition 8.1.23 *Let K be a simplicial complex. Then $|K|$ is a CW-complex.*

Proof $|K|$ is a subspace of some Euclidean space. Hence $|K|$ is a Hausdorff space. For each n-simplex s_n of K, let $\dot{s_n}$ denote the boundary of s_n and $\psi^n_{s_n} : (D^n, S^{n-1}) \to (s_n, \dot{s_n})$ be a relative homeomorphism. If A_n denotes the set of all n-simplexes of K, then the characteristic maps $\psi^n_{s_n}$ make $|K|$ into a CW-complex. This is because, since the properties **CW(3)** and **CW(5)** are obvious. The properties **CW(1)** and **CW(2)** and **CW(4)** follow from the properties of the simplicial complex K. ❑

Remark 8.1.24 Some authors prefer the following alternative definition of a CW-complex instead of Definition 8.1.17.

Definition 8.1.25 A CW-complex is an ordered triple (X, E, ψ), where X is a Hausdorff space, E is a family of cells in X, and $\psi = \{\psi_e : e \in E\}$ is a family of maps, such that

(i) $X = \cup\{e : e \in E\}$ (disjoint union);
(ii) for each k-cell $e \in E$, the map $\psi_e : (D^k, S^{k-1}) \to (e \cup X^{(k-1)}, X^{(k-1)})$ is a relative homeomorphism;
(iii) if $e \in E$, then its closure $\bar{e}$ is contained is a finite union of cells in E;
(iv) X has the weak topology determined by $\{\bar{e} : e \in E\}$.

If (X, E, ψ) is a CW-complex, then X is called a CW-space and or sometimes a CW-complex. The pair (E, ψ) is called a CW-decomposition of X, and $\psi_e \in \psi$ is called the characteristic map of e.

Remark 8.1.26 **(a)** One may consider a CW-complex space X as a generalized polyhedra and (E, ψ) as a generalized triangulation of X.
(b) 1. Axiom **(i)** indicates that the cells E partition X.
2. Axiom **(ii)** indicates that each k-cell e arises from attaching a k-cell to $X^{(k-1)}$ through the attaching map $\psi_e|_{S^{k-1}}$.
3. Axiom **(iii)** is called closure-finiteness.

Definition 8.1.27 Let (X, E, ψ) be a CW-complex. It is said to be finite if E is a finite set. A CW-complex X is regarded as a generalized polyhedron and a pair (E, ψ) as a generalized triangulation of X.

Definition 8.1.28 A pair (X, E) is a union of disjoint subspaces $e_\alpha (\alpha \in A)$ called cells satisfying the conditions **(i)** and **(ii)** of Definition 8.1.25.

Proposition 8.1.29 *Let X be a CW-complex, and let Y be any space. A function $f : X \to Y$ is continuous, iff each composite function $f \circ \psi^n_i : D^n \xrightarrow{\psi^n_i} X \xrightarrow{f} Y$ is continuous, for each $n \geq 0$ and $i \in A_n$.*

Proof Clearly, each $f \circ \psi_i^n$ is continuous if f is continuous. Conversely, let V be a closed subset of Y. Then each subset $(\psi_i^n)^{-1}(f^{-1}(V))$ is closed in D^n. Hence $f^{-1}(A)$ is closed in X by **CW(4)**. Consequently, f is continuous. ❑

Example 8.1.30 S^n is a CW-complex for $n \geq 0$. Consider S^n as a subspace of $\mathbf{R}^{n+1}$. For each $n \geq 1$ define

$$f : (D^n, S^{n-1}) \to (S^n, s_0),\ x \mapsto (2\sqrt{(1 - ||x||^2)}x, 2||x||^2 - 1),$$

where $s_0 = (0, 0, \ldots, 1) \in S^n$. Let e^i denote an i-cell. Then the map f shows that S^n is a CW-complex with $E = \{e^0, e^n\}$. Clearly, for $n = 0$, S^0 has a CW-decomposition with two 0-cells, which are $\{e_1^0, e_2^0\}$.

Example 8.1.31 S^n has a CW-decomposition with 2 i-cells in every dimension $0 \leq i \leq n$. Let E_+^n be the upper closed hemisphere and E_-^n be the lower closed hemisphere of S^n. Then $S^n = E_+^n \cup E_-^n$ and $E_+^n \cap E_-^n = S^{n-1}$(the equator). Hence there are two n-cells e_1^n and e_2^n with $\overline{e_1^n} = E_+^n$ and $\overline{e_2^n} = E_-^n$. This shows that S^n has a CW-decomposition with two i-cells in every dimension $0 \leq i \leq n$.

Example 8.1.32 The real number space $\mathbf{R}$ has the standard CW-structure with 0-skeletons the integers $\mathbf{Z}$ and as 1-cells the intervals $\{[n, n+1] : n \in \mathbf{Z}\}$.

Example 8.1.33 The space $\mathbf{R}^n$ has the standard CW-structure with cubical cells which are products of the 0-cells and 1-cells from $\mathbf{R}$.

Example 8.1.34 The torus T is a CW-complex. To show this consider T, formed from the square ABCD by identifying the edges AB, DC, and DA, CB as shown in Fig. 8.1.

Consider the maps f^0, f_1^1, f_2^1, f_1^2 defined as follows : $f^0 : D^0 \to T$ sends D^0 to the point where the four vertices A, B, C and D are identified; $f_1^1, f_2^1 : D^1 \to T$ are such that f_1^1 sends D^1 to DC and f_2^1 sends D^1 to DA, respectively, and $f_1^2 : D^2 \to T$ is defined by mapping D^2 homeomorphically onto the square ABCD and composing this map with the identification map onto T. Clearly, characteristic maps make T into a CW-complex with one 0-cell, two 1-cells, and one 2-cell.

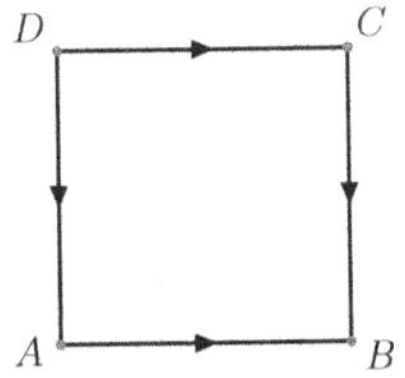

Fig. 8.1 Representation of the torus

8.1.3 *Examples of Spaces Which Are Neither CW-Complexes Nor Homotopy Equivalent to a CW-Complex*

This subsection now presents some examples of topological spaces which are neither CW-complexes nor homotopy equivalent to a CW-complex. As there exist spaces which are not Hausdorff, every space is not a CW-complex.

Example 8.1.35 **(i)** Let s_n $(n > 1)$ be an n-simplex. Then ∂s_n is regarded as a '0-dimensional cell complex': K^0, whose cells are the points of ∂s_n. This is closure finite but does not have the weak topology.

(ii) $s_n(n > 1)$ regarded as a complex $K^n = K^0 \cup e^n$, where $e^n = s_n - \partial s_n$ and $K^0 = \partial s_n$ as in (i). This has the weak topology, since $\overline{s}_n = K^n$, but is not closure finite.

(iii) Every cell complex is not a CW-complex. For example, consider a simplicial complex K, which has a metric topology but which is not locally finite (e.g., a complex covering coordinate axes in Hilbert space). The weak topology in such a complex K cannot be metricized, see [J.C Whitehead, 1939].

Remark 8.1.36 Every topological space is not homotopy equivalent to a CW-complex.

Example 8.1.37 Let X be a subspace of $\mathbf{R}^1$ consisting of points 0 and $1/n$ for all integers $n \geq 1$. The path components of X are just the single points (since each point $1/n$ is both open and closed). The topological space X is not homotopy equivalent to a CW-complex. If X were homotopy equivalent to a CW-complex Y, then Y would have an infinite number of path components. Suppose $f : X \to Y$ is a homotopy equivalence, then $f(X)$ is compact and hence $f(X)$ is contained in a finite subcomplex of Y. This shows that $f(X)$ is contained in the union of a finite number of path components, this contradicts the assumption that f is a homotopy equivalence. This implies that the topological space X is not homotopy equivalent to a CW-complex.

Example 8.1.38 An infinite-dimensional Hilbert space is not a CW complex. Since it is a Baire space, it cannot be expressed as a countable union of n-skeletons, each of which being a closed set with empty interior.

8.2 Cellular Spaces

This section conveys the concept of cellular spaces which reconciles the intuitive notion of a topological space built up by attaching cells with formal definition of a CW-complex. For this purpose the precise meaning of a 'space built up by attaching cells' is first given.

Definition 8.2.1 A cellular space is a topological space X, with a sequence of subspaces

$$X^0 \subset X^1 \subset X^2 \subset \cdots \subset X,$$

such that $X = \bigcup_{n=0} X^n$, with the following properties:

CS(1) X^0 is a discrete space.
CS(2) for each positive integer n, there is an indexing set A_n, and there exist continuous maps $\psi_i^n : S^{n-1} \to X^{n-1}$ for each $i \in A_n$. Moreover, X^n is the space obtained from X^{n-1} and (disjoint) copies D_i^n of D^n (one for each $i \in A_n$) by identifying the points x and $\psi_i^n(x)$ for each $x \in S_i^{n-1}$ and each $i \in A_n$.
CS(3) A subset Y of X is closed iff $Y \cap X^n$ is closed in X^n, for each $n \geq 0$.

Remark 8.2.2 If X is a finite dimensional CW complex, then **CS(3)** holds automatically, because all sets A_n are empty for sufficiently large n.

Example 8.2.3 Every CW-complex is a cellular space. Its converse is also true: every cellular space is a CW-complex (see Ex. 25. of Sect. 8.9)

8.3 Subcomplexes of CW-Complexes

We now introduce the concept of a subcomplex of a CW-complex which creates interest in many situations. We use the notation of Sect. 8.1.

Definition 8.3.1 Let X be a CW-complex and Y be a subspace of X. Then Y is said to be a subcomplex of X if, for each integer $n \geq 0$, there exists a subset C_n of A_n such that

(i) $Y = \cup \psi_i^n(e^n)$ for all $n \geq 0$ and $i \in C_n$;
(ii) $\psi_i^n(D^n) \subset Y$ for all $n \geq 0$ and $i \in C_n$.

If Y contains only a finite number of cells, Y is called a finite subcomplex.

Remark 8.3.2 **(i)** Arbitrary unions of subcomplexes of a CW-complex X is a subcomplex of X.
(ii) Arbitrary intersections of subcomplexes of a CW-complex X is also a subcomplex of X.

Example 8.3.3 Let X be a CW-complex. Its every n-skeleton $X^{(n)}$ is a subcomplex of X.

Proposition 8.3.4 *Let X be a CW-complex. For each integer $n \geq 0$ and $i \in A_n$, $\psi_i^n(D^n)$ is contained in a finite subcomplex of X.*

Proof As X is a CW-complex, it follows from condition **CW(5)** that $\psi_i^n(D^n)$ is contained in the union Y of a finite number of sets of the form $\psi_j^m(e^m)$. Since Y may not satisfy **CW(2)**, Y may not be a subcomplex of X. On the other hand, if $\psi_j^m(e^m)$ is a set of Y such that $\psi_j^m(e^m)$ is not contained in Y, then by using **CW(3)** and **CW(5)**, we can adjoin a finite number of sets $\psi_k^n(e^q)$ to Y with $q < m$ so as to include $\psi_j^m(S^{m-1})$. By going on working down in dimension, a finite number of sets $\psi_j^r(e^r)$ can be added to Y until Y becomes a finite subcomplex of X. ❑

8.4 Relative CW-Complexes

This section introduces the concept of relative CW-complexes which is a generalization of the notion of absolute CW-complexes.

Definition 8.4.1 A relative CW-complex (X, A) consists of a topological space X, a closed subspace A, and a sequence of closed subspaces $(X, A)^k$ for $k \geq 0$ such that

(i) $(X, A)^\circ$ is obtained from A by adjoining 0-cells;
(ii) For $k \geq 1$, $(X, A)^k$ is obtained from $(X, A)^{k-1}$ by adjoining k-cells;
(iii) $X = \cup(X, A)^k$;
(iv) X has the weak topology with respect $\{(X, A)^k\}_k$ (i.e., X has a topology coherent with $\{(X, A)^k\}_k$).

In this case, $(X, A)^k$ is called the k-skeleton of X relative to A. If $X = (X, A)^n$ for some n, we say that $\dim(X - A) \leq n$.

Remark 8.4.2 An absolute CW-complex X is a relative CW-complex $(X, \emptyset)$ and its k-skeleton is denoted by $X^{(k)}$.

Definition 8.4.3 A relative CW-structure on a pair (X, A) is a sequence

$$A = (X, A)^{-1} \subset (X, A)^0 \subset \cdots \subset (X, A)^n \subset (X, A)^{n-1} \subset \cdots \subset X$$

such that $(X, A)^n$ is obtained from $(X, A)^{n-1}$ by attaching n-cells, $n \geq 0$, $X = \bigcup\limits_{n \geq -1} (X, A)^{n-1}$ and X has the weak topology. We say $\dim(X, A) = n$ if $(X, A)^n = X$ and $(X, A)^{n-1} \neq X$.

Remark 8.4.4 If $A = \{x_0\}$ in (X, A), then X is a CW-complex. Conversely, if X is a CW-complex and $A \subset X$ is any subcomplex, then (X, A) a relative CW-complex. So one may replace the phrase 'relative CW-complex (X, A)' by the more restricted notion of a CW-complex pair (X, A), where X is a CW-complex and A is a subcomplex'.

Example 8.4.5 If (K, L) is a simplicial pair, then there is a CW-complex pair $(|K|, |L|)$, with $(|K|, |L|)^n = |K^{(n)} \cup L|$.

Example 8.4.6 I is a CW-complex, with $I^{(0)} = \{0, 1\} = \dot{I}$ and $I^{(n)} = I$ for some $n \geq 1$.

Definition 8.4.7 Let (X, A) and (Y, B) be relative CW-complexes and $f : (X, A) \to (Y, B)$ be a continuous map. Then f is said to be cellular if $f(\bar{X}^{(n)}) \subset \bar{Y}^{(n)}$, where $\bar{X} = X^{(n)} \cup A$ and $\bar{Y} = Y^{(n)} \cup B$ for every integer n.

8.5 Homotopy Properties of CW-Complexes, Whitehead Theorem and Cellular Approximation Theorem

This section conveys homotopy properties of CW-complexes, proves Whitehead theorem and cellular approximation theorem which play a key role in algebraic topology. Every continuous map between CW-complexes is homotopic to a cellular map and every two homotopic cellular maps are cellularly homotopic. This result leads to the CW-approximation theorem. On the other hand, Whitehead's theorem states that the continuous maps between CW-complexes that induce isomorphisms on all homotopy groups are actually homotopy equivalences.

8.5.1 Homotopy Properties of CW-Complexes

This subsection studies homotopy properties of CW-complexes. A CW-complex is a homotopy-theoretic generalization of the concept of a simplicial complex.

Theorem 8.5.1 *Let (X, A) and (Y, B) be relative CW-complexes and $f : (X, A) \to (Y, B)$ be continuous. Then $f \simeq g$ rel A, for some cellular map g.*

Proof Define homotopies $H^{(n)} : \bar{X}^{(n)} \times I \to Y$ inductively such that

(i) $H^{(n)}|_{X^{(n-1)} \times I} = H^{(n-1)}$;
(ii) $H^{(n)}(x, 0) = f(x)$;
(iii) $H^{(n)}(x, 1) \in \bar{Y}^{(n)}$;
(iv) $H^{(n)}(a, t) = f(a)$, $\forall a \in A$.

Then $\pi_n(Y, \bar{Y}^{(n)}, *) = 0$ by Ex.12 of Sect. 8.9 for any choice of $*$. Define $H : X \times I \to Y$ by $H|_{X^{(n)} \times I} = H^{(n)}$ and take $g(x) = H(x, 1)$. By (iii), $g(x) \in \bar{Y}^{(n)}$ and by (ii) and (iv), it follows that $H : f \simeq g$ rel A, where g is a cellular map. ❑

Corollary 8.5.2 *Let $f, g : (X, A) \to (Y, B)$ be cellular maps and $f \simeq g$ rel A. Then there is a cellular homotopy relative to A between them in the sense that there exists a continuous map $H : (X, A) \times I \to (Y, B)$ such that $H(\overline{X}^{(n)} \times I) \subset \overline{Y}^{(n+1)}$.*

Proof It is sufficient to show that there is a cellular homotopy $H : (X, A) \times I \to B$ such that $H(\overline{X}^{(n)} \times I) \subset \overline{Y}^{(n+1)}$. Clearly, $(X \times I, X \times \{0\} \cup A \times I \cup X \times 1)$ is a relative CW-complex with n-skeleton $X \times \{0\} \cup \overline{X}^{(n-1)} \times I \cup X \times 1$. Applying Theorem 8.5.1 to the given homotopy the required homotopy is obtained. ❑

Corollary 8.5.3 *Every continuous map between CW-complexes is homotopic to a cellular map and any two homotopic cellular maps between two CW-complexes are cellularly homotopic.*

Proof It follows from Theorem 8.5.1 and Corollary 8.5.2. ❑

Theorem 8.5.4 *Let (X, A) be a pair of spaces such that the inclusion map $i : A \hookrightarrow X$ is a weak homotopy equivalence. If K is a CW-complex, with a 0-cell as base point, then for any choice of base point in A, the induced map $i_* : [K, A] \to [K, X]$ is a bijection.*

Proof i_* **is onto**: Let $f : K \to X$ be a based continuous map. We show by induction on the skeletons of K that f can be deformed into A. The map f regarded as a map $K \times \{0\} \to X$ can be extended to a continuous map $f : K \times I \to X$ such that $f(K \times 1) \subset A$, and if L is a subcomplex of K, which is mapped by f into A, then $f(L \times I) \subset A$. Let L be such a subcomplex of K and $Y^{(n)} = K^{(n)} \cup L$. The map f is now extended as the constant homotopy to $(K \times \{0\}) \cup (L \times I)$. If x is any 0-cell of $K - L$, there exists a path $u : I \to X$ such that $u(0) = f(x)$ and $u(1) \in A$. Then we can continuously extend f to $Y^{(0)} \times I$ by setting $f(x, t) = u(t), 0 \leq t \leq 1$. We now start extension on n. We assume that f has been continuously extended to a map $f : (K \times \{0\}) \cup (Y^{(n-1)} \times I) \to X$, such that $f(Y^{(n-1)} \times 1) \subset A$. Then for each cell $\psi^n_\alpha(D^n)$ of $K - L$, the composite map

$$(D^n \times \{0\}) \cup (S^{n-1} \times I) \xrightarrow{\psi^n_\alpha \times 1_d} (K \times \{0\}) \cup (Y^{(n-1)} \times I) \xrightarrow{f} X,$$

sends $S^{n-1} \times 1$ to A. Define a homeomorphism

$$g : D^n \times I \to D^n \times I, \begin{cases} (x, 0) \mapsto (x/2, 0), \ \forall x \in D^n; \\ (x, t) \mapsto (\frac{1}{2}(1+t)x, 0), \ \forall x \in S^{n-1}, t \in I; \\ (x, 1) \mapsto \begin{cases} (x/||x||, 2 - 2||x||), & x \in D^n, ||x|| \geq \frac{1}{2} \\ (2x, 1), & x \in D^n : ||x|| \leq \frac{1}{2}. \end{cases} \end{cases}$$

Then, the map $f \circ (\psi^n_\alpha \times 1_d) \circ g^{-1} : (D^n, S^{n-1}) \to (X, A)$ represents an element of $\pi_n(X, A)$, with some base point. But by the exactness of the homotopy sequence, it follows that $\pi_n(X, A) = 0$. Consequently, $f \circ (\psi^n_\alpha \times 1_d) \circ g^{-1}$ can be continuously extended to a map of $D^n \times I$ that sends $D^n \times 1$ and $S^{n-1} \times I$ to A. Again applying the homeomorphism g, the map $f \circ (\psi^n_\alpha \times 1_d)$ can be extended to a continuous map of $D^n \times I$ that sends $D^n \times 1$ to A. This process defines a continuous extension $f : (K \times \{0\}) \cup (Y^{(n)} \times I) \to X$ such that $f(Y^{(n)} \times 1) \subset A$. Hence it gives a continuous extension $f : K \times I \to X$ such that $f(K \times 1) \subset A$. This shows that $i_* : [K, A] \to [K, X]$ is onto.

i_* **is injective**: Let $f, h : K \to A$ be based continuous maps such that $i \circ f \simeq i \circ h$ by a based homotopy $H : K \times I \to X$. Since $K \times I$ is a CW-complex and $(K \times \{0\}) \cup (k_0 \times I) \cup (K \times 1)$ is a subcomplex, H can be deformed to a map $F : K \times I \to A$ such that F coincides with H on $(K \times \{0\}) \cup (k_0 \times I) \cup (K \times 1)$.

Hence F is a based homotopy between f and h. This implies that i_* is injective. Consequently, i_* is a bijection. ❑

We now extend the Theorem 8.5.4 with the help of mapping cylinder when $f : A \to X$ be a weak homotopy equivalence.

Corollary 8.5.5 *Let $f : A \to X$ be a weak homotopy equivalence and K be a CW-complex. Then $f_* : [K, A] \to [K, X]$ is a bijection (where K has a 0-cell as base point, and A, X have any base points that correspond under f).*

Proof The map $f : A \to X$ may be considered as the composite $A \hookrightarrow M_f \xrightarrow{g} X$, where M_f is the mapping cylinder, $i : A \hookrightarrow M_f$ is an inclusion map and g is a homotopy equivalence by using Theorem 4.7.22 of Chap. 4. Since both f and g are weak homotopy equivalences, i is also so. Hence $i_* : [K, A] \to [K, M_f]$ is a bijection by Theorem 8.5.4. But as $g_* : [K, M_f] \to [K, X]$ is a bijection, f_* is also so and is such that $f_* = g_* \circ i_*$. ❑

Remark 8.5.6 This Corollary 8.5.5 leads to Whitehead theorem.

8.5.2 *Whitehead Theorem*

This subsection answers the question when the concepts of weak homotopy equivalence and homotopy equivalence coincide. Every homotopy equivalence is a weak homotopy equivalence. Is its converse true? Its answer is found in Whitehead theorem which asserts that a weak homotopy equivalence, for connected CW complexes, is a homotopy equivalence. So it has become necessary to introduce the concept of 'weak homotopy equivalence' at the beginning.

Definition 8.5.7 Let $f : (X, x_0) \to (Y, f(x_0))$ be a continuous map in the category of pointed topological spaces. Then f is called a weak homotopy equivalence if its induced map $f_* : \pi_0(X, x_0) \to \pi_0(Y, f(x_0))$ is a (1-1) correspondence and $f_* : \pi_m(X, x_0) \to \pi_m(Y, f(x_0))$ is an isomorphism for all $m \geq 1$ and all points $x_0 \in X$. The continuous map f is said to be an n-equivalence (for some $n \geq 1$) if $f_* : \pi_m(X, x_0) \to \pi_m(Y, f(x_0))$ is an isomorphism for $0 < m < n$ and an epimorphism for $m = n$ for all points $x_0 \in X$.

We are now in a position to present Whitehead Theorem, proved by J.H.C. Whitehead in his classical paper (Whitehead, Combinatorial Homotopy I, Bull Amer Math Soc **55**(1949), 213–245), where the concept of CW-complex was first defined.

Theorem 8.5.8 (Whitehead) *Let $f : K \to L$ be a weak homotopy equivalence of CW-complexes. Then f is a homotopy equivalence.*

Proof By Corollary 8.5.5, the induced function $f_* : [L, K] \to [L, L]$ is a bijection. Hence there exists a continuous map $g : L \to K$ such that $f \circ g \simeq 1_L$. Then, g is also a weak homotopy equivalence. Similarly, there exists a continuous map $f' : K \to L$ such that $g \circ f' \simeq 1_K$. Now,

$$f' \simeq (f \circ g) \circ f' = f \circ (g \circ f') \simeq f$$

shows that $g \circ f \simeq 1_K$. Hence f is a homotopy equivalence with g a homotopy inverse of f. ❑

There is another form of Whitehead theorem given below.

Theorem 8.5.9 (Alternative form of Whitehead theorem) *Let K and L be CW-complexes and $f : K \to L$ be a continuous map such that its induced homomorphisms*

$$f_* : \pi_n(K) \to \pi_n(L)$$

are isomorphism for all $n \geq 1$. Then f is a homotopy equivalence.

Proof Left as an exercise. ❑

Remark 8.5.10 **(i)** If X and Y are path-connected spaces, it is sufficient for f to be a weak homotopy equivalence that $f_* : \pi_m(X, x_0) \to \pi_m(Y, f(x_0))$ is an isomorphism for all $m \geq 1$ and for just one point $x_0 \in X$.

(ii) Every homotopy equivalence is a weak homotopy equivalence. Is its converse true? Whitehead theorem proves that the converse is also true if X and Y are both CW-complexes. Hence the concepts of weak homotopy equivalences and homotopy equivalences coincide for CW-complexes.

(iii) Whitehead theorem does not hold for general topological spaces or even for all subspaces of $\mathbf{R}^n$. For example, the Warsaw circle WS^1, which is a compact subset of Euclidean plane and obtained by closing up a topologist's sine curve with an arc, has all homotopy groups zero. On the other hand any continuous map $f : WS^1 \to A$ is not a homotopy equivalence, where A is a one-pointic space.

(iv) For possible generalizations of Whitehead theorem to more general topological spaces, one may study 'Shape theory.'

8.5.3 Cellular Approximation Theorem

This subsection proves a key theorem in algebraic topology: 'Cellular Approximation Theorem' which is an analogue for CW-complexes of the 'Simplicial Approximation Theorem 6.5.9 of Chap. 6 for simplicial complexes.

Theorem 8.5.11 (Cellular approximation theorem) *Let X and Y be CW-complexes, and $f : X \to Y$ be a continuous map such that $f|_A$ is cellular for some subcomplex A of X (possibly empty). Then there exists a cellular map $g : X \to Y$ such that $g|_A = f|_A$ and $g \simeq f$ rel A.*

Proof We use induction on the skeletons $X^{(n)}$ of X and can define a homotopy $H : X \times I \to Y$ that starts with f, ends with a cellular map, and is the constant homotopy on $A \times I$. Let x be a 0-cell of $X - A$. Then there is a path in Y from $f(x)$ to a point of $Y^{(0)}$. We can now define a map H on $X^{(0)} \times I \cup A \times I$. Suppose that H has been extended to $X^{(n-1)} \times I$, and $H(X^{(n-1)} \times 1) \subset Y^{(n-1)}$. Then H can be extended to each n-cell of $X - A$, since $\pi_n(Y, Y^{(n)}) = 0$. The result gives a continuous extension such that $H(X^{(n)} \times 1) \subset Y^{(n)}$. Then by inductive process, the required homotopy $H : X \times I \to Y$ is obtained. ❑

Remark 8.5.12 Whitehead theorem shows that despite every topological space is not a CW-complex, it is sufficient for many purposes to consider only CW-complexes instead of arbitrary topological spaces.

8.6 More on Homotopy Properties of CW-Complexes

This section proves an interesting property of CW-complexes which leads to the concept of Eilenberg–MacLane Spaces. Such spaces are discussed in details in Chap. 11. These spaces establish interlink between homotopy and cohomology theories (see Chaps. 15 and 17).

Theorem 8.6.1 *Let X be a CW-complex and $n \geq 0$ be an integer. Then there exists a CW-complex Y, having X as a subcomplex such that if $i : X \hookrightarrow Y$ is the inclusion, then*

(i) $i_* : \pi_m(X) \to \pi_m(Y)$ *are isomorphisms for* $0 < m < n$;
(i) $\pi_n(Y) = 0$.

Proof Let G be a set of generators for the group $\pi_n(X)$. For each $\alpha \in G$, take a based representative map $\psi_\alpha^n : S^n \to X$, which may be assumed to be cellular by Theorem 8.5.11. Let the space Y be obtained from X by attaching cells e_α^{n+1} by the maps ψ_α^n, one for each $\alpha \in G$. Then the space Y is a CW-complex and X is a cellular space. Hence Y is also so, since the maps ψ_α^n send S^n into $X^{(n)}$. Moreover, X is a subcomplex of Y and $i_* : \pi_m(X) = \pi_m(Y^{(n)} \cup X) \to \pi_m(Y)$ is an isomorphism for $0 < m < n$, and onto for $m = n$. But for each $\alpha \in G$, $i_*(\alpha) \in \pi_n(Y)$ is represented by the map $i \circ \psi_\alpha^n : S^n \to Y$ and this is homotopic to the constant map, since Y has an $(n + 1)$-cell attached by ψ_α^n. Consequently, $\pi_n(Y) = 0$. ❑

Remark 8.6.2 Given a CW-complex X and integer $n \geq 0$ the process described above can be iterated to "kill off" the higher homotopy groups $\pi_m(X)$ for all $m \geq n$.

8.7 Blakers–Massey Theorem and a Generalization of Freudenthal Suspension Theorem

This section presents Freudenthal suspension theorem obtained as a consequence of Blakers–Massey theorem (see Blakers and Massey 1952) and also a generalization of Freudenthal suspension theorem. One of the main problems of homotopy theory is to determine the homotopy groups $\pi_r(S^n)$ of spheres explicitly for $r \geq n$. Such search has discovered many techniques developing algebraic topology. Freudenthal invented in 1937 the concepts of the suspension ΣX of a pointed space X and the homotopy suspension map E. He defined a map E

$$E : \pi_r(X, x_0) \to \pi_{r+1}(\Sigma X, *), n \geq 0, [f] \mapsto [1_{S^1} \wedge f], \tag{8.1}$$

as a natural transformation from the functor π_r to the functor $\pi_{r+1} \circ \Sigma$ which is a homomorphism for all (X, x_0) and $n \geq 1$, called Freudenthal suspension homomorphism, where $S^{r+1} \approx S^1 \wedge S^r \xrightarrow{1_{S^1} \wedge f} S^1 \wedge X = \Sigma X$. In particular, for $X = S^r$ and $Y = S^n$, E gives group homomorphisms

$$E : \pi_r(S^r) \to \pi_{r+1}(S^{n+1}) \text{ for } r \geq 1, n \geq 1 \tag{8.2}$$

E is a bijection for $1 \leq r \leq 2n - 1$ and surjection for $r = 2n - 1$.

Remark 8.7.1 Blakers–Massey theorem is a fundamental result in algebraic topology. It is used to prove a general form of Freudenthal suspension theorem.

Remark 8.7.2 Given a triad $(X, A, *)$, let $X_2 = A \bigcup_{\alpha} e^n$, $X_1 = A \bigcup_{\beta} e^m$. If $X = X_1 \cup X_1$, such that $A = X_1 \cap X_2$ and $i : (X_1, A) \hookrightarrow (X, X_2)$ is the inclusion, then $X_1 - A \approx X - X_2 \approx \mathbf{R}^m$.

Theorem 8.7.3 (Blakers–Massey theorem) *Let X be a CW-complex and A be a subcomplex of X such that $X_2 = A \bigcup_{\alpha} e^n$, $X_1 = A \bigcup_{\beta} e^m$. If $X = X_1 \cup X_1$, $A = X_1 \cap X_2$ and $i : (X_1, A) \hookrightarrow (X, X_2)$ is the inclusion, then $i_* : \pi_r(X_1, A, *) \to \pi_r(X, X_2, *)$ is an isomorphism for $r < m + n - 2$ and is onto if $r = m + n - 2$.*

Proof See Blakers and Massey 1952 or Gray 1975. ❑

Definition 8.7.4 A pointed topological space (X, x_0) is said to be n-connected if $\pi_i(X, x_0) = 0$ for all $i \leq n$.

Remark 8.7.5 A 0-connected space means path-connected and 1-connected space means simply connected.

Theorem 8.7.6 (General Form of Freudenthal Suspension Theorem) *If X is an n-connected CW-complex $(n \geq 1)$, then the suspension homomorphism*

$$E : \pi_r(X, x_0) \to \pi_{r+1}(\Sigma X, *)$$

is an isomorphism for all $1 \leq r \leq 2n$, *and an epimorphism for* $r = 2n - 1$.

Proof It follows from Blakers–Massey theorem. ❑

Corollary 8.7.7 *For all* $n \geq 1$, $\pi_n(S^n)$ *is isomorphic to* $\mathbf{Z}$.

Proof As S^n is $(n-1)$-connected, the suspension map $\Sigma : \pi_r(S^n) \to \pi_{r+1}(S^{n+1})$ is an isomorphism for $r < 2n - 1$. For $n \geq 2$, $n + 1 < 2n$ and hence, in particular, $\Sigma : \pi_n(S^n) \to \pi_{n+1}(S^{n+1})$ is an isomorphism for $n \geq 2$. As $\pi_1(S^1)$ and $\pi_2(S^2)$ are both isomorphic to $\mathbf{Z}$, hence it follows that $\pi_n(S^n)$ is isomorphic to $\mathbf{Z}$ for all n. ❑

Remark 8.7.8 If we take $X = S^n$ in Theorem 8.7.6, then the Freudenthal suspension Theorem 7.10.3 of Chap. 7 follows. In this sense, the Theorem 8.7.6 is said to be a general form of Freudenthal suspension theorem.

8.8 Applications

This section presents some interesting applications by utilizing the main features of CW-complexes.

Theorem 8.8.1 $\pi_m(S^n) \cong \begin{cases} 0, & m < n \\ \mathbf{Z}, & m = n \end{cases}$

Proof Suppose $m < n$. Let $[f] \in \pi_m(S^n)$. By cellular approximation theorem we may assume that $f : S^m \to S^n$ is cellular. S^n is a CW-complex with a 0-cell e^0 and an n-cell e^n under the standard cellular structure. Hence its m-skeleton is just the base point e^0, because for $m < n$ there is only the 0-cell e^0. This shows that the map $f : S^m \to S^n$ is the trivial map. If $m = n$, then it follows from Freudenthal Suspension Theorem (Chap. 7) that for all n, $\pi_n(S^n) \cong \mathbf{Z}$. ❑

Remark 8.8.2 (Eilenberg–MacLane Space) It follows from Theorem 8.8.1 that there exists a CW-complex X such that $\pi_m(X) = 0$ for $m \neq n$, and $\pi_n(X) \cong \mathbf{Z}$. Such a CW-complex is called an Eilenberg–MacLane Space $K(\mathbf{Z}, n)$. Eilenberg–MacLane spaces are studied in the Chap. 11. The importance of such spaces is two-fold: the spaces develop homotopy theory as well as the cohomology theory of CW-complexes.

Theorem 8.8.3 *Let* X *be a* CW*-complex. Then*

(i) *every path component of* X *is a* CW*-complex, hence it is closed;*
(ii) *the path components of* X *are closed and open;*
(iii) *path components of* X *are the components of* X*;*
(iv) X *is connected iff* X *is path-connected.*

Proof **(i)** As X is a disjoint union of cells, each of which is path-connected, it follows that each path component Y of X is a union of cells. If e is an n-cell with $e \subset Y$, $\bar{e} = \psi_e(D^n)$ is also path- connected and hence $\bar{e} \subset Y$. Consequently, Y is a CW-complex and hence Y is closed.

(ii) Let Y be a path component of X and Z be the union of other path components. Since Z is a union of CW-subcomplexes, it is a CW-subcomplex and it is closed. Again since Z is the complement of Y in X, it follows that Y is open.

(iii) Let Y be a path component of X and Z be the complement of X containing Y. Since Y is closed and open, it is connected. Hence $Y = Z$.

(iv) follows from **(iii)**. ❑

Theorem 8.8.4 *Let (X, E) be a CW-complex.*

(i) *If $e \in E$ is an n-cell $(n > 0)$ with characteristic map ψ_e, then $\bar{e} = \operatorname{Im} \psi_e = \psi_e(D^n)$;*

(ii) *If E' is a finite subset of E, then $|E'|$ is a CW-subcomplex iff $|E'|$ is closed.*

(iii) *If $e \in E$, then its closure $\bar{e}$ is contained in a finite CW-subcomplex;*

(iv) *Every compact subset A of X lies in a finite CW-subcomplex and hence a CW-space X is compact iff (X, E) is a finite CW-complex for every CW-decomposition E;*

(v) *A subset A of X is closed iff $A \cap Y$ is closed in Y for every finite CW-subcomplex Y in X;*

(vi) *If E' (possibly infinite) is a subset of E, then $|E'|$ is a CW-subcomplex iff $|E'|$ is closed.*

Proof **(i)** Since ψ_e is continuous, it follows that

$$\psi_e(D^n) = \psi_e(\overline{D^n - S^{n-1}}) \subset \overline{\psi_e(D^n - S^{n-1})} = \bar{e} \implies \psi_e(D^n) \subset \bar{e}.$$

For the reverse inclusion, using the compactness of D^n, it follows that $\psi_e(D^n)$ is compact. Again as the space X is Hausdorff, $\psi_e(D^n)$ is a closed subset of X containing $e = \psi_e(D^n - S^{n-1})$ and hence $\bar{e} \subset \psi_e(D^n)$.

(ii) Let E' be a finite subset of E. Assume that $|E'|$ is closed and $e \in E'$. Then $e \subset |E'|$ and $\bar{e} \subset |E'|$. Consequently, $|E'|$ is a CW-subcomplex. Conversely, let $|E'|$ be a CW-subcomplex. Then $\bar{e} \subset |E'|$ for every $e \in E'$. Consequently, $|E'| = \cup\{e : e \in E'\} = \cup\{\bar{e} : e \in E'\}$ is a finite union of closed set, and hence $|E'|$ is closed.

(iii) Let dimension of e be n. Then for $n = 0$, the proof obvious. We now use induction on n. If $n > 0$, then **(i)** shows that $\bar{e} - e = \psi_e(D^n) - e \subset (e \cup X^{(n-1)}) - e \subset X^{(n-1)}$. Now by axiom **(iii)** of Definition 8.1.25, $\bar{e}$ meets only finitely many cells other than e; $e_1, e_2, \ldots, e_r$ (say). Then $\dim(e_i) \leq n - 1$ for all i. By induction, there is a finite CW-subcomplex X_i containing $\bar{e}_i$, for $i = 1, \ldots, r$ and each X_i is closed by (ii). But $\bar{e} \subset e \cup X_1 \cup \cdots \cup X_r$. Hence it is a union of finitely many cells and hence it is closed and so it is a finite CW-complex.

(iv) Suppose $A \cap e \neq \emptyset$ for every $e \in E$. Choose a point $a_e \in A \cap e$. Let Y be the set of all such points a_e. Then for each $e \in E$, by **(iii)** there exists a finite CW-complex X_e containing $\bar{e}$. Hence $Y \cap \bar{e} \subset Y \cap X_e$ is a finite set and hence is closed in $\bar{e}$. Again since X has the weak topology, Y is closed in X. Again the same reasoning implies that every subset of Y is closed in X and hence Y is discrete. Since Y is compact, as it is a closed subset of A. Consequently, Y is finite and hence A meets only finitely many $e \in E : e_1, \ldots, e_r$ (say). Then by using **(iii)**, there are finite CW-subcomplexes X_i with $\bar{e}_i \subset X_i$, for $i = 1, 2, \ldots, r$. Hence it follows that A is contained in the finite CW-subcomplex $\cup X_i$.

(v) Let A be closed in X. Then $A \cap Y$ is closed in Y. Conversely, for each $e \in E$, let X_e be a finite subcomplex containing $\bar{e}$. Then by hypothesis, $A \cap X_e$ is closed in X_e. Hence $A \cap \bar{e} = (A \cap X_e) \cap \bar{e}$ is closed in X_e and is also closed in the smaller set $\bar{e}$. Consequently, A is closed in X, since X has the weak topology determined by all $\bar{e}$.

(vi) Let $|E'|$ be closed and $e \in E'$. Then $\bar{e} \subset |E'|$ and $|E'|$ is a CW-subcomplex. Conversely, let $|E'| = Y$ be a CW-subcomplex. Then by **(v)** it is sufficient to prove that $Y \cap A$ is closed in A for every finite CW-subcomplex A of X. Clearly, $Y \cap A$ is a finite union of cells: $Y \cap A = e_1 \cup e_2 \cup \cdots \cup e_r$(say). As $Y \cap A$ is a CW-subcomplex, $\bar{e}_i \subset Y \cap Y$ for all i. Hence $Y \cap A = \bar{e}_1 \cup \bar{e}_2 \cup \cdots \cup \bar{e}_r$. This shows that $Y \cap A$ is closed in A (also in X). ❑

8.9 Exercises

1. Let X be a CW-complex. Show that

(i) the n-skeleton $X^n \subset X$ is closed for every n;
(ii) X is locally path-connected;
(iii) if X connected, then X is path-connected.
(iv) if A is a CW-subcomplex of X, then the inclusion map $i : A \subset X$ is a cofibration.

2. Let X be a CW-complex. Show that the following statements are equivalent:

(i) X is path-connected;
(ii) X is connected;
(iii) the 1-skeleton X' is connected;
(iv) the 1-skeleton X' is path-connected.

[Use the fact that every CW-complex is locally path-connected.]

3. Let X be a CW-complex and $f : S^n \to X^n \subset X$. Show that $X \bigcup_f e^{n+1}$ is a CW-complex.

4. Let X be a CW-complex and e_i^n be an n-cell. Show that $A = X^n - e_i^n$ is a skeleton of X and $X^n = A \bigcup_f e^n$ for some $f : S^{n-1} \to X^{n-1}$.

5. Let X be a CW-complex. Show that arbitrary unions and intersection of subcomplexes of X are again subcomplexes of X.

6. If X and Y are finite CW-complexes, show that the product space $X \times Y$ is also a CW-complex.
[Hint. Let $\{e_i\}$ and $\{e'_k\}$ be cellular decompositions of X and Y, respectively. Then the family $\{e_i \times e'_k\}$ is a cellular decomposition of $X \times Y$. Again, if f and g are characteristic maps of e_i and e'_k, respectively, then $f \times g$ is also a characteristic map $e_i \times e'_k$.]

7. If X is a CW-complex and A is a subcomplex of X, show that the quotient space X/A is also a CW-complex.

8. Let X be a CW-complex and $A \subset X$ be compact. Show that

(i) $A \subset X^{(n)}$ for some n;
(ii) $A \subset Y$ for a subcomplex of X, such that Y has only a finite number of cells.

9. Let X be a CW-complex. Prove that

(i) X is a T_1-space;
(ii) X is a normal space.

10. Let (X, A) be a relative CW-complex. Show that it has the absolute homotopy extension property.

11. Let (X, A) be a relative CW-complex and $\bar{X} = X^{(n)} \cup A$. Let $H : \bar{X}^{(n-1)} \times I \cup \bar{X}^{(n)} \times 0 \to Y$. If $\pi_n(Y, B, *) = 0$ for any choice of $*$ and $H(x, 1) \in B$. Then show that, H has a continuous extension $\tilde{\pi} : \bar{X}^{(n)} \times I \to Y$ such that $\tilde{\pi}(x, 1) \in B$.

12. Let (X, A) be a relative CW-complex with cells only in $X - A$ with dimensions $\geq n$. Show that $\pi_k(X, A, *) = 0$ for $k < n$.

13. Let X be an n-dimensional CW-complex, e_n be an n-cell in X and p be a point in e. Prove that $X - e$ is a strong deformation retract of $X - \{p\}$.
[Hint. For $n = 0$, it is trivial. For $n > 0$, show that there is a retraction $r : X - \{p\} \to X - e$ and a homotopy $H : (X - \{p\}) \times I \to X - \{p\}$ such that $H(x, 0) = x$, $H(x, 1 = r(x))$].

14. Let (X, E) be a CW-complex, for some fixed $n > 0$, let E' be a family of n-cells in E. Show that

(i) $X' = |E'| \cup X^{(n-1)}$ is closed in X;
(ii) every n-skeleton $X^{(n)}$ is closed in X for $n > 0$;
(iii) every n-cell e is open in $X^{(n)}$;
(iv) $X^{(n)} - X^{(n-1)}$ is an open subset of $X^{(n)}$;

15. Let X be a Hausdorff space. Show that there exists a CW-complex Y and a weak homotopy equivalence $f : Y \to X$, where Y is uniquely determined upto homotopy equivalence.

16. (Serre) Let $p : X \to B$ be a Serre fibration (i.e., p has the homotopy lifting property with respect to every cube $I^n, n \geq 0$, where I^0 is a singleton by definition). Show that it has the homotopy lifting property with respect to every CW-complex K.

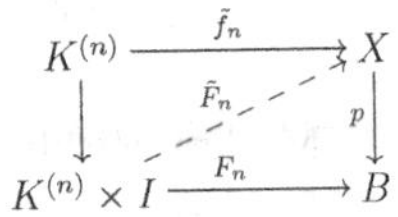

Fig. 8.2 HLP with respect to CW-complex K

[Hint. As every CW-complex has the weak topology determined by its skeletons, it is sufficient to show that there exists a map $\tilde{F}_n$, for every $n \geq 0$, making the diagram in Fig. 8.2 commutative, where $\tilde{f} : K \to X$ and $F : K \times I \to B$ are given and $\tilde{f}_n$ and F_n are appropriate restrictions. Then use induction on n.]

17. A space X is called compactly generated if X is a Hausdorff space and it has the weak topology determined by its compact subsets (see Appendix B). Show that every CW-complex is compactly generated.

18. If (X, E) is a CW-complex, show that $X^{(0)}$ is a discrete closed subset of X.

19. Show that Klein bottle has a decomposition of the form $\{e^0, e_1^1, e_2^1, e^2\}$, i.e., with one 0-cell, two 1-cells, and one 2-cell.

20. Define the dimension of a CW-complex (X, E) to be

$$\dim X = \sup\{\dim(e) : e \in E\}.$$

Show that $\dim X$ is independent of CW-decomposition of X.

21. Show that a CW-complex is connected iff its 1-skeleton $X^{(1)}$ is connected.

22. Let A be a subcomplex of the CW-complex X. If $\pi_n(X, A) = 0$ for all n, show that A is a string deformation retract of X.

23. If X has the homotopy type of a CW-complex and is a Lindelöf space, show that X has the homotopy type of a countable CW-complex.

24. Show that any continuous map $f : K \to L$ between CW-complexes is homotopic to a cellular map and any two cellular maps g_1, g_2 that are homotopic are also homotopic by a cellular homotopy.

25. Show that every CW-complex is a cellular space and every cellular space is a CW-complex.

26. Let X and Y be CW complexes. Show that the function space $F(X, Y)$ of all continuous maps from X to Y (with the compact-open topology) is not a CW complex in general, but if X is locally finite, then $F(X, Y)$ is homotopy equivalent to a CW complex.
[Hint: Milnor, 1959.]

8.10 Additional Reading

1. Adams, J.F., *Algebraic Topology: A student's Guide*, Cambridge University Press, Cambridge, 1972.
2. Aguilar, Gitler, S., Prieto, C., *Algebraic Topology from a Homotopical View Point*, Springer-Verlag, New York, 2002.

3. Arkowitz, Martin, *Introduction to Homotopy Theory*, Springer, New York, 2011.
4. Dieudonné, J., *A History of Algebraic and Differential Topology*, 1900–1960, Modern Birkhäuser, 1989.
5. Dold,A., *Lectures on Algebraic Topology*, Springer-Verlag, New York, 1972.
6. Dugundji, J., *Topology*, Allyn & Bacon, Newtown, MA, 1966.
7. Dyer, E., *Cohomology Theories*, Benjamin, New York, 1969.
8. Eilenberg, S., and Steenrod, N., *Foundations of Algebraic Topology*, Princeton University Press, Princeton, 1952.
9. Hilton, P.J., *An introduction to Homotopy Theory*, Cambridge University Press, Cambridge, 1983.
10. Hilton, P. J. and Wyle, S. *Homology Theory*, Cambridge University Press, Cambridge, 1960.
11. Husemöller, D., *Fibre Bundles*, Springer, New York, 1975.
12. Hu, S.T., *Homotopy Theory*, Academic Press, New York, 1959.
13. Hu, S.T., *Homology Theory*, Holden Day, Oakland CH, 1966.
14. Mayer, J. *Algebraic Topology*, Prentice-Hall, New Jersy, 1972.
15. Massey, W.S., *A Basic Course in Algebraic Topology*, Springer-Verlag, New York, Berlin, Heidelberg, 1991.
16. Milnor, J., On Spaces having the Homotopy Type a CW-complex, Trans. Amer. Maths. Soc. 90 (1959), 272–280.
17. Munkres, J.R., *Elements of Algebraic Topology*, Addition-Wesley-Publishing Company, 1984.
18. Singer, I.M., and Thrope, J.A., *Lecture Notes on Elementary Topology and Geometry*, Springer-Verlag, New York, 1967.
19. Steenrod. N., *The Topology of Fibre Bundles*, Prentice University Press, Prentice, 1955.
20. Steenrod, N., *A Convenient Category of Topological Spaces*, Mich. Math J 14 (1967), 133–152.
21. Whitehead, J.H,C. *A certain exact sequence,* Ann. of Math. 52 (1950), 51–110.

References

Adams, J.F.: Algebraic Topology: A Student's Guide. Cambridge University Press, Cambridge (1972)
Aguilar, M., Gitler, S., Prieto, C.: Algebraic Topology from a Homotopical View Point. Springer, New York (2002)
Arkowitz, Martin: Introduction to Homotopy Theory. Springer, New York (2011)
Blakers, A.L., Massey, W.S.: The homotopy groups of a triad II. Ann. Math. **55**(2), 192–201 (1952)
Dieudonné, J.: A History of Algebraic and Differential Topology, 1900–1960. Modern Birkhäuser, Boston (1989)
Dold, A.: Lectures on Algebraic Topology. Springer, New York (1972)
Dugundji, J.: Topology. Allyn & Bacon, Newtown (1966)
Dyer, E.: Cohomology Theories. Benjamin, New York (1969)
Eilenberg, S., Steenrod, N.: Foundations of Algebraic Topology. Princeton University Press, Princeton (1952)

Gray, B.: Homotopy Theory. An Introduction to Algebraic Topology. Academic Press, New York (1975)
Hatcher, A.: Algebraic Topology. Cambridge University Press, Cambridge (2002)
Hilton, P.J.: An Introduction to Homotopy Theory. Cambridge University Press, Cambridge (1983)
Hilton, P.J., Wyue, S.: Homology Theory. Cambridge University Press, Cambridge (1960)
Hu, S.T.: Homotopy Theory. Academic Press, New York (1959)
Hu, S.T.: Homology Theory. Holden Day, Oakland CH (1966)
Mayer, J.: Algebraic Topology. Prentice-Hall, New Jersy (1972)
Massey, W.S.: A Basic Course in Algebraic Topology. Springer, New York (1991)
Maunder, C.R.F.: Algebraic Topology. Van Nostrand Reinhhold, London (1970)
Milnor, J.: On spaces having the homotopy type a CW-compex. Trans. Am. Math. Soc. **90**, 272–280 (1959)
Munkres, J.R.: Elements of Algebraic Topology. Addition-Wesley-Publishing Company, Menlo Park (1984)
Rotman, J.J.: An Introduction to Algebraic Topology. Springer, New York (1988)
Singer, I.M., Thrope, J.A.: Lecture Notes on Elementary Topology and Geometry. Springer, New York (1967)
Spanier, E.: Algebraic Topology. McGraw-Hill, New York (1966)
Steenrod. N.: The Topology of Fibre Bundles. Prentice University Press, Prentice (1955)
Steenrod, N.: A convenient category of topological spaces. Mich. Math. J. **14**, 133–152 (1967)
Switzer, R.M.: Algebraic Topology-Homotopy and Homology. Springer, Berlin (1975)
Whitehead, G.W.: Elements of Homotopy Theory. Springer, New York (1978)
Whitehead, J.H.C.: Combinatorial homotopy. I. Bull. Am. Math. Soc. **55**, 213–245 (1949a)
Whitehead, J.H.C.: Combinatorial homotopy. II. Bull. Am. Math. Soc. **55**, 453–496 (1949b)
Whitehead, J.H.C.: A certain exact sequence. Ann. Math. **52**(2), 51–110 (1950)

Chapter 9
Products in Homotopy Theory

This chapter continues to study homotopy theory through different products defined between homotopy groups such as the Whitehead product introduced by J.H.C. Whitehead in 1941, the Samelson product introduced by H. Samelson in 1953 and the mixed product introduced by McCarty in 1964. Moreover, this chapter finds a generalization of Whitehead product and a relation between Whitehead and Samelson products. These products are used to solve several problems in algebraic topology. Computing the homotopy groups of even simple spaces is one of the basic problems in homotopy theory. The problem of computing the homotopy groups of n-sphere is not completely solved. In most cases, it is not known whether the homotopy groups are trivial or not. Different products are used to solve such problems. For example, Whitehead product provides methods for computing nonzero elements of homotopy groups of spheres. Throughout this chapter we consider topological spaces with base points. The base point is denoted by $*$ (unless otherwise stated), and not often explicitly mentioned.

For this chapter the books and papers Gray (1975), Hatcher (2002), Hu (1959), James (1971), Maunder (1980), Spanier (1966), Whitehead (1941, 1944) and some others are referred in Bibliography.

9.1 Whitehead Product Between Homotopy Groups of CW-Complexes

This section studies Whitehead product defined by J.H.C. Whitehead (1904–1960) in 1941 (Whitehead 1941) between two homotopy groups to study homotopy groups of pointed CW-complexes X. This product associates with each pair of elements $\alpha \in \pi_p(X, x_0)$ and $\beta \in \pi_q(X, x_0)$ an element denoted $[\alpha, \beta] \in \pi_{p+q-1}(X)$, called Whitehead product in his honor. This product provides a technique at least in some

M.R. Adhikari, *Basic Algebraic Topology and its Applications*,
DOI 10.1007/978-81-322-2843-1_9

cases for constructing nonzero elements of $\pi_{p+q-1}(X)$. He also defined generalized products involving the rotation groups. The concept of Whitehead product is utilized in this chapter as well as in Chap. 17 to solve several problems in algebraic topology.

Definition 9.1.1 Let X be a CW-complex and $\alpha \in \pi_{p+1}(X),\ \beta \in \pi_{q+1}(X)$. Then α and β can be represented by the maps

$$f : (E_1, \dot{E}_1) \to (X, *)$$

$$g : (E_2, \dot{E}_2) \to (X, *)$$

where E_1 and E_2 are oriented cells of dimensions $p+1$ and $q+1$, respectively, with boundaries $\dot{E}_1$ and $\dot{E}_2$. Hence $E_1 \times E_2$ is a cell, oriented by the product of the given orientations of E_1 and E_2; the base point of $E_1 \times E_2$ is the point $(*, *)$. Its boundary $S = (E_1 \times E_2)^{\bullet} = E_1 \times \dot{E}_2 \cup \dot{E}_1 \times E_2$ is an oriented $(p+q+1)$- sphere and the map $h : (S, *) \to (X, *)$ defined by

$$h(x, y) = \begin{cases} f(x), & \text{if } x \in E_1 \text{ and } y \in \dot{E}_2, \\ g(y), & \text{if } x \in \dot{E}_1 \text{ and } y \in E_2. \end{cases}$$

represents an element $[\alpha, \beta] \in \pi_{p+q+1}(X)$, called Whitehead product of α and β. This product depends only on the homotopy classes α, β of f, g, respectively.

Remark 9.1.2 The binary operation

$$\pi_{p+1}(X, x_0) \times \pi_{q+1}(X, x_0) \to \pi_{p+q+1}(X, x_0),\ \ (\alpha, \beta) \mapsto [\alpha, \beta]$$

is natural in the sense that if

$$\psi : X \to Y \text{ is continuous, and } \alpha \in \pi_{p+1}(X, x_0),\ \ \beta \in \pi_{q+1}(X, x_0), \text{ then}$$

$$\psi_*([\alpha, \beta]) = [\psi_*(\alpha), \psi_*(\beta)] \in \pi_{p+q+1}(Y).$$

If in particulars, $p = 0 = q$, then $[\alpha, \beta] = \alpha\beta\alpha^{-1}\beta^{-1} \in \pi_1(X, x_0)$ is the commutator of α and β (see Proposition 9.1.12). This justifies the notation '[,]' of the Whitehead product.

Remark 9.1.3 α and β can also be represented by maps

$$f : (S^{p+1}, *) \to (X, x_0)$$

$$g : (S^{q+1}, *) \to (X, x_0)$$

where S^{p+1} and S^{q+1} are oriented $(p+1)$-sphere and $(q+1)$-sphere, respectively.

Definition 9.1.4 A continuous map $h : S^p \times S^q \to X$ is said to have type (α, β) if $h|_{S^p \times \{*\}}$ represents $\alpha \in \pi_p(X)$ and $h|\{*\} \times S^q$ represents $\beta \in \pi_q(X, x_0)$.

We can characterize maps of types (α, β) with the help of Whitehead products.

Theorem 9.1.5 *Let $\alpha \in \pi_p(X, x_0)$, $\beta \in \pi_q(X, x_0)$. Then there exists a continuous map $S^p \times S^q \to X$ of type (α, β) if and only if the Whitehead product $[\alpha, \beta] = 0$*

Proof Let $f : (S^p, *) \to (X, *)$ and $g : (S^q, *) \to (X, *)$ be representatives of α and β, respectively. Clearly, $(S^p \times S^q, S^p \vee S^q)$ is a relative CW-complex with just one cell; a characteristic map for this cell is

$$\omega_{p,q} = \omega_p \times \omega_q : (E^p \times E^q, (E^p \times E^q)^\bullet) \to (S^p \times S^q, S^p \vee S^q),$$

the attaching map for this cell is a representative of the Whitehead product $[i_1, i_2]$ of the homotopy classes of the inclusion maps $i_1 : S^p \to S^p \vee S^q$, and $i_2 : S^q \to S^p \vee S^q$.

Let $k = (f, g) : S^p \vee S^q \to X$ be the map determined by f and g. Then there exists a continuous map $S^p \times S^q \to X$ of type (α, β) if and only if the map k can be extended over $S^p \times S^q$. Since $\omega_{p,q}$ is a relative homeomorphism, it is so iff the map $\psi = k \circ \omega_{p,q}|(E^p \times E^q)^\bullet$ can be extended over $E^p \times E^q$, i.e., iff ψ is nullhomotopic. But the homotopy class of ψ is

$$k_*[i_1, i_2] = [k \circ i_1, k \circ i_2] = [\alpha, \beta].$$

This proves the theorem. ❑

Corollary 9.1.6 *Let X be an H-space with continuous multiplication μ. Then $[\alpha, \beta] = 0$ for any $\alpha \in \pi_p(X, *)$ and $\beta \in \pi_q(X, *)$.*

Proof Let f, g be any representatives of α, β, respectively. Consider the map

$$h = \mu \circ (f \times g) = (S^p \times S^q, (*, *)) \to (X, *).$$

Then h has type (α, β) for all representative f, g of α, β, respectively. Consequently the corollary follows from Theorem 9.1.5.

Theorem 9.1.7 *Let $\alpha \in \pi_m(X, x_0)$, $\beta \in \pi_n(X, x_0)$ be such that their Whitehead product $[\alpha, \beta] = 0$. If $\xi \in \pi_p(S^m)$, $\eta \in \pi_q(S^n)$, then the Whitehead product $[\alpha \circ \xi, \beta \circ \eta] = 0$.*

Proof Since $[\alpha, \beta] = 0$, by Theorem 9.1.5, there is a map $h : (S^m \times S^n, (*, *)) \to (X, *)$ of type (α, β).

Let

$$f : (S^p, *) \to (S^m, *)$$

$$g : (S^q, *) \to (S^n, *)$$

be the representatives of ξ and η, respectively. Then the map $h \circ (f \times g) : (S^p \times S^q, (*, *)) \to (X, *)$ defined by $(h \circ (f \times g))(x, y) = h(f(x), g(y))$ is of type $(\alpha \circ \xi, \beta \circ \eta)$. Hence it follows by Theorem 9.1.5 that the Whitehead product $[\alpha \circ \xi, \beta \circ \eta] = 0$.

Definition 9.1.8 A bunch of spheres is a topological space which is homeomorphic to a union of spheres with a single point in common.

Theorem 9.1.9 *If X has the trivial Whitehead product and $\Sigma\Omega X$ has the homotopy type of a bunch of spheres, then ΩX is homotopy commutative.*

Proof Let Y denote the appropriate union of spheres and $g : Y \to \Sigma\Omega X$ be a homotopy equivalence. Let $h : Y \vee Y \to X$ be the map $d \circ g$ on each factor, where $d : \Sigma\Omega X \to X$ is defined by $d(t, \lambda) = \lambda(t), \ \forall\, \lambda \in \Omega X$. Since all Whitehead products vanish in X, we can extend $h|S^p \vee S^q$ to $S^p \times S^q$ for each pair of spheres S^p, S^q contained in Y. Hence we can extend h to $Y \times Y$. Using a homotopy inverse to g, we obtain the required map $f : \Sigma\Omega X \times \Sigma\Omega X \to X$.

An equivalent definition of Whitehead product is given in Definition 9.1.10 which is sometimes convenient for use.

Definition 9.1.10 Let X be a pointed topological space with base point $x_0 \in X$, $m \geq 1$ and $n \geq 1$ be given integers. Let $\alpha \in \pi_m(X, x_0)$ and $\beta \in \pi_n(X, x_0)$ be two given elements. We construct an element $[\alpha, \beta] \in \pi_{m+n-1}(X, x_0)$, called Whitehead product of α and β as follows:

Let α and β be represented by the continuous maps $f : (I^m, \partial I^m) \to (X, x_0)$ and $g : (I^n, \partial I^n) \to (X, x_0)$ respectively. Clearly, $I^{m+n} \approx I^m \times I^n, \partial I^{m+n} = (I^{m+n})^\bullet = (I^m \times \partial I^n) \cup (\partial I^m \times I^n)$ and the intersection of these two sets is $\partial I^m \times \partial I^n$.

Define a function

$$h : \partial I^{m+n} \to X, (s, t) \mapsto \begin{cases} f(s), & s \in I^m, t \in \partial I^n \\ g(t), & t \in I^n, s \in \partial I^m \end{cases} \tag{9.1}$$

Then h is well defined and continuous, since $f(s) = g(t) = x_0 \,\forall\, (s, t) \in \partial I^m \times \partial I^n$.

As the point $\mathbf{0} = (0, 0, \ldots, 0)$ of ∂I^{m+n} is in $\partial I^m \times \partial I^n$, it follows that $h(\mathbf{0}) = x_0$. Again since $\partial I^{m+n} \approx S^{m+n-1}$, h represents an element $\gamma \in \pi_{m+n-1}(X, x_0)$, which is the homotopy class of h. As γ depends only on the elements α and β define $[\alpha, \beta]$ by setting $[\alpha, \beta] = \gamma$.

Remark 9.1.11 $[\alpha, \beta] = \gamma$ depends only on the elements α, β. Hence $[\alpha, \beta]$ is well defined.

Proposition 9.1.12 *If $\alpha \in \pi_1(X, x_0)$ and $\beta \in \pi_1(X, x_0)$, then $[\alpha, \beta] = \alpha\beta\alpha^{-1}\beta^{-1}$ is the commutator of $\pi_1(X, x_0)$.*

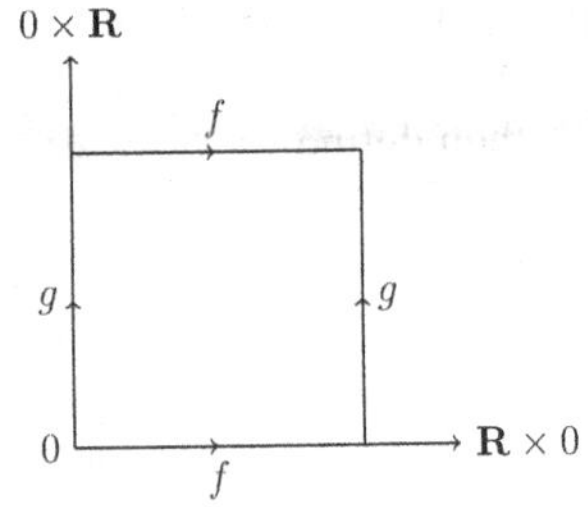

Fig. 9.1 Boundary of a square

Proof Suppose $m = 1 = n$. Then $(I \times I)^\bullet$ is the boundary of the unit square $I \times I$ in the plane $\mathbf{R}^2$, with clockwise orientation, and with the origin as base point as shown in Fig. 9.1.

The maps f, g are loops representing α, β, respectively. Hence it follows that $[\alpha, \beta] = \alpha\beta\alpha^{-1}\beta^{-1} \in \pi_1(X, x_0)$. ❑

Remark 9.1.13 The notation $[\alpha, \beta]$ for the Whitehead product is consistent with our standard notation for the commutator of two elements in a group.

Remark 9.1.14 Given base point preserving maps $f : S^m \to X$ and $g : S^n \to X$, let $[f, g] : S^{m+n-1} \to X$ be the composite

$$S^{m+n-1} \to S^m \vee S^n \xrightarrow{f \vee g} X,$$

where the first map is the attaching map of the $(m+n)$-cell of $S^m \times S^n$ with its usual CW-structure. Since homotopies of f or g give rise to homotopies $[f, g]$, we have a well-defined product $\pi_m(X, x_0) \times \pi_n(X, x_0) \to \pi_{m+n-1}(X, x_0)$. The notation $[f, g]$ is used since for $m = n = 1$, this is just the commutator product in $\pi_1(X, x_0)$. Clearly, for $m = 1$ and $n > 1$, $[f, g]$ is the difference between g and its image under the π_1-action of f.

9.2 Whitehead Products Between Homotopy Groups of H-Spaces

This section studies Whitehead products between the homotopy groups of H-spaces (Hopf's spaces) and topological groups. Let X be a given H-group and $x_0 \in X$ be a homotopy unit of X. Then the group operation in $\pi_n(X, x_0)$ is closely related to the multiplication in X.

Definition 9.2.1 Let X be a given H-group and $x_0 \in X$ be a homotopy unit of X. For $n > 0$, let $\alpha, \beta \in \pi_n(X, x_0)$ be represented by the maps $f, g : (I^n, \partial I^n) \to (X, x_0)$, respectively. Define $h : (I^n, \partial I^n) \to (X, x_0), t \mapsto f(t) \cdot g(t), t \in I^n$ (the right hand product is the usual multiplication in X). Then h represents the element $\alpha + \beta \in \pi_n(X, x_0)$. In particular, if X is a topological group and x_0 is the neutral element of X, then define

$$k : (I^n, \partial I^n) \to (X, x_0), t \mapsto f(t)\cdot[g(t)]^{-1}, t \in I^n,$$

where right hand multiplication is the usual multiplication in the topological group X. Then k represents the element $\alpha - \beta \in \pi_n(X, x_0)$.

Theorem 9.2.2 *Let X be a topological group. Then for every pair of elements $\alpha \in \pi_m(X, x_0)$ and $\beta \in \pi_n(X, x_0)$, their Whitehead product $[\alpha, \beta] = 0$.*

Proof Let X be topological group and $\alpha \in \pi_m(X, x_0)$ and $\beta \in \pi_n(X, x_0)$ be represented by $f : (I^m, \partial I^m) \to (X, x_0)$, and $g : (I^n, \partial I^n) \to (X, x_0)$, respectively. Define

$$h : I^{m+n} \to X, (s, t) \mapsto f(s)\cdot g(t), s \in I^m, t \in I^n,$$

where the right hand product is the usual multiplication in X. Then $h|_{\partial I^{m+n}}$ represents the Whitehead product $[\alpha, \beta]$. This shows that $[\alpha, \beta] = 0$. ❑

Theorem 9.2.3 *Let X be an H-space with multiplication μ. Then for every pair of elements $\alpha \in \pi_m(X, x_0)$ and $\beta \in \pi_n(X, x_0)$, their Whitehead product $[\alpha, \beta] = 0$.*

Proof ‘Let

$$f : (S^m, *) \to (X, x_0)$$

$$g : (S^n, x_0) \to (S^n, x_0)$$

be the representatives of α and β, respectively. Consider the map

$$h = \mu \circ (f \times g) : (S^m \times S^n, (*, *)) \to (X, x_0).$$

Then h is of type (α, β) for any representative f, g of α, β, respectively. This concludes by Theorem 9.1.5 that the Whitehead product $[\alpha, \beta] = 0$ ❑

An important characterization of a continuous map $h : S^m \times S^n \to X$ to be of type (α, β) with the help of Whitehead product.

Theorem 9.2.4 *Let X be an H-group. If $\alpha \in \pi_m(X, x_0)$, $\beta \in \pi_n(X, x_0)$, then there exists a map $S^m \times S^n \to X$ of type (α, β) iff $[\alpha, \beta] = 0$.*

Proof $(S^m \times S^n, S^m \vee S^n)$ is a relative CW-complex with just one cell. A characteristic map for this cell is $\psi_{m,n} = \psi_m \times \psi_n : (D^m \times D^n, (D^m \times D^n)^\bullet) \to (S^m \times S^n, S^m \vee S^n)$; the attaching map for the cell is a representative of the Whitehead product $[i_1, i_2]$ of the homotopy classes of the inclusion maps

$$S^m \to S^m \vee S^n, S^n \to S^m \vee S^n.$$

Let α, β be represented by f, g and $k = (f, g) : S^m \vee S^n \to X$ be the map determined by them. Then there is a continuous map $S^m \times S^n$ into X of type (α, β) iff k can be extended over $S^m \times S^n$. Since $\psi_{m,n}$ is a relative homeomorphism, it is true iff the map $k \circ \psi_{m,n}|_{(D^n \times D^n)^\bullet}$ can be extended over $D^m \times D^n$, i.e., it is nullhomotopic. But the homotopy class of the latter map is $k_*[i_1, i_2] = [k_* \circ i_1, k_* \circ i_2] = [\alpha, \beta]$. This proves the theorem. ❑

Corollary 9.2.5 *If X is an H-group, then $[\alpha, \beta] = 0$ for every $\alpha \in \pi_m(X, x_0)$, $\beta \in \pi_n(X, x_0)$.*

Proof Let X be an H-group with multiplication μ and α, β be represented by f and g respectively. Then $\mu \circ (f \times g) : S^m \times S^n \to X$ has type (α, β). Then by Theorem 9.2.4, it follows that $[\alpha, \beta] = 0$. ❑

9.3 A Generalization of Whitehead Product

This section studies the generalized Whitehead product which is obtained by a generalization of Whitehead product. The set $[\Sigma X, Y]$ has the group structure under the product of two continuous maps: for $f, g : \Sigma X \to Y$ their product denoted by $f.g : \Sigma X \to Y$, is defined by

$$(f \cdot g)(x, t) = \begin{cases} f(x, 2t), & 0 \leq t \leq 1/2, \\ g(x, 2t - 1), & 1/2 \leq t \leq 1, \end{cases} \tag{9.2}$$

where $x \in X$ and $t \in I$.

The inverse of a map $f : \Sigma X \to Y$ denoted by $f^{-1} : \Sigma X \to Y$ and defined by

$$f^{-1}(x, t) = f(x, 1 - t), \text{ where } x \in X \text{ and } t \in I. \tag{9.3}$$

Let A and B be polyhedra, X be any pointed topological space and $\alpha \in [\Sigma A, X]$ be represented by $f : \Sigma A \to X$ and $\beta \in [\Sigma B, X]$ be represented by $g : \Sigma B \to X$. If $p_1 : A \times B \to A$ and $p_2 : A \times B \to B$ are the respective projection maps, define

$$f' = f \circ \Sigma p_1 : \Sigma(A \times B) \to X$$

$$g' = g \circ \Sigma p_2 : \Sigma(A \times B) \to X$$

and the commutator $k' = (f'^{-1}.g'^{-1}).(f'.g') : \Sigma(A \times B) \to X$,
where the products and inverses come from the suspension structure of $\Sigma(A \times B)$ given by (9.2) and (9.3). Then $k'|_{\Sigma(A \vee B)} \simeq 0$, since $k'|_{\Sigma(A \times *)} \simeq 0$ and $k'|_{\Sigma(* \times B)} \simeq 0$. Using the homotopy extension property for the polyhedral pair $(\Sigma(A \times B), \Sigma A \vee B)$, there is a map $k : \Sigma(A \times B) \to X$ such that $k \simeq k'$ and $k|\Sigma(A \vee B) \simeq 0$. Hence k induces a map $\tilde{k} : \Sigma(A \wedge B) = \Sigma(A \times B)/\Sigma(A \vee B) \to X$ with the property $k =$

$\tilde{k} \circ \Sigma q$, where $q : A \times B \to A \wedge B$ is the projection. As $[\tilde{k}]$ does not depend on the choice of its representatives, it follows that $[\tilde{k}]$ is well defined.

Definition 9.3.1 The generalized Whitehead product of $\alpha = [f] \in [\Sigma A, X]$ and $\beta = [g] \in [\Sigma B, X]$ denoted by $[\alpha, \beta]_{GW}$ is defined by setting $[\alpha, \beta]_{GW} = [\tilde{k}] \in [\Sigma(A \wedge B), X]$

Remark 9.3.2 The classical definition of Whitehead product is obtained from the the generalized Whitehead product when $A = S^p$ and $B = S^q$. In this sense the above $[\alpha, \beta]_{GW}$ is called generalized Whitehead product.

Proposition 9.3.3 *If X is an H-group, then $[\alpha, \beta]_{GW} = 0, \quad \forall \alpha \in [\Sigma A, X]$ and $\beta \in [\Sigma B, X]$.*

Proof If X is an H-group, then the group $[\Sigma(A \times B), X]$ is abelian and hence the commutator map $k' = [f, g]$ is nullhomotopic. If $q : A \times B \to A \wedge B$ is the projection, then it follows that $\Sigma\tilde{k} \circ \Sigma^2 q \simeq 0$. Hence $k \circ \Sigma\tilde{k} \circ \Sigma q \simeq 0 : \Sigma(A \times B) \to \Omega\Sigma X$. Consequently, $k \circ \Sigma\tilde{k} \simeq 0$ and hence $\Sigma\tilde{k} \simeq 0$. This implies that $[\alpha, \beta]_{GW} = 0$. ❑

9.4 Mixed Products in Homotopy Groups

This section studies mixed products introduced by McCarty in 1964 (McCarty 1964) associated with pointed topological spaces and fibrations. It is basically a part of theory of Hopf construction.

9.4.1 Mixed Product in the Homotopy Category of Pointed Topological Spaces

This subsection defines mixed product in the homotopy category of pointed topological spaces.

Definition 9.4.1 Let X and Y be pointed topological spaces with base point denoted by $*$ and $h : X \to Y$ be a base point preserving continuous map. Let A be a pointed space and $m : A \times X \to Y$ be a continuous map such that

$$\left.\begin{array}{ll} m(a, *) = *, & \forall a \in A \\ m(*, x) = h(x), & \forall x \in X \end{array}\right\} \tag{9.4}$$

Then given $p, q \geq 1$, a product of $\pi_p(A)$ with $\pi_q(X)$ to $\pi_{p+q}(Y)$ is defined as follows:

$$\begin{array}{ccc} \pi_p(A) \times \pi_q(X) & \longrightarrow & \pi_{p+q}(Y) \\ \downarrow {\scriptstyle 1_d \times \Sigma_*} & & \downarrow {\scriptstyle \Sigma_*} \\ \pi_p(A) \times \pi_{q+1}(\Sigma X) & \longrightarrow & \pi_{p+q+1}(Y) \end{array}$$

Fig. 9.2 Anticommutativity of the mixed product

Let $f : S^p \to A, g : S^q \to X$ represent $\alpha \in \pi_q(A), \beta \in \pi_q(X)$, respectively. Consider $m \circ (f \times g) : S^p \times S^q \to Y$ satisfying (9.4) with h replaced by $h \circ g : S^q \to Y$. Then $m \circ (f \times g)$ agrees with $g \circ \rho$ on the subspace

$$S^p \vee S^q = S^p \times * \cup * \times S^q \subset S^p \times S^q,$$

where $\rho : S^p \times S^q \to S^q$ denotes the right projections. Thus the separation element $d(\mu \circ (f \times g), g \circ \rho) \in \pi_{p+q}(Y)$ is defined in the natural way. The element $d(\mu \circ (f \times g), g \circ \rho)$ denoted by $\langle \alpha, \beta \rangle_m$ is referred as the mixed product associated with the given map m.

Remark 9.4.2 It is convenient for formal reasons to define

$$\langle \beta, \alpha \rangle_m = (-1)^{pq+1} \langle \alpha, \beta \rangle_m.$$

Each element of A determines via m, a map of X into Y. If we suspend this map and reverse the process we obtain a map $k : A \times \Sigma X \to \Sigma Y$ satisfying (9.4) with h replaced by $\Sigma h : \Sigma X \to \Sigma Y$. Then $\Sigma_* \langle \alpha, \beta \rangle = -\langle \alpha, \Sigma_* \beta \rangle_m$, where Σ_* denotes the Freudenthal suspension. In otherwords, the diagram in Fig. 9.2 is anticommutative, where the upper row is given by the mixed product associated with m and the lower row by the mixed product associated with k.

Proof See G.W. Whitehead (1944). ❑

9.4.2 *Mixed Product Associated with Fibrations*

This subsection defines mixed product corresponding to a fiber space and a topological transformation group acting on it. F be a fiber with base point e, and let H be a topological transformation group acting on F. We denote the transformation of $x \in F$ under $g \in H$ by $g \cdot x$. Suppose that

$$g \cdot e = e, \ \forall g \in H \tag{9.5}$$

Let $\alpha \in \pi_p(H)$ and $\beta \in \pi_q(F)$. Their mixed product $\langle \alpha, \beta \rangle_m \in \pi_{p+q}(F)$ is defined as follows:

Take representatives $u : S^p \to H$, $v : S^q \to F$ of α, β, respectively, and let $h, k : S^p \times S^q \to F$ be the maps given by
$h(\xi, \eta) = u(\xi).v(\eta)$, where . represents the action of H on G.
$k(\xi, \eta) = v(\eta)$, where $\xi \in S^p$, $\eta \in S^q$. Since h and k agree on $S^p \vee S^q$ by (9.5), their separation element $d(h, k) \in \pi_{p+q}(F)$ is defined. We denote $d(h, k) \in \pi_{p+q}(F)$ by $\langle \alpha, \beta \rangle_m$, this is well defined because it is independent of the choice of representatives and is called the mixed product of α, β associated with fiber F.

9.5 Samelson Products

This section studies Samelson product, generalized samelson product and iterated Samelson product given by Hans Samelson (1916–2005).

9.5.1 *The Samelson Product*

This subsection presents Samelson product associated with Hopf groups.

Definition 9.5.1 Let X be an H-group or a topological group with multiplication μ and two-sided homotopy inversion ϕ. Define continuous maps $k : X \to X, x \mapsto x^{-1}$ and $\psi : X \times X \to X, (x, y) \mapsto (x, y)(x^{-1}y^{-1})$. Let ψ' be a map homotopic to ψ and restricts to the trivial group on $X \vee X$ and thus factors through $X \wedge X$. For the based spaces P and Q, define the generalized Samelson product

$$\langle -, - \rangle : [P, X] \otimes [Q, X] \to [P \wedge Q, X]$$

by $\langle \alpha, \beta \rangle = [\psi \circ f \wedge g]$, where α and β are represented by f and g, respectively. In particular, for $P = S^p$ and $Q = S^q$,$P \wedge Q$ can be identified with S^{p+q}, and the Samelson product becomes a map

$$\langle \alpha, \beta \rangle : \pi_p(X) \otimes \pi_q(X) \to \pi_{p+q}(X)$$

Remark 9.5.2 The map $S^p \times S^q \to S^q \times S^p$ which interchanges the coordinates induces a map from S^{p+q} into itself which has degree $(-1)^{p+q}$.

Theorem 9.5.3 *The map $(\alpha, \beta) \mapsto \langle \alpha, \beta \rangle$ is bilinear and so defines a pairing $\pi_p \otimes \pi_q \to \pi_{p+q}$. Moreover, $\langle \beta, \alpha \rangle = (-1)^{pq+1} \langle \alpha, \beta \rangle$.*

Proof Left as an exercise.

Remark 9.5.4 One important application of Samelson product is that it can be considered as an obstruction to homotopy commutativity.

9.5.2 *The Iterated Samleson Product*

This subsection presents the Samelson product and the generalized Samelson product the associated with topological groups. Let G be a topological group. Corresponding to each element $\alpha \in \pi_1(G)$, there exists an operator

$$\alpha_G : \pi_n(G) \to \pi_{n+1}(G), n = 1, 2, \dots$$

defined by taking the Samelson product with α. From the Jacobi identity, each of these operations constitutes a derivation with respect to the Samelson product in $\pi_*(G)$. For a Lie group G, $\pi_2(G) = 0$ and then these derivations form an anticommuting set of operations, and in particular $2\alpha_G^2 = 0$. Clearly, $\alpha_G = 0$ if α can be represented by a loop within the center of G.

Definition 9.5.5 Let $\mathbf{R}_n$ represent the group of rotations of Euclidean n-space, where $n = 1, 2, \dots$. Let $D : \pi_r(\mathbf{R}_n) \to \pi_{r+1}(\mathbf{R}_n)$ be the operator defined by taking Samelson product with the generator $\alpha \in \pi_1(\mathbf{R}_n)$. In general the operator $D^t : \pi_r(\mathbf{R}_n) \to \pi_{r+t}(\mathbf{R}_n)$ is defined in a similar way for $t \geq 1$.

Remark 9.5.6 For $n = 2$, D is trivial.

Theorem 9.5.7 *For $n > 2$ and $n \equiv 2 \bmod 4$, the operator $D^2 : \pi_r(\mathbf{R}_n) \to \pi_{r+2}(\mathbf{R}_n)$ is trivial.*

Proof See James (1971). ❑

Similar result holds for all values of n.

Theorem 9.5.8 *The operator $D^6 : \pi_r(\mathbf{R}_n) \to \pi_{r+6}(\mathbf{R}_n)$ is trivial.*

Proof See James (1971). ❑

Remark 9.5.9 The operator $D^4 : \pi_r(\mathbf{R}_n) \to \pi_{r+4}(\mathbf{R}_n)$ is trivial for $n = 3$ or 4 (see James (1971)).

Remark 9.5.10 If the topological group is homotopy commutative, then the Samelson product $\langle \alpha, \beta \rangle$ is trivial. Hence $\langle \alpha, \beta \rangle$ can be thought of an obstruction to homotopy commutativity. Samelson used this criterion to show that the unitary group $U(2.\mathbf{C})$ in two variables is not homotopy abelian. One of the results of Samelson asserts that if α is a generator of $\pi_3(U(2, \mathbf{C}))$, then $\langle \alpha, \beta \rangle$ is nonzero. The paper (James and Thomas 1959) asserts that among the classical compact groups G only the truely commutative ones are homotopy commutative. This method is utilized again to find elements α of $\pi_n(G)$ such that $\langle \alpha, \beta \rangle$ is nonzero.

9.6 Some Relations Between Whitehead and Samelson Products

This section gives certain relations between Samelson and Whitehead product in homotopy groups. We first compare the Whitehead product in homotopy groups of a pointed topological space X with the Samelson product in homotopy groups of ΩX. By using the adjointness relation, we have

$$\pi_{p+1}(X) = [S^{p+1}, X] = [\Sigma S^p, X] \cong [S^p, \Omega X] = \pi_p(\Omega X).$$

We utilize this isomorphism $\rho = \rho_p : \pi_{p+1}(X) \to \pi_p(\Omega X)$ to define ρ explicitly.

Proposition 9.6.1 *Let $f : (I^{p+1}, \partial I^{p+1}) \to (X, *)$ represent $\alpha \in \pi_{p+1}(X, *)$. Then the map*

$$\rho(f) : (I^p, \partial I^p) \to (\Omega X, *), (x_1, x_2, \ldots, x_p)(t) \mapsto f(t, x_1, \ldots, x_p)$$

represents $\rho(\alpha) \in \pi_p(\Omega X)$.

Proof It follows from the above discussion by using the relative homeomorphism $\gamma_p : (I^p, \partial I^p) \to (S^p, *)$.

Theorem 9.6.2 *If $\alpha \in \pi_{p+1}(X), \beta \in \pi_{q+1}(X)$, then $\rho[\alpha, \beta] = (-1)^p \langle \rho(\alpha), \rho(\beta) \rangle \in \pi_{p+q}(\Omega X)$.*

Proof See Whitehead (1978, pp. 476–478). ❑

Corollary 9.6.3 *If $\alpha_1, \alpha_2 \in \pi_{p+1}(X), \beta \in \pi_{q+1}(X)$ and $p > 0$, then*

$$[\alpha_1 + \alpha_2, \beta] = [\alpha_1, \beta] + [\alpha_2, \beta], \ [\beta, \alpha_1 + \alpha_2] = [\beta, \alpha_1] + [\beta, \alpha_2].$$

Corollary 9.6.4 *Given positive integers p, q, r, let $\alpha \in \pi_{p+1}(X), \beta \in \pi_{q+1}(X), \gamma \in \pi_{r+1}(X)$. Then*

$$(-1)^{r(p+1)}[\alpha, [\beta, \gamma]] + (-1)^{p(q+1)}[\beta, [\gamma, \alpha]] + (-1)^{q(r+1)}[\gamma, [\alpha, \beta]] = 0.$$

Theorem 9.6.5 *If $\alpha \in \pi_p(X), \beta \in \pi_q(X), \gamma \in \pi_m(S^p), \delta \in \pi_n(S^q)$, and if $[\alpha, \beta] = 0$, then $[\alpha \circ \gamma, \beta \circ \delta] = 0$.*

Proof By hypothesis, $[\alpha, \beta] = 0$. Hence there exists a continuous map $f : S^p \wedge S^q \to X$ of type (α, β). Let $g : S^m \to S^p, h : S^n \to S^q$ be representatives of γ and δ, respectively. Then $f \circ (g \times h) : S^m \times S^n \to X$ is a continuous map of the type $(\alpha \circ \gamma, \beta \circ \delta)$. Hence $[\alpha \circ \gamma, \beta \circ \delta] = 0$. ❑

Let G be a topological group with a subgroup H and quotient space $G/H = Y$(say). Consider the standard action of H on Y given by I.M. James in 1971. This is a pointed action and hence every H-bundle with fiber Y gives a canonical cross section. The classes of H-bundles over S^n correspond to an element $\alpha \in \pi_{n-1}(H)$. Consider the bundle E with fiber Y and base S^n which corresponds to an element $\alpha \in \pi_{n-1}(H)$. Let $i_* : \pi_*(Y) \to \pi_*(E)$ be the homomorphism induced by the inclusion $i : Y \hookrightarrow E$. Let $\xi \in \pi_n(E)$ denote the class of the canonical cross section.

Theorem 9.6.6 *Under the above notations, for any element $\beta \in \pi_q(Y)$, the relation*

$$i_*\langle\alpha, \beta\rangle = [\xi, i_*\beta] \text{ holds,} \tag{9.6}$$

where the brackets on the left denote the relative Samelson product corresponding to the standard action of H on G/H and those on the right denote the Whitehead product in $\pi_*(E)$.

Proof See James (1971). ❑

Remark 9.6.7 The relation (9.6) gives an interesting relation between the relative Samelson product and Whitehead product, since the existence of a cross section implies i_* is injective.

9.7 Applications

This section presents some interesting applications of different products in homotopy theory.

9.7.1 *Adams Theorem Using Whitehead Product*

This subsection conveys the most important application of Whitehead product appearing in the study of Hopf invariant (see Chap. 17). For example, if τ_{2n} is a generator of the group $\pi_{2n}(S^{2n})$ and $[\tau_{2n}, \tau_{2n}] \in \pi_{4n-1}(S^{2n})$ is the Whitehead product. then the its Hopf invariant $H([\tau_{2n}, \tau_{2n}]) = 2$, which shows the existence of nonzero Hopf invariant. There is a natural question: does there exist an element in $\pi_{4n-1}(S^{2n})$ with the Hopf invariant 1? This problem has several reformulations. One of them is: for what values of n the real vector space $\mathbf{R}^{n+1}$ admit a structure of real division algebra with a unit. Frank Adams solved this problem in 1960. Section 17.5 of Chap. 17 (The Hopf Invariant and Adams Theorem) of this book is referred. Again, if τ_n is a generator of the group $\pi_n(S^n)$ represented by the identity map $1_d : S^n \to S^n$, Adams proved in 1960 in his paper Adams (1960) that $[\tau_n, \tau_n] = 0$ only if $n = 1, 3, 7$, so that the n-sphere S^n is an H-space only for these values of n.

Remark 9.7.1 The importance of Whitehead product can be realized by the results given in Ex.16, Ex.18 and Ex.19 of Sect. 9.8.

9.7.2 *Homotopical Nilpotence of the Seven Sphere S^7*

This subsection studies homotopical nilpotence of S^7. Let $[X, Y]$ denote the set of all homotopy classes of base point preserving continuous maps from X to Y. We will not distinguish notationally between a map and its homotopy class. The multiplication and inversion in the unit Cayley numbers induce the standard multiplication $\mu : S^7 \times S^7 \to S^7$ and two-sided inverse $\phi : S^7 \to S^7$. Then $\mu \in [S^7 \times S^7, S^7]$ and $\phi \in [S^7, S^7]$. For the H-space (S^7, m, ϕ) define a commutator map $\psi : S^7 \times S^7 \to S^7, (x, y) \mapsto (xy)(x^{-1}y^{-1})$ using the multiplication μ and inversion ϕ.

Recall that Cayley multiplication is not associative but is disassociative in the sense that any two elements generate an associative subalgebra. Define inductively the n-fold commutator map $m : (S^7)^n \to S^7$ by $m_n = m \circ (m_{n-1} \times 1_d)$, where $m_1 = 1_d$, the identity map on S^7. Then m_n induces a unique homotopy class $k_n : \wedge^n S^7 \to S^7$ with $k_n \circ q_n = m_n$, where $\wedge^n S^7$ is the n-fold smash product of $S^7 (\approx S^{7n})$ and $q_n : (S^7)^n \to \wedge^n S^7$ is the projection map. The homotopical nilpotence of the H-space (S^7, μ, ϕ) denoted by nil (S^7, μ, ϕ) is the least integer n such that m_{n+1} (and hence k_{n+1}) is nullhomotopic.

Theorem 9.7.2 nil $(S^7, \mu, \phi) = 3$.

Proof See Gilbert (1972). ❑

9.8 Exercises

In this exercise, let $[\alpha, \beta]$ and $\langle \alpha, \beta \rangle$ denote Whitehead product and Samelson product of α and β respectively. Prove the following:

1. If $m > 1$, $\alpha \in \pi_m(X, x_0)$ and $\beta \in \pi_1(X, x_0)$, then $[\alpha, \beta]$ is the element $\beta\alpha - \alpha \in \pi_m(X, x_0)$.
2. If $m > 1$, then the assignment $\alpha \mapsto [\alpha, \beta]$ for a given $\beta \in \pi_n(X, x_0)$ defines a homomorphism $\beta_* : \pi_m(X, x_0) \to \pi_{m+n-1}(X, x_0)$.
3. If $m + n > 2$, given $\alpha \in \pi_m(X, x_0)$ and $\beta \in \pi_n(X, x_0)$, $[\beta, \alpha] = (-1)^{mn}[\alpha, \beta]$.
4. Let $\alpha \in \pi_m(X, x_0)$, $\beta \in \pi_n(X, x_0)$ and $\gamma \in \pi_p(X, x_0)$. Then $(-1)^{mp}[[\alpha, \beta], \gamma] + (-1)^{nm}[[\beta, \gamma], \alpha] + (-1)^{pn}[[\gamma, \alpha], \beta] = 0$.
   mra
5. Let X be an arbitrary homotopy abelian H- space and α be an arbitrary element in $\pi_3(X)$. Show that $\langle \alpha, \alpha \rangle = 0$ iff $2(\alpha \wedge \alpha)^*(\beta) = 0$ in $\pi_6(X)$ for any $\beta \in [X \wedge X.X]$.
6. Let τ_{2n} be a generator of the group $\pi_{2n}(S^{2n})$. Show that
   (i) the group $\pi_{4n-1}(S^{2n})$ is infinite for any $n \geq 1$;
   (ii) $[\tau_{2n}, \tau_{2n}] \in \pi_{4n-1}(S^{2n})$ is in the kernel of the suspension homomorphism
   $$\Sigma : \pi_{4n-1}(S^{2n}) \to \pi_{4n}(S^{2n+1}).$$

7. Let X be a topological group and $\alpha \in \pi_p(X)$ and $\beta \in \pi_q(X)$. Show that

(i) for a fixed α, the map $\beta \mapsto [\alpha, \beta]$ is a homomorphism of groups and hence for $n \geq 2$,

$$[\alpha, \beta_1 + \beta_2] = [\alpha, \beta_1] + [\alpha, \beta_2].$$

(ii) for $m + n \geq 3$, $[\beta, \alpha] = (-1)^{mn}[\alpha, \beta]$ and for $m + np \geq 4$, there is a Jacobi identity,

$$(-1)^{mp}[[\alpha, \beta], \gamma] + (-1)^{nm}[[\beta, \gamma], \alpha] + (-1)^{pn}[[\gamma, \alpha], \beta] = 0.$$

(iii) for any continuous map between two topological groups

$$\psi : (X, x_0) \to (Y, y_0), \psi_*([\alpha, \beta]) = [\psi_*(\alpha), \psi_*(\beta)].$$

8. For $n \geq 2$, show that the operator $D^4 : \pi_r(S^n) \to \pi_{r+4}(S^n)$ given in Definition 9.5.5 is trivial.

9. Let X be an H-space and A, B be polyhedra. Given $\alpha \in [\Sigma A, X]$ and $\beta \in [\Sigma \beta, X]$, show that $[\alpha, \beta] = 0$ iff there is a map $\mu : \Sigma A \times \Sigma B \to X$ such that $[\mu|_{\Sigma A}] = \alpha$ and $[\mu|_{\Sigma B}] = \beta$.

10. Show that all generalized Whitehead products vanish in a pointed space X iff $[\Sigma P, X]$ (equivalently, $[P, \Omega X]$) is abelian for all polyhedra P.

11. Let A be a polyhedron and τ denote the class of identity map of ΣA. If the generalized Whitehead product $[\tau, \tau] = 0$, show that ΣA is an H-space.

12. If X is an H-space, show that all Whitehead products are trivial on X.

13. If $q \leq 3n - 3$, homomoporphism

$$E : \pi_q(S^n) \to \pi_{q+1}(S^{n+1})$$

is generated by all Whitehead products $[\alpha, \beta]$ with $\alpha \in \pi_r(S^n)$, $\beta \in \pi_r(S^n)$, $r + s = q + 1$.
mra

14. (a) Let $\mu : S^3 \times S^3 \to S^3$ be a homotopy associative multiplication of the 3-sphere S^3 and $\beta \in \pi_3(S^3) \cong \mathbf{Z}$ be a generator. Show that Samelson (commutator) product $\langle \beta, \beta \rangle_\mu \in \pi_6(S^3) \cong \mathbf{Z}_{12}$.

(b) Let X be a finite CW-complex which is also an H-space with $\pi_3(X) \cong \mathbf{Z}$ and $\beta \in \pi_3(X)$ be a generator. Show that

(i) if m is any homotopy associative multiplication on X, then the Samelson product
$\langle \beta, \beta \rangle_m$ generates $\pi_6(X)$.

(ii) given a multiplication m on S^3, there exists a multiplication μ on the H-space X such that the generator $\beta : (S^3, m) \to (X, \mu)$ is an H-map.

(iii) given any multiplication μ on X, there exists a multiplication m on S^3 such that the generator $\beta : (S^3, m) \to (X, \mu)$ is an H-map.

[Hint: See Arkowitz and Curjel (1969) for **(a)** and Stephen (1978) for **(b)**.]

15. Examine the validity of following statements:

(i) Let $\mu : S^3 \times S^3 \to S^3$ be any given multiplication and X be a pointed space. Then there exists a multiplication $m : X \times X \to X$ such that $\beta : (S^3, \mu) \to (X, m)$ is an H-map;

(ii) Let $m : X \times X \to X$ be any given multiplication and $\mu : S^3 \times S^3 \to S^3$ be a multiplication such that such that $\beta : (S^3, \mu) \to (X, m)$ is an H-map.

16. Show that the n-sphere S^n ia an H-space iff the Whitehead product $[\tau_n, \tau_n] = 0$, where $\tau_n \in \pi_n(S^n)$ is represented by the identity map.

17. Using Samelson product prove that among the classical compact groups G, the only commutative ones are those which are homotopy abelian. Hence find elements $\alpha \in \pi_n(G)$ such that the Samelson product $\langle \alpha, \alpha \rangle \neq 0$ (James and Thomas 1959).

18. Let $X = S^{2n} \times S^{2n} / \sim$ be the quotient space obtained by identifying the points (x, x_0) and (x_0, x), where x_0 is the base point of S^{2n}. Show that the space $X = S^{2n} \times S^{2n} / \sim$ is homeomorphic to the space $S^{2n} \bigcup_f D^{4n}$, where f is the map defining the Whitehead $[\tau_{2n}, \tau_{2n}]$, where τ_{2n} is a generator of the group $\pi_{2n}(S^{2n})$.

19. Let $\tau_n \in \pi_n(S^n)$, $\alpha_k \in \pi_k(S^k)$ be the generators given by the identity maps $1_d : S^n \to S^n$ and $1_d : S^k \to S^k$ respectively. Show that the

(i) Whitehead product $[\tau_n, \alpha_k] \in \pi_{n+k-1}(S^n \vee S^k)$ has infinite order;

(ii) group $\pi_{n+k-1}(S^n \vee S^k)$ is infinite.

9.9 Additional Reading

[1] Adams, J.F., *Algebraic Topology: A student's Guide*, Cambridge University Press, Cambridge, 1972.

[2] Arkowitz, M., *Whitehead products as images of Pontrjagin products*, Trans. Amer. Math. Soc. **158**, 453–463, 1971.

[3] Blakers, A.L., and Massy, W.S., *Products in homotopy theory*, Ann. of Math. **58**, 295–324, 1953.

[4] Dieudonné, J., *A History of Algebraic and Differential Topology*, 1900-1960, Modern Birkhäuser, 1989.

[5] Hilton, P.J., *Homotopy Theory and Duality*, Nelson, London, 1965.

[6] Hilton, P.J., *Note on the Jacobi identity for Whitehead products*, Proc. Cambridge Philos. Soc., **57**, 180–182, 1961.

[7] James, I.M., and Thomas E., *Which Lie Groups are homotopy abelian*?, Proc. Nat. Acad. Sc (USA) **45**(5), 131–140, 1959.

[8] Massey, W.S., and Uehara,H., *The Jacobi identity for Whitehead products*, Princeton Univ. Press, 1957.

[9] May, J.P., *A Concise Course in Algebraic Topology*, University of Chicago Press, 1999.

[10] McCarty, G.S.,*Products between homotopy groups and J-morphism*, Quart.J. of Math. Oxford **15**(2), 362–370, 1964

[11] Miyazaki, H., *On realizations of some Whitehead products*, Tohoku Math. J., **12**, 1–30, 1960.

[12] Nakaoka, M., and Toda, H., *On Jacobi identity for Whitehead products*, J. Inst. Polytech, Osaka City Univ. Ser.A.5, 1–13, 1954.

[13] Samelson H., *A connection between the Whitehead product and Pontragin product*, Amer. J. Math **75**, 744–752,1953.

[14] Toda, H., *Generalized Whitehead products and homotopy groups of spheres*, J. Inst. Polytech. Osaka City Univ. Ser. A,**3**, 43-48, 1953.

[15] Whitehead, G.W., *Elements of Homotopy Theory*, Springer-Verlag, New York, Heidelberg, Berlin, 1978.

References

Adams, J.F.: On the non-existence of elements of Hopf invariant one. Ann. Math. **72**, 20–104 (1960)

Adams, J.F.: Algebraic Topology: A student's Guide. Cambridge University Press, Cambridge (1972)

Arkowitz, M.: Whitehead products as images of Pontrjagin products. Trans. Am. Math. Soc. **158**, 453–463 (1971)

Arkowitz, M., Curjel, C.R.: Some properties of the exotic multiplications on the three-sphere. Q. J. Math. Oxf. Ser. **20**(2), 171–176 (1969)

Blakers, A.L., Massy, W.S.: Products in homotopy theory. Ann. Math. **58**, 295–324 (1953)

Dieudonné, J.: A History of Algebraic and Differential Topology, pp. 1900–1960. Modern Birkhäuser, Boston (1989)

Gilbert, W.J.: Homotopical nilpotence of the seven sphere. Proc. Am. Math. Soc. **32**, 621–622 (1972)

Gray, B.: Homotopy Theory. An Introduction to Algebraic Topology. Acamedic Press, New York (1975)

Hatcher, A.: Algebraic Topology. Cambridge University Press, Cambridge (2002)

Hilton, P.J.: Homotopy Theory and Duality. Nelson, London (1965)

Hilton, P.J., Whitehead, J.H.C.: Notes on the Whitehead product. Ann. Math. **58**(2), 429–442 (1953)

Hilton, P.J.: Note on the Jacobi identity for Whitehead products. Proc. Camb. Philos. Soc. **57**, 180–182 (1961)

Hu, S.T.: Homotopy Theory. Academic Press, New York (1959)

James, I.M.: Products between homotopy groups. Compositio Mathematica **23**, 329–45 (1971)

James, I.M., Thomas, E.: Which Lie groups are homotopy abelian? Proc. Nat. Acc. Sc. USA **25**, 131–140 (1959)

Massey, W.S., Uehara, H.: The Jacobi Identity for Whitehead Products. Princeton University Press, Princeton (1957)

Maunder, C.R.F.: Algebraic Topology. Van Nostrand Reinhold Company, London (1980)

May, J.P.: A Concise Course in Algebraic Topology. University of Chicago Press, Chicago (1999)

McCarty, G.S.: Products between homotopy groups and J-morphism. Q. J. Math. Oxf. **15**(2), 362–370 (1964)

Miyazaki, H.: On realizations of some Whitehead products. Tohoku Math. J. **12**, 1–30 (1960)

Nakaoka, M., Toda, H.: On Jacobi identity for Whitehead products. J. Inst. Polytech, Osaka City Univ. Ser A **5**, 1–13 (1954)

Samelson, H.: A connection between the Whitehead product and Pontragin product. Am. J. Math **75**, 744–752 (1953)

Stephen, J.S.: A Samelson product and homotopy associativity. Proc. Am. Math. Soc **70**(2), 189–195 (1978)

Spanier, E.: Algebraic Topology. McGraw-Hill, New York (1966)

Toda, H.: Generalized Whitehead products and homotopy groups of spheres. J. Inst. Polytech. Osaka City Univ. Ser. A,**3**, 43-48 (1953)

Whitehead, G.W.: On products in homotopy groups. Ann. Math. **47**(2), 460–475 (1944)

Whitehead, G.W.: Elements of Homotopy Theory. Springer, Heidelberg (1978)

Whitehead, J.H.C.: On adding relations to homotopy groups. Ann. of Math. **42**(2), 409–428 (1941)

Chapter 10
Homology and Cohomology Theories

This chapter opens with homology and cohomology theories which play a key role in algebraic topology. Homology and cohomology groups are also topological invariants like homotopy groups and Euler characteristic. Homology (cohomology) theory is a sequence of covariant (contravariant) functors from the category of chain (cochain) complexes to the category of abelian groups (modules). A key feature of these functors is their homotopy invariance in the sense that homotopic maps induce the same homomorphism in homology (cohomology). In particular, topological spaces of the same homotopy type have isomorphic homology (cohomolgy) groups.

Homotopy groups are easy to define but very difficult to compute in general. For example, for spheres the computation of $\pi_m(S^n)$ for $m > n$ faces serious problems. Fortunately, there is a more computable alternative approach to homotopy groups, the so-called homology groups $H_n(X)$ of a topological space X. For example, for spheres, the homology groups $H_m(S^n)$ are isomorphic to the homotopy groups $\pi_m(S^n)$ for $1 \leq m \leq n$ and $H_m(S^n) = 0$ for all $m > n$, which is an advantage of homology groups. Historically, homology groups came earlier than homotopy groups. Homology invented by H. Poincaré in 1895 is one of the most fundamental influential invention in mathematics. Homology groups are refinements, in some sense, of Euler characteristic.

Chapter 12 presents another approach, known as an axiomatic approach to homology and cohomology theories defined on the category of spaces having homotopy type of finite CW-complexes. This approach is essentially due to Eilenberg and Steenrod and is the most important contribution to algebraic topology since the invention of the homology groups by Poincaré. Homotopy and homology groups have some close relations at least for a certain class of topological spaces.

The aim of homology theory is to assign a group structure to cycles that are not boundaries. The basic tools such as complexes and incidence numbers for constructing simplicial homology groups were given by Poincaré in 1895. The basic idea of his construction is that it starts with a geometric object (a space) which is given

M.R. Adhikari, *Basic Algebraic Topology and its Applications*,
DOI 10.1007/978-81-322-2843-1_10

by combinatorial data (a complex). Then the linear algebra and boundary relations determined by this data are used to construct homology groups. It took more than thirty years to develop homology theory (H_n) applicable to curvilinear polyhedra, embodying the notions given by Poincaré in 1895. The functor H_n measures the number of 'n-dimensional holes' in the simplicial complex (or in the the space), which means that the n-sphere S^n has exactly one n-dimensional hole and there is no m-dimensional hole if $m \neq n$. A 0-dimensional hole is a pair of points in different path components which asserts that H_0 measures path connectedness. The simplicial techniques in the simplicial homology theory prescribed by Poincaré were gradually generalized to singular homology using the algebraic properties of the singular complex. The cohomology groups of a topological space were not recognized until 1930.

After setting up the basic apparatus, H. Poincaré (1854–1912) constructed the homology groups of a polyhedron in 1895. These homology groups have several generalizations to singular homology groups of an arbitrary topological space made by S. Lefschetz (1884–1972) in 1933, S. Eilenberg (1915–1998) in 1944, E. Čech (1893–1960) in 1932, and for compact metric spaces by L. Vietoris (1891–2002) in 1927. Their approaches for constructing homology and cohomology theories and choice of methods are often dictated by the nature of the problems. For example, singular homology and cohomology theories are defined for all topological spaces.

The idea of Poincaré on homology theory was generalized in two directions

(i) from complexes to more general topological spaces where the homology groups are not characterized by numerical invariants;
(ii) from the group **Z** to arbitrary abelian groups.

There exist different homology theories such as simplicial homology, singular homology, Čech homology, cellular homology, etc., and their corresponding cohomology theories with different constructions. But Eilenberg–Steenrod theorem unified them by showing that any two homology theories with isomorphic coefficient groups on the category of all compact polyhedral pairs are isomorphic (see Chap. 12). The cohomology groups (modules) of a topological space were not recognized until S. Lefschetz formulated a simplified method of the duality theorem for manifolds in the 1930s.

Homology theory H_n and cohomology theory H^n are dual to each other in some sense: there is a bilinear pairing of chains and cochains and H_n is a covariant functor but H^n is a contravariant functor. The basic property of cohomology which distinguishes it from homology is the existence of a natural multiplication called cup product which makes the direct sum of all cohomology modules with coefficient in a ring R into a graded R-algebra. This extra structure is more subtle than the additive structure of homology module (group) of the space.

The most important homology theory in algebraic topology is the singular homology. The simplicial techniques are gradually modified until the creation of singular homolgy by S Eilenberg which is a topological invariant. Simplicial homology is the primitive version of singular homology. To inaugurate a simplicial homology theory, H. Poincaré started in 1895 with a geometric object (a space) which is given by combinatorial data (a simplicial complex). Then the linear algebra and boundary relations by these data are used to construct homology groups. Using these tools Poincaré defined directly the Betti numbers invented by E. Betti (1823–1892) and torsion numbers which are numerical invariants and characterize the homology groups having coefficient group **Z** of integers.

Attention for shift from numerical invariants to groups associated with homology theories was successfully made during the period 1925–1935. This shift is partly due to Emmy Noether (1882–1935). Her algebraic approach to homology conveys a major contribution to the geometrical approach to homology given in 1895 by H. Poincaré. There is a natural question: how to relate the groups $C_p(K; G)$, $Z_p(K; G)$, and $B_p(K; G)$ defined in Sect. 10.2 to the topological spaces whose triangulation is K? Is it possible for $C_p(K; G)$, to express any property which remains unchanged under homeomorphism? Homology groups provide the desired topological invariant. Cohomoogy theory invented by J.W. Alexander (1888–1971) and A. Kolmogoroff (1903–1987) independently in 1935 is dual to homology theory. E.Čech and H. Whitney (1907–1989) developed simplicial cohomology theory during 1935–1940. The terms 'coboundary', 'cocycle', 'cochain', and 'cohomology' were given by E.Čech.

More precisely, this chapter conveys constructions of simplicial, singular, Čech and cellular homology theories, and their dual cohomology theories. Moreover, this chapter studies basic properties of homology and cohomology theories, Euler characteristic (a numerical topological invariant) and Betti number from the viewpoint of homology theory, Hurewictz theorem, Mayer–Vietoris sequences, Jordan curve theorem, and universal coefficient theorem and also discusses cohomology theory.

For this chapter the books Croom (1978), Dold (1972), Gray (1975), Hatcher (2002), Maunder (1970), Rotman (1988), Spanier (1966) and some others are referred in Bibliography.

10.1 Chain Complexes

This section studies chain complexes with their basic properties needed for constructing homology groups. W. Mayer (1887–1947) studied in 1929 chain complex, boundary, cycle from purely algebraic viewpoint.

Definition 10.1.1 A sequence $C = \{C_n, \partial_n\}, n \in \mathbf{Z}$ of abelian groups and their homomorphisms

$$\partial_n : C_n \to C_{n-1}$$

such that $\partial_n \circ \partial_{n+1} = 0$, for all n, is called a chain complex and ∂_n is called a boundary homomorphism. More precisely,

$$C : \cdots \to C_{n+1} \xrightarrow{\partial_{n+1}} C_n \xrightarrow{\partial_n} C_{n-1} \to \cdots \tag{10.1}$$

is called a chain complex if for all $n \in \mathbf{Z}$, the equality $\partial_n \circ \partial_{n+1} = 0$ holds. The group C_n is called the n-dimensional chain group of the complex C and elements of C_n are called n-chains for C.

Definition 10.1.2 The elements of $Z_n = \ker \partial_n$ and elements of $B_n = \operatorname{Im} \partial_{n+1}$ in the sequence (10.1) are called n-cycles and n-boundaries for the complex C respectively.

Proposition 10.1.3 *For any chain complex C in the sequence (10.1), $B_n = \operatorname{Im} \partial_{n+1}$ is a subgroup of $Z_n = \ker \partial_n$.*

Proof It follows from the condition of a chain complex C that $\partial_n \circ \partial_{n+1} = 0$, for all $n \in \mathbf{Z}$. ❑

Definition 10.1.4 The quotient group Z_n/B_n for any chain complex C is called the n-dimensional homology group of the chain complex C, denoted by $H_n(C)$ or simply H_n. The complex C is said to be acyclic if $H_n(C) = 0$ for all n. The elements of $H_n = Z_n/B_n$ are called homology classes, denoted by $[z]$ for every $z \in Z_n$.

Remark 10.1.5 For an acyclic complex C, $H_n(C) = 0$ for all n implies that the sequence (10.1) is exact at C_n for all n and hence it makes the sequence (10.1) exact. This shows that the homology group of a chain complex measures its deviation from the exactness of the sequence (10.1).

Example 10.1.6 Consider the chain complex

$$C : \cdots \to 0 \to \mathbf{Z} \oplus \mathbf{Z} \xrightarrow{\partial_2} \mathbf{Z} \xrightarrow{\partial_1} 0 \to \cdots,$$

where the chain groups are given by

$$C_1 = \mathbf{Z},\ C_2 = \mathbf{Z} \oplus \mathbf{Z},\ C_n = 0 \text{ for } n \neq 1, 2,$$

and the homomorphism ∂_2 is defined by $\partial_2(x, y) = 3x + 3y$. The group Z_2 of 2-cycles is isomorphic to $\mathbf{Z}$ and the group Z_1 of 1-cycles is $C_1 = \mathbf{Z}$. All other groups of cycles are zero. On the other hand, the groups B_n of n-boundaries are zero except for $n = 1$ and B_1 is generated by the element 3 of the group $\mathbf{Z}$. Hence $H_1(C) = \mathbf{Z}_3$, $H_2(C) = \mathbf{Z}$ and $H_n(C) = 0$ for $n \neq 1, 2$.

Definition 10.1.7 Let $C = \{C_n, \partial_n\}$ and $C' = \{C'_n, \partial'_n\}, n \in \mathbf{Z}$ be two chain complexes of abelian groups. A sequence $f = \{f_n : C_n \to C'_n\}, n \in \mathbf{Z}$ of homomorphisms is called a chain map from C to C' if these homomorphisms commute with the boundary homomorphisms, i.e., if each square in the Fig. 10.1 is commutative, i.e.,

$$f_n \circ \partial_{n+1} = \partial'_{n+1} \circ f_{n+1},\ \forall n \in \mathbf{Z}.$$

We abbreviate the entire above collection to $f : C \to C'$ and call f a chain map.

Fig. 10.1 Chain map

$$\cdots \longrightarrow C_{n+1} \xrightarrow{\partial_{n+1}} C_n \xrightarrow{\partial_n} C_{n-1} \longrightarrow \cdots$$

$$\Big\downarrow f_{n+1} \qquad \Big\downarrow f_n \qquad \Big\downarrow f_{n-1}$$

$$\cdots \longrightarrow C'_{n+1} \xrightarrow{\partial'_{n+1}} C'_n \xrightarrow{\partial'_n} C'_{n-1} \longrightarrow \cdots$$

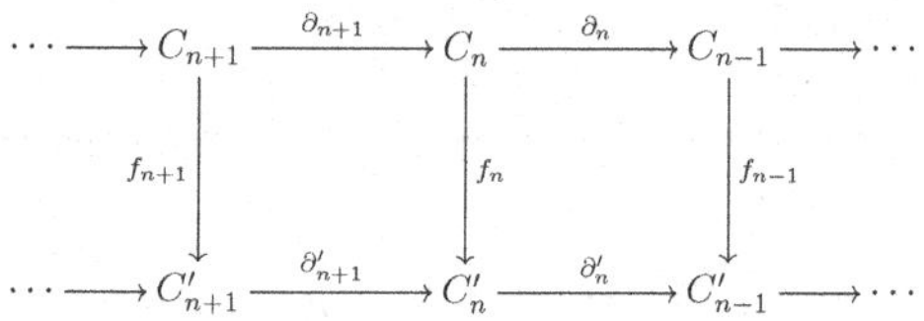

Proposition 10.1.8 *Let $C = \{C_n, \partial_n\}$ and $C' = \{C'_n, \partial'_n\}$ be two chain complexes of abelian groups and $f = \{f_n : C_n \to C'_n\}$ be a chain map. Then f_n maps n-cycles of C into n-cycles of C' and n-boundaries of C into n-boundaries of C' for all $n \in \mathbf{Z}$, i.e., $f_n(Z_n) \subset Z'_n$ and $f_n(B_n) \subset B'_n$ for every n.*

Proof The proof follows from the commutativity of each square in Fig. 10.1. ❏

Theorem 10.1.9 *Let $f : C \to C'$ be a chain map between two chain complexes $C = \{C_n, \partial_n\}$ and $C' = \{C'_n, \partial'_n\}$. Then for each integer n, $f_n : C_n \to C'_n$ induces a homomorphism.*

$$H_n(f_n) = f_{n*} : H_n(C) \to H_n(C'), [z] \mapsto [f_n(z)].$$

Proof Left as an exercise. ❏

Definition 10.1.10 $H_n(f_n) = f_{n*} : H_n(C) \to H_n(C')$ defined in Theorem 10.1.9 is called the homomorphism in homology induced by f_n for every integer n.

Remark 10.1.11 f and f_* are written in places of f_n and f_{n*} respectively, unless there is any confusion.

Proposition 10.1.12 **(a)** *Let $f : C \to C'$ and $g : C' \to C''$ be two chain maps. Then their composite $g \circ f : C \to C''$ is also a chain map such that $(g \circ f)_* = g_* \circ f_* : H_n(C) \to H_n(C'')$.*
(b) *If $1_C : C \to C$ is the identity chain map, then $(1_C)_* : H_n(C) \to H_n(C)$ is the identity automorphism.*

Proof Left as an exercise. ❏

Definition 10.1.13 Let $C = \{C_n, \partial_n\}$ and $C' = \{C'_n, \partial'_n\}$ be two chain complexes and $f, g : C \to C'$ be two chain maps. Then f is said to be chain homotopic to g, denoted by $f \simeq g$, if there is a sequence $\{F_n : C_n \to C'_{n+1}\}$ of homomorphism such that

$$\partial'_{n+1} F_n + F_{n-1} \partial_n = f_n - g_n : C_n \to C'_n, \forall n \in \mathbf{Z}$$

holds.
In particular, a chain map $f : C \to C'$ is called a chain homotopy equivalence if there exists a chain map $g : C' \to C$ such that $g \circ f \simeq 1_C$ and $f \circ g \simeq 1_{C'}$.

Proposition 10.1.14 *The relation of chain homotopy on the set* $\mathrm{S}(C, C')$ *of all chain maps from* C *to* C' *is an equivalence relation.*

Proof Left as an exercise. ❑

Theorem 10.1.15 *Two homotopic chain maps* $f, g : C \to C'$ *induce the same homomorphisms in the homology, i.e., if* $f \simeq g : C \to C'$*, then the homomorphisms* $f_* = g_* : H_n(C) \to H_n(C')$ *for every* n.

Proof Let $f \simeq g : C \to C'$. Then there exists a chain homotopy $\{F_n : C_n \to C'_{n+1}\}$. Let $[z] \in H_n(C)$. Then $\partial_n([z]) = 0$ shows that $f_n([z]) - g_n([z]) = \partial_{n+1} F_n([z])$ is a boundary. Hence $[f_n[z]] = [g, [z]]$ implies that $f_{n*}([z]) = g_{n*}([z])$, $\forall\, [z] \in H_n(C)$. Consequently, $f_{n*} = g_{n*}$ for all n. Hence $f_* = g_*$. ❑

Proposition 10.1.16 **(a)** *All chain complexes and chain maps form a category denoted by Comp.*
(b) *For each* $n \in \mathbf{Z}$, H_n *is a covariant functor from the category Comp of chain complexes and chain maps to the category* $\mathcal{Ab}$ *of abelian groups and homomorphisms.*

Proof **(a)** The objects here are taken chain complexes and the morphisms are taken chain maps. The composition of chain maps is defined coordinatewise: $\{g_n\} \circ \{f_n\} = \{g_n \circ f_n\}$. Hence they form a category written *Comp*.
(b) The object function is defined by assigning to each chain complex the sequence of its homology groups, and morphism function is defined by assigning to each chain map f between chain complexes the induced map f_* between their homology group. This shows that for each n, $H_n : Comp \to \mathcal{Ab}$ is a covariant functor by Proposition 10.1.12. ❑

10.2 Simplicial Homology Theory

This section begins with the simplicial homology theory invented by H.Poincaré in 1895 on the category of simplicial pairs starting with construction of the homology groups of a simplicial complex in two steps: first by assigning to each simplicial complex a certain complex, called chain complex followed by assigning to the chain complex its homology group. This theory stems from Poincaré's seminal mathematics paper 'Analysis situs' and five supplements to the paper around the turn of the nineenth century to the beginning of the twentieth century (between 1895 and 1904). This theory characterizes topological spaces which look like polyhedra. These can be used to cover a manifold by a process called triangulation. The most advantage of this theory is: it is easier to visualize geometrically than other homology theories.

10.2.1 *Construction of Homology Groups of a Simplicial Complex*

This subsection constructs homology groups of oriented simplicial complexes K and it is shown that they coincide with the homology groups of $|K|$. This construction assigns a group structure to cycles that are not boundaries. This subsection associates to every simplicial map a homomorphism on the corresponding simplicial homology groups and presents functorial properties of simplicial homology. The problem is: how to relate $C_p(K)$, $Z_p(K)$ and $B_p(K)$ to the topological properties of the spaces whose triangulation is K? This subsection gives its answer.

Definition 10.2.1 An oriented simplicial complex K is a simplicial complex and a partial order on vertex set $\text{Vert}(K)$ whose restriction to the vertices of any simplex in K is a linear order.

An oriented n-simplex for $n \geq 1$, is obtained from an n-simplex $s_n = \langle v_0 v_1 \dots v_n \rangle$ by assigning an ordering of its vertices. The two equivalence classes of permutations of the ordering of its vertices determine the orientations of s_n. The equivalence class of even permutations of the chosen ordering determines the positively oriented simplex $+s_n$. On the other hand the equivalence class of odd permutations determines the negatively oriented simplex $-s_n$. A vertex is considered a zero simplex positively oriented.

Example 10.2.2 In the 2-simplex $s_2 = \langle v_0 v_1 v_2 \rangle$, if the ordering $v_0 < v_1 < v_2$ is assigned, then $+s_2 = \langle v_0 v_1 v_2 \rangle$ and $-s_2 = \langle v_2 v_1 v_0 \rangle$.

Definition 10.2.3 Let K be an oriented simplicial complex with simplexes s_{p+1} and s_p, whose dimensions differ by 1. Then for each such pair (s_{p+1}, s_p) a number $[s_{p+1}, s_p]$, called incidence number is assigned as follows:
If s_p is not a face of s_{p+1}, then $[s_{p+1}, s_p] = 0$.
If s_p is a face of s_{p+1}, then vertices $v_0, v_1, \dots, v_p$ of s_p are labeled so that $+s_p = \langle v_0 v_1 \dots v_p \rangle$.
Let v denote the vertex of s_{p+1} which is not vertex of s_p. Then $s_{p+1} = \pm \langle v_0 v_1 \dots v_p \rangle$.
If $+s_{p+1} = +\langle v_0 v_1 \dots v_p \rangle$, then $[s_{p+1}, s_p] = +1$.
If $+s_{p+1} = -\langle v_0 v_1 \dots v_p \rangle$, then $[s_{p+1}, s_p] = -1$.

Remark 10.2.4 If $[s_{p+1}, s_p] = +1$, then s_p is a positively oriented face of s_{p+1} and if $[s_{p+1}, s_p] = -1$, then s_p is a negatively oriented face of s_{p+1}.

The choice of a positive ordering of vertices of s_{p+1} clearly induces a natural ordering of the vertices in each face of s_{p+1}. Thus an orientation of s_{p+1} induces a natural ordering of its vertices. Hence the Definition 10.2.3 implies that if s_p is a face of s_{p+1}, then the incidence number $[s_{p+1}, s_p]$ is positive or negative according as the chosen orientation of s_{p+1} agrees or disagrees with orientation of s_p respectively.

Example 10.2.5 If $+s_2 = \langle v_0 v_1 v_2 \rangle$, $\sigma_1 = \langle v_0 v_1 \rangle$, $\rho_1 = \langle v_0 v_2 \rangle$, as shown in Fig. 10.2, then $[s_2, \sigma_1] = +1$ but $[s_2, \rho_1] = -1$.

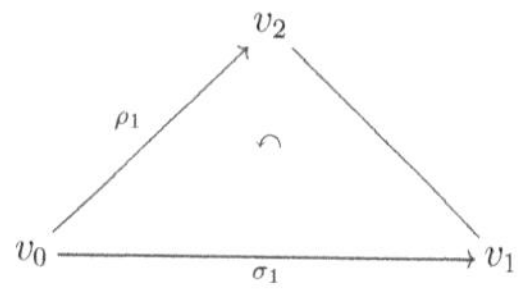

Fig. 10.2 Orientations involving $+s_2$, σ_1 and ρ_1.

Theorem 10.2.6 *Let K be an oriented complex, s_p be an oriented p-simplex of K and s_{p-2} be a $(p-2)$-face of s_p. Then*

$$\sum_{s_{p-1}\in K} [s_p, s_{p-1}][s_{p-1}, s_{p-2}] = 0.$$

Proof Order the vertices $v_0, v_1, \ldots, v_{p-2}$ so that $+s_{p-2} = \langle v_0 \ldots v_{p-2}\rangle$. Then s_p has two additional vertices a and b(say). Assume that $+s_p = \langle abv_0 \ldots v_{p-2}\rangle$. Every $(p-2)$-simplex in s_p is a face of exactly two $(p-1)$-faces of s_p, which are taken $+s^1_{p-1} = \langle av_0 \ldots v_{p-2}\rangle$ and $+s^2_{p-1} = \langle bv_0 \ldots v_{p-2}\rangle$ (say). Thus the nonzero terms in the sum

$$\sum_{s_{p-1}\in K} [s_p, s_{p-1}][s_{p-1}, s_{p-2}]$$

occur for only $+s^1_{p-1}$ and $+s^2_{p-1}$.

Case 1 Suppose $+s^1_{p-1} = +\langle av_0 \ldots v_{p-2}\rangle$ and $+s^2_{p-1} = +\langle bv_0 \ldots v_{p-2}\rangle$.

Then $[s_p, s^1_{p-1}] = -1$, $[s^1_{p-1}, s_{p-2}] = +1$, $[s_p, s^2_{p-1}] = +1$ and $[s^2_{p-1}, s_{p-2}] = +1$.

Hence $\sum_{s_{p-1}\in K} [s_p, s_{p-1}][s_{p-1}, s_{p-2}] = (-1)(+1) + (+1)(+1) = 0$.

Case 2 Suppose $+s^1_{p-1} = +\langle av_0 \ldots v_{p-2}\rangle$ and $+s^2_{p-1} = -\langle bv_0 \ldots v_{p-2}\rangle$.

Then $[s_p, s^1_{p-1}] = -1$, $[s^1_{p-1}, s_{p-2}] = +1$, $[s_p, s^2_{p-1}] = -1$ and $[s^2_{p-1}, s_{p-2}] = -1$.

Hence $\sum_{s_{p-1}\in K} [s_p, s_{p-1}][s_{p-1}, s_{p-2}] = 0$. The remaining other cases may be proved similarly. ❑

To define the homology groups of an oriented simplicial complex K, we need the concepts of p-chains, p-cycles and p-boundaries corresponding to K.

Definition 10.2.7 Let K be an oriented simplicial complex and p be a positive integer. A p-dimensional chain or a p-chain is a function c_p from the family of oriented p-simplicial of K to $\mathbf{Z}$ such that for each p-simplex s_p, $c_p(-s_p) = -c_p(+s_p)$. The family of p-chains forms a group called the p-dimensional chain group of K with coefficients in $\mathbf{Z}$ under usual functional addition:

$$(c^1_p + c^2_p)(s_p) = c^1_p(s_p) + c^2_p(s_p), \tag{10.2}$$

where the addition on right-hand side (RHS) of (10.2) is the usual addition in **Z**. This group denoted by $C_p(K, \mathbf{Z})$ is called the p-dimensional chain group of K with coefficients in **Z**.

The Definition 10.2.7 can be extended for an arbitrary abelian group G in place of **Z**.

Definition 10.2.8 A p-dimensional chain or a p-chain of an oriented simplicial complex K with coefficients in an additive abelian group G is a function c_p from the family of oriented p-simplexes of K to G such that if $c_p(+s_p) = g$ for some $g \in G$, then $c_p(-s_p) = -g$. The family of such p-chains forms a group denoted by $C_p(K; G)$ under usual functional addition

$$(c_p^1 + c_p^2)(s_p) = c_p^1(s_p) + c_p^2(s_p), \tag{10.3}$$

where the addition in RHS of (10.3) is the usual group operation in G. As G is an abelian group, $C_p(K; G)$ is also an abelian group. The group $C_p(K; G)$ is called the p-dimensional chain group of K with coefficients in G.

Remark 10.2.9 If the oriented complex K has no p-simplex for some p, then we take $C_p(K, G) = 0$, the trivial group consisting of the identity element 0 of G only.

Remark 10.2.10 If the oriented complex K is infinite, then $c_p(s_p) = 0$ for all but a finite number of p-simplexes of K.

Definition 10.2.11 An elementary p-chain on an oriented complex K is a p-chain c_p for which there is a p-simplex s_p in K such that $c_p(+s_p) = g$ and $c_p(t_p) = 0$, whenever, $s_p \neq \pm t_p$. Such a p-chain is denoted by a formal product $g \cdot s_p$, when $g = c_p(+s_p)$.

An arbitrary p-chain d_p on K can be written as a formal finite sum $d_p = \sum g_i \cdot s_p^i$ of elementary p-chains where $g_i = c_p(+s_p^i)$ and the index i ranges over all p-simplexes of K. This notation justifies the use of the word coefficient.

Proposition 10.2.12 *Let K be a finite oriented simplicial complex. Then*

(i) *If $c_p = \Sigma f_i \cdot s_p^i$ and $d_p = \Sigma g_i \cdot s_p^i$ are two p-simplexes on K, then*

$$c_p + d_p = \Sigma(f_i + g_i) \cdot s_p^i;$$

(ii) *The additive inverse of d_p in the group $C_p(K; G)$ is the chain $-d_p = \Sigma(-g_i) \cdot s_p^i$.*

Proof **(i)** and **(ii)** follow from the definition of p-chains. ❑

Theorem 10.2.13 *Let K be a finite oriented simplicial complex and n_p be the number of p-simplexes in K. Then the chain group $C_p(K; G)$ is isomorphic to the direct sum of n_p-copies of G.*

Proof Let $C_p(K; G)$ be the p-dimensional chain group of K with coefficients in G. Define a map

$$\psi : C_p(K; G) \to G \oplus G \oplus \cdots \oplus G, \sum_{s_p \in K} g_i \cdot s_p^i \mapsto (g_1, g_2, \ldots, g_{n_p}).$$

Then ψ is a group isomorphism. ❑

Remark 10.2.14 If the coefficient group G is taken to be a commutative ring or field, then $C_p(K; G)$ is a module or a vector space. Poincaré original definition was given in terms of integers.

To define the homology groups of a finite oriented simplicial complex K with coefficient group G, we introduce the following concepts. Let K be a finite oriented simplicial complex and G be an abelian group.

Definition 10.2.15 If $g \cdot s_p$ is an elementary p-chain with $p \geq 1$, then the boundary of $g \cdot s_p$, denoted by $\partial(g \cdot s_p)$, is defined by

$$\partial(g \cdot s_p) = \sum_{s_{p-1}^i \in K} [s_p, s_{p-1}^i] g \cdot s_{p-1}^i \tag{10.4}$$

The boundary operator ∂ is now extended by linearity to a homomorphism

$$\partial : C_p(K; G) \to C_{p-1}(K; G), c_p \mapsto \sum \partial(g_i \cdot s_{p-1}^i) \tag{10.5}$$

The boundary of a 0-chain is taken to be 0.

The boundary operator ∂ has an interesting property.

Theorem 10.2.16 *Let K be a finite oriented simplicial complex and $p \geq 2$. Then the composite homomorphism*

$$C_p(K; G) \xrightarrow{\partial_p} C_{p-1}(K; G) \xrightarrow{\partial_{p-1}} C_{p-2}(K; G)$$

is trivial, i.e., $\partial_{p-1} \circ \partial_p = 0$ for all $p \geq 2$.

Proof To prove the theorem it is sufficient to prove the result for an elementary p-chain $g \cdot s_p$ for $p \geq 2$. We claim that for such a p-chain $g \cdot s_p$, the composite $\partial_{p-1} \circ \partial_p = 0$. Now

$$\begin{aligned}(\partial_{p-1} \circ \partial_p)(g \cdot s_p) &= \partial_{p-1}(\partial_p(g \cdot s_p)) \\ &= \partial_{p-1}\left(\sum_{s_{p-1} \in K} [s_p, s_{p-1}^i] g \cdot s_{p-1}^i\right) \text{ by (10.4)}\end{aligned}$$

$$= \sum_{s^i_{p-1}\in K} \sum_{s^j_{p-2}\in K} \left([s_p, s^i_{p-1}][s^i_{p-1}, s^j_{p-2}]g\cdot s^j_{p-2}\right) \tag{10.6}$$

Reversing the order of summation and collecting coefficients of each simplex $s^j_{p-2} \in K$, we have

$$\partial_{p-1}(\partial_p(g\cdot s_p)) = \sum_{s^j_{p-2}\in K} \left(\sum_{s^i_{p-1}\in K} [s_p, s^i_{p-1}][s^i_{p-1}, s^j_{p-2}]g\cdot s^j_{p-2} \right) \tag{10.7}$$

$= 0$ by Theorem 10.2.6 for all elementary p-chains $g\cdot s_p$.

Hence from (10.7) it follows that $\partial_{p-1} \circ \partial_p = 0$. ❑

Corollary 10.2.17 *For any oriented simplicial complex K and an abelian group G, the groups $C_p(K; G)$ and the homomorphisms $\partial_p : C_p(K; G) \to C_{p-1}(K; G)$ form a chain complex, denoted by $C(K; G)$.*

Corollary 10.2.18 Im ∂_p *is a subgroup of* ker ∂_{p-1}.

Proof It follows from Theorem 10.2.16. ❑

Definition 10.2.19 Let K be an oriented simplicial complex and G be an an abelian group. For $p \geq 0$, a p-chain $b_p \in C_p(K; G)$ is called a p-dimensional boundary or a p-boundary on K if there is a $(p + 1)$-chain $c_{p+1} \in C_{p+1}(K; G)$ such that $\partial(c_{p+1}) = b_p$. The set of all p-boundaries is the homomorphic image $\partial(C_{p+1}(K; G))$ and is a subgroup of $C_p(K; G)$, called the p-dimensional boundary group of K, denoted by $B_p(K; G)$.

Remark 10.2.20 Let K be an oriented n-dimensional simplicial complex. Then there are no p-chains on K for $p > n$. As a result $C_p(K; G) = 0$ for $p > n$. Thus there exists no $(n + 1)$-chain on K and hence $C_{n+1}(K; G) = 0$; so $B_n(K; G) = 0$.

Definition 10.2.21 Let K be an oriented complex. If p is a positive integer, a p-dimensional cycle or p-cycle on K is a p-chain denoted by z_p such that $\partial_p(z_p) = 0$. The set of all p-cycles is denoted by $Z_p(K; G)$.

Proposition 10.2.22 *$Z_p(K; G)$ is a subgroup of $C_p(K; G)$, and $B_p(K; G)$ is a subgroup of $Z_p(K; G)$ for each p such that $0 \leq p \leq n$, where n is the dimension of K.*

Proof It follows trivially from Theorem 10.2.16. ❑

Remark 10.2.23 The subgroup $Z_p(K; G)$ is the kernel of the homomorphism

$$\partial_p : C_p(K; G) \to C_{p-1}(K; G).$$

Definition 10.2.24 The group $Z_p(K; G)$ is called the p-dimensional cycle group of K. The group $Z_0(K; G)$ of 0-cycles is the group $C_0(K; G)$ of 0-chains.

Remark 10.2.25 Intuitively, a p-cycle on K is a linear combination of p-simplexes which makes a complete circuit. The p-cycles enclosing 'holes' of K are those cycles which are not boundaries of $(p + 1)$-chains. A p-cycle which is the boundary of a $(p + 1)$-chain was called by Poincaré 'a cycle homologous to zero'.

Definition 10.2.26 Two p-cycles c_p and d_p on an oriented complex K are said to be homologous, denoted by $c_p \sim d_p$ if there is a $(p + 1)$-chain e_{p+1} such that

$$\partial(e_{p+1}) = c_p - d_p.$$

If a p-cycle f_p is the boundary of a $(p + 1)$-chain, f_p is said to be homologous to zero, denoted by $f_p \sim 0$.

Remark 10.2.27 As the relation of being homologous for p-cycles is an equivalence relation, it partitions $Z_p(K; G)$ into homology classes

$$[z_p] = \{c_p \in Z_p(K; G) : c_p \sim z_p\}.$$

The homology class $[z_p]$ is the coset

$$z_p + B_p(K; G) = \{z_p + \partial(c_{p+1}) : \partial(c_{p+1}) \in B_p(K; G)\}.$$

Definition 10.2.28 For $p \geq 1$, $Z_p(K; G) = \ker \partial_p$ is called the group of p-cycles of K with coefficients in G.

Remark 10.2.29 Since we take the boundary of every 0-chain to be 0, we define 0-cycle to be 0-chain. Thus for $p = 0$, $Z_0(K; G)$ of 0-cycles is the group $C_0(K; G)$ of 0-cycles.

Definition 10.2.30 If $p \geq 0$, the image $\mathrm{Im}\, \partial_{p+1}$ is a subgroup of $C_p(K; G)$ and is called the group of p-dimensional boundaries or p-boundaries of $C_p(K; G)$, denoted by $B_p(K; G)$.

Remark 10.2.31 For any chain complex $C_p(K; G)$, the group of boundaries B_p is a subgroup of the group of cycles Z_p by Theorem 10.2.16. For the converse if $B_p \subset Z_p$ for all p, then the corresponding sequence of groups and their homomorphisms is a chain complex, i.e., $\partial_p \circ \partial_{p+1} = 0$

Definition 10.2.32 Let K be an oriented simplicial complex and G be an abelian group. Then the simplicial homology group of the corresponding chain complex $C(K; G)$, denoted by $H_p(K; G)$ is the quotient group $\ker \partial_p / \mathrm{Im}\, \partial_{p+1}$, i.e., $H_p(K; G) = Z_p / B_p$.

Remark 10.2.33 The homology classes $[z_p]$ are actually members of the simplicial homology group $H_p(K; G)$.

Remark 10.2.34 For $p < 0$ or $p > \dim K$, we take $C_p(K; G) = 0$. Hence $H_p(K; G) = 0$ for all such p. This group $H_p(K; G)$ is sometimes called an 'absolute' simplicial homology group.

The following natural questions arise:
Do the homology groups $H_n(K; G)$ depend on the choice of an orientation of K? Is it possible for $C_p(K)$ to express any property which remains unchanged under homeomorphism.

To solve such problems consider two copies K_1 and K_2 of the given simplicial complex K endowed with distinct orientations.
Consider the map $\psi : C(K_1; G) \to C(K_2; G)$,

$$\psi(s) = \begin{cases} +s, & \text{if the two orientations of } s \text{ coincide} \\ -s, & \text{otherwise.} \end{cases}$$

Then ψ is an isomorphism. In other words, given K and an abelian group G, $C(K, G)$ is uniquely determined up to isomorphism and hence $H_n(K; G)$ is uniquely determined up to isomorphism.

More precisely, we prove that

Theorem 10.2.35 *Let K_1 and K_2 denote the same simplicial complex K endowed with distinct orientations. Then given an abelian group G, $H_p(K_1; G) \cong H_p(K_2; G)$ for all $p \geq 0$.*

Proof Let $p \geq 0$ be an integer. For a p-simplex s_p of K, let ${}^1s_p$ and ${}^2s_p$ denote the p-simplex in the oriented complex K_1 and K_2 respectively. Define a function α on the simplexes of K such that $\alpha(s_p) = \pm 1$ and ${}^1s_p = \alpha(s_p) \cdot {}^2s_p$. Then define a sequence $\psi = \{\psi_p\}$ of homomorphisms

$$\psi_p : C_p(K_1; G) \to C_p(K_2; G), \sum g_i \cdot {}^1s_p^i \mapsto \sum \alpha(s_p^i) g_i \cdot {}^2s_p^i,$$

where $\sum g_i \cdot {}^1s_p^i$ is a p-chain on K_1. Hence ψ_p is well defined. Clearly, the diagram in Fig. 10.3 is commutative. Hence it follows that if $z_p \in Z_p(K_1; G)$, then $\partial_p \psi_p(z_p) = \psi_{p-1} \partial_p(z) = 0$ shows that $\psi_p(z_p) \in Z_p(K_2; G)$. This means that ψ_p maps a p-cycle of K_1 into a p-cycle of K_2. Similarly, ψ_p maps a p-boundary of K_1 into a p-boundary of K_2. Consequently, ψ_p induces a homomorphism

$$\begin{array}{ccc} C_p(K_1; G) & \xrightarrow{\psi_p} & C_p(K_2; G) \\ \downarrow \partial_p & & \downarrow \partial_p \\ C_{p-1}(K_1; G) & \xrightarrow{\psi_{p-1}} & C_{p-1}(K_2; G) \end{array}$$

Fig. 10.3 Diagram involving ψ_p and ∂_p

$$\psi_{p*} : H_p(K_1; G) \to H_p(K_2; G), [z_p] \mapsto [\psi_p(z_p)],$$

for all homology classes $[z_p] \in H_p(K_1; G)$. Reversing the roles of K_1 and K_2 we have a sequence $\{\phi_p\}$ of homomorphisms

$$\phi_p : C_p(K_2; G) \to C_p(K_1; G)$$

such that

$$\phi_p \circ \psi_p = 1_d \text{ (identity automorphism of } C_p(K_1; G)) \tag{10.8}$$

and

$$\psi_p \circ \phi_p = 1_d \text{ (identity automorphism of } C_p(K_2; G)) \tag{10.9}$$

(10.8) shows that $(\phi_p \circ \psi_p)_* = \phi_{p*} \circ \psi_{p*} =$ identity 1_d and
(10.9) shows that $(\psi_p \circ \phi_p)_* = \psi_{p*} \circ \phi_{p*} =$ identity 1_d. Consequently, $\psi_{p*} : H_p(K_1; G) \to H_p(K_2; G)$ is an isomorphism of groups. ❑

Remark 10.2.36 The structure of 0-dimensional homology group $H_0(K; G)$ shows whether the polyhedron $|K|$ is connected: there is no torsion in dimension 0 and the rank of the free abelian group $H_0(K; G)$ is the number of components of $|K|$.

10.2.2 *Induced Homomorphism and Functorial Properties of Simplicial Homology*

This subsection associates to every simplicial map a homomorphism on the corresponding simplicial homology groups and presents functorial properties of simplicial homology.

Definition 10.2.37 Let K and L be oriented simplicial complexes and $f : K \to L$ be a simplicial map. For each $p \geq 0$, define

$$f_\square : C_p(K) \to C_p(L), \ \langle v_0, v_1, \ldots, v_p\rangle \mapsto \langle f(v_0), \ldots, f(v_p)\rangle.$$

If some $f(v_i)$ is repeated, then the term on the right is zero.

Proposition 10.2.38 *If $f : K \to L$ is a simplicial map, then*

$$f_\square : C_*(K) \to C_*(L)$$

is a chain map, which means that $f_\square \circ \partial = \partial \circ f_\square$.

Proof It follows from the definitions of $f_\square$ and ∂. ❑

Proposition 10.2.39 *If $f : K \to L$ is a simplicial map, then the induced homomorphism*

$$f_* : H_n(K) \to H_n(L),\ z + B_n(K) \mapsto f_\square(z) + B_n(L)$$

is a homomorphism of groups.

Proof Left as an exercise. ❑

Proposition 10.2.40 *Chain complexes and chain maps form a category under usual composition. This category is denoted by Comp.*

Proof The objects here are taken chain complexes and the morphisms are taken chain maps. The composition of chain maps is defined coordinatewise: $\{g_n\} \circ \{f_n\} = \{g_n \circ f_n\}$. Hence they form a category. ❑

Proposition 10.2.41 *H_n is a covariant functor from the category Comp of chain complexes and chain maps to the category $\mathcal{Ab}$ of abelian groups and homomorphisms for each $n \in \mathbf{Z}$. Moreover, H_n is a topological invariant.*

Proof The object function is here defined by assigning to each chain complex the sequence of its homology groups, and morphism function is defined by assigning to each chain map f between chain complexes the induced map f_* between their homology groups. This shows that for each n, $H_n : Comp \to \mathcal{Ab}$ is a covariant functor which is a topological invariant by Proposition 10.1.12. ❑

10.2.3 Computing Homology Groups of Polyhedra

The subsection considers the problem: how to relate $C_p(K;G)$, $Z_p(K;G)$, and $B_p(K;G)$ to the topological spaces whose triangulation is K? Let X be a polyhedron and G be an abelian group. For calculation of the homology groups we use the following steps:
Step 1: Triangulate X.
Step 2: Choose an orientation for the simplicial complex K thus obtained by triangulation.
Step 3: Calculate the chain group $C_n(K;G)$.
Step 4: Describe the boundary homomorphisms ∂_n.
Step 5: Calculate the groups of cycles $Z_n(K;G)$.
Step 6: Calculate the groups of boundaries $B_n(K;G)$.
Step 7: Calculate the quotient group $H_n(K;G) = Z_n(K;G)/B_n(K;G)$.

Example 10.2.42 Consider the simplicial complex K having only one vertex v with $\mathbf{Z}$ as the coefficient group. As there is only one possible orientation on K and with that orientation, $C_p(K;\mathbf{Z}) = 0$ for all $p \neq 0$ and $C_0(K;\mathbf{Z})$ is the free abelian group on the single generator v. Hence

$$H_p(K;\mathbf{Z}) = \begin{cases} 0, & p \neq 0 \\ \mathbf{Z}, & p = 0. \end{cases}$$

Example 10.2.43 Let S^2 be the 2-sphere. Then $H_2(S^2; G) = G$. Consider the simplicial complex K consisting of all 2-simplexes, 1-simplexes and 0-simplex that are faces of a single 3-simplex s_3, where s_3 is not in K. Geometrically, this is the surface of a tetrahedron and this surface is homeomorphic to S^2. The simplex S^2 is precisely the two skeletons of the complex K. We orient the complex K by chosing a fixed ordering of its vertices: $a_0 < a_1 < a_2 < a_3$. It will induce the positive orientation of the simplexes. In this way, we have the following oriented simplexes of K:

1-simplexes: $+s_1^1 = \langle a_2 a_3 \rangle, +s_1^2 = \langle a_1 a_3 \rangle, +s_1^3 = \langle a_0 a_3 \rangle, +s_1^4 = \langle a_1 a_2 \rangle, +s_1^5 = \langle a_0 a_2 \rangle$
2-simplexes: $s_2^1 = \langle a_1 a_2 a_3 \rangle, +s_2^2 = \langle a_0 a_2 a_3 \rangle, +s_2^3 = \langle a_0 a_1 a_3 \rangle, +s_2^4 = \langle a_0 a_1 a_2 \rangle$.

Then the only 2-cycles on S^2 are the chains of the form $g \cdot s_2^1 - g \cdot s_2^2 + g \cdot s_2^3 - g \cdot s_2^4, g \in G$. Hence $Z_2(S^2; G) \cong G$. Since there are no 3-simplexes in S^2, the chain $C_3(S^2; G) = 0$ and hence $\partial C_3(S^2; G) = 0$ gives $B_2(S^2; G) = 0$. Consequently, $H_2(S^2; G) = Z_2(S^2; G) \cong G$.

Remark 10.2.44 The homology groups of any polyhedron do not depend on any particular choice of its triangulation. Because if K and L are two triangulations of a polyhedron X, then given an abelian group G, $H_p(K; G) \cong H_p(L; G)$ for each $p \geq 0$.

Remark 10.2.45 The functor H_p measures the number of 'p-dimensional holes' in the simplicial complex, in the sense that the p-sphere S^p has exactly one p-dimensional hole and no m-dimensional hole if $m \neq p$. A 0-dimensional hole is a pair of points in different path components and hence H_0 measures path connectedness.

10.3 Relative Simplicial Homology Groups

This section extends the concept of absolute simplicial homology groups to the concept of relative simplicial homology groups. If K is an oriented simplicial complex and L is a subcomplex, then L is also oriented in the induced orientation, by the partial order on $\text{Vert}(L)$ inherited from that on $\text{Vert}(K)$.

Definition 10.3.1 Let L be a subcomplex of an oriented simplicial complex K. The relative chain group $C_p(K, L)$ of the pair (K, L) is defined to be the free abelian group freely generated by by all p-simplexes with interiors in $K - L$.

Definition 10.3.2 For $p \geq 1$, the boundary operator $\partial_p : C_p(K, L) \to C_{p-1}(K, L)$ is defined by the formula (10.5), where the summation is taken over all those $(p-1)$-simplexes s_{p-1}^i of K whose interiors do not intersect L.

Proposition 10.3.3 $\partial_p \circ \partial_{p+1} = 0$

Proof It is left as an exercise. ❑

Definition 10.3.4 The relative chain groups $C_p(K, L)$ and the operators ∂_p form a chain complex $C(K, L)$, called the relative chain complex of the pair (K, L).

Definition 10.3.5 A simplicial map of pairs $f : (K, L) \to (K_1, L_1)$ is a simplicial map $f : K \to K_1$ such that $f(L) \subset L_1$.

Definition 10.3.6 The group of cycles $Z_p(K, L)$ is the kernel of the homomorphism ∂_p and the group of boundaries $B_p(K, L)$ is the Im ∂_{p+1} in the chain complex $C(K, L)$.

Definition 10.3.7 The relative homology group $H_p(K, L)$ is the quotient group $Z_p(K, L)/B_k(K, L)$.

Remark 10.3.8 H_p satisfies the functorial properties.

Proposition 10.3.9 *Let (K, L) and (K_1, L_1) be two oriented simplicial pairs. Then any simplicial map $f : (K, L) \to (K_1, L_1)$ induces homomorphisms $f_* : H_p(K, L) \to H_p(K_1, L_1)$ such that*

(a) *if $g : (K_1, L_1) \to (K_2, L_2)$ is a simplicial map, then $(g \circ f)_* = g_* \circ f_* : H_p(K, L) \to H_p(K_2, L_2)$;*
(b) *if $1_{(K,L)} : (K, L) \to (K, L)$ is the identity map, then $1_{(K,L)*} : H_p(K, L) \to H_p(K, L)$ is the identity automorphism.*

Proof Left as an exercise. ❑

Corollary 10.3.10 *H_p is a covariant functor from the category of all relative simplicial chain complexes and chain maps to the category of abelian groups and homomorphisms.*

Proof It follows from Proposition 10.3.9. ❑

Remark 10.3.11 The relative homology groups $H_p(K, \emptyset)$ of an oriented simplicial complex K with respect to empty set $\emptyset$ coincide with the absolute ones, i.e., $H_p(K)$ and $H_p(K, \emptyset)$ are always isomorphic.

Theorem 10.3.12 *Let the simplicial complexes K_1 and K_2 intersect along a simplicial complex K_3, which is a subcomplex in both K_1 and K_2. Then the embedding i of the pair (K_1, K_3) to the pair $(K_1 \cup K_2, K_2)$ induces isomorphisms, for all $p \geq 0$,*

$$i_* = \psi_p : H_p(K_1, K_3) \to H_p(K_1 \cup K_2, K_2).$$

Proof The chain complexes of the pairs (K_1, K_3), $(K_1 \cup K_2, K_2)$ coincide, because they are constructed over the same set of simplexes. Hence the theorem follows. ❑

10.4 Exactness of Simplicial Homology Sequences

This section conveys the relations between absolute simplicial homology groups of simplicial chain complexes and the relative simplicial homology groups of relative simplicial chain complexes using the language of exact sequences and shows that the relative simplicial homology groups $H_p(K, L)$ for any pair (K, L) of simplicial complexes fit into a long exact sequence.

We recall that a sequence of groups and homomorphisms

$$\cdots \to C_{n+1} \xrightarrow{f_{n+1}} C_n \xrightarrow{f_n} C_{n-1} \to \cdots \tag{10.10}$$

is called exact if $\ker f_n = \operatorname{Im} f_{n+1}$ for all n.

This definition shows that any exact sequence of groups of the form (10.10) is a chain complex. Since $\ker f_n = \operatorname{Im} f_{n+1}$, all the homology groups of that complex are trivial. The converse result is also true: any chain complex with trivial homology groups is an exact sequence. This shows that the homology groups of a chain complex give a measure of its inexactness in some sense.

The definition of relative chain complex $C(K, L)$ of the pair (K, L) of simplicial complexes shows that the sequence

$$0 \longrightarrow C(L) \xrightarrow{i_*} C(K) \xrightarrow{j_*} C(K, L) \longrightarrow 0 \tag{10.11}$$

with homomorphism i_* induced by the embedding i of L in K and the homomorphism j_* obtained by forgetting those the simplexes of K that are contained in L, is exact.

Theorem 10.4.1 *Let C, C_1 and C_2 be chain complexes related by the short exact sequence*

$$0 \to C_1 \xrightarrow{i} C \xrightarrow{j} C_2 \to 0 \tag{10.12}$$

Then there are homomorphisms ∂ making the long sequence of homology groups

$$\cdots \to H_p(C_1) \xrightarrow{i_*} H_p(C) \xrightarrow{j_*} H_p(C_2) \xrightarrow{\partial} H_{p-1}(C_1) \to \cdots \tag{10.13}$$

exact.

Proof Left as an ecxercise. ❑

An immediate consequences of Theorem 10.4.1 is that the relative homology groups $H_p(K, L)$ for any pair (K, L) of simplicial complexes fit into a long exact sequence.

Theorem 10.4.2 (Exact sequence of the pair) *For any pair (K, L) of simplicial complexes, the sequence of the homology groups of these complexes and the relative homology groups of the pair (K, L)*

$$\cdots \to H_p(L) \xrightarrow{i_*} H_p(K) \xrightarrow{j_*} H_p(K, L) \xrightarrow{\partial} H_{p-1}(L) \to \cdots$$

is exact.

Proof Since the chain complexes $C(K)$, $C(L)$, $C(K, L)$ form a short exact sequence, the theorem follows from Theorem 10.4.1. ❑

Theorem 10.4.3 *If K_1 and K_2 are subcomplexes of a simplicial complex K such that $K = K_1 \cup K_2$, then there exist homomorphisms*

$$\partial : H_p(K_1 \cup K_2) \to H_{p-1}(K_1 \cap K_2)$$

such that the homology groups of the complexes $L = K_1 \cap K_2$, K_1, K_2 and K form the long exact sequence

$$\cdots \to H_p(L) \xrightarrow{i_*} H_p(K_1) \oplus H_p(K_2) \xrightarrow{j_*} H_p(K) \xrightarrow{\partial} H_{p-1}(L) \to \cdots$$

Proof It follows from Theorem 10.4.1. ❑

Remark 10.4.4 In addition to the long exact sequence of homology groups for the pair (K, L) of simplicial complexes, there is another long exact sequence, known as Mayer–Vietoris sequence, which is convenient at many situations.

Theorem 10.4.5 (Mayer–Vietoris) *If K_1 and K_2 are subcomplexes of a simplicial complex K such that $K_1 \cup K_2 = K$, then there is an exact sequence*

$$\cdots \to H_{p+1}(K) \xrightarrow{\partial} H_p(K_1 \cap K_2) \xrightarrow{i_*} H_p(K_1) \oplus H_p(K_2) \xrightarrow{j_*} H_p(K) \xrightarrow{\partial} H_{p-1}(K_1 \cap K_2)$$

Proof It follows from Theorem 10.4.3. ❑

10.5 Simplicial Cohomology Theory: Introductory Concepts

This section conveys the concept of simplicial cohomology theory. A cohomology group is the dual to homology group. Given an oriented simplicial complex K, cochain complex, cocycle, coboundary, cohomology class and the cohomology group are defined dually. In contrast with homology theory, a cohomology theory is a contravariant functor. In some sense, these two theories are adjoint (or dual) to each other. Historically, J.W. Alexander and A. Kolmogoroff invented simplicial cohomology in 1935 independently. E.Čech and H. Whitney developed it during 1935–1940. The terms 'coboundary', 'cocycle', 'cochain,' and 'cohomology' were given by E.Čech. The advantage of cohomology over homology is that the cohomology group admits an additional structure making it a ring.

Definition 10.5.1 A sequence $C^* = \{C^n, \delta^n\}, n \in \mathbf{Z}$ of additive abelian groups C^n together with a sequence of group homomorphisms $\delta^n : C^{n-1} \to C^n$, such that $\delta^{n+1} \circ \delta^n = 0, \forall n \in \mathbf{Z}$, is called a cochain complex and δ^n is called a coboundary homomorphism.

More precisely,

$$C^* : \cdots \longrightarrow C^{n-1} \xrightarrow{\delta^n} C^n \xrightarrow{\delta^{n+1}} C^{n+1} \longrightarrow \cdots \tag{10.14}$$

is called a cochain complex if $\delta^{n+1} \circ \delta^n = 0 \forall n \in \mathbf{Z}$.

Definition 10.5.2 The elements of $Z^n = \ker \delta^{n+1}$ are called n-cocycles and the elements of $B^n = \operatorname{Im} \delta^n$ are called n-coboundaries of the cochain complex C^* (10.14).

Proposition 10.5.3 *Z^n and B^n form groups for all n for the cochain complex C^* (10.14).*

Proof It follows from respective definitions. ❑

Proposition 10.5.4 *B^n is a subgroup of Z^n for all n for the cochain complex C^* (10.14).*

Proof It follows from the property of the cochain complex C^* (10.14) that $\delta^{n+1} \circ \delta^n = 0$. ❑

Definition 10.5.5 The quotient group Z^n/B^n for any cochain complex C^* (10.14), denoted by $H^n(C^*)$ (or simply H^n), is called the n-dimensional cohomology group of the cochain complex C^*.

For an oriented simplicial complex K, we define cohomology groups of K as follows:

Definition 10.5.6 Let K be an oriented simplicial complex. Then the cohomology groups of the corresponding cochain complex $C^*(K)$ are called the cohomology groups of K and are denoted by $H^n(K)$.

Definition 10.5.7 Let $C^* = \{C^n, \delta^n\}$ and $C'^* = \{C'^n, \delta'^n\}$ be two cochain complexes. Then a sequence $f = \{f^n : C^n \to C'^n\}, n \in \mathbf{Z}$ is called a cochain map from C^* to C'^* if the diagram in Fig. 10.4

commutes,.i.e., $f^{n+1} \circ \delta^n = \delta'^n \circ f^n$ holds for all n.

$$\begin{array}{ccccccccc}
\cdots & \longrightarrow & C^n & \xrightarrow{\delta^n} & C^{n+1} & \xrightarrow{\delta^{n+1}} & C^{n+2} & \longrightarrow & \cdots \\
 & & \downarrow{\scriptstyle f^n} & & \downarrow{\scriptstyle f^{n+1}} & & \downarrow{\scriptstyle f^{n+2}} & & \\
\cdots & \longrightarrow & C'^n & \xrightarrow{\delta'^n} & C'^{n+1} & \xrightarrow{\delta'^{n+1}} & C'^{n+2} & \longrightarrow & \cdots
\end{array}$$

Fig. 10.4 Cochain map

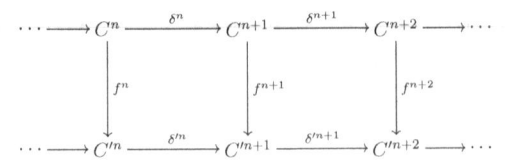

Each simplicial map $f : K \to K'$ between simplicial complexes induces homomorphisms:

$$f^* : H^n(K') \to H^n(K)$$

of the cohomology groups.

Proposition 10.5.8 *A simplicial cohomology theory $H^* = \{H^n, \delta^n\}$ is a sequence of contravariant functors from the category of simplicial complexes and the simplicial maps to the category of abelian groups and homomorphisms.*

Proof Each simplicial map $f : K \to K'$ between simplicial complexes induces homomorphisms

$$f^* : H^n(K') \to H^n(K)$$

of their cohomology groups and these homomorphisms act in the opposite direction. Hence it is easy to prove the theorem. ❑

Remark 10.5.9 The homology groups and cohomology groups of a simplicial complex are closely related. If we know homology groups of K, we can find the cohomology groups of K and conversely, provided the corresponding chain complex is finitely generated and free.

Recall that for a finitely generated abelian group G, if F and T represent free part and torsion part of G respectively, then G is always isomorphic to the group $F \oplus T$ (Adhikari and Adhikari 2014).

Theorem 10.5.10 *For the simplicial homology groups $H_n(C)$ and the cohomolgy groups $H^n(C)$ of any finitely generated free chain complex C,*

(i) *Free part of $H^n(C)$= Free part of $H_n(C)$;*
(ii) *Torsion part of $H^n(C)$= Torsion part of $H_{n-1}(C)$;*
(iii) *$H^n(C)$ is isomorphic to the direct sum*

$F \oplus T$, where F is the free part of $H_n(C)$ and T is the torsion group of $H_{n-1}(C)$.

Proof Left as an exercise. ❑

10.6 Simplicial Cohomology Ring

This section defines cup product of cochains with an eye to endow the direct sum of all the cohomolgy groups of a simplicial complex (with coefficients in a commutative ring) a ring structure. This algebraic structure given independently by Alexander and Kolmogroff in 1935 by defining a product, now called cup product of cochains, has wide applications in algebraic topology. But this does not fit for homolgy groups. This is the advantage of cohomolgy theory over homology theory.

Let K be simplicial complex and $\Delta : |K| \to |K| \times |K|$ be the diagonal map defined by $\Delta(x) = (x, x)$. It induces homomorphisms $\Delta^* : H^n(K \times K) \to H^n(K)$. We are now in a position to define the cohomology product.

Definition 10.6.1 Let $x \in H^n(K)$, $y \in H^m(K)$ be two elements of cohomology groups of a simplicial complex K. Then their cup product $x \cup y$ is defined by the rule $x \cup y = \Delta^* i(x \otimes y)$, where $x \otimes y$ defines embedding

$$i : \sum_{n+m} H^n(K) \otimes H^n(K) \to H^{n+m}(K \times K).$$

This definition shows that to find the product of two elements of cohomology groups we first consider their tensor product as an element of the corresponding cohomology group of the direct product $K \times K$ and take its image under the homomorphism Δ^* induced by the diagonal map.

Let R be a ring. The cup product is associative and distributive, it is natural to try to make the cup product the multiplication in a ring structure on the cohomology groups of a simplicial complex K. If we define $H^*(K; R)$ to be the direct sum of the groups $H^n(K; R)$, then the elements of $H^*(K; R)$ are the finite sum $\sum x_i$ with $x_i \in H^i(K; R)$ and the product of two such sums is defined to be

$$\left(\sum_i x_i\right)\left(\sum_j y_j\right) = \sum_{i,j} x_i y_j.$$

This makes $H^*(K; R)$ into a ring with identity if R has an identity.

Example 10.6.2 For the real projective plane $\mathbf{R}P^2$, $H^*(\mathbf{R}P^2, \mathbf{Z}_2)$ consists of all the polynomials $a_0 + a_1x + a_2x^2$ with coefficients $a_0, a_1, a_2 \in \mathbf{Z}_2$ and hence $H^*(\mathbf{R}P^2, \mathbf{Z}_2)$ is the quotient ring $\mathbf{Z}_2[x]/< x^3 >$.

Remark 10.6.3 Geometrically, the cup product on manifolds is interpreted as 'intersection numbers', see Dold (1972, Chap. 7), Munkres (1984, Chap. 8).

10.7 Singular Homology

This section presents singular homology using the algebraic properties of singular chain complexes. Singular homology theory generalizes the simplicial homology theory. The former is easier to work while the latter is easier to visualize geometrically. These two theories are related by the basic result that the singular homology of a polyhedron is isomorphic to the simplicial homology of any of its triangulated simplicial complexes. Again for any pointed topological space X, the homotopy groups $\pi_n(X)$ are very important invariants. It is easy to define $\pi_n(X)$ but difficult to compute them. Only for a few CW-complexes their homotopy groups are known. So this section defines singular homology groups $H_n(X)$ which are different invariants from simplicial homology groups. The algebraic properties of singular chain complexes lead to singular homology theory.

10.7.1 Singular Homology Groups

This subsection defines singular homology groups (modules). We have defined in Chap. 6 the standard n-simplex

$$\Delta^n = \{(t_0, t_1, \ldots, t_n) \in \mathbf{R}^{n+1} : 0 \le t_i \le 1, \Sigma t_i = 1\} \subset \mathbf{R}^{n+1}.$$

It has vertices $v_0 = (1, 0, \ldots, 0)$, $v_1 = (0, 1, 0, \ldots, 0), \ldots, v_n = (0, 0, \ldots, 0, 1)$ in the space $\mathbf{R}^{n+1}$. In particular, it defines orientation of Δ^n. The simplex Δ^n has the ith face

$$\Delta^{n-1}(i) = \{(t_0, t_n, \ldots, t_n) : t_i = 0\}.$$

It is a standard $(n-1)$-simplex in the space $\mathbf{R}^n$.

Definition 10.7.1 A singular n-simplex of the space X is a continuous map $f : \Delta^n \to X$.

Definition 10.7.2 A singular n-chain is a finite linear combination $\Sigma n_i f_i$, where each $f_i : \Delta^n \to X$ is a singular n-simplex, $n_i \in \mathbf{Z}$. The group of n-chains $C_n(X)$ is a free abelian group generated by all singular n-simplexes of X.

Definition 10.7.3 The boundary homomorphism is defined by

$$\partial : C_n(X) \to C_{n-1}(X),\ f \mapsto \Sigma_{i=0}^{n}(-1)^i e_i(f),$$

where e_i is given by $e_i(f) = f|_{\Delta^{n-1}(i)}$, the restriction of f on the ith face $\Delta^{n-1}(i)$.

Lemma 10.7.4 *The composite* $\partial^2 = \partial \circ \partial = 0$.

Proof It is sufficient to prove that the composite homomorphism

$$C_{n+1}(X) \xrightarrow{\partial_{n+1}} C_n(X) \xrightarrow{\partial_n} C_{n-1}(X)$$

is trivial, i.e., Im $\partial_{n+1} \subset$ ker ∂_n. Clearly it follows from the definition:

$$(e_i \circ e_j)(f) = e_i(e_j(f)) = \begin{cases} e_{j-1}(e_i(f)), & \text{for } j > 1, \\ e_j(e_{i+1}(f)), & \text{for } j \le 1 \end{cases}$$ ❑

The remaining part of the proof is left as an exercise.

Proposition 10.7.5 *The complex $C(X)$ defined by*

$$\cdots \longrightarrow C_{n+1}(X) \xrightarrow{\partial_{n+1}} C_n(X) \xrightarrow{\partial_n} C_{n-1}(X) \to \cdots \to C_1(X) \xrightarrow{\partial_1} C_0(X) \to 0. \tag{10.15}$$

is a chain complex.

Proof It follows from the Lemma 10.7.4. ❑

Definition 10.7.6 The group $Z(X) = \ker \partial_n$ is called the group of cycles, and the group $B_n(X) = \operatorname{Im} \partial_{n+1}$ is called the group of boundaries.

Proposition 10.7.7 *$B_n(X)$ is a subgroup of $Z_n(X)$.*

Proof It follows from the Lemma 10.7.4 that $B_n(X)$ is a subgroup of $Z_n(X)$. ❑

Definition 10.7.8 The nth homology group (module) $H_n(X)$ of X is defined by $H_n(X) = \ker \partial_n / \operatorname{Im} \partial_{n+1}$, i.e., $H_n(X) = Z_n(X)/B_n(X)$.

Remark 10.7.9 Convention: The group $H_0(X) = C_0(X)/\operatorname{Im} \partial_1$ and $H_n(X) = 0$ for $n < 0$.

Definition 10.7.10 Let $c_1, c_2 \in C_n(X)$. Then the chain c_1 is said to be chain homotopic to the chain c_2 if $c_1 - c_2 = \partial_{n+1}(d)$ for some $d \in C_{n+1}(X)$. The class $[c]$ (under the chain homotopy relation)$\in H_n(X)$ is called a homological class of the cycle $c \in C_n(X)$.

Remark 10.7.11 Chain homotopic maps induce the same homomorphism on homology.

Remark 10.7.12 The group $H_n(X)$ is an abelian group which is a module. If it is finitely generated, then

$$H_n(X) \cong \mathbf{Z} \oplus \mathbf{Z} \oplus \cdots \oplus \mathbf{Z} \oplus \mathbf{Z}_{n_1} \oplus \cdots \oplus \mathbf{Z}_{n_t} \tag{10.16}$$

Definition 10.7.13 The rank of $H_n(X)$ is the number of $\mathbf{Z}$'s in the decomposition (10.16) and it is called the Betti number of the space X. In symbol, $\beta_n(X) = rank(H_n(X))$.

Remark 10.7.14 The Betti numbers are named after E. Betti (1823–1892) and generalize the connectivity number which he used while studying curves and surfaces. Euler characteristic of X is closely related to its Betti number $\beta_n(X)$ (see Sect. 10.17).

Remark 10.7.15 $H_n(X) = H_n(C(X))$.

Definition 10.7.16 Let $g : X \to Y$ be a continuous map. Then g induces the homomorphism $g_\square : C_n(X) \to C_n(Y)$, which maps a singular simplex $f : \Delta^n \to X$ to a singular simplex $g \circ f : \Delta^n \to Y$. It defines a homomorphism $g_* : H_n(X) \to H_n(Y)$, called the homomorphism induced by g.

Theorem 10.7.17 *Let $f \simeq g : X \to Y$. Then $f_* = g_* : H_n(X) \to H_n(Y)$ for all $n \geq 0$.*

Proof As $f \simeq g : X \to Y$, there exists a homotopy $H : X \times I \to Y$ such that $H(x, 0) = f(x)$ and $H(x, 1) = g(x)$. Then for any singular simplex $s : \Delta^n \to X$, we have a map $H \circ (s \times I) : \Delta^n \times I \to Y$. The cylinder $\Delta^n \times I$ has a canonical simplicial structure. We divide $\Delta^n \times I$ into $(n+1)$-simplices $\widetilde{\Delta}^{n+1}(i), i = 0, 1, \ldots, n$ as follows:

$$\widetilde{\Delta}^{n+1}(i) = \{(t_0, t_1, \ldots, t_n, \alpha) \in \Delta^n \times I : t_0 + \cdots + t_{i-1} \leq \alpha \leq t_0 + \cdots + t_i\}.$$

Then the map $G : H \circ (s \times I) : \Delta^n \times I \to Y$ defines $(n+1)$-singular simplices of dimension $(n+1)$. We now define k as $k(s) = \sum_{i=0}^{n} (-1)^i G|_{\widetilde{\Delta}^{n+1}(i)}$.
Clearly, the homomorphisms

$$k_n : C_n(X) \to C_{n+1}(Y),\ \Sigma n_i s_i \mapsto \Sigma n_i k_n(s_i),\ n_i \in \mathbf{Z}$$

define a chain homotopy $k : C(X) \to C(Y)$. Hence the theorem follows. ❑

Corollary 10.7.18 *Homotopic maps induce the same homomorphism in homology groups.*

Proof It follows from Theorem 10.7.17. ❑

Corollary 10.7.19 *Homotopy equivalent spaces have isomorphic homology groups.*

Proof Let X and Y be two homotopy equivalent spaces. Then $H_n(X) \cong H_n(Y)$ for all $n \geq 0$ by Corollary 10.7.18. ❑

Corollary 10.7.20 *If X is contractable, then $H_n(X) = 0$ for all $n > 0$.*

Proof Since X has the homotopy type of one-point space. The corollary follows from Corollary 10.7.19 and dimension axiom (Exercises 6 of Sect. 10.21). ❑

Remark 10.7.21 The characteristic of homology groups is a homotopy invariant and hence it is a fundamental feature of homology groups.

Theorem 10.7.22 *Let X be a topological space and A be a subspace of X. Then the sequence of homology groups*

$$\cdots \to H_n(A) \xrightarrow{i_*} H_n(X) \xrightarrow{j_*} H_n(X, A) \xrightarrow{\partial} H_{n-1}(A) \xrightarrow{i_*} H_{n-1}(X) \to \cdots$$

is exact.

Proof Left as an exercise. ❑

10.7.2 Reduced Singular Homology Groups

This subsection defines reduced singular homology groups (modules) $\widetilde{H}_*(X)$ on a nonempty topological space X and shows that reduced homology modules are completely determined in terms of singular homology modules of X and vice-versa. Let $S_*(X)$ be the singular chain complex of X and C_p be an abstract chain complex such that $C_0 = R$, the R-module R and $C_p = 0$ for all $p \neq 0$. Then there is a chain map

$$f : S_*(X) \to C_*, \ \Sigma r_i x_i \mapsto \begin{cases} \Sigma r_i, & \text{if } p = 0 \\ 0, & \text{if } p > 0. \end{cases}$$

Then f is an onto map and $\ker f$ is a sub-chain complex of $S_*(X)$, written as $\widetilde{S}_*(X)$, called the reduced singular chain complex of X. As $S_*(X)$ is a chain complex of free R-modules, the augmented map f splits and hence $S_0(X) = \widetilde{S}_0(X) \oplus R$ and $S_p(X) = \widetilde{S}_p(X)$ for all $p > 0$.

Definition 10.7.23 The reduced p-dimensional homology module of a nonempty space X, denoted by $\widetilde{H}_p(X)$ is defined to be the homology module of the chain complex $\widetilde{S}_*(X)$.

As $f(\text{Im}\, \partial_1) = 0$, it follows that $\text{Im}\, \partial_1 \subset \ker f = \widetilde{S}_0(X)$. Hence

$$H_0(X) = \frac{S_0(X)}{\text{Im}\, \partial_1} = \frac{\widetilde{S}_0(X)}{\text{Im}\, \partial_1} \oplus R = \widetilde{H}_0(X) \oplus R,$$

$$H_p(X) = \widetilde{H}_p(X)$$

for all $p > 0$

Remark 10.7.24 The relation $\widetilde{H}_p = H_p$ except $\widetilde{H}_0 \oplus R = H_0$ shows that the reduced singular homology modules are completely characterized in terms of the reduced singular homology modules and play an important role in computing homology groups.

Remark 10.7.25 If G is the coefficient group of a homology theory H_*, then for the unique map f from a topological space X to a one-point space, the kernel of the homomorphism

$$f_* : H_0(X) \to G$$

is $\widetilde{H}_0(X)$

10.7.3 *Relative Singular Homology Groups*

This subsection generalizes the concept of singular homology groups of a nonempty space defined in Sect. 10.7.1 by introducing the concept of relative singular homology groups in the following way:

Let X be a space and A be a subspace of X. Then $C_n(A) \subset C_n(X)$, and $\partial_n(C_n(A)) \subset C_{n-1}(A)$ and each generator of the group $C_n(A)$ maps to a generator of the group $C_n(X)$. Let $C_n(X, A)$ be the quotient group $C_n(X)/C_n(A)$, which is a group of relative n-chains of the space X modulo the subspace A. The group $C_n(X, A)$ is a free abelian group with generators $g : \Delta^n \to X$, $f(\Delta^n) \cap (X \setminus A) \neq \emptyset$. Since the boundary operator $\partial_n : C_n(X) \to C_{n-1}(X)$ takes $C_n(A)$ to $C_{n-1}(A)$, it induces a quotient boundary map $\partial_n : C_n(X, A) \to C_{n-1}(X, A)$. Hence it produces a sequence of boundary maps by varying n

$$\cdots \to C_n(X, A) \xrightarrow{\partial_n} C_{n-1}(X, A) \xrightarrow{\partial_{n-1}} \cdots \xrightarrow{\partial_2} C_1(X, A) \xrightarrow{\partial_1} C_0(X, A) \to 0.$$

Just as before, a calculation shows that $\partial^2 = 0$. Hence $\{C_n(X, A), \partial_n\}$ forms a chain complex.

Definition 10.7.26 The resulting homology groups $H_n(X, A)$ are called relative singular homology groups of (X, A).

Remark 10.7.27 By considering the definition of the relative boundary map ∂, we see:

(i) elements of $H_n(X, A)$ are represented by relative cycles: n-chains $\alpha \in C_n(X)$ are such that $\partial\alpha \in C_{n-1}(A)$;
(ii) a relative cycle α is trivial in $H_n(X, A)$ off it is a relative boundary, i.e., $\alpha = \partial\beta + \gamma$ for some $\beta \in C_{n+1}(X)$ and $\gamma \in C_n(A)$.

These properties present the intuitive idea precisely that $H_n(X, A)$ is 'homology of X modulo A'.

Theorem 10.7.28 (Exact sequence of the pairs of spaces) *Let (X, A) be a pair of spaces. Then the sequence of homology groups*

$$\cdots \xrightarrow{\partial} H_n(A) \xrightarrow{i_*} H_n(X) \xrightarrow{j_*} H_n(X, A) \xrightarrow{\partial} H_{n-1}(A) \xrightarrow{i_*} \cdots$$

is exact, where $i : A \hookrightarrow X$ and $j : X \to (X, A)$ are inclusion maps.

Proof See Rotman (1988, pp. 96). ❑

Proposition 10.7.29 (Homotopy property for pairs of spaces) *Let $f, g : (X, A) \to (Y, B)$ be two maps homotopic through maps of pairs $(X, A) \to (Y, B)$ of spaces. Then $f_* = g_* : H_n(X, A) \to (Y, B)$.*

Proof See Rotman (1988, pp. 104). ❑

10.8 Eilenberg–Zilber Theorem and Künneth Formula

This section gives Eilenberg–Zilber theorem and Künneth formula which are used for computing homology or cohomolgy of product spaces.

10.8.1 Eilenberg–Zilber Theorem

This subsection gives Eilenberg–Zilber Theorem.

Theorem 10.8.1 *For topological spaces X and Y, there is a (natural) chain equivalence*

$$\psi : C_*(X \times Y) \to C_*(X) \otimes C_*(Y)$$

which is unique up to chain homotopy. Moreover,

$$H_n(X \times Y) \cong H_n(C_*(X) \otimes C_*(Y))$$

for all $n \geq 0$.

Proof See Rotman (1988, pp. 266). ❑

10.8.2 Künneth Formula

This subsection gives Künneth formula which gives a split exact sequence with middle term as given in Theorem 10.8.2.

Theorem 10.8.2 (Künneth formula) *For every pair of topological spaces X and Y and for every integer $n \geq 0$,*

$$H_n(X \times Y) \cong \sum_{i+j=n} H_i(X) \otimes H_j(Y) \oplus \sum_{p+q=n-1} \text{Tor}\,(H_p(X), H_q(Y))$$

Proof See Rotman (1988, pp. 270). ❑

Remark 10.8.3 (*Original version of Künneth formula*) If X and Y are compact polyhedra, then

$$b_n(X \times Y) = \sum_{i+j=n} b_i(X) b_j(Y),$$

where $b_i(X)$ is the ith Betti number of X. It follows from the Theorem 10.8.2. Because for any $f \cdot g$ (finitely generated) abelian groups A and B, the group Tor (A, B) is finite and hence it has no contribution to the calculation of the Betti numbers.

Example 10.8.4 For positive integers m, n and $m \neq n$,

$$H_p(S^m \times S^n; \mathbf{Z}) \cong \begin{cases} \mathbf{Z}, & \text{if } p = 0, m, n, m+n \\ 0, & \text{otherwise.} \end{cases}$$

If $m = n$, then

$$H_p(S^n \times S^n; \mathbf{Z}) \cong \begin{cases} \mathbf{Z}, & \text{if } p = 0, 2n \\ \mathbf{Z} \oplus \mathbf{Z}, & \text{if } p = n \\ 0, & \text{otherwise.} \end{cases}$$

For $m = n = 1$, this gives the homology groups of the torus $S^1 \times S^1$.

Example 10.8.5 If $X = S^1 \vee S^2 \vee S^3$ (wedge), then

$$H_p(X; \mathbf{Z}) \cong \begin{cases} \mathbf{Z}, & \text{if } p = 0, 1, 2, 3 \\ 0, & \text{otherwise.} \end{cases}$$

Remark 10.8.6 It follows from Examples 10.8.4 and 10.8.5 that the spaces $S^1 \times S^2$ and $S^1 \vee S^2 \vee S^3$ are the same homology groups, but they are not homotopy equivalent

10.9 Singular Cohomology

This section introduces the concept of singular cohomology which is an algebraic variant of homology. The basic difference between them is that cohomology groups are contravariant functors but homology groups are covariant functors. The homology groups determine the corresponding cohomology groups but its converse is true if the homology groups are finitely generated. The cohomology groups (modules or rings) of a topological space were not recognized until late 1930, when S. Lefschetz formulated a simplified proof of the duality theorem for orientable n-manifold with boundary.

Definition 10.9.1 Given a topological space X and an abelian group G, the singular n-cochain group $C^n(X; G)$ with coefficients in G is defined to be the dual group given by $C^n(X; G) = \text{Hom}\,(C_n(X; G), G)$ of the singular chain group $C_n(X; G)$.

Remark 10.9.2 An n-cochain $\alpha \in C^n(X; G)$ assigns to each n-simplex $\sigma : \Delta^n \to X$ a value $\alpha(\sigma) \in G$. Since the singular n-simplexes form a basis of $C_n(X)$, these values can be assigned arbitrarily. Hence n-cochains are precisely the functions from singular n-simplices to G. Again $C^n(X; G)$ is isomorphic to the direct product of as many copies of G as there are n-simplexes in X.

Definition 10.9.3 Let G be an abelian group. Given a cochain $\alpha \in C^n(X; G)$, the element $\alpha : C_n(X; G) \to G$ is a a homomorphism. Then the coboundary map $\delta : C_{n+1} \to G$ is defined by $\delta(\alpha) = \alpha \circ \partial$, which is the composite map

$$C_{n+1}(X; G) \xrightarrow{\partial} C_n(X; G) \xrightarrow{\alpha} G.$$

This implies that for a singular $(n+1)$-simplex

$$\sigma : \Delta^{n+1} \to X, \ \ \delta\alpha(\sigma) = \sum_i (-1)^{\alpha}(\sigma|_{\langle v_0 \cdots \hat{v}_i \cdots v_{n+1} \rangle}),$$

where 'hat' symbol $\hat{}$ over v_i indicates that this vertex is deleted from the sequence $v_0, v_1, \ldots, v_n$.

Clearly, $\delta^2 = \delta \circ \delta = 0$.

Definition 10.9.4 The cohomology group $H^n(X; G)$ with coefficient group G is defined to be the quotient group $\ker \delta / \operatorname{Im} \delta$ at $C^n(X; G)$ in the cochain complex

$$\cdots \leftarrow C^{n+1}(X; G) \xleftarrow{\delta^n} C^n(X; G) \xleftarrow{\delta^{n-1}} C^{n-1}(X; G) \leftarrow \cdots \leftarrow C^0(X; G) \leftarrow 0$$

The group of n-cocycles is $\ker \delta^n$ and is denoted by $Z^n(X; G)$ and the group of n-boundaries is $\operatorname{Im} \delta^{n-1}$ and is denoted by $B^n(X; G)$.

Definition 10.9.5 An element of $H^n(X; G)$ is a coset $\beta + B^n(X; G)$, where β is an n-cocycle. This is called the cohomology class of β, denoted by $[\beta] \in H^n(X; G)$.

Remark 10.9.6 For a cochain α, $\delta \circ \alpha = \alpha \circ \delta = 0$. Hence α vanishes on n boundaries.

10.10 Relative Cohomology Groups

The relative cohomology groups $H^n(X, A; G)$ for a pair (X, A) with coefficient group G are defined by dualizing the short exact sequence

$$0 \longrightarrow C_n(A; G) \xrightarrow{i} C_n(X; G) \xrightarrow{j} C_n(X, A; G) \longrightarrow 0$$

and by applying Hom $(-, G)$ functor to obtain

$$0 \leftarrow C^n(A; G) \xleftarrow{i^{\square}} C^n(X; G) \xleftarrow{j^{\square}} C^n(X, A; G) \leftarrow 0$$

where by definition $C^n(X, A; G) = \operatorname{Hom}(C_n(X, A), G)$. This sequence is exact.

Relative coboundary operators

$$\delta : C^n(X, A; G) \to C^{n+1}(X, A; G)$$

are defined by restriction of absolute δ's, and hence relative cohomology groups $H^n(X, A; G)$ are obtained.

The maps $i^{\square}$ and $j^{\square}$ commute with δ, since i and j commute with δ. Hence the short exact sequence of cochain groups is part of short exact sequence of cochain complexs, which give rise to an associated long exact sequence of cohomology groups

$$\cdots \to H^n(X, A; G) \xrightarrow{j^*} H^n(X; G) \xrightarrow{i^*} H^n(A; G) \xrightarrow{\delta} H^{n+1}(X, A; G) \to \cdots$$

10.11 Hurewicz Homomorphism

This section establishes a close connection between homotopy and homology groups of a certain class of topological spaces through Hurewicz homomorphism defined by Hurewicz (1904–1956) in 1935 in his paper (Hurewicz 1935). He first asserted that for a simplicial pair (K, L) if $\pi_r(K, L) = 0$ for $1 \leq r < n (n \geq 2)$, then $\pi_r(K, L) \to H_r(K, L)$ is an isomorphism (Original version of Hurewicz theorem). This paper cast light for the first time onto the relationship between homological and homotopical invariants. A series of four papers of Hurewicz published during 1935–1936, has greatly influenced the development the modern homotopy theory. In one sense homology is an approximation to homotopy. For more results see Sects. 17.1.2 and 17.1.3 of Chap. 17.

The original version of Hurewicz theorem is subsequently refined and is now studied.

Definition 10.11.1 Let X be a pointed topological space with a base point $x_0 \in X$ and s_n be the standard generator of $H_n(S^n), n = 1, 2, \ldots,$ If $f : S^n \to X$ represents an $\alpha \in \pi_n(X, x_0)$, then the induced homomorphism $f_* : H_n(S^n) \to H_n(X)$ defines an element $f_*(s_n) \in H_n(X)$. Set $h(\alpha) = f_*(s_n)$. Since $h(\alpha) \in H_n(X)$ does not depend on f, define a correspondence $h : \pi_n(X, x_0) \to H_n(X), \alpha \mapsto h(\alpha), n = 1, 2, \ldots$. Then h is a homomorphism, called the Hurewicz homomorphism.

Theorem 10.11.2 (Hurewicz) *Let (X, x_0) be a pointed topological space, such that $\pi_0(X, x_0) = 0, \pi_1(X, x_0) = 0, \ldots, \pi_{n-1}(X, x_0) = 0$, where $n \geq 2$. Then*

$$H_1(X) = 0, H_2(X) = 0, \ldots, H_{n-1}(X) = 0,$$

and the Hurewicz homomorphism $h : \pi_n(X, x_0) \to H_n(X)$ is an isomorphism.

$$\begin{array}{ccc} \pi_n(\vee S^n_\alpha) & \xrightarrow{h} & H_n(\vee S^n_\alpha) \\ \downarrow{\scriptstyle (\gamma_\beta)_*} & & \downarrow{\scriptstyle (\gamma_\beta)_*} \\ \pi_n(S^n_\beta) & \xrightarrow{h} & H_n(S^n_\beta) \end{array}$$

Fig. 10.5 Rectangular diagram related to h

Proof By Ex. 15 of sect. 8.9 of Chap. 8, there exists a CW-complex Y and a weak homotopy equivalence $f : Y \to X$. Then f induces an isomorphism in homology groups. So without loss of generality we may assume that X is a CW-complex. This means by the given conditions of the theorem that X is an $(n-1)$-connected CW-complex. Then up to homotopy equivalence X may be chosen so that X has a single 0-cell, and it does not have any cells of dimensions $1, 2, \ldots, n-1$. Hence $H_1(X) = 0, H_2(X) = 0, \ldots, H_{n-1}(X) = 0$. The nth skeleton $X^{(n)}$ is a wedge of spheres, i.e., $X^{(n)} = \bigvee\limits_{\alpha} S^n_\alpha$. Let $g_\alpha : S^n_\alpha \to \bigvee\limits_{\alpha} S^n_\alpha$ be the embedding of the αth sphere, and let $k_\beta : S^n \to \bigvee\limits_{\alpha}^{n}$ be the attaching maps of the $(n+1)$-cells e^{n+1}_β. Then the maps g_α determine the generators of the group $\pi_n(X^{(n)})$. Let $\gamma_\beta \in \pi_n(X^{(n)})$ be the element determined by the maps k_β.

The first nontrivial homotopy group $\pi_n(X)$ is given as the factor group of the homotopy group $\pi_n(X^{(n)}) \cong \mathbf{Z} \oplus \cdots \oplus \mathbf{Z}$ by the subgroup generated by γ_β. The cellular chain group

$$C(X) = H_n(X^{(n)}) = H_n(\bigvee_\alpha S^n_\alpha) \text{ and } H_n(X) = C(X)/\text{Im}\, \partial_{n+1}.$$

Since the Hurewicz homomorphism $h : \pi_n(S^n) \to H_n(S^n)$ is an isomorphism, the diagram in Fig. 10.5 is commutative, where the horizontal homomorphisms are isomorphisms. Hence h induces an isomorphism $\pi_n(X, x_0) \to H_n(X)$. ❑

Corollary 10.11.3 *Let X be a simply connected topological space, and $H_1(X) = 0, H_2(X) = 0, \ldots, H_{n-1}(X) = 0$. Then $\pi_1(X) = 0, \pi_2(X) = 0, \ldots, \pi_{n-1}(X) = 0$ and the Hurewicz homomorphism $h : \pi_n(X, x_0) \to H_n(X)$ is an isomorphism.*

Definition 10.11.4 A pointed topological space X with base point x_0 is said to be n-connected if $\pi_i(X, x_0) = 0$, for all $i \leq n$.

Remark 10.11.5 0-connected space means path-connected and 1-connected means simply connected.

Corollary 10.11.6 *If X is a path-connected topological space, then there is an epimorphism*

$$h_1 : \pi_1(X) \to H_1(X; \mathbf{Z})$$

which induces an isomorphism

$$h_{1*} : \pi_1(X)/\ker h_1 \to H_1(X; \mathbf{Z})$$

with ker h_1 *the commutator subgroup of* $\pi_1(X)$.

Corollary 10.11.7 *If* X *is path-connected, then* h_1 *is an isomorphism iff the group* $\pi_1(X)$ *is abelian.*

Example 10.11.8 If X is the figure-eight, then its fundamental group is the free group on two generators by van Kampen theorem. Hence by Corollary 10.11.7, $H_1(X; \mathbf{Z})$ is the free abelian group on 2 generators (i.e., $\mathbf{Z} \times \mathbf{Z}$).

Theorem 10.11.9 *If* X *is an* n*-connected topological space with* $n \geq 2$, *then* $\widetilde{H}_q(X) = 0$ *for all* $q \leq n$, *and the Hurewicz map* $h : \pi_{n+1}(X) \to \widetilde{H}_{n+1}(X)$ *is an isomorphism.*

Proof It follows from Theorem 10.11.2 and Corollary 10.11.3. ❑

Remark 10.11.10 A complete proof can be found in Spanier (1966), Maunder (1970).

Corollary 10.11.11 $\pi_n(S^n) \cong \mathbf{Z}$.

Proof Since S^n is $(n-1)$-connected, $\pi_n(S^n) \cong \mathbf{Z}$. ❑

Remark 10.11.12 An equivalent formulization of Hurewicz theorem with its generalization has been discussed in Chap. 17.

10.12 Mayer–Vietoris Sequences

This section introduces the concepts of Mayer–Vietoris sequences in singular and simplicial homology theories. Let A, B be subspaces of a topological space X. Mayer–Vietoris Sequence prescribes a method to compute the singular homology groups of $A \cup B$ if we know the homology groups of A, B, and $A \cap B$.

10.12.1 Mayer–Vietoris Sequences in Singular Homology Theory

This subsection displays Mayer–Vietoris sequences in singular homology theory.

Definition 10.12.1 Let X be topological space. For a pair of subspaces $A, B \subset X$ such that $X = \text{Int}\,(A) \cup \text{Int}\,(B)$, in addition to the long exact sequence of the homology groups of the pair (X,A) of spaces, there is an exact sequence of the form

$$\cdots \xrightarrow{\partial} H_n(A \cap B) \xrightarrow{i_*} H_n(A) \oplus H_n(B) \xrightarrow{j_*} H_n(X) \xrightarrow{\partial} H_{n-1}(A \cap B) \longrightarrow \cdots \longrightarrow H_0(X) \longrightarrow 0 \tag{10.17}$$

The sequence (10.17) is called a Mayer–Vietoris sequence in singular homology.

Let A, B be two subspaces of X such that $A \cap B \neq \emptyset$ and $A \cup B = \text{Int}\,(A) \cup \text{Int}\,(B)$. Then the sequence

$$\cdots \to \widetilde{H}_n(A \cap B) \to \widetilde{H}_n(A) \oplus \widetilde{H}(B) \to \widetilde{H}_n(A \cup B) \to \widetilde{H}_{n-1}(A \cap B) \to \qquad (10.18)$$

is exact and is called reduced singular Mayer–Vietoris exact sequence for reduced singular homology.

Remark 10.12.2 There is another form of Mayer–Vietoris sequence which is sometimes more convenient to apply.

Definition 10.12.3 Let X_1 and X_2 be two subspaces of X such that X is the union of interiors of X_1 and X_2. If $f_i : X_1 \cap X_2 \to X_i$ and $g_i : X_i \to X$ are inclusion maps for $i = 1, 2$, define

$$\phi : H_n(X_1 \cap X_2) \to H_n(X_1) \oplus H_n(X_2), \alpha \mapsto (f_{1*}(\alpha), f_{2*}(\alpha))$$

$$\psi : H_n(X_1) \oplus H_n(X_2) \to H_n(X), (\alpha_1, \alpha_2) \mapsto g_{1*}(\alpha) - g_{2*}(\alpha)$$

Then there exists a long exact sequence

$$\cdots \to H_{n+1}(X) \xrightarrow{\Delta} H_n(X_1 \cap X_2) \xrightarrow{\phi} H_n(X_1) \oplus H_n(X_2) \xrightarrow{\psi} H_n(X) \xrightarrow{\Delta} H_n(X_1 \cap X_2) \to \cdots \qquad (10.19)$$

The sequence (10.19) is also called the Mayer–Vietoris sequence and the homomorphisms Δ are called the connecting homomorphisms.

Remark 10.12.4 The sequence (10.18) can be viewed as an analog of the von Kampen theorem, since if $A \cap B$ is path-connected, then H_1 terms of this sequence gives an isomorphism

$$\psi : H_1(X) \to H_1(A) \oplus H_1(B)/\text{Im}\, f.$$

It is the abelianized statement of the von Kampen theorem, and H_1 is the abelianized of π_1 for the path-connected spaces.

10.12.2 *Mayer–Vietoris Sequences in Simplicial Homology Theory*

If K_1, K_2 are any two subcomplexes of a simplicial complex K, then the following sequence

$$\cdots \xrightarrow{\partial} H_n(K_1 \cap K_2) \xrightarrow{i_*} H_n(K_1) \oplus H_n(K_2) \xrightarrow{j_*} H_n(K_1 \cup K_2) \xrightarrow{\partial} H_{n-1}(K_1 \cap K_2) \rightarrow \cdots \tag{10.20}$$

is exact, and is called Mayer–Vietoris sequence in simplicial homology.

Example 10.12.5 If $X = S^n$ with $A = E_n^+$ and $B = E_n^-$, the north and south hemisphere of S^n, then $A \cap B = S^{n-1}$. Consequently, in the reduced Mayer–Vietoris sequence (10.18), the terms

$$\widetilde{H}_i(A) \oplus \widetilde{H}_n(B) = 0$$

give isomorphisms

$$\widetilde{H}_i(S^n) \cong \widetilde{H}_{i-1}(S^{n-1}).$$

Example 10.12.6 (*One-point union of n-cells*) Let $X = X_1 \cup X_2 \cup \cdots \cup X_n$ be the one-point union of n-spaces X_i, each of which is homomorphic to S^1. Then X has a triangulation, which is homeomorphic to the union of n triangle $T_i, T_2, \ldots, T_n$ all having one vertex v (say) is common. Then

$$H_p(X_1 \cup X_2 \cup \cdots \cup X_n; \mathbf{Z}) \cong \begin{cases} \mathbf{Z}, & \text{if } p = 0 \\ \oplus^n \mathbf{Z}, & \text{if } p = 1 \\ 0, & \text{otherwise.} \end{cases}$$

10.13 Computing Homology Groups

The section computes homology groups of some spaces.

10.13.1 Homology Groups of a One-Point Space

Let P be a one-point space. Then there is a unique map $f_n : \Delta^n \rightarrow P$ for any n. Hence the chain complex corresponding to the point P which is viewed as a 0-dimensional simplex, is $C_n(P) = \mathbf{Z}$ for all $n \geq 0$.

Clearly,

$$H_n(P; \mathbf{Z}) \cong \begin{cases} \mathbf{Z}, & \text{if } n = 0 \\ 0, & \text{otherwise} \end{cases}$$

Definition 10.13.1 A topological space X having the same homology groups of a one- point space is called an acyclic space.

Example 10.13.2 Every contractible space X is acyclic with

$$H_n(X; \mathbf{Z}) \cong \begin{cases} Z, & \textbf{if } n = 0 \\ 0, & \text{otherwise.} \end{cases}$$

10.13.2 *Homology Groups of CW-complexes*

The main aim of this subsection is to develop a technique to compute homology groups of CW-complexes. The singular chain complex is too large to compute. We construct here a cellular chain complex $\mathcal{E}(X)$ which is smaller than $C(X)$. We start into computations of homology groups of spheres. The reduced homology groups $\widetilde{H}_n(X)$ are defined viz the augmented chain complex with coefficients $\mathbf{Z}$:

$$\cdots \longrightarrow C_0(X) \xrightarrow{\mathcal{E}_S} \mathbf{Z} \longrightarrow 0$$

with $\mathcal{E}_S$ defined by summing coefficients.

Theorem 10.13.3 (Reduced Homology groups of spheres) $\widetilde{H}_p(S^n) = \begin{cases} \mathbf{Z}, & \textit{if } p = n \\ 0, & \textit{otherwise}. \end{cases}$

Proof Consider a long exact sequence of the pair of spaces (D^n, S^{n-1}):

$$\cdots \longrightarrow \widetilde{H}_p(D^n) \longrightarrow H_p(D^n, S^{n-1}) \longrightarrow \widetilde{H}_p(S^{n-1}) \longrightarrow \widetilde{H}_{p-1}(D^n) \cdots \longrightarrow .$$

Hence $\widetilde{H}_p(D^n) = 0$, $\widetilde{H}_{p-1}(D^n) = 0$. Consequently, $H_p(D^n, S^{n-1}) \cong \widetilde{H}_p(S^{n-1})$. The theorem follows by induction on n. ❑

Theorem 10.13.4 *Let X be a CW-complex. Then $\widetilde{H}_{n+1}(\Sigma X) \cong \widetilde{H}_n(X)$ for each n.*

Proof $\Sigma X = C^+X \cup C^-X$ described as in Fig. 10.6.

We now consider a long exact sequence in homology for the pair of spaces (C^+X, X):

$$\cdots \longrightarrow \widetilde{H}_n(C^+X) \longrightarrow H_n(C^+X, X) \longrightarrow \widetilde{H}_n(X) \longrightarrow \widetilde{H}_{n-1}(C^+X) \longrightarrow \cdots$$

Since the cone C^+X is contractible, $\widetilde{H}_*(C^+X) = 0$. Consequently, $H_n(C^+X, X) \cong \widetilde{H}_n(X)$. Again since (C^+X, X) is always a Borsuk pair, it follows that

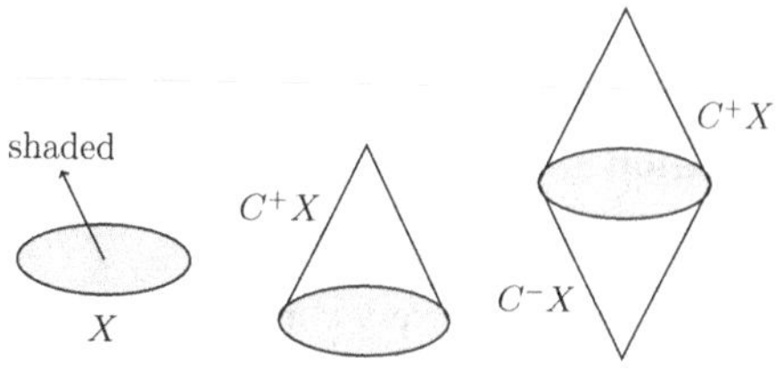

Fig. 10.6 Diagram for Suspension

$$H_n(C^+X, X) \cong \widetilde{H}_n(C^+X/X) \cong \widetilde{H}_n(C^+X \cup C^{-1}X) = \widetilde{H}_n(\Sigma X).$$

This proves the theorem. ❑

10.14 Cellular Homology

This section presents cellular homology which is a homology for the category of CW-complexes. It agrees with singular homology, and can provide an effective means for computing homology groups. This section introduces cellular homology theory that reflects the cellular structure of a CW-complex X. This theory is most suitable for computing homology groups of CW-complexes. Given a CW-complex decomposition E of a CW-complex X, a chain complex is defined whose group of n-chains, is a free abelian group for each $n \geq 0$, whose rank is the number of n-cells in E. The cellular chain complex is defined directly in terms of singular homology groups. Cellular homology is a homology functor from the category of CW-complexes and cellular maps to the category of abelian groups and homomorphisms. On the category of CW-complexes there is a natural equivalence from the cellular homology to the singular homology.

Definition 10.14.1 Let X be a cellular space and $n \geq 0$. Define $W_n(X) = H_n(X^{(n)}, X^{n-1})$(singular homology) and $d_n : W_n(X) \to W_{n-1}(X)$ as the composite $d_n = j_* \circ \partial$ as shown in Fig. 10.7.
where

$$j : (X^{(n-1)}, \emptyset) \hookrightarrow (X^{(n)}, X^{(n-1)})$$

is the inclusion map and ∂ is the connecting homomorphism arising from the long exact sequence of the pairs $(X^{(n)}, X^{(n-1)})$.

Proposition 10.14.2 *For a cellular space, $(W_*(X), d)$ is a chain complex.*

Proof It is sufficient to show that $d_n \circ d_{n+1} = 0$. Clearly, $d_n \circ d_{n+1}$ is the composite.

$$H_{n+1}(X^{(n+1)}, X^{(n)}) \to H_n(X^{(n)}) \to H_n(X^{(n)}, X^{(n-1)}) \to H_{n-1}(X^{(n-1)}) \to H_{n-1}(X^{(n-1)}, X^{(n-2)}).$$

This is zero, since the middle two arrows are adjacent arrows is the long exact sequence of the pair $(X^{(n)}, X^{(n-1)})$. ❑

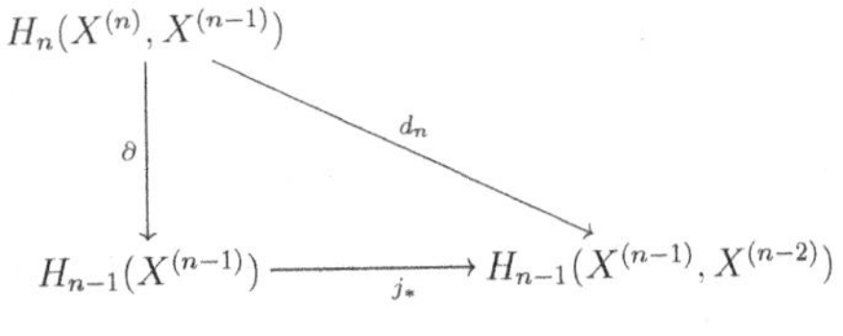

Fig. 10.7 Commutative triangle involving ∂, j_* and d_n

Definition 10.14.3 If X is a cellular space and $n \geq 0$, then its cellular homology group is defined as usual

$$H_n(W_*(X)) = \ker d_n \,/\, \mathrm{Im}\ d_{n+1}$$

Remark 10.14.4 Cellular homology in algebraic topology is a homology for the category of CW-complexes. It agrees with singular homology, and can provide an effective means for computing homology groups.

Remark 10.14.5 One sees from the cellular chain complex that the n-skeleton $X^{(n)}$ determines all lower-dimensional homology modules:
$H_k(X) \cong H_k(X^{(n)})$ for $k < n$.

Remark 10.14.6 An important consequence of this cellular chain complex is that if a CW-complex has no cells in consecutive dimensions, then all of its homology modules are free. For example, the complex projective space $\mathbf{CP}^n$ has a cell structure with one cell in each even dimension; it follows that for $0 \leq k \leq n$, $H_{2k}(\mathbf{CP}^n; \mathbf{Z}) \cong \mathbf{Z}$ and $H_{2k+1}(\mathbf{CP}^n; \mathbf{Z}) = 0$.

Remark 10.14.7 Generalization The Atiyah-Hirzebruch spectral sequence is the analogous method of computing the (co)homology of a CW-complex, for an arbitrary extraordinary (co)homology theory.

10.15 Čech Homology and Cohomology Groups

This section introduces Čech homology and cohomology groups. The homology theory constructed by Čech is called Čech homology theory after his name. Čech homology group of X with coefficient group G is denoted by $\check{H}_i(X; G)$. This theory defines homology invariants on topological spaces which are more general than polyhedra. The Čech cohomology theory is dual to the Čech homology theory.

In simplicial homology theory, certain topological properties of a polyhedra $|K|$ are expressed in algebraic terms. The algebraic information is obtained through the arrangement of the complex K into simplexes as faces of each other. In Čech homology theory corresponding to every finite open cover $\mathbf{U}$ of a topological space X, there is assigned a simplicial complex.

Definition 10.15.1 Let $\mathbf{U}$ be a finite open cover of a topological space X. The nerve of $\mathbf{U}$ denoted by $N(\mathbf{U})$, is the abstract complex whose vertices are members of $\mathbf{U}$ and whose simplexes are those subfamilies of $\mathbf{U}$ which have a nonempty intersection.

Example 10.15.2 The n-dimensional simplexes are those subfamilies of $\mathbf{U}$ with $(n+1)$ elements which have a nonempty intersection.

Remark 10.15.3 The term 'the complex $\mathbf{U}$' is used in place of 'the nerve defined by the finite open cover $\mathbf{U}$'.

A homology theory attains its full height only if it is defined for a pair of topological spaces. Let (X, A) be a topological pair and $\mathbf{U}$ be a cover of X. If $\mathbf{U}'$ is a cover of A, then $(\mathbf{U}, \mathbf{U}')$ is a pair of complexes but it may not be a simplicial pair, because $\mathbf{U}'$ is not in general a subcomplex of $\mathbf{U}$. So we assume that (X, A) is a compact pair.

Let (X, A) be a compact pair and $\mathbf{U}$ be an open cover of X. Then there is a simplicial pair $(\mathbf{U}, \mathbf{U}_A)$, where $\mathbf{U}_A$ is subfamily of $\mathbf{U}$ consisting of those sets which meet A. Then this pair defines a homology group $\check{H}((\mathbf{U}, \mathbf{U}_A); G)$, called Čech homology group of (X, A) with coefficient in G, where G is either a topological group or a vector space over a field.

Definition 10.15.4 Let $\mathbf{U} = \{U_\alpha\}$ be an open covering of X and $V = \{V_\beta\}$ be a refinement of $\mathbf{U}$ in the sense that V_β is contained in some U_α. Then these inclusions induce a simplicial map

$$N(V) \to N(V),$$

which is unique up to homotopy. The direct limit group $\lim_{\rightarrow} H^i(N(\mathbf{U}); G)$ with respect to finer and finer open cover $\mathbf{U}$ is called the Čech cohomologyindex $vCech$ cohomology group $\check{H}(X; G)$.

Remark 10.15.5 Relative Čech cohomology groups are defined in a way analogous to the Definition 10.15.4.

For full exposition of Čech homology and cohomology groups for an arbitrary pair (X, A) over a coefficient group G see Eilenberg and Steenrod (1952).

10.16 Universal Coefficient Theorem for Homology and Cohomology

This section studies universal coefficient theorem for homology and cohomology theories. The basic need for such study comes from the fact that homology and cohomolgy theories with coefficients in different abelian groups are frequently convenient than the corresponding theories in integral coefficients. For example, in the group $\mathbf{Z}_2$ the elements 1 and -1 coincide. This shows that there is no need to consider orientations of simplexes but it is simplicial to consider unoriented complexes. This makes many definition simpler. If the coefficient group is a field of characteristic 0 such as field $\mathbf{R}$ or field $\mathbf{Q}$, then there is no torsion and any homology group has the form $F \oplus F \oplus \cdots \oplus F$, which is completely determined by its rank.

10.16.1 Homology with Arbitrary Coefficient Group

This subsection discusses homology groups $H_n(X; G)$ with an arbitrary coefficient group G (an abelian group), which is a natural generalization of $H_n(X) = H_n(X; \mathbf{Z})$.

It is sometimes gained by this generalization. It has been working so far with homology groups of chain complexes in which the chain groups are free abelian groups. Thus given a topological space X, each element of $S(X)$ is a formal linear combination $\sum_i m_i s_i$, where each s_i is a singular simplex and $m_i \in \mathbf{Z}$. Given an abelian group G. it is sometimes helpful to make a generalization in which $m_i \in G$. The new born complex is as usual denoted $S(X; G)$ and the corresponding homology groups $H_n(X; G)$ are called the homology group of X with coefficients in G. In this sense $H_n(X; G)$ is a generaization of $H_n(X) = H_n(X; \mathbf{Z})$.

There is a natural question: how are the homology groups with coefficients in an arbitrary abelian group G and those with coefficients in $\mathbf{Z}$ related. Universal coefficient theorem gives its answer.

Theorem 10.16.1 (Universal coefficient theorem for homology)

(i) *For any space X and any abelain group G, there are exact sequences for all $n \geq 0$:*

$$0 \longrightarrow H_n(X) \otimes G \xrightarrow{\alpha} H_n(X; G) \longrightarrow \text{Tor}\,(H_{n-1}(X; G) \longrightarrow 0 \qquad (10.21)$$

where $\alpha : H_n(X) \otimes G \rightarrow H_n(X; G)$, $[z]' \otimes g \mapsto [z \otimes g]$.

(ii) *The sequence (10.21) splits:*

$$H_n(X; G) \cong H_n(X) \otimes G \oplus \text{Tor}\,(H_{n-1}(X); G) \qquad (10.22)$$

where '$\otimes$' denotes the usual tensor product of two groups and 'Tor' denotes the usual torsion product of two abelian groups.

Proof See Rotman (1988, pp. 262). ❑

Corollary 10.16.2 *If* Tor $(H_{n-1}(X), G)$ *vanishes in (10.22), then*

$$H_n(X) \otimes G \cong H_n(X; G).$$

Proof See Rotman (1988, pp. 264) ❑

Theorem 10.16.3 (Universal Coefficient Theorem for simplicial homology) *Let K be an oriented simplicial complex. Then for any integer n,*

$$H_n(K; G) \cong H_n(K; \mathbf{Z}) \otimes G \oplus \text{Tor}\,(H_{n-1}(K; \mathbf{Z}) * G.$$

Proof It follows from Theorem 10.16.1(ii). ❑

Remark 10.16.4 It is sometimes used to write simply $H_n(X)$ in place of $H_n(X; \mathbf{Z})$, which is a special case of $H_n(X; G)$.

As the groups $H_n(X; G)$ are completely determined by the groups $H_n(X)$ (see Universal Coefficient Theorem) this generalization cannot offer new information about X. But one can gain in this generalization: $H_*(X; G)$ may be easier sometimes to handle than $H_*(X)$. For example, if K is a simplicial complex and $G = F$ is a field, then $H_n(K; F)$ is a finite dimensional vector space for each n over F, and hence it is determined up to isomorphism by dimension of this vector space. Another gain in this case is that the homorphisms induced by continuous maps in the corresponding homology groups are linear maps of vector spaces which are utilized to obtain many interesting results.

Another convenience is for example, in the group $\mathbf{Z}_2$ the elements 1 and -1 coincide. This implies that there is no need of keeping track of orientations of simplexes and unoriented complexes need to be considered.

Remark 10.16.5 There is a natural relation between homology groups with coefficients in an arbitrary abelian group G and homolgy groups with integral coefficients given by Universal Coefficient Theorem.

10.16.2 Universal Cohomology Theorem for Cohomology

We now give the universal coefficient theorem for cohomology corresponding to the universal coefficient theorem of homology.

Definition 10.16.6 A chain complex C_* is said to be of finite type if each of its terms C_n is finitely generated.

Definition 10.16.7 A topological space X is said to be of finite type if each of its homology groups $H_n(X)$ is finitely generated.

Example 10.16.8 **(i)** Every compact polyhedra in a space of finite type;
(ii) Every compact CW-complex is a space of finite type;
(iii) $\mathbf{R}P^\infty$ is a space of finite type but it is not compact.

Theorem 10.16.9 (Universal coefficient theorem for cohomology)

(i) *Let X be a topological space of finite type and G be an abelian group. There is an exact sequence for every $n \geq 0$;*

$$0 \longrightarrow H^n(X) \otimes G \xrightarrow{\alpha} H^n(X; G) \longrightarrow \text{Tor}\,(H^{n+1}(X), G) \longrightarrow 0. \quad (10.23)$$

Here $\alpha : H^n(X) \otimes G \longrightarrow H^n(X; G)$, $[z] \times g \mapsto [zg]$, where $zg : \sigma \mapsto z(\sigma)g$ for an n-simplex σ in X (as $z(\sigma) \in \mathbf{Z}$, $z(\sigma)g$ has a meaning).

(ii) *The sequence (10.23) splits:*

$$H^n(X; G) \cong H^n(X) \otimes G \oplus \text{Tor}\,(H^{n+1}(X), G).$$

Proof See Rotman (1988, pp. 388). ❑

10.17 Betti Number and Euler Characteristic

This section studies Betti number and Euler characteristic of a polyhedron, which are closely related and revisits Euler characteristic from the viewpoint of homology theory. The Swiss mathematician Leonhard Euler gave a formula for comparing geometrical objects mathematically which relates the number of vertices V, the number of edges E, and the number of faces F of a polyhedron in an alternating sum $V - E + F = 2$ of a 3-dimensional polyhedron P. This result was given by Euler in 1752. Poincaré gave first application of his homology theory, which is a generalization of Euler formula to general polyhedra. The characteristic $\chi(P)$ of P is defined by $\chi(P) = V - E + F$.

Definition 10.17.1 If $G = \mathbf{R}$ in the homology group $H_n(K; G)$ of a finite simplicial complex K with coefficient group G, then the group $H_n(K; G)$ is a real vector space. If its dimension is q, then q is called the called the qth Betti number of K, denoted by β_q.

Euler characteristic of a topological space is also a topological invariant readily computable by 'polyhedronization' of the space. Homology groups are refinements of the Euler characteristic in some sense. Euler characteristic of a space X is an integer. It is different from other topological invariants such as compactness or connectedness which reflects geometrical properties of X. In this section we study Euler characteristic using vector spaces, graph theory, and algebraic topology.

10.17.1 Euler Characteristics of Polyhedra

Euler characteristic is a numerical invariant which can be used to distinguish topologically nonequivalent spaces. The search of other invariants has established a remarkable connection between two branches of modern mathematics: topology and modern algebra. There are many algebraic invariants associated with topological spaces: most commonly, we associate a group to a space, in such a way that topologically equivalent spaces have isomorphic groups. Historically, Euler's theorem asserts: if K is any polyhedron homeomorphic to S^2, with V vertices, E edges, and F 2-dimensional faces, then $V - E + F = 2$.

Definition 10.17.2 The Euler characteristic of a simplicial complex K of dimension n is defined by the alternative sum

$$\begin{aligned}\chi(K) &= \alpha_0(K) - \alpha_1(K) + \alpha_2(K) + \cdots + (-1)^n\alpha_n(K)\\ &= \sum_{i=0}^{n}(-1)^i\alpha_i(K),\end{aligned}$$

where $\alpha_i(K)$ is the number of simplexes of dimension i in K.

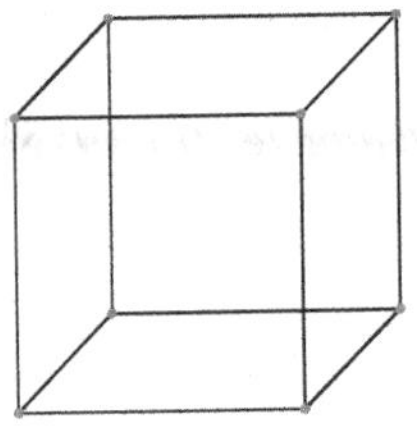

Fig. 10.8 Cube

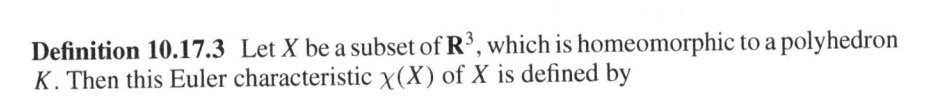

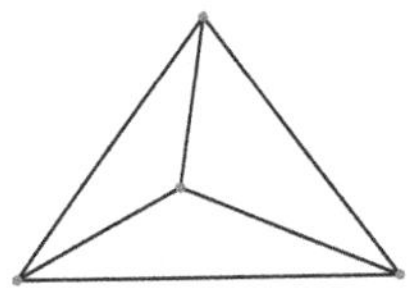

Fig. 10.9 Tetrahedron

Definition 10.17.3 Let X be a subset of $\mathbf{R}^3$, which is homeomorphic to a polyhedron K. Then this Euler characteristic $\chi(X)$ of X is defined by

$$\chi(X) = \text{(number of vertices in K)-(number of edges in K)} + \text{(number of faces in K)}.$$

Example 10.17.4 The Euler characteristic of the cube as shown in Fig. 10.8 is given by

$$\chi(\text{cube}) = 8 - 12 + 6 = 2.$$

Example 10.17.5 The Euler characteristic of the tetrahedron X as shown in Fig. 10.9 is given by

$$\chi(X) = 4 - 6 + 4 = 2.$$

Poincaré-Alexander theorem ensures that $\chi(X)$ is independent of the polyhedron K provided K is homeomorphic to X.

Theorem 10.17.6 (Pincaré-Alexander) *The Euler characteristic is independent of the polyhedron K as long as K is homeomorphic to X.*

Proof See Nakahara (2003). ❑

Example 10.17.7 **(i)** If X is a one-point space, then $\chi(X) = 1$.

(ii) If X is a line, then $\chi(X) = 2 - 1 = 1$, since a line has two vertices and an edge.

(iii) If X is a triangular disk, then $\chi(X) = 3 - 3 + 1 = 1$.

(iv) The simplest polyhedron which is homeomorphic to S^1 has three edges of a triangle. Then $\chi(S^1) = 3 - 3 = 0$.

(v) The simplest polyhedron which is homeomorphic to S^2 has the faces of a tetrahedron in Fig. 10.9. Then $\chi(S^2) = 4 - 6 + 4 = 2$.

Theorem 10.17.8 *Let X and Y be two geometrical objects in $\mathbf{R}^3$. If X is homeomorphic to Y, then $\chi(X) = \chi(Y)$. In particular, if $\chi(X) \neq \chi(Y)$, then X cannot be homeomorphic to Y.*

Proof It follows from Theorem 10.17.6. ❑

Example 10.17.9 S^1 cannot be homeomorphic to S^2, since $\chi(S^1) = 0$ but $\chi(S^2) = 2$.

Remark 10.17.10 Two figures, which are not homeomorphic to each other may have the same Euler characteristic.

For example, $\chi(X) = 1 = \chi(l)$, where l is a line and if $X = \mathbf{R}P^2$ is the real projective plane, then the Euler characteristic $\chi(\mathbf{R}P^2) = 1$. But l and $\mathbf{R}P^2$ are not of the same homotopy type. There is a general result which ensures that its converse is true.

Proposition 10.17.11 *If X and Y are two geometric objects such that they are of the same homotopy type, then $\chi(X) = \chi(Y)$.*

Proof It is left as an exercise. ❑

10.17.2 Euler Characteristic of Finite Graphs

This subsection studies Euler characteristic of a finite graph.

Definition 10.17.12 Let X be a finite graph. The Euler characteristic $\chi(X)$ is defined to be the number of vertices minus the number of edges.

Theorem 10.17.13 *Let X be a finite graph. Then $\chi(X) = 1$ if X is a tree and the rank of $\pi_1(X)$ is $1 - \chi(X)$ if X is connected.*

Proof It is left as an exercise. ❑

10.17.3 Euler Characteristic of Graded Vector Spaces

Definition 10.17.14 Let $\{V_n\}$ be a given graded vector space with $V_n \neq 0$ for only finitely many values of n. Euler characteristic $\chi(\{V_n\})$ of $\{V_n\}$ is defined to be the alternating sum $\Sigma(-1)^n \dim V_n$, i.e.,

$$\chi(\{V_n\}) = \sum_n (-1)^n \dim V_n.$$

Example 10.17.15 If $\{V_n, \partial_n\}$ is a chain complex with $\{V_n\}$ as above, then

$$\chi(\{V_n\}) = \chi(H(\{V_n, \partial_n\})).$$

10.17.4 Euler–Poincaré Theorem for Finite CW-complexes

This subsection gives the Euler–Poincaré theorem and studies Euler characteristic of a finite CW-complex by generalizing the celebrated formula: number of vertices − number of edges + number of faces for 2-dimensional complexes. Betti number and Euler characteristic have a close relation. For certain spaces such as spherical complexes, the homology groups with integral coefficients are finitely generated.

Definition 10.17.16 For any topological space X, if the homology group $H_n(X)$ is finitely generated, then its rank is called the nth 'Betti number' of X, denoted by $\beta_n(X)$.

Remark 10.17.17 Given a topological space X, its nth Betti number $\beta_n(X)$ is the number of free generators of $H_n(X)$. Thus $\beta_n(X)$ is the number of copies of $\mathbf{Z}$ in $H_n(X)$. The Euler characteristic $\chi(X)$ of X is defined as the alternating sum of its Betti numbers:

$$\chi(X) = \sum_{n=0}^{\infty}(-1)^n \beta_n(X).$$

Definition 10.17.18 Let X be a finite CW-complex. Then the Euler characteristic $\chi(X)$ is defined to be the alternating sum $\sum_n (-1)^n c_n$, where c_n is the number of n-cells of X.

Remark 10.17.19 This $\chi(X)$ can be defined purely in terms of homology of X and hence depends only on the homotopy type of X. This shows that $\chi(X)$ is independent of the choice of CW-structure of X. The rank of the finitely generated abelian group is the number of $\mathbf{Z}$ summands when the group is expressed as a direct sum of cyclic groups.

Theorem 10.17.20 *Let X be a finite CW-complex. Then the Euler characteristic*

$$\chi(X) = \sum_n (-1)^n \, rank \, H_n(X).$$

To prove this theorem we use the fact that if

$$0 \longrightarrow A \longrightarrow B \longrightarrow C \longrightarrow 0$$

is a short exact sequence of finitely generated abelian groups, then rank $B =$ rank $A+$ rank C.

Proof Let $0 \longrightarrow C_m \xrightarrow{d_m} C_{m-1} \longrightarrow \cdots \longrightarrow C_1 \xrightarrow{d_1} C_0 \longrightarrow 0$ be a chain complex of finitely generated abelian groups with cycles $Z_n = \ker d_n$, boundaries $B_n = \operatorname{Im} d_{n+1}$, and homology $H_n = Z_n/B_n$. Then we have short exact sequences

$$0 \longrightarrow Z_n \longrightarrow C_n \longrightarrow B_{n-1} \longrightarrow 0$$

$$\text{and } 0 \longrightarrow B_n \longrightarrow Z_n \longrightarrow H_n \longrightarrow 0.$$

Consequently, it follows that $\text{rank } C_n = \text{rank } Z_n + \text{rank } B_{n-1}$ and $\text{rank } Z_n = \text{rank } B_n + \text{rank } H_n$.
Hence it implies that

$$\sum_n (-1)^n \text{rank } C_n = \sum_n (-1)^n \text{rank } H_n.$$

Consequently, $\chi(X) = \sum_n (-1)^n \text{rank } H_n(X)$. ❑

Corollary 10.17.21 *Let X be a finite CW-complex. Then the Euler characteristic $\chi(X)$ is a homotopy invariant.*

Proof Let X be an n-dimensional CW-complex. Then by Theorem 10.17.20

$$\chi(X) = \sum_{r=0}^{n} (-1)^r \text{rank } H_r(X).$$

As each $H_r(X)$ is a homotopy invariant, it follows that $\chi(X)$ is a homotopy invariant. ❑

Theorem 10.17.22 (The Euler–Poincaré theorem) *Let K be an oriented simplicial complex of dimension n and for $q = 0, 1, 2, \ldots, n$, let α_q denote the number of q-simplexes of K. Then*

$$\sum_{q=0}^{n} (-1)^q \alpha_q = \sum_{q=0}^{n} (-1)^q \beta_q,$$

where β_q denotes the qth Betti number of K.

Proof The theorem follows from Theorem 10.17.20 and Definition 10.7.13. ❑

10.18 Cup and Cap Products in Cohomology Theory

This section conveys the basic concepts of cup and cap products. A basic property of cohomology which distinguishes from homology is the existence of a natural multiplication called cup product. This product makes the direct sum of all cohomology groups into a graded ring. This product is used to study 'duality' theorem on manifolds.

Recall that given a topological space X and an abelian group G, the singular n-cochain group $C^n(X; G)$ with coefficients in G is defined to be the dual group given by $C^n(X; G) = \text{Hom}\,(C_n(X), G)$ of the singular chain group $C_n(X; G)$. Instead of G, here we take commutative ring R with identity 1.

10.18.1 Cup Product

This subsection introduces the concept of cup product in cohomology theory.

Definition 10.18.1 Given a topological space X and a commutative ring R with 1, let $C^n(X; R) = \text{Hom}\,(C_n(X; R), R)$ and $C^*(X; R) = \bigoplus_{n \geq 0} C^n(X; R)$. If $\psi \in C^n(X; R)$ and $c \in C_n(X; R)$, their associated element in R denoted by $[c, \psi]$ is defined by setting $[c, \psi] = \psi(c) \in R$. In particular, if $c' \in c_{n+1}(X)$, then $[c', \delta(\psi)] = [\partial c', \psi]$.

Definition 10.18.2 Let σ be any singular $m + n$ simplex given by $\sigma = \langle v_0, \dots v_n \dots v_{m+n}\rangle$. Consider the affine maps

$$\lambda_n : \Delta_n \to \Delta_{m+n}, \ \rho_m : \Delta_m \to \Delta_{m+n}$$

given by $\lambda_n = \langle v_0 \dots v_n\rangle$ and $\lambda_m = \langle v_n \dots v_{m+n}\rangle$. Given $\psi \in [C^n(X; R)]$ and $\theta \in C^m(X; R)$, their associated element in R denoted by $[\sigma, \psi \cup \theta]$ is defined by setting

$$[\sigma, \psi \cup \theta] = [\sigma\lambda_n, \psi] \cdot [\sigma\rho_m, \theta] \tag{10.24}$$

where the right hand multiplication in (10.24) is the usual multiplication of scalars already defined in R.

Definition 10.18.3 Let X be a topological space and R be a commutative ring. If $\psi \in C^n(X; R)$ and $\theta \in C^m(X; R)$, then their cup product $\psi \cup \theta \in C^{m+n}(X; R)$ is defined by

$$[\sigma, \psi \cup \theta] = [\sigma\lambda_n, \psi][\sigma\rho_m, \theta]$$

for every singular $(m + n)$-simplex σ in X, where a right hand multiplication is the usual product of two elements in the ring R.

Remark 10.18.4 The relation in (10.24) shows that ψ operates on the front n-face of σ and θ operates on the back n-face of σ and the result is obtained by multiplication in R.

Remark 10.18.5 If $\psi = \sum\limits_i \psi_i$, $\theta = \sum\limits_j \theta_j$ are arbitrary elements of $C^*(X; R)$, then the cup product defines a function

$$C^*(X; R) \times C^*(X; R) \to C^*(X; R), \ (\sum_i \psi_i) \cup (\sum_j \theta_j) = \sum_{i,j} \psi_i \cup \theta_j,$$

where $\psi_i \in C^i(X; R)$ and $\theta_j \in C^j(X; R)$.

Proposition 10.18.6 *Let X be a topological space and R be a commutative ring. Then $C^*(X; R) = \bigoplus_{n\geq 0} C^n(X; R)$ is a graded ring under the cup product.*

Proof Let $\psi \in C^n(X; R), \theta \in C^m(X; R)$ and $\phi \in C^p(X; R)$. If σ is an $(n + m + p)$-singular simplex in X, then by definition of cup product it follows that $[\sigma, \psi \cup (\theta \cup \phi)] = [\sigma, (\psi \cup \theta) \cup \phi]$ for all σ. This shows that cup product is associative. Again, if $\psi \in C^n(X; R)$, $\theta \in C^m(X; R)$ and $\phi \in C^m(X; R)$ and σ is an $(n + m)$-singular simplex in X, then

$$\begin{aligned}[\sigma, \psi \cup (\theta + \phi)] &= [\sigma\lambda_n, \psi][\sigma\lambda_m, \theta + \psi] \\ &= [\sigma, \psi \cup \theta] + [\sigma, \psi\cup] \text{ for all } \sigma.\end{aligned}$$

This shows that the left distributivity holds. The right distributivity is similarly proved.

If R contains 1, define $c \in C^0(X; R)$ by $[x, c] = 1$ for all $x \in X$ (use the fact that 0-simplexes in X are identified with the points of X). Then c is a both-sided identity in $C^*(X; R)$. Consequently, it follows from the definition of cup product that $C^*(X; R)$ is a graded ring. ❑

Proposition 10.18.7 *The cup product in $C^*(X; R)$ is bilinear.*

Proof If follows from the definition of cup product and the above distributive laws. ❑

This shows that one may regard cup product as a map

$$\cup : C^*(X; R) \otimes C^*(X; R) \to C^*(X; R).$$

Remark 10.18.8 The coboundary operator is a derivation of the graded ring $C^*(X; R)$ in the sense that $\delta(\psi \cup \theta) = \delta\psi \cup \theta + (-1)^n \psi \cup \delta\theta$ for $\psi \in C^n(X; R), \theta \in C^m(X; R)$.

We summarize the above discussion in the basic and important result

Theorem 10.18.9 *Let X be a topological space and R be a commutative ring. Then*

$$H^*(X; R) = \bigoplus_{n\geq 0} H^n(X; R).$$

Proof The direct sum $Z^*(X; R)$ of the cocycles is a subring of $C^*(X; R)$ and the direct sum $B^*(X; R)$ of the coboundaries is a two-sided ideal in $Z^*(X; R)$. Hence passing of cup product to the quotient, the direct sum $H^*(X; R)$ of the cohomology rings becomes a graded ring. ❑

Definition 10.18.10 If X is a topological space and R is a commutative ring, then the cohomology ring with coefficients in R is

$$H^*(X; R) = \bigoplus_{n \geq 0} H^n(X; R).$$

Remark 10.18.11 The ring $C^*(X; R)$ has several demerits

(i) Its vast size makes it difficult to compute;
(ii) It does not satisfy the homotopy axioms;
(iii) It is not always commutative.

The ring structure on $C^*(X; R)$ inherited by $H^*(X; R)$ overcomes these defects.

Remark 10.18.12 For more properties of cup product see Ex.9 of Sect. 10.21.

10.18.2 Cap Product

This subsection discusses the cap product $\cap$, which is an adjoint operation of cup product $\cup$.

Definition 10.18.13 For an arbitrary topological space X, coefficient ring R, and each pair of integers m, n an R- bilinear cap product map

$$\cap : C_{n+m}(X : R) \times C^n(X; R) \to C_m(X; R).$$

is defined by the following rule:
For $\psi \in C^n(X; R)$, $z \in C_{m+n}(X; R)$, $z \cap \psi$ is the unique m-dimensional chain such that

$$[z \cap \psi, \theta] = [z, \psi \cup \theta] \tag{10.25}$$

for every m-cochains θ, i.e., for any singular $(m+n)$-simplex σ, set $\sigma \cap \psi = [\sigma\lambda_n, \psi]\sigma\rho_m$ and extend to arbitrary $(m+n)$-chain by linearity relation (10.25). Using a further extension by linearity there is a pairing:

$$\cap : C_*(X; R) \otimes C^*(X; R) \to C_*(X; R) \tag{10.26}$$

Proposition 10.18.14 *The pairing*

$$\cap : C_*(X; R) \otimes C^*(X; R) \to C_*(X; R)$$

makes $C_*(X; R)$ a graded commutative ring with identity 1.

Proof It follows from the definition of the pairing cap product map. ❑

Corollary 10.18.15 *There is a bilinear pairing*

$$\cap : H_m(X; R) \otimes H^n(X; R) \to H_{m-n}(X; R)$$

Proof It follows by passage to the quotients of the bilinear pairing in Proposition 10.18.14. ❑

10.19 Applications

This section presents some interesting applications such as Jordan curve theorem, Euler charact eristic of a cellular complex.

10.19.1 Jordan Curve Theorem

A homeomorphic image of a circle is called a Jordan curve. This section studies Jordan curve theorem with a homological proof of the theorem. A homeomorphic image of a circle is called a Jordan curve.This theorem is one of the most classical theorem on topology. It asserts that a subspace of $\mathbf{R}^2$ homeomorphic to S^1 separates $\mathbf{R}^2$ into two complementary components. This statement of 'Jordan curve theorem' appears to be an intuitive one, but Jordan asserted that intuition is not a proof. So it needs rigorous proof. The first rigorous proof was given by Veblem (1905). R. Maehara proved Jordan curve theorem using Brouwer fixed point theorem in 1984 Maehara (1984). But we prove this theorem in a different way.

Jordan curve theorem is one of the first problems of a purely topological nature, is related to connectedness, in particular to continuum theory. One version of this theorem says that a simple closed curve J in the Euclidean plane divides the plane into two regions and J is their common boundary. This result was announced and discussed by C. Jordan in 1887, but this proof was not rigorous.

Lemma 10.19.1 **(i)** *Let A be a subspace of S^n homomorphic to D^k for some $k \geq 0$. Then $\widetilde{H}_i(S^n - A) = 0$ for all i;*

(ii) *Let X is a subspace of S^n homomorphic to S^k for some k with $0 \leq k < n$. Then*

$$\widetilde{H}_i(S^n - X) \cong \begin{cases} \mathbf{Z}, & \text{if } i = n - k - 1 \\ 0, & \text{otherwise}. \end{cases}$$

Proof **(i)** We apply first induction on k. If $k = 0$, $S^n - A$ is homeomorphic to $\mathbf{R}^n$. So the proof follows trivially. Next let $h : I^k \to A$ be a homeomorphism. Consider the open sets $D = S^n - h(I^{k-1} \times [0, 1/2])$ and $S = S^n - h(I^{k-1} \times [1/2, 1])$, with $D \cap S = S^n - A$ and $D \cup S = S^n - h(I^{k-1} \times \{1/2\})$. By induction $\widetilde{H}_i(D \cup S) = 0$ for all i. Hence by Mayer–Vietoris sequence, there are isomorphisms $\psi : \widetilde{H}_i(S^n - A) \to \widetilde{H}_i(D) \oplus \widetilde{H}_i(S)$ for all i. The two components of ψ are induced by the inclusions $S^n - A \hookrightarrow D$ and $S^n - A \hookrightarrow S$. Hence if there exists an i-dimensional cycle α in $S^n - A$ which is not a boundary in $S^n - A$, then α is also not a boundary in at least one of D and S. If $i = 0$, 'cyclic' here is considered in the sense of augmented chain complexes, which are dealing with reduced homology. In a similar way, we can further subdivide the last I factor of I^k into quarters, eights, ... to obtain a nested sequence of closed subintervals $I_1 \supset I_2 \supset \cdots$ with intersection one point $p \in I$, such that α is not a boundary in $S^n - h(I^{k-1} \times I_m)$ for any m. By induction on k, α is the boundary of a chain β in $S^n - h(I^{k-1} \times \{p\})$. Hence β is a finite linear combination of singular simplices with compact image in $S^n - h(I^{k-1} \times \{p\})$. The union of these images is covered by the nested sequence of open sets $S^n - h(I^{k-1} \times I_m)$. Hence by compactness β must be a chain in $S^n - h(I^{k-1} \times I_m)$ for some m. This contradiction implies that α is a boundary in $S^n - A$. This completes the induction steps.

(ii) We prove it by induction on k, starting with the trivial case $k = 0$, when $S^n - X \approx S^{n-1} \times \mathbf{R}$. We represent X as a union of $A_1 \cup A_2$, where A_1 and A_2 homeomorphic to D^k and $A_1 \cap A_2$ is homeomorphic to S^{k-1}. We now use Mayer–Vietoris sequence for $A = S^n - A_1$ and $B = S^n - A_2$, both of which have trivial reduced homology groups by **(i)**. Hence there exist isomorphisms $\widetilde{H}_i(S^n - X) \cong \widetilde{H}_{i+1}(S^n - (A_1 \cap A_2))$ for all i. ❑

Theorem 10.19.2 (Jordan curve) *The complement in the plane $\mathbf{R}^2$ of a Jordan curve J consists of two open components, each of which as J as its boundary.*

Proof It follows from **(ii)** of Lemma 10.19.1 that a subspace of S^2 homeomorphic to S^1 separates S^2 into two complementary complements, i.e., into two path components since open subsets of S^n are locally path-connected. To complete the proof use now $\mathbf{R}^2$ in place of S^2, since deleting a point from an open set in S^2 does not affect its connectedness. ❑

Theorem 10.19.3 (Generalized Jordan curve theorem) *A subspace of S^n homeomorphic to S^{n-1} separates it into two components, and these components have the same homology group as a point. In particular, both complementary regions are homeomorphic to open balls.*

Proof It is left as an exercise. ❑

10.19.2 Homology Groups of $\bigvee_{\alpha \in a} S^n_\alpha$

Theorem 10.19.4 *Let $\mathbf{A}$ be an indexing set, and S^n_α be a copy of the n-sphere, $\alpha \in \mathbf{A}$. Then*

$$\widetilde{H}_p\left(\bigvee_{\alpha \in \mathbf{A}} S^n_\alpha\right) \equiv \begin{cases} \oplus \mathbf{Z}(\alpha), & \text{if } p = n \\ 0, & \text{otherwise}, \end{cases}$$

where $\bigoplus_{\alpha \in \mathbf{A}} \mathbf{Z}(\alpha)$ is a free abelian group with generators $\alpha \in \mathbf{A}$.

Proof The spaces $\Sigma(\bigvee_{\alpha \in \mathbf{A}} S^n_\alpha)$ and $\bigvee_{\alpha \in \mathbf{A}} \Sigma S^n_\alpha = \bigvee_{\alpha \in \mathbf{A}} S^{n+1}_\alpha$ are homotopy equivalent. Hence the theorem follows. ❑

Remark 10.19.5 For more generalization of the result in Sect. 10.19.2 see Ex. 9 of Sect. 10.21.

10.20 Invariance of Dimension

This section establishes the homotopy invariance of dimensions of spheres in the sense that the spheres S^m and S^n are not homotopy equivalent if $m \neq n$ and proves that the dimension of a Euclidean space is a topological invariant.

Proposition 10.20.1 *If m and n are two distinct nonnegative integers, then the spheres S^m and S^n are not homotopically equivalent.*

Proof Without loss of generality, we assume that $0 \leq m < n$. Let G be the nontrivial coefficient group of the ordinary homology theory. Then

$$H_n(S^m) = 0 \text{ and } H_n(S^n) \cong G \tag{10.27}$$

Since the group G is nontrivial, it follows from (10.27) that the spheres S^m and S^n cannot be homotopy equivalent. ❑

Corollary 10.20.2 *If $m \neq n$, then the spheres S^m and S^n are not homeomorphic.*

Proof It follows from Proposition 10.20.1. ❑

Proposition 10.20.3 *If m and n are two distinct nonnegative integers, then the Euclidean spaces $\mathbf{R}^m$ and $\mathbf{R}^n$ are not homeomorphic.*

Proof If possible, the Euclidean spaces $\mathbf{R}^m$ and $\mathbf{R}^n$ are homeomorphic. Then there exists a homeomorphism

$$f : \mathbf{R}^m \to \mathbf{R}^n \tag{10.28}$$

Hence the image point $f(0)$ of the origin of $\mathbf{R}^m$ is a point v in $\mathbf{R}^n$. Let g be the translation of the Euclidean space $\mathbf{R}^n$ defined by

$$g : \mathbf{R}^n \to \mathbf{R}^n, x \mapsto x - v.$$

Then the homeomorphism

$$k = g \circ f : \mathbf{R}^m \to \mathbf{R}^n \tag{10.29}$$

carries the origin of $\mathbf{R}^m$ into the origin of $\mathbf{R}^n$. This implies that there exists a homeomorphism $k' = k|_{\mathbf{R}^m - \{0\}} : \mathbf{R}^m - \{0\} \to \mathbf{R}^n - \{0\}$. Then using the fact that $\mathbf{R}^n - \{0\}$

is homotopy equivalent to S^{n-1}, we find that the spaces $\mathbf{R}^m - \{0\}$ and $\mathbf{R}^n - \{0\}$ are homotopy equivalent to the spheres S^{m-1} and S^{n-1} respectively. This shows that the spheres S^{m-1} and S^{n-1} are homotopy equivalent. This contradicts Proposition 10.20.1 as $m \neq n$. ❑

Remark 10.20.4 The dimension of a Euclidean space is a topological invariant.

10.21 Exercises

1. (Three utilities problem) Suppose there are three houses on a plane and each requires to be connected to the gas, water, and electricity lines. Show that there is no way to make all the nine connections without any of the lines crossing each other.
[Hint: Use Jordan curve theorem.]

2. Calculate the homology groups of the chain complex all of whose homology groups are 0 except the groups $C_0 = 2\mathbf{Z}$, $C_1 = 4\mathbf{Z}$, $C_2 = 3\mathbf{Z}$, $C_3 = \mathbf{Z}$, and the boundary homomorphism ∂_1 is given by the (2×4)-matrix composed of the rows $(1\ 1\ 1\ 1)$ and $(-1\ -1\ -1\ -1)$;

∂_2 is given by the the (4×3)-matrix composed of the rows $(1\ \ 1\ \ 1)$, $(1\ \ -1\ \ -1)$, $(-1\ \ -1\ 1)$, and $(-1\ 1\ -1)$ and ∂_3 maps the whole group C_3 to 0.
(Geometrically, the above chain complex corresponds to a cell decomposition of the closed orientable 3-manifold which is obtained by taking the quotient of the sphere S^2 by the linear action of the group $\{\pm 1, \pm i, \pm j, \pm k\}$ of the quaternion units. Hence the homology groups coincide with those of that manifold.)

3. For each fixed integer $n \geq 0$ and an abelian group G, show that cohomology is a contravariant functor

$$H^n(-; G) : \mathcal{T}op \to \mathcal{A}b.$$

4. Calculate the cohomolgy groups of the Klein bottle.

5. Show that every continuous map $f : (I^n, \partial I^n) \to (X, x_0)$ induces by passage to the quotient a singular n-simplex associated with an n-simplex, each of whose faces is the constant map on x_0. Hence deduce a homomorphism

$$h_n : \pi_n(X, x_0) \to H_n(X; \mathbf{Z}) \text{ for } n \geq 1.$$

Further, show that these homomorphisms are functorial in the sense that every continuous map $f : (X, x_0) \to (Y, y_0)$ induces a commutative diagram as shown in Fig. 10.10.

6. (Dimension axiom) If X is a one-point space,and G is an abelian group, show that

Fig. 10.10 Functional representation of h_n

$$\begin{array}{ccc} \pi_n(X, x_0) & \xrightarrow{h_n} & H_n(X; \mathbf{Z}) \\ \downarrow{\scriptstyle \pi_n(f)} & & \downarrow{\scriptstyle H_n(f)} \\ \pi_n(Y, y_0) & \xrightarrow{h_n} & H_n(Y; \mathbf{Z}) \end{array}$$

(i)

$$H_n(X; G) \cong \begin{cases} G, & \textbf{if } n = 0 \\ 0, & \textbf{if } n > 0.; \end{cases}$$

(ii)

$$H^n(X; G) \cong \begin{cases} G, & \textbf{if } n = 0 \\ 0, & \textbf{if } n > 0. \end{cases}$$

7. (Homotopy axiom) If $f, g : X \to Y$ are homotopic maps. show that they induce the same homomorphism

$$f^* = g^* : H^n(Y; G) \to H^n(X; G).$$

8. If n is a positive integer, show that

$$H_n(S^n) \cong \begin{cases} \mathbf{Z}, & \text{if } m = n \\ 0, & \text{otherwise.}; \end{cases}$$

[Hint. Let $X_1 = \{x \in S^n : x_n > -\frac{1}{2}\}$ and $X_2 = \{x \in S^n : x_n < \frac{1}{2}\}$. Then X_1 and X_2 are contractible spaces. Use induction on n and apply Mayer–Vietoris sequence.]

9. (X_α, x_α) be based spaces, where $\alpha \in \mathbf{A}$. Assuming that (X_α, x_α) is a Borsuk pair for each $\alpha \in \mathbf{A}$, show that

$$\widetilde{H}_n(\bigvee_{\alpha \in \mathbf{A}} X_\alpha) = \bigoplus_{\alpha \in \mathbf{A}} \widetilde{H}_n(X_\alpha).$$

10. (Cup Product) Let R be a ring such as $\mathbf{Z}$, $\mathbf{Z}_n$, and $\mathbf{Q}$. For cochains $\alpha \in C^k(X; R)$ and $\beta \in C^r(X; R)$, their cup product $\alpha \cup \beta \in C^{k+r}(X; R)$ is the cochain whose value on a singular simplex $\sigma : \Delta^{k+r} \to X$, $(\alpha \cup \beta)(\sigma) = \alpha(\sigma|\langle v_0, \cdots, v_k\rangle) \cdot \beta(\sigma|\langle v_k, \ldots, v_{k+r}\rangle)$, where the right-hand side is the usual product in R. Show that

(i) $\delta(\alpha \cup \beta) = \delta\alpha \cup \beta + (-1)^k \alpha \cup \delta\beta$;
(ii) the cup product of two cocycles is again a cocycle;
(iii) the cup product of a cocycle and a coboundary in either order, is a coboundary;
(iv) there is an induced cup product

$$H^k(X; R) \times H^r(X; R) \xrightarrow{\cup} H^{k+r}(X; R);$$

(v) the cup product '∪' is associative and distributive;
(vi) If R has an identity element, then there is an identity element for cup product;

11. Let $\mathcal{H}tp$ be the homotopy category of topological spaces and $\mathcal{GR}$ be the category graded rings. If R is a commutative ring, show that

$$H^*(\ ; R) = \bigoplus_{n \geq 0} H^n(\ ; R) : \mathcal{H}tp \to \mathcal{GR}$$

is a contravariant functor.

[Hint. Use $Z^*(X; R) = \bigoplus_{n\geq 0} Z^n(X; R)$ and $B^*(X; R) = \bigoplus_{n\geq 0} B^n(X; R)$.]

12. (Cohomology cross product) Given CW-complexes X and $Y \in \mathcal{C}_0$, define a cross product of cellular cochains $\alpha \in C^k(X; R)$ and $\beta \in C^r(Y; R)$ by setting

$$(\alpha \times \beta)(e_i^k \times e_j^r) = \alpha(e_i^k)\beta(e_j^r)$$

and letting $\alpha \times \beta$ take the value 0 on $(k+r)$-cells of $X \times Y$ which are not the product of a k-cell of X with an r-cell of Y.Prove that

(i) $\delta(\alpha \times \beta) = \delta\alpha \times \beta + (-1)^k \alpha \times \delta\beta$ for cellular cochains $\alpha \in C^k(X; R)$ and $\beta \in C^r(X; R)$ (coboundary formula);
(ii) given a definition of cross product there is a cup product (agreeing with the original definition) as the composite

$$H^k(X; R) \times H^r(X; R) \xrightarrow{\times} H^{k+r}(X \times X; R) \xrightarrow{\Delta^*} H^{k+r}(X; R),$$

where

$$\Delta : X \to X \times X, x \mapsto (x, x)$$

is the diagonal map.

13. (Exact homology sequence) Given CW-complexes X and $A \in \mathcal{C}_0$ and $(X, A) \in \mathcal{C}$ show that the sequence

$$\cdots \longrightarrow H_n(A) \xrightarrow{i_*} H_n(X) \xrightarrow{j_*} H_n(X, A) \xrightarrow{\partial} H_{n-1}(A) \xrightarrow{i_*} H_{n-1}(X) \longrightarrow$$

$$\cdots \longrightarrow H_0(X, A) \longrightarrow 0$$

is exact, when the boundary operator

$$\partial : H_n(X, A) \to H_{n-1}(A)$$

is defined as follows:
If an element $[f] \in H_n(X, A)$ is represented by a relative cycle f, then $\partial[f]$ is the class of the cycle $\partial\alpha$ in $H_{n-1}(A)$.

14. Show that for all $i > 0$,

$$H_i(D^n, \partial D^n) \cong \begin{cases} \mathbf{Z}, & \text{for } i = n \\ 0, & \text{otherwise .} \end{cases}$$

[Hint.Use the long exact sequence of reduced homology groups for the pairs $(D^n, \partial D^n)$, the homomorphisms $\partial : H_i(D^n, \partial D^n) \to \widetilde{H}_{i-1}(S^{n-1})$ are isomorphic for all $i > 0$, since the remaining terms $\widetilde{H}_i(D^n) = 0$ for all i.]

15. Show that

$$\chi(S^n) = \begin{cases} 2, & \text{if } n \text{ is even} \\ 0, & \text{if } n \text{ is odd.} \end{cases}$$

16. Let T^2 be the torus. Show that Euler characteristic $\chi(T^2) = 0$.

[Hint. $\chi(T^2) = \beta_0(T^2) - \beta_1(T^2) + \beta_2(T^2) = 1 - 2 + 1 = 0$.]

17. Let B be a finite CW-complex and $p : X \to B$ be an n-sheeted covering space. Show that $\chi(X) = n\chi(B)$.

18. Let $\mathbf{R}P^2$ be the real projective plane. Show that the Euler characteristic $\chi(\mathbf{R}P^2) = 1$.

19. Show that for any simplicial complex K the 0-dimensional homology group $H_0(K; \mathbf{Z})$ is the free abelian group whose rank is the same as the number of connected components of K.

20. For any connected closed triangulable manifold M of dimension n, show that

$$H_n(M; \mathbf{Z}) \cong \begin{cases} \mathbf{Z}, & \text{if } M \text{ is orientable} \\ 0, & \text{if } M \text{ is not so.} \end{cases}$$

21. Let K be a simplicial complex. Show that the groups $H_p(K; \mathbf{Z})$do not depend on the choice of an orientation of K.

22. Let X be a nonempty space and $x \in X$. Show that the inclusion map $X \hookrightarrow (X, \{x\})$ determines an exact homology sequence.

23. Show that

$$H^p(S^n \times S^m; \mathbf{Z}) \cong \begin{cases} \mathbf{Z}, & \text{if } p = 0, n, m, n, n+m \\ 0, & \text{otherwise.} \end{cases}$$

24. Show that

$$\widetilde{H}_p(S^n; \mathbf{Z}) \cong \begin{cases} \mathbf{Z}, & \text{if } p = n \\ 0, & \text{otherwise.} \end{cases}$$

25. Let X be a nonempty space. Show that $\widetilde{H}_{n+1}(\Sigma X) \cong \widetilde{H}_n(X)$ for each n.

26. Show that

(i) the Euler characteristic is additive:

for any cellular space and its finite cellular subspaces A and B,

$$\chi(A \cup B) = \chi(A) + \chi(B) - \chi(A \cap B)$$

(ii) Euler characteristic is multiplicative:

for any finite cellular spaces X and Y,

$$\chi(X \times Y) = \chi(X)\chi(Y).$$

27. Show that a finite connected cellular space X of dimension 1 is homotopy equivalent to the bouquet of $1 - \chi(X)$ circles.

28. Show that the fundamental group of S^2 with n points removed is a free group of rank $n - 1$.

29. Show that given any knot K, the homology groups $H_0(\mathbf{R}^3 - K; \mathbf{Z})$ and $H_1(\mathbf{R}^3 - K; \mathbf{Z})$ are isomorphic.

30. Let K be a simplicial complex. Then K is said to be connected if K is not the union of two nonempty subcomplexes of K which have no subcomplexes in common. Show that

(i) K is connected iff the polyhedra $|K|$ is connected;
(ii) if K is a connected complex, then $H_0(X; \mathbf{Z}) \cong \mathbf{Z}$.

31. Let K be a (finite) oriented simplicial complex of dimension m. Show that

(i) $H_n(K)$ is finitely generated $(f \cdot g)$ for every $n \geq 0$;
(ii) $H_n(K) = 0$ for all $n > m$;
(iii) $H_m(K)$ is free abelian, possibly zero.

[Hint. **(i)** $C_n(K)$ is $f \cdot g$ and hence its subgroup $Z_n(K)$ is also so.
(ii) $C_n(K) = 0$ for all $n > m$.
(iii) As $C_{m+1}(K) = 0$, $B_m(K) = 0$ and hence $H_m(K) = Z_m(K)$. Use the result that a subgroup of a free abelian group is also free abelian (see Sect. 14.8).]

32. Let $f, g : X \to Y$ be homotopic maps between two polyhedra. Show that the induced homomorphism

$$f_* = g_* : H_n(X) \to H_n(Y)$$

for all n. Hence show that the homotopy equivalence of polyhedra X and Y have isomorphic homology groups.

[Hint. As f and g are homotopic maps, any simplicial approximation to f is at the same time a simplicial approximation to g.]

33. Let K and L be two simplicial complexes. If $f : |K| \to |L|$ and $g : |L| \to |M|$ are continuous maps of polyhedra, show that

$$(f \circ h)_{n*} : H_n(K) \to H_n(M)$$

satisfying the relation

$$(f \circ h)_{n*} = f_{n*} \circ h_{n*}$$

in each dimension.

34. Let X be a topological space with base point x_0. Show that

(i) $H_n(X, x_0) \cong H_n(X)$ for all $n \geq 1$;
(ii) $H_n(X, x_0) \cong \widetilde{H}_n(X)$ for all $n \geq 0$.

35. Let (X, A) be a pair of topological spaces with X compact Hausdorff and A closed in X, where A is a strong deformation retract of some closed neighbourhood of A in X. If $p : (X, A) \to (X/A, y)$ is the identification map, show that its induced homomorphism

$$p_* : H_n(X, A) \to H_n(X/A, y)$$

is an isomorphism for all n.

36. If (X, A) is a compact Hausdorff pair of topological spaces for which A is a strong deformation retract of some compact neighbourhood of A in X. Show that $H_n(X, A) \cong \widetilde{H}_n(X/A)$ for all n.

37. If X is cellular space and $n \geq 0$, show that $H_n(W_*(X)) \cong H_n(X, X^{(n-1)})$, where $W_n(X) = H_n(X^{(n)}, X^{n-1})$.

38. If X is a cellular space with $X^{(-1)} = \emptyset$, then for all n, $H_n(X) \cong H_n(W_*(X))$, where $W_n(X) = H_n(X^{(n)}, X^{n-1})$.

39. Show that cellular homology is a homology functor from the category of CW-complexes and cellular maps to the category of R-modules and R-linear maps.

40. Show that there is a natural equivalence from the cellular homology theory to the singular homology theory.

10.22 Additional Reading

[**1**] Adams, J.F., *Algebraic Topology: A student's Guide*, Cambridge University Press, Cambridge, 1972.

[**2**] Adhikari, M.R., and Adhikari, Avishek, *Basic Modern Algebra with Applications*, Springer, New Delhi, New York, Heidelberg, 2014.

[**3**] Dieudonné, J., *A History of Algebraic and Differential Topology*, 19001960, Modern Birkhäuser, 1989.

[**4**] Eilenberg, S., and Steenrod, N., *Foundations of Algebraic Topology*, Princeton University Press, Princeton, 1952.

[**5**] Fulton, W., *Algebraic Topology, A First Course*, Springer-Verlag, New York, 1975.

[**6**] Greenberg, M.J., and Harper. J.R., *Algebraic Topology, A First Course*, Benjamin/Cummings Pub.Company, London, 1981.

[**7**] Hilton, P.J., and Wylie, S. *Homology Theory*, Cambridge University Press, Cambridge, 1960.

[**8**] Hu, S.T., *Homology Theory*, Holden Day, Oakland, CA, 1966.

[**9**] Lefschetz, S., *Algebraic Topology*, Amer. Math. Soc., New York, 1942.

[**10**] Matveev, Sergey V., *Lectures on Algebraic Topology*, European Math. Soc., 2006

[**11**] Mayer, J. *Algebraic Topology*, Prentice-Hall, New Jersy, 1972.

[**12**] Massey, W.S., *A Basic Course in Algebraic Topology*, Springer-Verlag, New York, Berlin, Heidelberg, 1991.

[**13**] Switzer, R.M., *Algebraic Topology-Homotopy and Homology*, Springer-Verlag, Berlin, Heidelberg, New York, 1975.

[**14**] Whitehead, G.W., *Elements of Homotopy Theory*, Springer-Verlag, New York, Heidelberg, Berlin, 1978.

References

Adams, J.F.: Algebraic Topology: A Student's Guide. Cambridge University Press, Cambridge (1972)

Adhikari, M.R., Adhikari A.: Basic Modern Algebra with Applications. Springer, Heidelberg (2014)

Čech, E.: Théorie génerale de l'homologie dans un espace quelconque'. Fundam. Math. **19**, 149–183 (1932)

Croom, F.H.: Basic Concepts of Algebraic Topology. Springer, Berlin (1978)

Dieudonné, J.: A History of Algebraic and Differential Topology, pp. 1900–1960. Modern Birkhäuser, Boston (1989)

Dold, A.: Lectures on Algebraic Topology. Springer, New York (1972)

Eilenberg, S., Steenrod, N.: Foundations of Algebraic Topology. Princeton University Press, Princeton (1952)

Fulton, W.: Algebraic Topology, A First Course. Springer, New York (1975)

Gray, B.: Homotopy Theory, An Introduction to Algebraic Topology. Acamedic Press, New York (1975)

Greenberg, M.J., Harper J.R.: Algebraic Topology, A First Course. Benjamin/Cummings Publisher Company, London (1981)
Hatcher, A.: Algebraic Topology. Cambridge University Press, Cambridge (2002)
Hilton, P.J., Wylie, S.: Homology Theory. Cambridge University Press, Cambridge (1960)
Hu, S.T.: Homology Theory. Holden Day, Oakland (1966)
Hurewicz, W.: Beitrage der Topologie der Deformationen. Proc. K. Akad. Wet. Ser. A **38**, 112–119, 521–528 (1935)
Lefschetz, S.: Algebraic Topology. American Mathematical Society, New York (1942)
Maehara, R.: The Jordan curve theorem via the Brouwer fixed point theorem. Am. Math. Mon. **91**, 641–643 (1984)
Mateev, S.V.: Lectures on Algebraic Topology. European Mathematical Society (2006)
Mayer, J.: Algebraic Topology. Prentice-Hall, Englewood Cliffs (1972)
Massey, W.S.: A Basic Course in Algebraic Topology. Springer, Heidelberg (1991)
Maunder, C.R.F.: Algebraic Topology. Van Nostrand Reinhhold, London (1970)
Munkres, J.R.: Elements of Algebraic Topology. Addision- Wesley, Reading (1984)
Nakahara, M.: Geometry, Topology and Physics. Taylor and Francis, Boca Raton (2003)
Rotman, J.J.: An Introduction to Algebraic Topology. Springer, New York (1988)
Spanier, E.: Algebraic Topology. McGraw-Hill, New York (1966)
Switzer, R.M.: Algebraic Topology-Homotopy and Homology. Springer-Verlag, Heidelberg (1975)
Veblem, O.: Theory of plane curves in nontrivial analysis situs. Trans. Am. Math. Soc. **6**, 83–98 (1905)
Whitehead, G.W.: Elements of Homotopy Theory. Springer, Heidelberg (1978)

Chapter 11
Eilenberg–MacLane Spaces

This chapter conveys homotopy theory through an important class of CW-complexes called Eilenberg–MacLane spaces introduced by S. Eilenberg (1915–1998) and S. MacLane (1909–2005) in 1945. An Eilenberg–MacLane space is a CW-complex having just one nonzero homotopy group G in dimension n (G is abelian if $n > 1$), denoted by $K(G, n)$. The spaces $K(G, 1)$ had been studied by W. Hurewicz (1904–1956) before Eilenberg and MacLane took up $K(G, n)$ as a general case. The importance of Eilenberg–MacLane spaces is twofold. First, they develop homotopy theory. Secondly, they are closely linked with the study of cohomology operations (see Chap. 15). This chapter proves that given an Eilenberg–MacLane space $K(G, n)$, and a CW-complex X, the group $[X, K(G, n)]$ is the cohomology group $H^n(X; G)$. This interesting result relates cohomology theory with homotopy theory.

More precisely, this chapter constructs Eilenberg–MacLane spaces, and finally studies their homotopy properties and relates cohomology theory with homotopy theory. The construction process of Eilenberg–MacLane spaces $K(G, n)$ for all possible (G, n) is very interesting and depends on a very natural class of spaces, called Moore spaces of type (G, n), denoted by $M(G, n)$. This chapter also studies Postnikov towers to meet the need for the construction of Eilenberg–MacLane spaces.

For this chapter the books and papers Eilenberg and MacLane (1945a), Gray (1975), Hatcher and Allen (2002), Hu (1959), Maunder (1980), Seree (1951), Spanier (1966), and some others are referred in Bibliography.

11.1 Eilenberg–MacLane Spaces: Introductory Concept

This section presents Eilenberg–MacLane spaces with some interesting examples. Such spaces interlink different concepts in algebraic topology and present some amazing results.

M.R. Adhikari, *Basic Algebraic Topology and its Applications*,
DOI 10.1007/978-81-322-2843-1_11

Definition 11.1.1 A pointed CW-complex is called an Eilenberg–MacLane space if it has only one nontrivial homotopy group. If G is a group and n is a positive integer, the Eilenberg–MacLane space of type (G, n) is a pointed CW-complex X whose homotopy groups vanish in all dimensions except n, where $G = \pi_n(X)$ and G is to be abelian for $n > 1$.

Remark 11.1.2 We use the notation $K(G, n)$ for a CW-complex which represents an Eilenberg–MacLane space of type (G, n). This is well defined, since there is only one space of type (G, n) up to homotopy equivalence. For a group G, $K(G, 0)$ is defined to be the group G with the discrete topology.

Definition 11.1.3 A path-connected space whose fundamental group is isomorphic to a given group G and which has a contractible universal covering space is called a $K(G, 1)$ space.

Example 11.1.4 We look at the following examples:

$$\pi_i(\mathbf{RP}^\infty) = \begin{cases} \mathbf{Z}_2, & \text{if } i = 1, \\ 0, & \text{if } i \neq 1. \end{cases}$$

$$\pi_i(\mathbf{CP}^\infty) = \begin{cases} \mathbf{Z}, & \text{if } i = 2, \\ 0, & \text{if } i \neq 2. \end{cases}$$

$$\pi_i(S^1) = \begin{cases} \mathbf{Z}, & \text{if } i = 1, \\ 0, & \text{if } i \neq 1. \end{cases}$$

where $\mathbf{RP}^\infty$, $\mathbf{CP}^\infty$, S^1 denote the infinite dimensional real projective space, infinite dimensional complex projective space, unit circle in $\mathbf{C}$ respectively. Consequently,

$$K(\mathbf{Z}_2, 1) = \mathbf{RP}^\infty(\text{infinite dimensional real projective space}).$$

$$K(\mathbf{Z}, 2) = \mathbf{CP}^\infty \text{ (infinite dimensional complex projective space)}$$

$K(\mathbf{Z}, 1) = S^1$ (unit circle in $\mathbf{C}$), but S^2 is not an Eilenberg–MacLane space of type $K(\mathbf{Z}, 2)$.

Example 11.1.5 $K(\mathbf{Z_m}, 1)$ is an infinite dimensional lens space $l^\infty(m) = S^\infty$ mod $\mathbf{Z}_m$, where $\mathbf{Z_m}$ acts on S^∞, regarded as the unit sphere in $\mathbf{C}^\infty$, by scalar multiplication by the mth roots of unity, this action being the map $(z_1, z_2, \ldots,) \longmapsto e^{2\pi i/m}(z_1, z_2, \ldots,)$.

Remark 11.1.6 The infinite dimensional lens space $K(\mathbf{Z_m}, 1)$ cannot be replaced by any finite dimensional complex.

Example 11.1.7 Given a closed connected subspace K of S^3 which is nonempty, the complement $S^3 - K$ is an Eilenberg–MacLane space. In particular, if K is the trivial knot, then $S^3 - K$ is an Eilenberg–MacLane space $K(\mathbf{Z}, 1)$.

11.2 Construction of Eilenberg–MacLane Spaces $K(G, n)$

This section prescribes a process of construction of Eilenberg–MacLane spaces $K(G, n)$ up to homotopy equivalence. An alternative proof of uniqueness of $K(G, n)$ up to homotopy equivalence is given in Theorem 15.11.14. Eilenberg–MacLane spaces are pointed CW-complexes X for which $\pi_r(X, x_0) = 0$ except for one value $n \geq 1$ of r. The existence of such spaces was shown by J.H.C. Whitehead (1904–1960) in 1949 by using the properties of CW-complexes. The proof is done by induction on $m > n$, the inductive assumption being that there exists a CW-complex $X^{(m)}$ such that

$$\pi_r(X^{(m)}, x_0) \cong \begin{cases} G \text{ for } r = m, \\ 0 \text{ for } 1 \leq r \leq m \text{ and } r \neq m \end{cases}$$

(for $m = n + 1$ the result of π_r is above).

11.2.1 A Construction of $K(G, 1)$

This subsection gives a construction of $K(G, 1)$. It can be obtained as an orbit space. Let G be for an arbitrary group G (not necessarily abelian). Let Δ_q be the q-simplex with ordered vertices $(g_0, g_1, \ldots, g_q)$ of elements of G and $C(G)$ be the complex obtained as a quotient space of the collection of disjoint simplices Δ_q by identifying their certain faces by canonical linear homeomorphism, preserving the ordering of the vertices. This attaches the q-simplex Δ_q to the $(q-1)$ simplexes $(g_0, \ldots, \widehat{g_i}, \ldots, g_n)$, where the notation $\widehat{g_i}$ indicates that this vertex is omitted. The group G acts on $C(G)$ by the left multiplication: $g.(g_0, g_1, \ldots, g_q) = (g.g_0, g.g_1, \ldots, g.g_q)$. Let $C(G)$mod G be its quotient space. This action of G on $C(G)$ is a covering space action. Hence the quotient map $p : C(G) \to C(G)$mod G is the universal covering of the orbit space $B_G = C(G)$mod G, which is a $K(G, 1)$ space.

Remark 11.2.1 The homotopy type of a CW-complex $K(G, 1)$ is uniquely defined by G.

11.2.2 A Construction of $K(G, n)$ for $n > 1$

This subsection conveys detailed process of construction of $K(G, n)$ for all possible (G, n) (group G is abelian for $n > 1$). This construction process is done on the following stages:

(i) Construction of 'Moore spaces' $M(G, n)$;
(ii) Using 'killing homotopy';

(iii) Applying 'Postnikov decomposition';
(iv) Construction of $K(G, n)$ uniquely determined up to homotopy equivalence by G and n.

11.2.3 Moore Spaces

This subsection introduces the concept of a Moore space, which is a generalization of Eilenberg–MacLane space. We claim that the Eilenberg–MacLane spaces $K(G, n)$ exist for all proper (G, n) and each one of them is unique up to homotopy type. The following class of spaces plays an important role in the construction process of $K(G, n)$.

Definition 11.2.2 A CW-complex X with one 0-cell, all of its other cells are in dimensions n and $n + 1$, and is such that $\pi_n(X) \cong G$, is called a Moore space of type (G, n), denoted by $M(G, n)$, G is abelian for $n > 1$.

Remark 11.2.3 For existence of Moore spaces see Exercise 8 of Sect. 11.4.

Example 11.2.4 (i) S^n is a Moore space $M(\mathbf{Z}, n)$.
(ii) $\mathbf{R}P^2$ is a Moore space $M(\mathbf{Z}_2, 1)$.
(iii) $\mathbf{C}P^1$ is a Moore space $M(\mathbf{Z}, 2)$.

Remark 11.2.5 There is a homological version of a Moore space.

Definition 11.2.6 Given an abelian group G and an integer $n \geq 1$, there is a CW-complex written $M(G, n)$ such that $H_n(X) \cong G$ and $\tilde{H}_i(X) = 0$ for $i \neq 0$. The space $M(G, n)$ is also called a Moore space of type (G, n).

11.2.4 Killing Homotopy Groups

This subsection conveys a process of construction for obtaining an Eilenberg–MacLane space $K(G, n)$ from a Moore space $M(G, n)$ by killing all homotopy groups above the nth.

Definition 11.2.7 The process of construction to obtain an Eilenberg–MacLane space $K(G, n)$ from a Moore space $M(G, n)$ by trivializing all homotopy groups beyond the nth is called 'killing homotopy groups'.

Theorem 11.2.8 (Killing homotopy theorem) *Given a CW-complex X and an integer $n > 0$, there exists a relative CW-complex (X', X) with cells in dimension $(n+1)$ only such that* *(i)* $\pi_n(X') = 0$;

(ii) $\pi_m(X') \cong \pi_m(X)$ *for* $m < n$,
here $X' = (X \bigcup \amalg_{\alpha \in \mathbf{A}} D_\alpha^{n+1})/\sim$, *where* x *is identified to* $f(x)$ *for* $x \in \partial D^{n+1}$, *i.e.,* $x \sim f(x)$, $\forall x \in \partial D_\alpha^{n+1}$.

Proof Let $\{f_\alpha : S^n \to X, \alpha \in \mathbf{A}\}$ represent a set of generators of $\pi_n(X)$. Then for each $\alpha \in \mathbf{A}$, take an $(n+1)$-ball D_α^{n+1} and attach it by f_α to X to obtain $X' = (X \bigcup \amalg_{\alpha \in \mathbf{A}} D_\alpha^{n+1})/\sim$, where x is identified to $f(x)$ for all $x \in \partial D^{n+1}$, i.e., $x \sim f(x)$, $\forall x \in \partial D_\alpha^{n+1}$. This shows that the relative CW-complex (X', X) has only these D_α^{n+1} as $(n+1)$-cells, which precisely make the generators of $\pi_{n+1}(X)$ inessential. Clearly, below dimension n, the space X' has the same homotopy groups as the space X by the inclusion $i : X \hookrightarrow X'$. ❑

Remark 11.2.9 X' resembles X below dimension n but at dimension n, $\pi_n(X') = 0$.

Example 11.2.10 A 4-ball D^4 may be attached to S^2 by Hopf map $S^3 \to S^2$ to kill the group $\pi_3(S^2)$.

11.2.5 Postnikov Tower: Its Existence and Construction

This subsection studies Postnikov tower (or Postnikov system) which gives a way of constructing a topological space given by M. Postnikov (1927–2004) in 1951 (Postnikov 1951).

Theorem 11.2.11 *Any CW-complex X admits a decomposition into a tower or a system of CW-complex pairs $(X^{[n]}, X)$, called a Postnikov tower or a system of X described in Fig. 11.1 with*

(i) cells in dimension $(n+1)$ and above only;

(ii) $\pi_m(X^{[n]}) = 0$ *for all* $m > n$;

(iii) $i_{n*}\pi_m(X) \cong \pi_m(X^{[n]})$ *for* $m \leq n$, *where* $i_n : X \hookrightarrow X^{[n]}$ *is the inclusion.*

Proof If $n \geq 0$ is a fixed integer, apply killing homotopy Theorem 11.2.8 to kill $\pi_{n+1}(X)$. Then $X_n^{(1)} = (X \bigcup_{\alpha \in \mathbf{A}} D_\alpha^{(n+2)})/\sim$, with $\pi_{n+1}(X_n^{(1)}) = 0$. Again apply the same procedure to $X_n^{(1)}$ to get $X_n^{(2)}$ with $\pi_{n+1}(X_n^{(2)}) = \pi_{n+2}(X_n^{(2)}) = 0$. Recursively, we obtain $X_n^{(r)}$. Then the required space $X^{[n]}$ is obtained by setting $X^{[n]} = \bigcup_{r \geq 1} X_n^{(r)}$ with weak topology. This gives (i) and (ii). (iii) follows from the preservation of π_m by direct limits for CW-complexes. ❑

Definition 11.2.12 $X^{[n]}$ is called the nth Postnikov section of X in the Postnikov tower which is defined uniquely up to homotopy equivalence.

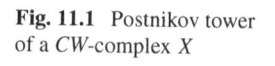

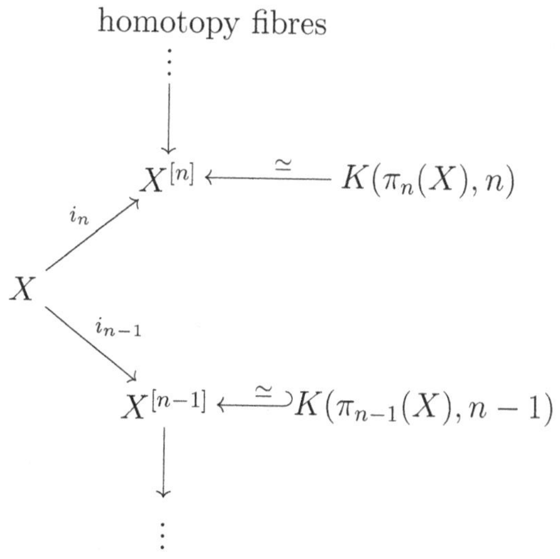

Fig. 11.1 Postnikov tower of a *CW*-complex X

Remark 11.2.13 Postnikov sections $X^{[n]}$ of X can be viewed as successive approximations to X and are considered dual to cellular skeletal approximations $X^{(n)}$.

Each inclusion $i_n : X \hookrightarrow X^{[n]}$ is convertible up to homotopy with a fibration.

$$\widetilde{X}^{(n)} \hookrightarrow \bar{X}^{(n)} \xrightarrow{p_n} X^{[n]},$$

where $j_n : \widetilde{X}^{(n)} \hookrightarrow \bar{X}^{(n)}$ is an inclusion and $\bar{X}^{(n)} = \{(x, f) \in X \times (X^{[n]})^I : f(0) = i_n(x)\} \simeq X$, $p_n : (x, f) \mapsto f(1)$ and fiber $\widetilde{X}^{(n)} = \{(x, f) \in \bar{X}^{(n)} : f(0) = f(1)\}$.

Definition 11.2.14 $\widetilde{X}^{(n)}$ is called the n-connected covering space of X, which is also usually denoted by $X^{(n)}$.

Using the above notations we have the following theorem:

Theorem 11.2.15 *Let X be a CW-complex. Then*

$$\pi_m(\widetilde{X}^{(n)}) \cong \begin{cases} 0 & \text{for } m \leq n \\ \pi_m(X) & \text{for } m > n. \end{cases}$$

Proof It follows from construction of $\widetilde{X}^{(n)}$ that the only nontrivial homotopy groups in $\pi_m(\widetilde{X}^{(n)})$ are those above n and are isomorphic to those of X by homomorphisms j_{n*} induced by $j_n : \widetilde{X}^{(n)} \hookrightarrow \bar{X}^{(n)}$. ❑

Remark 11.2.16 Postnikov tower is also given as follows:

For any CW-complex X, there is a sequence of fibrations

$$Y_n \to Y_{n-1} \to \cdots \to Y_2 \to Y_1 = K(\pi_1(X), 1)$$

with the fiber of $Y_m \to Y_{m-1}$ being $K(\pi_m(X), m)$; and connecting maps $f_m : X \to Y_m$ such that the homomorphisms

$$(f_m)_* : \pi_i(X) \to \pi_i(Y_m)$$

are isomorphism for $i \leq m$.

11.2.6 Existence Theorem

This subsection shows that given an abelian group G and an integer $n > 1$, there exists a CW-complex $K(G, n)$ determined uniquely up to homotopy equivalence by G and n.

Theorem 11.2.17 (Existence Theorem) *Given an abelian group G and an integer $n > 1$, there exists a CW-complex $K(G, n)$. The homotopy type of $K(G, n)$ is uniquely determined by G and n.*

Proof **Construction**: The construction of $K(G, n)$ is completed by setting $K(G, n) = M(G, n)^{[n]}$. **Uniqueness**: The CW-complex X of type (G, n) having all homotopy groups equal to 0 except for $\pi_r(X) = G$ is uniquely determined up to homotopy equivalence, because if A and B are both Eilenberg–MacLane spaces of type (G, n), then the identity homomorphism $1_d : G \to G$ induces a map $h : A \to B$ by Ex. 6 of Sect. 11.4. Then h is a homotopy equivalence. ❑

Remark 11.2.18 Let $G = \mathbf{Z}$. Then given an arbitrary n, a natural construction of $K(\mathbf{Z}, n)$ starts with S^n for which $\pi_n(S^n) \cong \mathbf{Z}$. Then attach $(n + 2)$-cells to kill $\pi_{n+1}(S^n)$ and iterate this process of attaching higher cells to kill higher homotopy groups. The resulting space is a $K(\mathbf{Z}, n)$.

11.3 Applications

This section presents some important applications of Eilenberg–MacLane spaces. Given an Eilenberg–MacLane space $K(G, n)$, and a CW-complex X, the group $[X, K(G, n)]$ is the cohomology group $H^n(X; G)$. This amazing result relates cohomology theory to homotopy theory by admitting a group structure on the set of homotopy classes of continuous maps from a CW-complex to an Eilenberg–MacLane space.

Theorem 11.3.1 (Whitehead theorem) *Let G be an abelian group and $K(G, n)$ be an Eilenberg–MacLane space. Then there is weak homotopy equivalence*

$$\alpha_n : K(G, n) \to \Omega K(G, n+1),$$

which is also a homotopy equivalence.

Proof Since $\pi_n(\Omega K(G, n+1)) \cong \pi_{n+1}(K(G, n+1) \cong G$ and $\pi_n(K(G, n)) \cong \pi_{n+1}(K(G, n+1) \cong G$, it follows that $\pi_n(\Omega K(G, n+1)) \cong \pi_n(K(G, n)$ for every $n \geq 1$. Consequently, there is a continuous map

$$\alpha_n : K(G, n) \to \Omega K(G, n+1)$$

such that its induced homomorphism

$$\alpha_{n*} : \pi_n(K(G, n)) \to \pi_n(\Omega K(G, n+1))$$

is an isomorphism. Again since all other groups are trivial, α_n is a weak homotopy equivalence. Moreover, $\Omega K(G, n+1)$ has the homotopy type of a CW-complex. Hence it follows that α_n is a homotopy equivalence. ❑

Proposition 11.3.2 *There is a natural group structure on $[X, K(G, n)]$.*

Proof The space $K(G, n)$ is homotopy equivalent to a loop space by Theorem 11.3.1. Since every loop space is an H-group and the set of homotopy classes of maps from any pointed space to an H-group admits a group structure (see Chap. 2), the proposition follows. ❑

Theorem 11.3.3 (Hopf theorem) *If X is a path-connected n-dimensional CW-complex, then $H^n(X; \mathbf{Z}) \cong [X, S^n]$ for an ordinary cohomology H^*.*

Proof Construct an Eilenberg–MacLane space $K(\mathbf{Z}, n)$ from S^n by attaching cells of dimensions $\geq n+2$. Then the inclusion $i : S^n \hookrightarrow K(\mathbf{Z}, n)$ induces a function $i_* : [X, S^n] \to [X, K(\mathbf{Z}, n)]$.
i_* **is injective**: Suppose $i_*(f) = i_*(g)$. Then there is a homotopy $H : X \times I \to K(\mathbf{Z}, n)$ between f and g. By cellular approximation, it can be made to have an image inside of $(n+1)$-skeleton of $K(\mathbf{Z}, n)$, which is S^n. This implies that $f \simeq g$.
i_* **is surjective**: Since the CW-complex X is n-dimensional, it follows by cellular approximation that i_* is surjective. ❑

Corollary 11.3.4 *Given an abelian group G, an integer $n \geq 1$ and a path-connected n-dimensional CW-complex X, the group $[X, K(G,n)]$ is the cohomology group $H^n(X; G)$.*

Proof The corollary follows likewise Theorem 11.3.3. ❑

Proposition 11.3.5

$$\pi_i(SP^\infty(S^n)) \cong \begin{cases} \mathbf{Z}, & \text{if } i = n, \\ 0, & \text{if } i \neq n. \end{cases}$$

Proof Since $SP^\infty(S^n)$ is the Eilenberg–MacLane space $K(\mathbf{Z}, n)$, the proposition follows. ❑

11.4 Exercises

1. Show that the Eilenberg–MacLane space $K(G, n)$ is an H-space iff the group G is abelian.
2. Show that for $n > 1$ the spaces $\Omega K(G, n)$ and $K(G, n-1)$ are homotopy equivalent.
3. Given a topological group G, show that the classifying space B_G for K_G (see Chap. 5) and the Eilenberg–MacLane space $K(G', n)$ are homotopy equivalent iff G and $K(G', n-1)$ are homotopy equivalent for $n > 1$ and in particular, B_G and $K(G, 1)$ are homotopy equivalent iff G is a discrete group.
4. Show that the Klein bottle is a $K(G, 1)$-space, where G is the group with two generators a, b and one relation given by $aba = b$. Is any other surface an Eilenberg–MacLane space $K(G, n)$? Justify your answer.
5. For any continuous map $f : X \to Y$, show that a weak decomposition exists. If $\pi_1(X)$ operates trivially on $\pi_n((M_f, X))$ for all n, show that f has a Moore–Postnikov decomposition, where M_f is the mapping cylinder of f.
6. Let G be an abelian group and $n \geq 1$. If $g : G \to H$ is a homomorphism of groups, show that there exists a map $h : K(G, n) \to K(H, n)$ such that $h_* = g : \pi_n(K(G, n)) \to \pi_n(K(H, n))$.
7. Let G be the free group on k generators. Show that the wedge sum of k unit circles $\bigvee_{i=1}^{k} S^i$ is a $K(G, 1)$ space.
8. Given an integer $n > 1$ and an abelian group G, show that there is a Moore space $M(G, n)$ which is a CW-complex with one 0-cell and all other cells in dimensions n and $n+1$ are such that $\pi_n(M(G, n)) \cong G$.
9. Show that the complement to any knot K in three-dimensional sphere S^3 is of type $K(G, 1)$, where G is a group depending on K.
10. (Whitehead tower) For any CW-complex X, there is sequence of fibrations:

$$\cdots \to X_n \to X_{n-1} \to \cdots \to X_1 \to X$$

where the fiber of $X_n \to X_{n-1}$ is $K(\pi_n(X), n-1)$, and X_1 is the universal cover of X. Show that $\pi_i(X_n) = 0$ for all $i \leq n$ and the map $f_n : X_n \to X$ induces isomorphisms $f_n* : \pi_i(X_n) \to \pi_i(X)$ for $i > n$.

11. Given Eilenberg–MacLane spaces $K(G, n)$ and $K(H, n)$ show that the product space $K(G, n) \times K(H, n)$ is a $K(G \times H, n)$. Hence show that the n-torus T^n is an example of $K(\mathbf{Z^n}, \mathbf{1})$.
12. Let X be a CW-complex of the form $\bigvee_{\alpha} S^n_\alpha \bigcup_{\beta} e^{n-1}_\beta$ for some $n \geq 1$. Show that for every homomorphism $\psi : \pi_n(X) \to \pi_n(Y)$ with Y path-connected, there exists a map $f : X \to Y$ such that $f_* = \psi$. Hence show that given an integer n and a group G (G is abelian if $n > 1$), the Eilenberg–MacLane space $K(G, n)$ is unique up to homotopy equivalence.

11.5 Additional Reading

[1] Adams, J.F., *Algebraic Topology: A student's Guide*, Cambridge University Press, Cambridge, 1972.
[2] Armstrong, M.A., *Basic Topology*, Springer-Verlag, New York, 1983.
[3] Aguilar, Gitler, S., Prieto, C., *Algebraic Topology from a Homotopical View Point*, Springer-Verlag, New York, 2002.
[4] Barratt, M.G., *Track groups I*, Proc. Lond. Math. Soc. 5(3), 71–106, 1955.
[5] Croom, F.H., *Basic Concepts of Algebraic Topology*, Springer-Verlag, New York, Heidenberg, Berlin, 1978.
[6] Dieudonné, J., *A History of Algebraic and Differential Topology*, 1900–1960, Modern Birkhäuser, 1989.
[7] Dold, A., *Lectures on Algebraic Topology*, Springer-Verlag, New York, 1972.
[8] Eilenberg, S., and MacLane, S., *Relations between homology and homotopy groups*, Proc. Natn. Acad. Sci. USA, **29**, 155–158, 1943.
[9] Eilenberg, S., and MacLane, S., *Relations between homology and homotopy groups of spaces I*, Ann. of Math., **46**, 480–509, 1945.
[10] Fulton, W., *Algebraic Topology, A First Course*, Springer-Verlag, New York, 1975.
[11] Hilton, P.J., *An introduction to Homotopy Theory*, Cambridge University Press, Cambridge, 1983.
[12] Hilton, P.J. and Wylie, S. *Homology Theory*, Cambridge University Press, Cambridge, 1960.
[13] Massey, W.S., *A Basic Course in Algebraic Topology*, Springer-Verlag, New York, Berlin, Heidelberg, 1991.
[14] Mayer, J., *Algebraic Topology*, Prentice-Hall, New Jersy, 1972.
[15] Munkres, J.R., *Elements of Algebraic Topology*, Addition-Wesley-Publishing Company, 1984.
[16] Rotman, J.J., *An Introduction to Algebraic Topology*, Springer-Verlag, New York, 1988.
[17] Switzer, R.M., *Algebraic Topology-Homotopy and Homology*, Springer-Verlag, Berlin, Heidelberg, New York, 1975.

References

Adams, J.F.: Algebraic Topology: A student's Guide. Cambridge University Press, Cambridge (1972)

Aguilar, M., Gitler, S., Prieto, C.: Algebraic Topology from a Homotopical View Point. Springer, New York (2002)

Arkowitz, M.: Introduction to Homotopy Theory. Springer, Berlin (2011)

Armstrong, A.: Basic Topology. Springer, New York (1983)

Barratt, M.G.: Track groups I. Proc. Lond. Math. Soc. **5**(3), 71–106 (1955)

Dieudonné, J.: A History of Algebraic and Differential Topology, 1900–1960, Modern Birkhäuser (1989)

Eilenberg, S., MacLane, S.: Relations between homology and homotopy groups. Proc. Nat. Acad. Sci. USA **29**, 155–158 (1943)

Eilenberg, S., MacLane, S.: General theory of natural equivalences. Trans. Amer. Math. Soc. **58**, 231–294 (1945a)

Eilenberg, S., MacLane, S.: Relations between homology and homotopy groups of spaces I. Ann. Math. **46**, 480–509 (1945b)

Eilenberg, S., Steenrod, N.: Foundations of Algebraic Topology. Princeton University Press, Princeton (1952)

Fulton, W.: Algebraic Topology, A First Course. Springer, New York (1975)

Gray, B.: Homotopy Theory, An Introduction to Algebraic Topology. Academic Press, New York (1975)

Hatcher, A.: Algebraic Topology. Cambridge University Press, Cambridge (2002)

Hilton, P.J.: An Introduction to Homotopy Theory. Cambridge University Press, Cambridge (1983)

Hu, S.T.: Homotopy Theory. Academic Press, New York (1959)

Massey, W.S.: A Basic Course in Algebraic Topology. Springer, New York (1991)

Maunder, C.R.F.: Algebraic Topology, Van Nostrand Reinhhold, London, 1970. Dover, Reprinted (1980)

Mayer, J.: Algebraic Topology. Prentice-Hall, New Jersy (1972)

Munkres, J.R.: Elements of Algebraic Topology. Addition-Wesley-Publishing Company, Reading (1984)

Postnikov, M.: Determination of the homology groups of a space by means of the homotopy invariants. Doklady Akademii Nauk SSSR **76**, 359–362 (1951)

Rotman, J.J.: An Introduction to Algebraic Topology. Springer, New York (1988)

Seree, J.-P.: Homologie singuliére des espaces fibrés. Ann. of Math. **54**, 425–505 (1951)

Spanier, E.: Algebraic Topology. McGraw-Hill, New York (1966)

Switzer, R.M.: Algebraic Topology-Homotopy and Homology. Springer, Berlin (1975)

Whitehead, G.W.: Elements of Homotopy Theory. Springer, New York (1978)

Chapter 12
Eilenberg–Steenrod Axioms for Homology and Cohomology Theories

This chapter presents an approach formulating axiomatization of homology and cohomology theories which makes the subject algebraic topology elegant and provides a quick access to further study. These axioms, now called Eilenberg and Steenrod axioms for homology and cohomology theories, were announced by S. Eilenberg (1915–1998) and N. Steenrod (1910–1971) in 1945 but first appeared in their celebrated book "The Foundations of Algebraic Topology" in 1952. This approach classifies and unifies different homology (cohomology) groups, and is the most important contribution to algebraic topology since the invention of the homology groups by Poincaré in 1895 and is called the axiomatic approach for homology theory given by a set of seven axioms by S. Eilenberg and N. Steenrod.

This axiomatic approach simplifies the proofs of many lengthy and complicated theorems and escapes the avoidable difficulty to motivate the students who are learning homology and cohomology theories for the first time as their systematic study. This approach gives the subject conceptual with coherence and elegance. It provides a quick approach for computing homology and cohomology groups. It unifies different homology groups (modules) on the category of compact triangulable spaces. It also inaugurates its dual theory called cohomology theory. This approach did not contain the term CW-complex whose definite study was first given by J.H.C Whitehead (1904–1960) in 1949. But it is thought today that this approach is considered on the category of finite CW-complexes.

Homology invented by Henry Poincaré in 1895 was studied by him during 1895–1904. This homology called simplicial homology is one of the most fundamental powerful inventions in mathematics. He started with a geometric object (a space) which is given by combinatorial data (a simplicial complex). Then the linear algebra and boundary relations by these data were used to construct homology groups. There are other homology theories:

(i) homology groups for compact metric spaces introduced by L. Vietoris (1891–2002) in 1927;

(ii) homology groups for compact Hausdorff spaces introduced by E. Cech (1893–1960) in 1932;

M.R. Adhikari, *Basic Algebraic Topology and its Applications*,
DOI 10.1007/978-81-322-2843-1_12

(iii) singular homology groups first defined by S. Lefschitz (1884–1972) in 1933.

Čech, Vietoris, Lefschitz constructed these homology theories in different methods and designed these tools for solving some specific problems. Initially, all these theories lived in isolation in the sense that no relation among them was established.

Algebraic topologists started around 1940 comparing various definitions of homology and cohomology given in the previous years. Eilenberg and Steenrod initiated a new approach by taking a small number of their properties (not focussing on machinery used for construction of homology and cohomology groups) as axioms to characterize a theory of homology and cohomology. The most interesting result is the proof that on the category of all topological pairs having homotopy type of finite CW-complex pairs all homology and cohomology theories satisfying these axioms have isomorphic groups. This result concludes that there is only one concept of homology and (cohomology) in that category.

For this chapter the books Eilenberg and Steenrod (1952), Gray (1975), Hu (1966), Maunder (1970), Rotman (1988), Spanier (1966) and some others are referred in the Bibliography.

12.1 Eilenberg–Steenrod Axioms for Homology Theory

This section conveys an axiomatic approach to homology theory announced by Eilenberg and Steenrod in 1945 but published in Eilenberg and Steenrod (1952). Let $\mathcal{C}_0$ be the category of all based topological spaces having homotopy type of finite CW-complexes and $\mathcal{C}$ be the category of all topological pairs having homotopy type of finite CW-complex pairs.

A homology theory $\mathcal{H}$ on the category $\mathcal{C}$ consists of three functions $\mathcal{H} = \{H, *, \partial\}$ which satisfy the following axioms:

(i): The first function H assigns to each topological pair (X, A) in $\mathcal{C}$ and each integer p, (positive, negative, or 0), an abelian group $H_p(X, A)$, called the p-dimensional homology group of the topological pair (X, A) in the homology theory $\mathcal{H}$. In particular, for $A = \emptyset$, it is called p-dimensional (absolute) homology group of the space X.

(ii): The second function $*$ assigns to each continuous map $f : (X, A) \to (Y, B)$ in $\mathcal{C}$ and each integer p a homomorphism

$$f_* = f_{p*} : H_p(X, A) \to H_p(Y, B),$$

called the homomorphism induced by the map f in the homology theory $\mathcal{H}$.

(iii): The third function ∂ assigns to each topological pair (X, A) in $\mathcal{C}$ and an integer p a homomorphism

$$\partial = \partial_p : H_p(X, A) \to H_{p-1}(A),$$

called the boundary operator on the group $H_p(X, A)$ in the homology theory $\mathcal{H}$.

Moreover, these functions satisfy the following seven axioms **H(1)-H(7)**, called the Eilenberg–Steenrod axioms for homology theory $\mathcal{H}$ on $\mathcal{C}$;

Axiom H(1)(Identity Axiom). If $1_X : (X, A) \to (X, A)$ is the identity map on a topological pair (X, A) in $\mathcal{C}$, then the induced homomorphism $1_{X*} : H_p(X, A) \to H_p(X, A)$ is the identity automorphism of the homology group $H_p(X, A)$ for every integer p.

Axiom H(2)(Composition Axiom). If $f : (X, A) \to (Y, B)$ and $g : (Y, B) \to (Z, C)$ are continuous maps in $\mathcal{C}$, then

$$(g \circ f)_{p*} = g_{p*} \circ f_{p*} : H_p(X, A) \to H_p(Z, C)$$

for every integer p.

Remark 12.1.1 The above axioms **H(1) and H(2)** show that for every fixed integer p, the functions H_p form a covariant functor from the category $\mathcal{C}$ to the category $\mathcal{Ab}$ of all abelian groups and their homomorphisms. We use the notation $H_p(f) = f_{p*}$. H_p is called the homology functor in the homology theory $\mathcal{H}$.

Axiom H(3)(Commutativity Axiom). If $f : (X, A) \to (Y, B)$ is a continuous map in $\mathcal{C}$ and if $g : A \to B$ is a continuous map in $\mathcal{C}$ defined by $g(x) = f(x)$ for all $x \in A$, then the diagram in Fig. 12.1 is commutative, i.e., $g_* \circ \partial = \partial \circ f_*$ for every integer p.

This axiom connects the homology functor in the homology theory $\mathcal{H}$ with boundary operator ∂ and induced homomorphisms.

Axiom H(4)(Exactness Axiom). If (X, A) is a topological pair in $\mathcal{C}$ and $i : A \hookrightarrow X$, $j : X \to (X, A)$ are the inclusion maps, then the sequence

$$\cdots \to H_p(A) \xrightarrow{i_*} H_p(X) \xrightarrow{j_*} H_p(X, A) \xrightarrow{\partial} H_{p-1}(A) \to \cdots$$

of groups and homomorphisms, called the homology sequence of (X, A), is exact.

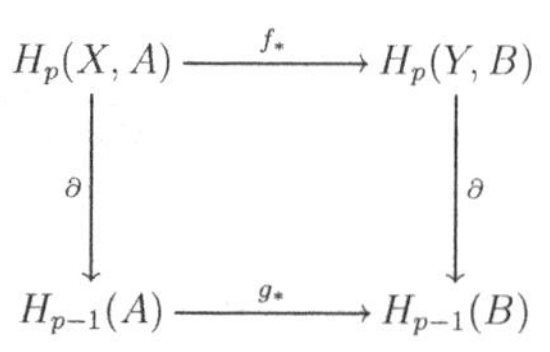

Fig. 12.1 Diagram connecting boundary operator ∂ and induced homomorphisms in $\mathcal{H}$

Remark 12.1.2 The above axioms **H(1)–H(4)** are algebraic axioms.

Axiom H(5)(Homotopy Axiom). If two continuous maps $f, g : (X, A) \to (Y, B)$ in $\mathcal{C}$ are homotopic in $\mathcal{C}$, then $f_{p*} = g_{p*}$ for every integer p.

Axiom H(6)(Excision Axiom). If U is an open set of a topological space X where closure $\overline{U}$ is contained in the interior $\mathring{A}$ of a subspace A of X (i.e., $\overline{U} \subset \mathring{A}$) and if the inclusion map

$$i : (X - U, A - U) \hookrightarrow (X, A)$$

is in $\mathcal{C}$, then the induced homomorphism

$$i_* : H_p(X - U, A - U) \to H_p(X, A)$$

is an isomorphism for every integer p.
The inclusion map i is called the excision of the open set U and i_* is called its p-dimensional excision isomorphism.

Axiom H(7)(Dimension Axiom). The p-dimensional homology group $H_p(X)$ of a one-point space $X = \{point\}$ in the homology theory $\mathcal{H}$ consists of a single element for every integer $p \neq 0$, in symbol, $H_p(point) = 0$, for $p \neq 0$.
This completes the definition of a homology theory $\mathcal{H}$ on the given category $\mathcal{C}$. If $\mathcal{H}$ satisfies only the first six axioms **H(1)–H(6)**, then $\mathcal{H}$ is called a generalized homology theory on the category $\mathcal{C}$.

The 0-dimensional homology group

$$G = H_0(point)$$

is called the coefficient group of the homology theory $\mathcal{H}$. Consequently, the dimension axiom locates the coefficient group at the right dimension.

Remark 12.1.3 The Eilenberg and Steenrod axioms for homology functors provide an elegant and quick access to the further study of algebraic topology.

Remark 12.1.4 The construction of simplicial homology theory and its development are given in Chap. 10. This homology theory applies to the category of pairs (X, A) of spaces, where X and A have triangulations K and L, respectively, for which L is a subcomplex of K. On the other hand the singular homology theory applies to all pairs of spaces (X, A), where X is a topological space and A is a subspace of X.

12.2 The Uniqueness Theorem for Homology Theory

This section gives the uniqueness theorem for the axiomatic approach to homology theory which deals with two homology theories in $\mathcal{C}$ with isomorphic coefficient groups. The most interesting result is the proof that on the category $\mathcal{C}$ of all topological pairs having homotopy types of finite CW-complex pairs all homology theories satisfying the Eilenberg–Steenrod axioms have isomorphic groups. This result concludes that there is only one concept of homology in that category. This uniqueness theorem is very important in the development of algebraic topology. Eilenberg and Steenrod proved that any two homology theories with isomorphic coefficient groups on the category of all compact polyhedral pairs are isomorphic.

Let $\mathcal{H} = \{H, *, \partial\}$ and $\mathcal{H}' = \{H', \square, \partial'\}$ be two arbitrary homology theories in $\mathcal{C}$. Suppose $G = H_0(point)$, $G' = H_0'(point)$ are their coefficient groups.

Definition 12.2.1 Let $\mathcal{H}$ and $\mathcal{H}'$ be two homology theories on $\mathcal{C}$. An isomorphism (natural) $\psi : \mathcal{H} \to \mathcal{H}'$ is a sequence of natural equivalences

$$\psi_n : H_n \to H_n',$$

for all $n \geq 0$ such that the diagram in Fig. 12.2 is commutative for all pairs (X, A) in $\mathcal{C}$ and for all $n \geq 0$.

Theorem 12.2.2 *Let G and G' be abelian groups and $h : G \to G'$ be a homomorphism. Then for every pair (X, A) in $\mathcal{C}$ and every integer n, there exists a unique homomorphism*

$$h_n : H_n(X, A) \to H_n'(X, A)$$

such that

(i) *$h_0 = h$ on $G = H_0(point)$;*
(ii) *for every map $f : (X, A) \to (Y, B)$ in $\mathcal{C}$ and every integer n, the diagram in Fig. 12.3 is commutative, i.e., $h_n \circ f_* = f_\square \circ h_n$*
(iii) *for every pair of spaces (X, A) in $\mathcal{C}$ and every integer n, the diagram in Fig. 12.4 is commutative, i.e., $h_{n-1} \circ \partial = \partial' \circ h_n$.*

Proof For proof see (Hu 1966, pp. 51). □

$$\begin{array}{ccc} H_n(X, A) & \xrightarrow{\partial} & H_{n-1}(A) \\ \downarrow{\scriptstyle \psi_n} & & \downarrow{\scriptstyle \psi_{n-1}} \\ H_n'(X, A) & \xrightarrow{\partial'} & H_{n-1}'(A) \end{array}$$

Fig. 12.2 Isomorphism of homology functors

$$\begin{array}{ccc} H_n(X, A) & \xrightarrow{f_*} & H_n(Y, B) \\ \downarrow{\scriptstyle h_n} & & \downarrow{\scriptstyle h_n} \\ H_n'(X, A) & \xrightarrow{f_\square} & H_n'(Y, B) \end{array}$$

Fig. 12.3 Diagram involving h_n and induced homomorphisms

$$\begin{array}{ccc} H_n(X, A) & \xrightarrow{\partial} & H_{n-1}(A) \\ \downarrow{\scriptstyle h_n} & & \downarrow{\scriptstyle h_{n-1}} \\ H'_n(X, A) & \xrightarrow{\partial'} & H'_{n-1}(A) \end{array}$$

Fig. 12.4 Diagram connecting boundary homomorphisms with h_n

Remark 12.2.3 The unique homomorphism

$$h_n : H_n(X, A) \to H'_n(X, A)$$

is an isomorphism of groups.

Theorem 12.2.4 (The Uniqueness Theorem) *Let G and G' be two abelian groups and $h : G \to G'$ be an isomorphism of groups. Then*

$$h_n : H_n(X, A) \to H'_n(X, A)$$

is also an isomorphism for every pair of spaces (X, A) in $\mathcal{C}$ and every integer n.

Proof Let $k : G' \to G$ be the isomorphism of groups defined by $k = h^{-1}$. Then by Theorem 12.2.2, there exists a unique homomorphism

$$k_n : H'_n(X, A) \to H_n(X, A)$$

satisfying the conditions **(i)**–**(iii)** of Theorem 12.2.2 for every pair of topological spaces (X, A) in $\mathcal{C}$ and every integer n. This shows that

(i) $k_n \circ h_n =$ Identity automorphism of the groups $H_n(X, A)$;
(ii) $h_n \circ k_n =$ Identity automorphism of the groups $H'_n(X, A)$.

Consequently, h_n is an isomorphism of groups. ❑

Remark 12.2.5 Given a coefficient group G, there exists only one homology theory in the category $\mathcal{C}$. Hence the name 'The Uniqueness Theorem' is justified.

12.3 Eilenberg–Steenrod Axioms for Cohomology Theory

This section presents an axiomatic approach to cohomology theory given by Eilenberg and Steenrod, which is dual (parallel) to their homology theory. In fact these two theories differ in only one point: homology functors are covariant; on the other hand, cohomology functors are contravariant. Hence one can expect a dual theorem in cohomology theory for every theorem established in homology theory. The Eilenberg–Steenrod axioms for homology and cohomology functors make the subject algebraic topology elegant and provides a quick access to further study.

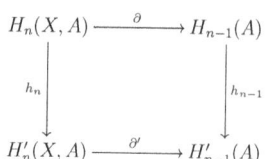

A cohomology theory $\mathcal{K}$ on the category $\mathcal{C}$ consists of three functions $\mathcal{K} = \{H, *, \delta\}$ satisfying the following axioms:

(i) The first function H assigns to each topological pair (X, A) in the category $\mathcal{C}$ and each integer p (positive, negative, or 0) an abelian group $H^p(X, A)$ which is called the p-dimensional cohomology group of the topological pair (X, A) in the cohomology theory $\mathcal{K}$. In particular, for $A = \emptyset$, it is called the p-dimensional (absolute) cohomology group of the space X.

(ii) The second function $*$ assigns to each $f : (X, A) \to (Y, B)$ in $\mathcal{C}$ and each integer p a homomorphism

$$f^* = f_p^* : H^p(Y, B) \to H^p(X, A),$$

called the homomorphism induced by the map f in the cohomology theory $\mathcal{K}$.

(iii) The third function δ assigns to each topological pair (X, A) in $\mathcal{C}$ and an integer p a homomorphism

$$\delta = \delta(X, A, p) : H^{p-1}(A) \to H^p(X, A),$$

called the coboundary operator on the group $H^{p-1}(A)$ in the cohomology theory $\mathcal{K}$.

Moreover, these three functions satisfy the following axioms **C(1)–C(7)**, called the Eilenberg–Steenrod axioms for cohomology theory on $\mathcal{C}$:

Axiom C(1)(Identity Axiom). If $1_X : (X, A) \to (X, A)$ is the identity map on a topological pair (X, A) in $\mathcal{C}$, then the induced homomorphisms

$$1_X^* : H^p(X, A) \to H^p(X, A)$$

is the identity automorphism of the cohomology group $H^p(X, A)$ for every integer p.

Axiom C(2)(Composition Axiom). If $f : (X, A) \to (Y, B)$ and $g : (Y, B) \to (Z, C)$ are maps in $\mathcal{C}$, then

$$(g \circ f)_p^* = f_p^* \circ g_p^* : H^p(Z, C) \to H^p(X, A)$$

for every integer p.

Remark 12.3.1 The above axioms **C(1) and C(2)** show that for every fixed integer p, the function H^p forms a contravariant functor from the category $\mathcal{C}$ to the category $\mathcal{Ab}$. We use notation $H^p(f) = f_p^*$. The functor H^p is called the p-dimensional cohomology functor in the cohomology theory $\mathcal{K}$.

Axiom C(3)(Commutativity Axiom). If $f : (X, A) \to (Y, B)$ is a map in $\mathcal{C}$ and if $g : A \to B$ is the map in $\mathcal{C}$ defined by $g(x) = f(x)$ for all $x \in A$, then the

$$\begin{array}{ccc} H^{p-1}(B) & \xrightarrow{g^*} & H^{p-1}(A) \\ \downarrow \delta & & \downarrow \delta \\ H^{p}(Y,B) & \xrightarrow{f^*} & H^{p-1}(X,A) \end{array}$$

Fig. 12.5 Diagram connecting coboundary operator δ with induced homorphisms in $\mathcal{K}$

diagram in Fig. 12.5 is commutative, i.e., $\delta \circ g^* = f^* \circ \delta$ for every integer p. This axiom connects the cohomology functors in the cohomology theory $\mathcal{K}$ with the coboundary operator δ and induced homomorphisms.

Axiom C(4)(Exactness Axiom). If (X, A) is a topological pair in $\mathcal{C}$ and $i : A \hookrightarrow X$, $j : X \hookrightarrow (X, A)$ are inclusion maps, then the cohomology sequence

$$\cdots \rightarrow H^{p-1}(A) \xrightarrow{\delta} H^p(X, A) \xrightarrow{j^*} H^p(X) \xrightarrow{i^*} H^p(A) \rightarrow \cdots$$

of the topological pair (X, A) is exact.

Remark 12.3.2 The above four axioms **C(1)–C(4)** are algebraic axioms.

Axiom C(5)(Homotopy Axiom). If two maps $f, g : (X, A) \rightarrow (Y, B)$ in $\mathcal{C}$ are homotopic in $\mathcal{C}$, then

$$f_p^* = g_p^*$$

for every integer p.

Axiom C(6)(Excision Axiom). If U is an open set of a topological space X whose closure $\overline{U}$ is contained in the interior $\mathring{A}$ of a subspace A of X (i.e., $\overline{U} \subset \mathring{A}$) and if the inclusion map $i : (X - U, A - U) \hookrightarrow (X, A)$ is in $\mathcal{C}$, then the induced homomorphism $i^* : H^p(X, A) \rightarrow H^p(X - U, A - U)$ is an isomorphism for every integer p.

The inclusion map i is called the excision of the open set U and i^* is called its p-dimensional excision isomorphism.

Axiom C(7)(Dimension Axiom). The p-dimensional cohomology group $H^p(X)$ of a one-point space $X = \{point\}$ consists of a single element for every integer $p \neq 0$, in symbol,

$$H^p(point) = 0, \text{ for } p \neq 0.$$

This completes the definition of a cohomology theory $\mathcal{K}$ on the given category $\mathcal{C}$. If $\mathcal{K}$ satisfies only the first six axioms **C(1)–C(6)**, then $\mathcal{K}$ is called a generalized cohomology theory on the category $\mathcal{C}$.

The 0-dimensional cohomology group $G = H^0(point)$ is called the coefficient group of the cohomology theory $\mathcal{K}$.

Remark 12.3.3 The Uniqueness Theorem for cohomology theory is similar to that of homology theory.

Remark 12.3.4 Section 12.3 conveys an axiomatic approach to cohomology theory announced by Eilenberg and Steenrod in 1945 but published in Eilenberg and Steenrod (1952). The most interesting result is the proof that on the category of all topological pairs having homotopy type of finite CW-complex pairs all cohomology theories satisfying these axioms have isomorphic groups. This result concludes that there is only one concept of cohomology in that category.

12.4 The Reduced 0-dimensional Homology and Cohomology Groups

This section conveys the concepts of 0-dimensional homology groups. Let P_0 denote a fixed reference point and also the space consisting of this single point in $\mathcal{C}_0$. The group $H_0(P_0)$ is as usual called the coefficient group of the given homology theory $\mathcal{H}$ and is denoted by G.

Definition 12.4.1 Let G be a coefficient group of a homology theory $\mathcal{H}$ on $\mathcal{C}_0$. Let X and P_0 be in $\mathcal{C}_0$. If $x \in X$ and $g \in G$, let $(Gx)_X$ denote the image of G in $H_0(X)$ under the homomorphism f_* induced by map $f : P_0 \to X$ defined by $f(P_0) = x$. The image of G in $H_0(X)$ under f_* is denoted by $(Gx)_X$.

Definition 12.4.2 If the unique map $f : X \to P_0$ is in $\mathcal{C}_0$, then space X is said to be collapsible. In such a case the kernel of the homomorphism $f_* : H_0(X) \to G$ is defined. It is called the reduced 0-dimensional homology group of X, denoted by $\widetilde{H}_0(X)$.

Definition 12.4.3 If a topological space X in $\mathcal{C}_0$ is collapsible in the sense that the unique map $f : X \to P_0$ is in $\mathcal{C}_0$, then the image of G in $H^0(X)$ under f^* is denoted by G_X. The factor group $\widetilde{H}^0(X) = H^0(X)/G_X$ is called the reduced 0-dimensional cohomology group of X.

Definition 12.4.4 Let $x \in X$, $h \in H^0(X)$ and $f : P_0 \to X$ be given by $f(P_0) = x$. Then $f^*(x) \in G$ is denoted by $h(x)$. The kernel of $f^* : H^0(X) \to G$ is denoted by $\widetilde{H}^0_x(X)$.

12.5 Applications

This section presents some interesting applications of homology theory $\mathcal{H} = \{H, *, \partial\}$ and cohomology theory $\mathcal{K} = \{H, *, \delta\}$ in the category $\mathcal{C}$ of all topological pairs having homotopy types of finite CW-complex pairs. Let $\mathcal{C}_0$ be the category of all based topological spaces having homotopy type of finite CW-complexes. Then $\mathcal{C}_0$ is a small subcategory of the category $\mathcal{C}$. For some direct consequences of Eilenberg–Steenrod axioms see Chap. 13 and for various applications of homology and cohomology theories see Chaps. 14 and 17.

12.5.1 *Invariance of Homology Groups*

This subsection proves invariance of homology groups in the sense that homeomorphic pairs of topological spaces in the category $\mathcal{C}$ have isomorphic homology groups.

Theorem 12.5.1 *A homeomorphism $f : (X, A) \to (Y, B)$ in the category $\mathcal{C}$ induces isomorphisms*

$$f_* : H_n(X, A) \to H_n(Y, B), \text{ for every integer } n.$$

Proof Since $f^{-1}f = 1_d$, $(f^{-1}f)_* = (f^{-1})_* f_* = 1_d$. Similarly, $f_*(f^{-1})_* = 1_d$. Consequently, f_* is an isomorphism with its inverse $(f_*)^{-1} = (f^{-1})_*$. ❑

Remark 12.5.2 $H_n(X, A)$ is a topological invariant. It is also a homotopy invariant in the sense that if $f : (X, A) \to (Y, B)$ in the category $\mathcal{C}$ is a homotopy equivalence, then it induces isomorphisms

$$f_* : H_n(X, A) \to H_n(Y, B), \text{ for every integer } n.$$

12.5.2 *Invariance of Cohomology Groups*

This subsection proves invariance of cohomology groups in the sense that homeomorphic pairs of topological spaces in the category $\mathcal{C}$ have isomorphic cohomology groups.

Theorem 12.5.3 *A homeomorphism $f : (X, A) \to (Y, B)$ in the category $\mathcal{C}$ induces isomorphisms*

$$f^* : H^n(Y, A) \to H^n(X, B), \text{ for every integer } n.$$

Proof The proof is similar to that of Theorem 12.5.1. ❑

Remark 12.5.4 $H^n(X, A)$ is a topological invariant. It is also a homotopy invariant.

12.5.3 *Mayer–Vietoris Theorem*

This subsection presents an application of "Excision Axiom" which provides a technique to compute homology groups. For example, Mayer–Vietoris theorem (see Sect. 10.12) leads in this respect.

$$\begin{array}{ccccccccc}
\cdots \longrightarrow H_n(A) & \xrightarrow{\alpha_1} & H_n(X_1) & \xrightarrow{\gamma_1} & H(X_1, A) & \xrightarrow{\delta_1} & H_{n-1}(A) \longrightarrow \cdots \\
\downarrow{\scriptstyle \alpha_2} & & \downarrow{\scriptstyle \beta_1} & & \downarrow{\scriptstyle \alpha} & & \downarrow{\scriptstyle \alpha_2} \\
\cdots \longrightarrow H_n(X_2) & \xrightarrow{\beta_2} & H_n(X) & \xrightarrow{\gamma_2} & H(X_1, X_2) & \xrightarrow{\delta_2} & H_{n-1}(X_2) \longrightarrow \cdots
\end{array}$$

Fig. 12.6 Four lemma diagram

Definition 12.5.5 A topological triad $(X; A, B)$ consists of a topological space X together with an ordered pair (A, B) of subspaces A and B of X. The topological triad $(X; A, B)$ is said to be proper with respect to a homology theory $\mathcal{H}$ if the inclusion maps

$$i : (A, A \cap B) \to (A \cup B, B)$$
$$j : (B, A \cap B) \to (A \cup B, A)$$

induce isomorphisms

$$i_* : H_n(A, A \cap B) \to H_n(A \cup B, B)$$
$$j_* : H_n(B, A \cap B) \to H_n(A \cup B, A)$$

in the homology theory $\mathcal{H}$ for every integer n.

Theorem 12.5.6 (Mayer–Vietoris Theorem) *Let X, X_1, X_2 and A be topological spaces in $\mathcal{C}_0$ such that $X = X_1 \cup X_2$, $A = X_1 \cap X_2$. If the inclusion $(X_1, A) \to (X, X_2)$ is an excision, then there is a long exact sequence in homology, called Mayer–Vietoris sequence of the proper topological triad $(X; X_1, X_2)$:*

$$\cdots \to H_n(A) \xrightarrow{\alpha} H_n(X_1) \oplus H_n(X_2) \xrightarrow{\beta} H_n(X) \xrightarrow{\Delta} H_{n-1}(A) \to \cdots$$

Proof Consider the commutative diagram with two long exact homology sequences provided by axiom **H(4)**, where by assumption $\alpha : H_n(X_1, A) \to H_n(X, X_2)$ is an isomorphism by Excision Axiom **H(6)**. Then use four lemma to complete the proof (Fig. 12.6). ❑

12.6 Exercises

In this section we use the notations described in Sect. 12.4.

1. If $f : X \to Y \in \mathcal{C}_0$, $x \in X$, $y = f(x)$ and $g \in G$, then show that f_* maps $(Gx)_X$ onto $(Gy)_Y$.

$$\begin{array}{ccccccccccc}
\cdots \longrightarrow H_n(A) & \longrightarrow & H_n(X) & \longrightarrow & H(X, A) & \longrightarrow & H_{n-1}(A) & \longrightarrow & H_{n-1}(X) \longrightarrow \cdots \\
\downarrow & & \downarrow & & \downarrow & & \downarrow & & \downarrow \\
\cdots \longrightarrow H_n(B) & \longrightarrow & H_n(Y) & \longrightarrow & H(Y, B) & \longrightarrow & H_{n-1}(B) & \longrightarrow & H_{n-1}(Y) \longrightarrow \cdots
\end{array}$$

Fig. 12.7 Five lemma diagram

2. If $f : X \to Y \in \mathcal{C}_0$ and Y is collapsible, show that X is collapsible. If $(X, A) \in \mathcal{C}$ and X is collapsible, show that A is collapsible and the map $(X, A) \to (P_0, P_0) \in \mathcal{C}$. If P is a space consisting of a single point, show that $\widetilde{H}_0(P) = 0$ and $H_0(P) = (G, P)_P$.

3. If X is collapsible and $x \in X$, show that $H_0(X)$ decomposes into the direct sum $H_0(X) = \widetilde{H}_0(X) \oplus (Gx)_X$ and the correspondence $g \mapsto (gx)_X$ maps G isomorphically onto $(Gx)_X$.

4. If X is a space consisting of a single point, show that $H^0(X) = G_X$ and $\widetilde{H}^0(X) = 0$.

5. Let $f : X \to Y, x \in X, y = f(x), h \in H^0(Y)$. Show that

(i) $(f^*h)(x) = h(y)$;
(ii) f^* maps $\widetilde{H}^0_y(Y)$ into $\widetilde{H}^0_x(X)$;
(iii) $\widetilde{H}^0_y(Y)$ contains the kernel of f^*.

6. If X is a collapsible space and x is in X, show that

(i) $H^0(X)$ decomposes into $H^0(X) = \widetilde{H}^0_x(X) \oplus G_X$;
(ii) The map $f : X \to P_0$ induces an isomorphism from G onto G_X.

7. Let G be a coefficient group of a homology theory $\mathcal{H}$ on $\mathcal{C}_0$. If $f : X \to Y$ in $\mathcal{C}_0$ is collapsible, $x \in X$, and $y = f(x)$, show that X is collapsible and f_* maps $\widetilde{H}_0(X)$ into $\widetilde{H}_0(Y)$ and maps $(Gx)_X$ isomorphically onto $(Gy)_Y$.

8. Let $f : (X, A) \to (Y, B)$ be a map of pair of spaces. If both $f : X \to Y$ and $g : f|_A : A \to B$ are homotopy equivalences, show that $f_* : H_n(X, A) \to H_n(Y, B)$ is an isomorphism for all n.
[Hint. Consider the commutative diagram of two rows of exact sequences as shown in Fig. 12.7, and use five lemma result]

12.7 Additional Reading

[1] Adams, J.F., *Algebraic Topology: A student's Guide*, Cambridge University Press, Cambridge, 1972.

[2] Dieudonné, J., *A History of Algebraic and Differential Topology*, 1900–1960, Modern Birkhäuser, 1989.

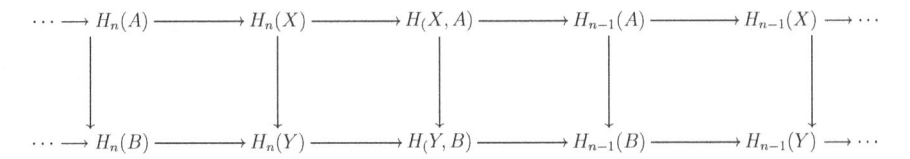

[3] Hatcher, Allen, *Algebraic Topology*, Cambridge University Press, 2002.
[4] Hilton, P.J. and Wylie, S. *Homology Theory*, Cambridge University Press, Cambridge, 1960.
[5] Switzer, R.M., *Algebraic Topology–Homotopy and Homology*, Springer-Verlag, Berlin, Heidelberg, New York, 1975.
[6] Whitehead, G.W., *Elements of Homotopy Theory*, Springer-Verlag, New York, Heidelberg, Berlin, 1978.

References

Adams, J.F.: Algebraic Topology: A student's Guide. Cambridge University Press, Cambridge (1972)
Dieudonné, J.: A History of Algebraic and Differential Topology, 1900–1960. Modern Birkhäuser Classics. Birkhäuser, Basel (1989)
Eilenberg, S., Steenrod, N.: Foundations of Algebraic Topology. Princeton University Press, Princeton (1952)
Gray, B.: Homotopy Theory: An Introduction to Algebraic Topology. Academic, New York (1975)
Hatcher, A.: Algebraic Topology. Cambridge University Press, Cambridge (2002)
Hilton, P.J., Wylie, S.: Homology Theory. Cambridge University Press, Cambridge (1960)
Hu, S.T.: Homology Theory. Holden Day, Oakland (1966)
Maunder, C.R.F.: Algebraic Topology. Van Nostrand Reinhold, London (1970)
Rotman, J.J.: An Introduction to Algebraic Topology, Springer, New York, 1988
Spanier, E.: Algebraic Topology. McGraw-Hill, New York (1966)
Switzer, R.M.: Algebraic Topology-Homotopy and Homology. Springer, Berlin (1975)
Whitehead, G.W.: Elements of Homotopy Theory. Springer, New York (1978)

Chapter 13
Consequences of the Eilenberg–Steenrod Axioms

This chapter continues the study of homology and cohomology theories by considering some immediate consequences of the Eilenberg–Steenrod axioms: **H(1)–H(7)** and **C(1)–C(7)** given by Eilenberg and Steenrod for homology and cohomology theories described in Chap. 12. Finally, this chapter establishes a close connection between cofibrations and homology theory, and computes the ordinary homology groups of S^n with coefficients in an abelian group G.

For this chapter, the books (Eilenberg and Steenrod 1952; Gray 1975; Hu 1966; Spanier 1966) and some others are referred in Bibliography.

13.1 Immediate Consequences

This section deals with some properties of homology and cohomology groups which directly follow from the Eilenberg and Steenrod axioms of homology and cohomology theories. Let $\mathcal{H} = \{H, *, \partial\}$ denote an arbitrary given homology theory on the category $\mathcal{C}$ of topological pairs having homotopy type of finite CW-complex pairs. Then $\mathcal{C}$ is a full subcategory of the category of pairs of topological spaces and maps of pairs and this admits the construction of mapping cones. Let $\mathcal{C}_0$ denote the category of pointed spaces having homotopy of finite CW-complexes. Throughout this chapter, it is assumed that (X, A) is in $\mathcal{C}$ and X is in $\mathcal{C}_0$ unless stated otherwise.

We first establish the homotopy invariance of the homology groups in $\mathcal{H}$.

Theorem 13.1.1 *Let $f : (X, A) \to (Y, B)$ in $\mathcal{C}$ be a homotopy equivalence. Then the induced homomorphism*

$$f_* : H_n(X, A) \to H_n(Y, B)$$

is an isomorphism for every n.

M.R. Adhikari, *Basic Algebraic Topology and its Applications*,
DOI 10.1007/978-81-322-2843-1_13

Proof By hypothesis, f is a homotopy equivalence. Hence there exists a map $g : (Y, B) \to (X, A)$ in $\mathcal{C}$ such that $g \circ f$ is homotopic to the identity map on the topological pair (X, A) and $f \circ g$ is homotopic to the identity map on the topological pair (Y, B).

Consequently, by Axioms **H(1)** and **H(2)** it follows that

$$g_* \circ f_* = (g \circ f)_* : H_n(X, A) \to H_n(X, A)$$

and

$$f_* \circ g_* = (f \circ g)_* : H_n(Y, B) \to H_n(Y, B)$$

are the identity automorphisms of the groups $H_n(X, A)$ and $H_n(Y, B)$, respectively. This shows that f_* is an isomorphism with g_* as the inverse of f_* for every integer n. ❑

Corollary 13.1.2 *Let X and Y be topological spaces in $\mathcal{C}_0$ such that X is homotopy equivalent to Y. Then the groups $H_n(X)$ and $H_n(Y)$ are isomorphic for every integer n.*

Proof The corollary follows from Theorem 13.1.1 by taking in particular $A = \emptyset$ and $B = \emptyset$. ❑

Remark 13.1.3 For every integer n, the n-dimensional homology groups $H_n(X)$ of space X in $\mathcal{C}_0$ and the n-dimensional homology groups $H_n(X, A)$ of (X, A) in $\mathcal{C}$ are both homotopy invariants.

Corollary 13.1.4 *Let the group G (abelian) be the coefficient group of the homology theory $\mathcal{H}$. If a topological space X is contractible, then*

$$H_0(X; G) \cong G,$$

and

$$H_n(X; G) = 0 \textit{ for } n \neq 0.$$

Proof By hypothesis X is contractible. Hence X is homotopically equivalent to the distinguished singleton space $\{*\}$. Hence the corollary follows from Corollary 13.1.2. ❑

We now consider some consequences of the Exactness Axiom: **H(4)** of Chap. 12.

Proposition 13.1.5 *Let X be a topological space in $\mathcal{C}_0$ and A be a subspace of X. If the inclusion map $i : A \hookrightarrow X$ is a homotopy equivalence, then $H_n(X, A) = 0$ for every integer n.*

Proof As $i : A \hookrightarrow X$ is a homotopy equivalence it follows from Theorem 13.1.1 that the induced homomorphism

$$i_* : H_n(A) \to H_n(X)$$

is an isomorphism for every integer n.

We now consider the homology sequence of the pair of topological spaces (X, A) in $\mathcal{C}$.

$$\cdots \to H_n(A) \xrightarrow{i_*} H_n(X) \xrightarrow{j_*} H_n(X, A) \xrightarrow{\partial} H_{n-1}(A) \xrightarrow{i_*} H_{n-1}(X) \to \cdots \tag{13.1}$$

Since the two homomorphisms i_* in (13.1) are isomorphisms, if follows from exactness of this sequence that $H_n(X, A)$ consists of a singleton element for every integer n. In other words, $H_n(X, A) = 0$ for every n. ❑

Corollary 13.1.6 *Given any topological space X in $\mathcal{C}_0$,*

$$H_n(X, X) = 0$$

for every integer n.

Proof The proof follows from Proposition 13.1.5 by taking in particular, $A = X$. ❑

For the pair (X, A) of topological spaces, we now establish some relations between homology groups of (X, A), X and A.

Proposition 13.1.7 *(a) If X is a topological space and A is a retract of X, then*

(i) the inclusion map $i : A \hookrightarrow X$ induces a monomorphism

$$i_* : H_n(A) \to H_n(X),$$

for every integer n;

(ii) the inclusion map $j : X \hookrightarrow (X, A)$ induces an epimorphism

$$j_* : H_n(X) \to H_n(X, A),$$

for every integer n;

(iii) the boundary operator

$$\partial : H_n(X, A) \to H_{n-1}(A)$$

is a trivial homomorphism for every integer n;

(b) $H_n(X) \cong H_n(A) \oplus H_n(X, A)$ for every integer n.

Proof (a) By hypothesis, A is a retract of X. Hence there exists a retraction $r : X \to A$ such that $r(x) = x$ for every $x \in A$. This shows that $r \circ i : A \to A$ is the identity map on A. By using axioms **H(1)** and **H(2)**, it follows that the composite homomorphism

$$H_n(A) \xrightarrow{i_*} H_n(X) \xrightarrow{r_*} H_n(A) \tag{13.2}$$

is the identity automorphism of the group $H_n(A)$ for every integer n. This implies from (13.2) that i_* is a monomorphism and r_* is an epimorphism and the abelian group $H_n(X)$ decomposes into the direct sum

$$H_n(X) = \operatorname{Im} i_* \oplus \ker r_*$$

for every integer n.

We now consider the homology sequence of the pair (X, A):

$$\cdots \to H_n(A) \xrightarrow{i_*} H_n(X) \xrightarrow{j_*} H_n(X, A) \xrightarrow{\partial} H_{n-1} \xrightarrow{i_*} H_{n-1}(X) \to \cdots \tag{13.3}$$

As $i_* : H_{n-1}(A) \to H_{n-1}(X)$ is a monomorphism, by using the exact sequence (13.3), we find that ∂ is a trivial homomorphism and j_* is an epimorphism.

(b) Let n be an arbitrary given integer. Since $i_* : H_n(A) \to H_n(X)$ is a monomorphism, it follows that $\operatorname{Im} i_* \cong H_n(A)$. Again from the exact sequence (13.3), it follows that $\ker j_* = \operatorname{Im} i_*$. Since $H_n(X) = \operatorname{Im} i_* \oplus \ker r_*$ and j_* is an epimorphism, it follows that

$$\ker r_* \cong H_n(X)/\operatorname{Im} i_* = H_n(X)/\ker j_*$$

by Isomorphism Theorem as shown in Fig. 13.1.
This shows that

$$H_n(X) = H_n(A) \oplus H_n(X, A)$$

for every integer n. □

Corollary 13.1.8 *Let G be the coefficient group of the homology theory $\mathcal{H}$. Then in $\mathcal{H}$, for every point x_0 of a topological space X,*

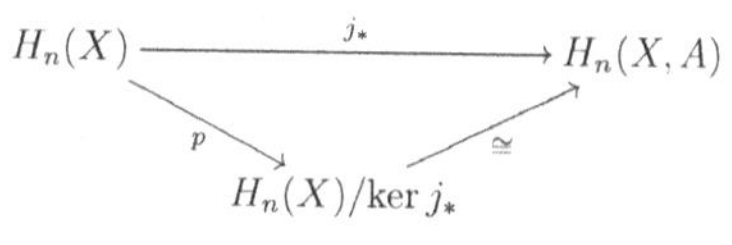

Fig. 13.1 Diagram involving isomorphism of homology groups

$$H_0(X) \cong G \oplus H_0(X, x_0),$$

$$H_n(X) \cong H_n(X, x_0) \text{ for } n \neq 0.$$

Proof Since every singleton subspace of a topological space X is a retract of X, the corollary follows from Propositions 13.1.7 and 13.1.4. ❑

We now study the effect of deformation retraction on the corresponding homology groups.

Proposition 13.1.9 *Let a topological X be deformable into a subspace A of X. Then for every integer n,*

(a) (i) the inclusion map $i : A \hookrightarrow X$ induces an epimorphism

$$i_* : H_n(A) \to H_n(X);$$

(ii) the inclusion map $j : X \hookrightarrow (X, A)$ induces trivial homomorphism

$$j_* : H_n(X) \to H_n(X, A);$$

(iii) the boundary operator

$$\partial : H_n(X, A) \to H_{n-1}(A)$$

is a monomorphism.

(b) $H_n(A) \cong H_n(X) \oplus H_{n+1}(X, A)$.

Proof (a) As X is deformable into its subspace A, there exists a homotopy

$$h_t : X \to X, \text{ for all } t \in I$$

such that $h_0 = 1_X$ and $h_1(X) \subset A$. Define a map

$$h : X \to A, x \mapsto h_1(x).$$

Then the composite map $i \circ h$

$$X \xrightarrow{h} A \hookrightarrow X$$

is homotopic to $h_0 = 1_X$. By using axioms **H(1), H(2),** and **H(5)**, it follows that the composite homomorphism $i_* \circ h_*$

$$H_n(X) \xrightarrow{h_*} H_n(A) \xrightarrow{i_*} H_n(X)$$

is the identity automorphism of the group $H_n(X)$ for every integer n. Consequently it implies that i_* is an epimorphism and h_* is a monomorphism and the abelian group $H_n(A)$ decomposes into the direct sum

$$H_n(A) = \operatorname{Im} h_* \oplus \ker i_*$$

for every integer n.

We now consider the homology sequence of the pair (X, A)

$$\cdots \longrightarrow H_{n+1}(X, A) \xrightarrow{\partial} H_n(A) \xrightarrow{i_*} H_n(X) \xrightarrow{j_*} H_n(X, A) \xrightarrow{\partial} H_{n-1}(A) \longrightarrow \cdots \quad (13.4)$$

As i_* is an epimorphism, it follows from the exact sequence (13.4) that j_* is a trivial homomorphism and ∂ is a monomorphism.

(b) Let n be an arbitrary given integer. Then the induced homomorphism

$$h_* : H_n(X) \to H_n(A)$$

is a monomorphism. Hence $H_n(X) \cong \operatorname{Im} h_*$. Again for the exact sequence (13.4), it follows that

$$\ker i_* = \operatorname{Im} \partial.$$

Again since $\partial : H_{n+1}(X, A) \to H_n(A)$ is a monomorphism, it implies that $\operatorname{Im} \partial \cong H_{n+1}(X, A)$. Hence it follows that

$$H_n(A) = \operatorname{Im} h_* \oplus \ker i_* \cong H_n(X) \oplus H_{n+1}(X, A)$$

for every integer n. ❑

Corollary 13.1.10 *Let G be the coefficient group of the homology theory $\mathcal{H}$ and X be a contractible space. If A is a nonempty subspace of X, then*
$H_0(A) \cong G \oplus H_1(X, A)$,
$H_n(A) \cong H_{n+1}(X, A), n \neq 0$.

Proof Since $A \neq \emptyset$ is a subspace of a contractible space X, it follows that X is deformable into A. Consequently the corollary follows from the Proposition 13.1.9 and Corollary 13.1.4. ❑

We are now in a position to study the Excision Axiom **H(6)**. Here $X - U$ is written as $X \setminus U$.

Theorem 13.1.11 *Let X be a topological space and U be an open set of X such that U is contained in a subspace A of X. Then in $\mathcal{C}$ the excision*

$$e : (X \setminus U, A \setminus U) \to (X, A)$$

induces an isomorphism

$$e_* : H_n(X \setminus U, A \setminus U) \to H_n(X, A)$$

for every integer n if there exists an open set V of X such that the closure $\overline{V}$ of V is contained in U and the inclusion map

$$i : (X \setminus U, A \setminus U) \hookrightarrow (X \setminus V, A \setminus V)$$

is a homotopy equivalence.

Proof By hypothesis $\overline{V} \subset U \subset A$. Hence $\overline{V} \subset \mathring{A}$ (interior of A). Then by Excision Axiom **H(6)**, it follows that the excision

$$\tilde{e} : (X \setminus V, A \setminus V) \to (X, A)$$

induces an isomorphism

$$\tilde{e}_* : H_n(X \setminus V, A \setminus V) \to H_n(X, A)$$

for every integer n.

Again as

$$i : (X \setminus U, A \setminus U) \hookrightarrow (X \setminus V, A \setminus V)$$

is a homotopy equivalence, it induces an isomorphism

$$i_* : H_n(X \setminus U, A \setminus U) \to H_n(X \setminus V, A \setminus V)$$

for every integer n. We now consider the composite map

$$e = \tilde{e} \circ i : (X \setminus U, A \setminus U) \hookrightarrow (X \setminus V, A \setminus V) \xrightarrow{\tilde{e}} (X, A).$$

Then by Axiom **H(2)**, it follows that

$$e_* = \tilde{e}_* \circ i_* : H_n(X \setminus U, A \setminus U) \to H_n(X, A)$$

is an isomorphism for every integer n. ❑

13.2 Applications

This section presents some applications derived as further consequences of Eilenberg and Steenrod axioms such as relation between cofibrations and homology. Moreover, this section computes the ordinary homology groups of S^n with coefficients in an arbitrary abelian group G. Let $\mathcal{C}_0$ be the full subcategory of $\mathcal{C}$, whose objects are topological spaces with base points. Throughout this section it is assumed that (X, A) is in $\mathcal{C}$ and X is in $\mathcal{C}_0$ unless stated otherwise.

13.2.1 *Cofibration and Homology*

There is close relation between cofibrations and homotopy theory. This subsection establishes some relations between cofibrations and homology theory $\mathcal{H} = \{H, *, \partial\}$.

Theorem 13.2.1 *If $i : A \hookrightarrow X$ is a cofibration and $a \in A$, then the projection*

$$p : (X, A) \to (X/A, \{a\})$$

induces isomorphisms in homology.

Proof Since i is a cofibration, its mapping cone C_i is homotopy equivalent to X/A. Again since $C_i = \mathring{X} \cup CA$, we have an inclusion $j : (X, A) \hookrightarrow (C_i, CA)$. This induces isomorphisms in homology. But $(CA, \{a\})$ is contractible and C_i/CA is homeomorphic to X/A. Hence it follows that the projection

$$p : (X, A) \to (X/A, \{a\})$$

induces isomorphisms

$$H_n(X, A) \cong H_n(C_i, CA) \cong H_n(X/A, \{a\}).$$

□

Theorem 13.2.2 *If $i : \{*\} \hookrightarrow X$ is a cofibration, then $H_n(X) \cong H_n(X, \{*\}) \oplus H_n(\{*\})$.*

Proof As $i : \{*\} \hookrightarrow X$ is a cofibration, there exists a map $p : X \to \{*\}$ such that $p \circ i = 1_d$. Hence the sequence

$$\cdots \longrightarrow H_n(\{*\}) \longrightarrow H_n(X) \longrightarrow H_n(X, \{*\}) \longrightarrow \cdots$$

splits for every n and hence the the desired result is proved. □

13.2.2 *Computing Ordinary Homology Groups of S^n*

This subsection computes the ordinary homology groups of S^n with coefficients in an abelian group G.

Theorem 13.2.3 *If $(X, \{x_0\}) \in \mathcal{C}$, then for all $n \in \mathbf{Z}$, there is an isomorphism*

$$\sigma_n = \widetilde{\Sigma}_n : H_n(X, \{x_0\}) \to H_{n+1}(\Sigma X, \{*\}).$$

Proof Consider the isomorphisms $H_{n+1}(CX, X) \cong H_n(X, \{x_0\})$, and compose these with those homomorphisms which are induced by the homeomorphisms $CX/X \approx \Sigma X$ and the projection

$$(CX, X) \to (CX/X, \{*\}).$$

This proves the theorem. ❑

Theorem 13.2.4 *For $m, n \in \mathbf{Z}$ with $m \geq 0$,*

$$H_n(S^m, \{*\}) \cong H_{n+1}(S^{m+1}, \{*\}) \cong H_{n-m}(S^\circ, \{*\}) \cong H_{n-m}(\{*\}).$$

Proof The first two isomorphisms follows from suspensions $S^{m+1} = \Sigma S^m = \Sigma^{m+1} S^\circ$. The other isomorphism follows from the triad $(S^\circ, \{-1\}, \{1\})$ and inclusion $(\{-1\}, \emptyset) \hookrightarrow (S^\circ, \{+1\})$. ❑

Theorem 13.2.5 *Let H_* be an ordinary homology with coefficient group G. Then*

$$H_n(S^m; G) \cong \begin{cases} G \oplus G, & \text{if } n = m = 0 \\ G, & \text{if } n = m \neq 0 \text{ or } n = 0, m \neq 0 \\ 0, & \text{otherwise.} \end{cases}$$

Proof Use Theorem 13.2.2 to obtain $H_n(S^m) \cong H_n(S^m, \{*\}) \oplus H_n(\{*\}) = H_n(S^m, \{*\})$ for $n \neq 0$. Again from Theorem 13.2.4, we have $H_n(S^m, \{*\}) \cong H_{n-m}(\{*\})$. Hence it follows that

$$H_n(S^m, \{*\}) \cong \begin{cases} G, & \text{if } n = m \\ 0, & \text{otherwise.} \end{cases}$$

Combining these, we have $H_0(S^\circ) \cong G \oplus G$, and for $m \neq 0$, $H_0(S^m) \cong H_m(S^m) \cong G$. ❑

13.3 Exercises

In this section $\mathcal{H} = \{H, *, \partial\}$ denotes an arbitrary given homology theory on the category $\mathcal{C}$ of topological pairs having homotopy type of finite CW-complex pairs.

1. If $f : X \to Y$ is a continuous map homotopic to a constant map, show that for every integer $n (n \neq 0)$ the induced homomorphisms

$$f_* : H_n(X) \to H_n(Y)$$

on the homology groups are trivial.

2. Let $f : (X, A) \to (Y, B)$ be deformable into the subspace B of Y. Show that for every integer n, the induced homomorphism

$$f_* : H_n(X, A) \to H_n(Y, B)$$

in the given homology theory is trivial.

3. By using homology prove Brouwer fixed point theorem: Every continuous map from the closed n-ball to itself, for $n \geq 1$, has a fixed point.
4. If $A \hookrightarrow X$ is a weak retract, prove that for all $n \in \mathbf{Z}$, $H_n(X) \cong H_n(A) \oplus H_n(X, A)$.
5. For a weak deformation retract A of X, show that $H_n(X, A) = 0$. Further show that $H_n(X, X) = 0$.
6. Show that for $n \neq m$, the spheres S^n and S^m cannot be homeomorphic. (Hint: For $n \neq m$, S^n and S^m have different homotopy types and hence they cannot be homeomorphic.)
7. Show that for $n \neq m$, $\mathbf{R}^n$ and $\mathbf{R}^m$ cannot be homeomorphic. (Hint: Add one point to compactify each $\mathbf{R}^n$ and $\mathbf{R}^m$ to obtain S^n and S^m, which would be homeomorphic if $\mathbf{R}^n$ and $\mathbf{R}^m$ were homeomorphic.)
8. Let G be the coefficient group of a given homology theory $\mathcal{H}$. Show that for any singleton space X

$$H_p(X; G) = \begin{cases} 0, & p \neq 0 \\ G, & p = 0. \end{cases}$$

9. Let X be a discrete space with n distinct points. Show that

$$H_p(X; G) \cong \begin{cases} G^n, & \text{if } p = 0 \\ 0, & \text{if } p \neq 0, \end{cases}$$

where G^n denotes the direct sum of n copies of the coefficient group G in the homology theory $\mathcal{H}$.

10. If a continuous map $f : X \to Y$ has a left (right) homotopy inverse, show that the induced homomorphism

$$f_* : H_n(X) \to H_n(Y)$$

in the homology theory $\mathcal{H}$ is a monomorphism (an epimorphism) for each n.

13.4 Additional Reading

[1] Adams, J.F., *Algebraic Topology: A student's Guide*, Cambridge University Press, Cambridge, 1972.

[2] Adhikari, M.R., and Adhikari, Avishek, *Basic Modern Algebra with Applications*, Springer, New Delhi, New York, Heidelberg, 2014.

[3] Dieudonné, J., *A History of Algebraic and Differential Topology*, 19001960, Modern Birkhäuser, 1989.

[4] Switzer, R.M., *Algebraic Topology-Homotopy and Homology*, Springer-Verlag, Berlin, Heidelberg, New York, 1975.

[5] Whitehead, G.W., *Elements of Homotopy Theory*, Springer-Verlag, New York, Heidelberg, Berlin, 1978.

References

Adams, J.F.: Algebraic Topology: A Student's Guide. Cambridge University Press, Cambridge (1972)

Adhikari, M.R., Adhikari, A.: Basic Modern Algebra with Applications. Springer, New Delhi (2014)

Dieudonné, J.: A History of Algebraic and Differential Topology, 19001960, Modern Birkhäuser (1989)

Eilenberg, S., Steenrod, N.: Foundations of Algebraic Topology. Princeton University Press, Princeton (1952)

Gray, B.: Homotopy Theory, An Introduction to Algebraic Topology. Academic Press, New York (1975)

Hu, S.T.: Homology Theory. Holden-Day Inc, San Francisco (1966)

Spanier, E.: Algebraic Topology. McGraw-Hill, New York (1966)

Switzer, R.M.: Algebraic Topology-Homotopy and Homology. Springer, Heidelberg (1975)

Whitehead, G.W.: Elements of Homotopy THeory. Springer, New York (1978)

Chapter 14
Applications

In earlier chapters some applications of algebraic topology have been discussed. This chapter conveys further applications to understand the scope and power of algebraic topology displaying the great beauty of the subject. Some concepts initially introduced in homology and homotopy theories to solve problems of topology have found fruitful applications to other areas of mathematics. More precisely, this chapter conveys some interesting applications of homotopy and homology theories. For example, Hopf classification theorem, Borsuk–Ulam theorem, Hairy Ball theorem, Ham Sandwich theorem, Lusternik–Schnirelmann theorem, Lefschetz fixed point theorem, van Kampen theorem are proved and also some results related to graphs, Mayer–Vietoris sequence, fixed points of continuous maps, vector fields and applications to algebra are studied in this chapter. Algebraic topology is now witnessing potential applications to various areas of science and engineering. This chapter also indicates some applications of algebraic topology to physics, chemistry, economics, biology, medical science, and engineering with specific references.

For this chapter the books Armstrong (1983), Croom (1978), Dodson and Parker (1997), Gray (1975), Hatcher (2002), Nakahara (2003), Spanier (1966) with some other books and papers are referred in Bibliography.

14.1 Degrees of Spherical Maps and Their Applications

This section introduces the concept of 'degree of a spherical map' and applies it to prove Brouwer degree theorem for an arbitrary degree, Hopf's classification theorem and Brouwer fixed point theorem. The degree of a spherical map $f : S^n \to S^n$ was defined and studied by L.E.J. Brouwer (1881–1967) during 1910–1912 to examine whether given two spherical maps are homotopic or not. He took the first step towards connecting the two basic concepts: homotopy and homology in topology by using his concept of degree of a spherical map which offers interesting applications. The

M.R. Adhikari, *Basic Algebraic Topology and its Applications*,
DOI 10.1007/978-81-322-2843-1_14

classical definition of the degree of a spherical map $f : S^n \to S^n$ given by Brouwer prior to the rigorous development of homology theory, is more intuitive than its definition from the view point of homology theory. The latter definition is more elegant but the geometric flavor is perhaps lost.

The concept of winding number of a curve with respect to a point in complex analysis or the concept of index of a vector field given by H. Poincaré (1854–1912) contained implicitly the idea of degree of a continuous map prior to Brouwer. This concept of degree of a spherical map is used to solve some problems. After having generalized Brouwer's result to an arbitrary dimension, H. Hopf (1895–1971) undertook a systematic study of the problem of classifying the continuous mappings of a polytope into a polytope (see Chap. 18).

14.1.1 Degree of a Spherical Map

This subsection introduces the concept of the degree function of spherical maps $f : S^n \to S^n$ through homology and characterizes homotopy property of spherical maps by their degrees. Recall that a group homomorphism $f : \mathbf{Z} \to \mathbf{Z}$ is completely determined by the image $f(1)$ of its generator $1 \in \mathbf{Z}$; i.e., f is simply multiplication by the integer $f(1)$. This leads to the following concept of the degree of a spherical map.

Definition 14.1.1 Given a continuous map $f : S^n \to S^n$ and a triangulation K of S^n for an integer $n \geq 1$, there is a homeomorphism $h : |K| \to S^n$; and a homomorphism $\psi : H_n(K; \mathbf{Z}) \to H_n(K; \mathbf{Z})$ defined by

$$\psi = (h^{-1} \circ f \circ h)_* : H_n(K; \mathbf{Z}) \to H_n(K; \mathbf{Z}).$$

Since $H_n(K; \mathbf{Z})$ is isomorphic to $\mathbf{Z}$, there exists a unique integer m with the property that $\psi(x) = m \cdot x$ for $x \in H_n(K; \mathbf{Z})$. This unique integer m is called the degree of f, and is denoted by $\deg f$. The function $d : f \mapsto \deg f$ is called degree function.

Remark 14.1.2 Given a continuous map $f : S^1 \to S^1$, if x moves around S^1, then its image f(x) moves around S^1, some integral number of times. This integer is called the degree of f. The degree function d sets up a (1-1) correspondence between the set $[S^1, S^1]$ of homotopy classes of continuous maps $f : S^1 \to S^1$ and the set $\mathbf{Z}$ of integers.

Let m be the degree of a spherical map $f : S^n \to S^n$ obtained by a triangulation K of S^n. A natural question arises: does this degree depend on a particular choice of the triangulation of S^n?.

To get its answer consider another triangulation L of S^n. Then there exists another homeomorphism $k : |L| \to S^n$. This shows that $\beta = k^{-1} \circ h : |K| \to |L|$ and $\beta^{-1} : h^{-1} o k : |L| \to |K|$ are both homeomorphisms. Hence $(k^{-1} o f o k) : |L| \to |L|$

is a continuous map and its induced homomorphism $(k^{-1} \circ f \circ k)_* : H_n(L; \mathbf{Z}) \to H_n(L; \mathbf{Z})$ is such that

$$\begin{aligned}(k^{-1} \circ f \circ k)_*(x) &= ((k^{-1} \circ h) \circ (h^{-1} \circ f \circ h) \circ (h^{-1} \circ k))_*(x) \\ &= (\beta_* \circ (h^{-1} \circ f \circ h)_* \circ \beta_*^{-1})(x) \\ &= (\beta_*(m(\beta_*^{-1}))(x)) \\ &= \beta_* \circ (\beta_*^{-1})(m \cdot x) \text{ for all } x \in H_n(L; \mathbf{Z}) \\ &= m \cdot x \text{ for all } x \in H_n(L; \mathbf{Z}).\end{aligned}$$

This shows that $\deg f$ of a map $f : S^n \to S^n$ does not depend on a particular choice of triangulations of S^n. Consequently, it follows that the $\deg f$ of a spherical map is well defined.

We now present some interesting properties of degree functions of spherical maps and classify such maps with the help of their degrees.

Theorem 14.1.3 (Brouwer's degree theorem) *Let $f, g : S^n \to S^n$ be two homotopic maps. Then* $\deg f = \deg g$., *i.e., homotopic spherical maps have the same degrees.*

Proof Let $f, g : S^n \to S^n$ be two homotopic maps. As their degrees do not depend on the triangulation $k : |K| \to S^n$, we fix a triangulation k. Then the maps $\psi = h^{-1} \circ f \circ k$ and $\phi = k^{-1} \circ g \circ k$ are homotopic. This shows by the homotopy axiom **H(5)** of Eilenberg–Steenrod that $\psi_* = \phi_* : H_n(K; \mathbf{Z}) \to H_n(K : \mathbf{Z})$ (see Chap. 12). This implies by Definition 14.1.1 that deg f = deg g. ❑

Proposition 14.1.4 **(i)** *The identity map $1_{S^n} : S^n \to S^n$ has degree $+1$;*

(ii) *If $f : S^n \to S^n$ and $g : S^n \to S^n$ are continuous maps, then* $\deg(g \circ f) = \deg g \cdot \deg f$;

(iii) *If $f : S^n \to S^n$ is a homeomorphism, then* $\deg f = \pm 1$.

Proof **(i)** The identity map $1_{S^n} : S^n \to S^n$ induces the identity homomorphism in homology groups. Hence (i) follows.

(ii) Given a triangulation K of S^n, there is a homeomorphism $k : |K| \to S^n$. If $\deg f = n_1$ and $\deg g = n_2$, then $k^{-1} \circ g \circ f \circ k : |K| \to |K|$ is a continuous map and its induced homomorphism

$$(k^{-1} \circ g \circ f \circ k)_* : H_n(K; \mathbf{Z}) \to H_n(K : \mathbf{Z})$$

is such that

$$\begin{aligned}(k^{-1} \circ g \circ f \circ k)_*(x) &= ((k^{-1} \circ g \circ k) \circ (k^{-1} \circ f \circ k))_*(x) \\ &= (k^{-1} \circ g \circ k)_*(n_1 \cdot x) \\ &= n_2 \cdot (n_1 \cdot x) \\ &= (n_2 n_1) \cdot x \text{ for all } x \in H_n(K; \mathbf{Z}).\end{aligned}$$

by using the definition of degree of a spherical map.

Hence $\deg(g \circ f) = \deg g \cdot \deg f$.

(iii) Let $h : S^n \to S^n$ be a homeomorphism. Then $h^{-1} \circ h = 1_{S^n} : S^n \to S^n$ implies by (i) and (ii) that

$$deg(h^{-1} \circ h) = 1 = \deg h^{-1} \cdot \deg h.$$

Since both $\deg h$ and $\deg h^{-1}$ are integers, it follows that $\deg h = \deg h^{-1}$ is either $+1$ or -1.

❑

Remark 14.1.5 The classical definition of the degree of a spherical map $f : S^n \to S^n$ given by Brouwer is more intuitive than its definition from the view point of homology theory. Brouwer defined $deg f$ as the number of times that the domain sphere wraps around the range sphere. His definition shows that if $f : S^1 \to S^1, z \mapsto z^n$, then $\deg f = n$; if $f : S^n \to S^n, n \geq 1$ is a constant map, then $\deg f = 0$ and if $f : S^n \to S^n, n \geq 1$ is the identity map, then $\deg f = +1$.

We are now in a position to study homotopy properties of spherical maps and their fixed points with the help of their degrees.

Proposition 14.1.6 *If $r_i : S^n \to S^n$ is the reflection map defined by*

$$r_i : (x_1, x_2, \ldots, x_i, \ldots, x_{n+1}) \mapsto (x_1, x_2, \ldots, -x_i, \ldots, x_{n+1}),$$

then $\deg r_i = -1$.

Proof We first consider the case $i = 1$. Let $g = r_1 : S^n \to S^n$, be the continuous map defined by

$$g(x_1, x_2, \ldots, x_i, \ldots, x_{n+1}) = (-x_1, x_2, \ldots, x_i, \ldots, x_{n+1})$$

Let $k : S^n \to S^n$ be the homeomorphism which interchanges the coordinates x_1 and x_i. Then $\deg k = \pm 1$. Hence it follows that $r_i = k \circ g \circ k$. This shows that $\deg r_i = \deg k \cdot \deg g \cdot \deg k = \deg g$. we now claim that $\deg g = -1$. For $n = 0$, it is trivial and for all $n \geq 1$, the result follows by iterated suspension. ❑

Theorem 14.1.7 (Antipodal degree) *The antipodal map $A : S^n \to S^n, x \mapsto -x$ has degree $(-1)^{n+1}$ for $n \geq 1$.*

Proof Using the notation of Proposition 14.1.6, we see that $A = r_1 \circ r_2 \circ \cdots \circ r_{n+1}$ and $\deg r_i = -1$ for each i. Hence it follows that $\deg A = (-1)^{n+1}$. ❑

Corollary 14.1.8 *If a continuous map $f : S^n \to S^n$ has no fixed point, then f has degree $(-1)^{n+1}$ for $n \geq 1$.*

Proof Let $f : S^n \to S^n$ be a continuous map having no fixed point. Then the line segment $(1 - t)f(x) - tx$ does not pass through the origin for any t in I and any x in S^n. Consider now the continuous map

$$F : S^n \times I \to S^n : (x, t) \mapsto \frac{(1-t)f(x) - tx}{||(1-t)f(x) - tx||}.$$

As the map $A : S^n \to S^n, x \mapsto -x$ is the antipodal, it follows that $F : f \simeq A$.. Hence $\deg f = \deg A = (-1)^{n+1}$. ❑

Corollary 14.1.9 *If a continuous map $f : S^{2n} \to S^{2n}$ is such that $f \simeq 1_{S^{2n}}$, then f has a fixed point.*

Proof By hypothesis $f \simeq 1_{S^{2n}}$. Then $\deg f = +1$. If possible, f has no fixed point, then Corollary 14.1.8 implies that $\deg f = (-1)^{2n+1} = -1$. Hence we have a contradiction. In other words, f has a fixed point. ❑

14.1.2 Hopf Classification Theorem

This subsection provides a complete homotopy classification of spherical maps with the help of their degrees which are integers. H. Hopf generalized this result of Brouwer to an arbitrary dimension n and made a systematic study of the classification problems of continuous mappings between certain class of spaces, called polytopes (see Chap. 18). H. Hopf proved in 1927 that the converse of Brouwer degree theorem is also true for arbitrary dimension n. These two combined results are known as 'Hopf Classification Theorem'; of course Brouwer proved a partial converse for $n = 2$: if f and g are continuous maps on the 2-sphere which have the same degree, then $f \simeq g$.

Definition 14.1.10 Two continuous maps f and g are said to belong to the same homology class if they induce identical homomorphisms of homology groups (for all dimensions and all coefficient groups), and they are said to belong to the same homotopy class if they can be embedded into a common one-parameter continuous family of mappings.

Theorem 14.1.11 (Hopf Classification Theorem) *Let $f, g : S^n \to S^n$ be two continuous maps. Then $f \simeq g$ iff $\deg f = \deg g$.*

Proof Suppose that $\deg f = \deg g$. We claim that $f \simeq g$. By using induction on n, we prove the theorem. For $n = 1$, a map with degree m is representable as a periodic real function on $\mathbf{R}$ which increases by m each time as its argument increases by 1. Clearly, any two such maps are homotopic. Next suppose that the result is valid for $n - 1$. Then the two maps f and g (where $\deg f = \deg g$) admit representation as the suspensions of maps of S^{n-1} to itself. As the suspension map preserves the degree, then $deg f = deg(\Sigma f)$ and $\deg g = \deg(\Sigma g)$ by Ex. 12 of Sect. 14.11. Hence it follows by induction hypothesis that the maps Σf and Σg are homotopic. Conversely, let f and g be homotopic maps. Then $\deg f = \deg g$ by Brouwer's degree Theorem 14.1.3. ❑

Hopf extended Brouwer's definition of degree to maps from polyhedra to spheres and extended his classification theorem in 1933 to such maps:

Theorem 14.1.12 (Extended Hopf classification theorem) *If X is a polyhedron of dimension not exceeding n and $f, g : X \to S^n$ are two given continuous maps. Then $f \simeq g$ iff $\deg f = \deg g$.*

Proof See Spanier (1966). ❑

14.1.3 *The Brouwer Fixed Point Theorem*

This subsection proves Brouwer fixed point theorem and its immediate consequences by using homology theory. L.E.J. Brouwer took the first step toward connecting homotopy and homology by demonstrating in 1912 that two continuous mappings of a two-dimensional sphere into itself can be continuously deformed into each other if and only if they have the same degree. The papers of H. Poincaré during 1895–1904 can be considered as blue prints for theorems to come. The results of Brouwer during 1910–1912 may be considered as the first one of the proofs in algebraic topology. He proved the celebrated theorem 'Brouwer fixed point theorem' by using the concept of degree of a continuous spherical map defined by Brouwer himself.

Proposition 14.1.13 *The n-sphere S^n is not contractible for any finite $n \geq 0$.*

Proof Let $1_{S^n} : S^n \to S^n$ be the identity map. If possible, S^n is contractible. Then $1_{S^n} \simeq c$ for some constant map c. But 1_{S^n} has degree 1 for $n \geq 1$, and any constant map $f : S^n \to S^n$ has degree 0. This contradicts Hopf classification theorem and hence S^n is not contractible for $n \geq 1$. Again for $n = 0$, $S^0 = \{-1, 1\}$ is a discrete space. Hence S^0 can not be contractible. Consequently, the proposition follows for every $n \geq 0$. ❑

Remark 14.1.14 The infinite dimensional sphere S^∞ is contractible (see Chap. 2) but S^n is not so for any finite $n \geq 0$.

Theorem 14.1.15 (Brouwer no retraction theorem) *There exists no continuous onto map $f : D^{n+1} \to S^n$ which leaves every point of S^n fixed for each integer $n \geq 0$.*

Proof If possible, there exists a continuous onto map $f : D^{n+1} \to S^n$ for every $n \geq 0$ such that $f(x) = x$ for all $x \in S^n$. Define a map.

$$H : S^n \times I \to S^n, (x, t) \mapsto f((1 - t)x)$$

Then H is a continuous map such that $H : f \simeq c$ for some constant map c. This implies that S^n is contractible, which is a contradiction. ❑

Theorem 14.1.16 (Brouwer fixed point theorem) *Every continuous map* $f : D^{n+1} \rightarrow D^{n+1}$ *has a fixed point for every integer* $n \geq 0$.

Proof If possible, $f : D^{n+1} \rightarrow D^{n+1}$ has no fixed point for every integer $n \geq 0$. This implies in this case that $f(x)$ and x are distinct points and hence $f(x) \neq x$ for all $x \in D^{n+1}$. If $n = 0$, it an immediate contradiction. Hence it is well assumed from now that $n \geq 1$. By assumption, for each $x \in D^{n+1}$, the points x and $f(x)$ are distinct. For any $x \in D^{n+1}$ we now consider the half-line in the direction from $f(x)$ to x. Let g(x) denote the point of intersection of this ray with S^n. Then we may consider $g : D^{n+1} \rightarrow S^n$ as a continuous map. Moreover, $g(x) = x$ for every $x \in S^n$. This contradicts "Brouwer no retraction theorem". This asserts that $f(x)$ has a fixed point. ❑

14.2 Continuous Vector Fields

This section studies nonvanishing vector fields on S^n. A vector field v on S^n is a continuous function which associates to each vector x of unit length in $\mathbf{R}^{n+1}$ a unit vector $v(x)$ in $\mathbf{R}^{n+1}$ such that x and $v(x)$ are orthogonal (for $n = 1$, it is geometrically described in Fig. 14.1). If we imagine that $v(x)$ begins at the point $x \in S^n$, then $v(x)$ must be tangent to the circle.

Definition 14.2.1 Let x be a point of S^n. If a vector v in $\mathbf{R}^{n+1}$ beginning at x is tangent to S^n at x and whose endpoint $v(x)$ varies continuously in $\mathbf{R}^{n+1}$ as x moves in S^n, then $v : S^n \rightarrow \mathbf{R}^{n+1}$ is called a continuous vector field on S^n. Moreover, if $v(x) \neq 0$ for all $x \in S^n$, the vector field v is said to be nonvanishing.

Remark 14.2.2 A vector field $v : S^n \rightarrow \mathbf{R}^{n+1}$ on S^n is a continuous map such that for each $x \in S^n$, the vector $v(x)$ is orthogonal to the vector x.

Theorem 14.2.3 *The n-sphere* S^n *admits a continuous nonvanishing vector field iff n is odd.*

Proof Suppose n is odd and $n = 2m - 1$ and $x = (x_1, x_2, \ldots, x_{2m})$ is a point of S^n. Define a map

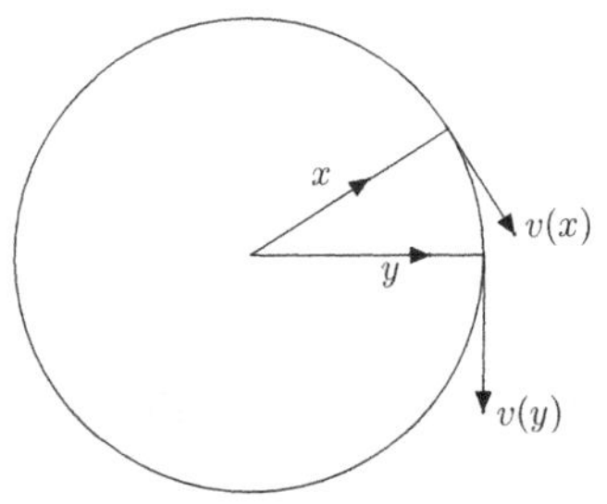

Fig. 14.1 Geometrical description of a vector field v

$$v : S^{2m-1} \to \mathbf{R}^{2m}, (x_1, x_2, \ldots, x_{2m-1}, x_{2m}) \mapsto (x_2, -x_1, \ldots, x_{2m}, -x_{2m-1}).$$

Then v is a vector field on $S^{2m-1} (= S^n)$ with the desired property.

For the converse, let v be a nonvanishing vector field on S^n. Define

$$H : S^n \times I \to S^n, (x, t) \mapsto \cos \pi t \cdot x + \sin \pi t \cdot v(x).$$

Then $H : 1_{S^n} \simeq A$, where map $1_{S^n} : S^n \to S^n$ is the identity map and

$$A : S^n \to S^n, \ x \mapsto -x.$$

is the antipodal map. This implies that the map A has degree 1. Consequently, by Theorem 14.1.7 n is odd. ❑

Remark 14.2.4 If n is odd, the difficult problem of determining the maximum number of linearly independent nowhere vanishing vector fields on S^n was solved by Adams (1962) by using K-theory.

Theorem 14.2.5 *There is a continuous nonvanishing tangent vector field on S^n iff the antipodal map*

$$A_n : S^n \to S^n, x \mapsto -x$$

is homotopic to the identity map 1_{S^n} on S^n.

Proof Let $v : S^n \to \mathbf{R}^{n+1}$ be a nonvanishing tangent vector field on S^n. Define a continuous map

$$H : S^n \times I \to S^n, (x, t) \mapsto (1 - 2t)x + 2\sqrt{t - t^2}\ v(x)/||v(x)||.$$

Then $H(x, 0) = x = 1_{S^n}(x)$ and $H(x, 1) = -x = A_n(x)$ for all $x \in S^n$. This shows that $H : 1_{S^n} \simeq A_n$ and hence $A_n \simeq 1_{S^n}$. Conversely, a homotopy $H : 1_{S^n} \simeq A_n$ can be approximated by a differentiable homotopy. This gives tangent curve elements and hence a nonzero tangent field of directions, because the tangent line to the curve $\alpha_x(t) = H(x, t)$ at $t = 0$ contains a unit vector pointing in the direction of increasing t, which is tangent to the sphere, and is nonvanishing. ❑

Theorem 14.2.6 *If the integer $n > 0$ is even, then there exists no continuous unit tangent vector field on the n-sphere S^n.*

Proof If possible, let $f : S^n \to S^n$ be a continuous unit tangent vector field on S^n (n is even). Define a homotopy

$$H_t : S^n \to S^n, x \mapsto x \cos(t\pi) + f(x) \sin(t\pi).$$

Then $||H_t(x)|| = 1$ for every $t \in I$ and every $x \in S^n$. Now $H_0(x) = x$ and $H_1(x) = -x$ for every $x \in S^n$ show that H_0 is the identity map and H_1 is the antipodal map of S^n which are homotopic. Hence $\deg(H_0) = 1, \deg(H_1) = (-1)^{n+1}$. As $H_0 \simeq H_1$, then $(-1)^{n+1} = 1$. This asserts that the integer n must be odd. This gives a contradiction. ❑

The absence of any nonvanishing vector field on S^2 presents a popular result called Hairy Ball Theorem.

Theorem 14.2.7 (Hairy Ball Theorem) *A hairy ball can not be combed flat.*

Proof If one imagines that he has a hair growing out from each point on the surface of a ball, then it is not possible to brush them flat. Otherwise, the tangent vectors to the hairs would show that S^2 would admit a continuous nonvanishing vector field. As it would contradict Theorem 14.2.3 for $n = 2$, we conclude that a hairy ball can not be combed flat. ❑

Remark 14.2.8 The hairy torus is the only orientable hairy surface that can be combed smoothly.

14.3 Borsuk–Ulam Theorem with Applications

This section proves Borsuk–Ulam Theorem theorem in a general form for all finite dimensions, which is a generalization of this theorem for 2-dimensional case proved in Chap. 3. This theorem was first formulated by S. Ulam (1909–1984) but its first proof was given by K. Borsuk (1905–1982) in 1933. Since then various alternative proofs have appeared in literature. Moreover, this section conveys Ham Sandwich Theorem and Lusternik–Schnirelmann Theorem as applications of Borsuk–Ulum theorem.

14.3.1 Borsuk–Ulam Theorem

Theorem 14.3.1 (Borsuk–Ulam Theorem) *Let m, n be integers such that $m > n \geq 0$. Then there does not exist a continuous map $f : S^m \to S^n$ preserving the antipodal points.*

Proof If possible, let there be a continuous map $f : S^m \to S^n$ such $f(x) = f(-x)$ for all $x \in S^m$. Suppose S^n is obtained from S^m by giving the last $m - n$ coordinates equal to zero and $i : S^n \hookrightarrow S^m$ is the usual inclusion map. Then the composite $i \circ f : S^m \to S^m$ also preserves the antipodal points. Hence $\deg(i \circ f)$ must an odd integer. Again the composite homomorphism

$$(i \circ f)_* = i_* \circ f_* : H_m(S^m; \mathbf{Z}) \xrightarrow{f_*} H_m(S^n; \mathbf{Z}) \xrightarrow{i_*} H_m(S^m; \mathbf{Z})$$

is the trivial homomorphism, since $H_m(S^n; \mathbf{Z}) = 0$ in this situation. This shows that $\deg(i \circ f) = 0$, which is a contradiction. ❑

Remark 14.3.2 Borsuk– Ulam theorem asserts that any continuous map $f : S^n \to \mathbf{R}^n$ must identify a pair of antipodal points of S^n.

Corollary 14.3.3 *If a continuous map $f : S^n \to \mathbf{R}^n$ preserves antipodal points, then there exists a point $x \in S^n$ such that $f(x) = 0$.*

Proof If possible, let $f(x) \neq 0$ for all $x \in S^n$. Define a continuous map

$$h : S^n \to S^{n-1}, x \mapsto \frac{f(x)}{||f(x)||}.$$

As f preserves antipodal points, it follows that h also preserves antipodal points. But this contradicts the Borsuk–Ulam Theorem 14.3.1. ❑

Corollary 14.3.4 *S^n cannot be embedded in $\mathbf{R}^n$.*

Proof S^n cannot be homeomorphic to a subspace of $\mathbf{R}^n$ by Remark 14.3.2. ❑

14.3.2 Ham Sandwich Theorem

This theorem is proved by applying Borsuk–Ulam Theorem.

Theorem 14.3.5 (Ham Sandwich Theorem) *Let $A_1, A_2, \ldots, A_n$ be n bounded convex subsets of $\mathbf{R}^n$. Then there exists a hyperplane which simultaneously bisects all of the A_i's.*

Proof For $n = 3$, given a three-layered ham sandwich, it can be divided with one cut that each of the three pieces is divided into two equal parts. To prove this result for $n = 3$, construct the continuous map $f : S^3 \to \mathbf{R}^3$ as follows:
for $x \in S^3$, take a hyperplane P_x perpendicular to x passing through the point $(0, 0, 0, 1/2)$. Let $v_i(x)$ be the volume of that part of A_i, which lies on the same side of the hyperplane P_x at the point x. Now construct the continuous map

$$f : S^3 \to \mathbf{R}^3, x \mapsto (v_1(x), v_2(x), v_3(x)).$$

Hence by Borsuk–Ulam Theorem, there exists a point $x_0 \in S^3$ such that $f(x_0) = f(-x_0)$. This implies that $v_1(x_0) = v_1(-x_0)$, $v_2(x_0) = v_2(-x_0)$ and $v_3(x_0) = v_3(-x_0)$. For $n > 3$ the same procedure is taken. ❑

14.3.3 Lusternik–Schnirelmann Theorem

This subsection proves the Lusternik–Schnirelmann theorem by using Borsuk–Ulam Theorem.

Theorem 14.3.6 (Lusternik–Schnirelmann) *Let S^n be covered by $n+1$ closed sets $A_1, A_2, \ldots, A_{n+1}$. Then one of them must contain a pair of antipodal points.*

Proof By hypothesis, $\bigcup_{j=1}^{n+1} A_j = S^n \subset \mathbf{R}^{n+1}$. Define a continuous map

$$f : S^n \to \mathbf{R}^n, x \mapsto (d(x, A_1), \ldots, d(x, A_n)),$$

where $d(x, A_i)$ is the distance of x from the closed set A_i. Then f must identity a pair of antipodal points by Borsuk–Ulam Theorem. Consequently, there is a point x_0 in S^n such that $d(x_0, A_j) = d(-x_0, A_j)$ for $0 \leq j \leq n$.

Now only two cases arise:

Case I: If $d(x_0, A_j) = 0$ for some j, then both the points $x_0, -x_0 \in A_j$, since each A_j is a closed set.

Case II: If $d(x_0, A_j) > 0$ for all $j = 1, 2, \ldots, n$, then $x_0, -x_0 \in A_{n+1}$, since the A_j's form a cover of S^n. □

14.4 The Lefschetz Number and Fixed Point Theorems

This section conveys the concept of Lefschetz number which is an integral homotopy invariant and generalizes the Euler characteristic. This number is closely related to the degree of a spherical map. S. Lefschetz (1884–1972) published the first version of his fixed point formula in 1923 which asserts that given a closed manifold M and a map $f : M \to M$, for each q there is an induced homomorphism on homology with rational coefficients $\mathbf{Q}$

$$f_{q*} : H_q(M; \mathbf{Q}) \to H_q(M; \mathbf{Q}).$$

For each q we may choose a basis for the finite-dimensional rational vector space $H_q(M; \mathbf{Q})$ and we write f_{q*} as a matrix with respect to this basis.

Lefschetz number is an important concept introduced by Lefschetz in 1923. It is a number associated with each continuous map $f : |K| \to |K|$ from a polyhedron into itself and the number is denoted by Λ_f. It is also closely related to the Euler characteristic formula. It proves a powerful fixed point theorem known as Lefschetz fixed point theorem, which is an important application of homology. This theorem generalizes Brouwer fixed point theorem. Moreover, some other results on fixed points follow as its applications.

Definition 14.4.1 (*Lefschetz number*) Let K be a fixed triangulation of a compact triangulable space X. Suppose n is the dimension of K and $f : X \to X$ is a continuous map. Then there is a homeomorphism $k : |K| \to X$ such that if n is the dimension of K, then each homology group $H_q(K; \mathbf{Q})$ with rational coefficients $\mathbf{Q}$ is vector space over $\mathbf{Q}$ and each homomorphism

$$f^k_{q*} = (k^{-1} \circ f \circ k)_* : H_q(K; \mathbf{Q}) \to H_q(K; \mathbf{Q})$$

is a linear transformation. The trace of the corresponding matrices does not depend on a particular choice of the basis. The alternating sum $\sum_{q=0}^{n} (-1)^q \text{trace } f^k_{q*}$ of the traces of f^k_{q*} of these linear transformations denoted by $\Lambda_f = \sum_{q=0}^{n} (-1)^q \text{trace } f^k_{q*}$, is called the Lefschetz number of f.

Remark 14.4.2 The number Λ_f does not dependent on the triangulation of X. Hence Λ_f is well defined.

Remark 14.4.3 If $f \simeq g$, then $f_* = g_*$ in homology. Hence it follows that $\Lambda_f = \Lambda_g$ whenever f is homotopic to g.

Definition 14.4.4 The rank of the free part of the abelian group $H_q(K; \mathbf{Z})$ of a finite complex K is called the Betti number of K, denoted by β_q.

Remark 14.4.5 The Lefschetz number is an integer and generalizes the Euler characteristics of an oriented complex.

Definition 14.4.6 (*Fixed Point Property*) If a topological space X is such that every continuous map $f : X \to X$ has a fixed point, then X is said to be a space with fixed - point property.

Example 14.4.7 The topological space $X = [0, 1]$ in the real line $\mathbf{R}$ is a space with the fixed point property. Every closed interval $[a, b]$ in the real line $\mathbf{R}$ has also the fixed-point property.

The following theorem shows that Λ_f is the 'obstruction' to f being fixed point free.

Theorem 14.4.8 (Lefschetz Fixed Point Theorem) *Let X be a compact triangulable space and $f : X \to X$ be a continuous map with Λ_f its Lefschetz number.*

(a) *If $\Lambda_f \neq 0$, then f has a fixed point;*
(b) *If X has the same rational homology groups as a point, then X has the fixed point property.*
(c) *(Brouwer Fixed Point Theorem): Any contractible compact triangulable space has the fixed point property.*

(d) *If the identity map 1_X of X is homotopic to a fixed point free map $f : X \to X$, then the Euler characteristic $\chi(X) = 0$;*
(e) *If $X = S^n$, then $\Lambda_f = 1 + (-1)^n \deg f$;*

Proof **(a)** **Case I:** Let K be a finite simplicial complex having X as its polyhedron. Suppose the simplicial map $f : |K| \to |K|$ has no fixed point. Then there exists a simplex σ in K such that $f(\sigma) \neq \sigma$. Now delete each q-simplex of K to obtain a basis over $\mathbf{Q}$ for the vector space $C_q(K; \mathbf{Q})$ in such a way that with respect to this basis the linear transformation $f_q : C_q(K; \mathbf{Q}) \to C_q(K; \mathbf{Q})$ will represent a matrix having zero along its diagonal, and hence having trace zero.

Hence

$$\sum_{q=0}^{n} (-1)^q \text{trace } f_q = \sum_{q=0}^{n} (-1)^q \text{trace } f_{q*}$$

gives that $\Lambda_f = 0$. In other words, if $\Lambda_f \neq= 0$, then f has a fixed point.

Remark 14.4.9 We may calculate the Lefschetz number of f at homology level or at chain level according to our convenience.

Case II: For general case see Gray (1975) or Armstrong (1983).

(b) By hypothesis, X has only one component. Hence the only nonzero rational homology group is $H_0(X; \mathbf{Q}) \cong \mathbf{Q}$, because $H_q(X; \mathbf{Q}) = 0$, for $q > 0$. By using the definition of Lefschetz number, for any map $f : X \to X$, the induced homomorphisms

$$f_{q*}^k : H_q(X; \mathbf{Q}) \to H_q(X; \mathbf{Q})$$

are all zero for $q > 0$. On the other hand, $f_{0*}^k : \mathbf{Q} \to \mathbf{Q}$ is the identity linear transformation. This proves that $\Lambda_f = 1(\neq 0)$ and hence f has a fixed pint.
(c) It follows from (b).
(d) Let f be fixed point free. Then $\Lambda_f = 0$. By hypothesis f is homotopic to 1_X. Then it induces identity homomorphisms on homology groups and the trace of an identity linear map is the dimension of its domain:

$$\Lambda_f = \sum_{q=0}^{n} (-1)^q \text{trace}(1_{q*}) = \sum_{q=0}^{n} (-1)^q \dim H_q(X; \mathbf{Q}) = \chi(X) = 0.$$

(e) The only nonzero rational homotopy groups of S^n are $\mathbf{Q}$ in dimensions 0 and n, i.e.,

$$H_0(S^n; \mathbf{Q}) \cong H_n(S^n; \mathbf{Q}) \cong \mathbf{Q}$$

and hence

$$\Lambda_f = \text{trace}\,(f_{0*}) + (-1)^n \text{tr}\,(f_{n*}) = 1 + (-1)^n \deg f.$$

❑

Remark 14.4.10 The Lefschetz number of $f : X \to X$ may be calculated at either homology level or at chain level according to our convenience.

14.5 Application of Euler Characteristic

This section coveys some interesting applications of Euler characteristic $\chi(X)$ of a finite CW-complex X. As it is defined in terms of homology of X, it depends only on the homotopy type of X.

Recall the definitions of Euler characteristic:

Definition 14.5.1 The Euler characteristic of a finite CW-complex X of dimension n is defined to be the alternating sum

$$\chi(X) = \sum_{q=0}^{n} (-1)^q \alpha_q,$$

where α_q denotes the number of q-cells of X, generalizing the formula of Euler characteristic: number of vertices − number of edges + number of faces for 2-dimensional complexes.

Definition 14.5.2 The Euler characteristic of a finite simplicial complex K of dimension n is defined to be the alternating sum

$$\chi(K) = \sum_{q=0}^{n} (-1)^q \beta_q,$$

where β_q denotes the number of q-dimensional simplices of K.

Remark 14.5.3 If $G = \mathbf{R}$ in the group $H_n(K; G)$, then the group $H_n(K; G)$ is a real vector space. If its dimension is q, then q is called the called the qth Betti number of K, denoted by β_q.

Theorem 14.5.4 *Let X be a finite CW-complex with $\chi(X) \neq 0$, and $\psi_t : X \to X$ be a flow. Then there exists a point $x_0 \in X$ such that $\psi_t(x_0) = x_0$, $\forall t \in \mathbf{R}$.*

Proof Under the given hypothesis, $\psi_t \simeq 1_X$ (see Chap. 2) and hence $\wedge_{\psi_t} = \wedge_{1_X} = \chi(X) \neq 0$. Consequently, there exists a fixed point $x_0^{(t)}$ of ψ_t for each $t \in \mathbf{R}$. Define $X_n = \{x \in X : \psi_{1/2^n}(x) = x\}$ for each natural number n. Then $X_n \supset X_{n+1}$, and each X_n is a nonempty closed set and hence $X_\infty = \bigcap_n X_n \neq \emptyset$. Let $x \in X_\infty$. Then x is a fixed point for any $\psi_{m/2^n}$. Since the numbers $m/2^n$ are dense in $\mathbf{R}$, x is a fixed point of ψ_t for any $t \in \mathbf{R}$. ❑

Remark 14.5.5 Given a finite CW-complex X and continuous map $f : X \to X$, the relation $\Lambda_f = \wedge_{1_X} = \chi(X) \neq 0$ (where Λ_f is the Lefschetz number of f and $\chi(X)$ is the Euler characteristic of X) shows that a flow $\psi_t : X \to X$ has a fixed point. We now claim that there exists a fixed point common to every ψ_t. We prove this by induction and the infinite intersection property as follows; corresponding to each rational n points, assign X_n to be the set of fixed points of $x_{1/2^n}$. Then X_n is a nonempty closed set. Since $X_{n+1} \subseteq X_n$, the set $X_\infty = \cap_n X_n \neq \emptyset$. This shows that X_∞ is a set of points fixed under all rational numbers of dyadic form $r/2^n$. Since the set of rational numbers are dense in the real number space $\mathbf{R}$, it follows that every element in X_∞ is fixed under ψ_t for all $t \in \mathbf{R}$.

Lefschetz Fixed Point Theorem 14.4.8 gives the following corollary.

Corollary 14.5.6 **(a)** $\wedge_{1_X} = \chi(X)$;
(b) *For the antipodal map* $A : S^n \to S^n, x \mapsto -x, \wedge_A = 0$.
(c) *If* $f : S^n \to S^n$ *is not a homeomorphism, then* f *must have a fixed point.*

Definition 14.5.7 A platonic solid is a special polyhedron having the property that its faces are congruent regular polygons and each vertex belongs to the same number of edges. It is sometimes called a regular simple polyhedron.

Theorem 14.5.8 *There are only five regular simple polyhedra.*

Proof Let P be a regular simple polyhedron with V number of vertices, E number of edges, and F number of faces and m be the number of edges meeting at each vertex and n be the number of edges of each face. For $n \geq 3$, by counting vertices by edges we obtain $2E = mV$. Again by counting faces by edges we obtain $2E = nF$. Then from Euler formula $V - E + F = 2$ for P it follows that $\frac{2}{m}E - E + \frac{2}{n}E = 2$. This shows that $E = 2(\frac{1}{m} + \frac{1}{n} - 1)^{-1}$, which must be a positive integer. Hence $\frac{1}{m} + \frac{1}{n} > 1$. The possibilities are only:

(i) If $m = 5, n = 3$, then $E = 30$, $V = 12$ and $F = 20$. Hence P is the regular icosahedron see Fig. 14.2.
(ii) If $m = 4, n = 3$, then $E = 12$, $V = 6$ and $F = 8$, show that P is the regular octahedron.
(iii) If $m = 3, n = 3$, then $E = 6$, $V = 4$ and $F = 4$ show that P is a regular tetrahedron.
(iv) If $m = 3, n = 4$, then $E = 12$, $V = 8$ and $F = 6$ show that P is the cube.
(v) If $m = 3, n = 5$, then $E = 30$, $V = 20$ and $F = 12$ show that P is the regular dodecahedron see Fig. 14.3.

Fig. 14.2 Icosahedron

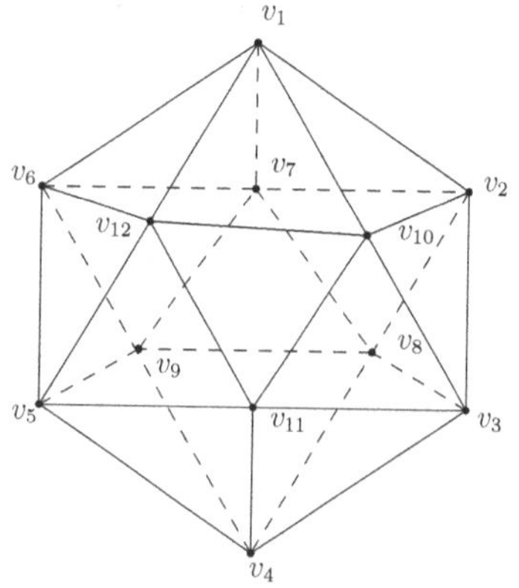

Fig. 14.3 Dodecahedron

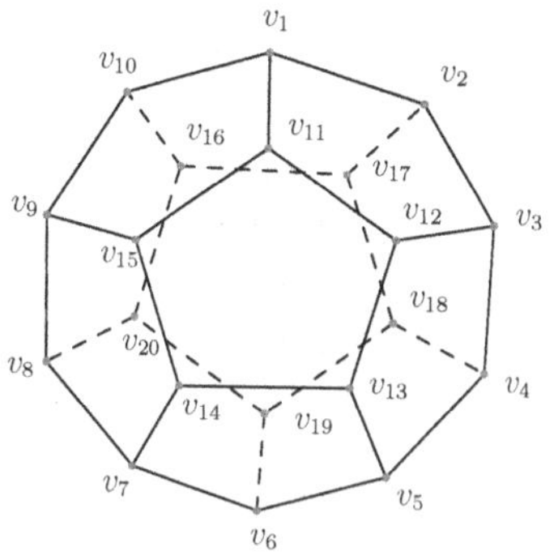

❑

Proposition 14.5.9 *Let B be a finite CW-complex and $p : X \to B$ be an m sheeted covering space. Then the Euler characteristic $\chi(X) = m\chi(B)$.*

Proof By hypothesis, it follows that X admits a CW-complex structure with the j-cells of X, which are the lifts to X of j-cells of B. As p is m-sheeted, the number of distinct such lifts is m. Hence it follows that the number of j-cells of X is exactly m times that of j-cells of B. This concludes that $\chi(X) = m\chi(B)$. ❑

14.6 Application of Mayer–Vietoris Sequence

This section conveys some interesting applications of Mayer–Vietoris sequence (see Chap. 10) defined by W.Mayer (1887–1947) and L. Vietoris (1891–2002) such as in computing homology groups of S^n. In addition to the long exact sequence of homology groups of a pair (X, A) of spaces, there is another long exact sequence of homology groups, called a Mayer–Vietoris sequence. This sequence is sometimes more convenient to apply.

Let U_1 and U_2 be two subspaces of X such that X is the union of the interiors of U_1 and U_2. If $f_i : U_1 \cap U_2 \to U_i$ and $g_i : U_i \to X$ are inclusion maps for $i = 1, 2$, define

$$\phi : H_n(U_1 \cup U_2) \to H_n(U_1) \oplus H_n(U_2), \quad \alpha \mapsto (f_{1*}(\alpha), f_{2*}(\alpha)),$$
$$\text{and} \quad \psi : H_n(U_1) \oplus H_n(U_2) \to H_n(X), \quad (\alpha_1, \alpha_2) \mapsto g_{1*}(\alpha_1) - g_{2*}(\alpha_2)$$

Under this situation one obtains an exact sequence

$$\cdots \to H_{n+1}(X) \xrightarrow{\triangle} H_n(U_1 \cup U_2) \xrightarrow{\phi} H_n(U_1) \oplus H_n(U_2) \xrightarrow{\psi} H_n(X) \xrightarrow{\triangle} H_n(U_1 \cap U_2) \to \cdots \tag{14.1}$$

The sequence (14.1) is called the Mayer–Vietoris sequence and the homomorphisms $\triangle$ are called connecting homomorphisms.

Mayer–Vietoris sequence is now applied to compute the homology groups of the n-sphere S^n.

Theorem 14.6.1 *Given a positive integer n,*

$$H_m(S^n; \mathbf{Z}) \cong \begin{cases} \mathbf{Z}, & \text{if } m = 0 \text{ or } n \\ 0, & \text{otherwise} \end{cases}$$

Proof Let $U_1 = \{x \in S^n : x_n > -\frac{1}{2}\}$ and $U_2 = \{x \in S^n : x_n < \frac{1}{2}\}$. Then U_1 and U_2 are contractible spaces. Use induction on n and apply Mayer–Vietoris sequence to compute $H_m(S^n)$. ❑

14.7 Application of van Kampen Theorem

This section presents an important application of van Kampen theorem which asserts that for $n > 1$, S^n is simply connected. The van Kampen Theorem has been studied in Chap. 6. An alternative form of van Kampen Theorem is now given.

Theorem 14.7.1 (van Kampen theorem) *Let $X = X_1 \cup X_2$ and X_1, X_2 and $A = X_1 \cap X_2$ are all open path-connected subsets of X. If $x_0 \in A$, then $\pi_1(X, x_0)$ is the free product:*

$$\pi_1(X, x_0) \cong \pi_1(X_1, x_0) *_{\pi_1(A, x_0)} \pi_1(X_2, x_0).$$

In other words, if $i_1 : A \hookrightarrow X_1$ *and* $i_2 : A \hookrightarrow X_2$ *are inclusions, then* $\pi_1(X, x_0)$ *is isomorphic to the free product* $\pi_1(X_1, x_0) *_{\pi_1(A, x_0)} \pi_1(X_2, x_0)$ *modulo the relations* $i_{1*}(\alpha) = i_{2*}(\alpha)$ *for every* $\alpha \in \pi_1(A, x_0)$.

Proof See [Gray, pp 40–41]. ❑

Corollary 14.7.2 *Under the hypothesis of van Kampen theorem,*

(i) *If* X_1 *and* X_2 *are simply connected, then* X *is also so;*
(ii) *If* A *is simply connected, then* $\pi_1(X, x_0) = \pi_1(X_1, x_0) * \pi_1(X_2, x_0)$.
(iii) *If* X_2 *is simply connected, then* $\pi_1(X, x_0) = \pi_1(X, x_0)/N(\pi_1(A, x_0))$

where $N(\pi_1(A, x_0))$ *is the normalizer of* $\pi_1(A, x_0)$.

Proof It is left as an exercise. ❑

Corollary 14.7.3 *For* $n > 1$, S^n *is simply connected.*

Proof Let $X_1 = S^n - N$, where $N = (0, 0, \ldots, 0, 1)$ and $X_2 = S^n - S$, where $S = (0, 0, \ldots, 0, -1)$. Clearly, X_1 and X_2 are both homeomorphic to $\mathbf{R}^n$. Consequently, they are path-connected spaces and simply connected. Again as $n > 1$, $X_1 \cap X_2$ is a path-connected space. Hence the corollary follows from Corollary 14.7.2(i). ❑

14.8 Applications to Algebra

Algebraic topology generally utilizes algebraic techniques to get topological information, but this direction is sometimes reversed in a convenient way. For example, it is proved by using the concepts of fundamental group and covering space that a subgroup of a free group is free. The other example is a proof of the algebraic result by using the concept of degree function that if n is an even integer, then $\mathbf{Z}_2$ is the only nontrivial group that can act freely on S^n.

Theorem 14.8.1 *Every subgroup of a free group is free.*

Proof Given a free group F, we can construct a graph B corresponding to a basis of F such that $\pi_1(B) \cong F$. This construction is possible. For example, one may take B to be the graph which is a wedge of circles corresponding to a basis for F for such a construction. Consequently, for each subgroup G of F, there exists a covering space $p : X \to B$ having induced monomorphism

$$p_* : \pi_1(X) \to \pi_1(B)$$

such that $p_*(\pi_1(X)) = G$ by Ex. 14 of Sect. 14.11. Since p_* is a monomorphism, $\pi_1(X) \cong G$. Again as X is a graph by Ex. 16 of Sect. 14.11, the group $G \cong \pi_1(X)$ is free by Ex. 15 of Sect. 14.11. ❑

Recall that the antipodal map

$$A : S^n \to S^n,\ x \mapsto -x$$

generates an action of $\mathbf{Z}_2$ on S^n with orbit space the real projection n-space $\mathbf{R}P^n$.

There is a natural question: does there exist any finite group that acts freely on S^n? If n is an even integer, $\mathbf{Z}_2$ is the only nontrivial group that can act freely on S^n (see Ex. 23 of Sect. 14.11).

Given a topological space X, if a topological group G acts on X (from the left), then for every $g \in G$, the map $\psi_g : X \to X$ defined by $\psi_g(x) = gx$ is a homeomorphism (see Appendix A). Hence for each $g \in G$, ψ_g is an element of the group Homeo(X) of homeomorphisms of X. This action is said to be free if the homeomorphism ψ_g in the group Homeo(X) corresponding to each nontrivial element g of G has no fixed points.

The above discussion may be summarized in an important result.

Proposition 14.8.2 *Under the above situation the antipodal map $A : S^n \to S^n, x \mapsto -x$ generates a free action of $\mathbf{Z}_2$.*

Remark 14.8.3 For more applications of algebraic topology to algebra see Sect. 14.11.

14.9 Application of Brown Functor

Brown functor defined in Appendix B plays a key role in the study of algebraic topology.

Definition 14.9.1 Let F be Brown functor and X be a topological space in the category $\mathcal{C}_0$. Then an element $u \in F(X)$ is said to be m-universal if the homomorphism

$$\psi_u : \pi_k(X) \to F(S^k),\ [f] \mapsto F([f])(u)$$

is an isomorphism for $k < m$ and an epimorphism for $k = m$. An m-universal element u is said to be a universal element if it is m-universal for all $m \geq 1$.

Proposition 14.9.2 *Let F be Brown functor in the category $\mathcal{C}_0$. Then it is representable in the sense that there is a pointed topological space $X \in \mathcal{C}_0$ which is determined up to homotopy equivalence such that there exists a natural equivalence of functors*

$$\psi : [-; X] \to F.$$

Proof For every pointed topological space $Y \in \mathcal{C}_0$, there is a bijection

$$\psi_Y : [Y; X] \to F(Y), [f] \mapsto F([f])(u),$$

where $u \in F(Y)$ is a universal element. ❑

Remark 14.9.3 For more results associated with Brown functor see Sect. 14.11.

14.10 Applications Beyond Mathematics

Algebraic topology has also interesting applications in some areas other than mathematics. The present book is beyond the scope of the study of these applications, except to give some references. However, the author is preparing a new book 'Topics in Topology with Applications' (unfinished).

14.10.1 Application to Physics

Algebraic topology has made a revolution in mathematical physics in the second half of the twentieth century. For example, fiber bundles and vector bundles constitute an extensive special class of manifolds, and play a key role in some theories of physics, general relativity, and gauge theories. Moreover, algebraic topology plays an important role in condensed matter physics, statistical mechanics, elementary particle theory, and some other branches of physics. Homotopy theory is specially used in the study of solitons, monopoles, and condensed system. Many interesting topological spaces appear in physics at different situations. For example, the phase space of a quantum system with n pure states can be considered as the complex projective space $\mathbf{C}P^{n-1}$. Each state (wave function) is a nonzero vector in $\mathbf{C}^n$, but the states that differ only by multiplicative factors are physically indistinguishable, and hence they are identified with each other. Another example: the phase space of a classical mechanical system with nondegenerate Lagrangian belongs to the same homotopy type of the configuration space.

For this subsection the following books are referred.

[1] M.F. Atiyah, The geometry and physics of knots, Cambridge University Press, Cambridge, 1990.
[2] M. Monastyrsky, Topology of Gauge Fields and Condensed Matter, Plenum, New York, 1993.
[3] M. Nakahara, Geometry, Topology and Physics, Taylor and Francis, 2003.
[4] C. Nash and S. Sen Topology and Geometry for Physicists, Academic Press, London, 1983.

[5] Schwartz, A. S., *Quantum field theory and topology*, Springer, Berlin, 1993.
[6] N. E. Steenrod, The Topology of Fibre Bundles, Princeton University Press, 1951

14.10.2 Application to Sensor Network

Algebraic topology is now used to solve coverage problems by integrating local data about sensor networks into global information and utilizes its strong tools to determine whether there is any hole in a sensor coverage. For example, certain topological invariants such as Euler characteristic, fundamental groups and higher homotopy groups, homology and cohomology groups play a key role in solving the coverage problems.

For this subsection the following papers are referred.

[1] E. W. Chambers, J. Erickson, and P. Worah, Testing contractibility in planar Rips complexes, in Proc. 24th Annu. Symp. Computat. Geom., College Park, MD, pp. 251–259. 2008.
[2] J. Cortes, S. Martinez, T. Karatas, and F. Bullo, Coverage control for mobile sensing networks, in Proc. IEEE Int. Conf. Robot. Autom., Washington, DC, Vol. **2**, pp. 1327–1332, 2002.
[3] de Silva and Robert Ghrist, Homological Sensor Networks, Notices of AMS **54** (1) pp 1–11, 2007.

14.10.3 Application to Chemistry

Topology and graph theory have strongly influenced the recent development of chemistry through their applications in nonroutine mathematical methods. For example, "chemical topology", "invariance of molecular topology", "chemical applications of topology and graph theory" and "topological methods in chemistry" are now outstanding developments of chemistry which are closely related to topology.

For this subsection the following books are referred:

[1] E. V. Babaev, The Invariance of molecular topology, Moscow State University, Moscow, 1994.
[2] Bonchev D., Rouvray R., (Eds) Chemical Topology: Introduction and Fundamentals, Gordon and Breach Publ., Reading, 1999.
[3] R.B. King, (Ed.)., Chemical Applications of Topology and Graph Theory; Studies in Physical and Theoretical Chemistry, Vol. **28**, Elsevier, Amsterdam, 1983.
[4] H. E. Simmons, Topological Methods in Chemistry, Wiley Interscience, New York, 1989.

14.10.4 Application to Biology, Medical Science and Biomedical Engineering

The fields of biological & medical physics and biomedical engineering are now emerging as a multidisciplinary area connecting topology with different areas of physics, biology, chemistry, medicine, and some of their closely related fields. For example, knot theory, a branch of topology, is used in biology to study the effects of certain enzymes on DNA. Algebraic topology addresses the growing need for this multidisciplinary research. For example, recent investigation in molecular biology, theory of protein and DNA involves application of algebraic topology, which is a stimulating feature.

For this subsection following books are referred.

[1] I. Darcy and D.Mners, Knot Theory, Polish Academy of Sciences, Warszawa, 1998

[2] M.I. Monastyrsky (Ed.) Topology in Molecular Biology, Springer-Verlag Berlin Heidelberg, 2007.

14.10.5 Application to Economics

The Brouwer fixed point theorem given by L.E.J. Brouwer in 1912 is one of the stimulating events in the history of topology. Since then this theorem has been extending its influence to diverse areas of mathematics, mathematical economics and related fields. For example, in economics Brouwer fixed point theorem plays a key role in studying general equilibrium theory and in the most basic and general models of economists. The other example, is the 'social choice' model which is a model for decision making in mathematical economics and social science which is closely related to homotopy problems. The author of this book earned an inspiration on the topic 'social choice and topology' by attending a lecture of Eckman (2003) at ETH, Zurich.

The following books and papers are referred for this subsection.

[1] Kim C.Border, ' Fixed Point Theorems with Applications to Economics and Game Theory,' Cambridge University Press, 1985.

[2] B. Eckman, 'Social Choice and Topology: A Case of Pure and Applied Mathematics,' 2003.

[3] A. Granas and J. Dugundji, Fixed Point Theory, Springer, New York, 2003.

14.10.6 Application to Computer Science

Algebraic topology has recently found some surprising fruitful results in computer science by establishing a close relation between the theory of concurrent computation and the theories of algebraic and combinatorial topology.

For this subsection the following references are given.

[1] Edelsbrunner, H. and Harer J. L., Computational Topology. An Introduction. Amer. Math. Soc., Providence, Rhode Island, 2009.
[2] Gyulassy A., Natarajan V., Pascucci V., Bremer P.T., and Ann B.H., A topological approach to simplification of three-dimensional scalar functions. IEEE Trans. Vis. Comput. Graph. Vol 12, 474–484. 2006.

14.11 Exercises

This section conveys some interesting results through different exercises.

1. Let K be a finite complex. Show that

(a) the qth Betti number β_q of K is the dimension of $H_q(K; \mathbf{Q})$ as a vector space over $\mathbf{Q}$;

(b) the complexes K whose polyhedra are homotopy equivalent have the same Euler characteristic

2. (Hopf Trace Theorem). Let K be a finite simplicial complex of dimension n, and $\psi : C(K; \mathbf{Q}) \to C(K; \mathbf{Q})$ be a chain map. Show that

$$\sum_{q=0}^{n} (-1)^q \text{trace } \psi_q = \sum_{q=0}^{n} (-1)^q \text{trace } \psi_{q^*},$$

where ψ_{q^*} is the homomorphism induced by chain map ψ_q in the corresponding homology groups.

3. Show that the only compact closed surfaces with Euler characteristic zero are the Klein bottle, which is non orientable, and the torus, which is orientable.

[Hint: Only these two compact closed surfaces admit a fixed-pint free map which is homotopic to the identity map. A fixed point free map on the torus can be obtained by a flow along a nowhere-zero (nonvanishing) tangent vector field.]

4. Use Euler characteristic to show that a sphere cannot be homotopy equivalent to a point.

5. Prove that the homotopy classes of maps of a sphere to itself can be characterized with the help of integers.

[Hint: Use Hopf classification theorem.]

6. Show that no subspace of $\mathbf{R}^n$ can be homeomorphic to S^n.

   [Hint: Use Borsuk–Ulam theorem to show that no continuous map from S^n to $\mathbf{R}^n$ is injective].
7. Show that while wrapping a soccer ball with three pieces of papers, one must contain a pair of antipodal points.

   [Use Theorem 14.3.6 (Lusternik–Schnirelmann).]
8. Show that the following statements are equivalent:

   (i) A continuous map $f : D^n \to D^n$ has a fixed point (Brouwer fixed point theorem).
   (ii) There does not exist a continuous map $r : D^n \to S^{n-1}$ which is identity on S^{n-1} (Retraction theorem).
   (iii) The identity map $1_d : S^{n-1} \to S^{n-1}$ is not nullhomotopic (Homotopy theorem).

9. Show that every nullhomotopic map $f : S^n \to S^n$ has at least one fixed point.
10. If $f : (D^{n+1}, S^n) \to (D^{n+1}, S^n)$ is a continuous map, show that $\deg f = \deg(f|_{S^n})$.

    [Hint: Consider the commutative diagram in Fig. 14.4]
11. Show that every continuous map $f : S^n \to S^n$ with $\deg f = d$ induces the homomorphism $f_* : \pi_n(S^n) \to \pi_n(S^n)$ which is a map multiple by d.
12. Let $f : S^n \to S^n$ be a continuous map for $n \geq 1$ and $\Sigma f : S^{n+1} \to S^{n+1}$ be its suspension map. Show that $\deg f = \deg(\Sigma f)$ for all $n \geq 1$.
13. Given a continuous map $f : S^n \to S^n$, show that $\Sigma f : \Sigma S^n \to \Sigma S^n$ has degree d iff the map $f : S^n \to S^n$ has degree d.
14. Let B be a path-connected, locally path-connected, and semi locally simply connected space. Show that for every subgroup G of $\pi_1(B, b_0)$, there is a covering space $p : X \to B$ such that $p_*(\pi_1(X, x_0)) = G$ for a suitable base point $x_0 \in X$.
15. Let X be a connected graph with maximal tree M. Show that $\pi_1(X)$ is a free group with basis the classes $[b_i]$ corresponding to the edges e_i of $X - M$.
16. Show that every covering space of a graph is also a graph, with vertices and edges the lifts of the vertices and edges in the base graph.
17. Let $X = S^n$ and E_n^+, E_n^- be the north and south hemispheres of S^n. Then $A \cap B = S^{n-1}$. Show that

$$\begin{array}{ccc} H_{n+1}(D^{n+1}, S^n) & \xrightarrow{f_*} & H_{n+1}(D^{n+1}, S^n) \\ \partial_* \downarrow \cong & & \partial_* \downarrow \cong \\ \widetilde{H}_n(S^n) & \xrightarrow{(f|_{S^n})_*} & \widetilde{H}_n(S^n) \end{array}$$

Fig. 14.4 Rectangle involving $(f|_{S^n})_*$

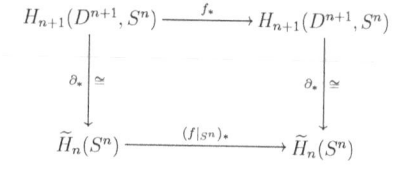

(i) $\widetilde{H}_i(E_n^+) \oplus \widetilde{H}_i(E_n^-) = 0$;
(ii) $\widetilde{H}_i(S^n) \cong \widetilde{H}_{i-1}(S^{n-1})$.

18. Let T be a Brown functor and $\{*\}$ denote the one-point space. Show that $T(\{*\})$ is a set consisting of a single element.
19. Let T be a Brown functor, X_0 be a pointed space. If $u_0 \in T(X_0)$, show that there is a pointed space X obtained from X_0 by attaching together with a universal element $u \in T(X)$ such that $u|_{X_0} = u_0$.
20. Let T be a Brown functor. If X and Y are pointed CW-complexes with universal element $u \in T(X)$ and $u' \in T(Y)$. Show that there exists a homotopy equivalence $f : X \to Y$ such that $T([f])(u') = u$.
21. Let T be a Brown functor and $u \in T(X)$ be a universal element. Show that if X is a pointed CW-complex, then the map

$$\psi_u : [Y, X] \to T(Y),\ [f] \to T([f])$$

is a bijection.
22. Show that $\mathbf{Z}_p \times \mathbf{Z}_p$ cannot act freely on S^n for a prime integer p.
23. If n is an even integer, show that $\mathbf{Z}_2$ is the only nontrivial group that can act freely on S^n.
24. Let $f : S^{2n} \to S^{2n}$ be any continuous map. Show that there is a point $x \in S^{2n}$ such that either $f(x) = x$ or $f(x) = -x$.
25. Show that every continuous map $f : \mathbf{R}P^{2n} \to \mathbf{R}P^{2n}$ has a fixed point.
26. Construct continuous maps $g : \mathbf{R}P^{2n-1} \to \mathbf{R}P^{2n-1}$ without fixed points from linear operators $\mathbf{R}^{2n} \to \mathbf{R}^{2n}$ without eigenvectors.
27. (Poincaré duality) Let K be a triangulation of a homology n-manifold. If $|K|$ is orientable, show that there exists an isomorphism

(i)

$$h : H^r(K; \mathbf{Z}) \to H_{n-r}(K'; \mathbf{Z})$$

for all r;

(ii)

$$h : H^r(K; \mathbf{Z}_2) \to H_{n-r}(K'; \mathbf{Z}_2)$$

for all r.

[Hint: See Maunder (1996).]

28. Let M be a compact manifold of odd dimension. Show that its Euler characteristic $\chi(M) = 0$.

[Hint: Since $H_r(M; \mathbf{Z}_2)$ is finite dimensional, the Euler characteristic $\chi(M)$ is well defined. Again since the dim $H_r(M; \mathbf{Z}_2) = dim H_{n-r}(M; \mathbf{Z}_2)$, it follows that

$$\chi(M) = \sum_r (-1)^r \dim H_r(M; \mathbf{Z}_2) = 0].$$

29. For any knot K, show that the homology groups $H_0(\mathbf{R}^3 - K; \mathbf{Z})$ and $H_1(\mathbf{R}^3 - K; \mathbf{Z})$ are isomorphic to $\mathbf{Z}$ and the remaining homology groups of the complement $\mathbf{R}^3 - K$ are trivial. Hence shows that the trefoil knot is nontrivial.

[Hint: Use the result that if two knots are equivalent, then their complements have isomorphic homology groups.]

30. T be an orthogonal transformation of $\mathbf{R}^m$ such that $T^n = 1_d$ and $T^k \neq 1_d$ for $0 < k < n$. Show that the group $G = \{T^0, T^1, \cdots, T^{n-1}\}$ acts freely on $S^{m-1} \subset \mathbf{R}^n$ if every eigenvalue λ of T satisfies $\lambda^k \neq 1$ for $0 < k < n$. Compute the fundamental group and homology groups of the orbit space S^{m-1} mod G, when $m = 4$ and G acts freely on S^3. (A orbit space of this type is called a lens space).

[Hint: For a free action, the group $\pi_1(S^{m-1} \bmod G) = G$. Use Hurewicz theorem and Poincaré duality for respective computations.]

31. Let $G_{n,k}$ be the Grassmann manifold of k-planes in $\mathbf{R}^n$. Show that

$$\pi_i(G_{n,k}) = \pi_{i-1}(O(k)) \text{ for } i < n - k.$$

[Hint: Use fibration $(V_{n,k}, G_{n,k}, O(k))$].

32. Apply the Lefschetz fixed point theorem to prove that

(i) if n is even, every continuous map $f : \mathbf{C}P^n \to \mathbf{C}P^n$ has a fixed point;
(ii) if n is odd, there is a fixed point unless $f^*(\beta) = -\beta$ for β, a generator of $H^2(\mathbf{C}P^n; \mathbf{Z})$.

[Hint: Use the results that $f^* : H^*(\mathbf{C}P^n; \mathbf{Z}) \to H^*(\mathbf{C}P^n; \mathbf{Z})$ is a ring homomorphism, each cohomology group is even degree $\leq n$ has rank 1, and each cohomology group in odd degree is 0.]

33. Using cup products, prove that every continuous map $S^{m+n} \to S^m \times S^n$ induces a trivial homomorphism $H_{m+n}(S^{m+n}) \to H_{m+n}(S^m \times S^n)$, for positive integers m and n.

[Hint: Use Künneth formula.]

34. Use homotopy groups to show that $\mathbf{R}P^m$ is not a retract of $\mathbf{R}P^m$ for $n > m > 0$.

35. Show that $\pi_i(V_{n,k}) = 0$, for $i < n - k$.

[Hint: Use the fibration $(V_{n,k+1}, V_{n,k}, S^{n-k-1})$.]

36. Show that for any closed surface other than the sphere and the projective plane, the homotopy groups of dimensions greater than 1 are trivial.

37. Show that $\pi_{n-2}(V_n, 2) \cong \begin{cases} \mathbf{Z}, & \text{for } n \text{ is even} \\ \mathbf{Z}_2, & \text{for } n \text{ is odd.} \end{cases}$

[Hint: Use the exact homotopy sequence of the fibration $(V_{n-2}, S^{n-1}, S^{n-2})$.]

14.12 Additional Reading

[1] Adams, J.F.,*Algebraic Topology: A student's Guide*, Cambridge University Press, Cambridge, 1972.

[2] Adhikari, M.R., and Adhikari, Avishek, *Basic Modern Algebra with Applications*, Springer, New Delhi, New York, Heidelberg, 2014.

[3] Dieudonné, J., *A History of Algebraic and Differential Topology*, 19001960, Modern Birkhäuser, 1989.

[4] Dold, A.,*Lectures on Algebraic Topology*, Springer-Verlag, New York, 1972.

[5] Eilenberg, S., and Steenrod, N., *Foundations of Algebraic Topology*, Princeton University Press, Princeton, 1952.

[6] Eschrig, H., *Topology and Geometry for Physics*, Springer-Verlag Berlin Heidelberg 2011.

[7] Fulton, W., *Algebraic Topology, A First Course*, Springer-Verlag, New York, 1975.

[8] Hilton, P.J., *An introduction to Homotopy Theory*, Cambridge University Press, Cambridge, 1983.

[9] Hilton, P. J. and Wylie, S. *Homology Theory*, Cambridge University Press, Cambridge, 1960.

[10] Mayer, J. *Algebraic Topology*, Prentice-Hall, New Jersy, 1972.

[11] Massey, W.S., *A Basic Course in Algebraic Topology*, Springer-Verlag, New York, Berlin, Heidelberg, 1991.

[12] Pearl Elliot (ed), Open Problems in Topology II, Elsevier, Amsterdam, 2007.

[13] Schwartz, A. S., *Quantum field theory and topology*, Springer, Berlin,, 1993.

[14] Switzer, R.M., *Algebraic Topology-Homotopy and Homology*, Springer-Verlag, Berlin, Heidelberg, New York, 1975.

[15] Whitehead, G.W., *Elements of Homotopy Theory*, Springer-Verlag, New York, Heidelberg, Berlin, 1978.

References

Adams, J.F.: Algebraic Topology: A student's Guide. Cambridge University Press, Cambridge (1972)

Adhikari, M.R., Adhikari, A.: Basic Modern Algebra with Applications, Springer, New Delhi, New York, Heidelberg (2014)

Armstrong, M.A.: Basic Topology. Springer, New York (1983)

Atiyah, M.F.: The geometry and physics of knots. Cambridge University Press, Cambridge (1990)

Babaev, E.V.: The Invariance of molecular topology. Moscow State University, Moscow (1994)

Bonchev D., Rouvray R. (eds.): Chemical Topology: Introduction and Fundamentals. Gordon and Breach Publ., Reading, (1999)

Border, K. C.: Fixed Point Theorems with Applications to Economics and Game Theory. Cambridge University Press, Cambridge (1985)

Chambers, E. W., Erickson, J., Worah, P.: Testing contractibility in planar Rips complexes. In: Proceedings of the 24th Annual Symposium of Computation Geometry. College Park, MD, pp. 251–259 (2008)

Cortes J., Martinez S., Karatas T., Bullo, F.: Coverage control for mobile sensing networks. In: Proceedings of the IEEE International Conference on Robotics Automation. Washington, DC, Vol.**2**, pp. 1327 –1332. (2002)

Croom, F.H.: Basic Concepts of Algebraic Topology. Springer, New York, Heidelberg, Berlin (1978)

Darcy, I., Mners, D.: Knot Theory. Polish Academy of Sciences, Warszawa (1998)

Dieudonné, J.: A History of Algebraic and Differential Topology, 1900–1960, Modern Birkhäuser (1989)

Dold, A.: Lectures on Algebraic Topology. Springer, New York (1972)

Dodson, C.T.J., Parker P.E.: A User's Guide to Algebraic Topology, Kluwer Academic Publishers, Berlin (1997)

Eckman, B.: "Social Choice and Topology A Case of Pure and Applied Mathematics", (2003). http://www.elsevier.de/expomath

Edelsbrunner, H., Harer, J.L.: Computational Topology. An Introduction. Amer. Math. Soc, Providence, Rhode Island (2009)

Eilenberg, S., Steenrod, N.: Foundations of Algebraic Topology. Princeton University Press, Princeton (1952)

Eschrig, H.: Topology and Geometry for Physics. Springer, Berlin (2011)

Fulton, W.: Algebraic Topology. A First Course. Springer, New York (1975)

Granas, A., Dugundji, J.: Fixed Point Theory. Springer, New York (2003)

Gray, B.: Homotopy Theory. An Introduction to Algebraic Topology. Acamedic Press, New York (1975)

Gyulassy, A., Natarajan, V., Pascucci, V., Bremer, P.T., Ann, B.H.: A topological approach to simplification of three-dimensional scalar functions. IEEE Trans. Vis. Comput. Graph **12**, 474–484 (2006)

Hatcher, A.: Algebraic Topology. Cambridge University Press, Cambridge (2002)

Hilton, P.J.: An Introduction to Homotopy Theory. Cambridge University Press, Cambridge (1983)

Hilton, P.J., Wylie, S.: Homology Theory. Cambridge University Press, Cambridge (1960)

King R.B.(ed.), Chemical Applications of Topology and Graph Theory; Studies in Physical and Theoretical Chemistry, Vol. 28, Elsevier, Amsterdam (1983)

Massey, W.S.: A Basic Course in Algebraic Topology. Springer, New York (1991)

Maunder, C.R.F.: Algebraic Topology, Van Nostrand Reinhold Company, London, 1970. Dover, Reprinted (1996)

Mayer, J.: Algebraic Topology. Prentice-Hall, New Jersy (1972)

Monastyrsky, M. (ed.): Topology in Molecular Biology. Springer, Berlin (2007)

Monastyrsky, M.: Topology of Gauge Fields and Condensed Matter. Plenum, New York (1993)

Nash, C., Sen, S.: Topology and Geometry for Physicists. Academic Press, London (1983)

Nakahara, M.: Geometry. Institute of Physics Publishing, Taylor and Francis, Bristol, Topology and Physics (2003)
Elliot, Pearl (ed.): Open Problems in Topology II. Elsevier, Amsterdam (2007)
Schwartz, A.S.: Quantum field theory and topology. Springer, Berlin (1993)
de Silva., Ghrist, R.: Homological Sensor Networks. Notices of AMS **54**(1), 1–11 (2007)
Simmons, H.E.: Topological Methods in Chemistry. Wiley Interscience, New York (1989)
Spanier, E.: Algebraic Topology. McGraw-Hill, New York (1966)
Steenrod N.E.: The Topology of Fibre Bundles. Princeton University Press, Princeton (1951)
Switzer, R.M.: Algebraic Topology: Homotopy and Homology. Springer, Berlin (1975)
Whitehead, G.W.: Elements of Homotopy Theory. Springer, New York (1978)

Chapter 15
Spectral Homology and Cohomology Theories

This chapter continues to study homology and cohomology theories through the concept of a spectrum and constructs its associated homology and cohomology theories, called spectral homology and cohomology theories. It also introduces the concept of generalized (or extraordinary) homology and cohomology theories. Moreover, this chapter conveys the concept of an Ω-spectrum and constructs a new Ω-spectrum $\underline{A}$, generalizing the Eilenberg–MacLane spectrum $K(G, n)$. It constructs a new generalized cohomology theory $h^*(\ ;\underline{A})$ associated with this spectrum $\underline{A}$, which generalizes the ordinary cohomology theory of Eilenberg and Steenrod. This chapter works in the category $\mathcal{C}$ whose objects are pairs of spaces having the homotopy type of finite CW-complex pairs and morphisms are continuous maps of such pairs. This is a full subcategory of the category of pairs of topological spaces and maps of pairs, and this admits the construction of mapping cones. Let $\mathcal{C}_0$ be the category whose objects are pointed topological spaces having the homotopy type of pointed finite CW-complexes and morphisms are continuous maps of such spaces. There exist the (reduced) suspension functor $\Sigma : \mathcal{C}_0 \to \mathcal{C}_0$ and its adjoint functor $\Omega : \mathcal{C}_0 \to \mathcal{C}_0$ which is the loop functor.

The idea of spectrum was originated by F. L. Lima (1929–) in 1958 and has been proved to be very useful. This chapter studies cohomology theories associated with different spectra, Brown representability theorem, stable homotopy groups, homotopical description of cohomology theory, and the cohomology operations. Around 1959, several topologists attempted to consider systems of covariant functors $h_n : \mathcal{C} \to \mathcal{Ab}$ (contravariant functors $h^n : \mathcal{C} \to \mathcal{Ab}$), where n takes all values in $\mathbf{N}$ or $\mathbf{Z}$ and $\mathcal{Ab}$ is the category of abelian groups and homomorphisms. These functors satisfy all the axioms of Eilenberg and Steenrod for homology (cohomology) with the exception of dimension axiom. The theory associated with such functors is now known as generalized (or extraordinary) homology and cohomology theories to distinguish them from ordinary homology and cohomology theories. There are several functors from (based) topological spaces to graded abelian groups such as stable homotopy groups or more generally, homology of a space with coefficients

M.R. Adhikari, *Basic Algebraic Topology and its Applications*,
DOI 10.1007/978-81-322-2843-1_15

in a spectrum. These are examples of generalized homology (cohomology) theories which are constructed in two equivalent methods: reduced and unreduced theories. The former one is a functor on $\mathcal{C}_0$ and the latter one is a functor on $\mathcal{C}$.

The cohomology groups $H^n(X; G)$ of a space $X \in \mathcal{C}_0$ with coefficients in an abelian group G can be characterized as the group of homotopy classes of maps of X into the Eilenberg–MacLane space $K(G, n)$. This asserts that the cohomology theory with coefficients in G can be described in this way. The spaces $K(G, n)$ are considered as the components of a spectrum. Given a cohomology theory, one may define the corresponding homology groups as the cohomology groups of the complement of X in a sphere in which X is embedded.

For this chapter, the books (Eilenberg and Steenrod 1952), (Gray 1975), (Hatcher 2002), (Maunder 1980), (Spanier 1966) and papers (Brown 1962), (Whitehead 1962) and some others are referred in the Bibliography.

15.1 Spectrum of Spaces

This section conveys the concept of a spectrum $\underline{E} = \{E_n, \alpha_n\}$ of spaces introduced by Lima in 1958. Let $\mathcal{Ab}$ be the category of abelian groups and homomorphisms. Special spectra, called Ω-spectra play a key role in algebraic topology. For example, Eilenberg–MacLane spectrum is an Ω-spectrum and it relates cohomology with homotopy (see Theorem 15.5.4). Each spectrum $\underline{E} = \{E_n, \alpha_n\}$ produces two different sequences of functors $\widetilde{h}_n(;\, , \underline{E})$ and $\widetilde{h}^n(; \underline{E})$ from the category $\mathcal{C}_0$ to the category $\mathcal{Ab}$, the first one is covariant and the second one is contravariant. These are called the spectral homology and cohomology functors associated with the spectrum $\underline{E}$, and in brief abbreviated $\widetilde{E_n}$ and $\widetilde{E^n}$.

Definition 15.1.1 A spectrum $\underline{E} = \{E_n, \alpha_n\}$ of spaces in $\mathcal{C}_0$ is a sequence $\{E_n\}$ of topological spaces in $\mathcal{C}_0$ together with a sequence of continuous maps

$$\alpha_n : E_n \to \Omega E_{n+1} \text{ in } \mathcal{C}_0$$

$$\text{(equivalently, } \widetilde{\alpha}_n : \Sigma E_n \to E_{n+1} \text{ in } \mathcal{C}_0\text{).}$$

.

Definition 15.1.2 A spectrum $\underline{E} = \{E_n, \alpha_n\}$ in $\mathcal{C}_0$ is is said to be an Ω-spectrum if $\alpha_n : E_n \to \Omega E_{n+1},\ n \in \mathbf{Z}$ is a base point preserving weak homotopy equivalence for every integer n.

Definition 15.1.3 (*Eilenberg–MacLane spectrum*) The spectrum $\underline{X} = \{X_n, \alpha_n\}$ given by $X_n = K(\mathbf{Z}, n)$ and $\alpha_n : K(\mathbf{Z}, n) \to \Omega K(\mathbf{Z}, n+1)$, a base point preserving weak homotopy equivalence, is called an Eilenberg–MacLane spectrum. In general, the Eilenberg–MacLane spectrum $\underline{X} = \{X_n, \alpha_n\}$, denoted by $\underline{HG}$, is defined by taking $X_n = K(G, n)$, where $K(G, n)$ is an Eilenberg–MacLane space of type (G, n).

$$\begin{array}{ccc} \Sigma E_n & \xrightarrow{\tilde{\alpha}_n} & E_{n+1} \\ \Sigma f_n \downarrow & & \downarrow f_{n+1} \\ \Sigma F_n & \xrightarrow{\tilde{\beta}_n} & F_{n+1} \end{array}$$

Fig. 15.1 Map between spectra

Definition 15.1.4 (*Suspension spectrum*) For $X \in \mathcal{C}_0$, the spectrum $\underline{X} = \{X_n, \tilde{\alpha}_n\}$ defined by taking $X_n = \Sigma^n X$, and $\tilde{\alpha}_n : \Sigma(\Sigma^n X) \to \Sigma^{n+1} X$ to be the natural homeomorphism is called a suspension spectrum. Any suspension spectrum is clearly of this form 'up to weak homotopy equivalence', where $X = X_0$. This spectrum is denoted by $\underline{X}$.

Definition 15.1.5 (*Sphere spectrum*) If $X_n = S^n$ in the suspension spectrum $\underline{X} = \{X_n, \alpha_n\}$, then the spectrum is called sphere spectrum and it is abbreviated as $\underline{S}$. Thus the sphere spectrum $\underline{S} = \{S^n, \alpha_n\}$, where

$$\alpha_n : \Sigma S^n \to S^{n+1}$$

is the identity map.

Definition 15.1.6 (*Unitary spectrum*) Let U be the infinite unitary group. There is a canonical homotopy equivalence $f : U \mapsto \Omega^2 U$. Suppose

$$E_n = \begin{cases} U, & \text{if } n \text{ is odd} \\ \Omega U, & \text{if } n \text{ is even.} \end{cases}$$

Let n be odd and $v_n : \Sigma U \to \Omega U$ be the map corresponding to f. Let n be even and $\tilde{v_n} : \Omega U \to \Omega U$ be the identity map. The resulting spectrum $\underline{U}$ is called the unitary spectrum.

Definition 15.1.7 Let $\underline{E} = \{E_n, \alpha_n\}$ and $\underline{F} = \{F_n, \beta_n\}$ be two spectra. A map $f : \underline{E} \to \underline{F}$ between spectra is a sequence of continuous maps

$$f_n : E_n \to F_n$$

such that the diagram in Fig. 15.1 is homotopy commutative for each n.

15.2 Spectral Reduced Homology Theory

This section constructs spectral reduced homology theory on $\mathcal{C}_0$ associated with a given spectrum $\underline{E}$.

Definition 15.2.1 A graded abelian group is a sequence $\{G_n\}$ of abelian groups, defined for each integer n. A homomorphism $f : \{G_n\} \to \{G'_n\}$ of graded groups is a sequence $\{f_n\}$ of homomorphisms $f_n : G_n \to G'_n$.

Remark 15.2.2 Sometimes, one writes G_* for $\{G_n\}$. Similar definitions are given for graded R-modules, or graded sets. Such objects and homomorphisms form a category denoted by $\mu_{\mathbf{Z}^*}$ (or $\mathcal{A}b$), μ_R, S_* in the cases of graded abelian groups, graded R-modules, and graded sets respectively.

Example 15.2.3 The sequence $G_n = \pi_n(X, *)$ for $n \geq 1$ and $G_n = 0$ if $n \leq 0$ is a graded abelian group if $G_1 = \pi_1(X, *)$ is abelian.

Definition 15.2.4 A reduced (spectral) homology theory on $\mathcal{C}_0$ associated with a spectrum $\underline{E} = \{E_n, \alpha_n\}$ is a sequence of covariant functors written $\{\widetilde{E}_n\} = \{h_n\}$ from $\mathcal{C}_0$ to $\mathcal{A}b$ satisfying:

RH(i)(Homotopy axiom): Let $f : X \to Y$ be in $\mathcal{C}_0$, and $f_* : \widetilde{E}_m(X) \to \widetilde{E}_m(Y)$ be the induced homomorphism. If $f \simeq g$ in $\mathcal{C}_0$ then $f_* = g_*$.

RH(ii)(Suspension axiom): There is a natural transformation $\sigma : \widetilde{E}_m(X) \to \widetilde{E}_{m+1}(\Sigma X)$, which is an isomorphism.

RH(iii)(Exactness axiom): If $(X, B) \in \mathcal{C}$ and $* \in B$, then the sequence

$$\widetilde{E}_n(B) \xrightarrow{i_*} \widetilde{E}_n(X) \xrightarrow{p_*} \widetilde{E}_n(X/B)$$

is exact at $\widetilde{E}_n(X)$ for each n, where $p : X \to X/B$ is the map collapsing B to a point and $i : B \hookrightarrow X$ is the inclusion map.

Each spectrum $\underline{E} = (E_n, \widetilde{\alpha}_n)$ gives rise to a reduced homology theory on $\mathcal{C}_0$, called spectral reduced homology theory.

Definition 15.2.5 Given $X \in \mathcal{C}_0$, consider the direct system

$$\cdots \longrightarrow \pi_{n+m}(X \wedge E_n) \xrightarrow{\beta_n} \pi_{n+m+1}(X \wedge E_{n+1}) \longrightarrow \cdots,$$

where the homomorphisms β_n are the composites

$$\pi_{n+m}(X \wedge E_n) \xrightarrow{E} \pi_{n+m+1}(X \wedge E_n \wedge S^1) \xrightarrow{(1_d \times \widetilde{\alpha}_n)_*} \pi_{n+m+1}(X \wedge E_{n+1})$$

for $n \geq m$, and E is the Freudenthal suspension homomorphism.

Define $\widetilde{E}_m(X) = \varinjlim\{\pi_{n+m}(X \wedge E_n), \beta_n\}$.

Theorem 15.2.6 *$\{\widetilde{E}_m\}$ constitutes a reduced homology theory on $\mathcal{C}_0$.*

Proof Let $f : X \to Y \in \mathcal{C}_0$. To define its induced homomorphism f_* consider the commutative diagram as shown in Fig. 15.2.

It gives rise to a map $f_* : \widetilde{E}_m(X) \to \widetilde{E}_m(Y)$ such that 1_{d^*}=1dentity, and $(f \circ g)_* = f_* \circ g_*$. We now verify **RH(i)–RH(ii)**. Since $f \simeq g$, $f \wedge 1_d \simeq g \wedge 1_d$ and hence

$$(f \wedge 1_d)_* = (g \wedge 1_d)_* : \pi_{n+m}(X \wedge E_n) \to \pi_{n+m}(Y \wedge E_n).$$

$$
\begin{array}{ccccc}
\pi_{n+m}(X \wedge E_n) & \xrightarrow{E} & \pi_{n+m+1}(X \wedge E_n \wedge S^1) & \xrightarrow{(1_d \times \tilde{\alpha}_n)_*} & \pi_{n+m+1}(X \wedge E_{n+1}) \\
\downarrow{\scriptstyle (f\wedge 1_d)_*} & & \downarrow{\scriptstyle (f\wedge 1_d)_*} & & \downarrow{\scriptstyle (f\wedge 1_d)_*} \\
\pi_{n+m}(Y \wedge E_n) & \xrightarrow{E} & \pi_{n+m+1}(Y \wedge E_n \wedge S^1) & \xrightarrow{(1_d \times \tilde{\alpha}_n)_*} & \pi_{n+m+1}(Y \wedge E_{n+1})
\end{array}
$$

Fig. 15.2 Diagram for reduced homology theory

This implies **RH(i)**.

Next define σ as follows. For any space $X \in \mathcal{C}_0$, define

$$\Sigma : \pi_n(X) \to \pi_{n+1}(S^1 \wedge X), [\theta] \mapsto [1_d \wedge \theta].$$

Consider the commutative diagram in Fig. 15.3.

Replacing $S^1 \wedge X$ by ΣX, it follows that Σ induces a natural transformation $\sigma : \widetilde{E}_m(X) \to \widetilde{E}_{m+1}(\Sigma X)$. Verify that σ is an isomorphism. For this verification consider the diagram in Fig. 15.4 which commutes up to sign, where $T_X : S^1 \wedge X \to X \wedge S^1 = \Sigma X$ is the natural homeomorphism. Hence **RH(ii)** follows. Verification of **RH(iii)** is left as an exercise. Consequently, $\{\widetilde{E}_m\}$ is a reduced homology theory on $\mathcal{C}_0$. ❑

Example 15.2.7 For any CW-complex X, the homology group $\widetilde{\underline{S}}_m(X)$ associated with the sphere spectrum $\underline{S}$ is given by the direct limit

$$\pi_{n+m}(X \wedge S^n) \xrightarrow{E} \pi_{n+m+1}(X \wedge S^{n+1}) \to \cdots .$$

These homology groups are also written $\pi_m^S(X)$ for any CW-complex X and are trivial if $m < 0$.

$$
\begin{array}{ccc}
\pi_{n+m}(X \wedge E_n) & \xrightarrow{\Sigma} & \pi_{n+m+1}(S^1 \wedge X \wedge E_n) \\
\downarrow{\scriptstyle E} & & \downarrow{\scriptstyle E} \\
\pi_{n+m+1}(X \wedge E_n \wedge S^1) & \xrightarrow{\Sigma} & \pi_{n+m+2}(S^1 \wedge X \wedge E_n \wedge S^1) \\
\downarrow{\scriptstyle (1_d \wedge \tilde{\alpha}_n)_*} & & \downarrow{\scriptstyle (1_d \wedge \tilde{\alpha}_n)_*} \\
\pi_{n+m+1}(X \wedge E_{n+1}) & \xrightarrow{\Sigma} & \pi_{n+m+2}(S^1 \wedge X \wedge E_{n+1})
\end{array}
$$

Fig. 15.3 Diagram involving suspension homomorphisms

$$
\begin{array}{ccc}
\pi_n(X) & \xrightarrow{\Sigma} & \pi_n(S^1 \wedge X) \\
\downarrow{\scriptstyle E} & \swarrow{\scriptstyle (T_X)_*} & \downarrow{\scriptstyle E} \\
\pi_n(X \wedge S^1) & \xrightarrow{\Sigma} & \pi_{n+1}(S^1 \wedge X \wedge S^1)
\end{array}
$$

Fig. 15.4 Diagram for construction of σ

15.3 Spectral Reduced Cohomology Theory

This section gives rise to spectral reduced cohomology theory which is dual to spectral reduced homology theory. A cohomology theory is not just a collection of cohomology functors. This needs connecting homomorphisms relating $h^n = [\ \ , A_n]$ with $h^{n+1} = [\ \ , A_{n+1}]$. Such a construction can be made using the map

$$\alpha_n : A_n \to \Omega A_{n+1}$$

for some Ω-spectrum $\underline{A} = \{A_n, \alpha_n\}$, where the spaces A_n are unique up to homotopy equivalence with $h^{n+1} = [-, A_{n+1}]$.

Definition 15.3.1 A reduced (spectral) cohomology theory on $\mathcal{C}_0$ associated with a spectrum $\underline{E} = \{E_n, \alpha_n\}$ is a sequence of contravariant functors written $\{\widetilde{E}^n\} = \{h^n\}$ from $\mathcal{C}_0$ to $\mathcal{Ab}$ together with a sequence

$$\sigma^n : h^{n+1} \, o \, \Sigma \to h^n$$

of natural transformations such that
(i) **Homotopy axiom**: If $f_0,\ f_1 \in \mathcal{C}_0$ and $f_0 \simeq f_1$ in $\mathcal{C}_0$, then $h^n(f_0) = h^n(f_1)$ (sometimes written as $f^* = g^*$) for all n;

(ii) **Suspension axiom**: $\sigma^n(X) : h^{n+1}(\Sigma X) \to h^n(X)$ is an isomorphism for all $X \in \mathcal{C}_0$; and

(iii) **Exactness axiom**: If $(X, B) \in \mathcal{C}$ and $* \in B$, then the sequence

$$h^n(X/B) \xrightarrow{h^n(p)} h^n(X) \xrightarrow{h^n(i)} h^n(B)$$

is exact at $h^n(X)$ for each n, where $p : X \to X/B$ is the map collapsing B to a point and $i : B \hookrightarrow X$ is the inclusion map.

Furthermore, the theory is called an ordinary cohomology theory (reduced), if the following axiom is also satisfied:

(iv) **Dimension axiom**: For the o-sphere S^0, $h^n(S^0) = 0$ if $n \neq 0$.

Definition 15.3.2 The graded group $\{h^n(S^0)\}$ is called the coefficient system of the cohomology theory $\{h^n\}$.

Definition 15.3.3 The cohomology group $h^n(X; \underline{A})$ of $X \in \mathcal{C}_0$ associated with the spectrum $\underline{A}$ is defined to be direct limit group of the sequence of groups and homomorphisms

$$\cdots \to [X, A_n] \xrightarrow{\alpha_n{}^*} [X, \Omega A_{n+1}] \xrightarrow{\Omega \alpha_n{}^*} [X, \Omega^2 A_{n+2}] \to \cdots$$

$$\text{i.e., } h^n(X; \underline{A}) = \lim_{k\to\infty} [X, \Omega^k A_{n+k}].$$

Theorem 15.3.4 *Let $\underline{A} = \{A_n, \alpha_n\}$ be an Ω-spectrum and X be a CW-complex. Then $h^n(X; \underline{A}) = [X, A_n]$.*

Proof As each $\alpha_k : A_k \to \Omega A_{k+1}$ in $\mathcal{C}_0$ is weak homotopy equivalence, it follows that $h^n(X; \underline{A}) = [X, A_n]$. ❑

15.4 Generalized Homology and Cohomology Theories

If a homology theory $\mathcal{H}$ satisfies only the first six axioms **H(1)–H(6)** (see Chap. 12), then $\mathcal{H}$ is called a generalized (or extraordinary) homology theory on the category $\mathcal{C}$. On the other hand, a cohomology theory which satisfies only the first six axioms **C(1)–C(6)** (see Chap. 12) is called a generalized (or extraordinary) cohomology theory on the category $\mathcal{C}$.

The 0-dimensional homology group

$$G = H_0(point)$$

is called the coefficient group of the homology theory $\mathcal{H}$. Consequently, the dimension axiom locates the coefficient group at the right dimension.

Remark 15.4.1 There is a natural question: whether there is a dual theory of cohomology theory for homology theory. The integral homology groups of a space X can be described by the Dold–Thom theorem, as the homotopy groups of the infinite symmetric product of X. However, the duality between homology and cohomology is not apparent from this description, nor is it clear how to generalize it. Examples of generalized homology theories are known; for instance, the stable homotopy groups (see Sect. 15.10).

15.5 The Brown Representability Theorem

This section studies the Brown representability theorem which relates homotopy theory with generalized cohomology theory. This theorem plays a key role in the applications of homotopy theory to other areas. Moreover, Brown proved that under certain conditions, any cohomology theory satisfying Eilenberg–Steenrod axioms can be obtained in the form [, Y] for some suitable space Y. More precisely, E.H. Brown (1962) proved in his paper that if H satisfies certain axioms, there is a space Y, unique up to homotopy type, such that H is naturally equivalent to the functor which assigns to each CW-Complex X with base point, the set of homotopy classes of maps of X into Y. More precisely, Brown representability theorem asserts that there exist connected CW-complexes A_n with base point and natural equivalences

$$\widetilde{h}^n(X) \cong [X, A_n],$$

where X runs over connected CW-complexes with base point (see Brown 1962). So we obtain a collection of spaces A_n $(n \in \mathbf{Z})$.

If we divert attention from the reduced $\widetilde{h}^n(X)$ to relative groups $h^n(X, Y)$ we should divert attention from suspension isomorphisms

$$\sigma : \widetilde{h}^n(X) \xrightarrow{\cong} \widetilde{h}^{n+1}(\Sigma X).$$

to the coboundary maps δ as a cohomology theory does not consists only of functors h^n; they are connected by coboundary maps.

Let H^n be a sequence contravariant functors from the category of pairs of finite CW-complexes to the category of abelian groups and $\delta^q : H^q(A) \rightarrow H^{q+1}(X, A)$ be a sequence of natural transformations. Furthermore, suppose H^q and δ^q satisfy all the Eilenberg–Steenrod axioms except the dimension axiom which is replaced by the condition that H^q on a point be countable.

E.H. Brown prescribed a very simple set of conditions on a functor H in his land mark paper (Brown 1959) that the functor H to be representable in the sense that H is naturally equivalent to $[\,, Y]$ for some space Y. This space Y is called a classifying space for the functor H. Every Ω-spectrum represents a generalized cohomology theory. Is its converse true? Do all cohomology theories arise in this way from an Ω-spectrum? E.H Brown gave necessary and sufficient conditions in 1962 under which a contravariant functor T has the form $[\,, Y]$ for some fixed space Y. It shows that there is a close relation between generalized cohomology theory and homotopy theory.

Theorem 15.5.1 *If $\underline{A} = \{A_n, \alpha_n\}$ is an Ω-spectrum, then the functors $X \mapsto h^n(X) = [X, A_n], n \in \mathbf{Z}$, define a reduced cohomology theory on the category of pointed CW-complexes and base point preserving maps.*

Proof See Brown (1962). ❑

Remark 15.5.2 Theorem 15.5.1 shows that every Ω-spectrum $\underline{A} = \{A_n, \alpha_n\}$ on the category $\mathcal{C}_0$ defines a cohomology theory given by $h^n(X) = [X, A_n], n \in \mathbf{Z}$, on the category $\mathcal{C}_0$. Is the converse of this theorem true? Brown proved in 1962 that all cohomology theories arise in this way from an Ω-spectrum $\underline{A} = \{A_n, \alpha_n\}$ on the category $\mathcal{C}_0$, where the spaces A_n are unique up to homotopy equivalence. In other words, Brown proved in 1962 that there is an Ω-spectrum $\underline{A} = \{A_n, \alpha_n\}$ such that $H^n(X)$ is naturally equivalent to the group of homotopy classes of maps of X into A_n.

Remark 15.5.3 There are natural isomorphisms between the groups $h^n(X; \mathbf{Z})$ and $H^n(X; \mathbf{Z})$ for the integral coefficients. In an analogous way, given an abelian group G, the singular cohomology group $H^n(X; G)$ can be defined as $H^n(X; G) = [X; K(G, n)]$. In particular,

(i) for $n = 1$, $H^1(X; \mathbf{Z}) = [X, S^1]$;
(ii) for $n = 2$, $H^2(X; \mathbf{Z}) = [X; \mathbf{C}P^\infty]$. This result implies that the complex line bundles are classified by the elements of $H^2(X; \mathbf{Z})$.

Theorem 15.5.4 (Brown Representability Theorem) *Every reduced cohomology theory h^* on the category $\mathcal{C}_0$ has the form $h^n(X) = [X, A_n]$ for some Ω-spectrum $\underline{A} = \{A_n, \alpha_n\}$, where the spaces A_n are unique up to homotopy equivalence.*

Proof See Brown (1962). ❑

Corollary 15.5.5 *Ordinary cohomology is representable as maps into Eilenberg–MacLane spaces.*

Proof Let $H^*(-; G)$ be an ordinary cohomology theory. Then $\pi_i(A_n) = [S^i, A_n] = H^n(S^i; G)$, where

$$H^n(S^i; G) = \begin{cases} G & \text{if } i = n \\ 0 & \text{otherwise} \end{cases}$$

This shows that A_n is an Eilenberg–MacLane space $K(G, n)$. This implies that $H^n(-; G) = [-, K(G, n)]$. ❑

Remark 15.5.6 Consider the singular ordinary cohomology group $H^n(X; G)$ for a CW-complex X with coefficients in an abelian group G. Then by Remark 15.5.3 $H^n(X; G)$ is the set of homotopy classes of maps from X to $K(G, n)$, the Eilenberg–MacLane space of type (G, n). The corresponding spectrum is the Eilenberg–MacLane spectrum $\underline{HG}$ has the nth space $K(G, n)$.

Remark 15.5.7 Let h^* be a generalized cohomology theory defined on CW-complex pairs. Then

$$h^n(X) = h^n(X, point) \oplus h^n(point)$$

and hence define $\widetilde{h}^n(X) = h^n(X, point)$. Now applying Brown representability theorem, there exists a connected CW-complex A_n with a base point and natural equivalences such that

$$\widetilde{h}^n(X) = [X, A_n]$$

where X runs over connected CW-complexes with base points. In this way, a sequence of spaces $\{A_n\}$ is obtained. However, a cohomology theory does not contain only the functors; also contains coboundary maps connecting them.

15.6 A Generalization of Eilenberg–MacLane Spectrum and Construction of Its Associated Generalized Cohomology Theory

In this section the author of the present book constructs a new Ω-spectrum generalizing the Eilenberg–MacLane spectrum and also presents its associated cohomology theory. The motivation of this construction comes from the Corollary 15.5.5 which asserts that the singular ordinary cohomology groups of a CW-complex can be identified with the groups of homotopy classes of continuous maps into Eilenberg–MacLane spectrum spaces. This section investigates a new generalized cohomology theory, constructed by replacing the Eilenberg–MacLane spectrum by a new Ω-spectrum.

This section works in the category $\mathcal{C}$ whose objects are pairs of spaces having the homotopy type of finite CW-complexes and morphisms are maps of such pairs. This is a full subcategory of the category of pairs of topological spaces and maps of pairs, and this admits the construction of mapping cones. In particular, there exists the (reduced) suspension functor $\Sigma : \mathcal{C} \to \mathcal{C}$ and its adjoint functor $\Omega : \mathcal{C} \to \mathcal{C}$ which is the loop functor. Let $\mathcal{C}_0$ be the full subcategory of $\mathcal{C}$, whose objects are spaces with base points, $\mathcal{Ab}$ be the category of abelian groups and homomorphisms.

15.6.1 *Construction of a New Ω-Spectrum $\underline{A}$*

This subsection constructs a new Ω-spectrum generalizing the Eilenberg–MacLane spectrum. Recall that an Ω-spectrum $\underline{A}$ is a sequence of spaces A_n, $n \in \mathbf{Z}$, in $\mathcal{C}_0$, together with a sequence of maps $\alpha_n : A_n \to \Omega A_{n+1}$, $n \in \mathbf{Z}$, in $\mathcal{C}_0$, where each α_n is a weak homotopy equivalence. Let $(X, *) \in \mathcal{C}_0$ and $X^n, n > 0$ be the n-fold cartesian product of X and S_n be the group of permutations of $\{1, 2, 3, \ldots, n\}$. Define a right action

$$X^n \times S_n \to X^n, \; (x_1, x_2, \ldots, x_n).\alpha = (X_{\alpha(1)}, X_{\alpha(2)}, \ldots, X_{\alpha(n)}).$$

The orbit space $SP^n(X) = X^n \bmod S_n$ is the n-fold symmetric product of X i.e., it is the quotient space of the n-fold product X^n obtained by identifying all n-tuples $(x_1, x_2, \ldots, x_n)$ that differ only by a permutation of their coordinates. Thus a typical point of $SP^n(X)$ is an unordered n-tuples $(x_1, x_2, \ldots, x_n)$, $x_i \in X$. There is a natural inclusion $X^n \subset X^{n+1}$, $(x_1, x_2, \ldots, x_n) \to (*, x_1, x_2, \ldots, x_n)$. This induces an inclusion $SP^n(X) \subset SP^{n+1}(X)$. This gives rise to an ascending sequence of spaces of $SP^n(X)$'s with the weak topology

$$SP^n(X) \subset SP^{n+1} \subset \cdots \subset \cdots$$

$SP^\infty(X)$ is defined by the union of this ascending sequences i.e., $SP^\infty(X) = \lim_{n\to\infty} SP^n(X)$.

Dold and Thom showed that $SP^\infty(S^n) = K(\mathbf{Z}, n)$ which is the Eilenberg–MacLane space with homotopy groups $\pi_r(K(\mathbf{Z}, n))$ are all zero except for $r = n$ (Dold and Thom 1958). We now utilize homotopy properties of infinite symmetric product (see Sect. 2.9 of Chap. 2): If $f : (X, *) \to (Y, *)$ is in $\mathcal{C}_0$, then $SP^n(f) : SP^n(X) \to SP^n(Y)$ is the map defined by passing to the quotient, and $SP^\infty(f) : SP^\infty(X) \to SP^\infty(Y)$ is the map defined by passing to the limit. SP^∞ (and also SP^n for each $n \geq 0$) is a covariant functor $\mathcal{C}_0 \to \mathcal{C}_0$ that if $f : (X, *) \to (Y, *)$ is a homotopy equivalence, so is $SP^\infty(f)$. Moreover, if X is a connected CW-complex, there is a weak homotopy equivalence

$$\rho : SP^\infty(X) \to \Omega SP^\infty(\Sigma X)$$

defined by

$$[\rho(x_1, x_2, \ldots, x_n)](t) = ((x_1, t), (x_2, t), \ldots, (x_n, t)), \ t \in I).$$

Using these facts, the author of the present book has constructed a new Ω-spectrum $\underline{A} = \{A_n, \alpha_n\}$ generalizing the Eilenberg–MacLane spectrum in the following way. Let Y be a connected CW-complex, define

$$A_n = \begin{cases} \Omega^{-n}\, SP^\infty(Y) & \text{if } n < 0 \\ \Omega\, SP^\infty(\Sigma^{n+1} Y) & \text{if } n \geq 0 \end{cases}$$

The homotopy equivalence $\alpha_n : A_n \to \Omega A_{n+1}$ is defined by

$$\alpha_n = \begin{cases} \text{identity} & \text{if } n < -1 \\ \Omega \circ \rho_n & \text{if } n \geq -1 \end{cases}$$

where $\rho_n : SP^\infty(\Sigma^{n+1} Y) \to \Omega SP^\infty(\Sigma^{n+2} Y)$, if $n \geq -1$, is a homotopy equivalence. Note that $\rho_{-1} = \rho$.

If we take in particular $Y = S^0$, the 0-sphere, then $A_n = K(\mathbf{Z}, n)$. Consequently, for $Y = S^0$, the above Ω-spectrum becomes an Eilenberg–MacLane spectrum. In this way, we obtain an Ω-spectrum $\underline{A} = \{A_n, \alpha_n\}$ which is a generalization of the Eilenberg–MacLane spectrum whose sequence of spaces are $K(\mathbf{Z}, n)$.

15.6.2 *Construction of the Cohomology Theory Associated with $\underline{A}$*

The author of this book constructs the cohomology theory associated with his Ω-spectrum $\underline{A} = \{A_n, \alpha_n\}$ defined in the previous subsection by generalizing the

Eilenberg–MacLane spectrum. Since α_k is weak homotopy equivalence

$$h^n(X; \underline{A}) = [X, A_n].$$

The coefficient system of the theory can be obtained in the following way: If $n < 0$,

$$\begin{aligned} h^n(S^0, \underline{A}) = [S^0, \Omega^{-n} SP^\infty(Y)] &= [\Sigma^{-n} S^0, SP^\infty(Y)] \\ &= [S^{-n}, SP^\infty(Y)] \\ &= \pi_{-n}(SP^\infty(Y)) \\ &= H_{-n}(Y; \mathbf{Z}), \end{aligned}$$

the $-n$th singular homology group of Y with coefficients in $\mathbf{Z}$.

If $n \geq 0$,

$$\begin{aligned} h^n(S^0, \underline{A}) = [S^0, \Omega SP^\infty(\Sigma^{n+1} Y)] &= [\Sigma S^0, SP^\infty(\Sigma^{n+1} Y)] \\ &= [S^1, SP^\infty(\Sigma^{n+1} Y)] \\ &= \pi_1(SP^\infty(\Sigma^{n+1} Y)) \\ &= H_1(\Sigma^{n+1} Y; \mathbf{Z}), \end{aligned}$$

the one-dimensional singular homology group of $\Sigma^{n+1} Y$ with coefficients in $\mathbf{Z}$.

In general, one may calculate the cohomology groups of the n-sphere S^n in the theory. If $n < 0$,

$$\begin{aligned} h^n(S^k, \underline{A}) = [S^k, \Omega^{-n} SP^\infty(Y)] &= [\Sigma^{-n} S^k, SP^\infty(Y)] \\ &= [S^{k-n}, SP^\infty(Y)] \\ &= \pi_{k-n}(SP^\infty(Y)) \\ &= H_{k-n}(Y; \mathbf{Z}), \end{aligned}$$

If $n \geq 0$,

$$\begin{aligned} h^n(S^k, \underline{A}) = [S^k, \Omega SP^\infty(\Sigma^{n+1} Y)] &= [\Sigma S^k, SP^\infty(\Sigma^{n+1} Y)] \\ &= [S^{k+1}, SP^\infty(\Sigma^{n+1} Y)] \\ &= \pi_{k+1}(SP^\infty(\Sigma^{n+1} Y)) \\ &= H_{k+1}(\Sigma^{n+1} Y; \mathbf{Z}). \end{aligned}$$

Remark 15.6.1 The relations displayed above show that our generalized cohomology theory has some close relations with the ordinary singular homology theory with integral coefficients.

15.7 K-Theory as a Generalized Cohomology Theory

This section presents an exposition of K-theory (instead its development), which is the first example of generalized cohomology theories. It plays a centrally important role in connecting algebraic topology to analysis and algebraic geometry. For example, an outstanding purely algebraic result in K-theory is that the only possible dimensions of a real (not necessarily associative) division algebra are 1, 2, 4, and 8 proved by J.F. Adams and M. Atiyah to solve 'Hopf invariant one problem'. This result asserts that **R**, **C**, **H** and Cayley numbers (an eight-dimensional non-associative algebra) are the only real division algebras (see Chap. 17).

Recall that the Grothendieck group $K(X)$ is defined in Chap. 5. Given a finite dimensional CW-complex X this group $K(X)$ has the natural structure of a commutative ring stemming from the tensor product of vector bundles. If $f : X \to Y$ is a continuous map, then f induces a ring homomorphism

$$f^* : K(Y) \to K(X),$$

where to each vector bundle ξ over X, there is the induced bundle $f^*(\xi)$ over Y. Moreover,

$$\text{if } f \simeq g : X \to Y, \text{ then } f^* = g^* : K(Y) \to K(X).$$

Hence the correspondence

$$X \mapsto K(X)$$

defines a homotopy invariant functor from category $\mathcal{C}_0$ to the category of rings.

Let (X, x_0) be a CW-complex. The natural inclusion $\{x_0\} \hookrightarrow X$ induces a homomorphism of rings

$$K(X) \to K(\{x_0\}) = \mathbf{Z},\ [\xi] - [\eta] \mapsto \dim \xi - \dim \eta \in \mathbf{Z} \tag{15.1}$$

Let $K^0(X, x_0)$ denote the kernel of the homomorphism (15.1)

$$K^0(X, x_0) = \ker(K(X) \to K(x_0)).$$

The elements of the subring $K^0(X, x_0)$ are represented by differences $[\xi] - [\eta]$ for which $\dim \xi = \dim \eta$.

Let (X, Y) be a finite CW-pair of spaces and $K^0(X, Y)$ denote the ring

$$K^0(X, Y) = K^0(X/Y).$$

For negative integer $-n$, let

$$K^{-n}(X, Y) = K^0(\Sigma^n X, \Sigma^n Y),$$

where $\Sigma^n X = (S^n \times X)/(S^n \vee X)$.

The above discussion can be summarized in the basic and important result.

Theorem 15.7.1 $K^*(-)$ *forms a generatized cohomology theory on* $\mathcal{C}$.

Remark 15.7.2 For deeper properties and applications of vector bundles, it is suggested to study Bott periodicity theorem as the main tool for calculation of K-theory, linear representations and cohomology operations in K-theory and Aitiyah–Singer formula for calculation of the indices of elliptic operators on compact manifolds and for this purpose the book (Luke and Mishchenko 1998) is referred.

15.8 Spectral Unreduced Homology and Cohomology Theories

This section generalizes the concepts of spectral homology and cohomology groups defined on $\mathcal{C}_0$ by introducing the concepts of spectral homology and cohomology groups defined on $\mathcal{C}$. The domain of our theories from the category $\mathcal{C}_0$ to $\mathcal{C}$ is transferred by a simple transformation. Homology and cohomology theories defined on pairs (X, A) are called unreduced homology and cohomology theories (sometimes the word 'unreduced' is dropped).

Definition 15.8.1 Let $\underline{E} = \{E_n, \widetilde{\alpha}_n\}$ be a spectrum on $\underline{\mathcal{C}}$. For $(X, A) \in \underline{\mathcal{C}}$, set

$$E_m(X, A) = \widetilde{E}_m(X \cup CA),\ E^m(X, A) = \widetilde{E}^m(X \cup CA),$$

where the vertex of the cone is taken as a base point. In particular, if $A \neq \emptyset$, we interpret $X \cup CA$ as the space X with a point added, which is used as base point. $\{E_m\}$ and $\{E^m\}$ are called the spectral unreduced homology and cohomology theories associated with the spectrum $\underline{E}$.

Definition 15.8.2 An unreduced homology theory on $\underline{\mathcal{C}}$ is a sequence of covariant functors $E_m : \mathcal{C} \to \mathcal{Ab}$ (category of abelian groups) for $m \in \mathbf{Z}$ satisfying the axioms

URH(i)(Homotopy axiom): Let $f, g : (X, A) \to (Y, B) \in \underline{\mathcal{C}}$, and $f \simeq g$. Then $f_* = g_* : E_m(X, A) \to E_m(Y, B)$;

URH(ii)(Excision axiom): If U is open and $\overline{U} \subset \text{Int}\, A$, then the inclusion map $e : (X - U, A - U) \to (X, A)$ induces isomorphisms in homology;

URH(iii)(Exactness axiom): There are natural transformations $\partial : E_m(X, A) \to E_{m-1}(A)$ which fit up to exact sequence

$$\cdots \to E_m(A) \to E_m(X) \to E_m(X, A) \xrightarrow{\partial} E_{m-1}(A) \to \cdots$$

An unreduced cohomology theory associated to a spectrum is dual to the corresponding unreduced homology theory.

Definition 15.8.3 An unreduced cohomology theory on $\underline{C}$ is a sequence of contravariant functors $E^m : \mathcal{C} \to \mathcal{Ab}$ for $m \in \mathbf{Z}$ satisfying the axioms

URC(i)(Homotopy axiom): Let $f, g : (X, A) \to (Y, B) \in \underline{C}$, and $f \simeq g$. Then $f^* = g* : E^m(Y, B) \to E^m(X, A)$;

URC(ii)(Excision axiom): If U is open and $\overline{U} \subset \text{Int}\, A$, then the inclusion map $e : (X - U, A - U) \to (X, A)$ induces isomorphisms in cohomology;

URC(iii)(Exactness axiom): There are natural transformations $\delta : E^m(A) \to E^{m+1}(X, A)$ which fit into exact sequence

$$\cdots \leftarrow E^m(A) \leftarrow E^m(X) \leftarrow E^m(X, A) \xleftarrow{\delta} E^{m-1}(A) \leftarrow \cdots$$

Remark 15.8.4 $E_m(X, \emptyset)$ is abbreviated $E_m(X)$ and $E^m(X, \emptyset)$ is abbreviated $E^m(X)$.

15.9 Cohomology Operations

This section studies cohomology operations which form an important topic in algebraic topology. The technique utilized for developing the algebraic structure of the cohomology ring has substantially enriched homotopy theory with some surprising results. Eilenberg–MacLane spaces are closely linked with the study of cohomology operations. The cohomology of an Eilenberg–Maclane space $K(G, n)$, depending on n and G has the surprising property. A cohomology operation is a natural transformation of cohomology functors

$$H^n(\,; G) \to H^n(-; G').$$

Given integers n, m and abelian groups G, G', in general, a cohomology operation is a natural transformation

$$\psi : H^n(X, Y; G) \to H^m(X, Y; G')$$

subject to one axiom only:

if $f : (X, Y) \to (X', Y')$ is continuous and $h \in H^n(X', Y'; G)$, then $\psi(f^*h) = f^*(\psi h)$.

On the other hand, a stable cohomology operation is a collection of cohomology operations

$$\psi_n : H^n(X, Y; G) \to H^{n+r}(X, Y; G').$$

Here n runs over $\mathbf{Z}$, but r, G and G' are fixed and each ψ_n is a natural transformation. Moreover, it is also necessary the following diagram in Fig. 15.5 to be commutative for each n.

Fig. 15.5 Diagram for stable cohomology operation

$$\begin{array}{ccc} H^n(X,Y;G) & \xrightarrow{\delta} & H^{n+1}(X,Y;G) \\ \downarrow{\scriptstyle \psi_n} & & \downarrow{\scriptstyle \psi_{n+1}} \\ H^{n+r}(X,Y;G') & \xrightarrow{\delta} & H^{n+r+1}(X,Y;G') \end{array}$$

Example 15.9.1 Let $G = G' = \mathbf{Z}_2$. Then ψ_n is called the Steenrod square denoted by S^r_q.

Remark 15.9.2 A cohomology operation is a concept which can be applied in any dimension.

Remark 15.9.3 Given a cohomology operation

$$\psi : H^n(X, Y; G) \to H^m(X, Y; G'),$$

it need not be necessary to appear as the nth term of any stable cohomology operation.

15.9.1 *Cohomology Operations of Type $(G, n; T, m)$ and Eilenberg–MacLane Spaces*

Ordinary cohomology of CW-complexes is representable by an Eilenberg–MacLane space. This section identifies the set of all cohomology operations of type $(G, n; T, m)$ with the mth ordinary cohomology group of the Eilenberg–MacLane space $K(T, n)$ with coefficient group G.

Definition 15.9.4 Let G, T be abelian groups and m, n be nonnegative integers. A cohomology operation θ of type $(G, n; T, m)$ is a natural transformation (Nat) $\theta : H^n(-; G) \to H^m(-; T)$ of functors defined on the category $\mathcal{C}_0$ of pointed CW-complexes.

Example 15.9.5 For each n, and each ring R the operation

$$H^n(X; R) \to H^{2n}; R), x \mapsto x \cup x = x^2$$

(cup product) is a cohomology operation. This is not generally a homomorphism.

The set of all cohomology operations $\mathrm{Nat}(H^n(-; G), H^m(-; T))$ of type $(G, n; T, m)$ is denoted by $\mathcal{O}(G, n; T, m)$. We now identify the set $\mathcal{O}(G, n; T, m)$ with the ordinary cohomology groups of the Eilenberg–MacLane spaces.

Theorem 15.9.6 $\mathrm{Nat}(H^n(-; G), H^m(-; T)) = H^m(K(T, n); G)$.

Proof Ordinary cohomology of CW-complexes is representable by an Eilenberg–MacLane space. Hence using Yoneda lemma (see Appendix B), a cohomology operation of type (n, m, G, T) is given by a homotopy class of maps $K(G, n) \to K(T, n)$.

Again, by representability, the cohomology operation is given by an element $H^m(K(G, n); T)$, because,

$$\begin{aligned}\mathrm{Nat}(H^n(-; G), H^m(-; T)) &= \mathrm{Nat}([-, K(G, n)], [-, K(T, m)]) \\ &= [K(G, n), K(T, m)] \\ &= H^m(K(G, n); T)(m\text{th ordinary cohomology group} \\ &\quad \text{of K(G, n) with coefficient group } T).\end{aligned}$$

❑

The above discussion is summarized in the basic surprising result.

Theorem 15.9.7 *Given an Eilenberg–MacLane space $X = K(G, n)$ there is a (1-1) correspondence between $H^m(K(G, n); T)$ and the set of all cohomology operations from $H^n(X; G)$ to $H^m(X; T)$ for some Eilenberg–MacLane space $K(T, m)$.*

Remark 15.9.8 For more study on cohomology operations (Mosher and Tangora 1968), (Steenrod and Epstein 1962) and (Spanier 1966) are referred.

15.9.2 Cohomology Operation Associated with a Spectrum

This subsection studies cohomology operations in the new cohomology theory $h^*(\ ;\underline{A})$ constructed in Sect. 15.6.2. There also exist relations between the cohomology operations and the general cohomology groups of some spaces in this general cohomology theory. This subsection establishes some such relations. More precisely, we prove that the abelian group of all cohomology operations of degree k for the cohomology theory $h^*(\ ;\underline{A})$ is isomorphic to the group $h^{n+k}(SP^\infty(\Sigma^n Y); \underline{A})$ and the graded abelian group of all stable cohomology operations of degree k for the cohomology theory $h^*(\ ;\underline{A})$ is isomorphic to the group $\varprojlim h^{n+k}(SP^\infty(\Sigma^n Y); \underline{A})$.

Definition 15.9.9 A cohomology operation in $h^*(\ ;\underline{A})$ of degree k is a natural transformation $\phi_m : h^m(\ ;\underline{A}) \to h^{m+k}(\ ;\underline{A})$.

Proposition 15.9.10 *Let ${O_m}^k$ be the set of all cohomology operations of degree k of the type ϕ_m for the cohomology theory $h^*(\ ;\underline{A})$. Then ${O_m}^k$ forms an abelian group.*

Proof We define an addition '+' on ${O_m}^k$ by the rule: $(\phi_m + \psi_m)(X)(x) = \phi_m(X)(x) + \psi_m(X)(x)$, for all $x \in h^m(X; \underline{A})$ and for all $X \in \mathcal{C}_0$, where the right-hand side addition is the addition in the additive abelian group $h^{m+k}(X; \underline{A})$. Then $({O_m}^k, +)$ is an abelian group. ❑

Remark 15.9.11 The group ${O_m}^k$ has some interesting properties. For example, in the ordinary singular cohomology theory $H^*(\ ;\mathbf{Z})$, the group ${O_m}^k$ is isomorphic to the group $H^{m+k}(K(\mathbf{Z}, m); \mathbf{Z})$, where $K(\mathbf{Z}, m)$ is an Eilenberg–MacLane space.

Theorem 15.9.12 *The homomorphism*

$$\lambda : {O_m}^k \to h^{m+k}(SP^\infty(\Sigma^m Y); \underline{A}) \text{ defined by } \lambda(\phi) = \phi[Id],$$

where $[Id]$ *is the homotopy class of the identity map*

$$I_d : SP^\infty(\Sigma^m Y) \to SP^\infty(\Sigma^m Y),$$

is an isomorphism of groups.

Proof Let $x \in h^m(X; \underline{A})$ be represented by a map $g : X \to SP^\infty(\Sigma^m Y)$ in $\mathcal{C}_0$. We now define a map

$$\mu : h^{m+k}(SP^\infty(\Sigma^m Y); \underline{A}) \to {O_m}^k \text{ by the rule } \mu(\alpha)(x) = [\alpha \circ g] = g^*(\alpha),$$

for all $\alpha \in h^{m+k}(SP^\infty(\Sigma^m Y); \underline{A})$. Now $\mu(\alpha + \beta)(x) = g^*(\alpha + \beta) = g^*(\alpha) + g^*(\beta)$. This implies that μ is a homomorphism. Moreover, $\lambda \circ \mu$ =identity and $\mu \circ \lambda$ = identity. Hence λ is an isomorphism with its inverse μ. ❑

Corollary 15.9.13 *For the ordinary singular cohomology theory* $H^*(\ ;\mathbf{Z})$, *the group* ${O_m}^k$ *is isomorphic to the group* $H^{m+k}(K(\mathbf{Z}, m); \mathbf{Z})$, *where* $K(\mathbf{Z}, m)$ *is an Eilenberg–MacLane space.*

Proof For $Y = S^0$ (0-sphere), the space $SP^\infty(\Sigma^m Y)$ becomes the Eilenberg–MacLane space $K(\mathbf{Z}, m)$ and the general cohomology theory reduces to the ordinary singular cohomology theory $H^*(\ ;\mathbf{Z})$. Hence the Corollary follows from Theorem 15.9.12. ❑

15.9.3 *Stable Cohomology Operations*

Definition 15.9.14 For the cohomology theory $h^*(\ ;\underline{A})$, a stable cohomology operation of degree k is a sequence $\phi_m : h^m(\ ;\underline{A}) \to h^{m+k}(\ ;\underline{A})$ of cohomology operations of degree k such that the following diagram commutes, i.e., $\sigma^{m+k}(\phi_m(x)) = \phi_{m+1}(\sigma^m(x))$, $\forall x \in h^m(X; \underline{A})$ and $\forall X \in \mathcal{C}_0$, where σ^m is the suspension isomorphism in $h^*(\ ;\underline{A})$.

Remark 15.9.15 Let $\{{\overline{O}_m}^k\}$ be the set of all stable cohomology operations of degree k for the cohomology of $h^*(\ ;\underline{A})$. We denote a sequence $\{\phi_m\} \in \{{\overline{O}_m}^k\}$ by a single letter ϕ (Fig. 15.6).

Fig. 15.6 Diagram for cohomology operation of degree k

$$\begin{array}{ccc} h^m(\ ;\underline{A}) & \xrightarrow{\phi_m} & h^{m+k+1}(\ ;\underline{A}) \\ \downarrow{\scriptstyle \sigma^m} & & \downarrow{\scriptstyle \sigma^{m+k}} \\ h^{m+k}(\Sigma,\underline{A}) & \xrightarrow{\phi_{m+1}} & h^{m+k+1}(\Sigma,\underline{A}) \end{array}$$

Proposition 15.9.16 $\{\overline{O}_m{}^k\}$ *forms an additive abelian group.*

Proof We define an addition '+' on $\{\overline{O}_m{}^k\}$ by the rule $(\phi+\psi)(x)=\phi(x)+\psi(x)$, $\forall x\in h^*(X;\underline{A})$ and $\forall X\in\mathcal{C}_0$. Then $\{\overline{O}_m{}^k\}$ becomes an additive abelian group. ❑

15.9.4 A Characterization of the Group $\{\overline{O}_m{}^k\}$

For this purpose, we define a sequence of homomorphisms

$$\gamma_{m+k}: h^{m+k+1}(SP^\infty(\Sigma^{m+1}Y);\underline{A})\to h^{m+k}(SP^\infty(\Sigma^m Y);\underline{A})$$

as the product of homomorphisms

$$h^{m+k+1}(SP^\infty(\Sigma^{m+1}Y);\underline{A})\xrightarrow{(\overline{\rho_m})^*}h^{m+k+1}(\Sigma SP^\infty(\Sigma^m Y);\underline{A})$$
$$\xrightarrow{(\sigma^{m+k})^{-1}}h^{m+k}(SP^\infty(\Sigma^m Y);\underline{A}),$$

i.e., $\gamma_{m+k}=(\sigma^{m+k})^{-1}\circ(\overline{\rho_m})^*$, where $(\overline{\rho_m})^*$ is the homomorphism induced by the adjoint map

$$\overline{\rho_m}\text{ of }\rho_m: SP^\infty(\Sigma^m Y)\to\Omega\, SP^\infty(\Sigma^{m+1}Y)$$

given by Spanier. Hence the following sequence of abelian groups and homomorphisms

$$\cdots\to h^{m+k+1}(SP^\infty(\Sigma^{m+1}Y);\underline{A})\xrightarrow{\gamma_{m+k}}h^{m+k}(SP^\infty(\Sigma^m Y);\underline{A})\to\cdots$$

form an inverse system of groups and homomorphisms. Let $\varprojlim h^{m+k}(SP^\infty(\Sigma^m Y);\underline{A})$ denote the inverse limit group of the above system.

Theorem 15.9.17 *The graded abelian group* $\{\overline{O}_m{}^k\}$ *of all stable cohomology operations of degree k for the cohomology theory $h^*(\ ;\underline{A})$ is isomorphic to the group* $\varprojlim h^{m+k}(SP^\infty(\Sigma^m Y);\underline{A})$.

Proof It follows from the definition of $\varprojlim h^{m+k}(SP^\infty(\Sigma^m Y);\underline{A})$ that an element of $\varprojlim h^{m+k}(SP^\infty(\Sigma^m Y);\underline{A})$ is a sequence of elements $x_m\in h^{m+k}(SP^\infty(\Sigma^m Y);\underline{A})$

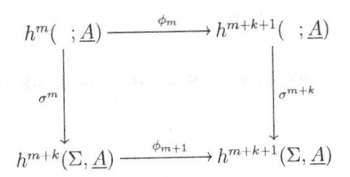

such that $\gamma_{m+k}(x_{m+1}) = x_m$. Hence $(\sigma^{m+k})^{-1} \circ (\overline{\rho_m})^*(x_{m+1}) = x_m$, i.e., $(\overline{\rho_m})^*(x_{m+1}) = \sigma^{m+k}(x_m)$.

We now show that to each sequence of elements $x_m \in h^{m+k}(SP^\infty(\Sigma^m Y); \underline{A})$, there corresponds a stable cohomology operation of degree k in $\{\overline{O_m}^k\}$ and conversely.

Let $\mu : h^{m+k}(SP^\infty(\Sigma^m Y); \underline{A}) \to O_m{}^k$ be the homomorphism defined in Theorem 15.9.12. Let $\mu(x_m) = \phi_m$. We now show that $\{\phi_m\}$ is a stable cohomology operation of degree k in $h^*(\ ; \underline{A})$.

Let $x \in h^m(X; \underline{A})$ be represented by a map $f : X \to SP^\infty(\Sigma^m Y)$. Then $\sigma^m(x)$ is represented by the composite map

$$\Sigma X \xrightarrow{\Sigma f} \Sigma\, SP^\infty(\Sigma^m Y) \xrightarrow{\overline{\rho_m}} SP^\infty(\Sigma^{m+1} Y).$$

Again $\phi_m(x) = \mu(x_m)(x) = f^*(x_m)$. Hence we have

$$\begin{aligned}\phi_{m+1}(\sigma^m(x)) &= \mu(x_{m+1})(\sigma^m(x)) = (\overline{\rho_m} \circ \Sigma f)^*(x_{m+1}) = (\Sigma f)^* \circ (\overline{\rho_m})^*(x_{m+1}) \\ &= (\Sigma f)^*(\sigma^{m+k}(x_m)) = \sigma^{m+k}(f^*(x_m)) = \sigma^{m+k}(\phi_m(x)),\ \forall x \in h^m(X; \underline{A}) \\ &\Rightarrow \phi_{m+1} \circ \sigma^m = \sigma^{m+k} \circ \phi_m \Rightarrow \{\phi_m\} \in \{\overline{O_m}^k\}.\end{aligned}$$

$$\begin{aligned}\phi_{m+1}(\sigma^m(x)) = \mu(x_{m+1})(\sigma^m(x)) &= (\overline{\rho_m} \circ \Sigma f)^*(x_{m+1}) \\ &= (\Sigma f)^* \circ (\overline{\rho_m})^*(x_{m+1}) \\ &= (\Sigma f)^*(\sigma^{m+k}(x_m)) \\ &= \sigma^{m+k}(f^*(x_m)) = \sigma^{m+k}(\phi_m(x)) \forall x \in h^m(X; \underline{A}) \\ \Rightarrow \phi_{m+1} \circ \sigma^m = \sigma^{m+k} \circ \phi_m &\Rightarrow \{\phi_m\} \in \{\overline{O_m}^k\}\end{aligned}$$

Conversely, let $\{\phi_m\} \in \{\overline{O_m}^k\}$. Then $\phi_{m+1}(\sigma^m(x)) = \sigma^{m+k}(\phi_m(x))$, $\forall x \in h^m(X; \underline{A})$. Let $\lambda(\phi_m) = x_m$, where λ is defined in Theorem 15.9.12. Then $x_m \in h^{m+k}(SP^\infty(\Sigma^m Y); \underline{A})$. Hence $\mu(x_m) = \phi_m$. Consequently,

$$\phi_{m+1}(\sigma^m(x)) = \mu(x_{m+1})(\sigma^m(x)) = (\overline{\rho_m} \circ \Sigma f)^*(x_{m+1}) = (\Sigma f)^* \circ (\overline{\rho_m})^*(x_{m+1}).$$

Again,

$$\sigma^{m+k}(\phi_m(x)) = \sigma^{m+k}(\mu(x_m)(x)) = \sigma^{m+k}(f^*(x_m)) = (\Sigma f)^* \circ \sigma^{m+k}(x_m).$$

Hence it follows that corresponding to each sequence ϕ_m, there exists a sequence of elements $x_m \in h^{m+k}(SP^\infty(\Sigma^m Y); \underline{A})$ such that $\sigma^{m+k}(x_m) = (\overline{\rho_m})^*(x_{m+1})$. ❑

Corollary 15.9.18 *If $Y = S^n$, then the graded abelian group $\{\overline{O}_m{}^k\}$ of all stable cohomology operations of degree k in $h^*(\ ;\underline{A})$ is isomorphic to $\varprojlim H^{m+k+n}$ $(K(\mathbf{Z}, m+n); \mathbf{Z})$, where $H^*(\ ;\mathbf{Z})$ is the ordinary cohomology theory with coefficients in $\mathbf{Z}$.*

Proof

$$h^{m+k}(SP^\infty(\Sigma^m S^n); \underline{A}) = h^{m+k}(K(\mathbf{Z}, m+n); \underline{A}) = H^{m+k+n}(K(\mathbf{Z}, m+n); \mathbf{Z}).$$

Hence the Corollary follows from Theorem 15.9.17. ❑

Corollary 15.9.19 *The graded abelian group $\{\overline{O}_m{}^k\}$ of all stable cohomology operations of degree k in $h^*(\ ;\underline{A})$ is isomorphic to $\varprojlim H^{m+k}(B(\mathbf{N}, \Sigma^m Y); \underline{A})$, where $B(\mathbf{N}, \Sigma^m Y)$ is the reduced monoid of the singular o-chain of $\Sigma^m Y$ with coefficients in $\mathbf{N}$ (multiplicative monoid of positive integers), i.e., $B(\mathbf{N}, \Sigma^m Y)$ is the set of all functions $(X, *) \to (\mathbf{N}, 1)$ with finite support.*

Proof In this case, $h^{m+k}(SP^\infty(\Sigma^m Y); \underline{A}) = h^{m+k}(B(\mathbf{N}, \Sigma^m Y); \underline{A})$. Hence the Corollary follows from Theorem 15.9.17. ❑

Definition 15.9.20 Let $\underline{E} = \{E_n, e_n\}$ and $\underline{E}' = \{E'_n, e'_n\}$ be two spectra. Let $\alpha_n : E_n \to E'_n$ be defined for $n \geq n_0$ such that the diagram in Fig. 15.7 is commutative. Such a sequence $\{\alpha_n\}$ is called a map of spectra of degree r.

Remark 15.9.21 Clearly, spectra form a category with this definition of morphism. Moreover, if $\alpha = \{\alpha_n\} : \underline{E} \to \underline{E}'$ is a map of spectra of degree r, then it induces natural homomorphisms of homology and cohomology theories

$$\alpha : \widetilde{E}_m(X) \to \widetilde{E}_{m-r}(X),\ \alpha : \widetilde{E}^n_m(X) \to \underline{\widetilde{E}}^{m+r}(X)$$

for all m commuting with the suspension isomorphism, i.e., $\alpha(\sigma(x)) = \sigma(\alpha(x))$.

Definition 15.9.22 The above transformation α is called a stable homology or cohomology operation associated with spectrum. The simplest examples of such operations are coefficient transformations.

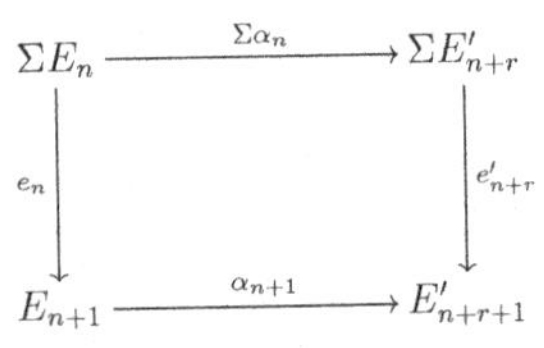

Fig. 15.7 Diagram for a map of spectra of degree r

15.10 Stable Homotopy Theory and Homotopy Groups Associated with a Spectrum

This section conveys the concept of 'stable homotopy groups' introduced in 1937 as a natural generalization of Freudenthal suspension theorem. In algebraic topology we use the word 'stable' when a phenomenon occurs essentially in the same way independent of dimension provided perhaps that the dimension is sufficiently large. The importance of stable homotopy theory was reinforced by two related developments in the late 1950s. One is the invention of spectral homology and cohomology theory and specially K-theory by Atiyah and Hirzebruch. The other one is the work of Thom which reduces the problem of classifying manifolds up to cobordism to a problem, a solvable problem in stable homotopy theory. Moreover, this section studies homotopy groups of a spectrum.

15.10.1 Stable Homotopy Groups

This subsection conveys a study of stable homotopy groups. If X is an n-connected complex, then the suspension map $\Sigma : \pi_r(X) \to \pi_{r+1}(X)$ is an isomorphism for $r < 2n + 1$. In particular, Σ is an isomorphism for $r \leq n$. This shows that ΣX is an $(n + 1)$-connected CW-complex. Hence it follows that $\Sigma^m X$ is $(n + m)$-connected.

Consider the sequence of groups

$$\pi_r(X) \longrightarrow \pi_{r+1}(\Sigma X) \longrightarrow \cdots \longrightarrow \pi_{r+m}(\Sigma^m X) \longrightarrow \cdots \qquad (15.2)$$

Since $\Sigma^m X$ is $(n + m)$-connected, the map $\pi_{r+m}(\Sigma^m X) \to \pi_{r+m+1}(\Sigma^{m+1} X)$ is an isomorphism for $r + m < 2(n + m) + 1$ (i.e., for $m \geq r - 2n$).

Hence for m sufficiently large, the sequence of maps in (15.2) are all isomorphisms. The resulting group is called the stable homotopy group denoted by $\pi_r^s(X)$. Since adding any finite number of terms to the beginning of (15.2) does not affect the resulting stable homotopy group, $\pi_r^s(S^n) \cong \pi_r^s(S^0)$. This shows that the only stable homotopy groups of spheres are the ones $\pi_r^s(S^0)$ for some value of r, which is simply denoted by π_r^s. These stable homotopy group classify the mapping of $(r + m)$-dimensional spheres onto m-dimensional spheres, for sufficiently large value of m.

Stable phenomena had of course appeared implicitly before 1937; reduced homology and cohomology are examples of functors that are invariant under suspension without limitations on dimension. Stable homotopy theory appeared as an important topic of algebraic homotopy with Adam's introduction of his spectral and conceptual use of the concept of stable phenomena in his solution to the Hopf invariant problems.

We recall that the suspension homomorphism asserts: $\pi_i(S^n) \to \pi_{i+1}(S^{n+1})$ is an isomorphism for $i < 2n - 1$ and an epimorphism for $i = 2n - 1$. More generally, this holds for the suspension $\pi_i(X) \to \pi_{i+1}(\Sigma X)$, whenever X is an $(n - 1)$-

connected CW-complex for $n \geq 1$. If q is small relative to n, then $\pi_{n+q}(S^n)$ is independent of n.

Example 15.10.1 Consider the homotopy groups $\pi_{n+r}(S^n)$ of spheres. We have the suspension homomorphism:

$$E : \pi_{n+r}(S^n) \to \pi_{n+r+1}(S^{n+1}).$$

The Freudenthal suspension theorem says that this homomorphism is an isomorphism for $n > r + 1$.

For example, $\pi_{n+1}(S^n)$ is isomorphic to $\mathbf{Z}_2$ for $n > 2$. The groups $\pi_{n+r}(S^n)$ $(n > r + 1)$ are called the stable homotopy groups of spheres.

More generally, let X and Y be two CW-complexes with base point which is assumed to be a 0-cell. The suspension ΣX is the reduced suspension: either $S^1 \wedge X$ or $X \wedge S^1$ which are homeomorphic. If $f : X \to Y$ is a map between CW-complexes with base point, its suspension Σf is to be $1_d \wedge f : S^1 \wedge X \to S^1 \wedge Y$ (or $f \wedge 1_d : X \wedge S^1 \to Y \wedge S^1$). Suspension defines a function

$$S : [X, Y] \to [\Sigma X, \Sigma Y]$$

Theorem 15.10.2 *Suppose Y is n- connected for $n \geq 1$. Then S is onto if* $\dim X \leq 2n + 1$ *and a bijection if* $\dim X < 2n$.

Proof See Spanier (1966, pp 458). ☐

Definition 15.10.3 An element of $[X, Y]$ (defined under the above situation) is called a stable homotopy class of maps.

Consider the exact homotopy sequence of the fibration

$$p : S^3 \to S^2 : \cdots \to \pi_3(S^1, s_0) \to \pi_3(S^3, s_0) \xrightarrow{p_*} \pi_3(S^2, s_0) \to \pi_3(S^1, s_0) \to \cdots$$

Since $\pi_3(S^1, s_0) = \pi_2(S^1, s_0) = 0$, $p_* : \pi_3(S^3, s_0) \to \pi_3(S^2, s_0)$ is an isomorphism. Consequently, $\pi_3(S^2, s_0) \cong \mathbf{Z}$, the first example, where $\pi_m(S^n, s_0) \neq 0$ for $m > n$. Since $\pi_3(S^3, s_0)$ is graded by $[1_{S^3}]$, it follows that $\pi_3(S^2, s_0)$ is generated by $[p]$. The map p is called the Hopf map.

For each q, consider

$$\pi_{2q+2}(S^{q+2}, s_0) \xrightarrow{\Sigma} \cong \pi_{2q+3}(S^{q+3}, s_0) \xrightarrow{\Sigma} \cong \cdots \xrightarrow{\Sigma} \cong \pi_{q+n}(S^n, s_0) \xrightarrow{\Sigma} \cong \cdots$$

We denote the common group $\pi_{n+q}(S^n, s_0)$, by π_q^S. It is called the kth stable homotopy group.

Consider the sequence of groups $\pi_{n+q}(\Sigma^n X, *)$ for $n = 0, 1, \ldots$, and the suspension homomorphisms between them. If X is a CW-complex, then $\Sigma^n X$ has no cells in dimension $< n$ except for a 0-cell. Hence it is $(n-1)$-connected, and

$$\pi_{n+q}(\Sigma^n X, *) \cong \pi_{n+q+1}(\Sigma^{n+1} X, *) \cong \cdots \text{ if } n > q+1 \tag{15.3}$$

Consequently, for large n, the sequence (15.3) stabilizes in the sense that all the groups in this sequence are isomorphic.

Definition 15.10.4 The stable value in the sequence (15.3) is called the qth stable homotopy group of X or q-stem of X denoted $\pi_q^S(X)$.

Remark 15.10.5 The importance of stable homotopy theory was reinforced by two related developments in the late 1950s. One is the introducing of spectral homology and cohomology theory and specially K-theory by Atiyah and Hirzebruch. The other one is the work of Thom which reduces the problem of classifying manifolds up to cobordism to a problem, a solvable problem in stable homotopy theory (see Gray pp. 324–357).

Remark 15.10.6 Higher algebraic K-theory introduced by Quillen in the early 1970 earns deep recognition by Segal and others.It can be viewed as a construction in stable homotopy.

The coefficient groups $\pi_q^S(S^0)$ are called the stable stems π_q^S. These groups are known only through a finite range of $n > 0$ (note $\pi_n^S = 0$ for $n < 0$, $\pi_0^S \cong \mathbf{Z}$). For details study see James (1995).

If X and Y are finite CW-complexes and $f : X \to Y$ induces the zero homomorphism: $\pi_q^S(X) \to \pi_q^S(Y)$, then $\Sigma^k f$ is nullhomotopic for some k.

15.10.2 *Homology Groups Associated with a Spectrum*

This subsection returns to homology theory associated with a given spectrum in $\mathcal{C}_0$. Such theories are closely related with stable homotopy theory and the study of spectra. Let $\underline{A} = \{A_n, \alpha_n\}$ be an arbitray spectrum. Then the spaces $X \wedge A_n$ form a spectrum $X \wedge \underline{A}$ and the resulting groups are given by

$$h_r(X; \underline{A}) = \lim_{n \to \infty} \pi_{n+r}(X \wedge A_n)$$

for a CW-complex X which form a homology theory, called homology theory associated with the spectrum $\underline{A}$ (Whitehead 1962).

15.10.3 Homotopy Groups Associated with a Spectrum

This subsection defines homotopy groups associated with a spectrum. We work in category $\mathcal{C}_0$. First we define the homotopy groups of a spectrum $\underline{A} = \{A_n, \widetilde{\alpha}_n\}$. These are really stable homotopy groups. We have the following homomorphisms:

$$\pi_{n+r}(A_n) \to \pi_{n+r+1}(\Sigma A_{n+1}) \xrightarrow{(\widetilde{\alpha}_n)_*} \pi_{n+r+1}(A_n) \tag{15.4}$$

Define $\pi_r(\underline{A}) = \lim\limits_{n\to\infty} \pi_{n+r}(\underline{A}_n)$; hence the homomorphisms of the direct system are those displayed in (15.4). If $\underline{A}$ is an Ω-spectrum then the homomorphism

$$\pi_{n+r}(\underline{A}) \to \pi_{n+r+1}(A_{n+1})$$

is an isomorphism for $n+r \geq 1$; the direct limit is obtained. Hence we have $\pi_r(\underline{A}) = \pi_{n+r}(A_n)$ for $n+r \geq 1$.

Example 15.10.7 For the Eilenberg–MacLane spectrum $\underline{A} = \{K(G,n), \alpha_n\}$,

$$\pi_r(\underline{A}) = \begin{cases} G, & \text{if } r = 0 \\ 0, & \text{if } r \neq 0. \end{cases}$$

Example 15.10.8 For $\underline{A} = BU$-spectrum,

$$\pi_r(\underline{A}) = \begin{cases} \mathbf{Z}, & \text{if } r \text{ is even} \\ 0, & \text{if } r \text{ is odd} \end{cases}$$

by the Bott periodicity theorem (Bott 1959).

Example 15.10.9 For the suspension spectrum $\underline{S} = \{A_n, \widetilde{\alpha}_n\}$,

$$A_n = \begin{cases} \Sigma^n X, & \text{if } n \geq 0 \\ \text{point}, & \text{if } n < 0. \end{cases}$$

Hence $\pi_r(\underline{S}) = \lim\limits_{n\to\infty} \pi_{n+r}(\Sigma^n X)$. The limit is attained for $n > r+1$. The homotopy groups of the spectrum $\underline{S}$ are stable homotopy groups of X.

15.11 Applications

This section presents some interesting applications associated with spectra.

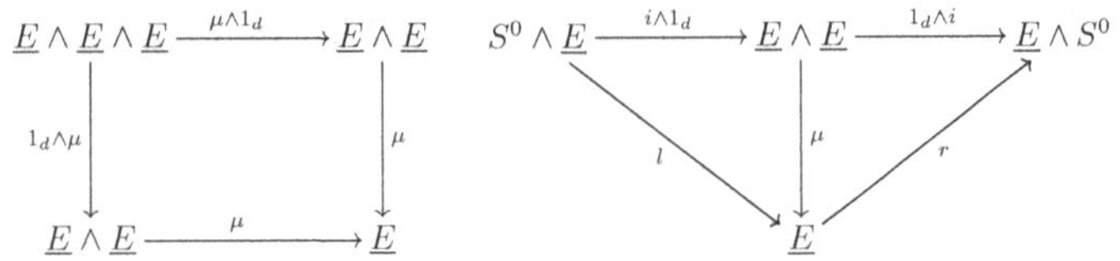

Fig. 15.8 Ring spectrum

15.11.1 Poincaré Duality Theorem

This subsection presents Poincaré duality theorem in the language of spectral homology $E_*(M; \underline{E})$ and spectral cohomology $E^*(M; \underline{E})$ associated with a ring spectrum $\underline{E}$ and also in some other forms. Manifolds generally arise in many problems of analysis. Historically, homology theory was first applied to manifolds by Poincaré, giving a key result, known as Poincaré duality theorem. He first stated this theorem in 1895 in terms of Betti numbers. At that time there was no concept of cohomology, which was invented in 1930s. Poincaré stated that the mth and $(n - m)$th Betti numbers of a closed (i.e., compact and without boundary) orientable n-manifold are equal. But there are at present its different forms given below.

Definition 15.11.1 A ring spectrum is a spectrum $\underline{E}$ with a product $\mu : \underline{E} \wedge \underline{E} \to \underline{E}$ (i.e., a map of spectra) and identity $i : S^0 \to \underline{E}$ such that the diagrams in Fig. 15.8 commute up to homotopy, where l and r are natural homotopy equivalences.

The product μ is said to be commutative if the diagram in Fig. 15.9 commutes up to homotopy.

Definition 15.11.2 An orientation of an n- manifold M at x with respect to a ring spectrum $\underline{E}$ is a choice of $E_0(P)$ module generators of $E_n(M, M - x)$. Given a collection $\{X\}$ of subsets of M, M is said to be consistently oriented along $\{X\}$ with respect to $\underline{E}$, if there is a chosen collection of classes $[X] \in E_n(M, M - x))$ satisfying

(i) $\left(\rho^X_{X\cap Y}\right)_* [X] = \left[\rho^Y_{X\cap Y}\right]_* [Y]$;
(ii) $\left(\rho^X_X\right)_* [X]$ is an orientation of X, where

$$\rho^A_B : (M, M - A) \to (M, M - B)$$

is the inclusion.

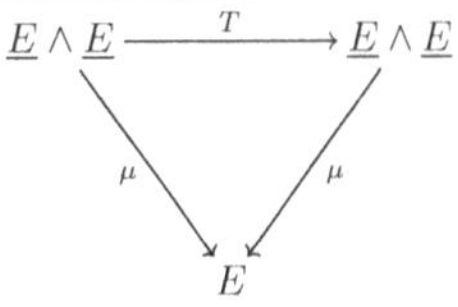

Fig. 15.9 Commutative multiplication

Definition 15.11.3 A manifold M is said to be oriented with respect to $\underline{E}$, if it can be consistently oriented along all compact subsets. A collection of such classes is called an $\underline{E}$-orientation.

Example 15.11.4 $\mathbf{R}P^n$ is orientable iff n is odd but $\mathbf{C}P^n$ is orientable for each n.

Theorem 15.11.5 (Poincaré duality theorem) *Let $\underline{E} = \{E_n, \alpha_n\}$ be a ring spectrum on $\mathcal{C}_0$. If M is a compact n-manifold oriented with respect to $\underline{E}$, then for the cohomology and homology associated with $\underline{E}$*

$$\tilde{E}^m(M; \underline{E}) \cong \tilde{E}_{n-m}(M; \underline{E}).$$

Proof See Gray (1975). ❑

Remark 15.11.6 If $\underline{E}$ is the Eilenberg–MacLane spectrum, then applying Poincaré duality theorem to ordinary homology, for any compact manifold M, it follows that $H_m(M; \mathbf{Z}_p) \cong H^{n-m}(M; \mathbf{Z}_p)$ and $H_m(M; \mathbf{Z}_p)$ is a finite dimensional vector space.

We now give another form Poincaré duality theorem connected with a finite simplicial complex K.

Definition 15.11.7 A path-connected space X is said to be a homology n-manifold if there exists a triangulation K of X such that for each point $x \in |K|$, and for each integer m, the homology groups $H_m(L_K(x))$ and $H_m(S^{n-1})$ are isomorphic, where $L_K(x)$ is the link of x in K.

Definition 15.11.8 A homolopy n-manifold X is said to be orientable if there exists a triangulation K of X, for which the n-simplexes of K can be identified with the elements of the chain group $C_n(K)$ in such a way that if σ is any $(n-1)$-simplex, and σ_1 and σ_2 are two n-simplexes that contain σ as a face, then σ occurs with opposite signs in $\partial(\sigma_1)$ and $\partial(\sigma_2)$.

Example 15.11.9 **(i)** A homology 0-manifold (a point) is orientable;
(ii) S^1 is orientable.

Theorem 15.11.10 (Another form Poincaré duality theorem) *Let K be a triangulation of a homology n-manifold and K' be the first barycentric subdivision of K. If $|K|$ is orientable, there is an isomorphism*

(i) $\psi_K : H^m(K) \to H_{n-m}(K')$ *for all integers m;*
(ii) *There is an isomorphism $\psi_K : H^m(K; \mathbf{Z}_2) \to H_{n-m}(K'; \mathbf{Z}_2)$ for all integers m.*

Proof See Maunder (1980). ❑

Remark 15.11.11 Poincaré duality theorem is not true for all homology n-manifolds, unless coefficient group is $\mathbf{Z}_2$ is used. Those homology manifolds for which the theorem is true for coefficient group $\mathbf{Z}$ are precisely those that are orientable.

One of the most important applications of cap product defined in Sect. 10.18 of Chap. 10 is the present form of Poincaré duality theorem.

Theorem 15.11.12 (An alternative form of Poincaré duality theorem) *If M is a compact connected oriented n-manifold with generator $z \in H_n(M; R)$ for the ordinary homology with coefficient ring R, then the map*

$$\psi_M : H^m(M; R) \to H_{n-m}(M; R), u \mapsto u \cap z$$

is an isomorphism for all integers n.

Proof See Vick (1994). ❑

Remark 15.11.13 H^m is a contravariant functor. On the other hand, H_{n-m} is a covariant functor. The family of isomorphisms

$$\psi_M : H^m(M; R) \to H_{n-m}(M; R)$$

is natural in following sense: if $f : M \to N$ is a continuous map between oriented n-manifolds which are compatible with orientation, then $\psi_N = f_* \circ \psi_M \circ f^*$, where f_* and f^* are homomorphisms induced by f in homology and cohomology respectively.

15.11.2 *Homotopy Type of the Eilenberg–MacLane Space $K(G, n)$*

We have discussed Eilenberg–MacLane spaces $K(G, n)$ in Chap. 11 and proved in Theorem 11.2.17 that it is uniquely determined by G and n. We prove the same result in an alternative way.

Theorem 15.11.14 *Let G be an abelian group. Then the homotopy type of the Eilenberg–MacLane space $K(G, n)$ is completely by the group G and the integer n.*

Proof Using the result of Ex. 4 of Sect. 15.12, any isomorphism $G \to G$ is induced by a continuous map

$$f : K(G, n) \to K(G, n).$$

Since all other groups are trivial, the map induces isomorphism in all homotopy groups. This asserts by Whitehead theorem (Theorem 8.5.9 of Chap. 8) that f is a homotopy equivalence. ❑

15.11.3 Application of Representability Theorem of Brown

Let h^* be a generalized cohomology theory associated with an Ω-spectrum $\underline{A} = \{A_n, \alpha_n\}$. We now observe that there are the following natural equivalences, at least if X is connected.

$$\begin{aligned}[X, A_n] &\cong \widetilde{h}^n(X) \cong \widetilde{h}^{n+1}(\Sigma X). \\ &\cong [\Sigma X, A_{n+1}] \cong [X, \Omega A_{n+1}].\end{aligned}$$

This natural equivalence is induced by a weak homotopy equivalence

$$\alpha_n : A_n \rightarrow \Omega A_{n+1}.$$

This shows that the above sequence of spaces forms a spectrum.

Example 15.11.15 Let H^* be ordinary cohomology : $H^n(X, Y) \cong H^n(X, Y; G)$. The corresponding spectrum $\underline{A}$ is the Eilenberg–MacLane spectrum for the group G; the nth space is the Eilenberg–MacLane space of type (G, n). That is, we have

$$\pi_r(A_n) = [S^n, A_n] \cong \widetilde{H}^n(S^r; G) = \begin{cases} G, & \text{if } r = n \\ 0, & \text{if } r \neq n. \end{cases}$$

Example 15.11.16 **(a)** Let K^* be complex K-theory. The corresponding spectrum is called the BU-spectrum. Each even term A_{2n} is the space BU or $\mathbf{Z} \times BU$, depends on whether we choose to work with connected spaces or not. Each odd term A_{2n+1} is the space U.

(b) Let K^* be real K-theory. The corresponding spectrum is called the BO-spectrum. Every eighth term A_{8n} is the space BO or $\mathbf{Z} \times BO$, depends on whether we chose to work with connected spaces or not. Each term A_{2n+1} is the space U.

Remark 15.11.17 All spectra are not Ω-spectra.

Example 15.11.18 Given a CW-complex X, let

$$A_n = \begin{cases} \Sigma^n X, & \text{if } n \geq 0 \\ \text{point}, & \text{if } n < 0 \end{cases}$$

with the obvious map. Define a spectrum $\underline{S}$ to be a suspension spectrum or S-spectrum of X if

$$\psi_n : \Sigma A_n \rightarrow A_{n+1}$$

is a weak homotopy equivalence for n sufficiently large. Then this spectrum is called the 'suspension spectrum' $\underline{S}$ is usually not an Ω-spectrum. In particular, let sphere spectrum $\underline{S}$ is the suspension spectrum of S^0; it has nth term S^n for $n \geq 0$.

15.11.4 More Applications of Spectra

This subsection presents more connections of homotopy theory with cohomology theory through spectra.

Theorem 15.11.19 *Let $\underline{E} = \{E_n, \widetilde{\alpha}_n\}$ be a spectrum and X be a CW-complex. Then there exists a natural isomorphism $\widetilde{E}^m(X) \cong [X, E_m]$.*

Proof Consider the diagram in Fig. 15.10 where $\alpha_n : E_n \to \Omega E_{n+1}$ is the adjoint to $\widetilde{\alpha}_n : E_n \to E_{n+1}$ and ψ is a bijective correspondence. Hence

$$\widetilde{E}^m(X) = [\Sigma^{n-m}X, E_n] \cong [X, \Omega^{n-m}E_n] \cong [X, E_m],$$

as Σ and Ω are adjoint functors (see Chap. 2). ❑

Remark 15.11.20 Theorem 15.11.19 interlinks homotopy theory with cohomology theory.

Theorem 15.11.21 *For any spectrum $\underline{E}$, $\widetilde{E}_m(S^k) \cong \widetilde{E}^k(S^m)$.*

Proof By property **RH(ii)**, it is sufficient to show that $\widetilde{E}_m(S^\circ) \cong \widetilde{E}^{-m}(S^\circ)$. Consider the direct limit of the sequence as shown in Fig. 15.11:

Hence the theorem follows ❑

Definition 15.11.22 The direct limit in Theorem 15.11.21 is called the *m*th homotopy group of the spectrum $\underline{E}$ and is sometimes abbreviated $\pi_m(\underline{E})$. It is also called the group of coefficients for the theories $\widetilde{\underline{E}}_*$ and $\widetilde{E}^*$ on $\mathcal{C}_0$.

Definition 15.11.23 Given an Ω-spectrum $\underline{E} = (E_n, \alpha_n)$ and a CW-complex pair (X, A), the cohomology groups of (X, A) associated with $\underline{E}$ are defined by $E^n(X, A; \underline{E}) = [X/A, E_n]$ with multiplication induced by α_n.

We write $E^*(X, A; \underline{E}) = \oplus E^m(X, A; \underline{E})$. The corresponding reduced groups are given by $A = \emptyset$; $E^n(X; \underline{E}) = E^n(X, \emptyset; \underline{E})$ and $\widetilde{E}(X; \underline{E}) = E^n(X, x_0; \underline{E})$, where x_0 is the base point(assumed to be a 0-cell).

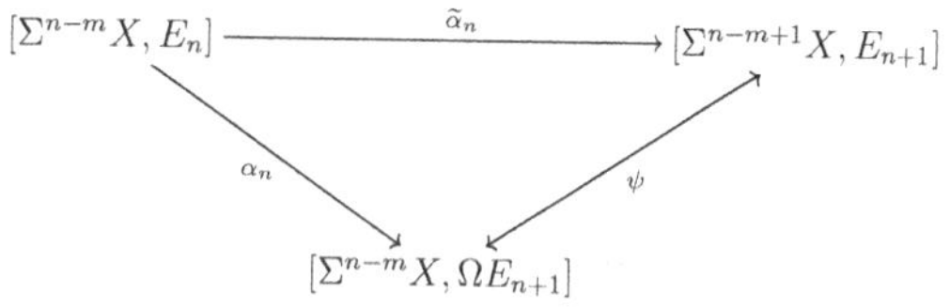

Fig. 15.10 Diagram associated with a spectrum $\underline{E}$

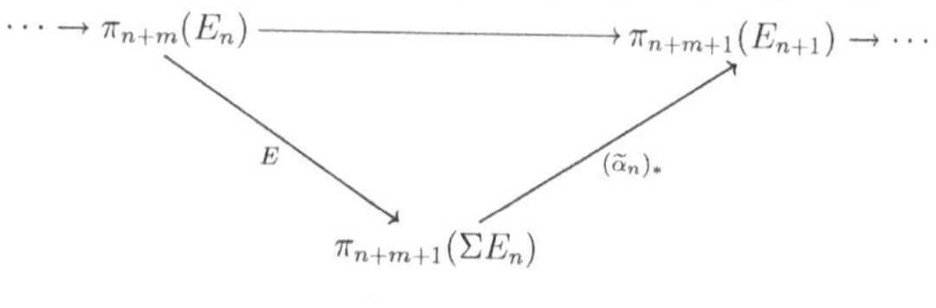

Fig. 15.11 Triangular diagram for groups and homomorphism

Remark 15.11.24 $E^n(X; \underline{E}) = [X^+, E_n]$, the set of unbased homotopy classes of maps X into E_n and $\widetilde{E}^n(X; \underline{E}) = [X; E_n]$.

Theorem 15.11.25 *The groups $\widetilde{E}^n(X, A; \underline{E})$(and hence also $E^n(X; \underline{E})$, $\widetilde{E}^n(X; \underline{E})$) are all abelian.*

Proof $E^n(X, A; \underline{E}) \cong [X/A, \Omega E_{n+1}] \cong [X/A, \Omega(\Omega E_{n+2})]$, which is an abelian group. ❑

The exactness of the cohomology sequence proves the following:

Theorem 15.11.26 *Let $\underline{E} = \{E_n, \alpha_n\}$ be an Ω-spectrum on $\mathcal{C}_0$. Then*

$$\widetilde{E}^n(X; \underline{E}) \cong \widetilde{E}^n(X; \underline{E}) \oplus \widetilde{E}^n(S^\circ; \underline{E}),$$

where X is a CW-complex.

Proof Let x_0 be the base point of X. Now consider the exact cohomology sequence of the pointed space (X, x_0):

$$\cdots \to E^n(X, x_0; \underline{E}) \xrightarrow{j_*} E^n(X; \underline{E}) \xrightarrow{i_*} \widetilde{E}^n(x_0; \underline{E}) \to \cdots$$

The above sequence splits, because, there is a map $p : X \to x_0$ such that $p \circ i = 1_d : x_0 \to x_0$. Hence $i^* \circ p^* = 1_d$. Consequently,

$$E^n(X; \underline{E}) \cong E^n(X, x_0; \underline{E}) \oplus E^n(x_0, \underline{E}).$$

But

$$E^n(X, x_0; \underline{E}) = \widetilde{E}^n(X; \underline{E}) \text{ and } E^n(x_0; \underline{E}) = \widetilde{E}^n(S^0; \underline{E}).$$

❑

15.11.5 *Homotopical Description of Singular Cohomology Theory*

Eilenberg–MacLane spaces are used in giving homotopical description of singular cohomology theory. The Eilenberg–MacLane spaces $K(\mathbf{Z}, n)$ fit together via the loop functor Ω

$$K(\mathbf{Z}, 0) \xleftarrow{\Omega} K(\mathbf{Z}, 1) \xleftarrow{\Omega} K\mathbf{Z}, 2) \xleftarrow{\Omega} \cdots$$

to form a spectrum, called the Eilenberg–MacLane spectrum. Given a connected CW-complex X define cohomology theory by setting

$$H^n(X; \mathbf{Z}) = [X, K(\mathbf{Z}, n)].$$

In general, given a CW-complex X, there is a bijective correspondence between the cohomology group $H^n(X; G)$ and the homotopy classes of maps X to $K(G, n)$. Using this fact it can be shown that the cohomology operations are completely classified by the cohomology groups of $K(G, n)$'s.

15.12 Exercises

1. Let $X = \cup X_\alpha = \underset{\rightarrow}{\lim} X_\alpha$ has the weak topology and assume

(i) for all $\alpha, \beta \in \mathbf{A}$, there exists $\delta \in \mathbf{A}$ such that $X_\alpha \cap X_\beta = X_\delta$;
(ii) for all $\alpha \in \mathbf{A}$, $\{\beta \in \mathbf{A} : \beta \leq \alpha\}$ is finite (ordering $\beta \leq \alpha$ is defined by $\beta \leq \alpha$ iff $X_\beta \subset X_\alpha$). Then show that $\widetilde{E}_m(\underset{\rightarrow}{\lim} X_\alpha) \cong \underset{\rightarrow}{\lim} \widetilde{E}_m(X_\alpha)$.

2. Show that for a spectrum $\underline{E}$, the homology and cohomology groups satisfy the following properties:

(i) $\widetilde{E}_m(X_1 \vee \cdots \vee X_n) \cong \widetilde{E}_m(X_1) \oplus \cdots \oplus \widetilde{E}_m(X_n)$;
(ii) $\widetilde{E}^m(X_1 \vee \cdots \vee X_n) \cong \widetilde{E}^m(X_1) \oplus \cdots \oplus \widetilde{E}^m(X_n)$.

3. Let $(X_\alpha, *_\alpha) \in \mathcal{C}_0$ be indexed by a set $\mathbf{A}$ and $\bigvee_{\alpha \in \mathbf{A}} X_\alpha$ denote the quotient space $\bigsqcup_{\alpha \in \mathbf{A}} X_\alpha / *_\alpha \sim *_\beta$. If the set $\mathbf{A}$ is finite, this is the one-point union $X_{\alpha_1} \vee \cdots \vee X_{\alpha_n}$.
Show that

(i) $\widetilde{E}_m(\bigvee_{\alpha \in \mathbf{A}} X_\alpha) \cong \oplus \widetilde{E}_m(X_\alpha)$;
(ii) if $(X_\alpha, *_\alpha)$ are CW-complexes, and $\underline{E} = \{E_n, \widetilde{\alpha}_n\}$ is an Ω-spectrum, them

$$\widetilde{E}^m(\bigvee_{\alpha \in \mathbf{A}} X_{\alpha \in \mathbf{A}}) \cong \Pi_{\alpha \in \mathbf{A}} \widetilde{E}^m(X_\alpha).$$

(This result is sometimes called the wedge axiom.)

4. Given abelian groups G and T, show that there is a bijection

$$\psi : [K(G, n), K(T, n)] \to Hom(G, T).$$

5. Given an abelian group G, show that there exists an Ω-spectrum $\underline{E}$ such that

$$E_n = \begin{cases} K(G,n), & n \geq 0 \\ \text{point}, & \text{otherwise.} \end{cases}$$

6. Let $H_*(\ ;G)$ denote the singular homology with coefficients in the abelian group G and $\{G_r\}_r \in \mathbf{Z}$ be a sequence of abelian groups. Show that $h_n(X,A) = \bigoplus_{m+r=n} H_m(X,A;G_r)$ defines a homology theory with coefficient group $h_r(\text{point}) = G_r$ for all r.

7. Let

$$0 \to G \xrightarrow{f} H \xrightarrow{g} K \to 0$$

be a short exact sequence of abelian groups. Construct natural long exact sequences

$$\cdots \to \widetilde{E}^m(X;G) \xrightarrow{c} \widetilde{E}^m(X;H) \xrightarrow{d} \widetilde{E}^m(X,K) \xrightarrow{\alpha} \widetilde{E}^{m+1}(X;G) \to \cdots,$$

$$\cdots \to \widetilde{E}_m(X;G) \xrightarrow{c} \widetilde{E}_m(X;H) \xrightarrow{d} \widetilde{E}_m(X;K) \xrightarrow{\alpha} \widetilde{E}_{m-1}(X;G) \to \cdots$$

This is called the Bockstein sequence and α is called the Bockstein homomorphism.

8. A spectrum $\underline{E}$ is called properly convergent if $\alpha_n : \Sigma E_n \to E_{n+1}$ is a $(2n+1)$-isomorphism for each n. Show that

(i) if E_n is a pointed topological space for each n and $\underline{E}$ is properly convergent, then E_n is $(n-1)$-connected;

(ii) if E is an Ω-spectrum and E_n is connected for each n, then $\underline{E}$ is properly convergent.

9. If X is a CW-complex, show that the reduced ordinary homology and cohomology groups $\widetilde{H}_m(X;G)$ and $\widetilde{H}^m(X;G)$ with coefficient groups G are trivial for $m < 0$.

[Hint: $X \wedge K(G,n)$ is a CW-complex with all cells in dimension n and larger, except for 0-cells. Hence $\pi_{m+n}(X \wedge K(G,n)) = 0$ for $m < 0$ and $n > 1$. This shows that $\widetilde{H}_m(X;G) = 0$ for $m < 0$. Since $\Sigma^{n-m}X$ has all cells in dimension $n-m$ and larger, except for 0-cells, $[\Sigma^{n-m}X, K(G,n)] = 0$, if $m < 0$.]

10. Show that the sphere spectrum $\underline{S}$ is a ring spectrum and every spectrum $\underline{E}$ is a module over $\underline{S}$.

11. If $\underline{E}$ is a ring spectrum, show that $\widetilde{E}^*(S^0)$ is a graded commutative ring with unit and $\widetilde{E}^*(X)$ is a module over $\widetilde{E}^*(S^0)$ for every $X \in \mathcal{C}_0$.

15.13 Additional Reading

[1] Adams, J.F., *Algebraic Topology: A student's Guide*, Cambridge University Press, Cambridge, 1972.
[2] Adams, J.F., *Stable Homotopy and Generalized Homology*, University of Chicago, 1974.
[3] Adhikari, M.R., and Adhikari, Avishek, *Basic Modern Algebra with Applications*, Springer, New Delhi, New York, Heidelberg, 2014.
[4] Bott, R., *The stable homotopy of the classical groups*, Ann. Math. Second Series **70**: 313–337, 1959.
[5] Croom, F.H., *Basic Concepts of Algebraic Topology*, Springer-Verlag, New York, Heidelberg, Berlin, 1978.
[6] Dieudonné, J., *A History of Algebraic and Differential Topology*, 1900–1960, Modern Birkhäuser, 1989.
[7] Dold, A., *Lectures on Algebraic Topology*, Springer-Verlag, New York, 1972.
[8] Dold, A., *Relations between ordinary and extraordinary homology*, Algebraic Topology Colloquium Aarhus, 2–9, 1962.
[9] Fulton, W., *Algebraic Topology, A First Course*, Springer-Verlag, New York, 1975.
[10] Hilton, P.J., *An introduction to Homotopy Theory*, Cambridge University Press, Cambridge, 1983.
[11] Hilton, P. J. and Wylie, 1945, 1942 S. *Homology Theory*, Cambridge University Press, Cambridge, 1960.
[12] Huber, P. J., *Homotopy theory in general categories*, Math. Ann. **144**, 361–385, 1961.
[13] Mayer, J., *Algebraic Topology*, Prentice-Hall, New Jersey, 1972.
[14] Massey, W.S., *A Basic Course in Algebraic Topology*, Springer-Verlag, New York, Berlin, Heidelberg, 1991.
[15] Mitra, Shibopriya, *A study of some notions of algebraic topology through homotopy theory*, PhD Thesis, University of Calcutta, 2007.
[16] Mosher, R. and Tangora, M.C., *Cohomology Operations and Applications in Homotopy Theory*, Harper and Row, New York, 1968.
[17] Steenrod, N.E. and Epstein, D. B. A., *Cohomology operations*, Princeton University Press, Princeton, 1962.
[18] Switzer, R.M., *Algebraic Topology-Homotopy and Homology*, Springer-Verlag, Berlin, Heidelberg, New York, 1975.
[19] Whitehead, G.W., *Elements of Homotopy Theory*, Springer-Verlag, New York, Heidelberg, Berlin, 1978.

References

Adams, J.F.: Algebraic Topology: A student's Guide. Cambridge University Press, Cambridge (1972)
Adams, J.F.: Stable Homotopy and Generalized Homology. University of Chicago, Chicago (1974)
Adhikari, M.R., Adhikari, A.: Basic Modern Algebra with Applications. Springer, New Delhi (2014)
Brown, E.H.: Cohomology theories. Ann. Math. **75**, 467–484 (1962)
Bott, R.: The stable homotopy of the classical groups. Ann. Math. Second Series **70**, 313–337 (1959)
Croom, F.H.: Basic Concepts of Algebraic Topology. Springer, New York (1978)
Dieudonné, J.: A History of Algebraic and Differential Topology, 1900-1960. Modern Birkhäuser, Boston (1989)
Dold, A.: Relations between ordinary and extraordinary homology. Algebraic Topology Colloquium, Aarhus, pp. 2–9 (1962)
Dold, A.: Lectures on Algebraic Topology. Springer, New York (1972)
Dold, A., Thom, R.: Quasifaserungen und unendliche symmetrische Produkte. Ann. Math. **67**(2), 239–281 (1958)
Eilenberg, S., Steenrod, N.: Foundations of Algebraic Topology. Princeton University Press, Princeton (1952)
Fulton, W.: Algebraic Topology, A First Course. Springer, New York (1975)
Gray, B.: Homotopy Theory, An Introduction to Algebraic Topology. Acamedic Press, New York (1975)
Hatcher, A.: Algebraic Topology. Cambridge University Press, Cambridge (2002)
Hilton, P.J.: An Introduction to Homotopy Theory. Cambridge University Press, Cambridge (1983)
Hilton, P.J., Wylie, S.: Homology Theory. Cambridge University Press, Cambridge (1960)
Huber, P.J.: Homotopy theory in general categories. Math. Ann. **144**, 361–385 (1961)
James, I.M. (ed.): Handbook of Algebraic Topology. North Holland, Amsterdam (1995)
Luke, G., Mishchenko, A.S.: Vector Bundles and Their Applications. Kluwer Academic Publishers, Boston (1998)
Massey, W.S.: A Basic Course in Algebraic Topology. Springer, New York (1991)
Maunder, C.R.F.: Algebraic Topology. Cambridge University Press, Cambridge (1980)
Mayer, J.: Algebraic Topology. Prentice-Hall, New Jersey (1972)
Mitra, S.: A study of some notions of algebraic topology through homotopy theory. Ph.D. thesis, University of Calcutta (2007)
Mosher, R., Tangora, M.C.: Cohomology Operations and Applications in Homotopy Theory. Harper and Row, New York (1968)
Spanier, E.: Algebraic Topology. McGraw-Hill, New York (1966)
Steenrod, N.E., Epstein, D.B.A.: Cohomology Operations. Princeton University Press, Princeton (1962)
Switzer, R.M.: Algebraic Topology-Homotopy and Homology. Springer, Berlin (1975)
Vick, J.W.: Homology Theory: Introduction to Algebraic Topology. Springer, New York (1994)
Whitehead, G.W.: Generalized homology theories. Trans. Am. Math. Soc. **102**, 227–283 (1962)
Whitehead, G.W.: Elements of Homotopy Theory. Springer, New York (1978)

Chapter 16
Obstruction Theory

This chapter studies a theory known as "Obstruction Theory" by applying cohomology theory to encounter two basic problems in algebraic topology such as extension and lifting problems. Obvious examples are the homotopy extension and homotopy lifting problems. The homotopy classifications of continuous maps together with the study of extension and lifting problems, play a central role in algebraic topology. Obstruction theory leads to make an attempt to find a general solution. This theory was originated in the classical work of H. Hopf (1894–1971), S. Eilenberg (1915–1998), N. Steenrod (1910–1971) and M. Postnikov (1927–2004) around 1940. The term "obstruction theory" refers to a technique for defining a sequence of cohomology classes that are obstructions to finding solution to the extension, lifting, or relative lifting problems. More precisely, this chapter studies certain sets of cohomology elements, called obstructions which are associated with both a single map in the case of extension and with a pair of maps in the case of homotopies. These are invariants depending only on the spaces and mappings. In polyhedra, these are the characteristics for the existence or nonexistence of the desired extensions and homotopies. The underlying idea of associating cohomology elements with mappings was implicitly used by H. Whitney (1907–1989) and first explicitly formulated by S. Eilenberg. Let X be a compact triangulable space and $f : X \to X$ be a continuous map. Then f has a fixed point if the Lefschetz number Λ_f of f is a nonzero integer. This implies that Λ_f is the "obstruction" to f being fixed point free. Such an example displays the basic objective of obstruction theory.

Extension problems play a central role in topology. Most of the basic theorems of topology together with their successful applications in other areas in mathematics are solutions of particular extension problems. The most successful results have been obtained using the tools of algebraic topology, which offers a conversion of the geometric problem into an algebraic problem. The homotopy classification problem is closely related to extension and lifting problems. Many extension and lifting problems are still unsolved.

M.R. Adhikari, *Basic Algebraic Topology and its Applications*,
DOI 10.1007/978-81-322-2843-1_16

N.E. Steenrod wrote in his excellent paper published in Steenrod (1972):

"Many of the basic theorems of topology, and some of its most successful applications in other areas of mathematics, are solutions of particular extension problems. The deepest results of this kind have been obtained by the method of algebraic topology. The essence of the method is a conversion of a geometric problem into an algebraic problem which is sufficiently complex to embody the essential features of the geometric problem, yet sufficiently simple to be solvable by standard algebraic methods. Many extension problems remain unsolved, and much of the current development of algebraic topology is inspired by the hope of finding a truly general solution".

For this chapter the books and papers Arkowitz (2011), Davis and Kirk (2001), Dodson and Parkar (1997), Eilenberg and Steenrod (1952), Gray (1975), Hatcher (2002), Hu (1959), Maunder (1970), Spanier (1966), Steenrod (1951, 1972) and some others are referred in Bibliography.

16.1 Basic Aim of Obstruction Theory

This section conveys the aim of obstruction theory and describes a technique for studying various homotopy problems such as extension problems, lifting problems, and relative lifting problems which are basic problems in algebraic topology. To earn the basic objective of obstruction theory we start with a simple example: given a group homomorphism f, $\ker f$ is an algebraic indicator which is an obstacle to f for being injective. We normally use homotopy theory to yield algebraic indicators for obstacles to extension and lifting problems of continuous maps. For example, a continuous map $f : S^n \to X$ has a continuous extension over the $(n + 1)$-ball D^{n+1} bounded by S^n iff f is nullhomotopic (see Theorem 2.10.1 of Chap. 2). Hence in this case the obstacle for extension of f is precisely $[f] = 0 (\in \pi_n(X))$. Again the obstacle to lifting problems in a principal fibration is a constant map. Hence this problem can be expressed as the homotopy class of the map into the classifying space.

There are several techniques to develop obstruction theory to extension problems using the tools of algebraic theory. Obstructions are built step by step using the tools of cohomology theory. The most useful technique is to associate certain sets of cohomology elements with a single map in case of extension, and with a pair of maps in case of homotopies. These cohomology elements are called obstructions. This idea of associating cohomolgy elements with mappings was first found implicitly in the work of Whitney and explicitly in the work of S. Eilenberg. The latter theory is traditionally called Eilenberg obstruction theory in his honor. It involves cohomology groups with coefficients in certain homotopy groups to define obstructions to extension problems. But there are several cohomology theories such as cellular, singular, simplicial, Čech cohomology, etc. The uniqueness theorem of cohomology theory asserts that any two cohomology groups having the same coefficient group coincide on finite CW-complexes, which implies that the cohomology groups of finite

CW-complexes are completely determined by the coefficient group. We study here obstruction theory for CW-complex (see (Hu 1959)) but Olum studied this theory for polyhedra (see (Olum 1950)) in almost identical techniques in view of the fact that polyhedron admits a CW structure.

Recall the concepts of cochain and cochain group which are applied in obstruction theory:

Definition 16.1.1 Given a topological space X and an abelian group G, the singular n-cochain group $C^n(X; G)$ with coefficients in G is defined to be the dual group given by $C^n(X; G) = \text{Hom}\,(C_n(X; G), G)$ of the singular chain group $C_n(X; G)$. An element of $C^n(X; G)$ is called a cochain.

Remark 16.1.2 An n-cochain $\alpha \in C^n(X; G)$ assigns to each n-simplex $\sigma : \Delta^n \to X$ a value $\alpha(\sigma) \in G$. Since the singular n-simplexes form a basis of $C_n(X; G)$, these values can be assigned arbitrarily. Hence n-cochains are precisely the functions from singular n-simplexes to G. Again $C^n(X; G)$ is isomorphic to the direct product of as many copies of G as there are n-simplexes in X.

The basic aim of obstruction theory is to study mainly the following four types of problems:

(i) The extension problem;
(ii) The lifting problem;
(iii) The relative lifting problem;
(iv) Cross section problem.

16.1.1 The Extension Problem

This subsection explains the extension problem. Given a CW-complex pair (X, A) with inclusion map $i : A \hookrightarrow X$, and a continuous map $f : A \to E$, does there exist a continuous map $h : X \to E$ (represented by dotted arrow) such that the triangle in Fig. 16.1 is commutative? This is called an extension problem. If such h exists, then h is called an extension of f. For understanding the technique of obstruction theory in extension problem, we first consider the Example 16.1.3.

Example 16.1.3 Let X be a finite CW-complex. If we need construction of a continuous function on X, we use induction: if the function is defined on r-skeleton $X^{(r)}$, we attempt to extend it over the $(r + 1)$-skeleton $X^{(r+1)}$. Then the obstruction to extending it over an $(r + 1)$-cell is an element of π_r.

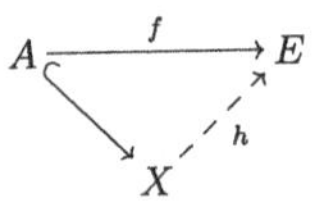

Fig. 16.1 Representing extension problem

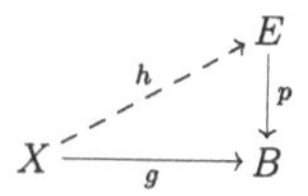

Fig. 16.2 Representing lifting problem

16.1.2 The Lifting Problem

This subsection explains the lifting problem: given a fibration $p : E \to B$ and a continuous map $g : X \to B$, does there exist a continuous map $h : X \to E$ (represented by dotted arrow) called a lift or lifting of g, such that the triangle in Fig. 16.2 is commutative? This problem is called a lifting problem. If such h exists, then h is called a lift or lifting of g. Thus an extension problem is the question of finding a criterian to make the triangle in Fig. 16.2 commutative.

Example 16.1.4 (*Path lifting Property*) Let $f : (I, \dot{I}) \to (S^1, 1)$ be continuous. Then there exists a unique continuous map $\tilde{f} : I \to \mathbf{R}$ with $p\tilde{f} = f$ and $\tilde{f}(0) = 0$. Hence $\tilde{f}$ is the unique lifting of f (see Chap. 3).

16.1.3 Relative Lifting Problem

This subsection explains the relative lifting problem which combines both the extension and lifting problems into a single problem described in the diagram in Fig. 16.3 with commutative square, called the extension-lifting square. In other words, a relative lifting problem is the question of finding a criterian to make the diagram in Fig. 16.3 commutative.

Thus given a CW-complex pair (X, A) if $i : A \hookrightarrow X$ is the inclusion map, then the problem is to determine a continuous map $h; X \to E$ (it exists) such that $hi = f$ and $ph = g$. This h (if it exists) is called a solution of the extension-lifting problem, called a relative lift. If we take in particular, $B = \{*\}$, we obtain the extension problem, and for $A = \{*\}$ we obtain the lifting problem.

Remark 16.1.5 In most cases, these obstructions are in the cohomology groups that are all zero, which gives a solution. On the contrary, if the obstructions are nonzero, it can be used to encounter the problem expressed in cohomology terms.

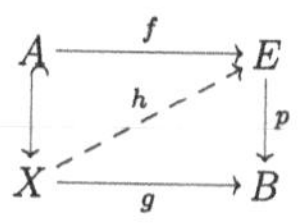

Fig. 16.3 Representing relative lifting problem

16.1.4 Cross Section Problem

It is a particular case of the relative lifting problem when $X = B$ and $f : X \to B$ is the identity map on B in the relative lifting case. Again the relative lifting problem reduces to determining a cross section of the induced bundle $f^*\xi$ of the bundle $\xi : p : E \to B$ under a continuous map $f : X \to B$. For step-by-step extension of cross section of fiber bundle see Sect. 16.4.2.

Remark 16.1.6 The above observation asserts that the relative lifting and cross section problems are equivalent.

16.2 Notations and Abbreviations

In obstruction theory, for simplicity we consider only finite cell complexes X of dimension n and continuous maps from X into a path-connected n-simple space Y. The standard notations and abbreviations used in obstruction theory are given with their meanings.

X	: finite CW-complex
A	: subcomplex of X
$X^{(n)}$	: n-skeleton of X
$\widetilde{X}^n$	: $A \cup X^{(n)}$
Y	: path-connected n-simple space
$\pi_n(Y)$	: nth homotopy group of Y
$C_{n+1}(X, A; \pi_n(Y))$	: $(n+1)$th chain group of X modulo A with coefficient group $\pi_n(Y)$
$C^{n+1}(X, A; \pi_n(Y))$	: $(n+1)$th cochain group of X modulo A with coefficient group $\pi_n(Y)$
$c^{n+1}(f)$	: obstruction of cochain of f in $C^{n+1}(X, A; \pi_n(Y))$
∂	: boundary homomorphism
δ	: coboundary homomorphism
fg	: $f \circ g$ (composite map)
$\partial\partial$	: $\partial \circ \partial = 0$ (0-homomorphism)
$Z^n(X, A; \pi_n(Y))$	: group of n-cycles of X modulo A with coefficient group $\pi_n(Y)$
$B^n(X, A; \pi_n(Y))$	: group of n-boundaries of X modulo A with coefficient group $\pi_n(Y)$
$H^n(X, A; \pi_n(Y))$	: nth cohomology group of X modulo A with coefficient group $\pi_n(Y)$
$[c^{n+1}(f)]$	: $\gamma^{n+1}(f)$ (cohomology class of $c^{n+1}(f)$) in $H^{n+1}(X, A; \pi_n(Y))$
$\psi^\#$	: cochain homomorphism induced by ψ
ψ^*	: homomorphism induced by ψ in cohomology
σ	: cell of the CW-complex X
$\|K\|$	: underlying space of a simplicial complex K
ξ	: fiber bundle.

16.3 The Obstruction Theory: Basic Concepts

This section presents the basic concepts of obstruction theory and uses the notations of Sect. 16.2. Before conveying the basic concepts of 'obstruction theory' we recall that a path-connected topological space is said to n-simple if there is point $x_0 \in X$ such that $\pi_1(X, x_0)$ acts trivially on $\pi_n(X, x_0)$ in the sense that each element of $\pi_1(X, x_0)$ acts on $\pi_n(X, x_0)$ as the identity. For example, every H-space is n-simple for all n. If X is n-simple, then the homoptopy groups $\pi_n(X, x_0)$ do not depend on its base point. Hence if X is n-simple, then the groups $\pi_n(X, x_0)$ is simply denoted by $\pi_n(X)$.

This section considers extension problem: Given a subcomplex A of a finite cell complex X, a path-connected n-simple space Y and a continuous map $f : A \to Y$, the extension problem for f over the whole X is to determine whether or not f can be continuously extended over X. In obstruction theory, an attempt is made to extend the given map f step by step over the subcomplexes $\tilde{X}^n, n = 0, 1, 2, \cdots$. of X, where $\tilde{X}^n = A \cup X^{(n)}$. This process is continued till some obstruction for further extension is met. Then the traditional technique is to measure this obstruction and to change the previously constructed partial extension of f so that this obstruction vanishes and hence further extension of f might be possible.

Definition 16.3.1 Given an integer $n \geq 0$, a subcomplex A of a finite cell complex X, and a path-connected n-simple space Y, a continuous map $f : A \to Y$ is said to be n-extensible over X if f has a continuous extension over the subcomplex $\tilde{X}^n$ of X.

Example 16.3.2 Every continuous map $f : X \to Y$ is 1-extendable.

Definition 16.3.3 The supremum of n for which f is n-extensible is called the extension index of f over X.

Proposition 16.3.4 *Homotopic maps have the same extension index.*

Proof Let $f, g : A \to Y$ be two maps such that $f \simeq g$. Let $\tilde{f}$ be an extension of f. Define $\tilde{g}$ which coincides with $\tilde{f}$ on $X^n - A$ and coincides with g on A. Then there is a homotopy between $\tilde{g}$ and $\tilde{f}$, so there is a homotopy between g and f on $X^n - A$. Hence the index of $g \leq$ index of f. Similarly, index of $f \leq$ index of g. This asserts that index of f = index of g. ❑

Corollary 16.3.5 *The extension index of a continuous map is a homotopy invariant in the sense that homotopic maps have the same extension index.*

Proof The corollary follows from Proposition 16.3.4. ❑

16.3.1 The Obstruction Cochains and Cocycles

This subsection conveys the concepts of obstruction cochain and cocycle. It is important that cycles and coboundaries are primary concepts of a cohomology group. These are linked to an obstruction to extension problems. Consider a given continuous map $f : \tilde{X}^n \to Y$. Then f determines a cochain $c^{n+1}(f)$ in $C^{n+1}(X, A; \pi_n(Y))$ as follows.

Let σ be an arbitrary $(n+1)$-cell of X. Then σ is in $X^{(n+1)} - X^{(n)}$ and its boundary $\partial\sigma$ is an oriented n-sphere. As $\partial\sigma \subset X^{(n)} \subset \tilde{X}^n$, the restriction map f_σ of f over $\partial\sigma$ determines an element $[f_\sigma] \in \pi_n(Y)$. Define a map

$$c^{n+1}(f) : X^{(n+1)} - X^{(n)} \to \pi_n(Y), \sigma \mapsto [f_\sigma].$$

Definition 16.3.6 The cochain $c^{n+1}(f) : X^{(n+1)} - X^{(n)} \to \pi_n(Y), \sigma \mapsto [f_\sigma]$. is called the obstruction cochain of $f : \tilde{X}^n \to Y$.

We claim that $c^{n+1}(f)$ is a cocycle i.e., it vanishes on boundaries.

Theorem 16.3.7 *Let X be a finite CW-complex and A be a subcomplex of X. If Y is a path-connected n-simple space, then given a continuous map $f : \tilde{X}^n \to Y$, its cochain $c^{n+1}(f)$ is a cocycle of X modulo A.*

Proof To prove the theorem it is sufficient to prove that $c^{n+1}(f) \in Z^{n+1}(X, A; \pi_n(Y))$. Let σ be an arbitrary $(n+2)$-cell of X. Then $\partial\sigma \in \tilde{X}^{n+1}$. We claim that $[\delta c^{n+1}(f)](\sigma) = 0$. Let B be the subcomplex $\partial\sigma$ of X and $B^{(n)}$ be the n-skeleton of B. Then we have the following results:

(i) there is a homomorphism $\partial : C_{n+1}(B; \pi_n(Y)) \to Z_n(B : \pi_n(Y)) = Z_n(B^{(n)} : \pi_n(Y))$, since there is no n-cycle in the $(n+1)$-skeleton of B;
(ii) there is an isomorphism $g : Z_n(B^{(n)}; \pi_n(Y)) \to H_n(B^{(n)}; \pi_n(Y))$ by quotient group definition of homology group;
(iii) there is a natural homomorphism $h : \pi_n(B^{(n)}) \to H_n(B^{(n)}; \pi_n(Y))$ by Hurewicz theorem;
(iv) there is a homomorphism $k_* : \pi_n(B^{(n)}) \to \pi_n(Y)$ induced by $k = f|_{B^{(n)}}$.

Combining the above homomorphisms, we have the homomorphisms

$$\begin{aligned} C_{n+1}(B; \pi_n(Y)) \xrightarrow{\partial} Z_n(B, \pi_n(Y)) &= Z_n(B^{(n)}, \pi_n(Y)) \\ &= H_n(B^{(n)}; \pi_n(Y)) \xleftarrow{h} \pi_n(B^{(n)}) \xrightarrow{k_*} \pi_n(Y) \end{aligned}$$

If $n > 1$ then $B^{(n)}$ is $(n-1)$-connected and hence h is an isomorphism by Hurewicz theorem. Again if $n = 1$, h is an epimorphism and $\ker h$ is contained in $\ker k_*$, since the group $\pi_n(Y))$ is abelian. Consequently, in either case, there is a well defined homomorphism

$$\psi : k_* h^{-1} : Z_n(B; \pi_n(Y)) \to \pi_n(Y).$$

Since $C_{n-1}(B, \pi_n(Y))$ is a free abelian group, the kernel $Z_n(B; \pi_n(Y))$ of $\partial : C_n(B, \pi_n(Y)) \to C_{n-1}(B, \pi_n(Y))$ can be expressed as a direct summand of

$C_n(B, \pi_n(Y))$. This asserts that the homomorphism

$$\psi : Z_n(B; \pi_n(Y)) \to \pi_n(Y)$$

has an extension

$$\tilde{\psi} : C_n(B; \pi_n(Y)) \to \pi_n(Y).$$

Again for every $(n+1)$-cell σ' in B, the element $[c^{n+1}(f)](\sigma')$ is represented by the partial map $k|\partial\sigma'$. Hence it follows that $[c^{n+1}(f)](\sigma') = k_* h^{-1}(\partial\sigma') = \tilde{\psi}(\partial\sigma')$. This asserts that $[\delta c^{n+1}(f)](\sigma) = [c^{n+1}(f)](\partial\sigma) = \tilde{\psi}(\partial\partial(\sigma')) = 0$. This implies that $c^{n+1}(f) \in c^{n+1}(f) \in Z^{n+1}(X, A; \pi_n(Y))$. ❑

Remark 16.3.8 Since $c^{n+1}(f) \in Z^{n+1}(X, A; \pi_n(Y))$, $c^{n+1}(f)$ is called the obstruction cocyle of f. The reason for naming obstruction cocycle is given in Proposition 16.3.9.

We assume that somehow an extension of a map over $\tilde{X}^n = A \cup X^n$ has been achieved for some n. We now consider its extension problem over $\tilde{X}^{n+1}$.

Proposition 16.3.9 *A continuous map $f : \tilde{X}^n \to Y$ has a continuous extension over $\tilde{X}^{n+1}$ iff $c^{n+1}(f) = 0$.*

Proof Let σ be an arbitrary $(n+2)$-cell of X and $\tilde{f} : \tilde{X}^{n+1} \to Y$ be a given continuous extension of f. Then $c^{n+1}(f) = c^{n+1}(\tilde{f}|\partial\sigma) = [\tilde{f}|\partial(\partial\sigma)] = 0$. Conversely, suppose $c^{n+1}(f) = 0$. Then f is a coboundary on X. Hence there is a continuous map $f : \tilde{X}^{n+1} \to Y$ such that $f = \delta\tilde{f} = \tilde{f}\delta$. This asserts that for any $(n+1)$-cell σ of X, $f(\sigma) = \tilde{f}(\partial\sigma)$. This shows that $\tilde{f}$ agrees on its boundary. Consequently, $\tilde{f}$ is an extension of f. ❑

Example 16.3.10 Let M be the Möbius band and f be the continuous map which defines the complex structure of M. If σ_1 is the boundary circle and σ_2 is the interior, then $[c^1(f)](\sigma_1) = 0$ shows that the extension of all 1-cells is admitted. The second obstruction $[c^2(f)](\sigma_2) = [f|_{\partial\sigma_2}] = [f|_{S^1}]$. It shows that there is an obstruction to the attaching discs to M. This result shows that the Möbius band M is non-orientable.

Proposition 16.3.11 *Let $f, g : \tilde{X}^n \to Y$ be two homotopic maps. Then $c^{n+1}(f) = c^{n+1}(g)$.*

Proof Let σ be an $(n+1)$-cell of X and $f, g : \tilde{X}^n \to Y$ be two homotopic maps. Then $f|\partial\sigma \simeq g|\partial\sigma$. Hence it shows that $[f|\partial\sigma] = [g|\partial\sigma]$ in $\pi_n(Y)$. This implies that $c^{n+1}(f) = c^{n+1}(g)$. ❑

Remark 16.3.12 The obstruction cocycle $c^{n+1}(f)$ of $f : \tilde{X}^n \to Y$ is a homotopy invariant in the sense that if $f, g : \tilde{X}^n \to Y$ are homotopic maps, then $c^{n+1}(f) = c^{n+1}(g)$.

Definition 16.3.13 Given a continuous map $f : \tilde{X}^n \to Y$, its obstruction cocycle $c^{n+1}(f)$ determines the cohomolgy class $[c^{n+1}(f)] \in H^{n+1}(X, A; \pi_n(Y))$. This cohomology class $[c^{n+1}(f)]$ is abbreviated as $\gamma^{n+1}(f)$ and $c^{n+1}(f)$ is said to represent it.

16.3.2 *The Deformation and Difference Cochains*

This subsection conveys the concepts of deformation and difference cochains and considers the problem of constructing homotopies between two given continuous maps $f, g : \tilde{X}^n \to Y$ which are assumed to be homotopic on $\tilde{X}^{n-1}$ (called partial homotopic on $\tilde{X}^{n-1}$). We claim that the difference of the obstruction cocycles $c^{n+1}(f)$ and $c^{n+1}(g)$ is a coboundary.

First we consider a partial homoptopy $H_t : \tilde{X}^{n-1} \to Y$ such that $H_0 = f|_{\tilde{X}^{n-1}}$ and $H_1 = g|_{\tilde{X}^{n-1}}$. Again the topological product $P = X \times I$ is also a cell complex. Let $P^{(n)}$ be the n-dimensional skeleton of P. As usual notation

$$\begin{aligned}\widetilde{P}^{(n)} &= (A \times I) \cup P^{(n)} \\ &= (\widetilde{X}^n \times \{0\}) \cup (\widetilde{X}^{n-1} \times I) \cup (\widetilde{X}^n \times \{1\}).\end{aligned}$$

Define a map

$$h : \widetilde{P}^n \to Y, (x, t) \mapsto \begin{cases} f(x), & x \in \widetilde{X}^n, t = 0 \\ H_t(x), & x \in \widetilde{X}^{n-1}, t \in I \\ g(x), & x \in \widetilde{X}^n, t = 1 \end{cases}$$

Then h is continuous and determines an obstruction cocycle $c^{n+1}(h)$ of the complex P modulo $A \times I$ with coefficient group $\pi_n(Y)$. Then it follows from the definition of h that $c^{n+1}(h)$ agrees with $c^{n+1}(f) \times 0$ on $X \times \{0\}$ and with $c^{n+1}(g) \times 1$ on $X \times \{1\}$.

Let B be the subcomplex $(X \times \{0\}) \cup (A \times I) \cup (X \times \{1\})$ of $X \times I = P$. Hence $c^{n+1}(h) - c^{n+1}(f) \times 0 - c^{n+1}(g) \times 1$ is a cochain of $X \times I$ modulo B with coefficient group $\pi_n(Y)$. Again the map $\sigma \mapsto \sigma \times I$ establishes a bijective correspondence between the n-cells of $X - A$ and $(n+1)$-cells of $P - B$. This correspondence defines an isomorphism

$$\psi : C^{n+1}(X, A; \pi_n(Y)) \to C^{n+1}(P, B; \pi_n(Y)).$$

This determines a unique cochain, denoted by $d^n(f, g; H_t)$ in $C^n(X, A; \pi_n(Y))$ such that

$$\psi d^n(f, g; H_t) = (-1)^{n+1}(c^{n+1}(h) - c^{n+1}(f) \times 0 - c^{n+1}(g) \times 1).$$

Definition 16.3.14 The unique cochain $d^n(f, g : H_t) \in C^n(X, A : \pi_n(Y))$ is called the deformation cochain of f and g. In particular, if $f|_{\tilde{X}^{n-1}} = g|_{\tilde{X}^{n-1}}$ and $H_t(x) = f(x) = g(x)$ for all $x \in \tilde{X}^{n-1}$ and for all $t \in I$, then $d^n(f, g : H_t)$ is abbreviated in brief as $d^n(f, g)$, and is called the difference cochain of f and g.

Remark 16.3.15 As $d^n(f, g) = c^{n+1}(f) - c^{n+1}(g)$, $d^n(f, g)$ is called the difference cochain of f and g.

Proposition 16.3.16 *Given two continuous maps $f, g : \tilde{X}^n \to Y$, and a partial homotopy $H_t : \tilde{X}^{n-1} \to Y$ has a continuous extension $\tilde{H}_t : \tilde{X}^n \to Y$ over $\tilde{X}^n$ with the property that $\tilde{H}_0 = f$ and $\tilde{H}_1 = g$ iff $d^n(f, g : H_t) = 0$.*

Proof It needs extensions of f and g to have an extension of H_t over $\tilde{X}^{n+1}$. This is possible iff $c^{n+1}(f) = 0 = c^{n+1}(g)$, and $c^{n+1}(H_t) = 0$. Then $\psi d^n(f, g; H_t) = (-1)^n(0 - 0 + 0) = 0$. Again since ψ is an isomorphism, the only element of ker ψ is the zero element. This implies that H_t has a continuous extension over $\tilde{X}^{n+1}$ iff $d^n(f, g : H_t) = 0$. ❑

Remark 16.3.17 $d^n(f, g : H_t)$ plays a key role in the study of obstruction theory because of its coboundary formula Proposition 16.3.18.

Proposition 16.3.18 (Coboundary formula) $\delta d^n(f, g : H_t) = c^{n+1}(f) - c^{n+1}(g)$.

Proof As $\delta I = 0$, isomorphism ψ commutes with δ, and hence it follows that

$$\psi\delta d^n(f, g : H_t) = \delta\psi d^n(f, g : H_t) \tag{16.1}$$

Again, as $c^{n+1}(h), c^{n+1}(f), c^{n+1}(g)$ are all cocycles and $\delta 0 = -I$ and $\delta 1 = I$, applying δ to both sides of (16.1) we have $\delta\psi\delta d^n(f, g : H_t) = c^{n+1}(f) \times I - c^{n+1}(g) \times I$. This shows that $\psi\delta d^n(f, g : H_t) = \psi(c^{n+1}(f) - c^{n+1}(g))$. Since ψ is an isomorphism, the proposition follows. ❑

16.3.3 *The Eilenberg Extension Theorem*

This subsection proves Eilenberg extension theorem which is a key result in obstruction theory when stepwise extension process faces an obstruction. Recall that given a continuous map $f : \tilde{X}^n \to Y$, its obstruction cocycle $c^{n+1}(f)$ determines an element $\gamma^{n+1}(f) \in H^{n+1}(X, A; \pi_n(Y))$.

Theorem 16.3.19 (Eilenberg extension theorem) *Given a continuous map $f : \tilde{X}^n \to Y$, the element $\gamma^{n+1}(f) \in H^{n+1}(X, A; \pi_n(Y))$ vanishes iff there exists a continuous map $\tilde{h} : X^{n+1} \to Y$ such that $\tilde{h}|_{\tilde{X}^{n-1}} = f|_{\tilde{X}^{n-1}}$.*

Proof First suppose that element $\gamma^{n+1}(f) = 0$. If $c^{n+1}(f) \in C^{n+1}(X, A : \pi_n(Y))$ is a representative of $\gamma^{n+1}(f)$, then $c^{n+1}(f)$ is homotopic to a constant map rel A. Hence there exists a continuous map $h : \tilde{X}^n \to \pi_n(Y)$ such that $f|_{\tilde{X}^{n-1}} = h|_{\tilde{X}^{n-1}}$ and $c^{n+1}(h) = 0$. This implies that h has a continuous extension $\tilde{h} : \tilde{X}^{n+1} \to Y$ by Proposition 16.3.9. Conversely, assume that there is a continuous extension $\tilde{h} : \tilde{X}^{n+1} \to Y$ of $h = \tilde{h}|_{\tilde{X}^n}$. Then $c^{n+1}(h) = 0$. Again since $f|_{\tilde{X}^{n-1}} = h|_{\tilde{X}^{n-1}}$, the difference cochain $d^n(f, h)$ is defined. Hence by Proposition 16.3.18, $c^{n+1}(h) = 0$. It implies that $c^{n+1}(f)$ is the coboundary of $d^n(f, h)$. This shows that $\gamma^{n+1}(f) = 0$. ❑

Remark 16.3.20 conveys the significance of Eilenberg extension theorem.

Remark 16.3.20 Assume that $\tilde{f}$ is a continuous extension of a given map $f : A \to Y$. If the obstruction cocycle $c^{n+1}(f) \neq 0$, then it follows from Proposition 16.3.9 that $\tilde{f}$ cannot be extended over $\tilde{X}^{n+1}$. This gives an obstruction in stepwise extending process. The importance of the Eilenberg extension theorem is that if $c^{n+1}(f)$ is nullhomotopic rel A, then the obstruction can be removed by modifying the values of f on the open n-cells in $X - A$ only.

16.3.4 The Obstruction Set for Extension

This subsection conveys the concept of 'obstruction set' which plays a key role in extensibility of a map $f : A \to Y$ over the whole space X.

Definition 16.3.21 Given a finite CW-complex pair and a path-connected n-simple space, a continuous map $f : \tilde{X}^n \to Y$ determines a cocycle $c^{n+1}(f)$ up to homotopy and hence determines an element $\gamma^{n+1}(f) \in H^{n+1}(X, A; \pi_n(Y))$, called an $(n+1)$-dimensional obstruction element of f.

Definition 16.3.22 The set of all $(n+1)$-dimensional obstruction elements of $f : \tilde{X}^n \to Y$ forms a subset of $H^{n+1}(X, A; \pi_n(Y))$, called an obstruction set of f, denoted by $\mathbf{O}^{n+1}(f)$.

Definition 16.3.23 A continuous map $f : A \to Y$ is said to be n-extensible over X if there exists a continuous extension $\tilde{f} : \tilde{X}^n \to Y$ of f.

Proposition 16.3.24 *If $f, g : \tilde{X}^n \to Y$ are two homotopic maps, then* $\mathbf{O}^{n+1}(f) = \mathbf{O}^{n+1}(g)$.

Proof Let $f, g : \tilde{X}^n \to Y$ be two maps such that $f \simeq g$. We claim that $\mathbf{O}^{n+1}(f) = \mathbf{O}^{n+1}(g)$. As $f \simeq g$, $c^{n+1}(f) = c^{n+1}(g)$ by Proposition 16.3.11. Hence they define the same equivalence class in $H^{n+1}(X, A; \pi_n(Y))$. This implies that $\mathbf{O}^{n+1}(f) = \mathbf{O}^{n+1}(g)$. ❑

Recall that if (X, A) and (X', A') be two cellular pairs and $\psi : (X, A) \to (X', A')$ is a cellular map, then given a continuous map $f : \tilde{X}^n \to Y$, the composite map $g = f\psi : \tilde{X}'^n \to Y$ induces a unique cochain homomorphism

$$\psi^{\#} : C^{n+1}(X, A; \pi_n(Y)) \to C^{n+1}(X', A'; \pi_n(Y))$$

such that $c^{n+1}(g) = \psi^{\#} c^{n+1}(f)$. Hence ψ induces a homomorphism

$$\psi^{*} : H^{n+1}(X, A; \pi_n(Y)) \to H^{n+1}(X', A'; \pi_n(Y))$$

Proposition 16.3.25 *Let $\psi : (X, A) \to (X', A')$ be a cellular map and $f : A \to Y$ be a continuous map. If $g = f\psi : A' \to Y$, then the induced homomorphism*

$$\psi^* : H^{n+1}(X, A; \pi_n(Y)) \to H^{n+1}(X', A'; \pi_n(Y))$$

sends $\mathbf{O}^{n+1}(f)$ into $\mathbf{O}^{n+1}(g)$.

Proof Using the relation $c^{n+1}(g) = \psi^{\#}c^{n+1}(f)$, it follows that for each obstruction of f, there is a corresponding obstruction of g, which is the image of $\psi^{\#}$. Consequently, $\psi^{\#}$ induces a map between the cohomology groups and hence in particular, in obstruction sets. ❑

Proposition 16.3.26 *A continuous map $f : A \to Y$ is $(n+1)$-extensible over X iff $\mathbf{O}^{n+1}(f) = \{0\}$.*

Proof $f : A \to Y$ is $(n+1)$-extensible over X iff $c^{n+1}(f) = 0$. This asserts that the nullity class of $c^{n+1}(f)$ and the cohomology class $\gamma^{n+1}(f)$ are equal. This shows that $\mathbf{O}^{n+1}(f) = \{0\}$ in $H^{n+1}(X', A'; \pi_n(Y))$. ❑

Theorem 16.3.27 *If Y is n-simple and $H^{n+1}(X, A; \pi_n(Y)) = \{0\}$, then for every integer m such that $m \leq n < r$, then the m-extensibility of $f : A \to Y$ over X implies its r-extensibility over X.*

Proof The theorem follows by repeated applications of the results of Exs. 12 and 13 of Sect. 16.5. ❑

Theorem 16.3.28 *Let Y be n-simple and $H^{n+1}(X, A; \pi_n(Y)) = \{0\}$ for every integer $n \geq 1$. Then every continuous map $f : A \to Y$ has a continuous extension over X.*

Proof Clearly, if $X - A$ is of dimension not exceeding r given in Theorem 16.3.27. Then it asserts that every continuous map $f : A \to Y$ has a continuous extension over X iff it is n-extensible over X. This implies the theorem. ❑

16.3.5 *The Homotopy Index*

This subsection gives the concept of homotopy index which is important in obstruction theory. Homotopy problem is a special case of extension problem. So the techniques of obstruction theory can be naturally applied.

Definition 16.3.29 Two maps f and g are said to be n-homotopic rel A if $f|_{\bar{X}^n}$ and $g|_{\bar{X}^n}$ are homotopic relative to A. If $f \simeq g$ rel A, then they are automatically n-homotopic.

Definition 16.3.30 The supremum of n such that f and g are n-homotopic is called the homotopy index of the pair of maps (f, g) rel A.

Proposition 16.3.31 *Every pair of maps $f, g : X \to Y$ such that $f|_A = g|_A$ are 0-homotopic*

Proof By assumption Y is path-connected. Hence the proposition follows. ❑

Remark 16.3.32 If (X, A) is a simplicial pair, then the homotopy index of any pair of maps $f, g : X \to Y$ rel A is a topological invariant. For more results see exercises in Sect. 16.5.

16.4 Applications

This section applies obstruction theory to solve some problems of algebraic topology and proves some key results.

16.4.1 A Link between Cohomolgy and Homotopy with Hopf Theorem

Theorem 16.4.1 *Given a CW-complex X and an abelian group G, there is a bijection*

$$\psi : [X, K(G, n)] \to H^n(X; G),\ [f] \mapsto f^*(\tau_n).$$

Proof *ψ is surjective:* Let $\beta \in H^n(X; G)$. Choose a cocycle $g : C^n(X) \to G$ which represents β. Then g assigns an element $g(\sigma_i^n) \in \pi_n(K(G, n)) = G$. Let $h_i : S^n \to K(G, n)$ be representatives of the elements of $g(\sigma_i^n)$. Let $f^{(n)}|_{X^{(n-1)}}$ be a constant map. Define a map $f^{(n)} : X^{(n)} \to K(G, n)$ as the composite

$$f^{(n)} : X^{(n)} \to X^{(n)}/X^{(n-1)} = \bigvee_i S_i^n \xrightarrow{\vee h_i} K(G, n).$$

such that $f^{(n)}|_{X^{(n-1)}}$ is a constant map. Since g is a cocycle and g coincides with the distinguishing cochain $d(*, f^{(n)})$, it follows that

$$0 = \delta g = \delta d(*, f^{(n)}) = c(f^{(n)}) - c(*) = c(f^{(n)}).$$

This shows that there is an extension of the map $f^{(n)} : X^{(n)} \to K(G, n)$ to a map $f^{(n+1)} : X^{(n+1)} \to K(G, n)$. Then the further obstructions to extend the map

$$f^{(n+1)} : X^{(n+1)} \to K(G, n)$$

to the skeletons $X^{(n+q)}$ are in the corresponding groups

$$C^{n+q}(X; \pi_{n+q-1}K(G, n)) = 0 \text{ for } q \geq 2.$$

This shows that ψ is a surjection.

ψ is injective: Let $f, g : X \to K(G, n)$ be two continuous maps such that $f^*(\tau_n) = g^*(\tau_n)$ in the cohomology group $H^n(X; G)$. Then by cellular approximation theorem, we assume that $f|_{X^{(n-1)}} = g|_{X^{(n-1)}} = *$. Hence the element $f^*(\tau_n)$ coincides with the cohomology class of the distinguishing cocycle $d(*, f)$. Then $f^*(\tau_n) = [d(*, f)]$ and $g^*(\tau_n) = [d(*, g)]$ give

$$[d(f, g)] = [d(f, *)] + [d(*, g)] = -f^*(\tau_n) + g^*(\tau_n) = 0.$$

Consequently, there exists a homotopy

$$H_t : f|_X(n) \simeq g|_{X^{(n)}}$$

relative to the skeleton $X^{(n-2)}$. Hence all obstructions to extend this homotopy to the skeletons $X^{(n+q)}$ are all zero. ❑

Remark 16.4.2 If X is a CW-complex of infinite dimension, using the intervals

$$\left[\frac{2^p - 1}{2^p}, \frac{2^{p+1} - 1}{2^{p+1}}\right] = \left[1 - \frac{1}{2^p}, 1 - \frac{1}{2^{p+1}}\right],$$

we construct a homotopy between $f|_{X^{(n+p)}}$ and $g|_{X^{(n+p)}}$.

Theorem 16.4.3 (Hopf) *Let X be a CW-complex of dimension n. Then there is a bijection $\psi : H^n(X; \mathbf{Z}) \to [X, S^n]$.*

Proof It follows from Theorem 16.4.1. ❑

Remark 16.4.4 The Theorem 16.4.1 asserts that the cohomology groups of a CW-complex can be identified with the groups of homotopy classes of continuous maps into Eilenberg-MacLane spaces.

16.4.2 *Stepwise Extension of A Cross Section*

This subsection considers the problems of constructing a cross section of a fiber bundle. Throughout this subsection it is assumed that for the fiber bundle $\xi : F \subset E \to X$, the base space X is a finite CW-complex. Suppose that A is a subcomplex of X and the fiber space F is path-connected and n-simple. We use the notations of Sect. 16.2. If A does not contain all of the 0-dimensional skeleton $X^{(0)}$ of X, then we assume that we have a partially defined cross section $f : \tilde{X}^n \to E$, the problem is to extend it over $\tilde{X}^{n+1}$. In such a problem, obstruction may appear. Indeed, if σ is an $(n+1)$-cell of $X - A$, the cross section $f|\partial\sigma$ might describe a nontrivial element in $\pi_n(F)$ and in this case f will not have a continuous extension over σ. Consider the

$$\psi_f : \{(n+1)\text{-cells } \sigma \text{ of } X\} \to \pi_n(F), \sigma \mapsto [f|\partial\sigma].$$

As by hypothesis, F is n-simple, and $f|\partial\sigma$ is a topological n-sphere, the function ψ_f, is well defined and ψ_f sends an $(n+1)$-cell σ to an element of $\pi_n(F)$ which is determined by $f|\partial\sigma$ through some random trivialization $\xi|_\sigma \approx \sigma \times F$. Then ψ_f can be extended by linearity and ψ_f can be regarded as a $\pi_n(F)$-valued cochain, abbreviated $c^{n+1}(f) \in C^{n+1}(X, A : \pi_n(F))$. Again for every $(n+2)$-cell σ, we have $(\delta\psi_f)(\sigma) = \psi_f(\partial\sigma) = [f|\partial\partial\sigma] = 0$. This shows that $c^{n+1}(f)$ is a cocycle. Hence its cohomology class $[c^{n+1}(f)]$ is an element of $H^{n+1}(X, A : \pi_n(F))$, the element is usually abbreviated $\gamma^{n+1}(f)$.

Remark 16.4.5 Because $c^{n+1}(f)$ being the zero indicator that all of these elements of $H^{n+1}(X, A : \pi_n(F))$ vanish, it asserts that the given partially cross section can be extended to $\tilde{X}^{n+1}$ using the homotopy between $f|\partial\sigma$ and the constant map.

If we start with a different partially defined cross section g that agrees with $f|\tilde{X}^{n-1}$, then the resulting cocycle $c^{n+1}(g)$ would differ from $c^{n+1}(f)$ by a coboundary. This asserts that there is a well-defined element of the cohomology group $H^{n+1}(X, A : \pi_n(F))$ such that if a partially defined cross section on $\tilde{X}^{n+1}$ exists that agrees with the given choice on $\tilde{X}^{n-1}$, then the cohomology class $\gamma^{n+1}(f)$ must be trivial. Its converse is also true as homotopy section is in the sense $pf \simeq 1_d$.

Definition 16.4.6 Given an $(n+1)$-cell σ of X, a function $c(f, \sigma)$ is defined by the rule $c(f, \sigma) = \psi_f(\sigma) = [f|\partial\sigma] \in \pi_n(F)$.

Definition 16.4.7 The function ψ_f of σ given $\psi_f(\sigma) = [f|\partial\sigma] = c(f, \sigma) \in \pi_n(F)$ is called the obstruction cocycle of f and is sometimes denoted by $c(f)$.

The above discussion with corresponding notations leads to the the following important results.

Theorem 16.4.8 *A cross section over $A \cup X^{(n)}$ can be extended over $A \cup X^{(n+1)}$ iff $c(f, \sigma)$ vanishes for each $(n+1)$-cell σ of X.*

Proof Let x_σ be a base point of a given $(n+1)$-cell σ of X and F_σ be the fiber over x_σ. Choose an orientation of σ. Consider σ the oriented cell and $\partial\sigma$ oriented boundary. If f is a cross section $f : \tilde{X}^n \to E$, define $c(f, \sigma) = \psi_f(\sigma) = [f|\partial\sigma] \in \pi_n(F)$. As F is n-simple, the theorem follows. ❑

Corollary 16.4.9 *Given a fiber bundle $p : E \to X$ whose fiber F is a path-connected n-simple space, then the element $\gamma^{n+1}(f) \in H^{n+1}(X : \pi_n(F))$ vanishes iff there are cross sections of $p : E \to X$ defined over the n-skeleton of X that extend over the $(n+1)$-skeleton.*

Remark 16.4.10 For more results see Exercises in Sect. 16.5.

16.4.3 *Homological Version of Whitehead Theorem*

This subsection presents a homological version of the Whitehead theorem (see Theorem 8.5.8) which asserts that if X and Y are connected abelian CW-complexes and if $f : X \to Y$ induces isomorphisms on all homotopy groups, then f is a homotopy equivalence.

Theorem 16.4.11 (Homological version of Whitehead theorem) *Let X and Y be simply connected CW-complexes. If a continuous map $f : X \to Y$ induces isomorphisms*

$$f_* : H_n(X) \to H_n(Y)$$

on all homology groups, then f is a homotopy equivalence.

Proof It can be proved by applying the Hurewicz theorem and obstruction theory to extend the homological version of Whitehead theorem to CW-complexes with trivial action of π_1 on all homotopy groups. ❑

Remark 16.4.12 For an alternative proof see Theorem 17.2.1 of Chap. 17.

16.4.4 *Obstruction for Homotopy Between Relative Lifts*

This subsection uses obstruction theory to obtain obstruction for homotopy between relative lifts. The extension problems have closed connection with homotopy problems. We now want to define obstruction for homotopy of maps. Given continuous two maps $f_0, f_1 : X \to Y$ a homotopy between them is an extension of the map $X \times \partial I \cup \{*\} \times I \to Y$ determined by f_0 and f_1 to the space $X \times I$. Hence we can use obstruction theory to obtain obstruction for homotopy.

Let (X, A) be a CW-complex pair with $i : A \hookrightarrow X$ inclusion map and $p : E \to B$ be a fiber map with fiber F. If $h_0, h_1 : X \to E$ are two relative lifts of f as shown in diagram Fig. 16.4, then h_0, h_1, g and f determine continuous maps

$$H : X \times \partial I \cup A \times I \to E,\ (x, t) \mapsto \begin{cases} h_t(x) \text{ if } t = 0, 1 \text{ and } x \in X \\ g(x) \text{ if } x \in A \end{cases}$$

and

$$G : X \times I \to B, (x, t) \mapsto f(x).$$

Then there is a commutative square shown in Fig. 16.5 where j is inclusion. A relative lift $L : X \times I \to E$ for this diagram is a homotopy between h_0 and h_1 such that $L(a, t) = g(a)$ and $(pL)(x, t) = f(x)$, $\forall a \in A$ and $\forall x \in X$ and $\forall t \in I$. If C_j is the mapping cone of j, then the obstructions to this relative lift are in

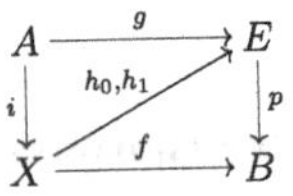

Fig. 16.4 Diagram for relative lifting

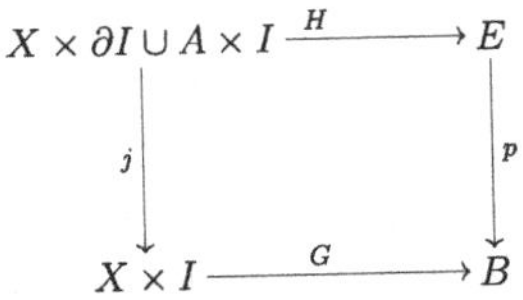

Fig. 16.5 Commutative square involving H and G

$$\begin{aligned} H^{n+1}(C_j; \pi_n(F)) &\cong H^{n+1}(X \times I/(X \times \partial I \cup A \times I); \pi_n(F)) \\ &\cong H^{n+1}(\textstyle\sum X/\sum A; \pi_n(F)) \\ &\cong H^{n+1}(\textstyle\sum X/A; \pi_n(F)) \\ &\cong H^n(C_i; \pi_n(F)), \end{aligned}$$

where C_i is the mapping cone of the inclusion map $i : A \hookrightarrow X$. This gives an obstruction for homotopy between relative lifts.

16.5 Exercises

1. Let $f : X \to Y$ and $g : Y \to Z$ be two continuous maps in $\mathcal{C}_0$. Show that $gf : X \to Z$ is inessential iff g extends to $\tilde{g} : C_f \to Z$.
2. Let X be a simple space (i.e., n-simple for every $n \geq 1$). Show that $\pi_r(X) = 0$ for $r \leq n-1$, where $n \geq 1$, iff for every CW-complex K of dimension $\leq n-1$, any continuous map $f : K \to X$ is nullhomotopic.
3. Let $f : X \to Y$ be a map, where X is the homotopy type of a CW-complex. Taking a Postnikov system for Y, the map f is said to be the n-trivial if the composite map
$$i_n f : X \xrightarrow{f} Y \xrightarrow{i_n} Y^{[n]}$$
is nullhomotopic. This is well defined, since it does not depend on the choice of $Y^{[n]}$. f is said to be trivial if it is n-trivial for all n.
If X is an n-dimensional CW-complex, show that f is n-trivial iff f is trivial.
4. Let X be a compact triangulable space and $f : X \to X$ be a continuous map. Show that Lefschetz number Λ_f of f is the "obstruction" to f being fixed point free.
[Hint. Use Lefschetz fixed point theorem.]
5. Let $f : X \to Y$ be $(n-1)$-trivial. For the exact sequence in $\underline{S}$

$$[X, K(\pi_n(Y), n)] \xrightarrow{\alpha_*} [X, Y^{[n]}] \longrightarrow [X, Y^{[n-1]}],$$

let the set

$$\mathrm{O}_n(f) = \alpha_*^{-1}[(i_n f)] \subset \widetilde{H}^n(X; \pi_n(Y)).$$

Show that the set $\mathrm{O}_n(f)$ is nonempty iff f is n-trivial.

($\mathrm{O}_n(f)$ is called the n-dimensional obstruction set to f being essential).

6. Let K be a finite CW-complex. If $\pi_n(X) = 0$ for all $n \geq 0$, show that any two continuous maps $f, g : |K| \to X$ are homotopic.

7. Show that

(i) a polyhedron $|K|$ is contractible iff $\pi_n(|K|) = 0$ for all $n \geq 0$.
(ii) Let K be an m-dimensional simplicial complex. If n is a positive integer and $\pi_n(Y)$ is zero for all $m \neq n$ (and if $n = 1$, Y is 1-simple), then there is a bijective correspondence between the homotopy classes of maps of $|K|$ into Y and the elements of $H^n(K; \pi_n(Y))$.

8. Let Y be an n-connected space and K be a CW-complex of dimension n. Show that $[K, Y] = \{*\}$.

9. Show that the extension index of a continuous map $f : A \to Y$ is a topological invariant.

10. Let (X, A) be a relative CW-complex and $f, g : X^{(n)} \to Y$ be two continuous maps such that $f|_{X^{(n-1)}} \simeq g|_{X^{(n-1)}}$ rel A. Show that a choice of homotopy defines a difference cochains $d \in C^n(X, A : \pi_n(Y))$ such that $\delta d = \theta^{n+1}(f) - \theta^{n+1}(g)$.

11. Let (X, A) be a relative CW-complex, $n \geq 1$, and Y be an $(n-1)$-connected space ($\pi_1(Y)$ is assumed to be abelian for $n = 1$). Let $f, g : X \to Y$ be two continuous maps such that $f|_A = g|_A$. Show that $f|_{X^{(n-1)}} \simeq g|_{X^{(n-1)}}$ rel A and the obstruction in $H^n(X, A; \pi_n(Y))$ to extending this homotopy to $X^{(n)}$ is independent of the choice of homotopy on $X^{(n-1)}$ and depends only on the homotopy classes of f and g relative to A.

12. Show that a continuous map $f : A \to Y$ is n-extensible over X iff $\mathbf{O}^{n+1}(f)$ is nonempty.

13. Show that a continuous map $f : A \to Y$ is $(n+1)$-extensible over X iff $\mathbf{O}^{n+1}(f)$ contains the zero element of $H^{n+1}(X, A; \pi_n(Y))$.

14. Let Y be a path-connected and n-simple space. If $H^{n+1}(X, A; \pi_n(Y)) = 0$ for every integer $n \geq 1$, show that every continuous map $f : A \to Y$ has a continuous extension over X.

15. (Poincaré-Hopf theorem) Show that a closed oriented n-manifold has a nowhere zero vector field iff its characteristic is zero.

16. Given a fiber bundle $\xi : F \subset E \to X$, the base space X is a finite CW-complex. Suppose that A is a subcomplex of X and the fiber space F is path-connected and n-simple. If ξ admits a cross section $f : \tilde{X}^n \to F$, using the notations used in Sect. 16.4.2 show that

(i) f has a continuous extension to a cross section $\tilde{f} : \tilde{X}^{n+1} \to F$ iff $c^{n+1}(f) = 0$;
(ii) If $f, g : \tilde{X}^n \to F$ are homotopic cross sections, then $c^{n+1}(f) = c^{n+1}(g)$;
(iii) $c^{n+1}(f)$ is a cocycle and its cohomology class $\gamma^{n+1}(f) \in H^{n+1}(X, A; \pi_n(F)$;
(iv) $\gamma^{n+1}(f)$ is a topological invariant.

17. If (X, A) is a simplicial pair, show that the homotopy index of any pair of maps $f, g : X \to Y$ rel A is a topological invariant.

18. Let X be a CW-complex, and Y be a path-connected $(n-1)$-simple space. Given a continuous map $f : X^{(n-1)} \to Y$, define

$$c(f) : \text{Hom}\,(C_n(X), \pi_{n-1}(Y)) \to \pi_{n-1}(Y), \ \sigma \mapsto [f\psi^n_\alpha],$$

where the characteristic map ψ^n_α is regarded as map of S^{n-1} to $X^{(n-1)}$. Show that

(i) $c(f)$ depends only on the homotopy class of f;
(ii) $c(f) = 0$ iff f has a continuous extension to a map $\tilde{f} : X^{(n)} \to Y$;
(iii) $\delta c(f) = 0$;
(iv) if $\gamma(f)$ represents the homotopy class of $c(f)$ in $H^n(X; \pi_{n-1}(Y))$, then $\gamma(f) = 0$ iff there exists a continuous map $g : X^{(n)} \to Y$ such that $f = g$ on $X^{(n-2)}$.
(v) Let B be a CW-complex and $A \subset B$ be its subcomplex. Suppose $\tilde{X}^n = B^{(n)} \cup A$, where $B^{(n)}$ is the nth skeleton of B. Let $\sigma = e^{n+1}$ be an $(n+1)$-cell of B, which does not belong to A and $\psi_\sigma : S^n \to X^n$ be the attaching map corresponding to A and $\psi_\sigma : S^n \to X^n$ be the attaching map corresponding to the cell σ. Any map $f : \tilde{X}^n \to Y$, where Y is homotopically simple (in the sense that $\pi_1(Y)$ acts trivially on $\pi_n(Y)$ for each n), defines a cochain $c(f)$ by taking the value $c(f)$ on the generator σ, given by $c(f)(\sigma) = [f\psi_\sigma] \in \pi_n(Y)$, where the composite is given by

$$f\psi_\sigma : S^n \xrightarrow{\psi_\sigma} \tilde{X}^n \xrightarrow{f} Y.$$

Show that the cochain $c(f)$ is a cocycle, i.e., $\delta c(f) = 0$.
(vi) Using the notation of the above Ex. 18 (v), show that a continuous map $f : \tilde{X}^n \to Y$ has a continuous extension $\tilde{f} : X^{n+1} \to Y$ iff $c(f) = 0$.
(vii) If $f \simeq f'$ and $g \simeq g'$ rel A, then the pair (f', g') has the same homotopy index rel A as the pair (f, g).
(viii) Let Y be a homotopically simple space, (X, A) a CW-pair and $\tilde{X}^n = X^{(n)} \cup A$ for $n = 0, 1, \ldots$. If $f : \tilde{X}^n \to Y$ is a continuous map, show that there exists a continuous map $g : \tilde{X}^{n+1} \to Y$ such that $g|_{\tilde{X}^{n-1}} = f|_{\tilde{X}^{n-1}}$ iff $[c(f)] = 0$ in $H^{n+1}(B, A; \pi_n(Y))$.

16.6 Additional Reading

[1] Adams, J.F., *On the non-existence of elements of Hopf invariant one*, Ann of Math. **72**(2), 20–104, 1960.
[2] Adams J.F., *Algebraic Topology: A student's Guide*, Cambridge University Press, Cambridge, 1972.
[3] Dieudonné J., *A History of Algebraic and Differential Topology*, 1900–1960, Modern Birkhäuser, 1989.
[4] Dold A., *Lectures on Algebraic Topology*, Springer-Verlag, New York, 1972.
[5] Hilton P.J., *An introduction to Homotopy Theory*, Cambridge University Press, Cambridge, 1983.
[6] Hilton P. J. and Wylie S., *Homology Theory*, Cambridge University Press, Cambridge, 1960.
[7] Massey W.S., *A Basic Course in Algebraic Topology*, Springer-Verlag, New York, Berlin, Heidelberg, 1991.
[8] Mayer J. *Algebraic Topology*, Prentice-Hall, New Jersy, 1972.
[9] Olum P., *Obstructions to extensions and homotopies*, Ann of Math. **25**, pp 1–25, 1950.
[10] Switzer R.M., *Algebraic Topology-Homotopy and Homology*, Springer-Verlag, Berlin, Heidelberg, New York, 1975.
[11] Whitehead, G.W., *On mappings into group like spaces*, Comment. Math. Helv. **28**, 320–328. 1954.
[12] Whitehead G.W., *Elements of Homotopy Theory*, Springer-Verlag, New York, Heidelberg, Berlin, 1978.

References

Adams, J.F.: On the non-existence of elements of Hopf invariant one. Ann. Math. **72**(2), 20–104 (1960)
Adams, J.F.: Algebraic Topology: A Student's Guide. Cambridge University Press, Cambridge (1972)
Arkowitz, M.: Introduction to Homotopy Theory. Springer, New York (2011)
Davis, J.F., Kirk, P.: Lecture Notes in Algebraic Topology. Indiana University, Bloomington (2001). http://www.ams.org/bookstore-getitem/item=GSM-35
Dieudonné, J.: A History of Algebraic and Differential Topology, 1900–1960. Modern Birkhäuser Classics. Birkhäuser, Basel (1989)
Dodson, C.T.J., Parkar, P.E.: User's Guide to Algebraic Topology. Kluwer Academic Publishers, Dordrecht (1997)
Dold, A.: Lectures on Algebraic Topology. Springer, New York (1972)
Eilenberg, S., Steenrod, N.: Foundations of Algebraic Topology. Princeton University Press, Princeton (1952)
Gray, B.: Homotopy Theory: An Introduction to Algebraic Topology. Academic, New York (1975)
Hatcher, A.: Algebraic Topology. Cambridge University Press, Cambridge (2002)
Hilton, P.J.: An Introduction to Homotopy Theory. Cambridge University Press, Cambridge (1983)
Hilton, P.J., Wylie, S.: Homology Theory. Cambridge University Press, Cambridge (1960)

Hu, S.T.: Homotopy Theory. Academic Press, New York (1959)
Massey, W.S.: A Basic Course in Algebraic Topology. Springer, New York (1991)
Maunder, C.R.F.: Algebraic Topology. Van Nostrand Reinhold Company, London (1970)
Mayer, J.: Algebraic Topology. Prentice-Hall, New Jersy (1972)
Olum, P.: Obstructions to extensions and homotopies. Ann. Math. **25**, 1–25 (1950)
Samelson, H.: Groups and spaces of loops. Comment. Math. Helv. **28**, 278–286 (1954)
Spanier, E.: Algebraic Topology. McGraw-Hill, New York (1966)
Steenrod, N.E.: The Topology of Fibre Bundles. Princeton University Press (1951)
Steenrod, N.E.: Chohomology operations and obstructions to extending continuous functions. Adv. Math. **8**, 371–416 (1972)
Switzer, R.M.: Algebraic Topology-Homotopy and Homology. Springer, Berlin (1975)
Whitehead, G.W.: On mappings into group like spaces. Comment. Math. Helv. **28**, 320–328 (1954)
Whitehead, G.W.: Elements of Homotopy Theory. Springer, New York (1978)

Chapter 17
More Relations Between Homology and Homotopy

This chapter displays some similarities and further interesting relations between homology and homotopy groups of topological spaces in addition to some relations between these theories discussed earlier. The concept of homotopy presents a mathematical formulation of the intuitive idea of a continuous transition between two geometrical configurations. On the other hand, the concept of homology presents a mathematical precision to the intuitive idea of a curve bounding an "area" or a surface bounding a "volume." L.E.J. Brouwer (1881–1967) first connected these two basic concepts of algebraic topology in 1912 by proving that two continuous maps of a two-dimensional sphere into itself can be continuously deformed into each other if and only if they have the same degree (i.e., if and only if they are equivalent from the view point of homology theory). Hopf's classification theorem generalizes Brouwer's result to an arbitrary dimension.

The homotopy groups resemble the homology groups in many respects under suitable situations as shown by Hurewicz in his celebrated "Equivalence Theorem." Homotopical and homological versions of Whitehead theorem are similar. Since homology groups are in general more computable than homotopy groups, the homological version of Whitehead theorem is often convenient to apply. Cohomology groups of a CW-complex are dual to homotopy groups in the sense that cohomology groups of a CW-complex can be identified with the groups of homotopy classes of continuous maps into Eilenberg–MacLane spaces. By replacing the Eilenberg–MacLane spaces by suitable spaces, "generalized cohomology theories" are constructed in Chap. 15.

There is also a lack of similarities between these two theories essentially due to absence in higher homotopy groups the excision property for homology and also absence in higher homotopy groups a theorem analogous to van Kampen theorem for fundamental group. This chapter continues to study Eilenberg–MacLane spaces, Moore spaces, Dold–Thom theorem, Hopf invariant and Adams classical theorem on Hopf invariant.

M.R. Adhikari, *Basic Algebraic Topology and its Applications*,
DOI 10.1007/978-81-322-2843-1_17

In this chapter, $\mathcal{C}_0$ denotes the category of pointed topological spaces having homotopy type of finite pointed CW-complexes and $\mathcal{C}$ denotes the category of topological pairs of spaces having homotopy type of finite CW-complex pairs.

For this chapter, the books Adams (1972), Gray (1975), Hatcher (2002), Maunder (1980), Spanier (1966), the papers Eilenberg and MacLane (1945), Steenrod (1949) and some others are referred in the Bibliography.

17.1 Some Similarities and Key Links

Higher homotopy groups which are the natural higher-dimensional analogue of the fundamental groups carry certain similarities and key links with homology groups.

17.1.1 Some Similarities

This subsection shows that the homotopy groups resemble the homology groups in many aspects.

Example 17.1.1 The fundamental groups $\pi_1(X)$ are not always abelian but the groups $\pi_n(X)$ are always abelian for $n \geq 2$. On the other hand, homology groups $H_n(X)$ are always abelian for $n \geq 1$.

Example 17.1.2 The relative homotopy groups give a long exact sequence like long exact sequence of homology groups.

17.1.2 Hurewicz Homomorphism Theorem: A Key Link

This subsection establishes a key link between homotopy and homology groups with the help of Hurewicz homomorphism given by Withold Hurewicz (1904–1956) during 1934–1936. His classical result known as Hurewicz theorem says that for $n \geq 1$ the first nonzero homotopy group $\pi_n(X)$ of a simply connected space X is isomorphic to the first nonzero ordinary homology group $H_n(X)$ for $n > 1$. Their relative version is also similar.

Recall that for any topological space X and positive integer m there exists a group homomorphism

$$h_*\colon \pi_m(X) \to H_m(X)$$

called the Hurewicz homomorphism from its m-th homotopy group to its m-th homology group (with integer coefficients). For $m = 1$, the fundamental group is not abelian in general but its abeliazation is the first homology group

$$H_1(X) \cong \pi_1(X)/[\pi_1(X), \pi_1(X)],$$

where $[\pi_1(X), \pi_1(X),]$ is the commutator subgroup of $\pi_1(X)$.

Definition 17.1.3 (*Hurewicz homomorphism*) Let $m > 0$ and H_* be the ordinary homology theory. The homomorphism $h : \pi_m(X) \to \widetilde{H}_m(X) \cong H_m(X)$ defined as the composite

$$\pi_m(X) \xrightarrow{E} \pi_{m+1}(X \wedge S^1) \cong \pi_{m+1}(X \wedge K(\mathbf{Z}, 1)) \longrightarrow \widetilde{H}_m(X),$$

(use the result that $S^1 \simeq K(\mathbf{Z}, 1)$) is a natural homomorphism, called the Hurewicz homomorphism, where E is the Freudenthal suspension homomorphism

Remark 17.1.4 An equivalent formulation of Hurewicz theorem given Theorem 10.11.2 of Chap. 10 is now presented.

Theorem 17.1.5 (Hurewicz) *Let X be a simply connected pointed topological space. Then the following statements are equivalent:*

(i) $\pi_i(X) = 0$, *if* $1 \leq i < n (n \geq 2)$;
(ii) $\widetilde{H}_i(X) = 0$, *if* $1 \leq i < n (n \geq 2)$.

Either implies that $h : \pi_r(X) \to \widetilde{H}_r(X)$ is an $(n + 1)$-isomorphism.

Proof Since h is natural, and X is well pointed we may assume that X is a CW-complex. First suppose that $\pi_i(X) = 0$ for $i < n$. Then $E : \pi_r(X) \to \pi_{r+1}(\Sigma X)$ is an $(n + 1)$-isomorphism, since $n > 1$. Consider the composite map β_m:

$$\pi_{r+m}(X \wedge K(\mathbf{Z}, m)) \xrightarrow{E} \pi_{r+m+1}(X \wedge K(\mathbf{Z}, m) \wedge S^1) \xrightarrow{(1_d \wedge h_m)_*} \pi_{r+m+1}(X \wedge K(\mathbf{Z}, m + 1)).$$

Again since $X \wedge K(\mathbf{Z}, m)$ is $(m + n - 1)$-connected, E is an isomorphism for $r < m + 2n - 1$ and is onto if $r = m + 2n - 1$. Let $f : X \to \Omega Y$ be continuous and $\tilde{f} : \Sigma X \to Y$ be adjoint to f. Consider the commutative diagram as shown in Fig. 17.2, Exercise 1 of Sect. 17.6.

Assume that $(K(\mathbf{Z}, m + 1), \Sigma K(\mathbf{Z}, m))$ is a relative CW-complex with cells in dimensions greater than $2m + 1$. This implies that $(X \wedge K(\mathbf{Z}, m + 1), X \wedge K(\mathbf{Z}, m) \wedge S^1)$ is a relative CW-complex with cells in dimensions $> n + 2m + 1$. Hence it follows that $(1_d \wedge h_m)_*$ is an isomorphism if $r + m + 1 < n + 2m + 1$ and is onto if $r + m + 1 = n + 2m + 1$. Hence, β_m is an isomorphism if $r < m + n$ and is onto if $r = m + n$. Consequently, h is an $(n + 1)$-isomorphism. This h is an $(n + 1)$-isomorphism under condition **(i)**. Hence it follows that the statement **(i)** is equivalent to the statement **(ii)**. ❑

Corollary 17.1.6 *Let X be a simply connected space. If $\widetilde{H}_i(X) = 0$ for all $i < n$, then $\pi_i(X, *) = 0$ for $i < n$ and the Hurewicz homomorphism $h : \pi_n(X, *) \to H_n(X)$ is an isomorphism for every integer $n \geq 2$.*

Proof It follows from Theorem 17.1.5 that $h : \pi_m(X) \cong H_m(X)$ for the smallest m such that $\pi_k(X) = 0$ for $1 \leq k < m$. ❑

Remark 17.1.7 The first nontrivial homotopy group of a simply connected space X and the first nontrivial homology group of the same space X occur in the same dimension and they are isomorphic under Hurewicz homorphism h.

Remark 17.1.8 For $n = 1$, the Hurewicz homomorphism $h : \pi_1(X) \to H_1(X)$ has as kernel the commutator subgroup of $\pi_1(X)$.

17.2 Relative Version of Hurewicz Homomorphism Theorem

This section conveys relative version of Hurewicz homomorphism theorem. Let $(X, A) \in \mathcal{C}$ and $A \neq \emptyset$. Define $k : \pi_i(X, A) \to H_i(X, A)$ to be the composite

$$\pi_i(X, A) \xrightarrow{(p_A)_*} \pi_i(X/A, *) \xrightarrow{h} \widetilde{H}_i(X/A) = H_i(X, A),$$

where h is the Hurewicz homomorphism.

Theorem 17.2.1 (Relative Hurewicz theorem) *Let A be simply connected, and $\pi_1(X, A) = 0$. Then the following statements are equivalent:*

(i) *$\pi_i(X, A) = 0$ for $1 \leq i < n(n \geq 2)$;*
(ii) *$H_i(X, A) = 0$ for $1 \leq i < n(n \geq 2)$.*

Either implies that $k : \pi_i(X, A) \to H_i(X, A)$ is an isomorphisms for $i \leq n$ and onto for $i = n + 1$.

Proof We may assume that (X, A) is a CW-pair. As in the case of Hurewicz Theorem 17.1.5, it follows that

(i) implies the final condition. But $\pi_i(X, A) \xrightarrow{(p_A)_*} \pi_i(X/A)$ is an $(n + 1)$-isomorphism. Hence the theorem follows. ❑

Definition 17.2.2 A map $f : X \to Y$ between CW-complexes is said to be an h-equivalence if its induced homomorphisms

$$f_* : \pi_m(X) \to \pi_m(Y)$$

are isomorphism for each $m \geq 1$ and X is said to be h-equivalent to Y if there exists an h-equivalence $f : X \to Y$.

Theorem 17.2.3 *Let X be simply connected CW-complex such that $H_m(X) \cong H_m(S^n)$ for $n \geq 2$, with integral coefficients. Then X is h-equivalent to S^n.*

Proof By Corollary 17.1.6, $\pi_n(X) \cong H_n(X)$, and this group is assumed to be isomorphic to $\mathbf{Z}$. Let $f : S^n \to X$ represent a generator. Then $f_* : \pi_n(S^n) \to \pi_n(X)$ is an isomorphism in homotopy and also $f_* : H_n(S^n) \to H_n(X)$ is so in homology. Then by Whitehead theorem, f is a homotopy equivalence and hence is h-equivalent to S^n ❑

Remark 17.2.4 Theorem 17.2.3 has interesting applications. For example, a closed connected n-manifold of the homotopy type of the n-sphere S^n is homeomorphic to the n-sphere (compare with Poincaré conjecture given in Sect. 18.1). Consequently, these spheres are characterized by invariants of algebraic topology.

Example 17.2.5 The spaces $S^n \vee S^n \vee S^{2n}$ and $S^n \times S^n$ are simply connected for $n \geq 2$ and have isomorphic homology groups. As their cohomology rings are different, they cannot be h-equivalent.

17.3 Alternative Proof of Homological Version of Whitehead Theorem

This section conveys an alternative proof of Theorem 16.4.11 of homological version of the well known classical Whitehead theorem given by J.H.C. Whitehead (1904–1960) in homotopy theory saying that a continuous map between CW-complexes which induces isomorphisms on all homotopy groups is a homotopy equivalence. Since homology groups are easier to compute in general than homotopy groups, the homological version of Whitehead theorem is often convenient to use.

Theorem 17.3.1 (Whitehead theorem in homological form) *Let X and Y be both simply connected CW-complexes and $f : X \to Y$ be a continuous map. If the induced homomorphism $f_* : H_m(X) \to H_m(Y)$ is an isomorphism for each m, then f is a homotopy equivalence.*

Proof Let $C_f = Z$ be the mapping cylinder of f. We may consider f to be an incusion $X \hookrightarrow Z$. Again since the spaces X and Y are both simply connected, it follows that $\pi_1(Z, X) = 0$. The relative version of Hurewicz homomorphism Theorem 17.2.1 asserts that the first nonzero homotopy group $\pi_m(Z, X)$ is isomorphic to the first nonzero homology group $H_m(Z, X)$ for $m > 1$. All the groups $H_m(Z, X)$ are zero from the long exact sequence of homology. This shows that all the groups $\pi_m(Z, X)$ also vanish. This implies that the inclusion $X \hookrightarrow Z$ induces isomorphisms on all homotopy groups. Consequently, this inclusion is a homotopy equivalence. Hence it follows from the diagram in Fig. 17.1 that $f = p \circ i$ is a homotopy equivalence. ❑

Remark 17.3.2 Theorem 17.3.1 gives an alternative proof of Theorem 16.4.11.

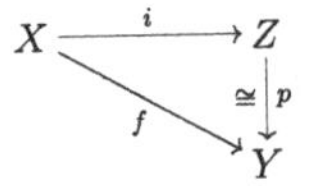

Fig. 17.1 Diagram for Whitehead theorem

17.4 Dold–Thom Theorem

This section studies Dold–Thom Theorem given by Albrecht Dold (1928–2011) and René Thom (1923–2002) in 1958 and its immediate consequences. This theorem presents a homotopy theoretic definition of homolgy. More precisely, let $\mathcal{C}_0$ be the category of base pointed finite CW-complexes and SP^∞ be the infinite symmetric product functor defined in Sect. B.2.5 of Appendix B on $\mathcal{C}_0$. The functor SP^∞ has the interesting property that it can be used to define Eilenberg–MacLane spaces: $SP^\infty(S^n)$ is an Eilenberg–MacLane space $K(\mathbf{Z}, n)$ and for a Moore space $M(G, n)$, the space $SP^\infty(M(G, n))$ is a $K(G, n)$. In particular, $SP^\infty(S^2) = \mathbf{C}P^\infty$ is an Eilenberg–MacLane space $K(\mathbf{Z}_2, 2)$. Dold–Thom theorem establishes a surprising close connection between $\pi_i(SP^\infty(X))$ and $H_i(X; \mathbf{Z})$ for every space X in $\mathcal{C}_0$.

Theorem 17.4.1 (Dold–Thom Theorem) *The functor $X \mapsto \pi_i(SP^\infty(X))$ for $i \geq 1$ coincides with the functor $X \mapsto H_i(X; \mathbf{Z})$ on $\mathcal{C}_0$.*

Proof See Dold and Thom (1958). ❑

Corollary 17.4.2 *For a connected CW-complex X, there is a natural isomorphism*

$$\psi : \pi_n(SP^\infty(X)) \to H^n(X; \mathbf{Z}).$$

for every $n \geq 1$.

Corollary 17.4.3 **(i)** *$SP^\infty(S^n)$ is a $K(\mathbf{Z}, n)$;*
(ii) *For a Moore space $M(G, n)$, $SP^\infty(M(G, n))$ is a $K(G, n)$.*

Proof **(i)** It follows from Dold–Thom Theorem by taking in particular, $X = S^n$.
(ii) It follows from Dold–Thom Theorem by taking $X = M(G, n)$. ❑

Corollary 17.4.4 *A path-connected, commutative, associative H-space X with a strict identity element has the weak homotopy type of a product of Eilenberg–MacLane spaces.*

Proof Left as an exercise.

Corollary 17.4.5 *The functor SP^∞ gives Eilenberg–MacLane spaces.*

Proof If X is a CW-complex, then $SP^\infty(X)$ is path-connected and has the weak homotopy type of $\prod_n K(H_n(X), n)$. Hence the corollary follows from Corollary 17.4.4. ❑

Remark 17.4.6 The map $\pi_n(X) \to \pi_n(SP^\infty(X)) = H_n(X; \mathbf{Z})$ induced by the inclusion $X = SP^1(X) \hookrightarrow SP^\infty X$ is the Hurewicz homomorphism. Using the Hurewicz homomorphism and naturality this reduces to the case $X = S^1$, where the map $SP^n(S^n) \hookrightarrow SP^\infty(S^n)$ induces on π_1 a homomorphism $\mathbf{Z} \to \mathbf{Z}$, which is an isomorphism. The suspension isomorphism makes a further definition to the case $n = 1$, where the inclusion $SP^1(S^1) \hookrightarrow SP^\infty(S^1)$ is a homotopy equivalence and hence it induces an isomorphism on π_1.

Remark 17.4.7 Dold–Thom theorem asserts that $\pi_n(SP^\infty)(X) \cong H_n(X)$ on $\mathcal{C}_0$ for all $n \geq 1$. Hence $\pi_n(SP^\infty(S^2)) = 0$ for all $n > 2$. On the other hand

$$\pi_3(\Omega^2\Sigma^2 S^2) \cong \pi_5(S^4) \cong \mathbf{Z}_2,$$

which is generated by the double suspension of the Hopf map $p : S^3 \to S^2$ (Dold and Thom 1958).

17.5 The Hopf Invariant and Adams Theorem

This section defines Hopf invariant using cup product and discusses Adams theorem which provides a solution of vector field theorem.

17.5.1 Hopf Invariant

H. Hopf (1894–1971) introduced in 1935 the concept of an invariant, now called Hopf invariant. The Hopf invariant has been generalized by G.W. Whitehead in 1950 to a homomorphism

$$H : \pi_m(S^n) \to \pi_m(S^{2n-1}) \ (m \leq 4n - 4)$$

and by P.J. Hilton in 1951 to a homomorphism

$$H : \pi_m(S^n) \to \pi_{m+1}(S^{2n}) \ (m > 0)$$

Homology is used to define the degree of a spherical map $f : S^n \to S^n$ which distinguishes different homotopy classes of maps f. Cup products can be used to define something similar concept for maps $S^{2n-1} \to S^n$. Hopf did this using more geometric constructions prior to the invention of cohomolgy and cup products. There are several definitions of the Hopf invariant $H(f)$ for a continuous map $f : S^{2n-1} \to S^n$. We define $H(f)$ here as one of the most remarkable applications of cup product in topology.

Recall that S^n and S^{2n-1} may be given the structure of finite CW-complexes, each having only two cells. Given a continuous map $f : S^{2n-1} \to S^n$ for $n \geq 2$, let S^n_f denote the space obtained by attaching a $2n$-cell to S^n via f. Then S^n_f is a finite CW-complex with these cells: one is of dimension 0, one is of dimension n and one is of dimension $2n$. Since $n > 1$, the cohomology of S^n_f is given by

$$H^m(S^n_f) \cong \begin{cases} \mathbf{Z}, & \text{if } m = 0, n, 2n \\ 0, & \text{otherwise.} \end{cases}$$

Definition 17.5.1 Let $a \in H^{2n}(S^n_f; \mathbf{Z})$ and $b \in H^n(S^n_f; \mathbf{Z})$ be a chosen pair of generators. The integer $H(f)$, called Hopf invariant is defined to be the integer for which the cup product $b \cup b = b^2 = H(f) \cdot a$ in $H^{2n}(S^n_f; \mathbf{Z})$.

Example 17.5.2 If n is odd then, $H(f) = 0$.

Example 17.5.3 Consider the exact homotopy sequence of the fibration $p : S^3 \to S^2$

$$\cdots \to \pi_3(S^1, s_0) \longrightarrow \pi_3(S^3, s_0) \xrightarrow{p_*} \pi_3(S^2, s_0) \longrightarrow \pi_2(S^1, s_0) \to \cdots$$

Since $\pi_3(S^1, s_0) = \pi_2(S^1, s_0) = 0$, $p_* : \pi_3(S^3, s_0) \to \pi_3(S^2, s_0)$ is an isomorphism. Consequently, $\pi_3(S^2, s_0) \cong \mathbf{Z}$, the first example, where $\pi_m(S^n, s_0) \neq 0$ for $m > n$. Since $\pi_3(S^3, s_0)$ is generated by $[1_{S^3}]$, it follows that $\pi_3(S^2, s_0)$ is generated by $[p]$. The map p is called the Hopf map.

For each q, consider the isomorphisms Σ

$$\pi_{2q+2}(S^{q+2}, s_0) \xrightarrow{\Sigma} \cong \pi_{2q+3}(S^{q+3}, s_0) \xrightarrow{\Sigma} \cong \cdots \xrightarrow{\Sigma} \cong \pi_{q+n}(S^n, s_0) \xrightarrow{\Sigma} \cong \cdots$$

For each $q > 1$ the common group $\pi_{n+q}(S^n, s_0)$ is denoted by by π^S_q. It is called the kth stable homotopy group. For example, $\pi^S_1 \cong \pi_4(S^3, s_0) \cong \mathbf{Z}_2$ and is generated by $\Sigma[p]$, where $p : S^3 \to S^2$ is the Hopf map.

The exceptional case $\pi_{4n-1}(S^{2n}, s_0)$ invites attraction in many respects. The homomorphism

$$H : \pi_{4n-1}(S^{2n}, s_0) \to \mathbf{Z},\ f \mapsto H(f)$$

defined by Hopf is now called the Hopf invariant.

Remark 17.5.4 If n is odd, then $H(f) = 0$, because of anticommutativity of the cup product.

Definition 17.5.5 Let X be the CW-complex obtained by attaching a $(2n+2)$-cell to S^n using f as the attaching map. Then

$$H^m(X) \cong \begin{cases} \mathbf{Z}, & \text{if } m = 0, n+1 \text{ and } 2n+2 \\ 0, & \text{otherwise} \end{cases}$$

Remark 17.5.6 The elements of $\pi_3(S^2)$ may be given a geometrical interpretation by assigning an integer (Hopf integer) to each element of $\pi_3(S^2)$.

An alternative definition of Hopf invariant given by N. Steenrod in 1949 is now conveyed.

Definition 17.5.7 (*Steenrod*) Given an element $\alpha \in \pi_{2n-1}(S^n)(n \geq 1)$, the Hopf invariant of α is also defined as follows: Represent α by a map $f : S^{2n-1} \to S^n$ and let $Y = C_f$. Then $H^n(Y) \cong \mathbf{Z}$, $H^{2n}(Y) \cong \mathbf{Z}$, the generators being τ_n, τ_{2n}, where $f_1^*(\tau_n) = s_n, \tau_{2n} = f_2^*(s_{2n})$, and s_n, s_{2n} are the generators of $H^n(S^n), H^{2n}(S^{2n})$ respectively. The Hopf invariant of α denoted by $H(\alpha)$ is the defined by

$$\tau_n^2 = \tau_n \cup \tau_n = H(\alpha) \cdot \tau_{2n}.$$

$H(\alpha)$ is an integer, called the Hopf invariant of α. This integer $H(\alpha)$ depends only on α in the sense that it does not depend on the choice of its representative f and hence it is well defined.

Remark 17.5.8 Consider a continuous map $f : S^{2n-1} \to S^n$ for $n \geq 1$. Then there exists a unique integer integer $H(f)$, called the Hopf invariant of f.It depends only on the homotopy class of f. The assignment

$$H : \pi_{2n-1}(S^n) \to \mathbf{Z},\ f \mapsto H(f)$$

is a homomorphism such that for $n = 2, 4, 8,\ H(f) = 1$. Its converse is also true: up to homotopy, the Hopf maps are the only ones of Hopf invariant 1. This proves the purely algebraic theorem that **C**, **H**,, and Cayley numbers are the only nontrivial real division algebra. (see Remark 17.5.11).

Remark 17.5.9 The Definition 17.5.7 is due to Steenrod given in 1947. It is some what different from Hopf original definition.

17.5.2 Vector Field Problem and Adams Theorem

The problem for which n there exists a continuous map $f : S^{2n-1} \to S^n$ with Hopf invariant $H(f) = 1$ was solved by J.F. Adams (1930–1989) in his papers (Adams 1958, 1960). This theorem, also called Adams' theorem, is a deep theorem in homotopy theory which states that the only n-spheres which are H-spaces are S^0, S^1, S^3, and S^7. This relates to the existence of division algebra structure on Euclidean space $\mathbf{R}^n$. A division algebra is a finite dimensional real vector space together with a bilinear multiplication having both-sided identity and such that each nonzero element has a both-sided multiplicative inverse. The real numbers, complex numbers, the real quaternions, and the Cayley numbers are examples of real division algebras. J.F. Adams proves that there are no other examples. Corresponding to each continuous map $f : S^{2n-1} \to S^n$ one can associate an integer $H(f)$, called Hopf invariant. This means that to each element of $\pi_{2n-1}(S^n)$, one can assign an integer which is its Hopf invariant.

Theorem 17.5.10 (Adams) *There exists a continuous map* $f : S^{2n-1} \to S^n$ *with Hopf invariant one only when* $n = 2, 4, 8$.

Proof See Adams (1958, 1960). ❑

Remark 17.5.11 The Definition 17.5.7 is due to Steenrod given in 1947. It is some what different from Hopf original definition. For any integer $n > 0$, that there exists continuous S^{4n-1} to S^{2n} of arbitrary even Hopf invariant. Does there exist maps having odd Hopf invariant? Hopf maps $S^3 \to S^2$ and $S^7 \to S^4$ are each of Hopf invariant one. Using Cayley numbers, one can define an analogous map from $S^{15} \to S^8$ of Hopf invariant one. Adams showed in 1952 that there exist maps $f : S^{4n-1} \to S^{2n}$ of odd Hopf invariant using cohomology operations only when n is a power of 2. If $\pi_{2n-1}(S^n)$ contains no element of Hopf invariant one, then there is no real division algebra of dimension n. Adams proved in 1960 that such elements exist precisely for $n = 1, 2, 4, 8$, whose simpler proof is given by (Atiyah 1967) using K-theory.This result asserts purely algebraic theorem that $\mathbf{R}$, C, H and Cayley numbers are the only real division algebras (see Steenrod and Epstein 1962).

17.6 Exercises

1. Let $f : X \to \Omega Y$ be a continuous map and $\tilde{f} : \Sigma X \to Y$ be adjoint to f. Show that the diagram in Fig. 17.2 commutes.
2. Show that the diagram in Fig. 17.3 commutes up to sign.
3. If X is path-connected, show that there is an epimorphism

$$h : \pi_1(X) \to H_1(X; \mathbf{Z}).$$

Fig. 17.2 Commutative square involving $\tilde{f}_*$ and f_*

$$\begin{array}{ccc} \pi_{m+1}(\Sigma X) & \xrightarrow{\tilde{f}_*} & \pi_{m+1}(Y) \\ \downarrow E & & \downarrow \cong \\ \pi_m(X) & \xrightarrow{f_*} & \pi_m(\Omega Y) \end{array}$$

Fig. 17.3 Diagram involving E and σ

$$\begin{array}{ccc} \pi_n(X) & \xrightarrow{E} & \pi_{n+1}(\Sigma X) \\ \downarrow h & & \downarrow h \\ \widetilde{H}_n(X) & \xrightarrow[\cong]{\sigma} & \widetilde{H}_{n+1}(\Sigma X) \end{array}$$

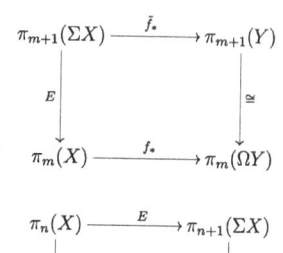

4. If X is path-connected, show that the homomorphism

$$h : \pi_1(X) \to H_1(X; \mathbf{Z})$$

induces an isomorphism

$$h_* : \pi_1(X)/kerh \to H_1(X; \mathbf{Z})$$

with ker h, the kernel of h the commutator subgroup of $\pi_1(X)$.

5. If X is a simply connected CW-complex, and

$$\widetilde{H}_n(X) = \begin{cases} \mathbf{Z}, & \text{if } n = m \\ 0, & \text{if } n \neq m, \end{cases}$$

show that $X \simeq S^m$.

6. Given an element $\alpha \in \pi_{2n-1}(S^n)(n \geq 1)$, let $H(\alpha)$ be the Hopf invariant of α. Prove that

- **(i)** $H(\alpha)$ depends only on α, and not on the choice of its representative $f : S^{2n-1} \to S^n$;
- **(ii)** $H : \pi_{2n-1}(S^n) \to \mathbf{Z}$, $f \mapsto H(f)$ is a homomorphism;
- **(iii)** $H(f) = 0$ if n is odd;
- **(iv)** if n is even, $H[\beta_n, \beta_n] = \pm 2$, where β_n is the generator of $\pi_n(S^n)$ represented by the identity map;
- **(v)** Deduce that if n is even, then $2 \in$ Im H and $\pi_{2n-1}(S^n)$ has an element of infinite order, and S^n can not be a Hopf space.

7. For any integer $n > 0$, show that there exists a continuous map f from S^{4n-1} to S^{2n} of arbitrary even Hopf invariant $H(f)$.

8. Show that two continuous maps from S^3 to S^2 are homotopic iff they have the same Hopf invariant.

9. Let $f : X \to Y$ be a continuous map of simply connected CW-complexes. Show that f is a homotopy equivalence if and only if $f_* : H_i(X) \cong H_i(Y)$ for all i.

10. Let (X, A) be a pair of simply connected CW-complexes. If $H_m(X, A) = 0$ for $m < n, n \geq 2$, then show that $\pi_m(X, A) = 0$ for $m < n$ and the Hurewicz homomorphism

$$h : \pi_n(X, A) \to H_n(X, A)$$

is an isomorphism.

11. (Homological description of Moore space) Given an abelian group G and an integer $n \geq 2$, a pointed CW-complex X is said to be a Moore space of type (G, n) if X is n-connected and

$$\widetilde{H}_i(X) = \begin{cases} G, & \text{if } i = n \\ 0, & \text{if } i \neq n \end{cases}.$$

A Moore space of type (G, n) or any space homeomorphic to it is denoted by $M(G, n)$. For example, S^n is a Moore space of type (G, n).
Show that

(i) The Hurewicz homomorphism

$$h_n : \pi_n(M(G, n)) \to H_n(M(G, n))$$

is an isomorphism.

(ii) $\pi_n(M(G, n)) = G$.

(iii) If $f : G \to H$ is a homomorphism of abelian groups, then there exists a continuous map

$$\psi : M(G, n) \to M(H, n)$$

such that its induced homomorpshim

$$\psi_* : H_n(M(G, n)) \to H_n(M(H, n))$$

coincides with f.

12. Let $f : X \to Y$ be a map between simply connected spaces. If $f_* : H_m(X) \to H_m(Y)$ is bijective if $m < n$ and surjective if $m = n(n \geq 2)$, show that

$$f_* : \pi_m(X) \to \pi_m(Y)$$

is bijective for $m < n$ and surjective for $m = n$.

13. Show that for an integer $n \geq 1$, there exist maps

$$f : S^{4n-1} \to S^{2n}$$

of arbitrary even Hopf invariant $H(f)$.

14. Show that the image of Hopf invariant $H : \pi_{4n-1}(S^{2n}) \to \mathbf{Z}$ is either $\mathbf{Z}$ or $2\mathbf{Z}$.

15. (Hopf) For each continuous map $f : S^3 \to S^2$, show that there exists an integer $H(f)$ with the properties:

(i) if $f, g : S^3 \to S^2$ are homotopic, then $H(f) = H(g)$;

(ii) If $h : S^2 \to S^3$ is a continuous map with degree n, then $H(foh) = n.H(f)$;

(iii) There exists a map $g : S^3 \to S^2$ such that $H(g) = 1$;

(iv) If there exists a Hopf invariant $H(f)$ with the above properties, then the map

$$\psi : \pi_3(S^2) \to \mathbf{Z}, [f] \mapsto H(f)$$

is surjective.

16. Let τ_{2n} be a generator of the group $\pi_{2n}(S^{2n})$ and $[\tau_{2n}, \tau_{2n}] \in \pi_{4n-1}(S^{2n})$ be the Whitehead product. Show that

(i) the Hopf invariant $H([\tau_{2n}, \tau_{2n}]) = 2$;
(ii) the element $[\tau_{2n}, \tau_{2n}]$ has infinite order;
(iii) the group $\pi_{4n-1}(S^{2n})$ is infinite for every $n \geq 1$.

17.7 Additional Reading

[1] Dieudonné, J., *A History of Algebraic and Differential Topology*, 1900–1960, Modern Birkhäuser, 1989.
[2] Eilenberg, S., and Steenrod, N., *Foundations of Algebraic Topology*, Princeton University Press, Princeton, 1952.
[3] Switzer, R.M., *Algebraic Topology:Homotopy and Homology*, Springer-Verlag, Berlin, Heidelberg, New York, 1975.

References

Adams, J.F.: On the nonexistence of elements of hopf invariant one. Bull. Am. Math. Soc. **64**, 279–282 (1958)
Adams, J.F.: On the nonexistence of elements of hopf invariant one. Ann. Math. **72**, 20–104 (1960)
Adams, J.F.: Algebraic Topology: A student's Guide. Cambridge University Press, Cambridge (1972)
Atiyah, M.F.: K-Theory. Benjamin, Elmsford (1967)
Dieudonné, J.: A History of Algebraic and Differential Topology, 1900–1960. Modern Birkhäuser, Basel (1989)
Dold, A., Thom, R.: Quasifaserungen und unendliche symmetrische Produkte. Ann. Math. **67**(2), 239–281 (1958)
Eilenberg, S., MacLane, S.: Relations between homology and homotopy groups of spaces, Ann. Math. **46**(2) (1945)
Eilenberg, S., Steenrod, N.: Foundations of Algebraic Topology. Princeton University Press, Princeton (1952)
Gray, B.: Homotopy Theory, An Introduction to Algebraic Topology. Acamedic Press, New York (1975)
Hatcher, A.: Algebraic Topology. Cambridge University Press, Cambridge (2002)
Maunder C.R.F: Algebraic Topology, Van Nostrand (1970). Cambridge University Press, Cambridge (1980)
Spanier, E.: Algebraic Topology. McGraw-Hill, New York (1966)
Steenrod, N.: Cohomology Invariants Mapp. **50**, 954–988 (1949)
Steenrod, N., Epstein, D.B.A.: Cohomolgy operations, Annals of Mathematics Studies, vol. 50. Prineton University Press, Princeton (1962)
Switzer, R.M.: Algebraic Topology-Homotopy and Homology. Springer, Berlin (1975)

Chapter 18
A Brief History of Algebraic Topology

This chapter focuses the history on the emergence of the ideas leading to new areas of study in algebraic topology and conveys the contributions of some mathematicians who introduced new concepts or proved theorems of fundamental importance or inaugurated new theories in algebraic topology starting from the creation of homotopy, fundamental group, and homology group by H. Poincaré (1854–1912) in 1895, which are the first most profound and far reaching inventions in algebraic topology. This subject arose through the study of the problems in mathematical analysis and geometry in Euclidean spaces, particularly, through Poincaré's work in the classification of algebraic surfaces. An important feature in the history of algebraic topology is that the concepts initially introduced in homology and homotopy theories for applications to problems of topology have found fruitful applications to other areas of mathematics and have become the starting points of various theories: category theory, homological algebra and K-theory are outstanding examples. The term "topology" was given by J.B. Listing (1802–1882) in 1862 instead of previously used "Analysis situs". The subject "topology" was studied by C. Felix Klein (1849–1925) in his "Erlangen Program" in 1872 and considered the invariants of arbitrary continuous transformation, a kind of geometry. He classified geometries by their underlying symmetry groups, and this classification greatly influenced the synthesis of the mathematics. J.W. Alexander (1888–1971) used 'topological' in the titles of his research papers in the twenties.

The basic problem in algebraic topology is to devise ways to assign various algebraic objects such as groups, rings, modules to topological spaces and homomorphisms to the corresponding structures in a functorial way. The literature on algebraic topology is very vast. Properties and characteristics which are shared by homeomorphic spaces are called topological properties and topological invariants; on the other hand those by homotopy equivalent spaces are called homotopy properties and homotopy invariants. The Euler characteristic invented by L. Euler (1707–1783) in 1752 is an integral invariant, which distinguishes non-homeomorphic spaces. The search of other invariants has established connections between topology and modern

M.R. Adhikari, *Basic Algebraic Topology and its Applications*,
DOI 10.1007/978-81-322-2843-1_18

algebra in such a way that homeomorphic spaces have isomorphic algebraic structures. Historically, the concept of fundamental group introduced by Poincaré in 1895 is the first important invariant of homotopy theory which came from such a search. His work explained the difference between curves deformable to one another and curves bounding a larger space. The first one led to the concepts of homotopy and fundamental group and the second one led to homology. Some concepts studied now in algebraic topology had been found in the work of B. Riemann (1826–1866), C. Felix Klein and H. Poincaré. But the foundation of algebraic (combinatorial topology) was laid in the decade beginning 1895 by H. Poincaré through publication of his famous series of memoirs 'Analysis Situs' during the years 1895–1904. His motivation was to solve specific problems involving paths, surfaces, and geometry in Euclidean spaces. His vision of the key role of topology in all mathematical theories began to materialize from 1920.

For this chapter the books and papers Adams (1960, 1962, 1972), Atiyah (1967), Barratt (1955), Brown (1962), Dieudonné (1960), Dold and Thom (1958), Dold (1962), Eilenberg and MacLane (1942, 1945a), Eilenberg and Steenrod (1952), Freudenthal (1937), Hopf (1935), Hurewicz (1935), James (1999), Maunder (1970), Poincaré (1895, 1900, 1904), Whitehead (1941, 1949, 1950), Whitehead (1950, 1953, 1962) and some others are referred in the Bibliography.

18.1 Poincaré and his Conjecture

H. Poincaré born in France is the first mathematician who systemically attacked the problems of assigning topological invariants to topological spaces. He also first introduced the basic concepts and invariants of combinatorial topology, such as Betti numbers and the fundamental group. He proved a formula relating the number of edges, vertices and faces of n-dimensional polyhedron (the Euler-Poincaré theorem) and gave the first precise formulation of the intuitive notion of dimension. The monumental work of Poincaré embodied in "Analysis situs", Paris, 1895 introduced the concepts of homotopy, fundamental group and homology. He is the first mathematician who applied algebraic objects in homotopy theory. His work organized the subject algebraic topology for the first time which has been discussed in earlier chapters. At the beginning his work did not attract mathematicians to a great extent but his promising work with his vision attracted mathematical community since 1920. This subject is an interplay between topology and algebra and studies algebraic invariants provided by homotopy and homology theories. The twentieth century witnessed its greatest development.

Poincaré remarked in 1912 "Geometers usually distinguish two kinds of geometry, the first of which they qualify as metric and the second as projective.· · · . But it is a third.· · · ; this is analysis situs. In this discipline, two figures are equivalent whenever one can pass from one to the other by a continuous deformation; whatever else the law of this deformation may be, it must be continuous. Thus a circle is equivalent to an ellipse or even to an arbitrary closed curve, but it not equivalent to a straight line segment since this segment is not closed. A sphere is equivalent to a convex surface; it is not equivalent to a torus since there is a hole in a torus and in a sphere there is not."

The idea of homotopy for the continuous maps of unit interval was originated by C. Jordan (1838–1922) in 1866 and that of for loops was introduced by H. Poincaré in 1895 to define an algebraic invariant called the fundamental group. H. Poincaré may be regarded as the father of algebraic topology. The concept of fundamental groups invented by H. Poincaré in 1895 conveys the first transition from topology to algebra by assigning an algebraic structure on the set of relative homotopy classes of loops in a functorial way. Fundamental group is a basic very powerful invariant in algebraic topology and is the first of a series of algebraic invariants π_n associated with a topological space with a base point. Historically, the concept of fundamental group introduced by Poincaré is the first important algebraic invariant of homotopy theory. This group provides information about the basic shape, or holes, of the topological space. His work in algebraic topology is mainly in geometric terms.

Poincaré posed in 1904 a conjecture which is called Poincaré conjecture. This conjecture asks whether a simply connected compact n-manifold having the same homology groups as S^n is homeomorphic to S^n? It is not hard to show that the conjecture is true for $n = 2$.

(i) For $n = 4$ the conjecture was proved to be true by M. Freedman (1951-) in 1982 and he was awarded the 1986 Fields medal for this work.

(ii) For $n = 5$ the conjecture was demonstrated by Christopher Zeeman (1925–2016) in 1961.

(iii) For $n = 6$ the conjecture was proved to be true by John R. Stallings (1935–2008) in 1962.

(iv) For $n \geq 7$ the conjecture was established by Stephen Smale (1930-) in 1961. He subsequently extended his proof for all $n \geq 5$ independently. He was awarded the Fields Medal in 1966 for this work.

(v) For $n = 3$, (its equivalent statement asserts that a compact 3-manifold homotopically equivalent to S^3 is homeomorphic to S^3) the problem has been solved by Grigori Yakovlevich Perelman (1966-) in year 1994. He was offered the 2006 Fields Medal for his contributions to geometry and his revolutionary insights into the analytical and geometric structure of the Ricci flow. But he declined to accept the award or to appear at the Spain ICM 2006. The scientific journal "Science" declared on 22nd December 2006 Perelman's proof of the Poincaré conjecture as the scientific "Breakthrough of the Year 2006", such recognition is possibly the first in the area of mathematics.

Poincaré also made significant contributions in algebra, differential equations, complex analysis, algebraic geometry, celestial mechanics, mathematical physics, philosophy of mathematics and popular science through his publication of 30 books and over 500 papers. Of course, some of the ideas which Poincaré developed had of course their origins prior to him, with L. Euler, and B. Riemann above all.

18.2 Early Development of Homotopy Theory

This section conveys the early development of homotopy theory arising through the work H. Poincaré, L.E.J. Brouwer (1881–1967), H. Hopf (1894–1971), W. Hurewicz (1904–1956), H. Freudenthal (1905–1990) and some others. Topologists regard Poincaré as the founder and H. Hopf and Hurewicz as the cofounders of algebraic topology. The concept of homotopy presents a mathematical formulation of the intuitive idea of a continuous transition between two geometrical configurations. On the other hand, the concept of homology presents a mathematical precision to the intuitive idea of a curve bounding an 'area' or a surface bounding a 'volume'. Algebraic topology attempts to measure degrees of connectivity by using homology and homotopy groups.

The concept of homotopy, at least for maps of the unit interval I was given by C. Jordan in 1886. The word 'homotopy' was first given by Max Dehn (1878–1952) and Paul Heegaard (1871–1948) in 1907. The Jordan Curve Theorem, a classical theorem, was first stated by Jordan in 1892. His paper contained some gaps. Its first rigorous proof given by Oswald Veblen (1880–1960) in 1905 is one of the greatest developments of algebraic topology.

The importance of homotopy theory was realized in 1930 with the discovery of the Hopf map with his striking result $\pi_3(S^2) \neq 0$. Prior to him homotopy theory was used as a secondary tool for the homology theory except for the fundamental group. Hopf fibering given by H. Hopf through his work during 1935–1941 plays an important role in the study of algebraic topology. The Hopf maps $p : S^{2n-1} \to S^n$ for $n = 2, 4, 8$ introduced by Hopf in 1935 are utilized to study certain homotopy groups of spheres such as $\pi_3(S^2) \neq 0$, $\pi_7(S^4) \neq 0$ and $\pi_{15}(S^8) \neq 0$. Homotopy theory is used in solving many of the old problems of classical topology. The fundamental groups are deeply connected with covering spaces. Historically, the systemic study of covering spaces appeared during the late 19th century and early 20th century through the theory of Riemann surfaces. But its origin was found before the invention of the fundamental groups by H. Poincaré in 1895. Poincaré introduced the concept of universal covering spaces in 1883 to prove a theorem on analytic functions.

Some analytical geometric tools are required for development of algebraic topology. These are: simplex, complex, subcomplex, simplicial map, triangulation, polyhedron and simplicial approximation. Simplicial complexes introduced by J.W. Alexander provide useful tools in computing fundamental groups of simple compact spaces as well as for the study of manifolds. For example, Poincaré duality theorem is one of the earliest theorems in topology. Simplicial complexes form building blocks of homology theory. The simplicial approximation theorem given by Brouwer and Alexander around 1920 by utilizing a certain good feature of simplicial complexes plays a key role in the study of homotopy and homology theories. The combinatorial device, now called abstract complex was systematically used by W. Mayer (1887–1947) in 1923.

It is easy to define algebraic invariants such as fundamental groups, higher homotopy groups and homology groups, etc., but difficult to compute them for different classes of topological spaces. To facilitate such computation, Chap. 6 conveys a special class of topological spaces that can be broken up into pieces which fit together in a nice way. Such spaces are called triangulable spaces. The concept of triangulation is also utilized to compute homology groups of triangulable spaces and to solve extension problems. The concept of edge-group $E(K, v)$ (which is isomorphic to the fundamental group $\pi_1(|K|, v)$ for any simplicial complex K) is applied to graph theory. For example, van Kampen theorem for fundamental groups given by E. van Kampen (1908–1942) in 1933 is proved by using graph-theoretic results.

The extension of 'Euler characteristic' was given by A.L. Cauchy (1789–1857) in 1813 and H. Poincaré in 1895. The study of 1-dimensional and 2-dimensional simplicial complexes dates back at least to Euler and that of higher-dimensions first appeared in the work of J. Listing (1808–1882) in 1862. Listing is the first mathematician who used the word 'topology' in his famous article published in 1847, although he had used the term in some of his previous correspondence.

Except for fundamental group, the subject of homotopy was first found in the work of L.E.J. Brouwer who made the first step in 1912 towards connecting homology and homotopy groups of certain spaces which are the two basic concepts of algebraic topology by showing that two continuous maps of a two-dimensional sphere into itself can be continuously deformed into each other if and only if they have the same degree (that is, if they are equivalent from the view point of homology theory). His definition of the degree of a spherical map is more intuitive than its definition from the view point of homology theory. He defined $degf$, the degree of f as the number of times of the domain sphere wraps around the range sphere and proves its homotopy invariance. He showed that for self maps of S^n, the homotopy class of a continuous map is characterized by its degree. His definition shows that if $f : S^1 \to S^1, z \mapsto z^n$, then $\deg f = n$. The most celebrated results of Brouwer include the proof of the topological invariance of dimension of $\mathbf{R}^n$, Brouwer fixed point theorem (the theorem is named after his name) and the simplicial approximation theorem. His results are important in the foundations of algebraic topology.

After having generalized Brouwer's result of degree mapping of a continuous map of a two-dimensional sphere into itself to an arbitrary dimension, H. Hopf made a systematic study of the problem of classifying the continuous maps of polytopes (a polytope is the union of finitely many simplices, with the additional property that, for any two simplices that have a nonempty intersection, their intersection is a vertex, edge, or higher-dimensional face of the two). His work is mainly based on highly geometric intuitions like Brouwer. He continued the work of Brouwer by using the degree mapping and the homotopy class of a map as tools. Hopf characterized homotopy class of self maps of a sphere by their degrees. His definition of the degree of a spherical map $f : S^n \to S^n$ is more intuitive than its definition from the view point of homology theory. His definition shows that if $f : S^n \to S^n, n \geq 1$ is a constant map, then $\deg f = 0$ and if $f : S^n \to S^n, n \geq 1$ is the identity map, then $\deg f = +1$.

He proved its homotopy invariance and gave a complete homotopy classification of mappings of n-dimensional polytopes into the n-dimensional sphere S^n.

W. Hurewicz made remarkable contributions to algebraic topology. His invention of the higher homotopy groups π_n in 1935–1936 is a natural generalization of the fundamental group to higher-dimensional analogue of the fundamental group. π_n is a sequence of functors given by W. Hurewicz in 1935 from topology to algebra by extending the concept of fundamental group. Historically, Hurewicz introduced higher homotopy groups by defining a group structure in $\pi_n(X) = [S^n, X]$. He showed that the higher homotopy groups $\pi_n(X)$ are all abelian for $n \geq 2$ though $\pi_1(X)$ is in general not abelian. By an action of π_1 on π_n it is also proved that for a simply connected space X, the group $\pi_n(X, x_0) \cong [S^n, X]$.

Hurewicz establishes a close connection between homotopy and homology groups of a certain class of spaces through Hurewicz homomorphism defined by Hurewicz in 1935 in his paper Hurewicz (1935). He first asserted that for a simplicial pair (K, L) if $\pi_r(K, L) = 0$ for $1 \leq r < n (n \geq 2)$, then $\pi_n(K, L) \to H_n(K, L)$ is an isomorphism. This work cast light for the first time onto the relationship between homological and homotopical invariants. A series of four papers of Hurewicz published during 1935–1936, has greatly influenced the development of the modern homotopy theory. More precisely, Hurewicz is the first mathematician who established a connection in 1935 between homology groups $H_n(X)$ and homotopy groups $\pi_n(X)$ for $(n-1)$-connected spaces X, when $n \geq 2$. by defining group homomorphisms $h : \pi_n(X) \to H_n(X)$ for all n. This result known as Hurewicz theorem asserts that the first nonzero homotopy groups $\pi_n(X)$ of a simply connected space X is isomorphic to the first nonzero homology group of $H_n(X)$ for $n \geq 2$. Their relative version is also similar. The one-dimensional case of this theorem was already given by Poincaré, who established that the homology relation appears after abelianizing the homotopy relation. His celebrated result is the "Equivalence theorem": if a space X has homotopy groups $\pi_n(X) = 0$ for $1 \leq n \leq n_0$ for some $n_0 > 1$, then $H_n(X) = 0$ for $1 \leq n \leq n_0$ and $\pi_{n_0+1}(X) \cong H_{n_0+1}(X)$. He also introduced the concepts of homotopy equivalence and homotopy equivalent spaces.

For the study of spaces X of low dimension, the fundamental group $\pi_1(X)$ is very useful. But it needs refined tools for the study of higher-dimensional spaces. For example, fundamental group can not distinguish spheres S^n with $n \geq 2$. Such limitation of low dimension can be removed by considering the natural higher-dimensional analogues of $\pi_1(X)$ given by Hurewicz. His another invention is the long exact homotopy sequence for fibrations in 1941, and the fundamental theorem, known as Hurewicz theorem which connects homotopy and homology groups in 1935. His work led to homological algebra. The homotopy extension property (HEP) and its dual the homotopy lifting property (HLP) play critical role in homotopy theory. There are various notions of fibrations in algebraic topolgy but the work of Hurewicz shows that a fiber map is simply a continuous map which has the HLP for arbitrary topological spaces.

The homotopy equivalent relation is much coarser than the relation of homeomorphism and hence is more accessible to classification. For example, the disk D^n is of the same homotopy type of a single point $\{p\} \subset D^n$ but D^n is not homeomorphic to $\{p\}$. The higher homotopy groups and homology groups are invariants of the the homotopy equivalence class of a space. This concept has offered a new foundation for the development of combinatorial invariants of spaces and manifolds. His contributions made breakthrough in the field of topology.

By a synthesis of the work of Hopf and Hurewicz, H. Freudenthal proved the completeness of Hopf 's classification and discovered the suspension map in around 1935. Since then the study of homotopy of spheres comes up a challenging field of research of many topologists including Hopf. Freudenthal made a breakthrough in algebraic topology by establishing a theorem in 1937 known as Freudenthal suspension theorem while investigating the homotopy groups $\pi_m(S^n)$ for $0 < m < n$. J.H.C. Whitehead (1904–1960) introduced the concept of simple homotopy theory, which has developed through algebraic K-theory.

Freudenthal also studied the nth cohomotopy set $\pi^n(X, A)$ on which K. Borsuk (1905–1982) endowed in 1936 the abelian group structure under certain conditions on (X, A) (Borsuk 1936). For each integer $p > \frac{n+1}{2}$, the cohomotopy groups of a compact pair (X, A) of finite dimension n, the pth cohomotopy group $\pi^p(X, A)$ is defined by $\pi^p(X, A) = [(X, A), (S^p, s_0)]$, which is an abelian group. The set $\pi^p(X, A)$ is defined for $p \geq 0$, but addition operation in it is defined for $p > \frac{n+1}{2}$. If $f : (X, A) \to (Y, B)$ is continuous, it induces maps

$$f^* : \pi^p(Y, B) \to \pi^p(X, A)$$

for all p, which are homomorphisms when both sides are groups. Again a mapping $\delta : \pi^{p-1}(A) \to \pi^p(X)$ is defined for $p > 0$, which is a homomorphism when both sides are groups. The basic difference between cohomology and cohomotopy is the lack of a group structure in $\pi^p(X, A)$ for $p \leq \frac{n+1}{2}$. So it is not possible to compute cohomotopy group by an induction on p starting with $p = 0$. The cohomotopy groups defined by K Borsuk in 1936 was deeply studied by Spanier in his paper Spanier (1949). Spanier has shown that with the induced homomorphism and the coboundary operator, the cohomotopy groups satisfy all the Eilenberg–Steenrod axioms for cohomology theory, and emphasized the importance of the cohomotopy groups. His investigation, however, has been restricted to the case of compact spaces.

Lens spaces defined by H. Tietze (1880–1964) in 1908 form an important class of 3-manifolds in the study of their homotopy classification. Subsequently, K. Reidemeister (1893–1971) established their topological classification in 1935 and J.W.C. Whitehead gave their homotopical classification in 1941. Tietze gave a finite presentation for the fundamental group and proved the topological invariant of fundamental groups. He also contributed to the development of knot theory, Jordan curves, cell complexes and even general topology which has now entered in the premises of analysis.

18.3 Category Theory and CW-Complexes

This section conveys the concepts of category theory and CW-complexes which play a key role in the development of algebraic topology. Category theory is very important in mathematics to unify different concepts in mathematics. It conveys a key language in algebraic topology. The concepts of categories, functors, natural transformations and duality introduced and studied during 1942–1945 by S. Eilenberg (1913–1998) and S. Mac-Lane (1909–2005) form category theory (Eilenberg and MacLane 1942, 1945). Originally, the purpose of these notions was to provide a technique for classifying certain concepts such as that of natural isomorphism. The whole idea of category theory arose through the field of algebraic topology. The first and the simplest realization of this idea is the fundamental group (or Poincaré group) of a pointed space. Many concepts of algebraic topology are unified and explained by category theory, and it plays a key role for the study of homotopy, homology and cohomology theories which constitute the basic text of this book, in addition to adjoint functor, representable functor, abelianization functor, Brown functor, and infinite symmetric product functor which are important functors in the study of algebraic topology.

J.H.C. Whitehead constructed a new category in 1949 in his paper Whitehead (1949), which is now called the category of CW-complexes. The concept of CW-complexes is a natural generalization of the concept of polyhedra, where cells are attached by arbitrary continuous maps starting with a discrete set, whose points are each regarded as a 0-cell. CW-complexes built up by successive adjunctions of cells of dimensions $1, 2, 3, \ldots$. CW-complexes form an extensive class of topological spaces for the study of algebraic topology, where a weak equivalence is necessarily a homotopy equivalence. CW-complexes carry excellent combinatorial properties which are flexible than simplicial complexes. The existence of Eilenberg and MacLane spaces was shown by J.H.C. Whitehead in 1949 by using the properties of CW-complexes.

CW-complexes give a convenient setting for homotopy theory. It is surprising that the homotopy groups of CW-complexes supply a vast information. For example, Whitehead theorem asserts that if a continuous map $f : X \to Y$ between connected CW-complexes induces isomorphisms $f_* : \pi_n(X) \to \pi_n(Y)$ for all n, then f is a homotopy equivalence. Again every space is not a CW-complex but for many purposes it is sufficient to consider only CW-complexes. This conclusion follows from a theorem of Whitehead established in 1950 that says that given any topological space X, there exists a CW-complex K and a weak homotopy equivalence $f : K \to X$. The cellular approximation theorem is an analogue result of simplicial approximation theorem.

The origin of CW-complexes closely relates to the birth of many concepts and development of algebraic topology in general. There are many advantages of CW-complexes over polyhedra. Since all CW-complexes are paracompact and all open coverings of a paracompact space are numerable, the results on the homotopy classification of principal G-bundles hold for locally trivial principal G-bundles over a

CW-complex. Algebraic topologists feel that the category of CW-complexes is a good category for homotopy and homology theories.

J.H.C. Whitehead defined a product between two groups in 1941 to study homotopy groups of pointed topological spaces X. This product associates with elements $\alpha \in \pi_p(X, x_0)$ and $\beta \in \pi_q(X, x_0)$ an element $[\alpha, \beta] \in \pi_{p+q-1}(X)$, called Whitehead product in his honor. This product provides a technique at least in some cases for constructing nonzero elements of $\pi_{p+q-1}(X)$. He also defined generalized products involving the rotation groups. Whitehead product is used to solve several problems proving many amazing results in algebraic topology.

18.4 Early Development of Homology Theory

This section conveys early development of homology theory starting from simplicial homology defining the homology groups of a polyhedron by Poincaré in 1895, followed by several generalizations of his homology beyond polyhedra during the period 1927–1944. The extensions were made by several topologists such as L. Vietoris (1891–2002) in 1927 (for compact metric spaces), E. Čech (1893–1960) in 1932 (for compact Hausdorff spaces), S. Lefschetz (1884–1972) in 1933, S. Eilenberg in 1944 (for arbitrary topological spaces). Singular homology of an arbitrary topological space constructed by Eilenberg in 1944 is the most powerful homology.

Homology theory plays a key role in algebraic topology. The basic tools such as complexes and incidence numbers are necessary for constructing simplicial homology groups as defined by Poincaré. To inaugurate a homology theory, Poincaré started in 1895 with a geometric object (a space) which is given by combinatorial data (a simplicial complex). Then the linear algebra and boundary relations by these data are used to construct homology groups, called simplicial homology groups. This theory stemmed from his 'Analysis Situs'. Using these tools Poincaré defined directly the Betti numbers invented by E. Betti (1823–1892) and torsion numbers which are numerical invariants and characterized the homology groups based on the coefficient group $\mathbf{Z}$ of integers. The concept of relative homology (modulo a subcomplex) was given by Lefschetz in 1927 and the operator ∂ was used by Lefschetz. Attention for shift from numerical invariants to groups associated with homology theories was successfully made during the period 1925–1935. This shift is partly due to Emmy Noether (1882–1935). The algebraic approach of Noether to homology is a fruitful contribution to the geometrical approach of Poincaré. Inspired by the above approach of Noether, P. Alexandroff (1896–1982) and H. Hopf gave jointly the first detailed study of homology theory from the view point of algebra in 1935.

There are two directions of generalizations of simplicial homology invented by Poincaré in 1895:

(i) from complexes to more general spaces where the homology groups are not characterized by numerical invariants;
(ii) from the group **Z** to arbitrary abelian groups.

Several homology theories are constructed other than homology invented by Henry Poincaré in 1895. They include

(i) Homology groups for compact metric spaces introduced by L. Vietoris (1891–2002) in 1927;
(ii) Homology groups for compact Hausdorff spaces introduced by E.Čech in 1932;
(iii) Singular homology groups are first defined by S. Lefschitz in 1933.
(iv) Cellular homology groups for CW-complexes (see Sect. 10.14).

All these homology theories and their dual theories called cohomology theories lived in isolation in the sense that their interrelations were not established for a long time. Cohomology is dual to homology and it arises from the algebraic dualization of the construction of homology. The origin of the concept of cohomology groups is the duality theorem given J.W. Alexander in 1935. Cohomology theory given by S. Lefschetz in 1930 was further developed by J.W. Alexander in 1936, H. Whitney in 1938 and Lefschetz himself in 1942. L. Pontryagin (1908–1988) proved in 1934 the complete group invariant form of the duality theorem in his paper Pontryagin (1934). Alexander gave the first formal definition of the cohomology groups in 1936 at the Moscow conference.

The cohomotopy groups defined by K. Borsuk in 1936 resemble to cohomology groups in some sense. E. Spanier (1921–1996) deeply studied cohomotopy groups in his paper (Spanier 1949) and has shown that the cohomotopy groups satisfy analogues of all the cohomology axioms under situations when they are meaningful.

Simplicial cohomology constructed by J.W. Alexander and A. Kolmogoroff (1903–1987) in 1935 was developed by E. Čech and H. Whitney during 1935–1940. Čech cohomology is a cohomology theory based on the intersection properties of open covers of a topological space. It is named after E.Čech. De Rham cohomolgy defined for smooth manifolds has many deep results including direct relationships to solutions of differential equations on manifolds. The homology and cohomology groups for CW-complexes can be directly calculated from the cellular structure like simplicial structure in the simplicial homology and cohomology groups of a polyhedron and they provide the most useful tools in algebraic topology. Cohomology has a multiplicative structure making it a ring (algebra). This advantage of cohomolgy over homology facilitates more development of cohomology than homology.

Historically, homology theory was first applied to manifolds by Poincaré, giving a result known as Poincaré duality theorem. He first stated this theorem in 1895 in terms of Betti numbers. At that time there was no concept of cohomology, which was invented in 1930s. Poincaré stated that the m-th and $(n-m)$ th Betti numbers of a closed (i.e., compact and without boundary) orientable n-manifold are equal. But there are at present its different forms.

Lefschetz number is an important concept introduced by Lefschetz in 1923. It is a number associated with each continuous map $f : |K| \to |K|$ from a polyhedron into itself and the number is denoted by Λ_f. It is also related to the Euler characteristic formula and proves a powerful fixed point theorem known as Lefschetz fixed point theorem, which is a classical application of homology and generalizes Brouwer fixed point theorem. Moreover, some other results on fixed points follow as its applications.

The proof of homotopy type invariance of homology was given by J.W. Alexander in 1915 and 1926 and by O. Veblen in 1922 through their work in terms of simplicial homology groups of a polyhedron. The concept of induced homomorphisms f_* was used since the time of Poincaré but it had neither any name nor any status for at least next 35 years. On default of formal recognition of boundary operator ∂ and f_* for such period the homology groups earned no formal status for a long period. The first formal recognition of the homology sequence and its exactness was found in 1941 in the work of Hurewicz.

The excision property does not hold in general for homotopy groups. This failure makes homotopy groups so much harder for computing than homology groups. However, Fredenthal suspension theorem shows that in some special cases there is a range of dimensions in which excision property holds. This leads to the concept of stable homotopy groups, which begins with stable homotopy theory. Computation of these groups even for simple spaces is a difficult problem. An interesting conjecture posed by Freyd also seems to be very hard (Gray, pp. 145). The stable homotopy groups of spheres are fundamental objects in algebraic topology and attempts are going on for their calculation. Stable homotopy groups ${\pi_n}^s(X)$ define a reduced homology theory on the category of pointed CW-complexes.

Higher homotopy groups have certain similarities with homology groups. For example, $\pi_n(X)$ are always abelian for $n \geq 2$ and there are relative homotopy groups which give a long exact sequence like long exact sequence of homology groups. The higher homotopy groups are easier to define but harder to compute than either homology groups or fundamental groups essentially due to absence in higher homotopy groups the excision property for homology and also absence in higher homotopy groups a theorem analogous to van Kampen theorem for fundamental group. In spite of these computational difficulties, homotopy groups are of great importance. For example, Whitehead theorem given by J.H.C. Whitehead which says that a continuous map between CW-complexes which induce isomorphisms on all homotopy groups is a homotopy equivalence. The homological version of Whitehead theorem is similar.

18.5 Hopf Invariant

This section presents Hopf invariant which is an important concept invented by Hopf to solve the problem when a map $f : S^m \to S^n$ for $m > n > 1$ is necessarily nullhomotopic. This problem was resolved by Hopf with the discovery of his famous map $f : S^3 \to S^2$. Hopf in his celebrated paper (Hopf 1931) studied the space of homotopically nontrivial continuous mappings of spheres: $S^3 \to S^2$. He showed in 1931 that $\pi_3(S^2, s_0)$ is nonzero. He later solved the general problem when a continuous mapping $f : S^m \to S^n$ for $m > n > 1$ is necessarily nullhomotopic.

The basic problem which led to the discovery of homotopy groups was to classify homotopically the maps of an n-sphere S^n into a given space. Hopf introduced the concept of an invariant $H(f)$ of f, now called Hopf invariant of f, which depends only on the homotopy class of f. Hopf invariant for certain class of mappings and Hopf group (which is a generalization of topological group) are two important inventions of H. Hopf. It is proved that two continuous maps from S^3 to S^2 are homotopic iff they have the same Hopf invariant. Hopf developed vector field theory. His work has earned a permanent place in the history of algebraic topology.

The main thrust in homotopy theory appears to centralize on the problems of determining homotopy groups of spheres. The basic tools used in this search are of algebraical nature, like "generalized Hopf invariants" studied by G. Whitehead, or "cup products" introduced by N. Steenrod. Homology is used to define the degree of spherical maps $f : S^n \to S^n$ which distinguish different homotopy classes of maps f. Cup products can be used something similar for maps $f : S^{2n-1} \to S^n$, i.e., for elements of $\pi_{2n-1}(S^n)$, Hopf did this by using more geometric constructions prior to the invention of cohomolgy and cup products. The Hopf invariant $H(f)$ defined now by using cup product is as one of the most remarkable applications of cup product in topology.

The problem for which n there exists a continuous map $f : S^{2n-1} \to S^n$ with Hopf invariant $H(f) = 1$ was solved by Adams (Adams 1960, 1962). This theorem, also called Adams theorem, is a deep theorem in homotopy theory which states that the only n-spheres which are H-spaces are S^0, S^1, S^3, and S^7. This relates to the existence of division algebra structure on Euclidean space $\mathbf{R}^n$. The real numbers, complex numbers, the real quaternions, and the Cayley numbers are examples of real division algebras. J.F. Adams proves that there are no other examples. Corresponding to each continuous map $f : S^{2n-1} \to S^n$ one can associate an integer $H(f)$ its Hopf invariant. This means that to each element of $\pi_{2n-1}(S^n)$, one can assign an integer which is its Hopf invariant. If $\pi_{2n-1}(S^n)$ contains no element of Hopf invariant one, then there is no real division algebra of dimension n. Adams theorem shows that such elements exist precisely for $n = 1, 2, 4, 8$, whose simpler proof is given by Atiyah (1967) by using K-theory introduced by M.F. Atiyah (1929-) and F.E. Peter Hirzebruch (1927–2012) in 1961.

18.6 Eilenberg and Steenrod Axioms

This section presents the axiomatic approach of homology and cohomolgy theories given by S. Eilenberg and N.E. Steenrod (1910–1971) in 1945 (Eilenberg and Steenrod 1945) as axioms to characterize a theory of homology and cohomology (see Chap. 12). The usual approach to homology arises through the complicated notion of a complex. Many of the ideas used in constructions, such as orientation, chain and algebraic boundary seem to be artificial (see Chap. 10). The motivation of these concepts appears only in retrospect. Several homology theories were constructed during 1927–1933 which are different from simplicial homology theory invented by Poincaré in 1895. Since their constructions are complicated and different, for greater logical simplicity, algebraic topologists started around 1940 comparing various definitions of homology and cohomology given in the previous years. The construction of a homology theory and proofs of its main properties are extremely complicated. To avoid these problems, S. Eilenberg and N.E. Steenrod initiated a new approach in 1945 by taking a small number of their properties (not focusing on machinery used for construction of homology and cohomology groups) as axioms to characterize a theory of homology and cohomology. This axiomatic approach has greatly influenced later developments of algebraic topology. The axioms reveal that the first six axioms carry a very general character, while the seventh axiom, which is the "Dimension Axiom" is very specific. There exist many such theories such as stable homotopy, various K-theories and bordism theories. The author of the present book has constructed a new generalized cohomology theory (see Chap. 15).

This axiomatic approach given by a set of seven axioms of S. Eilenberg and N. Steenrod, announced in 1945 and published in their book in 1952 (Eilenberg and Steenrod 1952) with the proof of their uniqueness is the most important contribution to algebraic topology since the invention of the homology groups by Poincaré. The uniqueness theorem of homology theory asserts that any two homology groups having the same coefficient group coincide on finite CW-complexes, which implies that the homology groups of finite CW-complexes are completely determined by the coefficient group, and hence are computable from the axioms. This approach classifies and unifies different homology groups on the category of compact triangulated spaces. An analogous approach given by them also inaugurated its dual theory called cohomology theories.

The exactness of the homology sequence of a pair of topological spaces was formalized by Eilenberg and Steenrod in 1945 while giving "Axiomatic Approach to Homology Theory". On the other hand, the form of the Mayer–Vietoris sequence (exact) of a triad which is now used was also given by Eilenberg and Steenrod in 1952, although formulae for the homology groups of the union of two polyhedra were prescribed by W. Mayer (1887–1948) in 1929 and L. Vietoris in 1930.

18.7 Fiber Bundle, Vector Bundle, and K-Theory

This section conveys the early development of fiber bundles, vector bundles and K-theory. Fiber bundles and vector bundles are special bundles with additional structure and are closely related to the homotopy theory. The recognition of bundles in mathematics was realized during 1935–1940 through the work of H. Whitney, H. Hopf and E. Stiefel and some others. Since then the subject has created a general interest. There is a link-up between the study of vector bundles and homotopy theory. The K-theory studied in 1959 by Atiyah and Hirzebruch (1959) connects vector bundles with homotopy theory and is a generalized cohomolgy theory.

One of the most important notions in topology is the notion of fiber spaces which is the most fruitful generalization of covering spaces. Although this notion had appeared in the literature before 1955, the definition introduced by Hurewicz in 1955 (Hurewicz 1955) is much more general and useful. The concept of fiber bundles arose through some problems in topology and geometry of manifolds around 1930. Its first general definition was given by H.Whitney. His work and that of H. Hopf, E. Stiefel (1909–1978), J. Feldbau (1914–1945), and many others displayed the importance of the subject for the application of topology to different areas of mathematics and to other fields also (see Chaps. 14 and 17). This subject also marks a return of algebraic topology to its origin.

Covering spaces provide tools to study the fundamental groups. Fiber bundles provide likewise tools to study higher homotopy groups (which are generalization of fundamental groups). The importance of fiber spaces was realized during 1935–1950 to solve several problems relating to homotopy and homology. The motivation of the study of fiber bundles and vector bundles came from the distribution of signs of the derivatives of the plane curves at each point.

The concept of fiber bundle arose through some problems in topology and geometry of manifolds around 1930. Fiber bundles form a nice class of maps in topology, and many naturally emerging maps are fiber bundles. Fiber bundles are fibrations and fibrations are a natural class of maps in algebraic topology. The notion of fiber bundles plays a central role to study spaces up to homotopy. A fiber bundle is a bundle with an additional structure derived from the action of a topological group on the fibers. On the other hand, a vector bundle is a bundle with an additional vector space structure on each fiber.

The concept of a vector bundle came from the study of tangent vector fields to smooth manifolds, such as spheres, projective spaces etc. A fiber bundle is a locally trivial fibration and has covering homotopy property. Theory of fiber bundles including classifying theorem, with a special attention to vector bundles with fibers of different dimensions and K-theory (which is generalized cohomology theory) interlinks vector bundles with homotopy theory.

The concept of fibration plays a key role in the study of homotopy theory, which appeared implicitly in 1937 in the work of Borsuk but explicitly in the work of Whiteney during 1935–1940, first on sphere bundles. This concept led to general fiber bundles. Hurewicz and Steenrod made the first attempt in 1940 to formulate the homotopy-theoretic properties latent in the notion of fiber bundles and gave a set of sufficient conditions to establish that a large class of homotopy lifting problems always has a solution. More precisely, if $p : X \to B$ is a continuous map, the condition for a homotopy lifting problem consists of a map $f : Y \to X$ and a homotopy $G : Y \times I \to B$ of its projection $p \circ f$. A solution of this problem is a homotopy $H : Y \times I \to X$ of f such that $p \circ H = G$.

J.P. Serre (1926-) studied fibrations and showed in 1950 that a continuous map to be a fibration iff every homotopy lifting problem with X a finite complex has a solution. This result characterizes a map to be a fibration and may be considered as a definition of a fibration. Hurewicz modified Serre's definition in 1955 by removing all restrictions on X. Hurewicz established that the projection of every fiber bundle with paracompact base space is in particular a fibration according to his definition.

J.W. Milnor (1931-) invented a new method in 1956 for giving a classifying space and a universal principal fiber space associated with principal fiber bundle. The most celebrated published work of Milnor is his proof in 1956 of the existence of 7-dimensional sphere with nonstandard differential structure. He constructed a universal fiber bundle for any topological group G and homotopy classification of principal G-bundles. The relations between G and a classifying space B_G can be readily displayed using a geometric analogue of the resolution of homological algebra. The above homotopy classification of vector bundles, Milnor's construction of a universal fiber bundle for any topological group G with homotopy classification of numerable principal G-bundles and corresponding to the set of the isomorphism classes of F-vector bundles over a paracompact space B, the group $K_F(B)$ called the K-theory introduced by M.F. Atiyah (1929-) and Hirzebruch in 1961 are very powerful results.

18.8 Eilenberg–MacLane Spaces and Cohomology Operations

This section conveys the concept of Eilenberg and MacLane spaces introduced by S. Eilenberg and S. MacLane during 1942–1943 which plays a central role in algebraic topology. The importance of Eilenberg–MacLane spaces is twofold. First, they are important in homotopy theory. Second, they are closely linked with the study of cohomology operations (invented by Serre). They carry close connection with cohomology. The cohomology classes of a CW-complex have a bijective correspondence with the homotopy classes of continuous maps from the complex into an Eilenberg–MacLane space. This gives a strict homotopy-theoretic interpretation of cohomology. In this sense cohomology groups may be considered 'dual' to homotopy groups for CW-complexes. Moreover, every topological space has the homotopy type of an iterated fibration of Eilenberg–MacLane spaces (called a Postnikov system).

Given an abelian group G and an integer $n > 0$, Eilenberg and MacLane constructed a space $X = K(G, n)$ with nth homotopy group G and all other homotopy groups vanish. Such spaces $K(G, n)$ are now called an Eilenberg–MacLane spaces. The homotopy sets $[X, Y]$ were first systematically studied by Barratt in 1955 while studying 'Track groups'. Eilenberg and MacLane studied the homological and cohomological structures of the complex $K(G, n)$. These complexes were defined in a purely algebraic fashion for every abelian group G and any integer $n = 1, 2, \ldots$. The topological significance of these complexes $X = K(G, n)$ are on the fact that homotopy groups $\pi_n(X) \cong G$ and $\pi_i(X) = 0$ for $i \neq n$. There are many other important topological and also algebraic applications of these complexes.

The concept of cohomology operations introduced by Serre is a natural transformation of functors

$$H^n(\,;G) \to H^n(-;G').$$

Steenrod defined operations from one cohomology group to another (the so-called Steenrod squares) that generalized the cup product. The additional structure made cohomology a finer invariant. More precisely, Steenrod defined a family of new operations $S_q^i : H^n(\,;\mathbf{Z}_2) \to H^{n+1}(\,;\mathbf{Z}_2)$ which is a sequence of operations, one for each dimension, and behaves well with respect to suspension and they are the components of a stable operation. These operations form a (noncommutative) algebra under composition, known as the Steenrod algebra.

The method of cohomology operations can be used to study homology groups of spheres. Let $\alpha \in \pi_n(S^m)$ and $\psi : H^m(\,;G) \to H^{n+1}(\,;L)$ be a cohomology operation. The mapping cone C_α of α is a complex with one 0-cell, the base point (may be ignored), one m-cell, and one $(n+1)$-cell. We say that α is determined by ψ iff the operator

$$\psi : H^m(C_\alpha;G) \to H^{n+1}(C_\alpha;L)$$

is nonzero. For example, if α is a Hopf map, then C_α is $\mathbf{CP^2}$ and the operation $S_q^2 : H^2 \to H^4$ is the cup square, which is nonzero in C_α.

18.9 Generalized Homology and Cohomology Theories

This section presents certain functors which satisfy all the axioms of Eilenberg and Steenrod for homology (resp. cohomology) with the exception of dimension axiom. The theory of such functors is known as the generalized (or extraordinary) homology (resp. cohomology). These theories first appeared in print in 1952 (Eilenberg and Steenrod 1952). Several such functors have been found to be very useful. For example, K-theory, various forms of bordism and cobordism theories, stable homotopy and cohomotopy theories are their outstanding examples.

Around 1959 several algebraic topologists, working in different directions, considered systems of covariant functors

$$h_n : \mathcal{C}_0 \to \mathcal{A}b$$

(resp. contravariant functors)

$$h^n : \mathcal{C}_0 \to \mathcal{A}b$$

from the category $\mathcal{C}_0$ whose objects are pointed topological spaces having the homotopy type of pointed finite CW-complexes and morphisms are maps of such spaces to the category $\mathcal{A}b$ of abelian groups and their homomorphisms. These functors satisfy all the axioms of Eilenberg and N. Steenrod for homology (resp. cohomology) with exception of dimension axiom. The notions initially introduced in homology and homotopy theories for applications to problems of topology have found fruitful applications to other parts of mathematics such as algebra, analysis, geometry, graph theory. Homological algebra and K-theory are their outstanding examples. Among the various homology theories, ordinary homology theory H_* is the most useful, it is usually much easier in most of the cases, to compute the ordinary homology groups of a given space X than computing $h_*(X)$ for some other homology theory h_*. The first step to computing $h_*(X)$ usually consists of computing $H_*(X)$. In this sense, ordinary homology theory is the fundamental homology theory.

18.10 Ω-Spectrum and Associated Cohomology Theories

This section conveys the concept of Ω-spectrum and its associated cohomology theories. The notion of spectrum introduced by F. L. Lima (1929-) in 1958 has proved to be very useful.

A spectrum $\underline{A} = \{A_n, \alpha_n\}$ in $\mathcal{C}_0$ is a sequence $\{A_n\}$ of spaces in $\mathcal{C}_0$ together with a sequence of continuous maps

$$\alpha_n : A_n \to \Omega A_{n+1} \text{ in } \mathcal{C}_0$$

$$(\text{equivalently, } \widetilde{\alpha}_n : \Sigma A_n \to A_{n+1} \text{ in } \mathcal{C}_0).$$

and it is said to be an Ω-spectrum if $\alpha_n : A_n \to \Omega A_{n+1},\ n \in \mathbf{Z}$ is a base point preserving weak homotopy equivalence for every integer n. There is a special sequence $\{A_n\}$ of spaces, $A_n = K(G, n)$ together with a sequence $\{\alpha_n\}$ of homotopy equivalences, $\alpha_n : A_n \to \Omega A_{n+1}$, relating the cohomology groups of Eilenberg–MacLane spaces to the homotopy groups of spaces by the relation $H^n(X; G) = [X, K(G, n)]$, the homotopy classes of continuous maps from X to $K(G, n)$, which admits a natural group structure. For example, the spectrum $\underline{A} = \{A_n, \alpha_n\}$ given by $A_n = K(\mathbf{Z}, n)$ and $\alpha_n : K(\mathbf{Z}, n) \to \Omega K(\mathbf{Z}, n+1)$, a base point preserving weak homotopy equivalence, is called an Eilenberg–MacLane spectrum. In general, the Eilenberg–MacLane spectrum $\underline{A} = \{A_n, \alpha_n\}$, denoted by $\underline{HG}$, is defined by taking $A_n = K(G, n)$, where $K(G, n)$ is an Eilenberg–MacLane space of type (G, n).

Each spectrum $\underline{A}=\{A_n, \alpha_n\}$ produces two different sequences of functors $\widetilde{h}_n(;\ , \underline{A})$ and $\widetilde{h}^n(; \underline{A})$ from the category $\mathcal{C}_0$ to the category $\mathcal{Ab}$, the first one is covariant and the second one is contravariant. These are called the spectral homology and cohomology functors associated with the spectrum $\underline{A}$, and in brief abbreviated $\widetilde{A_n}$ and $\widetilde{A^n}$.

The homology theory associated with a spectrum was first defined by G.W. Whitehead (1918–2004) in 1962, now called a spectral homology theory associated with spectrum. Its dual theory is called a spectral cohomology theory. For example, the Eilenberg–MacLane space $K(G, n)$ form a spectrum $\underline{K(G)}$, such that its associated cohomology group $H^n(Y; \underline{K(G)}) \cong H^n(Y; G)$ for a finite CW-complex Y. In particular, as the infinite symmetric product $SP^\infty(S^n)$ of the n-sphere S^n is the Eilenberg–MacLane space $K(\mathbf{Z}, n)$, it follows that $H^n(X, \mathbf{Z}) = [X, K(\mathbf{Z}, n)]$. Using this fact the author of the book has constructed in Chap. 15 a new Ω-spectrum

$\underline{A}$, generalizing the Eilenberg–MacLane spectrum $K(G, n)$ and also constructed its associated cohomology theory $h^*(\ ; \underline{A})$ which generalizes the ordinary cohomology theory of Eilenberg and Steenrod.

Instead of defining products axiomatically, G W Whitehead defined products directly from the Ω-spectrum in 1962. For example, for a finite CW-complex X, if $\underline{X} = \{X_n, f_n\}$, where $X_n = \Sigma^n X$ and f_n is the natural homeomorphism $\Sigma X_n \to X_{n+1}$, then it is a spectrum and if Y is a complex, let $H^n(Y; \underline{X})$ be the distinct limit of the groups $[\Sigma^m Y, X_{n+m}]$ under the composite maps

$$[\Sigma^m Y, X_{n+m}] \xrightarrow{E} [\Sigma^{m+1} Y, \Sigma X_{n+m}] \xrightarrow{f_{n*}} [\Sigma^{m+1} Y, X_{n+m+1}].$$

The functors $H^n(\ ; \underline{X})$ behave very much like cohomology group; indeed, they satisfy the Eilenberg–Steenrod axioms with the exception of the 'Dimension Axiom', which says that the cohomology groups of point vanish except in dimension zero.

There are more interesting examples of cohomology theories derived from spectrum: for example, if $\underline{S}$ is the suspension functor, the cohomolgy theory associated this spectrum is $H^n(Y; \underline{S})$ which is just the stable cohomotopy groups. Other important cohomology theories are various Bordism and K-theories. Atiyah and Hirzebruch made a study of the group $K(X)$ in 1959 from the category of complex vector bundles over a finite dimensional CW-complex. They developed their study of $K(X)$ in 1961 into a generalized cohomology. K-theory carries many similarities to ordinary cohomology theory and plays a key role in many areas of mathematics such as modern algebra and number theory.

18.11 Brown Representability Theorem

This section presents a surprising theorem proved by E.H. Brown (1926-) in 1962, now known as Brown representability theorem. This theorem solves the problem: every Ω-spectrum defines a cohomology theory. Is its converse true? Brown proves all cohomology theories on the category of CW-complexes arise from Ω-spectra. This theorem relates homotopy theory with generalized cohomology theory and plays a key role in the applications of homotopy theory to other areas. Brown proved that under certain conditions, any cohomology theory satisfying Eilenberg–Steenrod axioms can be obtained in the form $[\ , Y]$ for some suitable space Y. Brown representability theorem presents necessary and sufficient conditions under which a contravariant functor on X has the form [X,Y] for some fixed Y. This shows that there is a close relation between generalized cohomology theory and homotopy theory, which plays a key role in the later development of algebraic topology.

More precisely, E.H. Brown proved in his paper Brown (1962) that, if H satisfies certain axioms, there is a space Y, unique up to homotopy type, such that H is naturally equivalent to the functor which assigns to each CW-Complex X with base point the set of homotopy classes of maps of X into Y. Thus Brown representability theorem asserts that there exist connected CW-complexes A_n with base point and natural equivalences

$$h^n(X) \cong [X; A_n],$$

where X runs over connected CW-complexes with base point. In this way, Brown constructed a new contravariant functor in his paper (Brown 1962). In other words, he proved that every reduced cohomology theory on the category of CW-complexes with base points and base point preserving maps has the form

$$h^n(X) = [X; A_n]$$

for some Ω-spectrum $\underline{A} = \{A_n, \alpha_n\}$. This functor now called Browns functor. This theorem in homotopy theory presents a necessary and sufficient condition for a contravariant functor on the homotopy category of pointed connected CW-complexes, to the category of sets, to be a representable functor. The representability theorem of Brown shows that the set of all cohomology operations of above type is in bijective correspondence with the group $H^m(K(G, n); G')$. These groups were studied intensively by Eilenberg and Steenrod during 1950–1952 and determined by Henri Paul Cartan (1904–2008) in 1953. This theorem has made a turning point in algebraic topology. Brown showed in 1963 that many of the most important functors in algebraic topology are essentially homotopy functors and hence accessible to the methods of homotopy theory. It is proved that cohomology theory is to a large extent a branch of stable homotopy theory.

18.12 Obstruction Theory

This section conveys the early development of obstruction theory by using cohomology theory which describes a technique for studying various homotopy problems such as extension problems, lifting problems and relative lifting problems. These are basic problems in algebraic topology. The origin of obstruction theory is found in the classical works of H. Hopf, S. Eilenberg, N.E. Steenrod and M. Postnikov (1927–2004). It appears in most textbooks on algebraic topology with different approaches. For example, Steenrod studied obstruction on fiber bundles in his book (Steenrod 1951), Spanier in his book (Spanier 1966) and Whitehead in his book (Whitehead 1978) used the concept of obstruction to solve extension and classification problems for continuous maps of a CW-complex into a topological space. One application of obstruction theory is to define characteristic classes. N.E. Steenrod wrote in his excellent paper (Steenrod 1972):

"Many of the basic theorems of topology, and some of its most successful applications in other areas of mathematics, are solutions of particular extension problems. The deepest results of this kind have been obtained by the method of algebraic topology. The essence of the method is a conversion of a geometric problem into an algebraic problem which is sufficiently complex to embody the essential features of the geometric problem, yet sufficiently simple to be solvable by standard algebraic methods. Many extension problems remain unsolved, and much of the current development of algebraic topology is inspired by the hope of finding a truly general solution".

18.13 Additional Reading

[1] Adams J.F., *Stable Homotopy Theory*, Springer-Verlag, Berlin, 1964.
[2] Alexander, J.W., *On the chains of a complex and their duals*, Proc. Nat. Acad. Sci., USA **21**, 509–511, 1935.
[3] Aull, C.E., and Lowen R., (Eds) *Handbook of the History of General Topology*, Volume 3, Kluwer Academic Publishers, 2001.
[4] Bott, R., *The stable homotopy of the classical groups*, Ann. of Math., **70**, 313–337 1959.
[5] Čech,E., *Théorie génerale de l'homologie dans un espace quelconque'*, Fund. Math. **19**, 149–183, 1932.
[6] Eilenberg, S., and MacLane, S., *Relations between homology and homotopy groups of spaces*, Ann. of Math., **46**(2), 1945.
[7] Freedman, M., *The topology of four-dimensional manifolds*, Journal of Differential Geometry**17**, 357–453, 1982.
[8] Glenys, G.L., and Mishchenko, A.S., *Vector bundles and their applications*, Mathematics and its Applications, 447, Kluwer Academic Publishers, Dordrecht, 1998.
[9] Hatcher, Allen, *Algebraic Topology*, Cambridge University Press, 2002.
[10] Hilton, P.J., *An introduction to Homotopy Theory*, Cambridge University, 1983.
[11] Hilton, P.J., and Wylie, S., *Homology Theory*, Cambridge University Press, Cambridge, 1960.
[12] Kampen van. E., *On the connection between the fundamental groups of some related spaces*, American Journal of Mathematics, **55**, 261–267, 1933.
[13] Mayer, J., *Algebraic Topology*, Prentice-Hall, New Jersey, 1972.
[14] Milnor, J.W., *Spaces having the homotopy type of a CW-complex*, Trans Amer. Math. Soc., **90**, 272–280, 1959.
[15] Olum, P., *Obstructions to extensions and homotopies*. Ann. of Math., **25**, 1–50,1950.
[16] Poincaré, H., *Papers on Topology: Analysis Situs and its Five Supplements*, Translated by J. Stillwell, History of Mathematics, **37**, Amer. Math. Soc, 2010.
[17] Smale, S., *Generalized Poincaré's conjecture in dimensions greater than 4*, Ann. of Math., **74** (2), 391–406. 1961.
[18] Spanier, E., *Algebraic Topology*, McGraw-Hill, 1966.
[19] Spanier, E., *Infinite symmetric products, function spaces and duality*, Ann. of Math., **69**, 142–198, 1959.

[**20**] Stallings, J.R., *On fibering certain 3-manifolds, Topology of 3-manifolds and related topics*, Proc. The Univ. of Georgia Institute, 1961, Prentice Hall, 95–100, 1962.

[**21**] Steenrod, N.E., *The Topology of Fibre Bundles*, Princeton, 1951.

[**22**] Switzer, R.M., *Algebraic Topology-Homotopy and Homology*, Springer-Verlag, Berlin, Heidelberg, New York, 1975.

[**23**] Whitehead, G.W., *Elements of Homotopy Theory*, Springer-Verlag, New York, Heidelberg, Berlin, 1978.

[**24**] Zeeman, C., *The generalised Poincaré conjecture*, Bull. Amer. Math. Soc **67**, 270, 1961.

References

Adams, J.F.: On the non-existence of elements of Hopf invariant one. Ann. Math. **72**, 20–104 (1960)

Adams, J.F.: Vector fields on spheres. Ann. Math. **75**, 603–632 (1962)

Adams, J.F.: Stable Homotopy Theory. Springer, Berlin (1964)

Adams, J.F.: Algebraic Topology: A Student's Guide. Cambridge University Press, Cambridge (1972)

Alexander, J.W.: On the chains of a complex and their duals. Proc. Natl. Acad. Sci., USA **21**, 509–511 (1935)

Atiyah, M.F.: K-theory. Benjamin, New York (1967)

Atiyah, M.F., Hirzebruch, F.: Riemann-Roch theorems of differentiable manifolds. Bull. Am. Math. Soc. **65**, 276–281 (1959)

Atiyah, M.F., Hirzebruch F.: Vector bundles and homogeneous spaces. Proc. Symp. Pure Math., Am. Math. Soc. 7–38 (1961)

Aull, C.E., Lowen R. (eds.): Handbook of the History of General Topology, vol. 3. Kluwer Academic Publishers, Berlin (2001)

Barratt, M.G.: Track groups I. Proc. Lond. Math. Soc. **5**(3), 71–106 (1955)

Borsuk, K.: Sur les groupes des classes de transformations continues, pp. 1400–1403. C.R. Acad. Sci, Paris (1936)

Bott, R.: The stable homotopy of the classical groups. Ann. Math. **70**, 313–337 (1959)

Brown, E.H.: Cohomology theories. Ann. Math. **75**, 467–484 (1962)

Čech, E.: Théorie génerale de l'homologie dans un espace quelconque'. Fund. Math. **19**, 149–183 (1932)

Dieudonné, J.: A History of Algebraic and Differential Topology, 1900–1960. Modern Birkhäuser, Boston (1989)

Dold, A.: Relations between ordinary and extraordinary homology. Algebr. Topol. Colloq. Aarhus **2–9** (1962)

Dold, A., Thom, R.: Quasifaserungen und unendliche symmetrische Produkte. Ann. Math. Second Series **67**, 239–281 (1958)

Eilenberg, S., MacLane, S.: Natural isomorphism in group theory. Proc. Natl. Acad. Sci. USA **28**, 537–544 (1942)

Eilenberg, S., MacLane, S.: General theory of natural equivalence. Trans. Am. Math. Soc. **58**, 231–294 (1945a)

Eilenberg, S., MacLane, S.: Relations between homology and homotopy groups of spaces. Ann. Math. **46**(2) (1945b)

Eilenberg, S., Steenrod, N.: Axiomatic approach to homology theory. Proc. Natl. Acad. Sci. USA **31**, 117–120 (1945)

Eilenberg, S., Steenrod, N.: Foundations of Algebraic Topology. Princeton University Press, Princeton (1952)

Freedman, M.: The topology of four-dimensional manifolds. J. Differ. Geom. **17**, 357–453 (1982)

Freudenthal, H.: Über die Klassen von Sphärenabbildungen. Compositio Mathematica **5**, 299–314 (1937)

Hatcher, A.: Algebraic Topology. Cambridge University Press, Cambridge (2002)

Hilton, P.J.: An Introduction to Homotopy Theory. Cambridge University Press, Cambridge (1983)

Hilton, P.J., Wylie, S.: Homology Theory. Cambridge University Press, Cambridge (1960)

Hopf, H.: Über die Abbildungen der 3-Sphäre auf die Kugelfleche. Math. Ann. **104**, 637–665 (1931)

Hopf, H.: Über die Abbildungen von Sphären auf Sphären niedrigerer Dimension. Fund. Math. **25**, 427–440 (1935)

Hurewicz, W.: Beitrage der Topologie der Deformationen. Proc. K. Akad. Wet., Ser. A **38**, 112–119, 521–528 (1935)

Hurewicz, W.: On duality theorems. Bull. Am. Math. Soc. **47**, 562–563 (1941)

Hurewicz, W.: On the concept of fibre space. Proc. Natl. Acad. Sci., USA **41**, 956–961 (1955)

James, I.M.: (ed.) History of Topology. North-Holland, Amsterdam (1999)

Kampen van. E.: On the connection between the fundamental groups of some related spaces. Am. J. Math. **55**, 261–267 (1933)

Maunder, C.R.F.: Algebraic Topology. Van Nostrand Reinhold Company, London (1970)

Mayer, J.: Algebraic Topology. Prentice-Hall, New Jersey (1972)

Milnor, J.W.: Spaces having the homotopy type of a CW-complex. Trans. Am. Math. Soc. **90**, 272–280 (1959)

Olum, P.: Obstructions to extensions and homotopies. Ann. Math. **25**, 1–50 (1950)

Poincaré, H.: Analysis situs. J. Ecole polytech **1**(2), 1–121 (1895)

Poincaré, H.: Second complément *à* l'analysis situs. Proc. Lond. Math. Soc. **32**, 277–308 (1900)

Poincaré, H.: Cinquiéme, complément *à* l'analysis situs. Rc. Cir. Mat. Palermo. **18**, 45–110 (1904)

Poincaré, H.: Papers on topology: analysis situs and its five supplements (Translated by J. Stillwell, History of Mathematics). Am. Math. Soc. **37** (2010)

Pontryagin, L.: The general topological theorem of duality for closed sets. Ann. Math. **35**(2), 904–914 (1934)

Smale, S.: Generalized Poincaré's conjecture in dimensions greater than 4. Ann. Math. **74**(2), 391–406 (1961)

Spanier, E.H.: Borsuk's cohomotopy groups. Ann. Math. **50**, 203–245 (1949)

Spanier, E.: Infinite symmetric products, function spaces and duality. Ann. Math. **69**, 142–198 (1959)

Spanier, E.: Algebraic Topology. McGraw-Hill, New York (1966)

Stallings, J.R.: On fibering certain 3-manifolds, topology of 3-manifolds and related topics. In: Proceedings of The University of Georgia Institute, 1961, pp. 95–100. Prentice Hall, Englewood Cliffs (1962)

Steenrod, N.E.: The Topology of Fibre Bundles. Princeton University Press, Princeton (1951)

Steenrod, N.E.: Chohomology operations and obstructions to extending continuous functions. Adv. Math. **8**, 371–416 (1972)

Switzer, R.M.: Algebraic Topology-Homotopy and Homology. Springer, Berlin (1975)

Whitehead, J.H.C.: On adding relations to homotopy groups. Ann. Math. **42**, 409–428 (1941)

Whitehead, J.H.C.: Combinatorial homotopy: I. Bull. Am. Math. Soc. **55**, 213–245 (1949)

Whitehead, J.H.C.: A certain exact sequence. Ann. Math. **52**(52), 51–110 (1950)

Whitehead, G.W.: A generalization of the Hopf invariant. Ann. Math. **51**, 192–237 (1950)

Whitehead, G.W.: On the Freudenthal theorem. Ann. Math. **57**, 209–228 (1953)

Whitehead, G.W.: Generalized homology theories. Trans. Am. Math. Soc. **102**, 227–283 (1962)

Zeeman, C.: The generalised Poincaré conjecture. Bull. Am. Math. Soc. **67**, 270 (1961)

Appendix A
Topological Groups and Lie Groups

This appendix studies topological groups, and also Lie groups which are special topological groups as well as manifolds with some compatibility conditions. The concept of a topological group arose through the work of Felix Klein (1849–1925) and Marius Sophus Lie (1842–1899). One of the concrete concepts of the theory of topological groups is the concept of Lie groups named after Sophus Lie. The concept of Lie groups arose in mathematics through the study of continuous transformations, which constitute in a natural way topological manifolds. Topological groups occupy a vast territory in topology and geometry. The theory of topological groups first arose in the theory of Lie groups which carry differential structures and they form the most important class of topological groups. For example, $GL\,(n, \mathbf{R})$, $GL\,(n, \mathbf{C})$, $GL\,(n, H)$, $SL\,(n, \mathbf{R})$, $SL\,(n, \mathbf{C})$, $O(n, \mathbf{R})$, $U(n, \mathbf{C})$, $SL\,(n, H)$ are some important classical Lie Groups. Sophus Lie first systematically investigated groups of transformations and developed his theory of transformation groups to solve his integration problems.

David Hilbert (1862–1943) presented to the International Congress of Mathematicians, 1900 (ICM 1900) in Paris a series of 23 research projects. He stated in this lecture that his Fifth Problem is linked to Sophus Lie theory of transformation groups, i.e., Lie groups act as groups of transformations on manifolds. A translation of Hilbert's fifth problem says "It is well-known that Lie with the aid of the concept of continuous groups of transformations, had set up a system of geometrical axioms and, from the standpoint of his theory of groups has proved that this system of axioms suffices for geometry".

For this appendix, the books Bredon (1993), Chevelly (1957), Pontragin (1939), Sorani (1969), Switzer (1975) and some others are referred in Bibliography.

A.1 Topological Groups: Definitions and Examples

This section introduces the concept of topological groups with illustrative examples. A topological group is simply a combination of two fundamental concepts: group and topological space and hence the axiomatization of the concept of topological groups

M.R. Adhikari, *Basic Algebraic Topology and its Applications*,
DOI 10.1007/978-81-322-2843-1

is a natural procedure. O. Schreior (1901–1929) gave a formal definition of modern concept of topological groups in 1925 and F. Leja (1885–1979) in 1927 in terms of topological spaces. Topological groups are groups in algebraic sense together with continuous group operations. This means that the topology of a topological group must be compatible with its group structure.

Definition A.1.1 A topological group G is a Hausdorff topological space together with a group multiplication such that

TG(1) group multiplication $m : G \times G \to G, (x, y) \mapsto xy$ is continuous;
TG(2) group inversion $\text{inv} : G \to G, x \mapsto x^{-1}$ is continuous.

The continuity in **TG(1)** and **TG(2)** means that the topology of G must be compatible with the group structure of G. The conditions **TG(1)** and **TG(2)** are equivalent to the single condition that the map

$$G \times G \to G, (x, y) \mapsto xy^{-1}$$

is continuous.

Remark A.1.2 Some authors do not assume 'Hausdorff property' for a topological group.

Example A.1.3 $\mathbf{R}^n$ (under usual addition) and $S^1 = \{z \in \mathbf{C} : |z| = 1\}$ (under usual multiplication of complex numbers) are important examples of topological groups.

We now describe some classical topological groups $\text{GL}(n, \mathbf{R})$, $\text{SL}(n, \mathbf{R})$, $\text{O}(n, \mathbf{R})$, $\text{SO}(n, \mathbf{R})$ and their complex analogues.

Definition A.1.4 (*General linear group*) $\text{GL}(n, \mathbf{R})$ is the set of all $n \times n$ non-singular matrices with entries in $\mathbf{R}$. It is a group under usual multiplication of matrices, called general linear group over $\mathbf{R}$.

Definition A.1.5 (*Special linear group*) $\text{SL}(n, \mathbf{R})$ defined by $\text{SL}(n, \mathbf{R}) = \{A \in \text{GL}(n, \mathbf{R}) : \det A = 1\}$ is a subgroup of $\text{GL}(n, \mathbf{R})$.

Definition A.1.6 (*Orthogonal group*) $\text{O}(n, \mathbf{R})$ defined by $\text{O}(n, \mathbf{R}) = \{A \in \text{GL}(n, \mathbf{R}) : AA^t = I\}$ is a subgroup of $\text{GL}(n, \mathbf{R})$.

Definition A.1.7 (*Special orthogonal group*) $\text{SO}(n, \mathbf{R})$ defined by $\text{SO}(n, \mathbf{R}) = \{A \in \text{O}(n, \mathbf{R}) : \det A = 1\}$ is a group.

Theorem A.1.8 *The general real linear group* $\text{GL}\,(n, \mathbf{R})$ *of all invertible* $n \times n$ *matrices over* $\mathbf{R}$ *is a topological group. This group is neither compact nor connected.*

Proof Let $M_n(\mathbf{R})$ be the set of all $n \times n$ real matrices. Let $A = (a_{ij}) \in M_n(\mathbf{R})$. We can identify $M_n(\mathbf{R})$ with the Euclidean space $\mathbf{R}^{n^2}$ by the mapping

$$f : M_n(\mathbf{R}) \to \mathbf{R}^{n^2}, (a_{ij}) \mapsto (a_{11}, a_{12}, \ldots, a_{1n}, a_{21}, a_{22}, \ldots, a_{2n}, \ldots, a_{n1}, a_{n2}, \ldots, a_{nn}).$$

This identification defines a topology on $M_n(\mathbf{R})$ such that the matrix multiplication

$$m : M_n(\mathbf{R}) \times M_n(\mathbf{R}) \to M_n(\mathbf{R})$$

is continuous.

Let $A = (a_{ij})$ and $B = (b_{ij}) \in M_n(\mathbf{R})$. Then the ijth entry in the product $m(A, B)$ is $\sum_{k=1}^{n} a_{ik}b_{kj}$. As $M_n(\mathbf{R})$ has the topology of the product space $\mathbf{R}^1 \times \mathbf{R}^1 \times \ldots, \times \mathbf{R}^1$ (n^2 copies), and for each pair of integers i, j satisfying $1 \leq i, j \leq n$, we have a projection $p_{ij} : M_n(\mathbf{R}) \to \mathbf{R}^1$, which sends a matrix A to its ijth entry. Then m is continuous if and only if the composite maps

$$M_n(\mathbf{R}) \times M_n(\mathbf{R}) \xrightarrow{m} M_n(\mathbf{R}) \xrightarrow{p_{ij}} \mathbf{R}^1$$

are continuous. But $p_{ij}m(A, B) = \sum_{k=1}^{n} a_{ik}b_{kj}$, which is a polynomial in entries of A and B. Hence the composite maps $p_{ij} \circ m$ are continuous.

GL $(n, \mathbf{R})$ topologized as a subspace of the topological space $M_n(\mathbf{R})$ is such that the matrix multiplication

$$\mathrm{GL}\,(n, \mathbf{R}) \times \mathrm{GL}\,(n, \mathbf{R}) \to \mathrm{GL}\,(n, \mathbf{R}), (A, B) \mapsto AB$$

is continuous. We next claim that the inverse map

$$\mathrm{inv} : \mathrm{GL}\,(n, \mathbf{R}) \to \mathrm{GL}\,(n, \mathbf{R}), A \mapsto A^{-1}$$

is continuous. The map

$$\mathrm{inv} : \mathrm{GL}\,(n, \mathbf{R}) \to \mathrm{GL}\,(n, \mathbf{R}) \subset \mathbf{R}^1 \times \mathbf{R}^1 \times \ldots \times \mathbf{R}^1 (n^2 \text{ copies})$$

is continuous if and only if all the composite maps

$$\mathrm{GL}\,(n, \mathbf{R}) \xrightarrow{\mathrm{inv}} \mathrm{GL}\,(n, \mathbf{R}) \xrightarrow{p_{jk}} \mathbf{R}^1, 1 \leq j, k \leq n$$

are continuous. But each composite map $p_{jk} \circ \mathrm{inv}$ sends a matrix A to the jkth element of A^{-1}, which is $(1/\det A)$ (kjth cofactor of A), where $\det A \neq 0 \;\forall\, A \in$ GL $(n, \mathbf{R})$. Hence the composite maps $p_{jk} \circ \mathrm{inv}$ are continuous. Consequently, GL $(n, \mathbf{R})$ is a topological group.

The group GL $(n, \mathbf{R})$ is not compact: Clearly, GL $(n, \mathbf{R})$ is the inverse image of nonzero real numbers under the determinant function

$$\det : M_n(\mathbf{R}) \to \mathbf{R}.$$

The determinant function is continuous, since it is just a polynomial in the matrix coefficient. Hence the inverse image of $\{0\}$ is a closed subset of $M_n(\mathbf{R})$. Its complement is the set of all nonsingular $n \times n$ real matrices is an open subset of $M_n(\mathbf{R})$. Hence $\mathrm{GL}\,(n, \mathbf{R})$ is not compact.

The group $\mathrm{GL}\,(n, \mathbf{R})$ is not connected: Clearly, the matrices with positive and negative determinants give a partition of $\mathrm{GL}\,(n, \mathbf{R})$ into two disjoint nonempty open sets. Hence $\mathrm{GL}\,(n, \mathbf{R})$ is not connected. ❑

Definition A.1.9 $\mathrm{GL}\,(n, \mathbf{C})$ is the set of all $n \times n$ nonsingular matrices with complex entries. It is a group under usual multiplication of matrices, called the general complex linear group.

Theorem A.1.10 $\mathrm{GL}\,(n, \mathbf{C})$ *is a topological group. It is not compact.*

Proof Every element $A \in \mathrm{GL}\,(n, \mathbf{C})$ is a nonsingular linear transformation of $\mathbf{C}^n$ over $\mathbf{C}$. If $\{z_1, z_2, \ldots, z_n\}$ is a basis of $\mathbf{C}^n$, then $\{x_1, y_1, \ldots, x_n, y_n\}$ is a basis of $\mathbf{R}^{2n}$, where $z_i = x_i + iy_i$. Every element $A \in \mathrm{GL}\,(n, \mathbf{C})$ determines a linear transformation $\tilde{A} \in \mathrm{GL}\,(2n, \mathbf{R})$ into a subset of $\mathrm{GL}\,(n, \mathbf{R})$. Since $\mathrm{GL}\,(n, \mathbf{C})$ is an open subset of a Euclidean space, it is not compact. ❑

Corollary A.1.11 *The set* $\mathrm{U}(n, \mathbf{C}) = \{A \in \mathrm{GL}\,(n, \mathbf{C}) : AA^* = I\}$, *is a compact subgroup of* $\mathrm{GL}\,(n, \mathbf{C})$, *where* A^* *denotes the transpose of the complex conjugate of* A.

Remark A.1.12 $\dim_{\mathbf{C}} \mathrm{GL}\,(n, \mathbf{C}) = n^2$.

Definition A.1.13 A homomorphism $f : G \to H$ between two topological groups G and H is a continuous map such that f is a group homomorphism. An isomorphism $f : G \to H$ between two topological groups is a homeomorphism and is also a group homomorphism between G and H.

Example A.1.14 The special orthogonal group $\mathrm{SO}\,(2, \mathbf{R})$ and the circle group S^1 are isomorphic topological groups under an isomorphism f of topological groups given by

$$f : \mathrm{SO}\,(2, \mathbf{R}) \to S^1, \begin{pmatrix} \cos\theta & -\sin\theta \\ \sin\theta & \cos\theta \end{pmatrix} \mapsto e^{i\theta}.$$

Remark A.1.15 For quaternionic analogue see the sympletic group $SU(n, \mathbf{H}) = \{A \in \mathrm{GL}\,(n, \mathbf{H}) : AA^* = I\}$ (Ex. 10 of Sect. A.4).

A.2 Actions of Topological Groups and Orbit Spaces

This section introduces the concept of actions of topological groups and studies some important orbit spaces (thus obtained) with an eye to compute their fundamental groups. Real and complex projective spaces, torus, Klein bottle, lens spaces, and figure-eight are important objects in geometry and topology and they can be represented as orbit spaces.

Definition A.2.1 Let G be a topological group with identity element e and X a topological space. An action of G on X is a continuous map $\sigma : G \times X \to X$, with the image of (g, x) being denoted by gx such that

(i) $(gh)x = g(hx)$;
(ii) $ex = x$,

$\forall\, g, h \in G$ and $\forall\, x \in X$. The pair (G, X) with the given action σ is called a topological transformation group and in brief we call X a G-space.

Remark A.2.2 If we change any one G, X or σ, then we get a different transformation group. If we forget the topologies from the space, then the group G and the action σ give together the concept of G-set.

Orbit spaces are closely related with G-spaces. Let X be a G-space. Two elements $x, y \in X$ are said to be G-equivalent if $\exists$ an element $g \in G$ such that $gx = y$. The relation of being G-equivalent is an equivalence relation and the set $\{gx : g \in G\}$ denoted by Gx, the equivalence class determined by x, is called the orbit of x. If the group G is compact and the space X is Hausdorff, then the orbits are closed sets of X and in this case, the coset space G/Gx is homeomorphic to the orbit Gx. The action of G on X is said to be free if $Gx = \{e\}$, $\forall\, x \in X$. Two orbits in X are either identical or disjoint. The set of all distinct orbits of X, denoted by $X \mod G$, with the quotient topology induced from X, is called the orbit space of the transformation group.

Proposition A.2.3 *Let X be a G-space. For every $g \in G$, the map $\psi_g : X \to X$ defined by $\psi_g(x) = gx$ is a homeomorphism.*

Proof Since the action of G on X given by $x \mapsto gx$ is continuous, ψ_g is continuous for each $g \in G$. Moreover, $\psi_g \circ \psi_g^{-1} = \psi_{gg^{-1}} = \psi_e = 1_X$ and $\psi_{g^{-1}} \circ \psi_g = \psi_{g^{-1}g} = \psi_e = 1_X \Rightarrow$ for each g the map $\psi_g : X \to X$ is a homeomorphism. ❑

Let Homeo (X) denote the group of all homeomorphisms of X under usual composition of mappings.

Proposition A.2.4 *Let X be a G-space. Then the action of G on X induces a homomorphism $f : G \to$* Homeo (X).

Proof For each $g \in G$, the map $f : G \to$ Homeo (X) defined by $f(g) = \psi_g$ is a homeomorphism by Proposition A.2.3 $\Rightarrow f$ is well defined. Moreover, for $g, h \in G$, $f(gh) = f(g)f(h) \Rightarrow f$ is a homomorphism. ❑

Definition A.2.5 Let X be a G-space. The action of G on X is said to be effective if the map $f : G \to$ Homeo (X), $g \mapsto \psi_g$ is a monomorphism i.e., $g \neq e \Rightarrow \psi_g \neq 1_X$.

Proposition A.2.6 *Every effective action of a topological group G on a topological space X induces an embedding* $f : G \to \text{Homeo}\,(X)$.

Proof Let $\sigma : G \times X \to X$ be an effective action. Then $g \neq e \Rightarrow \psi_g \neq 1_X$. Consider the homomorphism $f : G \to \text{Homeo}\,(X)$ defined by $f(g) = \psi_g$, where $\psi_g : X \to X$ is given by $\psi_g(x) = \sigma(g, x) = gx$. Then for $g \neq h (g, h \in G) \Rightarrow h^{-1}g \neq e \Rightarrow \psi_{h^{-1}g} \neq 1_X \Rightarrow \psi_{h^{-1}g}(x) \neq x, \forall x \in X \Rightarrow (h^{-1}g)x \neq x, \forall x \in X \Rightarrow gx \neq hx, \forall x \in X \Rightarrow \psi_g(x) \neq \psi_h(x), \forall x \in X \Rightarrow \psi_g \neq \psi_h \Rightarrow f(g) \neq f(h) \Rightarrow f : G \to \text{Homeo}\,(X)$ is an embedding. ❑

Corollary A.2.7 *Every topological group G can be viewed as a group of homeomorphisms of a topological space X. This is an analogue of Caley's Theorem for group.*

Remark A.2.8 **Geometrical interpretation of free action**: The action of G on X is free if and only if each $g(\neq e) \in G$ moves every point of X. Given a group G of homeomorphisms of X, we can always define an action σ of G on X by taking $\sigma(g, x) = g(x)$. What is the topology of G? Since $G \subseteq \text{Homeo}\,(X)$, any topology on Homeo (X)(viz. the compact open topology) will induce a topology on G. Now the problem is: whether or not the map $\sigma : G \times X \to X$ defined above is continuous. If X is a locally compact Hausdorff space, then the compact open topology gives σ a continuous action (see Dugundji, pp 259).

Example A.2.9 (*Real Projective Space* $\mathbf{R}P^n$) Let $f : S^n \to S^n (n \geq 1)$ be the antipodal map i.e., $f(x) = -x$. Then $f \circ f = f^2 = 1_{S^n}$. Thus $G = \{1_{S^n}, f\} \cong \mathbf{Z}_2$ is a group of homeomorphisms of S^n and so it acts on S^n. The action is free and the orbit space $S^n \mod G$ denoted by $\mathbf{R}P^n$, is called the real projective n-space. The space $\mathbf{R}P^n$ is a compact, connected manifold of dimension n.

Example A.2.10 (*Complex Projective Space* $\mathbf{C}P^n$) Let $S^1 = \{z \in \mathbf{C} : |z| = 1\}$ be the circle group. Then S^1 is a topological group. Let $S^{2n+1} = \{(z_0, z_1, \ldots, z_n) \in \mathbf{C}^n : \sum_{i=0}^{n} |z_i|^2 = 1\}$, be the $(2n+1)$-dimensional unit sphere. Then S^1 acts S^{2n+1} continuously under the action defined by $z \cdot (z_0, z_1, \ldots, z_n) = (zz_0, zz_1, \ldots, zz_n)$. This action is free. The orbit space $S^{2n+1} \mod S^1$, denoted by $\mathbf{C}P^n$ is called the complex projective n-space. This space is a compact , connected manifold of real dimension $2n$.

Example A.2.11 (*Torus*) Let $\mathbf{R}$ be the real line and $f : \mathbf{R} \to \mathbf{R}$ be a homeomorphism defined by $f(x) = x + 1$, which is a translation. Then for each integer $n, h^n : \mathbf{R} \to \mathbf{R}, x \mapsto x + n$ is also a homeomorphism. These are just translations by integer amounts. Then the cyclic group $< f >$ generated by the homeomorphism h of $\mathbf{R}$ is the infinite cyclic group $\mathbf{Z}$. Endowing $\mathbf{Z}$ the discrete topology, we find that $\mathbf{Z}$ acts as a group of homeomorphisms on $\mathbf{R}$. Again the action is free and the orbit space $\mathbf{R} \mod \mathbf{Z} = \mathbf{R}/\mathbf{Z}$ is homeomorphic to the circle group S^1. We now consider the product action of the discrete group $\mathbf{Z} \times \mathbf{Z}$ on $\mathbf{R} \times \mathbf{R}$. Let f, g be two homeomorphisms

$\mathbf{R} \to \mathbf{R}$. Define the action of (f, g) on $\mathbf{R} \times \mathbf{R}$ by $(f, g)(x, y) = (x + 1, y + 1)$. Then $(f^m, g^n)(x, y) = (x + m, y + n)$ for every pair of integers. Again the action is free and the orbit space $S^1 \times S^1$ is the 2-torus. An n-torus is obtained similarly as an orbit space of $\mathbf{R}^n$ for $n \geq 2$.

A.3 Lie Groups and Examples

This section introduces the concept of Lie groups with illustrative examples. Lie groups play an important role in geometry and topology. A Lie group is a topological group having the structure of a smooth manifold for which the group operations are smooth functions. Such groups were first considered by Sophus Lie in 1880 and are named after him. He developed his theory of continuous maps and used it in investigating differential equations. The fundamental idea of his Lie theory was published in his paper Lie (1880) and his later book with F. Engel published in 1893 (Lie and Engel 1893). Lie classified infinitesimal groups acting in dimensions 1 and 2 up to analytic coordinate changes. Lie displayed the key role of his Lie theory as a classifying principle in geometry, mechanics and ordinary and partial differential equations. This theory made a revolution in mathematics and physics.

A.3.1 Lie Group: Introductory Concepts

Definition A.3.1 A topological group G is called a real Lie group if

(i) G is a differentiable manifold;
(ii) the group operations $(x, y) \mapsto xy$ and $x \mapsto x^{-1}$ are both differentiable.

Definition A.3.2 A topological group G is called a complex Lie group if

(i) G is a complex manifold;
(ii) the group operations $(x, y) \mapsto xy$ and $x \mapsto x^{-1}$ are both holomorphic.

Definition A.3.3 The dimension of a Lie group is defined to be its dimension as a manifold.

Remark A.3.4 A Lie group is not necessarily connected. Given a Lie group, let G° denote the connected component of G which contains the identity element of G. Then G° is a closed subgroup of G. Any other connected component of G is homeomorphic to G°. This shows that if G is a Lie group, the (real or complex) dimension of G is well defined and it is the dimension of the manifold G°.

Definition A.3.5 Let G and H be Lie groups. A differentiable map $f : H \to G$ is called a homomorphism if f is a group homomorphism and a regular analytic map. $f(H)$ is a subgroup of G and a submanifold of G.

Definition A.3.6 Let G be a Lie group and $f : \mathbf{R} \to G$ be a homomorphism of Lie groups. Then $f(\mathbf{R})$ is called a one parameter subgroup of G.

Remark A.3.7 It was believed before 1956 that a topological space may admit only one differentiable structure. Examples show that differentiable structure of a topological space may not be unique. For example, John Willard Milnor (1931-) proved in 1956 that the 7-sphere S^7 admits 28 different differentiable structures (Milnor 1956). Milnor was awarded the Fields Medal in 1962 for his work in differential topology. He was also awarded the Abel Prize in 2011. For another example, Sir Simon Kirwan Donaldson (1957-) proved in 1983 that $\mathbf{R}^4$ admits an infinite number of different differentiable structures (Donalson 1983). He was awarded a Fields Medal in 1986.

A.3.2 *Some Examples of Lie Groups*

This subsection presents important examples of Lie groups.

(i) The real line $\mathbf{R}$ is a Lie group under usual addition.

(ii) The classical Lie groups:
$GL(n, \mathbf{R}), SL(n, \mathbf{R}), O(n, \mathbf{R}), SO(n, \mathbf{R})$ are all manifolds, because for each point x of any one of these groups, there exists an open neighborhood homeomorphic to a Euclidean space. All of them are real Lie groups, called classical Lie groups. $GL(n, \mathbf{R})$ is a real Lie group of dimension n^2, $SL(n, \mathbf{R})$ is a real Lie group of dimension $n^2 - 1$, $O(n, \mathbf{R})$ is a real Lie group of dimension $n(n-1)/2$, $SO(n, \mathbf{R})$ is a real Lie group of dimension $n(n-1)/2$. Their complex and quaternionic analogues are also Lie groups.

A.4 Exercises

1. Show that the circle group S^1 in the complex plane is a Lie group. (This group is denoted by $U(1, \mathbf{C})$ or by simply $U(1)$).

2. Prove that the general linear group $GL(n, \mathbf{H})$ over the quaternions $\mathbf{H}$ is a topological group but it is not compact.
[Hint: In absence of a determinant function in this case, use the result that $GL(n, \mathbf{H})$ is an open subset of an Euclidean space.]

3. Show that the special real linear group $SL(n.\mathbf{R})$ defined by $SL(n, \mathbf{R}) = \{X \in GL(n, \mathbf{R}) : \det X = 1\}$ is a noncompact connected topological group and is a real Lie group of dimension $n^2 - 1$.
[Hint: $SL(n, \mathbf{R})$ is a subgroup of $GL(n, \mathbf{R})$. It is a hypersurface of $GL(n, \mathbf{R})$.]

4. Prove that the special complex linear group $SL(n, \mathbf{C})$ given by $SL(n, \mathbf{C}) = \{X \in GL(n, \mathbf{C}) : \det X = 1\}$ is a noncompact connected topological group and is a complex Lie group of dimension $n^2 - 1$.

[Hint: SL $(n, \mathbf{C})$ is a subgroup of GL $(n, \mathbf{C})$.]

5. Show that the orthogonal group given by $O(n, \mathbf{R}) = \{A \in \mathrm{GL}\,(n, \mathbf{R}) : AA^t = I = A^tA\}$ is a compact non-connected topological group and is a real Lie group of dimension $\frac{n(n-1)}{2}$.
[Hint: $O(n, \mathbf{R})$ is a closed subspace of GL $(n, \mathbf{R})$. It contains matrices $A \in$ GL $(n, \mathbf{R})$ with $\det A = \pm 1$. Thus $O(n, \mathbf{R})$ is a bounded closed subset of the Euclidean space $\mathbf{R}^{n^2}$. Hence it is compact. Thus $O(n, \mathbf{R})$ is a compact subgroup of GL $(n, \mathbf{R})$. Since it contains matrices with determinant equal to 1 and matrices with determinant equal to -1, it non-connected.]

6. Prove that the special orthogonal group SO $(n, \mathbf{R})$ given by SO $(n, \mathbf{R}) = O(n, \mathbf{R}) \cap \mathrm{SL}\,(n, \mathbf{R})$ is a real compact connected topological group and is a real Lie group of dimension $\frac{n(n-1)}{2}$.

7. Prove that the general (complex) linear group GL$(n, \mathbf{C})$ is a topological group and is a connected, noncompact complex Lie group of dimension n^2.

8. Show that the unitary group $U(n, \mathbf{C})$ defined by $U(n, \mathbf{C}) = \{A \in \mathrm{GL}\,(n, \mathbf{C}) : AA^* = A^*A = I\}$ is a connected compact topological group, and is a real Lie group of dimension n^2, where A^* denotes the conjugate transpose of A (conjugate means reversal of all the imaginary components).
[Hint: It is a subgroup of GL $(n, \mathbf{C})$. It is not a complex submanifold of GL $(n, \mathbf{C})$. As a subspace of GL $(n, \mathbf{C})$ it can be embedded as a subgroup of GL $(2n, \mathbf{R})$.]

9. Let $SU(n, \mathbf{C})$ denote the special unitary group defined by SL $(n, \mathbf{C}) = U(n, \mathbf{C}) \cap$ SL $(n, \mathbf{C})$. Show that the group $SU(2, \mathbf{C}) = \{A = \begin{pmatrix} z & w \\ -\bar{w} & \bar{z} \end{pmatrix} : z, w \in \mathbf{C}$ and $|z|^2 + |w|^2 = 1\}$ is isomorphic to S^3.
[Hint. Use the form of A.]

10. The quaternionic analogue of orthogonal unitary groups is the sympletic group $SU(n, \mathbf{H}) = \{A \in \mathrm{GL}\,(n, \mathbf{H}) : AA^* = I\}$, where A^* denotes the quaternionic conjugate transpose of A. Prove that it is a compact topological group.

11. Show that the 3-dimensional projective space $\mathbf{R}P^3$ and SO $(3, \mathbf{R})$ are homeomorphic.

12. Show that

(i) Let M be a manifold. A flow $\psi_t : M \to M$ is an action of $\mathbf{R}$ on M. If a flow is periodic with period p, it may be regarded as an action of U $(1, \mathbf{C})$ or SO $(2, \mathbf{R})$ on M.

(ii) Let $A \in \mathrm{GL}\,(n, \mathbf{R})$ and $x \in \mathbf{R}^n$. The action σ of GL $(n, \mathbf{R})$ on $\mathbf{R}^n$ is defined by the usual matrix action on a vector

$$\sigma : \mathrm{GL}\,(n, \mathbf{R}) \times \mathbf{R}^n \to \mathbf{R}^n, (M, x) \mapsto M \cdot x$$

13. Show that the right translation $R : (a, g) \mapsto R_a g$ and left translation $L : (a, g) \mapsto L_a g$ of a Lie group are free and transitive actions.

14. Let a Lie group G act on a manifold M. Show that

(i) the isotropy group G_x of any $x \in M$ is a Lie subgroup;

(ii) if G acts on M freely, then the isotropy group G_x of any $x \in M$ is trivial.

15. Show that the orthogonal group O $(n+1, \mathbf{R})$ acts on $\mathbf{R}P^n$ transitively from left.

16. Show that orthogonal group O $(n, \mathbf{R})$ acts transitively on the Grassmann manifold (see Chap. 5) $G_{n,r} (r \leq n)$.

17. Show that the special orthogonal group SO $(n, \mathbf{R})$ acts transitively on the Stiefel manifold (see Chap. 5) $V_{n,r} = V_r(\mathbf{R}^n)$, $(r \leq n)$.

A.5 Additional Reading

[**1**] Adhikari, M.R., and Adhikari, Avishek, *Basic Modern Algebra with Applications*, Springer, New Delhi, New York, 2014.

[**2**] Borel, A., *Topology of Lie groups and characteristic classes*, Bull. Amer. Math. Soc.**61** (1955), 397–432.

[**3**] Dieudonné, J., *A History of Algebraic and Differential Topology*, 1900–1960, Modern Birkhäuser, 1989.

[**4**] Dupont, J., *Fibre Bundles*, Aarhus Universitet, 2003.

[**5**] Nakahara, M., *Geometry, Topology and Physics*, Institute of Physics Publishing, Taylor and Francis, Bristol, 2003.

[**6**] Samelson, H., *Topology of Lie groups*, Bull. Amer. Math. Soc. **52** (1952), 2–37.

[**7**] Spanier, E., *Algebraic Topology*, McGraw-Hill Book Company, New York, 1966.

References

Adhikari, M.R., Adhikari, A.: Basic Modern Algebra with Applications. Springer, New Delhi (2014)

Bredon, G.: Topology and Geometry. Springer, Heidelberg (1993) (GTM 139)

Borel, A.: Topology of Lie groups and characteristic classes. Bull. Am. Math. Soc. **61**, 397–432 (1955)

Chevelly, C.: Theory of Lie Groups I. Princeton University Press, Princeton (1957)

Dieudonné, J.: A History of Algebraic and Differential Topology, 1900–1960. Modern Birkhäuser, Boston (1989)

Donalson, S.: Self-dual connections and the topology of smooth 4-manifolds. Bull. Am. Math. Soc. **8**(1), 81–83 (1983)

Dugundji, J.: Topology. Allyn & Bacon, Newtown (1966)

Dupont, J.: Fibre Bundles. Aarhus Universitet, Denmark (2003)

Lie, S.: Theorie der Transformations gruppen. Math. Annalen **16**, 441–528 (1880)

Lie, S., Engel, F.: Theorie der Transformations gruppen, Bd III. Teubner, Leipzig (1893)

Milnor, J.: On manifolds homeomorphic to the 7-sphere. Ann. Math. **64**(2), 399–405 (1956)

Nakahara, M.: Geometry, Topology and Physics, Institute of Physics Publishing. Taylor and Francis, Bristol (2003)

Pontragin, L.: Topological Groups. Princeton University Press, Princeton (1939)

Samelson, H.: Topology of Lie groups. Bull. Am. Math. Soc. **52**, 2–37 (1952)

Sorani, G.: An Introduction to Real and Complex Manifolds. Gordon and Breech, Science Pub., Inc, New York (1969)

Spanier, E.: Algebraic Topology. McGraw-Hill Book Company, New York (1966)
Switzer, R.M.: Algebraic Topology-Homotopy and Homology. Springer, Berlin (1975)

Appendix B
Categories, Functors and Natural Transformations

This appendix conveys category theory through the study of categories, functors, and natural transformations with an eye to study algebraic topology which consists of the constructions and use of functors from certain category of topological spaces into an algebraic category. Algebraic topology studies techniques for forming algebraic images by mechanisms that create these images which are known as functors. They have the characteristic feature that they form algebraic images of spaces and project continuous maps into their corresponding algebraic images. Category theory plays an important role for the study of homotopy, homology and cohomology theories which constitute the basic text of algebraic topology. So the readers of algebraic topology can not escape learning the concepts of categories, functors and natural transformations.

The present book uses category theory and conveys a study of some important functors such as homotopy, homology and cohomology functors in addition to *adjoint functor, representable functor, abelianization functor, Brown functor, and infinite symmetric product functor.* All constructions in algebraic topology are in general functorial. Fundamental groups, higher homotopy groups, homology and cohomology groups are not only algebraic invariants of the underlying topological space, in the sense that two topological spaces which are homeomorphic have the isomorphic associated groups (or modules) but also they are homotopy invariants in the sense that homotopy equivalent spaces have isomorphic algebraic structures. Moreover, corresponding to a continuous mapping of topological spaces the induced group (or module) homomorphism on the associated groups (modules) can be used to show the non-existence (or much more deeply, existence) of a continuous mapping of the spaces.

Historically, the whole idea of category theory arose through the field of algebraic topology. The first and the simplest realization of this idea is the fundamental group (or Poincaré group) of a pointed topological space. The concepts of categories, functors, natural transformations and duality were introduced during 1942–1945 by S. Eilenberg (1913–1998) and S. MacLane (1909–2005).[1] Originally, the purpose

---

[1](i) Natural isomorphism in group theory, *Proc. Nat. Acad Sc.*, USA **28** (1942), 537–544.

M.R. Adhikari, *Basic Algebraic Topology and its Applications*,
DOI 10.1007/978-81-322-2843-1

of these notions was to provide a technique for classifying certain concepts, such as that of natural isomorphism. Many concepts of algebraic topology are unified and explained by category theory which is a very important branch of modern mathematics. This branch has been quite rapidly growing both in contents and applicability to other branches of mathematics.

For this chapter, the books and papers Adhikari and Adhikari (2014), Eilenberg and MacLane (1942, 1945), Eilenberg and Steenrod (1952), Gray (1975), Hatcher (2002), MacLane (1972), Rotman (1988), Spanier (1966), Steenrod (1967) and some others are referred in Bibliography.

B.1 Categories: Introductory Concepts

This section introduces the concept of 'category' to specify a class of objects for their study. It is observed that to define a new class of mathematical objects in modern mathematics, it becomes necessary to specify certain types of functions between the objects such as topological spaces and continuous maps, groups and homomorphisms, modules and module homomorphism. A formulation of this observation leads to the concept of 'categories'. A 'category' may be thought roughly as consisting of sets, possibly with additional structures, and functions, possibly preserving additional structures. More precisely, a category can be defined with the following characteristics.

Definition B.1.1 A category $\mathcal{C}$ consists of

(a) a class of objects $X, Y, Z, \dots$ denoted by $ob(\mathcal{C})$;

(b) for each ordered pair of objects X, Y, a set of morphisms with domain X and range Y denoted by $\mathcal{C}(X, Y)$ or simply $\text{mor}\,(X, Y)$; i.e., if $f \in \text{mor}\,(X, Y)$, then X is called the domain of f and Y is called the co-domain (or range) of f: one also writes $f : X \to Y$ or $X \xrightarrow{f} Y$ to denote the morphism from X to Y;

(c) for each ordered triple of objects X, Y and Z and a pair of morphisms $f : X \to Y$ and $g : Y \to Z$, their composite denoted by $gf : X \to Z$ i.e., if $f \in \text{mor}\,(X, Y)$ and $g \in \text{mor}\,(Y, Z)$, then their composite $gf \in \text{mor}\,(X, Z)$ and satisfies the following two axioms:

 (i) *associativity*: if $f \in \text{mor}\,(X, Y)$, $g \in \text{mor}\,(Y, Z)$ and $h \in \text{mor}\,(Z, W)$, then $h(gf) = (hg)f \in \text{mor}\,(X, W)$;

 (ii) *identity*: for each object Y in $\mathcal{C}$, there is a morphism $1_Y \in \text{mor}\,(Y, Y)$ such that if $f \in \text{mor}\,(X, Y)$, then $1_Y f = f$ and if $h \in \text{mor}\,(Y, Z)$, then $h 1_Y = h$. Clearly, 1_Y is unique.

If the class of objects is a set, the category is said to be *small*.

(ii) General theory of natural equivalence, *Trans Amer. Math. Soc.*, **58** (1945), 231–294.

Example B.1.2 **(i)** Sets and functions form a category denoted by $\mathcal{S}et$.
(Here the class of objects is the class of all sets and for sets X and Y, mor (X, Y) equals the set of functions from X to Y and the composition has the usual meaning, i.e., usual composition of functions).

(ii) Finite sets and functions form a category denoted by $\mathcal{S}et_F$.

(iii) Groups and homomorphisms form a category denoted by $\mathcal{G}rp$.
(Here the class of objects is the class of all groups and for groups X and Y, mor (X, Y) equals the set of homomorphisms from X to Y and the composition has the usual meaning).

(iv) Abelian groups and homomorphisms form a category denoted by $\mathcal{A}b$.

(v) Rings and homomorphisms form a category denoted by $\mathcal{R}ing$.

(vi) R-modules and R-homomorphisms form a category denoted by $\mathcal{M}od_R$.

(vii) Exact sequences of R-modules and R-homomorphisms form a category.

(viii) Topological spaces and continuous maps form a category denoted by $\mathcal{T}op$.

(ix) Topological spaces and homotopy classes of maps form a category denoted by $\mathcal{H}tp$.
(Here the class of objects is the class of all topological spaces and for topological spaces X and Y, mor (X, Y) equals the set of homotopy classes of maps from X to Y and the composition has the usual meaning.)

Definition B.1.3 A subcategory $\mathcal{C}' \subset \mathcal{C}$ is a category such that

(a) the objects of $\mathcal{C}'$ are also objects of $\mathcal{C}$, i.e., $ob(\mathcal{C}') \subset ob(\mathcal{C})$;

(b) for objects X' and Y' of $\mathcal{C}'$, $\mathcal{C}'(X', Y') \subset \mathcal{C}(X', Y')$;

(c) if $f' : X' \to Y'$ and $g' : Y' \to Z'$ are morphisms of $\mathcal{C}'$, their composite in $\mathcal{C}'$ equals their composite in $\mathcal{C}$.

Definition B.1.4 A subcategory $\mathcal{C}'$ of $\mathcal{C}$ is said to be a full subcategory of $\mathcal{C}$ if for objects X' and Y' in $\mathcal{C}'$, $\mathcal{C}(X', Y') = \mathcal{C}'(X', Y')$.

The category in Example B.1.2(ii) is a full subcategory of the category in Example B.1.2(i).

Remark B.1.5 The category $\mathcal{S}et_F$ in Example B.1.2(ii) is a subcategory of the category $\mathcal{S}et$ in Example B.1.2(i). On the other hand, the categories in Example B.1.2(iii)–(vii) are not subcategories of the category in Example B.1.2(i), because each object of one of the former categories consists of a set, endowed with an additional structure (hence different objects in these categories may have the same underlying sets).

Remark B.1.6 In category Example B.1.2(ix), the morphisms are not functions and so this category is not a subcategory of the category in Example B.1.2(i).

Let $\mathcal{C}$ be a category and $A, B, C, \ldots$ be objects of $\mathcal{C}$.

Definition B.1.7 A morphism $f : A \to B$ in $\mathcal{C}$ is called a *coretraction* if there is a morphism $g : B \to A$ in $\mathcal{C}$ such that $gf = 1_A$. In this case g is called a left inverse of f and f is called a right inverse of g and A is called a *retract* of B.

Dually we say that f is a retraction if there is a morphism $g' : B \to A$ such that $fg' = 1_B$ in $\mathcal{C}$. In this case g' is called a right inverse of f.

Definition B.1.8 A two-sided *inverse* (or simply an inverse) of f is a morphism which is both a left inverse of f and a right inverse of f.

Lemma B.1.9 *If $f : A \to B$ in $\mathcal{C}$ has a left inverse and a right inverse, they are equal.*

Proof Let $g' : B \to A$ be a left inverse of f and $g'' : B \to A$ a right inverse of f, then $g'f = 1_A$ and $fg'' = 1_B$. Now $g' = g'1_B = g'(fg'') = (g'f)g'' = 1_A$ $g'' = g''$. ❑

Definition B.1.10 A morphism $f : A \to B$ is called an *equivalence* (or an *isomorphism*) in a category $\mathcal{C}$ denoted by $f : A \approx B$ if there is a morphism $g : B \to A$ which is a two-sided inverse of f.

Example B.1.11 An equivalence in the category $\mathcal{T}op$ of topological spaces and their continuous maps is a homeomorphism and that in category $\mathcal{H}tp$ of topological spaces and their homotopy classes of maps is a homotopy equivalence.

Remark B.1.12 An equivalence $f : A \approx B$ has a unique inverse denoted by $f^{-1} : B \to A$ and f^{-1} is also an equivalence.

Definition B.1.13 Two objects A and B in $\mathcal{C}$ are said to be equivalent denoted by $A \approx B$ if there is an equivalence $f : A \approx B$ in $\mathcal{C}$.

Remark B.1.14 As the composite of equivalences is an equivalence, the relation of being equivalent is an equivalence relation on any set of objects of a category $\mathcal{C}$.

B.2 Functors: Introductory Concepts and Examples

This section introduces the concept of functors and studies functors of different nature such as *covariant and contravariant functors, adjoint functor, representable functor, abelianization functor, Brown functor, and infinite symmetric product functor which play a key role in algebraic topology.* The main interest in category theory is in the maps from one category to another. Those maps which have the natural properties of preserving identities and composites are called *functors* (covariant or contravariant). An algebraic representation of topology is a mapping from topology to algebra. Such a representation, formally called a functor, converts a topological problem into an algebraic one. The concept of algebraic functors is very important in algebraic topology. For example, homotopy and homology theories provide a sequence of covariant (algebraic) functors. On the other hand, cohomology theory provides a sequence of contravariant (algebraic) functors.

B.2.1 Functors: Introductory Concepts

Definition B.2.1 Let $\mathcal{C}$ and $\mathcal{D}$ be categories. A *covariant functor* (or *contravariant functor*) T from $\mathcal{C}$ to $\mathcal{D}$ consists of

(i) an object function which assigns to every object X of $\mathcal{C}$ an object $T(X)$ of $\mathcal{D}$; and

(ii) a morphism function which assigns to every morphism $f : X \to Y$ in $\mathcal{C}$, a morphism $T(f) : T(X) \to T(Y)$ (or $T(f) : T(Y) \to T(X)$) in $\mathcal{D}$ such that

  (a) $T(1_X) = 1_{T(X)}$;

  (b) $T(gf) = T(g)T(f)$ (or $T(gf) = T(f)T(g)$) for $g : Y \to W$ in $\mathcal{C}$.

Example B.2.2 **(i)** **(Forgetful functor)** There is a covariant functor from the category of groups and homomorphisms to the category of sets and functions which assigns to every group its underlying set. This functor is called a *forgetful functor* because it forgets the structure of a group.

(ii) (Hom $_R$ **functor)** Let R be a commutative ring. Given a fixed R-module M_0, there is a covariant functor π_{M_0} (or *contravariant functor* π^{M_0}) from the category of R-modules and R-homomorphisms to itself which assigns to an R-module M the R-module $\text{Hom}_R\,(M_o, M)$ (or $\text{Hom}_R(M, M_o)$) and if $\alpha : M \to N$ is an R-module homomorphism, then
$\pi_{M_0}(\alpha) : \text{Hom}\,_R(M_0, M) \to \text{Hom}\,_R(M_0, N)$ is defined by $\pi_{M_0}(\alpha)(f) = \alpha f$ $\forall\, f \in \text{Hom}\,_R(M_0, M)$
($\pi^{M_0}(\alpha) : \text{Hom}\,_R(N, M_0) \to \text{Hom}\,_R(M, M_0)$ is defined by
$\pi^{M_0}(\alpha)(f) = f\alpha\, \forall\, f \in \text{Hom}\,_R(N, M_0)$).

(iii) **(Dual functor)** Let $\mathcal{C}$ be any category and $C \in ob(\mathcal{C})$. Then there is a *covariant functor* $h_C : \mathcal{C} \to \mathcal{S}et$ (category of sets and functions), where the object function is defined by $h_C(A) = \mathcal{C}(C, A)$ (set of all morphisms from the object C to the object A in $\mathcal{C}$) $\forall$ objects $A \in ob(\mathcal{C})$ and for $f : A \to B$ in $\mathcal{C}$, the morphism function $h_C(f) : h_C(A) \to h_C(B)$ is defined by $h_C(f)(g) = fg\, \forall\, g \in h_C(A)$ (the right hand side is the composite of morphisms in $\mathcal{C}$).
Its dual functor h^C defined in a usual manner is a contravariant functor.

Remark B.2.3 A functor from a category $\mathcal{C}$ to itself is sometimes called a functor on $\mathcal{C}$. Any contravariant functor on $\mathcal{C}$ corresponds to a covariant functor on $\mathcal{C}^0$ and vice versa. Thus any functor can be regarded as a covariant (or contravariant) functor on a suitable category. In spite of this, we consider covariant as well as contravariant functors on $\mathcal{C}$.

Definition B.2.4 A functor $T : \mathcal{C} \to \mathcal{D}$ is called

(i) faithful if the mapping $T : \mathcal{C}(A, B) \to \mathcal{D}(T(A), T(B))$ is injective;

(ii) full if the mapping $T : \mathcal{C}(A, B) \to \mathcal{D}(T(A), T(B))$ is surjective; and

(iii) an embedding if T is faithful and $T(A) = T(B) \implies A = B$.

Definition B.2.5 A category $\mathcal{C}$ is called concrete if there is a faithful functor $T : \mathcal{C} \to \mathcal{S}et$.

Theorem B.2.6 *Let $F : \mathcal{C}_1 \to \mathcal{C}_2$ be a functor from a category $\mathcal{C}_1$ to a category $\mathcal{C}_2$. If two objects X and Y are isomorphic in $\mathcal{C}_1$ then the objects $F(X)$ and $F(Y)$ are isomorphic in $\mathcal{C}_2$.*

Proof Let $f : X \to Y$ be a isomorphism in $\mathcal{C}_1$. Then there exists a morphism $g : Y \to X$ in $\mathcal{C}_1$ such that $g \circ f = 1_X$ and $f \circ g = 1_Y$. If F is a covariant functor, then $F(g) \circ F(f) = 1_{F(X)}$ and $F(f) \circ F(g) = 1_{F(Y)}$, which imply that the objects $F(X)$ and $F(Y)$ are isomorphic in $\mathcal{C}_2$. If F is a contravariant functor, then the proof is similar. ❑

Remark B.2.7 Consider a functor from the category $\mathcal{T}op$ to another category, say the category $\mathcal{G}rp$ of groups. Let X and Y be objects in $\mathcal{T}op$. Then $F(X)$ and $F(Y)$ are objects in $\mathcal{G}rp$. If $F(X)$ and $F(Y)$ are not isomorphic, then X and Y can not be homeomorphic. To the contrary if $F(X)$ and $F(Y)$ are isomorphic, then X and Y may not be homeomorphic. For example, for the covariant functor π_m, $\pi_m(S^2) \cong \pi_m(S^3)$ for $m \geq 3$ but S^2 and S^3 are not homeomorphic (see Hopf fibering of spheres, Chap. 7).

Definition B.2.8 (*Bifunctor*) A bifunctor is a mapping $T : \mathcal{C}_1 \times \mathcal{C}_2 \to \mathcal{C}$, defined on the Cartesian product of two categories $\mathcal{C}_1$ and $\mathcal{C}_2$ with values in the category $\mathcal{C}$, which assigns to each pair of objects in $A_1 \in \mathcal{C}_1$ and $A_2 \in \times\mathcal{C}_2$, some object $A \in \mathcal{C}$, and to each pair of morphisms

$$\alpha : A_1 \to A_1', \ \beta : A_2 \to A_2'$$

the morphism

$$T(\alpha, \beta) : T(A_1', A_2) \to (A_1, A_2')$$

The following conditions

$$T(1_{A_1}, 1_{A_2}) = 1_{T(A_1, A_2)}$$

must also be satisfied. In such a case one says that the functor is contravariant with respect to the first argument and covariant with respect to the second.

Remark B.2.9 Similarly one can define bifunctors contravariant in both arguments and covariant in the first and contravariant in the second argument. Thus a bifunctor is a functor whose domain is product category.

Example B.2.10 Hom functor is of the type $\mathcal{C}^{op} \times \mathcal{C} \to \mathcal{S}et$. It is a bifunctor in two arguments. The Hom functor is a natural example; it is contravariant in one argument, covariant in the other.

B.2.2 Abelianization Functor

This subsection defines abelianization functor. The concept of abelianized groups is utilized to establish a connection between the fundamental group and first homology group of a pointed space. If G is a group (not necessarily abelian), then the commutator subgroup $[G, G]$ is a normal subgroup of G and the quotient group $G/[G, G]$ is called the abelianized group of G, and denoted by G^{ab} : If $f : G \to H$ is a homomorphism, it induces a homomorphism

$$f_* : G/[G, G] \to H/[H, H].$$

This gives a functor $G \mapsto G^{ab}$, $f \mapsto f_*$ which is called 'abelianization functor' from the category of groups and homomorphisms to the category of abelian groups and homomorphisms.

B.2.3 Adjoint Functor

This subsection discusses adjoint functors which are very important in the study of homotopy theory. Recall that the loop functor Ω is a covariant functor from the category of pointed topological spaces and continuous maps to the category of H-groups and continuous homomorphisms such that the functor Ω also preserves homotopies (see Chap. 2). In the same chapter we have also described suspension space which is dual to the loop space. Let X be a pointed topological space with base point x_0. Then the suspension space of X, denoted by ΣX, is defined to be the quotient space of $X \times I$ in which $(X \times \{0\}) \cup \{x_0\} \times I) \cup (X \times 1)$ is identified to a single point. If $(x, t) \in X \times I$, $[x, t]$ denotes the corresponding points of ΣX under the quotient map $X \times I \to \Sigma X$. The point $[x_0, 0] \in \Sigma X$ is also denoted by x_0 and ΣX becomes a pointed space with base point x_0. If $f : X \to Y$ is a continuous map, then $\Sigma f : \Sigma X \to \Sigma Y$ is defined by $(\Sigma f)([x, t]) = [f(x), t]$. Hence Σ is a covariant functor from the category of pointed topological spaces and continuous maps to itself. Recall that Σ is also a covariant functor from the category of pointed topological spaces and continuous maps to the category of H-cogroups and homomorphisms (see Chap. 2).

Definition B.2.11 The functors Ω and Σ defined from the category of pointed topological spaces and continuous maps to itself form a pair of functors, called an adjoint pair in the sense that for pointed topological spaces X and Y, there is an equivalence between mor $(\Sigma X, Y) \approx$ mor $(X, \Omega Y)$, where both sides are interpreted as the set of morphisms in the category of pointed topological spaces and continuous maps. If $f : X \to \Omega Y$ is a morphism, then the corresponding morphism $\tilde{f} : \Sigma X \to Y$ is defined by $\tilde{f}([x, t]) = f(x)(t)$ for all $x \in X$ and $t \in I$.

Definition B.2.12 Corresponding to the adjoint pair of functors Ω and Σ, the continuous maps $f : X \to \Omega Y$ and $\tilde{f} : \Sigma X \to Y$ in the category pointed topological spaces and continuous maps are said to be adjoint to each other if the morphism $\tilde{f} : \Sigma X \to Y$ is defined by $\tilde{f}([x, t]) = f(x)(t)$ for all $x \in X$ and $t \in I$.

Remark B.2.13 The adjoint relation mor $(\Sigma X, Y) \approx$ mor $(X, \Omega Y)$ holds, because base point preserving maps $\Sigma X \to Y$ are exactly the same as the base point preserving maps $X \to \Omega Y$, the correspondence is given by assigning to $f : \Sigma X \to Y$ the family of loops by restricting f to the images of the segment $\{x\} \times I$ in ΣX.

B.2.4 *Brown Functor*

This subsection defines Browns functor which is used to prove Brown representability theorem.

Definition B.2.14 Let $T : \mathcal{H}tp_* \to \mathcal{S}et_*$ be a contravariant functor from the homotopy category of pointed topological spaces and homotopy classes of their maps to the category of pointed sets and their maps. If $i : X \hookrightarrow Y$ in $\mathcal{H}tp_*$, $u \in T(Y)$ and $u|_X$ denotes the element $T([i])(u) \in T(X)$, then T is called a Brown functor if it satisfies following two axioms:

B(1): Wedge axiom If $\{X_j\}$ is a family of pointed topological spaces and $i_j : X_j \hookrightarrow \vee X_j$ is the inclusion, then $T(i_j) : T(\vee_j X_j) \to \Pi_j T(X_j)$ is an equivalence of sets;

B(2): Mayer–Vietoris axiom For any excisive triad $(X; A, B)$ (i.e., $X =$ Int $(A) \cup$ Int (B)) and for any $u \in T(A)$ and $v \in T(B)$ such that $u|_{A\cap B} = v|_{A\cap B}$, there exists an element $w \in T(X)$ such that $w|_A = u$ and $w|_B = v$.

B.2.5 *Infinite Symmetric Product Functor*

This subsection defines 'infinite symmetric product functor'. This functor is important to prove Dold–Thom theorem, which is a very key result of algebraic topology. Let X be a pointed topological space with base point x_0 and $X^n = X \times \times \ldots \times X$ be its nth cartesian product for $n \geq 1$. If S_n denotes the symmetric groups of the set $\{1, 2, \ldots, n\}$, then there is a right action on X^n, which permutes the coordinates, i.e., for $\sigma \in S_n$, we define $(x_1, x_2, \ldots, x_n)\cdot\sigma = (x_{\sigma(1)}, x_{\sigma(2)}, \ldots, x_{\sigma(n)}), x_i \in X$. The orbit space $X^n \mod S_n$ of this action denoted by $SP^n X$ and is called the nth symmetric product of X. The equivalence class of $(x_1, x_2, \ldots, x_n)$ is denoted by $[x_1, x_2, \ldots, x_n]$.

Define inclusions $SP^n X \to SP^{n+1} X, [x_1, x_2, \ldots, x_n] \mapsto [x_0, x_1, \ldots, x_n]$ for $n \geq 1$ and form the union $SP^\infty X = \bigcup_n SP^n X$ equipped with the union topology

$$\begin{array}{ccccc} \cdots \longrightarrow & SP^nX & \longrightarrow & SP^{n+1}X \cdots & \longrightarrow \\ & \downarrow f_*^{(n)} & & \downarrow f_*^{(n+1)} & \\ \cdots \longrightarrow & SP^nY & \longrightarrow & SP^{n+1}Y \cdots & \longrightarrow \end{array}$$

Fig. B.1 Commutative diagram for symmetric product

(weak topology), which means that a subset $A \subset SP^\infty X$ is closed iff $A \cap SP^n X$ is closed for each $n \geq 1$.

Definition B.2.15 For a pointed space X the space $SP^\infty X$ is called the infinite symmetric product of X.

Remark B.2.16 For a pointed space X the elements of $SP^n X$ may be considered as unordered n-tuples $[x_1, x_2, \ldots, x_n]$, where $n \geq 1$. Then $SP^\infty X$ is a pointed space with the base point $0 = [x_0]$. There exists a natural inclusion $i : X \hookrightarrow SP^\infty X$, where $X = SP^1 X$.

Let $f : X \to Y$ be a base point preserving continuous map. Then f induces maps $f_*^{(n)} : SP^n X \to SP^n Y$, which are compatible with the action of the group S_n. Moreover, these maps make the diagram in Fig. B.1 commutative and hence induce a map $\tilde{f}_* : SP^\infty X \to SP^\infty Y$.

Remark B.2.17 SP^∞ satisfies the functorial properties.

Proposition B.2.18 *Let $\mathcal{Top}_*$ be the category of pointed topological spaces.*

(i) *If $f = 1_X$ in $\mathcal{Top}_*$, then $f_* = 1_{SP^\infty(X)}$;*
(ii) *If $f : X \to Y$ and $g : Y \to Z$ are in $\mathcal{Top}_*$, then $(g \circ f)_* = g_* \circ f_* : SP^\infty X \to SP^\infty Z$.*

Hence it follows that SP^∞ is a covariant functor. It proves the following proposition.

Proposition B.2.19 *$SP^\infty : \mathcal{Top}_* \to \mathcal{Top}_*$ is a covariant functor.*

Example B.2.20 **(i)** For each integer $n \geq 1$, $SP^\infty S^n$ is an Eilenberg–MacLane space $K(\mathbf{Z}, n)$, which is a CW-complex having just one nontrivial homotopy group $\pi_n(K(\mathbf{Z}, n)) \cong \mathbf{Z}$.
(ii) $SP^n S^2 \approx \mathbf{C}P^n$.
(iii) $SP^\infty S^2 \approx \mathbf{C}P^\infty$.

B.3 Natural Transformations

This section introduces the concept of natural transformations. In some occasions we have to compare functors with each other. We do this by means of suitable maps between functors, called natural transformations which are very important in algebraic topology. For example, this concept is used in homology and cohomology theories, to compare homotopy and homology groups and applies to Yoneda lemma and 'representable functor'.

B.3.1 Introductory Concepts

Definition B.3.1 Let $\mathcal{C}$ and $\mathcal{D}$ be categories. Suppose T_1 and T_2 are functors of the same variance (either both covariant or both contravariant) from $\mathcal{C}$ to $\mathcal{D}$. A *natural transformation* ϕ from T_1 to T_2 is a function from the objects of $\mathcal{C}$ to morphisms of $\mathcal{D}$ such that for every morphism $f : X \to Y$ in $\mathcal{C}$ the appropriate one of the following conditions hold:

$\phi(Y)T_1(f) = T_2(f)\phi(X)$ (when T_1 and T_2 are both covariant functors)
or $\phi(X)T_1(f) = T_2(f)\phi(Y)$ (when T_1 and T_2 are both contravariant functors).

Definition B.3.2 Let $\mathcal{C}$ and $\mathcal{D}$ be categories and T_1, T_2 be functors of the same variance from $\mathcal{C}$ to $\mathcal{D}$. If ϕ is a natural transformation from T_1 to T_2 such that $\phi(X)$ is an equivalence in $\mathcal{D}$ for each object X in $\mathcal{C}$, then ϕ is called a natural equivalence.

Example B.3.3 Let R be a commutative ring and $\mathcal{M}od$ be the category of R-modules and R-homomorphism, M and N be objects in $\mathcal{M}od$. Suppose $g : M \to N$ is a morphism in $\mathcal{M}od$. So by Example B.2.2(ii), π_M, π_N are both covariant functors and π^M, π^N are both contravariant functors from $\mathcal{M}od$ to itself. Then there exists a natural transformation $g^* : \pi_N \to \pi_M$, where $g^*(X) : \pi_N(X) \to \pi_M(X)$ is defined by $g^*(X)(h) = hg$ for every object X in $\mathcal{M}od$ and for all $h \in \pi_N(X)$; and a natural transformation

$g_* : \pi_M \to \pi_N$, where $g_*(X)$ is defined in an analogous manner.
If g is an equivalence in $\mathcal{M}od$, then both the natural transformations g_* and g^* are natural equivalences.

Example B.3.4 If $f : X \to \Omega Y$ is a morphism in the category of pointed topological spaces and continuous maps, then its corresponding adjoint morphism $\tilde{f} : \Sigma X \to Y$ is defined by $\tilde{f}([x, t]) = f(x)(t)$ for all $x \in X$ and $t \in I$. The equivalence $f \leftrightarrow \tilde{f}$ comes from a natural equivalence from the functor mor $(\Sigma-, -)$ to the functor $(-, \Omega-)$.

B.3.2 Yoneda Lemma

This lemma provides important tools in algebraic topology. Moreover, it defines a new functor called 'representable functor'.

Lemma B.3.5 (Yoneda) *Let $\mathcal{C}$ be any category and T be a covariant functor from $\mathcal{C}$ to $\mathcal{S}et$ (category of sets and set functions). Then for any object C in $\mathcal{C}$, there is an equivalence $\theta = \theta_{C,T} : (h_C, T) \to T(C)$, where (h_C, T) is the class of natural transformations from the set valued functor h_C to the set valued functor T such that θ is natural in C and T.*

Proof Let $\eta : h_C \to T$ be a natural transformation. Define $\theta : (h_C, T) \to T(C)$, $\eta \mapsto \eta(C)(1_C) \in T(C)$ and $\rho : T(C) \to (h_C, T)$, given by $\rho(x)(X)(f) = T(f)(x)$

$\forall x \in T(C)$, $X \in \mathcal{C}$ and $f \in \text{mor}\,(C, X)$. Then $\rho(x) \in (h_C, T)$. Moreover, $\rho \circ \theta =$ identity and $\theta \circ \rho =$ identity show that θ is an equivalence. it is left as an exercise to show that θ is natural in C and T. ❑

Remark B.3.6 For detailed proof of Yoneda lemma see Adhikari and Adhikari (2014). For dual of the result of Yoneda lemma see Ex. 11 of Sect. B.5.

Definition B.3.7 Let $T : \mathcal{C} \to \mathcal{S}et$ be a representable (contravariant) functor and C is an object in $\mathcal{C}$. If $\eta : h^C \to T$ is natural equivalence then under the equivalence θ given by Yoneda lemma, $\theta : (h^C, T) \to T(C)$ the associated element $\theta(\eta)$ of η is defined by $\theta(\eta) = \eta(C)(1_C) \in T(C)$.

Example B.3.8 Let $\mathcal{G}rp$ be the category of groups and homomorphisms; $\mathcal{S}et$ be the category of sets and functions and $S : \mathcal{G}rp \to \mathcal{S}et$ be the forgetful functor which assigns to each group G its underlying set SG. Then

(i) there is a natural equivalence from the covariant functor $h_{\mathbf{Z}}(= Hom(\mathbf{Z}, -))$ to the covariant functor $\mathcal{S}et$; and
(ii) there is an equivalence $\theta : (S, S) \to S\mathbf{Z}$.

B.3.3 Representable Functor

This functor is defined to prove Brown Representability Theorem. Let $\mathcal{C}$ be a category. Then each object C of $\mathcal{C}$ defines a contravariant functor $h^C : \mathcal{C} \to \mathcal{S}et$ which assigns to an object X of $\mathcal{C}$, the set mor (X, C) (the set of all morphisms for the object X to the object C in $\mathcal{C}$) and for the morphism $f : X \to Y$ in $\mathcal{C}$, $h^C(f) = f^* : h^C(Y) \to h^C(X)$ is defined by $f^*(g) = g \circ f$, $\forall g \in h^C(Y)$.

Definition B.3.9 Let $\mathcal{C}$ be a category. A contravariant functor $T : \mathcal{C} \to \mathcal{S}et$ is said to be representable if there is an object C in $\mathcal{C}$ and a natural equivalence $\psi : h^C \to T$. The object C is called a classifying space for T and C is said to be representable.

Definition B.3.10 If T is a representable functor and $\theta : h^C \to T$ is a natural equivalence, then the associated element $\theta(\eta) = \eta(C)(1_C) \in T(C)$ is called the universal element for T and C is called the representable object.

B.4 Convenient Category of Topological Spaces

This section describes an important category called the category of compactly generated Hausdorff spaces convenient for the study of homology and cohomology theories. This category introduced by N. Steenrod (1910–1971) in 1967, includes almost all important spaces in topology. For this section the paper Steenrod (1967) and the book Gray (1975) are referred.

Definition B.4.1 A Hausdorff space is said to be a compactly generated Hausdorff space if each subset which intersects every compact set in a closed set is itself closed.

The category of compactly generated Hausdorff spaces and their continuous maps is denoted by $\mathcal{CG}$.

Proposition B.4.2 *Let X be a Hausdorff space and if for each subset S and each limit point x of S, there exists a compact set C in X such that x is a limit point of $S \cap C$, then $X \in \mathcal{CG}$.*

Proof To prove this proposition it is sufficient to show that if each limit relation in X lies in some compact subset of X, then $X \in \mathcal{CG}$. Let S meet each compact set in a closed set, and let x be a limit point of S. Then by hypothesis, there exists a compact set C such that x is a limit point of $S \cap C$. Since $S \cap C$ is closed, $x \in S \cap C$ and hence $x \in S$. This shows that S is closed and hence $X \in \mathcal{CG}$. ❑

Definition B.4.3 A topological space X is said to satisfy the first axiom of countability if there exists a countable open base about every point in X and X is then called a first countable space.

Example B.4.4 Every metrizable space is a first countable space.

Proposition B.4.5 *The category $\mathcal{CG}$ contains all locally compact spaces and all topological spaces satisfies the first axiom of countability.*

Proof If X is locally compact, we take C to be the compact closure of a neighborhood of $x \in X$. If X is first countable, we take C to consist of x and a sequence in S converging to x. ❑

Corollary B.4.6 *The category $\mathcal{CG}$ includes all metrizable spaces.*

Remark B.4.7 All Hausdorff spaces are not in $\mathcal{CG}$.

Example B.4.8 Let X be the set of all ordinal numbers preceding and including the first noncountable ordinal Ω. Endow X the topology defined by its natural order. Let S be the subspace of the topological space X obtained by deleting all limit ordinals except Ω. Since each infinite set contains a sequence converging to a limit ordinal of second kind, the only compact subsets of S are finite sets. This shows that the set $S - \Omega$ meets each compact set in a closed set, but is not closed in S, because it has Ω as a limit point.

Remark B.4.9 The Example B.4.8 shows that a subspace S of a compactly generated topological space X need not be compactly generated.

Proposition B.4.10 *Let $X \in \mathcal{CG}$ and Y be a Hausdorff space. If $f : X \to Y$ is a quotient map, then $Y \in \mathcal{CG}$.*

Proof Let $B \subset Y$ meet each compact set of Y in a closed set. If C is a compact set in X, then $f(C)$ is compact and hence $B \cap f(C)$ is closed. This shows that $f^{-1}(B \cap f(C))$ is closed and thus $f^{-1}(B \cap f(C)) \cap C = f^{-1}(B) \cap C$ is closed. This implies that $f^{-1}(B)$ meets each compact set of X in a closed set. As $X \in \mathcal{CG}$, $f^{-1}(B)$ is closed. Again since f is a quotient map, B must be closed in Y. Consequently, $Y \in \mathcal{CG}$. ❑

Remark B.4.11 It follows from the preceding results that the category $\mathcal{CG}$ is larger in the sense that it contains almost all important spaces in topology.

For example, it contains all continuous maps between any two of its spaces.

The following proposition facilitates to examine the continuity of a function.

Proposition B.4.12 *Let $X \in \mathcal{CG}$ and Y be a Hausdorff space and $f : X \to Y$ be continuous on each compact subset of X. Then f is continuous.*

Proof Let $B \subset Y$ be closed and A be compact in X. Since Y is a Hausdorff space and $f|_A$ is continuous, $f(A)$ is compact and hence $f(A)$ is closed in Y. This shows that $B \cap f(A)$ is closed, and hence also

$$(f|_A)^{-1}(B \cap f(A)) = (f^{-1}(B)) \cap A.$$

Since $X \in \mathcal{CG}$, $f^{-1}(B)$ is closed in X. This proves the continuity of f. ❑

Definition B.4.13 Let X be a Hausdorff space. The associated compactly generated topological space $k(X)$ is the set X with the topology defined as follows:
a closed set of $k(X)$ is a set that meets each compact set of X in a closed set. If $f : X \to Y$ is a mapping of Hausdorff spaces, $k(f)$ denotes the same function $k(X) \to k(Y)$.

Theorem B.4.14 (Steenrod) *Let X be a Hausdorff space and $k(X)$ be its associated compactly generated space. Then*

- **(i)** *the identity function $1_d : k(X) \to X$ is continuous;*
- **(ii)** *$k(X)$ is a Hausdorff space;*
- **(iii)** *$k(X)$ and X have the same compact sets;*
- **(iv)** *$k(X) \in \mathcal{CG}$;*
- **(v)** *if $X \in \mathcal{CG}$, then $1_d : k(X) \to X$ is a homeomorphism;*
- **(vi)** *if $f : X \to Y$ is continuous on compact sets, then $k(f) : k(X) \to k(Y)$ is continuous;*
- **(vii)** *$1_{d*} : \pi_n(k(X), *) \to \pi_n(X, *)$ establishes a 1-1 correspondence for all n and all $*$, where $\pi_n(X, *)$ is the nth homotopy group of $(X, *)$.*

Proof **(i)** Let $A \subset X$ be closed in X and C be compact in X. Then C is closed in X and hence $A \cap C$ is also closed in X. This implies that A is also closed in $k(X)$.

(ii) Since X is a Hausdorff space, **(ii)** follows from **(i)**.

(iii) Let A be a compact subset in $k(X)$. Then by **(i)**, A is also compact in X. If C is compact in X, and $\tilde{C}$ is the same set C with its relative topology from $k(X)$. Then the identity map $\tilde{C} \to C$ is continuous by **(i)**. We claim that its inverse is also continuous. Let B be a closed set of $\tilde{C}$. Then by definition B meets every compact set of X in a closed set. This shows that $B \cap C = B$ is closed in C and hence the identity map $C \to \tilde{C}$ is also continuous. Consequently, $\tilde{C}$ is compact.

(iv) Let a set A meet each compact set of $k(X)$ in a closed set. Then by **(iii)**, A meets each compact set of X in a compact set and hence in a closed set. This shows that A is closed in $k(X)$.

(v) If follows from **(iv)**.

(vi) It is sufficient to prove by Proposition B.4.12 that $k(f)$ is continuous on each compact set of $k(X)$. Let $\tilde{C}$ be a compact set in $k(X)$. Then by **(iii)**, the set $\tilde{C}$ endowed with the topology induced from X, which is C say, is compact and the identity map $\tilde{C} \to C$ is a homeomorphism. Since $f|_C$ is continuous, $f(C)$ is compact, and hence by **(iii)**, $f(C)$ is the same set $f(\tilde{C})$ with its topology in $k(Y)$. Consequently, the map $k(f)|_{\tilde{C}} : \tilde{C} \to f(C)$ factors into the composition of $f|_C$ and the two identity maps

$$\tilde{C} \to C \to f(C) \to f(\tilde{C}).$$

Consequently, $k(f)|_{\tilde{C}} : \tilde{C} \to f(C)$ is continuous and hence $k(f) : k(X) \to k(Y)$ is continuous.

(vii) Since the maps of closed cells into X coincide with the maps into closed cells into $k(X)$ by **(vi)**, then **(vii)** follows from **(vi)** as the sets under consideration are derived from such mappings. ❑

Definition B.4.15 Let X and Y be in $\mathcal{CG}$ and their product $X \times Y$ (in $\mathcal{CG}$) be $k(X \times_c Y)$, where '$\times_c$' denotes the product under the usual product topology.

Remark B.4.16 Given $X, Y \in \mathcal{CG}$, $X \times_c Y$ may not be in $\mathcal{CG}$ but this product satisfies the universal property given in Fig. B.2.

There are continuous projections $p_1 : X \times Y \to X$ and $p_2 : X \times Y \to Y$ such that if $f : Z \to X$ and $g : Z \to Y$ are continuous, and Z is in $\mathcal{CG}$, then there exists a unique map $F : Z \to X \times Y$ such that $f = p_1 \circ F$ and $g = p_2 \circ F$.

Theorem B.4.17 *Let X and Y be in $\mathcal{CG}$. There are continuous projections $p_1 : X \times Y \to Y$ and $p_2 : X \times Y \to Y$ such that if $f : Z \to X$ and $g : X \to Y$ are continuous, and Z is in $\mathcal{CG}$, there exists a unique map $F : Z \to X \times Y$ with $f = p_1 \circ F$ and $g = p_2 \circ F$ as shown in Fig. B.2.*

Proof Since the identity function

$$X \times Y \to X \times_c Y$$

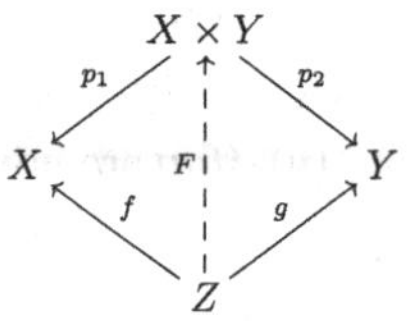

Fig. B.2 Existence of the unique map $F : Z \to X \times Y$

is continuous by Theorem B.4.14 and the projections $X \times_c Y$ into X and Y are also continuous, their compositions projecting $X \times Y$ into X and Y are continuous. Consequently they are in $\mathcal{CG}$. Let $Z \in \mathcal{CG}$ and $f : Z \to X, g : Z \to Y$ be in $\mathcal{CG}$. Then f and g are the components of the unique mapping $(f, g) : Z \to X \times_c Y$. Since $k(Z) = Z$ and $k(X \times_c Y) = X \times Y$, there exists a unique mapping $k(f, g) : Z \to X \times Y$, satisfying the required properties. ❑

B.5 Exercises

1. Let V_F be the category of vector spaces over a field F and their linear transformations. If $D : V_F \to V_F$ is given by $D(V) = V^*$ and $D(T) = T^*$, where V^* is the dual space of V and T^* is the adjoint of T, show that D is a contravariant functor.
2. Let $\mathcal{Grp}$ be the category of groups and their homomorphisms and $C(G)$ be the commutator subgroup of G (i.e., the subgroup generated by $[g_1, g_2] = g_1 g_2 g_1^{-1} g_2^{-1}$). Show that $C : \mathcal{Grp} \to \mathcal{Grp}$ is a covariant functor.
3. Let $\mathcal{Ab}$ denote the category of abelian groups and their homomorphisms.

   (a) For an abelian group G, let $T(G)$ denote its torsion group. Show that $T : \mathcal{Ab} \to \mathcal{Ab}$ defines a functor if $T(f)$ is defined by $T(f) = f|T(G)$ for every homomorphism f in $\mathcal{Ab}$ such that

   (i) f is a monomorphism is $\mathcal{Ab}$ implies that $T(f)$ is also so;
   (ii) f is an epimorphism in $\mathcal{Ab}$ does not always imply that $T(f)$ is also so.

   (b) Let p be a fixed prime integer. Show that $T : \mathcal{Ab} \to \mathcal{Ab}$ defines a functor, where the object function is defined by $T(G) = G/pG$ and the morphism function $T(f)$ is defined by $T(f) : G/pG \to H/pH, x + pG \mapsto f(x) + pH$ for every homomorphism $f : G \to H$ in $\mathcal{Ab}$ such that

   (i) f is an epimorphism in $\mathcal{Ab}$ implies that $T(f)$ is also so;
   (ii) f is a monomorphism in $\mathcal{Ab}$ does not always imply that f is also so.

4. Show that the equivalences in the category

   (i) $\mathcal{Set}$ are bijections of sets;
   (ii) $\mathcal{Grp}$ are isomorphisms of groups;

(iii) $\mathcal{Ring}$ are isomorphisms of rings;
(iv) $\mathcal{Mod}_R$ are isomorphisms of modules;
(v) $\mathcal{Top}$ are homeomorphisms of topological spaces.

5. Let X, Y be Hausdorff spaces and $C(X, Y)$ be the topological space of all continuous maps $X \to Y$ with compact open topology. Define $Y^X = kC(X, Y)$, where $kC(X, Y)$ is the associated compactly generated topological space of $C(X, Y)$. Show that the evaluation map $E : C(X, Y) \times_C X \to Y, (f, x) \mapsto f(x)$ is continuous on compact sets. Moreover, if X and Y are in $\mathcal{CG}$, show that E is continuous as a mapping $Y^X \times X \to Y$.

6. Let X, Y and Z be in $\mathcal{CG}$. Show that $Z^{Y\times X} \approx (Z^Y)^X$.

7. If $\alpha : A \to B$ is a retraction and also a monomorphism in a category $\mathcal{C}$, prove that α is an isomorphism.

8. Show that if $\alpha : A \to B$ is an epimorphism in $\mathcal{Set}$, then $\alpha^0 \in \mathcal{Set}^0$ is a monomorphism.

9. Let X and Y be objects of a category $\mathcal{Set}$ and let $g : X \to Y$ be a morphism in $\mathcal{Set}$. Show that there is a natural transformation g^* from the covariant functor h_Y to the covariant functor h_X and a natural transformation g_* from the contravariant functor h^X to the contravariant functor h^Y. Further show that if g is an equivalence in $\mathcal{Set}$, both these natural transformations g_* and g^* are natural equivalences.

10. Let A and C be objects of a category $\mathcal{C}$. Using Yoneda Lemma, show that $(h_C, h_A) \approx \mathcal{C}(A, C)$.
[Hint. Take $T = h_A$. Then by Yoneda's Lemma $(h_C, h_A) \approx h_A(C) = \mathcal{C}(A, C)$.]

11. **(Dual of Yoneda lemma)** Let $\mathcal{C}$ be any category and T a contravariant functor from $\mathcal{C}$ to $\mathcal{Set}$. Show for any object $C \in \mathcal{C}$, there is an equivalence $\theta : (h^C, T) \to T(C)$ such that θ is natural in C and T, where (h^C, T) is the class of natural transformations from the set valued functor h^C to the set valued functor T such that θ is natural C and T.

12. **(a)** Let $\mathcal{C}$ be a category and $T : \mathcal{C} \to \mathcal{Set}$ be a contravariant functor. Show that for each object C in $\mathcal{C}$ there is a one-to-one correspondence θ between natural transformations $\eta : h^C \to T$ and elements $x \in T(C)$.
[Hint: Use dual result Yoneda lemma.]
(b) Let $\mathcal{C}$ be a category and $T : \mathcal{C} \to \mathcal{Set}$ be a representable functor. If C and C' are representable objects for T with universal elements x, x' respectively. Show that there is an isomorphism $\psi : C \to C'$ such that $T(\psi)(x') = x$.

13. **(a)** If T is a Brown functor and $*$ is a one-point space, show that $T(*)$ is a set which also consists of a single element.
(b) Let T be a Brown functor and $X = \Sigma Y$ be the suspension for some topological space Y. Show that $T(X)$ admits a group structure with the distinguished element in the pointed set $T(X)$ as identity element.

B.6 Additional Reading

[1] Aguilar, Gitler, S., Prieto, C., *Algebraic Topology from a Homotopical View Point*, Springer-Verlag, New York, 2002.
[2] Arkowitz, Martin, *Introduction to Homotopy Theory*, Springer, New York, 2011.
[3] Atiyah, M. F., *K-Theory*, Benjamin, New York, 1967.
[4] Switzer, R.M., *Algebraic Topology-Homotopy and Homology*, Springer-Verlag, Berlin, Heidelberg, New York, 1975.
[5] Whitehead, G.W., *Elements of Homotopy Theory*, Springer-Verlag, New York, Heidelberg, Berlin, 1978.

References

Adhikari, M.R., Adhikari, A.: Basic Modern Algebra with Applications. Springer, New Delhi (2014)
Aguilar, M., Gitler, S., Prieto, C.: Algebraic Topology from a Homotopical View Point. Springer, New York (2002)
Arkowitz, M.: Introduction to Homotopy Theory. Springer, New York (2011)
Atiyah, M.F.: *K-Theory*. Benjamin, New York (1967)
Eilenberg, S., Mac Lane, S.: Natural isomorphism in group theory. Proc. Natl. Acad. Sci., USA **28**, 537–544 (1942)
Eilenberg, S., Mac Lane, S.: General theory of natural equivalence. Trans. Am. Math. Soc. **58**, 231–294 (1945)
Eilenberg, S., Steenrod, N.: Foundations of Algebraic Topology. Princeton University Press, Princeton (1952)
Gray, B.: Homotopy Theory, An Introduction to Algebraic Topology. Acamedic Press, New York (1975)
Hatcher, A: Algebraic Topology. Cambridge University Press, Cambridge (2002)
MacLane, S.: Categories for the Working Mathematician. Springer, Heidelberg (1972)
Rotman, J.J.: An Introduction to Algebraic Topology. Springer, New York (1988)
Spanier, E.: Algebraic Topology. McGraw-Hill Book Company, New York (1966)
Steenrod, N.: A Convenient Category of Topological Spaces. Michigan Math J. **14**, 133–152 (1967)
Switzer, R.M.: Algebraic Topology-Homotopy and Homology. Springer, Berlin (1975)
Whitehead, G.W.: Elements of Homotopy Theory. Springer, New York (1978)

List of Symbols

$\emptyset$	: empty set
$X \subset Y$ or $Y \supset X$	: set-theoretic containment (not necessarily proper)
$\mathbf{N}$	: set of natural numbers (or positive integers)
$\mathbf{Z}$	: ring of integers (or set of integers)
$\mathbf{R}$	: field of real numbers (or set of real numbers)
$\mathbf{Q}$	: field of rational numbers (or set of rational numbers)
$\mathbf{C}$	: field of complex numbers (or set of complex numbers)
$\mathbf{H}$	: division ring of quaternions (or set of quaternions)
pp(or p.)	: particular page of reference
$\times, \Pi$	: product of sets, groups, modules, or spaces
$\cong$	: isomorphism between groups
$\approx$	: homeomorphism between topological spaces
iff	: if and only if
$\lvert X \rvert$	: cardinal of a set X
$\mathbf{Z}_n$	: ring of integers modulo n (or residue classes of integers modulo n), 2
$\mathrm{GL}(V)$	: general linear group on V, 7
$G = \prod_{i \in I} G_i$	: direct product of family $\{G_i : i \in I\}$ of groups, 9
$G = \oplus_{i \in I} G_i$	: direct sum of family $\{G_i : i \in I\}$ of groups, 10
$G \oplus H$	: direct sum of groups, 10
$G * H$	: free product of groups, 13
$G_x = \{g \in G : g \cdot x = x\}$	: isotropy group or the stabilizer group of x, 16
$G \otimes H$	: tensor product of modules ,18
$\mathbf{C}^n$	: complex n-space, 22

M.R. Adhikari, *Basic Algebraic Topology and its Applications*,
DOI 10.1007/978-81-322-2843-1

I	: closed unit interval [0,1], 22
I^n	: n-cube $= \{x \in \mathbf{R}^n : 0 \leq x_i \leq 1 \text{ for } 1 \leq i \leq n\}$ for $x = (x_1, x_2, \ldots, x_n)$, 22
D^n	: n-disk or n-ball $= \{x \in \mathbf{R}^n : \|x\| \leq 1\}$, 22
S^n	: n-sphere $= \{x \in \mathbf{R}^{n+1} : \|x\| = 1\} = \partial D^{n+1}$ (the boundary of the $(n+1)$-disk D^{n+1}), 22
$\mathbf{R}P^n$	: real projective space = quotient space of S^n with x and $-x$ identified, 22
$\mathbf{C}P^n$	: complex projective space = space of all complex lines through the origin in the complex space $\mathbf{C}^{n+1}$, 22
$\partial I^n = \dot{I}^n = I^n - \text{Int } I^n$	: boundary of I^n, 26
$\Vert\ \Vert : X \to \mathbf{R}$	: norm function, 27
X/A	: quotient space (endowed with quotient topology), 33
M_f	: mapping cylinder of f, 35
C_f	: mapping cone of f, 35
$\bigsqcup$	: disjoint union of sets or spaces, 36
$X \vee Y$	: wedge (one-point union), 36
$X \wedge Y$	: smash product, 36
ΣX	: suspension space, 36
$Y^X = F(X, Y)$	: function space (endowed with compact open topology), 37
$V_r(\mathbf{R}^n)$	: Stiefel manifold of orthonormal r-frames in $\mathbf{R}^n$, 39
$G_r(\mathbf{R}^n)$	: Grassmann manifold of r-planes of $\mathbf{R}^n$ through the origin, 39
$\text{GL}(n, \mathbf{R})$	: general real linear group, 41
$F : f \simeq g$	: F is a homotopy between f and g, 47
$f \simeq g$	: homotopic maps, 47
$F : f \underset{p}{\simeq} g$	: F is a path homotopy between f and g, 48
$[X, Y]$	: set of homotopy classes of maps from X to Y, 53
$\mathcal{H}tp_*$	: homotopy category of pointed spaces, 53
$\mathcal{H}tp$	: homotopy category of spaces, 54
(X, A)	: topological pair, 55
$X \simeq Y$	: homotopy equivalent spaces, 59
H-space	: Hopf space, 66
H-group	: Hopf group, 67
$\Omega(X)$	: loop space of pointed space X, 75
$\Omega^n X$	: iterated loop space of pointed space X, 75
$\nabla : X \vee X \to X$	: folding map, 77
ΣX	: suspension space of a pointed space X, 78
Ω	: loop functor, 79

BS_p	: $G_n(\mathbf{H}^\infty)$, 228
$GL(n, F)$	: group of nonsingular $n \times n$ matrices over F, 232
$Vect_n(B)$	: B-isomorphic classes of n-dimensional vector bundles over B, 234
$G_n(F^\infty) = \bigcup_{n \leq m} G_n(F^\infty)$	: Grassmannian manifold, 235
γ_n^∞	: n-dimensional vector bundle over $G_n(F^\infty)$, 235
$\xi \oplus \eta$	: Whitney sum of vector bundles over B, 236
$K_F(B)$	: K-theory on F-vector bundles over B, 237
$K(B)$	: Grothendieck group of vector bundles over B, 245
Δ^n	: standard n-dimensional simplex, 250
$\|K\|$	: polyhedron of simplicial complex K, 253
$\mathrm{st}(v)$	: star of v, 255
$\mathrm{ost}(v)$	: open star of v, 255
$N_K(x)$	: simplicial neighbourhood of $x \in \|K\|$, 256
$L_K(x)$	: link of $x \in \|K\|$, 256
(K, L)	: simplicial pair, 257
K'	: barycentric subdivision of K, 258
$B(s_p)$	: barycenter of s_p, 258
$K^{(n)}$	: nth barycentric subdivision of K, 258
$\mu(K)$	: mesh of simplicial complex K, 260
$E(K, v)$	: edge group of K based at v, 264
$\pi_n(X, x_0)$	: nth homotopy groups of (X, x_0), 274
$\pi_n(X, A, x_0)$	: nth relative homotopy group of (X, A, x_0), 282
∂I^n	: boundary of I^n, 274
$f \simeq g$ rel ∂I^n	: f is homotopic to g rel ∂I^n, 275
∂	: boundary operator in homotopy theory, 283
$\mathcal{H}tp^2$	: homotopy category of triplets, 285
π_n	: homotopy functor, 285
π^n	: cohomotopy functor, 296
$PCHP$	: polyhedra covering homotopy property, 298
e^n	: open n-cell, 306
$X^{(n)}$	: n-skeleton of cell complex, 307
CW-complex	: cell complex having closure finite and weak topology, 307
ψ_i^n	: characteristic maps for CW-complex, 309
WS^1	: Warsaw circle, 318
$[\alpha, \beta]$	: Whitehead product of α and β, 330
$[\alpha, \beta]_{GW}$	: generalized Whitehead product of α and β, 336
$\langle \alpha, \beta \rangle_m$	: mixed product of α and β, 338

$\langle\alpha, \beta\rangle$	: Samelson product of α and β, 338
τ_n	: generator of $\pi_n(S^n)$, 341
$\text{nil}\,(S^7, \mu, \phi)$	: homotopical nilpotence of the H-space (S^7, μ, ϕ), 342
Z_n	: group of n-cycles, 350
B_n	: group of n-boundaries, 350
H_n	: homology functor, 352
$\mathcal{Comp}$	: category of chain complexes, 352
$\mathcal{Ab}$	: category of abelian groups, 352
$c_p \sim d_p$	: c_p and d_p are homologous p-cycles, 358
$Z_p(K; G)$	: group of p-cycles of K with coefficients in G for $p \geq 1$, 358
$B_p(K; G)$	: group of p-boundaries of K with coefficients in G for $p \geq 1$, 358
$H_p(K; G)$	: p-dimensional simplicial homology group of K with coefficients in G for $p \geq 1$, 359
$Z^n = \ker \delta^{n+1}$	: group of n-cocycles, 366
$B^n = \text{Im}\,\delta^n$	: group of n-coboundaries, 366
H^n	: n-dimensional cohomology, 366
$H^* = \{H^n, \delta^n\}$	: cohomology theory, 367
Δ^n	: standard n-simplex, 369
$\beta_n(X)$	: $\beta_n(X) = rank(H_n(X))$ = nth Betti number of X, 370
$\widetilde{H}_*$	: reduced singular homology, 372
$H_n(X, A)$	: relative singular homology groups of (X, A), 373
$C^n(X; G) = Hom(C_n(X; G), G)$	: singular n-cochain group, 375
$H^n(X, A)$	: relative singular chomology groups of (X, A), 376
$H^n(X, A; G)$	: relative singular chomology group with coefficient group G, 377
$W_*(X)$	: cellular chain complex, 383
$H_n(W_*(X))$	: cellular homology group, 383
$\check{H}^*(X; G)$	: Čech cohomology group, 385
$\chi(K)$	: Euler characteristic of a simplicial complex K, 388
$\psi \cup \theta$	: cup product of $\psi \in C^n(X; R)$ and $\theta \in C^m(X; R)$, 393
$\sigma \cap \psi$	: cap product of $\sigma \in C_{m+n}(X; R)$ and $\psi \in C^n(X; R)$, 395
$\mathcal{GR}$	: category of graded rings, 401
$\alpha \times \beta$	: cohomology cross product of cellular cochains, 401
$l^\infty(m) = S^\infty \bmod \mathbf{Z}_m$	: infinite dimensional lens space, 408

$c(f)$	: obstruction cocycle of f, 525
$H(f)$	: Hopf invariant of $f : S^{2n-1} \to S^n$, 540
$\underline{A} = \{A_n, \alpha_n\}$	: spectrum of spaces, 563
$\mathrm{SL}(n, \mathbf{R})$	: special linear group, 570
$\mathrm{O}(n, \mathbf{R})$	: orthogonal group, 570
$\mathrm{SO}(n, \mathbf{R})$	: special orthogonal group, 570
$\mathrm{GL}(n, \mathbf{C})$	: general complex linear group, 572
$X \mod G$	: orbit space of G-space X, 573
$\mathrm{SU}(n, \mathbf{H})$	: sympletic group, 577
$\mathcal{C}$	: category, 582
$\mathcal{G}rp$	: category of groups, 583
$\approx$	: equivalence between objects in category theory, 584
$A \approx B$	: A and B are equivalent objects in a category, 584
$[G, G]$	: commutator subgroup of G, 587
$G/[G, G]$	: abelianized group of G, 587
Ω & Σ	: pair of adjoint functors, 587
SP^nX (or $SP^n(X)$)	: nth symmetric product of X, 588
$\mathcal{CG}$	: category of compactly generated Hausdorff spaces, 592

Author Index

M.R. Adhikari, *Basic Algebraic Topology and its Applications*,
DOI 10.1007/978-81-322-2843-1

Subject Index

M.R. Adhikari, *Basic Algebraic Topology and its Applications*,
DOI 10.1007/978-81-322-2843-1

D

E